AF685004

The Apennines: Tectonics, Sedimentation, and Magmatism from the Palaeozoic to the Present

The Apennines: Tectonics, Sedimentation, and Magmatism from the Palaeozoic to the Present

Editors

Domenico Liotta
Giancarlo Molli
Angelo Cipriani

MDPI • Basel • Beijing • Wuhan • Barcelona • Belgrade • Manchester • Tokyo • Cluj • Tianjin

Editors
Domenico Liotta
Università di Bari Aldo Moro
Italy

Giancarlo Molli
Università di Pisa
Italy

Angelo Cipriani
Università degli Studi di Roma
"La Sapienza"
Servizio Geologico
d'Italia—ISPRA
Italy

Editorial Office
MDPI
St. Alban-Anlage 66
4052 Basel, Switzerland

This is a reprint of articles from the Special Issue published online in the open access journal *Geosciences* (ISSN 2076-3263) (available at: https://www.mdpi.com/journal/geosciences/special_issues/Apennines).

For citation purposes, cite each article independently as indicated on the article page online and as indicated below:

LastName, A.A.; LastName, B.B.; LastName, C.C. Article Title. *Journal Name* **Year**, *Volume Number*, Page Range.

ISBN 978-3-0365-2251-7 (Hbk)
ISBN 978-3-0365-2252-4 (PDF)

© 2023 by the authors. Articles in this book are Open Access and distributed under the Creative Commons Attribution (CC BY) license, which allows users to download, copy and build upon published articles, as long as the author and publisher are properly credited, which ensures maximum dissemination and a wider impact of our publications.
The book as a whole is distributed by MDPI under the terms and conditions of the Creative Commons license CC BY-NC-ND.

Contents

About the Editors vii

Preface to "The Apennines: Tectonics, Sedimentation, and Magmatism from the Palaeozoic to the Present" ix

Eugenio Turco, Chiara Macchiavelli, Giulia Penza, Antonio Schettino and Pietro Paolo Pierantoni
Kinematics of Deformable Blocks: Application to the Opening of the Tyrrhenian Basin and the Formation of the Apennine Chain
Reprinted from: *Geosciences* **2021**, *11*, 177, doi:10.3390/geosciences11040177 1

Giovanni Luca Cardello, Giuseppe Vico, Lorenzo Consorti, Monia Sabbatino, Eugenio Carminati and Carlo Doglioni
Constraining the Passive to Active Margin Tectonics of the Internal Central Apennines: Insights from Biostratigraphy, Structural, and Seismic Analysis
Reprinted from: *Geosciences* **2021**, *11*, 160, doi:10.3390/geosciences11040160 37

Fabrizio Piana, Luca Barale, Carlo Bertok, Anna d'Atri, Andrea Irace and Pietro Mosca
The Alps-Apennines Interference Zone: A Perspective from the Maritime and Western Ligurian Alps
Reprinted from: *Geosciences* **2021**, *11*, 185, doi:10.3390/geosciences11050185 87

Enrico Capezzuoli, Amalia Spina, Andrea Brogi, Domenico Liotta, Gabriella Bagnoli, Martina Zucchi, et al.
Reconsidering the Variscan Basement of Southern Tuscany (Inner Northern Apennines)
Reprinted from: *Geosciences* **2021**, *11*, 84, doi:10.3390/geosciences11020084 113

Francesco Brozzetti, Daniele Cirillo and Lucina Luchetti
Timing of Contractional Tectonics in the Miocene Foreland Basin System of the Umbria Pre-Apennines (Italy): An Updated Overview
Reprinted from: *Geosciences* **2021**, *11*, 97, doi:10.3390/ geosciences11020097 137

Stefano Conti, Claudio Argentino, Chiara Fioroni, Aura Cecilia Salocchi and Daniela Fontana
Miocene Seep-Carbonates of the Northern Apennines (Emilia to Umbria, Italy): An Overview
Reprinted from: *Geosciences* **2021**, *11*, 53, doi:10.3390/geosciences11020053 159

Mario Costa, Jessica Chicco, Chiara Invernizzi, Simone Teloni and Pietro Paolo Pierantoni
Plio–Quaternary Structural Evolution of the Outer Sector of the Marche Apennines South of the Conero Promontory, Italy
Reprinted from: *Geosciences* **2021**, *11*, 184, doi:10.3390/geosciences11050184 179

Aurélie Labeur, Nicolas E. Beaudoin, Olivier Lacombe, Laurent Emmanuel, Lorenzo Petracchini, Mathieu Daëron, et al.
Burial-Deformation History of Folded Rocks Unraveled by Fracture Analysis, Stylolite Paleopiezometry and Vein Cement Geochemistry: A Case Study in the Cingoli Anticline (Umbria-Marche, Northern Apennines)
Reprinted from: *Geosciences* **2021**, *11*, 135, doi:10.3390/geosciences11030135 199

Giacomo Prosser, Giuseppe Palladino, Dario Avagliano, Francesco Coraggio, Eleonora Maria Bolla, Marcello Riva and Daniele Enrico Catellani
Stratigraphic and Tectonic Setting of the Liguride Units Cropping Out along the Southeastern Side of the Agri Valley (Southern Apennines, Italy)
Reprinted from: *Geosciences* **2021**, *11*, 125, doi:10.3390/geosciences11030125 229

Stefano Vitale, Ernesto Paolo Prinzi, Maria Monda, Francesco D'Assisi Tramparulo and Sabatino Ciarcia
Structural and Stratigraphic Setting of Campagna and Giffoni Tectonic Windows: New Insights on the Orogenic Evolution of the Southern Apennines (Italy)
Reprinted from: *Geosciences* **2020**, *10*, 405, doi:10.3390/geosciences10100405 . . . . **259**

Stefano Vitale, Ernesto Paolo Prinzi, Francesco D'Assisi Tramparulo, Claudio De Paola, Rosa Di Maio, Ester Piegari, et al.
Late Miocene-Early Pliocene Out-of-Sequence Thrusting in the Southern Apennines (Italy)
Reprinted from: *Geosciences* **2020**, *10*, 301, doi:10.3390/geosciences10080301 . . . . **283**

Caterina Bianco
The Capo Castello Shear Zone (Eastern Elba Island): Deformation at the Contact between Oceanic and Continent Tectonic Units
Reprinted from: *Geosciences* **2020**, *10*, 361, doi:10.3390/geosciences10090361 . . . . **307**

Andrea Brogi, Alfredo Caggianelli, Domenico Liotta, Martina Zucchi, Amalia Spina, Enrico Capezzuoli, et al.
The Gavorrano Monzogranite (Northern Apennines): An Updated Review of Host Rock Protoliths, Thermal Metamorphism and Tectonic Setting
Reprinted from: *Geosciences* **2021**, *11*, 124, doi:10.3390/geosciences11030124 . . . . **333**

Massimiliano Ghinassi, Mauro Aldinucci, Valeria Bianchi, Andrea Brogi, Enrico Capezzuoli, Tsai-Luen Yu and Chuan-Chou Shen
Lifecycle of an Intermontane Plio-Pleistocene Fluvial Valley of the Northern Apennines: From Marine-Driven Incision to Tectonic Segmentation and Infill
Reprinted from: *Geosciences* **2021**, *11*, 141, doi:10.3390/geosciences11030141 . . . . **363**

Ilaria Isola, Francesco Mazzarini, Giancarlo Molli, Leonardo Piccini, Elena Zanella, Giovanni Zanchetta, et al.
New Chronological Constraints from Hypogean Deposits for Late Pliocene to Recent Morphotectonic History of the Alpi Apuane (NW Tuscany, Italy)
Reprinted from: *Geosciences* **2021**, *11*, 65, doi:10.3390/geosciences11020065 . . . . **391**

Giancarlo Molli, Isabelle Manighetti, Rick Bennett, Jacques Malavieille, Enrico Serpelloni, Fabrizio Storti, et al.
Active Fault Systems in the Inner Northwest Apennines, Italy: A Reappraisal One Century after the 1920 Mw ~6.5 Fivizzano Earthquake
Reprinted from: *Geosciences* **2021**, *11*, 139, doi:10.3390/geosciences11030139 . . . . **409**

About the Editors

Domenico Liotta

Domenico Liotta got the Ph.D. in Geology in 1990 at University of Siena. From 1998 is associate professor at Bari University. He carried out researches on the relationships between geological structures and geothermal fluid flow in Iceland, Turkey, Mexico, and Northern Apennines. His expertise in the field mapping activity and collection of structural and kinematic data in brittle and ductile deformation domains. His papers concern structural and filed mapping works, with emphasis in geothermal areas. He was partner and coordinator of several national and international joint programs. He is author of more than 100 peer reviewed scientific journals.

Giancarlo Molli

Giancarlo Molli got the diploma in Geological Sciences (1990) and the Ph.D. in Structural Geology (1995) at the University of Pisa. From 1995 to 1997 he was post-Doc Researcher at the University of Pisa and Researcher since 1999 where he is now Associate Professor. Visiting researcher at the University of Manchester (UK); Basel (CH) and CNRS Geosciences Montpellier. His research interest focus on the following subjects: (1) orogenic processes and their geometric and kinematic records; (2) zone of localized deformation, their microstructures and relationships with tectonic setting; (3) ciclicity of deformation and record of seismicity in exhumed fault zone; (4) recent and active tectonics of the northern Apennines. The research areas he is currently involved with, and his past studies include: Apennines, Corsica, Alps, Himalaya, Taiwan, Andes. Teaching activity in Tectonics and Applied Structural Geology at the University of Pisa. Authors of more than 90 papers in international peer reviewed scientific journals.

Angelo Cipriani

Angelo Cipriani got the Master Degree in Exploration Geology (2013) and the Ph.D. in Earth Sciences (2017) at "Sapienza" University of Rome. In 2018, he was a Post-Doctoral Researcher at the Department of Earth Sciences of "Sapienza" University of Rome. Since 2020, he works as Researcher at the Geological Survey of Italy (ISPRA) for the National geological mapping project (CARG Project at 1:50,000 scale). His field-based expertise and research interest concerns the geological mapping, the carbonate sedimentology of pelagic and shallow-water environments, Meso-Cenozoic stratigraphy of Tethyan realm, the interaction between syn-depositional tectonics and sedimentation, the paleotectonic reconstruction of rifted basins, carbonate platform-basin settings, the analysis of structurally complex areas. Angelo Cipriani has received numerous awards (i.e., "Emilio Cortese", "David Giuntini", and "Secondo Franchi" awards) and has published more than 20 papers and geological maps in international scientific journals.

Preface to "The Apennines: Tectonics, Sedimentation, and Magmatism from the Palaeozoic to the Present"

This Special Issue of Geosciences consists of a selection of papers, some of them presented at the International Meeting EGU 2020 in the session: "Apennines tectonics, magmatism and sedimentation: from Permian to Present". This session emphasized the necessity of a multidisciplinary and integrated approach to the study of the Apennines and the related crustal structures.

Papers included in this Special Issue deal with structural, stratigraphic, geochemical and geophysical methods, and cover arguments from local to regional scale, investigating different sectors of the Apennines (Figure 1) and their role from pre-orogenic to late- and post-orogenic stages during belt's evolution.

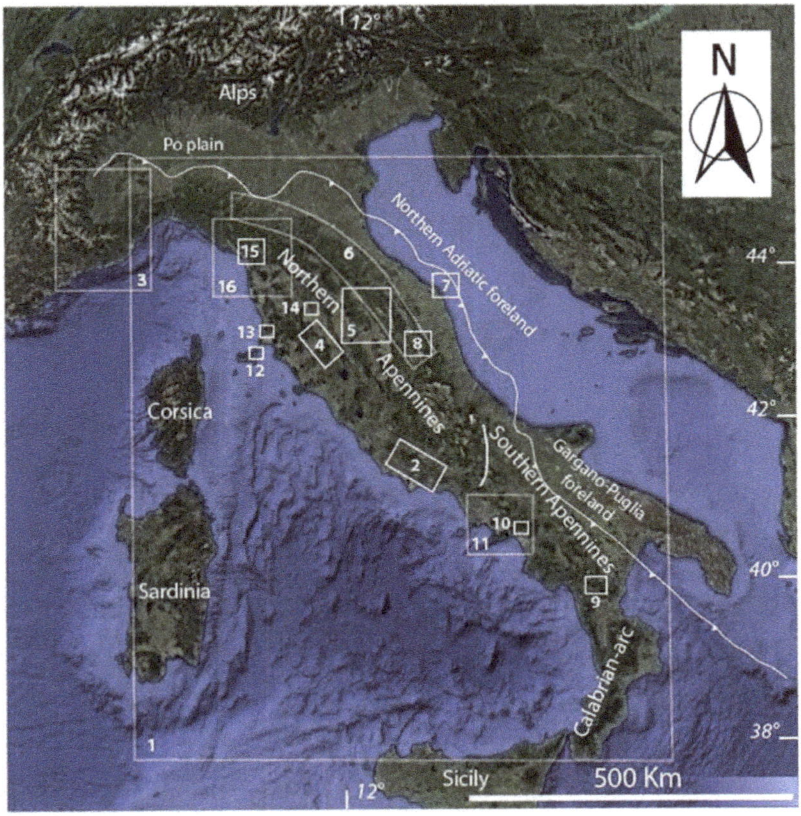

Figure 1. Numbers are the same of the references list, indicating the areas where the papers of this special issue are related. Modified after Google Earth 2018.

The completeness of the information collected in this book, makes it an optimal revision and implementation of the knowledge for the entire Apennines, as also shown by the location of the different study areas, distributed throughout the entire belt.

The geodynamic context is investigated by Turco et al. [1], proposing a reconstruction on the coeval development of the Tyrrhenian Basin and Apennines, from Oligocene to present. This is based on the assumption that different crustal blocks, delimited by shear zones and involved in the tectonic evolution, rotated under the same Eulero pole. Values at the base of this study are interestingly obtained from balanced crustal geological sections and stratigraphic data.

A multidisciplinary approach to unravel the evolution of the Apennines from Cretaceous to present is then presented by Cardello et al. [2]. These Authors deal with central Apennines, with emphasis on thrust evolution and inheritance of extensional structures, a crucial theme in the frame of the Alpine-Apenninic orogeny.

The relationships between Alps and Apennines are faced by Piana et al. [3]. These Authors conclude accounting for the role of a regional Oligocene–Miocene transfer zone which favored the Adria indentation and drastically influenced kinematics and tectono-metamorphic evolution of the involved tectonic units.

Regarding the northern Apennines, a long-lasting issue on the age and evolution of the Paleozoic basement cropping out in Tuscany is discussed by Capezzuoli et al. [4]. The new investigations on palynoflora and sporomorphs, joined with original structural studies, relate these Paleozoic rocks to the Middle Mississippian and exclude their involvement in the Hercynian deformation.

Several contributions are concerned with tectonics and sedimentation relationships in the outer zone of Apennines.

Brozzetti et al. [5] illustrate the Miocene foredeep evolution, providing paleontological, sedimentary, and structural data to constrain the timing and reconstruct the evolution of the compressional deformations in the central part of the northern Apennines.

The same time-span is investigated by Conti et al. [6], discussing the seep-carbonates cropping out in the outer zone of the northern Apennines. These Authors integrate sedimentary, geochemical, and paleontological data, providing a unique dataset for cold seepage systems, with fallouts for the understanding of the modern carbonates, located in comparable tectonic settings.

Costa et al. [7] discuss the relationships between sedimentation and tectonics during the Pliocene–Quaternary in the outermost (i.e., eastern) part of northern Apennines. This work integrates field and geophysical data, and highlights the relationships between negative vs. positive reactivation of inherited structures in this part of the Apennine fold and thrust belt.

A study, reconstructing the burial-deformation history of sedimentary rocks in outer zones of the Apenninic Chain (Cingoli anticline), is then provided by Labeur et al. [8]. The work is based on the combination of geochemical and microstructural analysis, allowing to identify the effects of the regional tectonics from layer-parallel shortening to fold growth.

Three different contributions describing crucial issues in the tectono-stratigraphic evolution of the southern Apennines are also included in this Special Issue.

Based on a multidisciplinary approach involving field mapping, structural geology, biostratigraphy, and geophysical methods, Prosser et al. [9] shed light on the age and structural setting of crucial deposits belonging to the Tethyan oceanic cover (i.e., Albidona formation). Surface geology, coupled with subsurface data, has allowed the Authors to better define the stratigraphic features of the Albidona Formation, and to provide an updated review on the tectonic setting of the well-known Agri Valley.

In the same line of innovation, Vitale et al. [10,11] present two contributions on the inner part of southern Apennines. The Authors analyze compressional deformation structures in several key-areas of the western part of the southern Apennines. A detailed kinematic, structural and stratigraphic approach permitted to define the timing of in- and out-of-sequence thrusts, thus better explaining the relationships among the geological bodies of southern Apennines.

Geological features related to the extensional tectonics affecting the inner zone of the northern Apennines are reported by Bianco [12], who studied a Miocene extensional shear zone, precursor of other regional low angle normal faults. The microstructural study highlighted the role of the strain partitioning and of the fluid assisted deformation during this extensional event.

Inland Pliocene extension is accompanied by magmatism. The relationships between magma emplacement and structures is discussed in Brogi et al. [13], who describe the role of the transfer zones and their internal deformation to favor up flow of magmatic, deep fluids in localized, fracture-controlled, permeable sectors.

Effects of extensional tectonics in recent times are then investigated by Ghinassi et al. [14], who applied a multidisciplinary approach based on stratigraphy, dating analyses and structural geology to depict the tectonic control on fluvial tracks, by studying a key area where one of the main Italian river changed its flow direction.

The morphogenetic stage of the evolution of the inner NW zone of the Northern Apennines is described by Isola et al. [15]. They constrained the uplift of the Alpi Apuane region, joining structural and sedimentological analyses in caves with speleothem dating. Hence, indications for the erosion rate during the Late Neogene uplift of the inner NW Apennines belt is provided.

Finally, Molli et al. [16], one century after the 1920 Mw 6.5 Fivizzano earthquake, presented a reappraisal of the potentially seismogenic faults and fault systems of this sector of the inner northern Apennines, reviewing existing data along with new observations. The result is an integrated approach, leading to a catalog of active faults and a new tectonic scenario accounting for their development.

Domenico Liotta, Giancarlo Molli, Angelo Cipriani
Editors

geosciences

Article

Kinematics of Deformable Blocks: Application to the Opening of the Tyrrhenian Basin and the Formation of the Apennine Chain

Eugenio Turco [1], Chiara Macchiavelli [2], Giulia Penza [1,*], Antonio Schettino [1] and Pietro Paolo Pierantoni [1]

1 Geology Division—School of Science and Technology, University of Camerino, Via Gentile III da Varano, 62032 Camerino (MC), Italy; eugenio.turco@unicam.it (E.T.); antonio.schettino@unicam.it (A.S.); pietropaolo.pierantoni@unicam.it (P.P.P.)
2 Group of Dynamics of the Lithosphere (GDL), Geosciences Barcelona, Geo3Bcn—CSIC, Lluís Solé Sabarís s/n, 08028 Barcelona, Spain; cmacchiavelli@geo3bcn.csic.es
* Correspondence: giulia.penza@unicam.it

Abstract: We describe the opening of back-arc basins and the associated formation of accretionary wedges through the application of techniques of deformable plate kinematics. These methods have proven to be suitable to describe complex tectonic processes, such as those that are observed along the Africa–Europe collision belt. In the central Mediterranean area, these processes result from the passive subduction of the lithosphere belonging to the Alpine Tethys and Ionian Ocean. In particular, we focus on the opening of the Tyrrhenian basin and the contemporary formation of the Apennine chain. We divide the area of the Apennine Chain and the Tyrrhenian basin into deformable polygons that are identified on the basis of sets of extensional structures that are coherent with unique Euler pole grids. The boundaries between these polygons coincide with large tectonic lineaments that characterize the Tyrrhenian–Apennine area. The tectonic style along these structures reflects the variability of relative velocity vectors between two adjacent blocks. The deformation of tectonic elements is accomplished, allowing different rotation velocities of lines that compose these blocks about the same stable stage poles. The angular velocities of extension are determined on the basis of the stratigraphic records of syn-rift sequences, while the rotation angles are obtained by crustal balancing.

Keywords: Tyrrhenian–Apennine system; non-rigid plate kinematics; rotation models

check for updates

Citation: Turco, E.; Macchiavelli, C.; Penza, G.; Schettino, A.; Pierantoni, P.P. Kinematics of Deformable Blocks: Application to the Opening of the Tyrrhenian Basin and the Formation of the Apennine Chain. *Geosciences* **2021**, *11*, 177. https://doi.org/10.3390/geosciences11040177

Academic Editors: Jesus Martinez-Frias, Domenico Liotta, Giancarlo Molli and Angelo Cipriani

Received: 31 January 2021
Accepted: 9 April 2021
Published: 14 April 2021

Publisher's Note: MDPI stays neutral with regard to jurisdictional claims in published maps and institutional affiliations.

Copyright: © 2021 by the authors. Licensee MDPI, Basel, Switzerland. This article is an open access article distributed under the terms and conditions of the Creative Commons Attribution (CC BY) license (https://creativecommons.org/licenses/by/4.0/).

1. Introduction

The peri-Tyrrhenian orogenic belt, which is formed by the Apennine Chain, the Calabrian arc, and the Sicily Chain, is the most recent expression of the geodynamic process that created the western Mediterranean basin after the Europe–Africa collision (Figure 1). Large-scale extensional tectonics, coupled with orogenic processes, formed the Tyrrhenian basin, while the thrust belt–foredeep system of the Apennine chain continued migrating towards the present-day Adriatic–Ionian foreland. The Tyrrhenian margin of the Apennine chain experienced widespread extensional tectonics, characterized by formation of several marine basins, intramontane troughs, and intense magmatism. The Tyrrhenian Sea, which developed since Middle Tortonian times, is the youngest basin of the western Mediterranean region [1]. It has been extensively studied since the 1960s. In spite of the huge amount of available data, the geodynamic evolution of the Tyrrhenian basin and surrounding regions are not yet coherently described and have been subject to controversial interpretations [2–17]. In particular, the kinematic relationships between extension in the Tyrrhenian Sea, basin formation along the Tyrrhenian margin of the Apennine chain, migration of the Apennine arcs, and volcanism still remain to be determined.

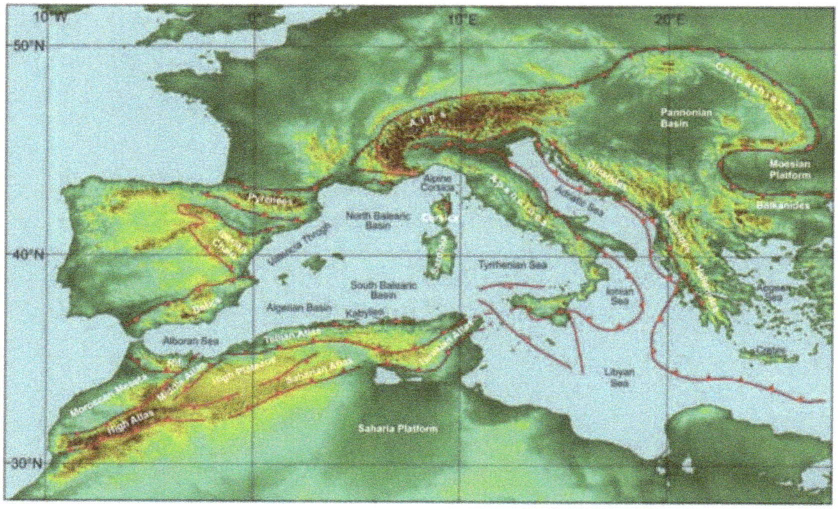

Figure 1. Digital terrain model (ASTER images, 1.5 s) of the western Mediterranean region with major, simplified, tectonic lineaments; modified after [18].

The main reason for the existence of controversial interpretations is due to the complexity of the geodynamic processes that generated the Tyrrhenian–Apennine system. It belongs to the Africa–Europe collision belt along which the fragmentation of the Adriatic plate started [19] since the upper Cretaceous, followed by upper Oligocene slab-retreat events. All these processes have produced an articulated Africa–Europe collision front, which includes back-arc basins and the Apennine chain. Most models proposed so far for the description of the evolution of the Tyrrhenian–Apennine system are based on stratigraphic and structural analyses of transects, not always correctly oriented along the flow lines of relative motion. Tectonic reconstructions obtained with such a 2D method often neglect important 3D kinematic constraints expressed by structures that are transversal to the chain [4,20–22]. Only a few authors have followed an approach based on the laws of plate kinematics [15,23–26].

In this work, we propose a quantitative method for describing the evolution of a system of deformable tectonic elements in the context of a back-arc extension and associated building of an accretionary wedge mountain belt. The technique is then applied to the kinematic reconstruction of the Tyrrhenian–Apennine region. Previous works [6,27,28] have described the tectonic evolution of this area in the rigid plate kinematics approximation. Here, we use the same kinematic framework but allow internal deformation of the blocks during their motion. This approach provides a better representation of the geological processes associated with the formation of back-arc basins, in particular the existence of transverse structures along the axis of the accretionary wedge. Finally, we will show that the resulting model supports the formation of at least three STEP (Subduction-Transform-Edge-Propagator) faults along the subducting slabs.

2. Geological Setting

2.1. The Apennine Chain

The Apennine–Maghreb chain is considered a Neogene thrust belt, which comprises Mesozoic to Palaeogene sedimentary rocks, derived from different basins and shelf located in paleogeographic domains of the Adria continental margin [27] (and references therein).

According to Turco et al. [18], when the Sardinian-Corsica block (SCB) started separating from the European plate, a long trench was present in the Central-Western Mediterranean region, where Liguride-Tethys lithosphere was subducting. We know that the polarity of subduction flipped from NW below the Calabrian Arc-Kabylies to SE below Alpine

Corsica (AlC) and the Western Alpine Arc (WAA). Such a structure always determines the formation of a strike-slip fault that links the two branches of the subduction zone [29]. The rotation of the Sardinian–Corsican block accompanied this process, favouring the sinistral transpressional character of the plate boundary between Adria and Sardinia-Corsica block (yellow line A, Figure 2). At the end of the Sardinia–Corsica rotation, such a lithosphere fault reached its maximum length of ~500 km (Figure 3).

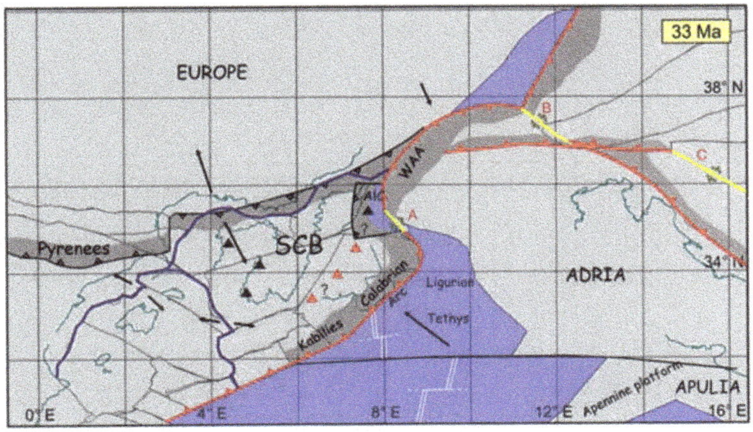

Figure 2. Plate reconstruction of the western Mediterranean region at 33 Ma. The distribution of the continental lithosphere is shown in (**gray**). Present-day coastlines are shown for reference. (**Black**) arrows represent direction and magnitude of relative motion. Strike–slip faults are shown in (**yellow**), labeled A, B, C. (**Red lines**) are convergent boundaries. Blue lines are divergent boundaries. (**White lines**) represent extinct spreading centers. (**Red triangles**) are volcanoes associated to the Tethys subduction; black triangles are volcanoes probably associated with the pyrenaic subduction. AlC: Alpine Corsica; SCB: Sardinian-Corsican block; WAA: Western Alpine Arc; modified after [18].

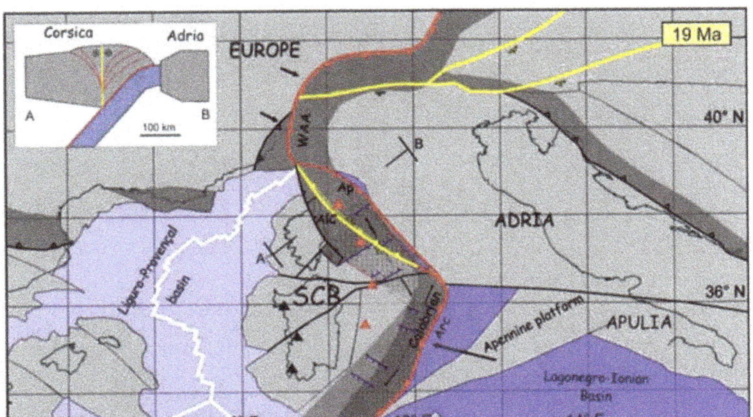

Figure 3. Plate reconstruction of the western Mediterranean region at 19 Ma (late Aquitanian). Dotted areas indicate wedge-top basins. Lower Miocene Chains are shown in (**dark gray**), the Africa-Adria continental lithosphere is in (**light gray**), the oceanic crust is in (**blue**). (**Red lines**) are active boundaries. (**Black lines**) are inactive boundaries. (**Yellow lines**) are strike-slip faults. (**Red triangles**) are volcanoes associated to the Tethys subduction; (**black triangles**) are volcanoes probably associated with the pyrenaic subduction. AlC: Alpine Corsica; Ap: proto-Apennine chain; SCB: Sardinian-Corsican block; WAA: Western Alpine Arc. Other symbols are the same from Figure 2; modified after [18].

At the end of the rotation, the Liguride slab was juxtaposed to the Corsica block and had dragged with it the deepest parts of the Calabrian accretion wedge, thereby a mélange of rocks belonging to the accretion wedge formed along the transpressive boundary. Complex structures, probably associated to the transpression, formed top-wedge basins, filled by sediments of the external Liguride flysch, today outcropping along the Tyrrhenian margin from Liguria to northern Calabria [30]. During the Burdigalian, the extension jumped from the western to the eastern margin of the Sardinian-Corsican block (future Tyrrhenian area) and the construction of the Apennine Chain continued further East. Therefore, the process of Tyrrhenian extension is strictly connected to the formation of the Apennine Chain. In current literature, the Apennine Chain extends from the Sestri-Voltaggio line to the Sangineto line (northern Calabria) [31–34] (Figure 4) and, on the basis of the paleogeographic domains involved in its structuring, it is subdivided in: (1) Northern Apennine, where the Ligurian allochthonous units extensively crop out [7,35–40], (2) Umbria–Marche Apennine, where sediments of the homonymous continental paleogeographic basin outcrop [7,38,41–43], (3) Lazio–Abruzzi Apennine, characterized by the presence of the homonymous Cretaceous carbonatic platform [44–49], and (4) Southern Apennine, which resulted from the deformation of the Campania–Lucania platform and the Lagonegro Basin [11,50–54]. The lateral continuity of the Apennine units is interrupted by the Calabrian arc along the Sangineto line and finds again its lateral continuity in the Maghrebide belt, which starts outcropping in Sicily, west of the Taormina line [33,55,56] (Figure 4).

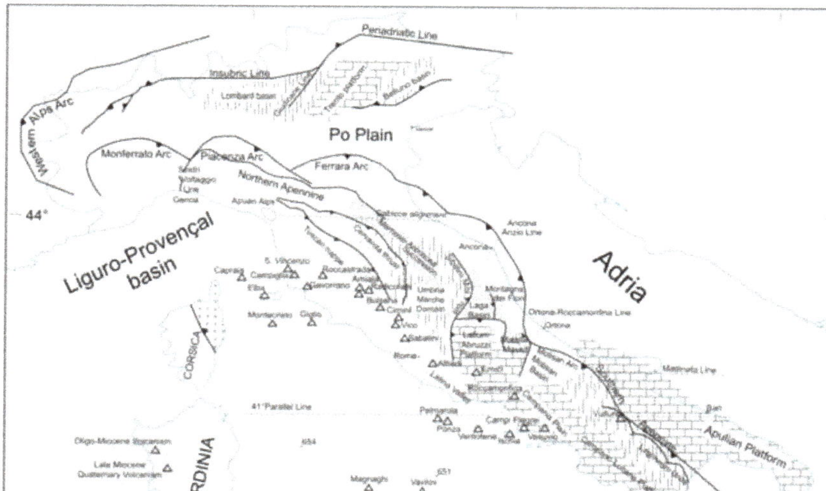

Figure 4. Location map of the main paleogeographic and structural units, modified after [28].

It is important to note the following key peculiarities of the Apennine chain: (a) the basement of the Adriatic margin is never involved in the structuring of the chain; (b) although the chain is exclusively made up of carbonate rocks deriving from the Adriatic domain, the foredeep and top-wedge sediments have a composition mainly referable to continental basement rocks [57]; (c) during its structuring, while the Apennine chain grows along the Adriatic front, it is subject to extension along the Tyrrhenian side.

2.2. The Tyrrhenian Sea

The Tyrrhenian Sea has a triangular shape, and its northern vertex is located in the proximity of the Elba Island. Starting from the northern tip of the triangle, the Ligurian-Provençal Sea extends westwards. From the end of the 1960s and up to the entire decade of the 1980s, the Tyrrhenian basin has been the subject of many scientific cruises (DSDP, ODP, and various seismic explorations). Its formation started after the cessation of sea floor spreading in the Ligurian–Provencal basin and, according to Malinverno and Rayan and Faccenna et al. [4,10], it was due to rollback of the subducting Ionian lithosphere and migration of the Calabrian Arc towards the southeast.

An E–W lineament extending from northern Sardinia to the Campania margin, known as the 41st parallel line, is suggested to be a lithospheric left-lateral transform fault that separates the Tyrrhenian Sea in two sectors [58]. The amount and directions of extension, as well as crustal and lithospheric thicknesses, are different to the north of the line with respect to the southern region (Figure 5).

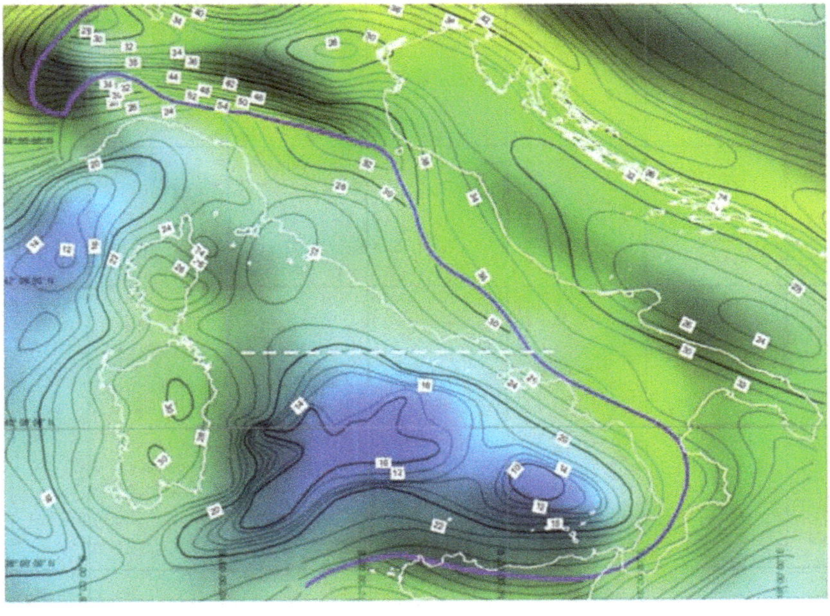

Figure 5. Moho map showing difference of crustal thickness to the north and south of the 41° parallel line. White dashed line is the 41°parallel line, violet line is the Moho discontinuity. [59].

In the sector north of the 41st parallel line, the continental crust is 20–25 Km thick and the lithosphere thickness is ~50–60 Km [60–62]. ODP site 654 [5] shows conglomerates covered by Tortonian, Messinian, and Plio-Pleistocene deposits. Conversely, the southern sector includes the Vavilov and Marsili basins, more than 3500 m deep, which have thin crust (25–10 Km or less) and lithosphere (30–50 Km) [60–62]. The Vavilov basin is characterized by a triangular shape, while the Marsili basin is almost squared. Both basins show magnetic lineaments comparable with the structural lineaments and the geometric shape of the basins (Figure 6).

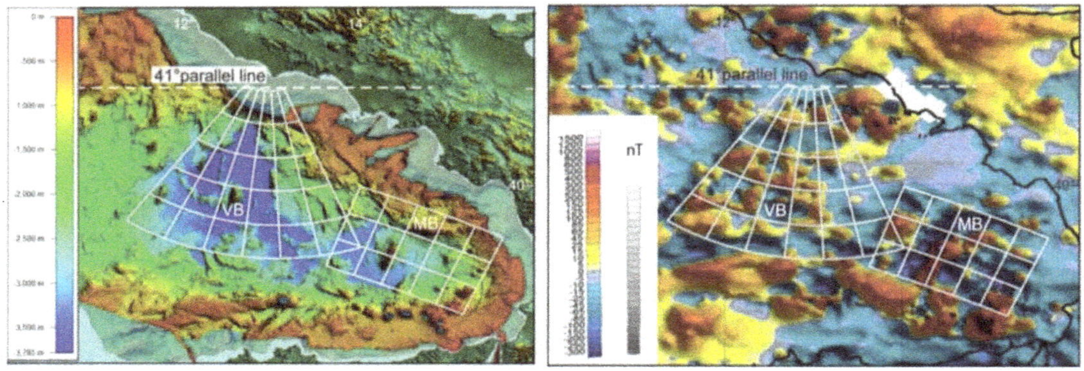

(a) (b)

Figure 6. (**a**) Bathymetry [63] and (**b**) magnetic anomalies [64] for the Tyrrhenian basin. White lines show the grid of the rotation pole for: Vavilov (VB) and Marsili (MB) basins. Both basins show magnetic anomalies comparable with the structural lineaments and the geometric shape of the basins.

According to some authors [23,65–67], rifting processes in this sector started along the Sardinia margin at ~10 Ma and then migrated eastward forming the Vavilov basin first (at 5.5 Ma) and later the Marsili basin (2.0–1.8 Ma). Despite the remarkable extension, the presence of oceanic crust is likely to be restricted within these two basins [1,68,69], as testified by the ODP sites 651 and 650. The first one shows, from the top, a 388 m thick Pliocene–Pleistocene succession (bio- zone MPL6/NN18, 2 Ma) above 39-m thick succession of dolostones, lying on 29 m of highly serpentinized peridotites covered by a 134 m thick succession of basalts (lava flows and breccia). The second borehole displays, starting from the bottom, 32 m of vesicular basalts followed by dolostones and a 602-m thick succession of Plio-Pleistocene deposits (bio zone MPL6/NN18, 2.0 Ma) (Figure 7).

Despite the great commitment in the exploration of the Tyrrhenian basin, the start time of the Tyrrhenian extension process is not well constrained. According to some authors [6,70], the time of cessation of sea floor spreading in the Ligure–Provençal basin is 19 Ma. Conversely, Speranza et al. [71] proposed that the rotation of the Sardinian-Corsican block ended at 15 Ma. These ages are important because they put a constrain on the timing of roll-back for the Adriatic slab. The end of sea-floor spreading in the Ligure–Provençal basin and the Algerian basin was determined by Marani [68] on the basis of marine magnetic anomalies. The end of rotation of the Sardinian–Corsica block can be also constrained by paleomagnetic data. While Marani [68] confirmed the age obtained by the analysis of marine magnetic anomalies (19 Ma) on the basis of a compilation of quality paleopoles, Gattacceca et al. [72] proposed a younger age of 15 Ma on the basis of new paleomagnetic data and 40Ar/39Ar dating. For the beginning of the Tyrrhenian extension, several authors used the ages of the stratigraphic successions from wells and seismic correlations to suggest that it took place during the Middle Tortonian times (~12 Ma) [5,66,73]. Therefore, the time elapsed between the end of the rotation of the Sardinian–Corsican block and the beginning of the Tyrrhenian rift is not sufficiently determined so far. However, it is important to note that the elapsed time between the onset of Tyrrhenian rifting and the beginning of sedimentation in the basin depends on the speed of extension in the rift and on the original thickness of the crust.

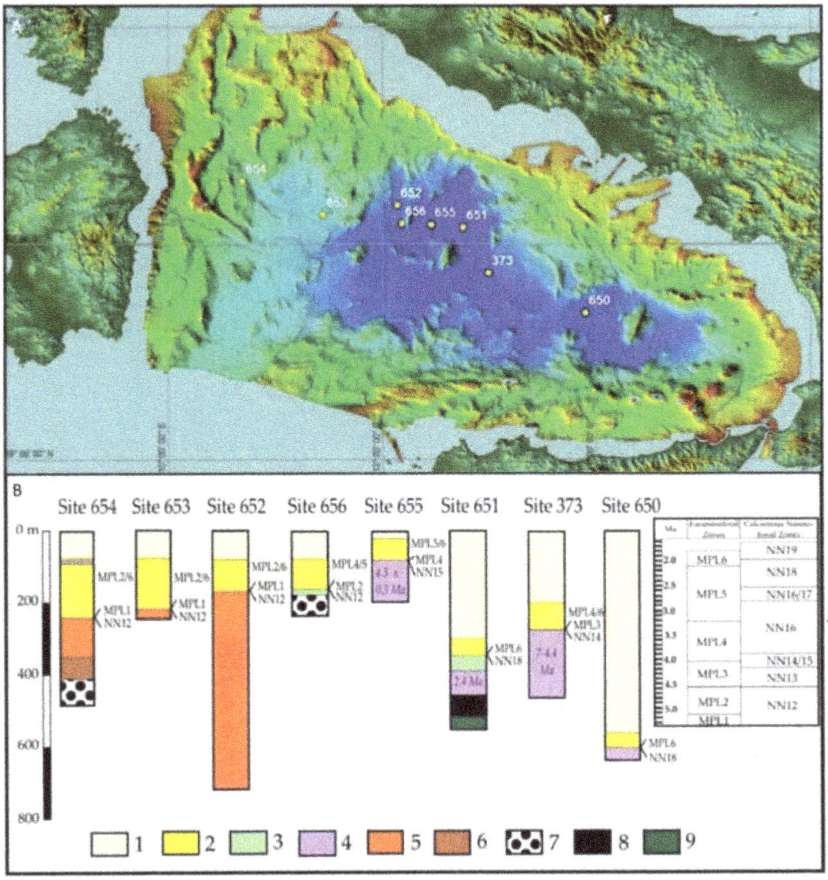

Figure 7. (**A**) Index map of Italy and Vavilov basin and DSDP and ODP well sites (**B**) Stratigraphic logs of the DSDP and ODP sites. (**1**) Pleistocene deposits; (**2**) Pliocene deposits; (**3**) dolostones; (**4**) basalts; (**5**) Messinian deposits; (**6**) Tortonian deposits; (**7**) conglomerates; (**8**) breccias; (**9**) serpentinized peridotites; after [69].

3. Methods

The fundamental tool for describing and measuring the deformation of the Earth's lithosphere, on a global scale, is the kinematics of tectonic plates. This allows to determine the path between the initial and final position of a plate with respect to another through the rotations of spherical caps about Eulerian poles [29,74] (Figure 8).

Plate kinematics has always encountered major obstacles in being used to represent deformation processes at the scale of mountain ranges, continental rifts and trascurrent boundaries, in the absence of magnetic isochrons. For this reason, palinspastic reconstructions along cross-sections at the macro-scale rely on the methods of structural geology. Retro-deformation (or crustal balancing) methods are well-known tools for the analysis of single transects, but are hardly applicable in complex areas characterized by the presence of triple junctions or polyphase systems. In these situations, the techniques of plate kinematics provide a convenient set of tools for describing the tectonic evolution and quantify the deformation of complex areas. On the other hand, structural geology can contribute to determine some fundamental parameters that must be specified to apply the plate kinematics approach, such as plate boundaries, kinematic indicators useful to determine rotation poles, and angles of rotation. It is important to specify which kind of structural data can be used to constrain the kinematics of large tectonic blocks. For example, single slip vectors observed along the Dead Sea fault zone cannot be used for determining the rotation pole of a microplate like Sinai. In fact, due to the heterogeneity of the rocks, these data record only strains generated by local stresses. We argue instead that the structural parameters that can be useful for determining rotation poles must be observed on geological structures at the same scale of the model that is being constructed.

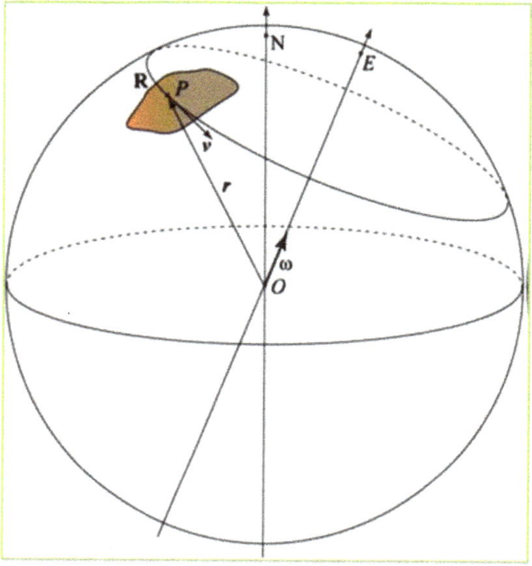

Figure 8. Geometry of the instantaneous motion of a tectonic plate **R**. *E* is the Euler pole, N is the North Pole. *P* is a representative point on **R**, whose instantaneous linear velocity is v. ω is the Euler vector of **R**; from [29].

For example, normal faults associated with an active rift such as the Tyrrhenian Sea represent a record that reflects the kinematics of extension. They exhibit sharp morpho-structural lineaments, easily observable on high-resolution DTMs or multibeam bathymetry. These features are generally devoid of "structural noise" generated by local paleo-stresses [75]. A shuttle radar topography mission (SRTM) image of the Tyrrhenian region shows lineaments that can be grouped into fan-shaped sets characterizing specific sectors of the Tyrrhenian and Apennine areas (Figure 9).

Thus, considering that these lineaments are expressions of several systems of normal faults associated with the Tyrrhenian extension, we can determine a set of Euler poles that describes the relative motion of different sectors of the Apennine chain. The classic rules of plate kinematics allow to assign these Euler poles on the basis of an estimate of the location of the convergence point of the observed lineaments [76]. All the Euler poles identified using this method describe the opening of the Tyrrhenian Sea in a European (Corsica–Sardinia) reference frame.

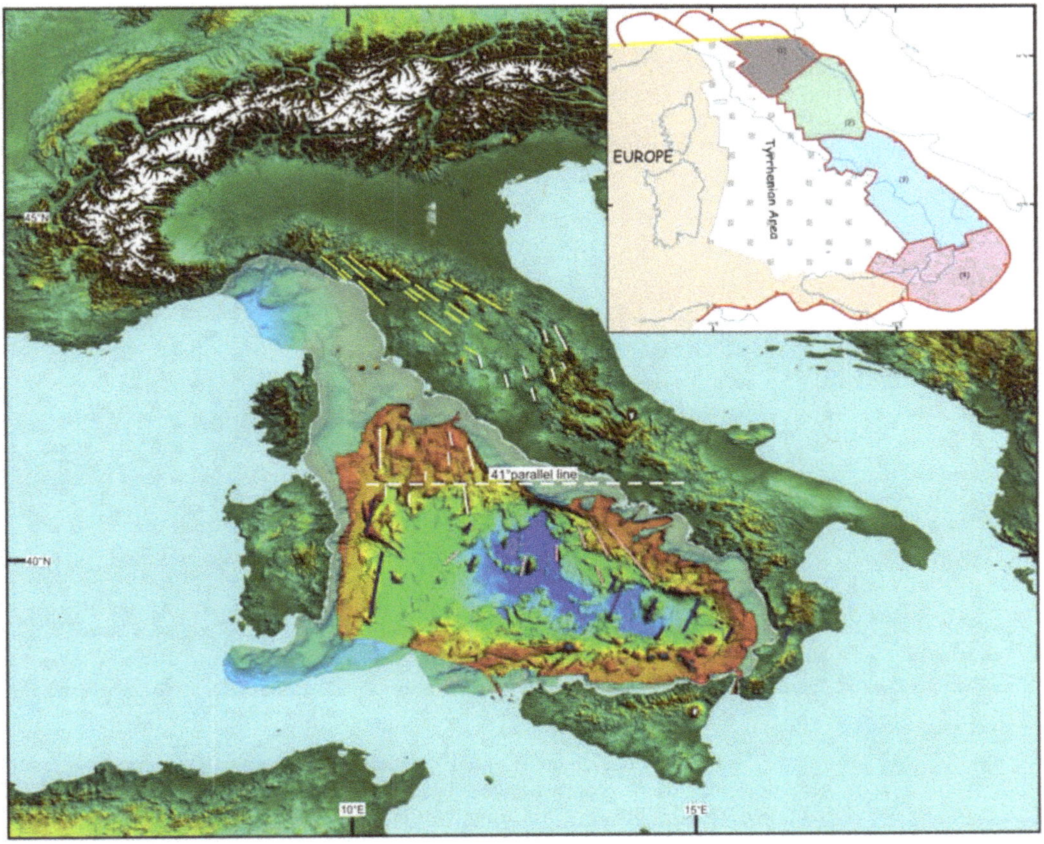

Figure 9. Morpho-tectonic map of the Tyrrhenian–Apennine system showing lineaments that can be grouped into fan-shaped sets characterizing specific sectors. Lineaments of the Northern Apennine sector are in (**yellow**; **white lines**) are for northern Tyrrhenian basin and Central Apennine sector; (**blue** and **pink lines**) are for Calabrian and Southern sectors respectively. The multibeam bathymetry is after [63]. The insert box shows the deformable blocks identified: (1) Northern sector in (**dark beige**), (2) Central Sector in (**green**), (3) southern sector in (**blue**); (4) Calabrian Arc Sector in (**pink**). Dotted areas are in extension; (**white dotted**) area is the extended Tyrrhenian area.

3.1. How to Identify Tectonic Elements in the Apennine Domain

Tectonic elements are semi-rigid crustal blocks, bounded by faults, which had an independent kinematic history in the geologic past [77]. Every single block of the Apennine area is characterized by boundaries that are divergent along the Tyrrhenian side and convergent along the Adriatic–Ionian side. They result from rotations about fixed Euler poles in the Corsica–Sardinia reference frame. The boundary between two adjacent blocks is represented by structural systems, usually transversal to the chain, that are the expression of continuously changing Euler poles in the same reference frame. The faults associated with the Tyrrhenian side boundaries represent eastward migrating fan-shaped extensional systems. On the Adriatic side, the frontal segments of the Apennine chain represent the eastern boundaries of the tectonic elements. On the basis of the lineaments identified for the Tyrrhenian area, we defined the deformable blocks shown in Figure 9, which extend along the Tyrrhenian side and grow by accretion along the Adriatic margin. Boundaries between adjacent blocks are not easily identified, because the Euler pole of relative motion is always changing and depends on the ratio of angular velocities (in the Sardinia–Corsica reference system) between pairs of tectonic elements. Boundaries with extensional kinematics produce systems of faults that break the continuity of the chain, forming transversal basins that are simultaneously included in the Apennine orogenic system and TTR-like and FTR-like triple junctions (Figure 10). Boundaries with strike-slip kinematics produce more complex structures and can form pull-apart basins or depressions. In some cases, for example the Ancona–Anzio line (Figure 4), the boundary can have compressive or transpressive kinematics. On the Tyrrhenian side, RRR-like triple junctions form at the intersection of rift axes (Figure 10).

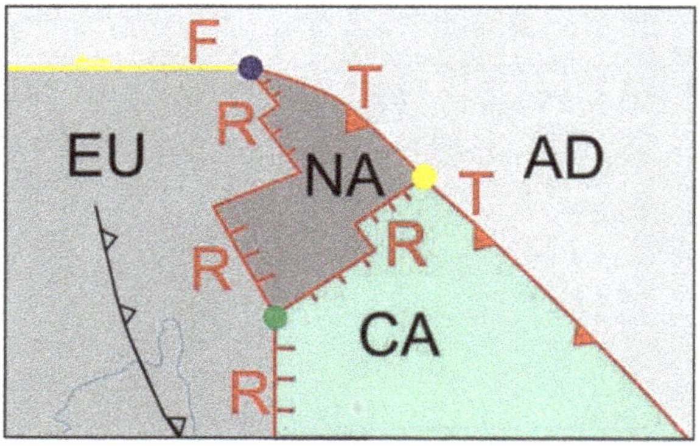

Figure 10. Scheme showing TTR-like (Trench-Trench-Ridge) and FTR-like (Fault-Trench-Ridge) triple junctions of the Apennine and RRR-like (Ridge-Ridge-Ridge) triple junction of the Tyrrhenian side. AD: Adria, CA: Central Apennine, EU: Europe, NA: North Apennine, T = Trench, R = Ridge, F: Fault; (**yellow dot**) is the TTR-like triple junction; (**green dot**) is the RRR-like triple junction, (**blue dot**) is the FTR triple junction.

The blocks identified in this way are highly deformable and it is not possible to identify any portion inside them that can be considered rigid. In general, the eastward migrating extensional axes overlap with some delay to the compressional structures, determining their collapse. Conversely, in global tectonics the deformed areas along plate margins represent bands with negligible surface compared to that of the entire rigid lithospheric plate. Therefore, the tectonic elements considered here behave like portions of large plate margins devoid of their rigid counterpart. However, considering the scale of the tectonic reconstructions, the internal deformation of the tectonic elements is not relevant, in our opinion, for the application of the method. The boundaries of the Apennine blocks cannot therefore be considered fixed over time. We can say that tectonic elements do not represent rigid polygons but are characterized by the rotation pole about which the lines and points contained in them rotate.

3.2. *How to Determine Rotation Poles of Apennine Blocks*

The stage poles between conjugate oceanic plates are determined building magnetic isochrons [29]. In continental tectonics, where magnetic lineaments are not available, the rotation pole between two divergent plates is generally determined qualitatively tracing central meridians of at least two parallels passing from conjugate points on the two undeformed margins [76]. Then, the Euler pole is the central point of a cloud of intersection points between meridians. Unfortunately, this method cannot be applied to the Tyrrhenian area, as it is not possible to recognize conjugate points belonging to the margins of Corsica–Sardinia and the Apennine domain, while there are no clear magnetic lineaments in the Tyrrhenian Sea. Therefore, the finite poles of rotation of the Apennine blocks can only be estimated by taking intersections of great circle arcs associated with morpho-structural lineaments of the Tyrrhenian rift. Finally, specialized software for plate kinematics (e.g., PCME-Paleo Continental Map Editor [78] or GPlates [79]) can be used to build Euler pole grids that show parallels and meridians about every single rotation pole. Parallel circles represent flow lines of relative motion, while meridians represent the trend of extensional structures. These grids can be superimposed on the observed morpho-structures to assess the correctness of the Euler poles.

3.3. How to Determine an Angle of Finite Rotation and the Start and End Times of Rotation

The angle of rotation about a Euler pole is a function of finite strain and can be determined through crustal balancing of transects along flow lines of relative motion, granted that an estimate of the initial thickness is available. The technique is described in [29] and can be easily applied to the restoration of rifted continental margins and the reconstruction of pre-rift configurations. In the case of migrating mountain belts, the angle of rotation does not need to be determined through crustal balancing, because it can be observed directly measuring the angular distance between the present day location of a point along the thrust front and a homologous point along the estimated location of the thrust at an earlier age. Two points along the thrust front at different ages are considered homologous when the more recent point can be obtained rotating the older one about the Euler pole. In this instance, crustal balancing can be applied a posteriori to obtain the initial thickness of the chain. For example, we estimated an initial thickness of ~42 km for the central Apennine chain through the crustal balancing procedure illustrated in Figure 11. We considered a reference point in the Sardinian–Corsican block (blue point FPL in Figure 11), which was considered at the western end of the extensional area, and a homologous point along the Sibillini mountains (red point ML00 in Figure 11), representative of the eastern end the rift. The angular distance between these two points is 13°, corresponding to 180 km along the small circle arc linking the two points. The crustal balancing procedure was performed using the Moho depths of [59] (Figure 5), which have an uncertainty of 2–4 km in this area.

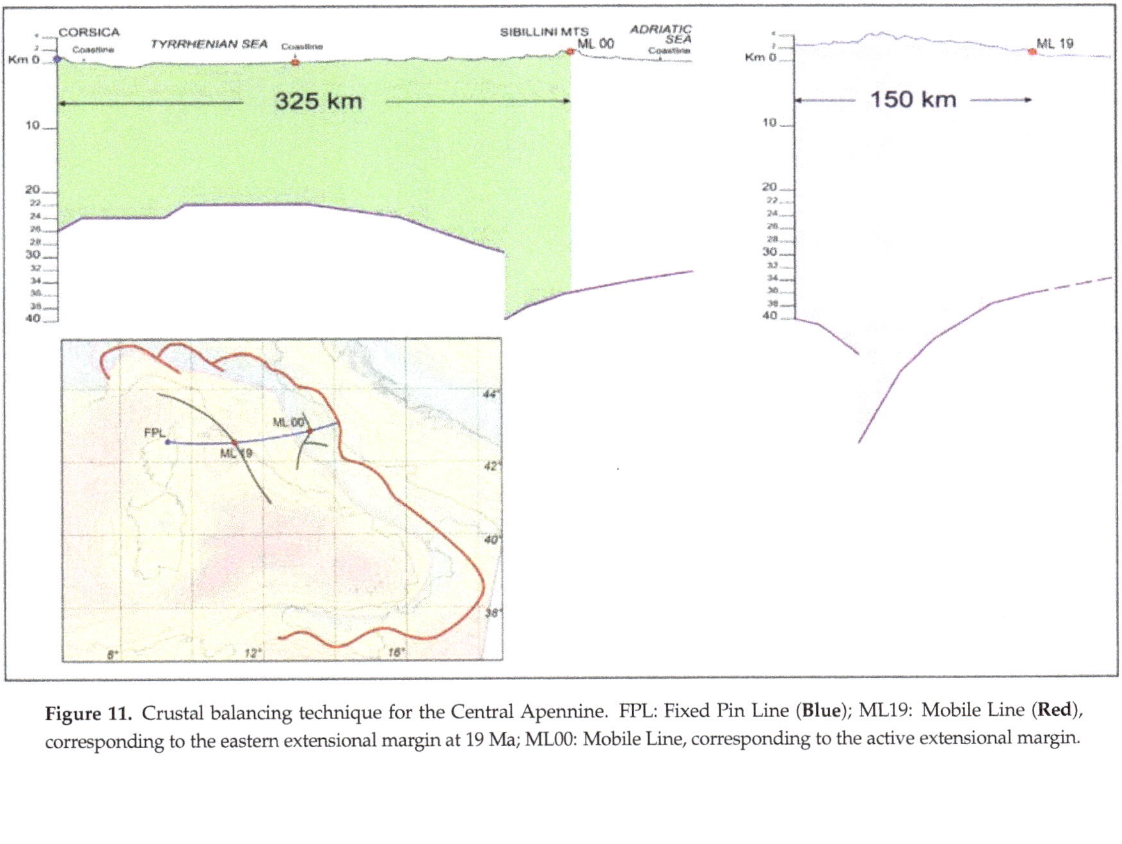

Figure 11. Crustal balancing technique for the Central Apennine. FPL: Fixed Pin Line (**Blue**); ML19: Mobile Line (**Red**), corresponding to the eastern extensional margin at 19 Ma; ML00: Mobile Line, corresponding to the active extensional margin.

To determine the velocity vectors, it is also necessary to know the timing of deformation. This parameter can be estimated by the analysis of stratigraphic successions that record the syn-rift tectonic activity. There is a wide array of literature for the Tyrrhenian basin with data deriving from well stratigraphy, seismic surveys, and dredging (see Figure 7) [1,67,80–82]. Furthermore, there are many other data on successions of the Apennine and foredeep basins that have recorded the time of tectonic activity of the chain [39,83].

3.4. Apennine Chain and Tyrrhenian Sea Sectors

We distinguished six groups of homogeneous structural systems, which characterize chain sectors. From north to south, they are: (1) Northern Apennine; (2) Umbria–Marche Apennine Arc; (3) Southern Apennine; (4) Calabrian Arc; (5) Sicilian Chain [27].

3.4.1. The Northern Sector

The Northern sector extends north of Elba Island and includes eastern Ligurian Sea, Northern Apennine and the area of the Tuscan rift. We consider the Northern Apennine as the northern limb of the Umbria–Marche Apennine Arc, running from the Sestri–Voltaggio line to the Gabicce alignment (Figure 4) and formed by chaotic sediments deriving from oceanic covers and boudins-like structures of Liguride ophiolites and their thick terrigenous top-wedge successions. The chain front of this sector is buried under Po river valley sediments. In the Corsica basin, syn-rift sediments are aged between Messinian and Oligocene [1,82,83]. The rift area is also characterized by the well-known Tyrrhenian-Tuscan magmatism; the age of the magmatism grows from east to west. From the structural point of view, this sector shows a fan of lineaments converging towards NW. The resulting Euler pole is located at 45.45° N, 8.40° E (Figure 12).

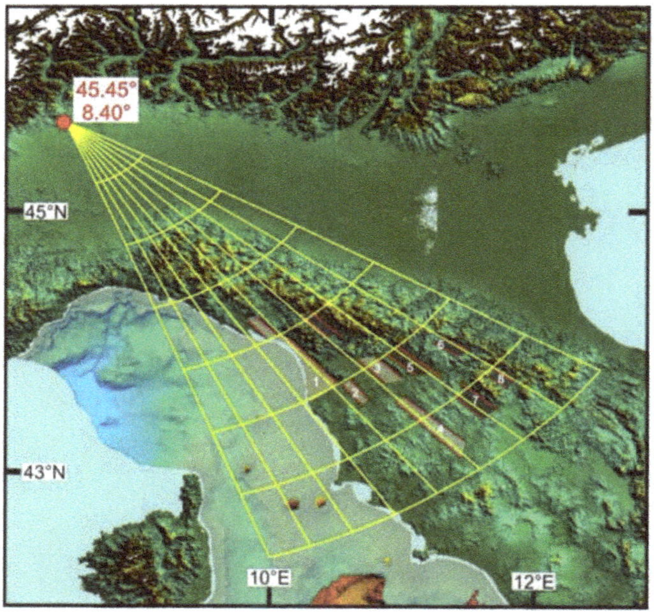

Figure 12. Morpho-structural lineaments of the northern Apennine and Euler pole of relative motion with respect to Sardinia-Corsica (**red dot**). The Euler pole grid shows the goodness of fit between kinematic model and geological structures. Some basins associated with the Tyrrhenian extension are also shown. Mio-Pliocene basins are in (**light gray**), Plio-Pleistocene basins are in (**dark gray**). (1) Viareggio basin, (2) Volterra basin, (3) Elsa basin, (4) Siena basin, (5) Firenze basin, (6) Mugello basin, (7) Valdarno basin, (8) Casentino basin.

3.4.2. Central Sector

The Central Sector extends from Elba Island to the 41° parallel and includes the northern Tyrrhenian Sea, the Umbria–Marche-Abruzzi Apennine Arc and the areas of Umbria-Lazio and Southern Tuscan rifts. The Umbria–Marche Apennine Arc is located between the Gabicce alignment (Figure 4) and the Ancona–Anzio line and is composed of deposits from the Adriatic margin (the well-known Umbria–Marche succession). The Umbria-Marche succession consists of Jurassic-Middle Miocene basinal sediments. The eastern part of the arc is the Ancona-Anzio line during the first phase of evolution and lately the Maiella front, also called "Ortona-Roccamonfina line" [34,51,84] (Figure 4). Between the two lines there is a thick Mesozoic carbonatic platform succession known as Lazio–Abruzzi platform. Tyrrhenian rift structures in this sector form lineaments converging toward north (Lat. 48.9°; Long. 10.4°; Figure 13). The Lazio margin of this sector is also affected by a Pleistocene–Oligocene volcanism.

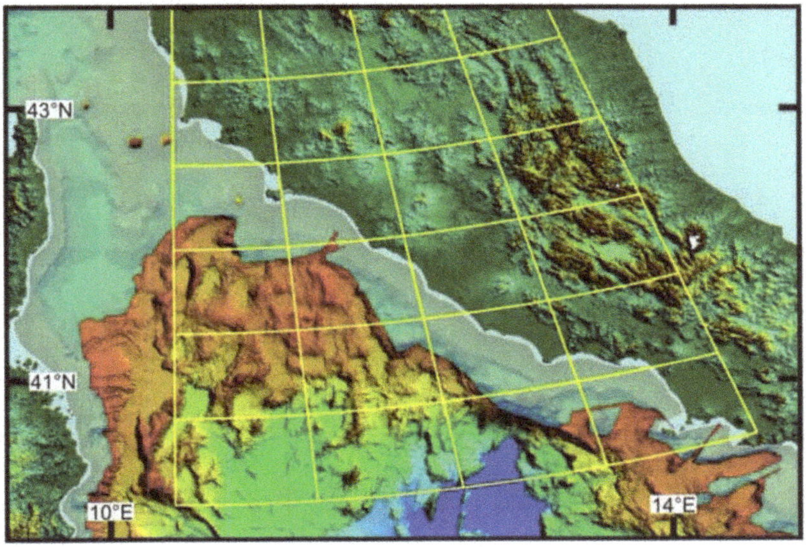

Figure 13. Morpho-structural lineaments of the central Apennine and Euler pole grid of relative motion with respect to Sardinia-Corsica. The Euler pole grid shows the goodness of fit between kinematic model and geological structures. The Euler pole has coordinates: 48.9° − 10.4°.

The interaction between the northern and central sectors is responsible for the formation of transversal lineaments (orthogonal features) that created rectangular basins and depressions (chocolate bar), e.g., the Trasimeno Lake [85] (Figure 14).

3.4.3. Southern Sector

The southern sector runs from south of the 41° parallel to the Sangineto line (northern Calabria) and includes Vavilov basin and the entire Southern Apennine. The Southern Apennine is composed, from the bottom to the top, of Middle Triassic—Lower Cretaceous Lagonegro units and of carbonatic Upper Triassic—Eocene Panormide units. In some places, the Panormide units are overlapped by ophiolites and oceanic sediments of the Liguride basin, covered by lower Miocene top-wedge sediments [86,87]. On the external chain front, at the boundary with the Central sector, there is a minor arc known as "Molise arc", extended from the Maiella front to the Vulture volcano. It is made of Jurassic-Cretaceous pelagic successions and of carbonatic platforms, overlapped by Mio-Pliocene chaotic and top-wedge successions [50]. Structures of the Tyrrhenian rift of this sector form a fan of lineaments that draw the triangular shape of the Vavilov basin, whose vertex is located at 41.23° N, 13.01° E (Figures 9 and 15).

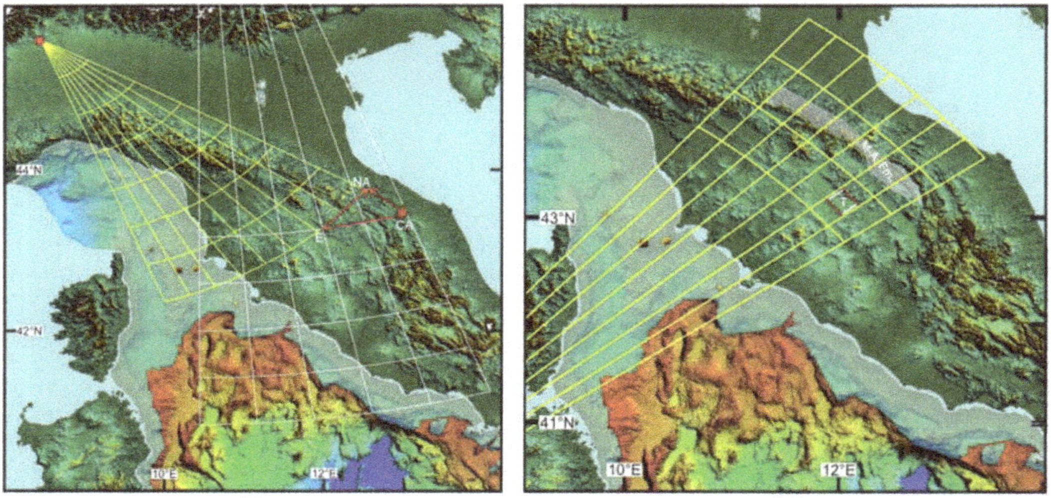

(a) (b)

Figure 14. (**a**) Interaction between northern (**yellow grid**) and central (**white grid**) sectors. Red arrows are velocity vectors of the relative motion between the two sectors. E: Europe, NA: Northern Apennine, CA: Central Apennine. (**b**) Comparison between morpho-structures and the Stage pole grid valid from 5 Ma to present, between Northern Apennine and Central Apennine sectors. T: Trasimeno Lake, M-A Fm: Marnoso-Arenacea Formation. The Euler pole has coordinates: 38.95° −5.61°.

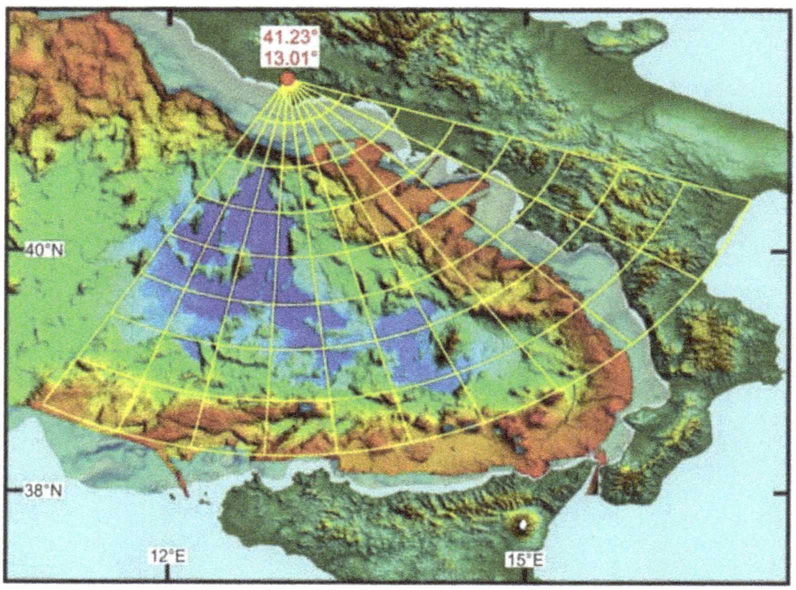

Figure 15. Morpho-structural lineaments of the southern sector and Euler pole of relative motion with respect to Corsica (**red dot**). The Euler pole grid shows the goodness of fit between kinematic model and geological structures.

The interaction between the southern and central sectors is responsible for the formation of transversal lineaments (ex. Gran Sasso front, Ancona–Anzio line) and basins (Laga basin) (Figure 16a). The southern sector is also characterized by complex extensional structures expressed by transverse features that cut, on the Apennine margin, the fan of the Vavilov basin. The transverse structures are the result of the interaction with the Calabrian arc sector and form basins and depressions since the Pleistocene. Some of these contain considerable thicknesses of sediment (ex. Sant'Arcangelo basin) [88,89] (and reference therein) (Figure 16b). The southern sector also shows a strong magmatism, both on the Tyrrhenian portion and on the Campania margin. A volcano located in the Apulian foredeep of the chain (Mount Vulture) belongs to this sector.

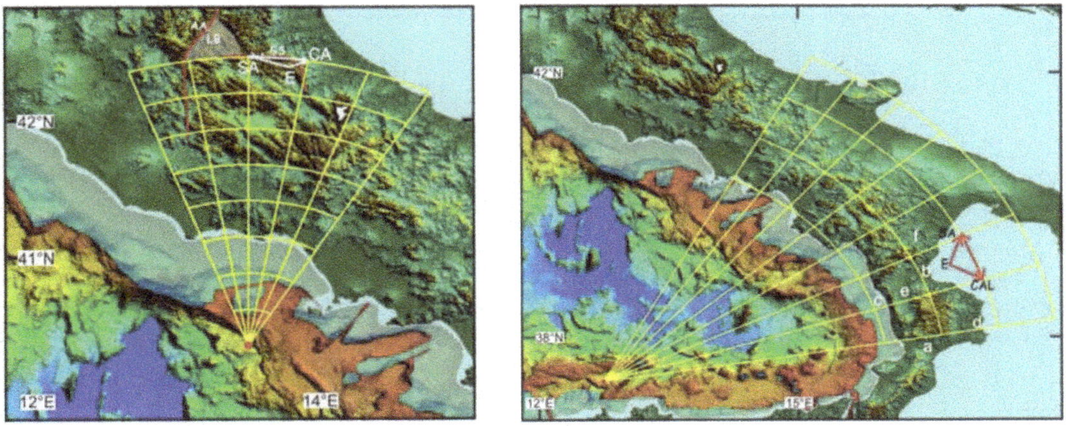

(a) (b)

Figure 16. Comparison between morpho-structures and the Stage pole grid valid from 5 Ma to present, between: (**a**) Central Apennine and southern sectors. The Euler pole has coordinates: 40.48°−13.52° (**b**) southern and Calabrian arc sectors. The Euler pole has coordinates: 38.43°−12.80°. a: Catanzaro trough, b: Sibari basin, c: Paola basin, d: Crotone basin, e: Crati valley, f: Sant' Arcangelo basin. Red and white arrows are velocity vectors of the relative motion between sectors. AA: Ancona-Anzio line, CA: Central Apennine, CAL: Calabrian Arc, E: Europe, LB: Laga Basin, GS: Gran Sasso front, SA: Southern Apennine.

3.4.4. Calabrian Arc Sector

The Calabrian Arc sector is bordered to the north by the Sangineto line and to the south by the Taormina line and includes the southern Tyrrhenian Sea (Marsili Basin). Structural lineaments converge towards SW. The Calabrian Arc includes the Coastal Chain, the Sila Massif, Le Serre, Aspromonte and Paleoritani mountains in Sicily. This segment of the chain is an accretion wedge formed, from bottom to top, by: Apennine carbonatic units, ophiolites (Liguride units), units consisting of low-grade metamorphic rocks with high-grade rocks on top, derived from continental crust with local flaps of Upper Trias–Upper Cretaceous sedimentary deposits. The Tyrrhenian extensional structures are represented by the lineaments that characterize the Marsili basin, the southern Calabria and the Strait of Messina. These lineaments form a mild fan with vertex towards SW (Lat. 21.85°; Long. 6.28°; Figure 17), which can be superimposed to the magnetic lineaments of the Marsili basin (Figure 6b). The Calabrian arc is also characterized by important transverse features that express extensional structures, essentially present in Northern Calabria. The most evident have generated the Catanzaro trough, the Sibari basin and also the Paola basin, the Crotonese Basin, and the Crati Valley, which are delimited by structures transversal to the arc and by high-angle N-S trending structures (Figure 16b).

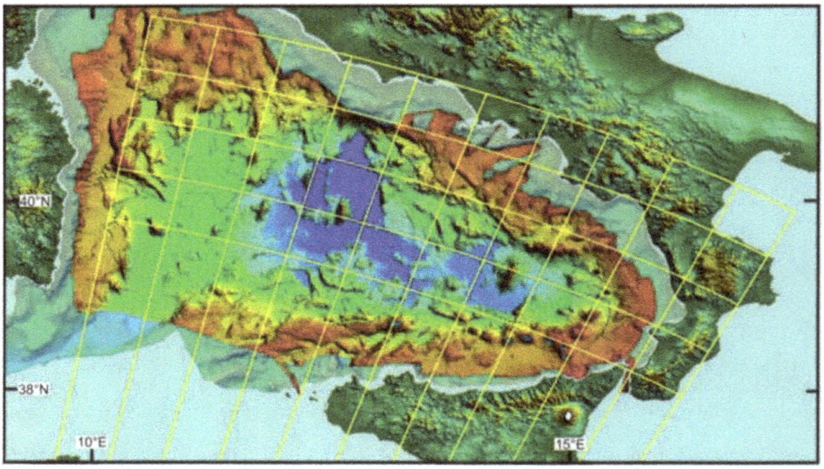

Figure 17. Morpho-structural lineaments of the southern Tyrrhenian and Euler pole grid of relative motion with respect to Sardinia-Corsica (**red dot**). The Euler pole grid shows the goodness of fit between kinematic model and geological structures. The Euler pole has coordinates: 21.85° 6.28°.

3.4.5. Sicilian Sector

The Sicilian sector is represented by the chain located west of the Taormina line. It is made up of Meso-Cenozoic basin and carbonate platform deposits, belonging to the African margin, overlapped by: sediments of the Middle Triassic-Jurassic/Cretaceous Imerese basin, sediments belonging to the Late Triassic-Eocene Panormide carbonatic platform, and Sicilidi Units covered by Oligo-Miocene sediments [90–93].

The Sicilian chain, with southern vergence is sharply divided in a northern internal sector and a southern external one by a long E-W trending lineament, known as "Monte Kumeta- Alcantara fault". This is considered a high-angle strike-slip structure with right-lateral kinematics [94]. The internal sector includes Nebrodi Mounts, Madonie and Palermo Mounts. The external sector includes Trapanese and Saccense Units, covered by sediments that filled the great depression of the Caltanissetta basin that ends south on the Gela foredeep. On the west side the foredeep is sharply interrupted in the proximity of Sciacca Mounts. To the East it passes through the Ragusa foreland and ends to the Etna Mount. The Caltanissetta basin is filled by thick Messinian and Pliocene successions and is highlighted by a strong gravimetric anomaly. This evidence suggests the presence of extensional structures that delimited the basin to the West and East [92].

The Sicilian sector, despite the Apennine sectors that show very visible and clear Tyrrhenian extensional structures, is characterized by complex structures. The Tyrrhenian offshore includes post-Messinian basins interposed between the Elimi chain and the Sicilian northern margin. The chain, E-W oriented, ends to the East with Ustica Island, where volcanic islands (Alicudi, Filicudi, Salina) mark the continuation of the lineament towards the East (Figure 4). Morpho-structures along the chain show characteristics similar to transpressive structures. Seismic profiles [92] and compressive earthquakes confirm this hypothesis. Therefore, the Tyrrhenian margin of Sicily, unlike the Apennine one, is characterized by a second phase with transpressive kinematics. To the south, the boundary with the Africa plate coincides with the Sicily channel, characterized by right transcurrent structures [95,96]. The two features delimiting Sicily converge towards west. Considering the dextral kinematic in the Sicily channel, is possible to deduce that the entire Sicily, compressed between Africa and Europa plates, is extruded towards East. The eastern side of the triangle forms the third boundary of the Sicily micro-plate. This boundary is divided in two segments: north of the Etna Mount the boundary is between Sicily and Calabrian Arc, to the south instead the boundary is between Sicily and African plate and is represented by the Malta escarpment. Along the Malta escarpment, the Ionian slab starts to break [28], interrupting the continuity between Sicilian and Ionian lithosphere. The complexity and the numerous implications of this process deserve to be dealt in a separate paper.

3.4.6. Rotation Model

The tectonic reconstructions proposed here can be considered as a refinement of the regional kinematic model of the western Tethys proposed by [18,19,27,28]. They were made using an interactive computer software for plate kinematic modelling [78]. This software allows to create and maintain data sets of plate boundaries and relative plate positions, and to test the predicted plate motions through the generation of velocity fields and inferred tectonic structures.

As explained above, the classic techniques of rigid plate kinematics are not applicable in the Tyrrhenian–Apennine area. To release the block rigidity constraint, we did not use polygons to represent the different Apennine blocks. The classic polygons were replaced by linear elements that allowed to implicitly define areas with changing geometry. Therefore, each sector in the rotational model includes two separate objects that rotate about the same pole but with different angles of rotation. The western side of a block rotates with a velocity defined by the local amount of Tyrrhenian extension, while the eastern side of the arc front rotates by a higher angle that represents the accretion rate of the corresponding sector. Euler poles for the Apennine sectors are shown in Figure 18 and can be found in Table 1.

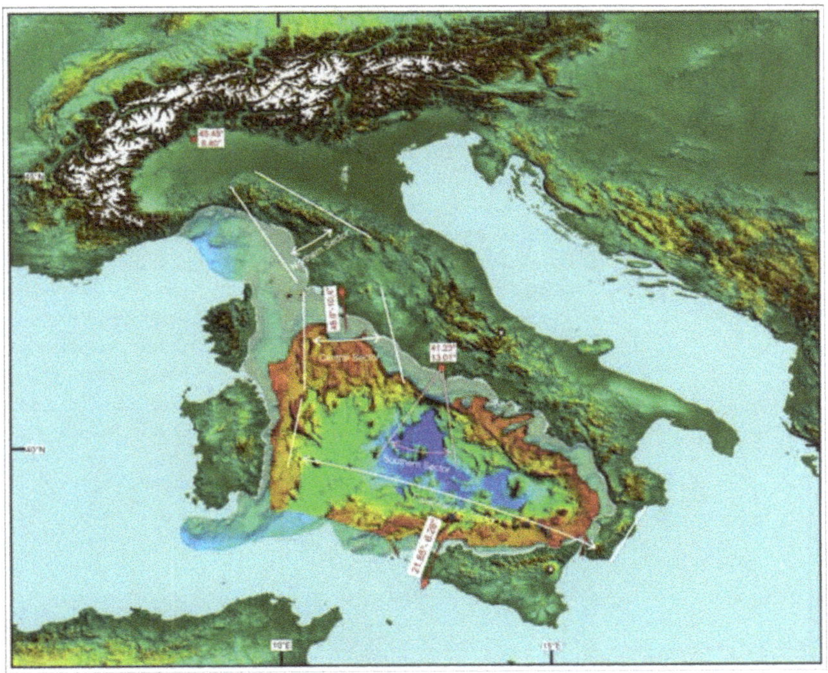

Figure 18. Morpho-tectonic map and multibeam bathymetry of the Tyrrhenian–Apennine system, showing extension directions in the Tyrrhenian Basin. (**White** and **pink lines**) show extension started respectively 19 Ma and 7 Ma. (**Red points**) are Euler poles. (**Red arrows**) indicate the Euler poles out of the figure, coordinates of these poles are shown. Real angle of rotation is shown for each pole. Modified after [27]. The multibeam bathymetry is after [63].

Table 1. Finite reconstruction parameters, from [27].

NAME	PLATE ID	TIME [Ma]	LAT [deg]	LON [deg]	ANG [deg]	REF. ID
Northern Sector Rear	384	19	45.45	8.40	−20	301(Europe)
North. Sect. Front	387	19	45.45	8.40	−38	301
Central Sector Rear	360	19	48.90	10.40	−13	301
Cent. Sect. Front	357	19	48.90	10.40	−18	301
Southern Sector Rear	383	7	41.23	13.01	−45.00	301
South. Sect. Front	390	7	41.23	13.01	−54	301
Calabrian Arc Rear	381	1	21.85	6.28	+0.46	301
Calabrian Arc Rear	381	7	21.85	6.28	+7.0	301
Calabrian Arc Rear	381	19	21.85	6.28	+12.00	301
Calab. Arc Front	388	1	21.85	6.28	+0.507	301
Calab. Arc Front	388	7	21.85	6.28	+8.75	301
Calab. Arc Front	388	19	21.85	6.28	+15	301
Sicilian Sector	394	2	0.00	0.00	+0.00	301
Sicilian Sector	394	2	−21.85	−173.72	1.55	381
Sicilian Sector	394	4	−16.43	−169.662	2.523	381
Sicilian Sector	394	7	49.408	33.075	1.398	381
Sicilian Sector	394	12	60.813	136.123	+0.773	381

The construction of the rotation model of Table 1 (rotation model.rot in Supplementary Material) was performed as follows. Although the Euler poles associated with each tectonic stage could be determined by fitting the observed morpho-structures, we did not know a priori the corresponding angles with the exception of the rotation angle of the central Apennine with respect to Sardinia-Corsica. Therefore, we first determined the ratios between the stage angles fitting the predicted relative motion between adjacent sectors to the observed geologic structures.

Therefore, with the exception of the central sector, the rotation angles of the other tectonic elements forming the Apennine chain were determined considering the velocity ratios between adjacent blocks, which control the kinematics and the trend of the structural associations that characterize the boundaries. In practice, we chose the rotation angles in such a way that the resulting velocity ratios were compatible with the observed geological structures along the boundaries. Figures 19 and 20 summarize the timing of deformation and amount of extension on the basis of the available literature and the proposed kinematic model.

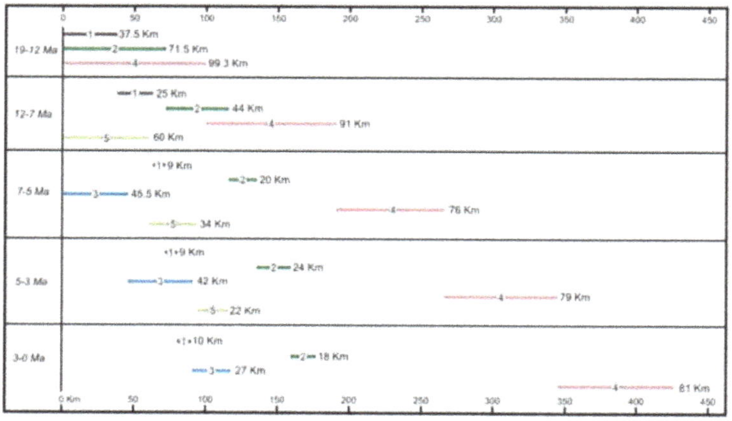

Figure 19. Estimated extension and timing along representative small circle arcs about the Euler poles of sectors 1–5 with respect to Sardinia-Corsica. Brown line (**1**): Northern sector, dark green line (**2**): Central sector, blue line (**3**): Southern sector, pink line (**4**): Calabrian arc sector, light green line (**5**): Sicilian sector.

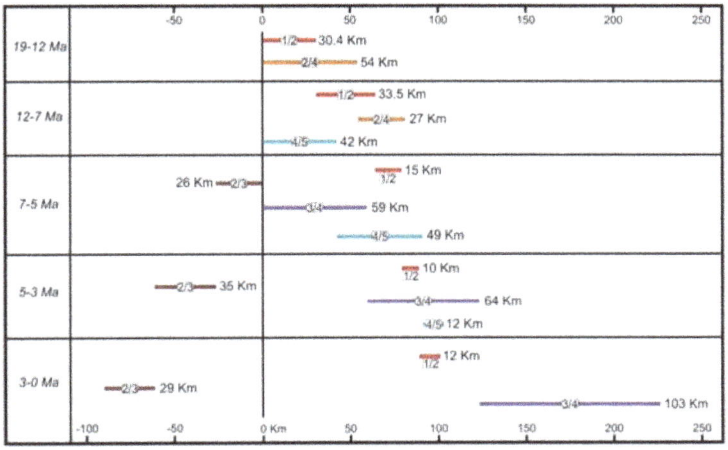

Figure 20. Estimated total shortening/extension and timing between adjacent sectors of Figure 19.

4. Results: Kinematic Evolution of the Tyrrhenian–Apennine System

4.1. How the Apennine Chain Evolves

The process of evolution of the Apennine Chain is originated by two big tectonic events: the rotation of the Sardinian–Corsican block and the formation of the Tyrrhenian basin.

4.1.1. First Apennine Event

According to [18,70], the first phase of the Apennine evolution (33–19 Ma) generated a long left-lateral transpression along the boundary between the Adriatic Plate and the Sardinian-Corsican block. This transpression, with a major trascurrent component, drove apart the Western Alpine Arc and the Calabrian–Kabylide Arc, which aligned with opposite vergence when the rotation started (Figure 3).

At the end of this phase the transpressive structure reached its maximum length [18]. The tectonic mélange that resulted from the deformation of the accretion wedges of the two arcs, locally covered by top-wedge successions (external Liguride flysch), was the proto-Apennine Chain (Figure 3). Rocks belonging to this proto-chain outcrop in the actual Apennine chain along the Tyrrhenian margin from Liguria to northern Calabria (Falda Toscana, Apuane Alps, Argentario, Zannone island, and Cetraro and Verbicaro Units in Calabria). During this phase Adria was migrating toward NNW, while its slab was sinking into the upper mantle. As a consequence, at the end of the rotation phase the Adriatic slab was juxtaposed to the Corsica block, together with the deep portions of the Calabrian accretion wedge. The upper portion of the Calabrian wedge was still attached to the Calabrian arc.

On the southern side, the rotation phase of the Sardinian–Corsican block produced a huge stretch that was thinning the Calabrian–Kabylides Arc in its central portion. Moreover, considering the huge amount of volcanic sediments present in the internal Apennine flysch and external Ligurids successions [97], we suppose that during the rotation of the Sardinian–Corsican block a second volcanic arc was forming due to the slab of Adriatic Plate. This arc was located on the eastern side of the block attached to the first embryo of the Apennine chain.

4.1.2. Second Apennine Event

We describe below the Tyrrhenian phase that results from the implications of the kinematic model described above. According to this model, this phase starts at the end of the Sardinia-Corsica block rotation, with an eastward jump of the axis of extension in the future Tyrrhenian area (Figure 21 and Plate 1). The beginning time of the Tyrrhenian rifting process is not well-constrained. The oldest stratigraphic record, made up of marine sediments in the western Tyrrhenian margin, is middle Tortonian in age [67,73] (and reference therein). Thus, the onset of rifting could be a few million years older. We assume the ending age of the Sardinia–Corsica block rotation (19 Ma) [70] as the starting age of the rifting phase, although early Burdigalian sediments can be found in the Corsica Basin, thereby a basin already existed at that time to the east of Corsica. In our interpretation, the sediments associated with the N6 biozone (18 Ma) represent the first syn-rift deposits in the Corsica basin [82], unconformably placed on older wedge-top basin sediments of the left-lateral transcurrent structure separating Corsica from Adria. However, it is important to note that the uncertainty about the starting age of Tyrrhenian extension does not affect the rotation model proposed here, because the structuring of the Apennine chain depends on the rotation speed ratios between the kinematic elements of the puzzle and not on their absolute speeds.

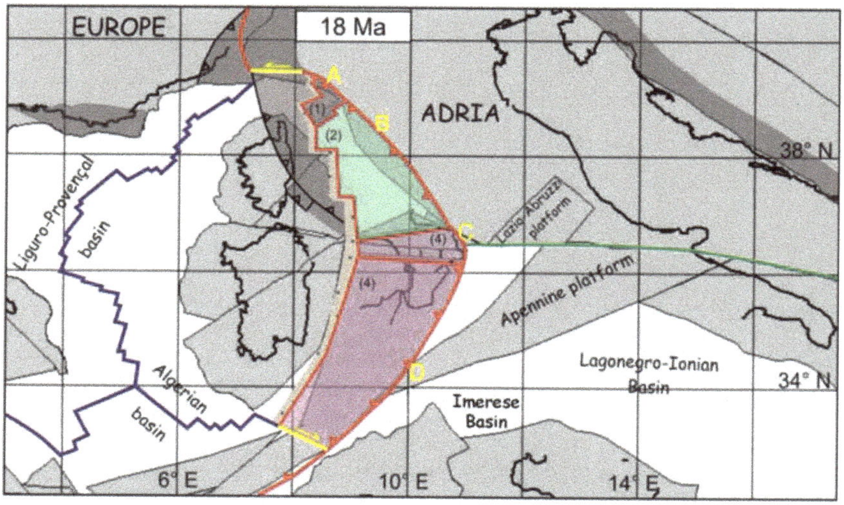

Figure 21. Plate reconstruction of the western Mediterranean region at 18 Ma. The distribution of the continental lithosphere is shown in (**gray**). Present-day coastlines are shown for reference. Strike–slip faults are shown in (**yellow**). (**Green line**) is an E-W directed strike-slip structure that separated Adria from Apulia from the late Cretaceous to the Eocene. (**Red**) lines are divergent boundaries, red lines with triangles are convergent boundaries. A: northern Apennine arc, B: central Apennine arc, C: northern Calabrian arc, D: southern Calabrian arc. The Northern Sector (**1**) is shown in (**dark beige**); the Central Sector (**2**) is in (**green**); the Calabrian Arc Sector (**4**) is in (**pink**). (**Light brown dotted**) area is the extended area.

The Tyrrhenian rifting presumably started behind the existing Apennine Chain, following the trend of the eastern volcanic arc (Figure 3). In the first phase of rifting, the Sardinia-Corsica margin necessarily underwent an important uplift, which inevitably caused strong erosion and further disruption of the volcanic chain already dismembered by the rifting. The Tyrrhenian extension also determined a separation between the Apennine chain and the Calabrian arc, which moved about different Euler poles with respect to the Sardinia-Corsica block. During this phase, the front of the Apennine Chain is no longer a transpressive structure, but acquires the character of an accretionary wedge. The subducting slab was made up of Adriatic lithosphere, probably teared along an E-W directed strike-slip structure that separated Adria from Apulia from the late Cretaceous to the Eocene [19] (Figure 21).

Later, the slab-retreat process produced important changes in the evolution of the Apennine chain and of the Tyrrhenian basin. We attribute these changes to new lithospheric tears that formed within the Ionian and Adriatic slabs. We can distinguish four phases for the evolution of the Apennine–Tyrrhenian system: (1) Beginning of rifting and chain segmentation (19–12 Ma); (2) Separation of the Sicilian sector (12–7 Ma); (3) Vavilov basin formation (7–3 Ma); and (4) Marsili basin formation (3–0 Ma).

4.2. Phases of the Evolution

4.2.1. Phase 1. Start of Rifting (19–12 Ma)

The Tyrrhenian rift ends in the north against the left-lateral transform fault that links the Apennine chain from the Western Alps (Figure 21). To the south, the rift ends with the right-lateral strike-slip fault that separates the Calabrian arc from Kabylides (Figure 21). At the end of this first phase, the rifting process did not form important marine basins. Only on the Corsica side of the northern Tyrrhenian some marine successions can be attributed to this phase [82]. However, there are no known successions related to this phase in the southern Tyrrhenian area. As mentioned above, during this phase the Apennine chain was made of two arcs separated by a wide extension area that hosted the Sannite basin. The successions of this basin, described by Patacca & Scandone [50], form the actual Molise Arc. The model implies that the basin in this phase is located on the lithospheric flexure, generated by the tear fault between the two main slabs, which increased its subsidence (Figure 22 and Plate 1).

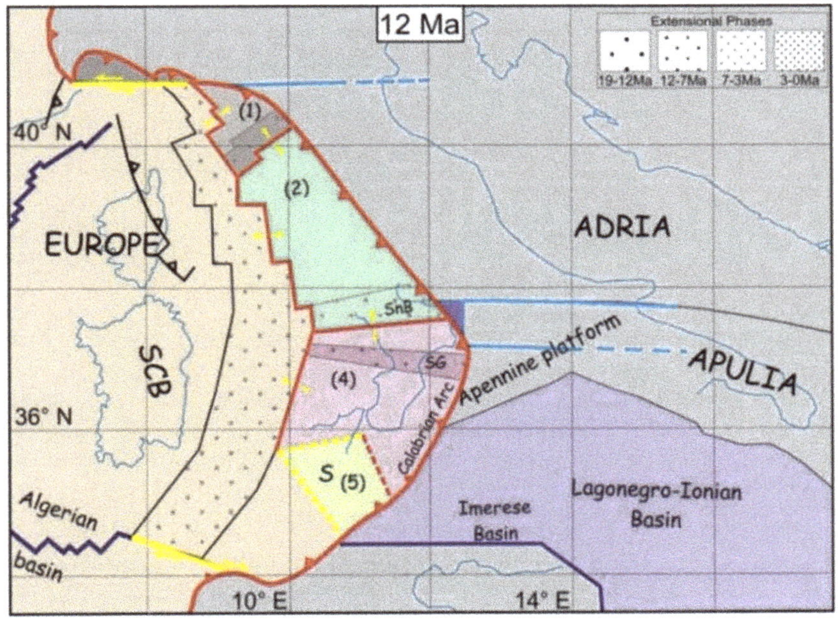

Figure 22. Plate reconstruction of the western Mediterranean region at 12 Ma. The Northern Sector (**1**) is shown in (**dark beige**); the Central Sector (**2**) is in (**green**); the Calabrian Arc Sector (**4**) is in (**pink**), the Sicilian Sector in light (**green**) (**5**). Dotted areas are in extension (see upper right legend). Dashed lines are incipient boundaries. (**Red lines**) are active boundaries. (**Black lines**) are inactive boundaries. (**Yellow lines**) are strike-slip faults. (**Light blue lines**) are transfer faults on the continental lithosphere and (**light blue dotted lines**) are the Jurassic COB (Continent Ocean Boundary). (**Dark blue lines**) are the Middle-Triassic COB. (**Yellow arrows**) indicate direction of extension. S: Sicily, SCB: Sardinian-Corsican block, SG: Squillace Gulf, SnB: Sannio basin.

Apennine Arc

During the first phase the arc was delimited to the north by a strike-slip fault that drove it apart from the Western Alpine arc. The fault was generated by the STEP fault of the Adriatic slab that at the same time formed an important eastward-migrating lithospheric flexure with NW-SE trend (Figure 23), visible today in the Po river Valley. Along the flexure, an important left-lateral transpression was forming the northern Apennine sector. Geometries of the buried north Apennine structures that characterize the Po river valley are the expression of this complex process. At the beginning of the rifting, the Apennine chain was further divided in two sectors: a northern Apennine arc (A) and a central Apennine arc (B) (Figure 21), separated in transverse direction by the extensional area covered by the Marnoso–Arenacea fm [98] (Figure 16b). The central sector of the Apennine chain is associated with the Adriatic slab-retreat, accompanied by the deformation of the Umbria–Marche succession, which in this phase includes the Lazio–Abruzzi platform. Probably the transversal basins that segmented the Apennine chain were fed by sediment coming from the Liguride units, that is, units of the previous top-wedges that covered the early Apennine chain. The part of the chain affected by this process today separates the two sectors of the Apennine arc and coincides essentially with the area where the Marnoso–Arenacea formation is exposed.

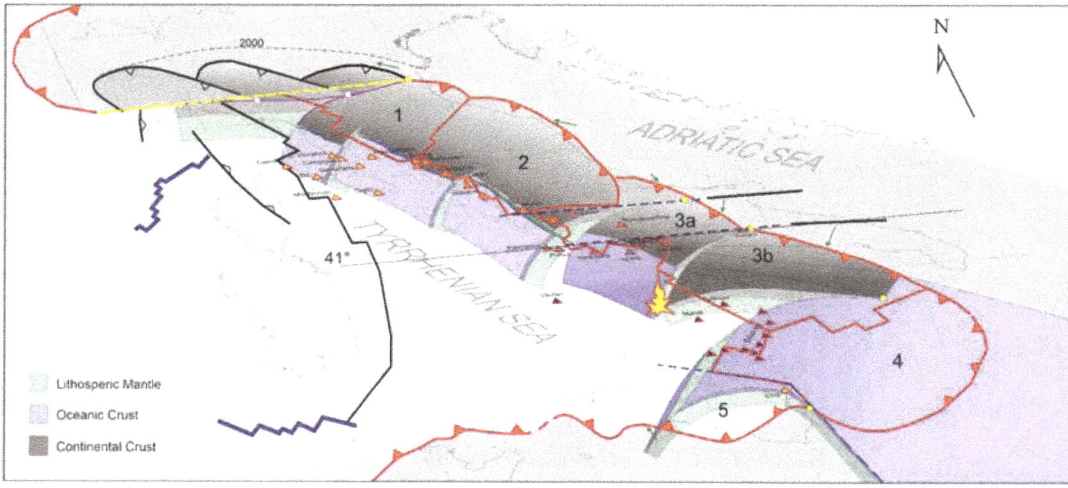

Figure 23. Proposed Ligurian-Ionian slab geometric reconstruction. The kinematic sectors are numbered from 1 to 5. (**Black dotted lines**) are tear faults. (**Yellow dot**) is active tip point of STEP fault (**white dot is paleo-tip point**). (**Triangles**) are volcanoes. The entire slabs length is not represented. Green arrows are the velocity vectors between Adria and the adjacent sector; modified after [28].

Calabrian Arc

The Calabrian arc is divided in two sectors: north Calabria (C) and south Calabria–Peloritani mounts (D), divided by the Catanzaro Trough (Figure 21). The transversal extension between these two sectors is recorded by late Tortonian successions in the Gulf of Squillace [73] (Figure 22). To the west, the Calabrian arc is separated by Kabylies by a large right-lateral transtensional fault that thinned the chain and delimited the Tyrrhenian rift to SW. After the subduction of the last remnant of Liguride ocean, the Calabrian wedge thrusted onto the Panormide and Imerese domains, incorporating them in the accretionary wedge. During this phase the trench of the Calabrian arc still provided an access path to the Numidian sands, which continue to fill the entire trench.

4.2.2. Phase 2. Sicilian Sector Separation (12–7 Ma)

The second phase of Tyrrhenian extension is marked by the separation of the Western Sicily chain from the Calabrian arc. The separation takes place along a right-lateral strike-slip fault that transfers the extension between Calabrian Arc and Kabylies further to the East, in the Gela foredeep. Such large transform structure that cuts the Sicilian upper plate is known as the Monte Kumeta–Alcantara fault. It separates the Sicilian Internal Chain from the External one [90]. The western Tethys reconstructions proposed in our earlier works (e.g., [19]) imply the existence of a STEP fault along the northern margin of Sicily until the Tortonian, where Ionian lithosphere was cut away from the African margin. Here, we propose that main feature of phase 2 of Tyrrhenian extension is related to the activity of this STEP fault. The associated flexure of African lithosphere determined a southward migration of the External Sicilian chain. The external chain migrated southwards, while the Internal chain continued to migrate toward SE together with the Calabrian Arc (Figure 24 and Plate 1). The differential motion between these two sectors formed the Monte Kumeta-Alcantara fault and caused a large E-W extension in the External Chain. In the section of the chain corresponding to the Gela foredeep, the decrease in thickness associated with the lithospheric flexure of the African plate was not compensated by the thrusting of the External Chain, thus generating a strong tectonic subsidence that started to form the Caltanissetta Basin (Figure 25). The kinematic relations in the other sectors of the Chain and Tyrrhenian Rift remain unchanged in this phase. The Apennine trench passed the Carbonatic platforms and Calabrian trench reached the Ionian–Lagonegrese basin.

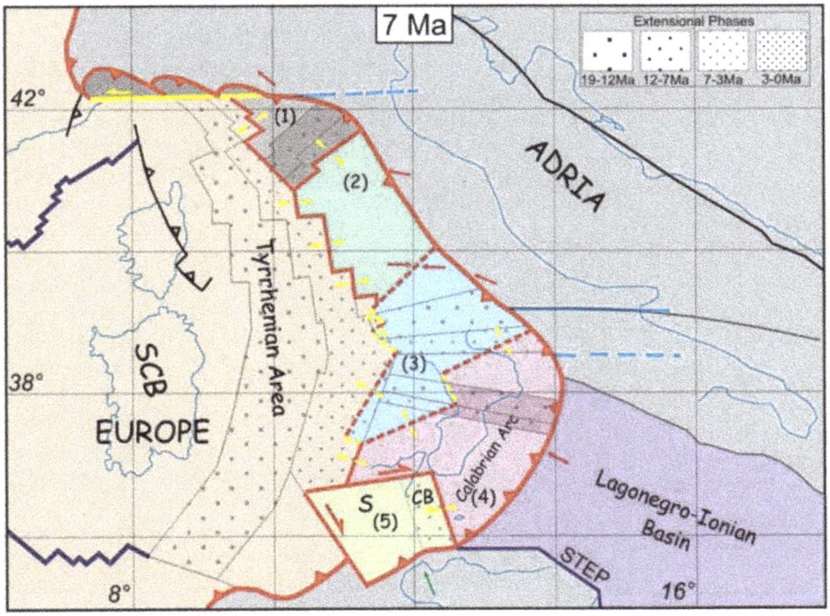

Figure 24. Plate reconstruction of the western Mediterranean region at 7 Ma. The Northern Sector (**1**) is shown in (**dark beige**); the Central Sector (**2**) is in (**green**); the Southern Sector (**3**) is in (**light blue**); the Calabrian Arc Sector (**4**) is in (**pink**), the Sicilian Sector in (**light green**) (**5**). (**Dotted areas**) are in extension (see upper right legend). (**Dotted lines**) are incipient boundaries. (**Red lines**) are active boundaries. (**Black lines**) are inactive boundaries. (**Yellow lines**) are strike-slip faults. (**Light blue lines**) are transfer faults on the continental lithosphere. (**Dark blue lines**) are the Middle-Triassic COB. (**Red arrows**) are the velocity vectors between Adria and the adjacent sector. CB, Caltanissetta Basin; S: Sicily, SCB: Sardinian-Corsican block.

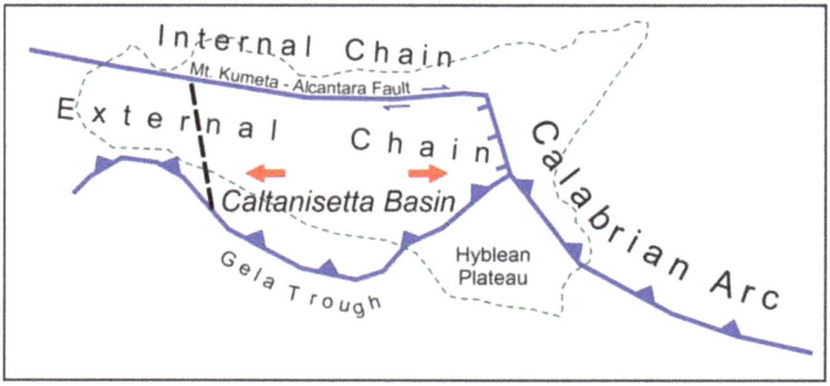

Figure 25. Schematic tectonic evolution of the Caltanissetta Basin. (**Red arrows**) show the direction of extension.

4.2.3. Phase 3. Vavilov Basin Formation (7–3 Ma)

Phase 3 of Tyrrhenian extension was characterized by the formation of the Vavilov basin, which records major changes in the Ionian slab-retreat process. At the beginning of this phase, the entire Imerese Basin and the promontory of the Panormide Carbonatic Platform were accreted to the Calabrian Arc accretionary wedge and the trench entered the narrow Ionian–Lagonegrese ocean corridor (Figure 24). The kinematic model proposed here implies the tearing of the Ionian slab along the Apulian continental margin through a STEP fault, while the eastward propagation of the 41st parallel STEP fault slowed down. On the Sicilian margin, the Ionian tear fault propagated following the continental margin and assumed a different strike. Consequently, during this phase, the units of the Lagonegro basin were also accreted to the SE migrating Calabrian wedge.

Figure 23 illustrates a possible present day 3D reconstruction of the Ionian slab and Apulian lithospheric flexure, based on kinematic considerations and seismic tomography [99–101]. This configuration suggests that during phase 2 the flexure of the Apulian lithosphere induced a rotation of the future southern Apennine domain determining the opening of the Vavilov Basin (Figures 26 and 27, Plate 1). The particular geometry of the three slabs shown in Figure 23 [28], whose flexure lines retreat rotating about different poles, gives the new nascent sector a striking rotation that is recorded by the simultaneous formation of the Vavilov Basin. Consequently, the Southern Apennine Sector rotated around a very close pole, located at the northern tip of the Vavilov Basin. At this stage, the Southern Apennine sector includes the Lazio–Abruzzi platform, separated from the Apennine Arc by the Ancona–Anzio Line. On the southern side, the boundary with the Calabrian Arc consists of an articulated low-angle fault that exhumes the Southern Apennines.

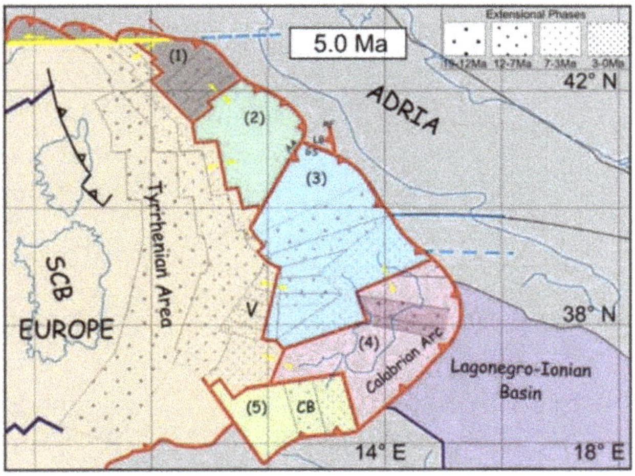

Figure 26. Plate reconstruction of the western Mediterranean region at 5 Ma. The Northern Sector (**1**) is shown in (**dark beige**); the Central Sector (**2**) is in (**green**); the Southern Sector (**3**) is in (**light blue**); the Calabrian Arc Sector (**4**) is in (**pink**), the Sicilian Sector in (**light green**) (**5**). (**Dotted areas**) are in extension (see upper right legend). (**Red lines**) are active boundaries. (**Black lines**) are inactive boundaries. (**Yellow lines**) are strike-slip faults. (**Light blue lines**) are transfer faults on the continental lithosphere. (**Dark blue lines**) are the Middle-Triassic COB. (**Red arrows**) are the velocity vectors between Adria and the adjacent sector. AA: Ancona-Anzio line; CB: Caltanissetta Basin; GS: Gran Sasso Front; LB: Laga Basin; MF: Montagna dei Fiori, SCB: Sardinian-Corsican block; V: Vavilov basin.

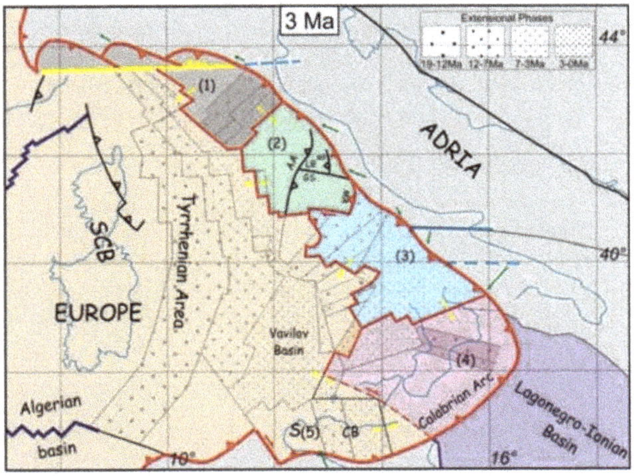

Figure 27. Plate reconstruction of the western Mediterranean region at 3 Ma. The Northern Sector (**1**) is shown in (**dark beige**); the Central Sector (**2**) is in (**green**); the Southern Sector (**3**) is in (**light blue**); the Calabrian Arc Sector (**4**) is in (**pink**), the Sicilian Sector in (**light green**) (**5**). (**Dashed lines**) are incipient boundaries (**Dotted areas**) are in extension (see upper right legend). (**Red lines**) are active boundaries. (**Black lines**) are inactive boundaries. (**Yellow lines**) are strike-slip faults. (**Light blue lines**) are transfer faults on the continental lithosphere. (**Dark blue lines**) are the Middle-Triassic COB. (**Green arrows**) are the velocity vectors between Adria and the adjacent sector. A-A: Ancona-Anzio line; CB: Caltanissetta Basin, GS: Gran Sasso Front; LB: Laga Basin; MA: Maiella Front; MF: Montagna dei Fiori; S: Sicily; SCB: Sardinian-Corsican block, TL: Taormina Line.

Vavilov Basin

The Phase 3 marks the beginning of the differentiation of the Tyrrhenian basin in two areas, roughly separated by the 41st parallel lineament [58]. As mentioned above, the formation of the Vavilov Basin, located south of the lineament, arises from the rotation of the Southern Apennine Sector. The extension begins on the eastern margin and propagates with the rotation, significantly increasing the extension of the Southern Tyrrhenian Sea. The abyssal plain, more than 3500 m deep, probably made up of exhumed mantle or oceanic crust, is covered by limited thicknesses of Pliocene sediments [5]. Therefore, the abyssal area of the basin began to form in post-Messinian age. The Vavilov basin, with its triangular shape and the fan of extensional structures placed on the Apennine margin, provides the best kinematic record of the rotation of the Southern Apennine Sector. The magnetic features of the abyssal plain, on the other hand, appear less clear (Figure 6a).

Ancona-Anzio Line

The kinematics of the Ancona-Anzio line, which represents the boundary between the Umbria-Marche Apennine and the Lazio–Abruzzi Platform, was the facies line between the Carbonate Platform and Basin during the Mesozoic. The role of this structure has been much debated [102–104]. The Southern Apennine Sector represents, according to our rotation model, a portion of the upper plate placed above three slabs. This boundary assumes an important kinematic role for the evolution of the Apennine Arc. The rotation of the two sectors around different poles generates a right-lateral transpression that cuts the Adriatic front of the chain, forming a triple junction between three transpressive trenches (Figures 22 and 23). The northern section of the trench forms the northern segment of the Ancona-Anzio line and coincides with the front of the Sibillini Mountains with eastern vergence. It represents the boundary between Adria and the Central Sector of the Apennine arc. The southern section of the line represents the boundary between the Central Sector and the Southern Apennine Sector. In this section, the transpression with western vergence is confirmed by the Antrodoco1 well [105]. The opposite vergence between the two lines gives the Ancona–Anzio line its typical flexure. The third section is outlined by the front of the Gran Sasso, which represents the boundary between the Southern Apennine Sector and Adria foreland and is characterized by right transpressive kinematics.

Laga Basin and Montagna dei Fiori

The Ancona-Anzio line, which cut the Adriatic front of the chain above the flexure of the Adriatic slab, forms a deep depression at the triple junction, not compensated by the essentially right-lateral Gran Sasso front. The Laga Basin formed at this location, fed by the pre-Tyrrhenian Apennine flysches. At the same time, the front of the Sibillini Mountains was migrating eastwards, forming the Montagna dei Fiori with an axis orthogonal to the Gran Sasso front.

Southern Apennine

The formation of the Southern Apennines is perhaps the most interesting result of the kinematic model proposed here. At the beginning of phase 3, the Apennine Arc and the Calabrian Arc were separated by the Sannio Basin. The tectonic units that form the actual Southern Apennines (carbonatic platform and Lagonegrese Units) represented the basal part of the accretionary wedge of the Calabrian arc. The boundary between the Southern Apennine and the Calabrian arc sectors is characterized by an articulated low-angle fault, generated by the rotations between the two sectors (Figure 16b). The activity of this fault produced a large extension that exhumed the carbonatic and Lagonegrese units from the accretionary wedge of the Calabrian Arc. At the same time, the exhumed units rotated with the Southern Apennine Sector, thereby migrating towards the Apulian flexure.

In summary, the Southern Apennine was generated by a sequence of events that took place in the following order: (1) accretion of the carbonate and Lagonegrese units in the Calabrian arc wedge; (2) exhumation from the accretionary wedge; (3) accretion of the exhumed units to the Southern Apennine Wedge. This singular process created the space occupied by the present chain, whose front developed along a NW-SE direction.

The direction of extension that generated the exhumation of the Apennine units, according to the kinematic model, varies from NW-SE to N-S as a function of the position of the instantaneous pole of rotation between the two sectors (Figures 19 and 23). In addition, the extension is also evidenced by the formation of top-wedge basins present both in the Southern Apennines and in the Calabrian Arc (Irpino Basin, Gulf of Sibari, Crati Valley, Crotone Basin, etc.).

Taormina Line

At the end of this phase, the Internal and External Sicilian chains started moving together with respect to the Calabrian arc along the right-lateral Taormina fault. Consequently, the activity of Monte Kumeta-Alcantara fault and the extension along the External Sicilian Chain ended. The latter was previously responsible for the formation of the Gela foredeep and the Caltanissetta basin.

Northern Tyrrhenian and Apennine Arc

Along the Apennine Arc continues the rotation of the two sectors with the previous kinematics. Crustal thinning in the Northern Tyrrhenian area and between the two sectors of the Apennine Arc also continue. Tuscan magmatism is starting near the triple junction of the two rifts.

4.2.4. Phase 4. Marsili Basin Formation (3–0 Ma)

At 3 Ma extension jumped eastwards of the Vavilov basin and the southern branch of the triple junction of the Southern Tyrrhenian rifts started to form the Marsili basin in an area already thinned by transversal extension associated with the exhumation of the southern Apennines (Figure 27). At the same time, exhumation in the Southern Apennine the Lagonegro Units continued. During this phase, the Ancona-Anzio line stops its activity while, due to the propagation of the flexure of Adriatic lithosphere, the Adriatic trench jumps east of the Montagna dei Fiori and the Maiella Massif, up to the Ortona-Roccamonfina line. The Lazio–Abruzzi segment is again incorporated into the Central Sector of the Apennine Arc. The Ortona–Roccamonfina line from this moment forms the new boundary between the Apennine Arc and the Southern Apennine. The line ends North of the Roccamonfina volcano, from where it transfers a convergent movement on the southern tip of the Ancona-Anzio line through an articulated dextral transtensive structure. This E-W oriented structure runs along the Latina Valley, where the eruptive centers of the Ernici Mounts are located. The southern Apennine continues its exhumation from below the Calabrian Arc while migrating towards the Apulian flexure and generating transpessive structures in the exhumed Lagonegrese Units. At the same time, further away from the Apulian front, basins filled with marine sediments (Potenza, Santarcangelo) formed on the exhumed units still in extension. On the Calabrian Arc, the same extensional event formed the Sibari–Corigliano Basin and the Paola Basin, while affecting the existing Crati Valley and Crotonese Basin [106] (and reference therein) (Figure 16b). Further south the Catanzaro Trough was reactivated. In Sicily, the Ionian STEP fault intersected the Malta Escarpment and the tear fault began to propagate along this structure. Our model suggests that this event started widening the narrow Ionian Slab and interrupted the continuity between the Sicilian and Ionian lithosphere. All this causes dramatic changes in Sicily. During the upper Pleistocene, the triple junction of the Southern Tyrrhenian abandons the area of the Marsili and jumps near the Aeolian Arc (Figure 28 and Plate 1). The slab-retreat process from this moment is exclusively guided by the Ionian slab.

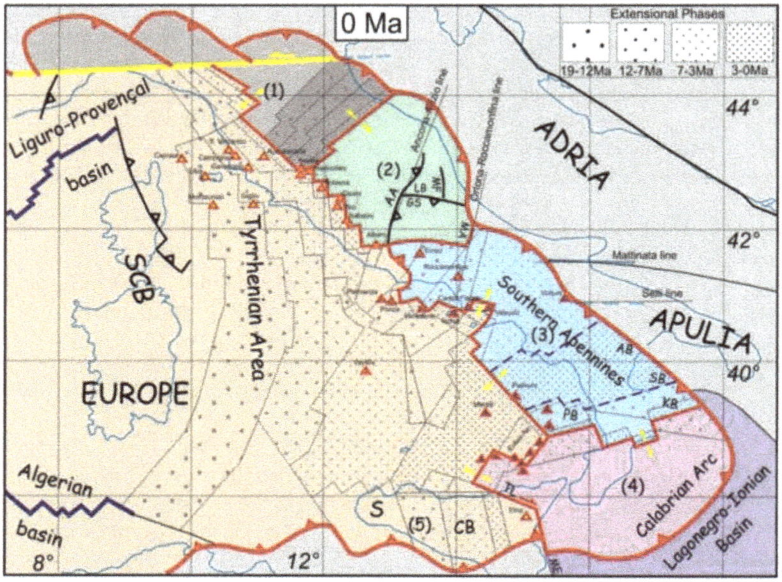

Figure 28. Plate reconstruction of the western Mediterranean region at 0 Ma. Triangles are volcanoes: the Tuscan province is in filled in (**yellow**), the Roman province is in orange, the Campanian province is in (**blue**), the Aeolian Arc with Marsili and Palinuro is in (**black**). The Northern Sector (**1**) is shown in (**dark beige**); the Central Sector (**2**) is in (**green**); the Southern Sector (**3**) is in (**light blue**); the Calabrian Arc Sector (**4**) is in (**pink**), the Sicilian Sector in (**light green**) (**5**). (**Dotted areas**) are in extension (see upper right legend). (**Red lines**) are active boundaries. (**Black lines**) are inactive boundaries. (**Yellow lines**) are strike-slip faults. (**Light blue lines**) transfer faults on the continental lithosphere. (**Dark blue lines**) are the Middle-Triassic COB and transfer faults on the continental lithosphere. Blue dashed lines identify the area of the Southern Apennine exhumed from 7 Ma. A-A: Ancona-Anzio line; AB: St. Arcangelo Basin; CB, Caltanissetta Basin, KB: Crotone Basin; GS: Gran Sasso Front; LB: Laga Basin; MA: Maiella Front; MF: Montagna dei Fiori; PB: Paola Basin; S: Sicily; SB:Sibari Basin; SCB: Sardinian-Corsican block, TL: Taormina Line.

5. Discussion

5.1. General Considerations

The method described in this work is based on the laws of plate kinematics and takes into account deformation of blocks through time. The proposed model describes quantitatively the complex evolution of the system of deformable tectonic elements belonging to the Tyrrhenian–Apennine region. The main tectonic consequences of this model shed new light on the geodynamic processes that generated the Tyrrhenian extension and the corresponding formation of the Apennine chain. The slab retreat process, which is generally recognized as responsible for the origin of the Tyrrhenian rift, has complex geodynamic implications, not always sufficiently considered. It arises when the speed between two converging plates is lower than the retreat speed of the slab inflection point, caused by the lithospheric sinking [107]. The process is often illustrated by a lithospheric profile that shows the kinematics of extension in the upper plate along the line of the profile. However, a correct description of the retreat of a trench should be done in 3D, considering the rotation of the trench line about the Euler pole of the upper plate. Besides the kinematic aspects, the slab-retreat process is associated with important geodynamic implications, such as the formation of STEP faults and the elastic rebound that follows slab detachment, and requires a description that takes into account of all the boundary conditions. For example, STEP faults determine a downward flexure that partly affects the unsubducted plate, whose margin is dragged to some extent in the asthenosphere. Such a flexure propagates together with the rip between the slab and the continental margin. In the crust, the effects of this process are strictly related to the extent of the flexure, which in turn depends on the rigidity and strength of the lithosphere. In general, the marginal flexure of the continental lithosphere occurs along a hinge line that is oblique with respect to the STEP fault strike. The subsequent rebound determines the formation of crustal structures that are pairwise oblique with respect to the direction of slab retreat. In the Apennine area this effect is quite evident. In addition to this, the lithospheric flexure along a STEP fault produces important transpressive structures at crustal scale that simulate normal mountain chains. On the basis of these considerations, it is possible to describe the evolution of some sectors of the Apennine and Sicilian chains.

5.2. STEP Fault Evolution Within the Northern Apennine

In previous models, the northern and central Apennine are part of a unique arc that extends from the Sestri Voltaggio line to the Ortona–Roccamonfina line [108]. These models are based on the hypothesis that a continuous Adriatic slab exists. Our kinematic model suggests instead that the northern Apennine segment represents a transform structure of the upper plate with left-lateral motion, which links the western Alpine arc with the Apennine chain with opposite vergence. The presence of a mountain chain along this fault is a consequence of the geodynamic complexity of the processes involved in STEP faults. The flexure of the continental margin of Adria in the Padania Valley originated a submerged chain that was later exhumed by elastic rebound. A series of arcs formed, with a symmetry axis aligned with the hinge line of the flexure, thereby oblique with respect to the strike of the STEP fault. These arcs propagated eastwards and were eventually exhumed (the Monferrato arc) following the tear migration (Figure 4 and [109]).

5.3. STEP Fault Evolution Within the Sicilian Chain

Our model proposes that, differently from the northern Apennine, the western Sicilian chain developed along three distinct STEP faults. The oldest (and most important) one had E-W trend (in present day coordinates) and was placed along the northern continental Hyblean paleomargin. The next, very short STEP coincided with the modern Taormina line. Finally, the modern STEP fault associated with the rip of the Ionian oceanic lithosphere from the African margin coincides with the Malta Escarpment. The eastward propagation of the tearing determined flexure along the Iblei and SSE migration of the external chain while the internal chain moved together with the Calabrian Arc (Figure 4). This process separated the two chains along the Monte Kumeta fault, which is characterized by right-lateral kinematics. The eastern termination of this fault is represented by the northeastern tip of the Caltanissetta extensional system. During the early Pliocene slab tearing was transferred to the Taormina line, thereby starting from this time tectonic activity along the external and internal chains was essentially driven by the convergence between Africa and Eurasia. Finally, during the Pleistocene, slab tearing started along the Malta Escarpment, determining an eastern break in the continuity between the African continental margin and the Ionian lithosphere. We hypothesize that this event originated an eastward escape of Sicily with respect to stable Africa. Such a hypothesis is supported by the dextral kinematics along the Sicily Channel and by the submerged Elimi Chain [110,111]. A successive paper will address this point.

5.4. STEP Fault Evolution Within the Southern Apennine

The southern Apennine is aligned with an important STEP fault that separates the continental margin of Apulia from the Ionian slab. Initially, the southwestern margin of Apulia bended downwards following the subduction of the Ionian oceanic lithosphere, while its northern margin separated by Adria along the 41° parallel STEP fault. Starting from the late Tortonian formation of the Vavilov basin, a tear formed that separated Apulia from the Ionian oceanic slab. As soon as Apulia was not anymore pulled downwards by the Ionian slab, the flexure line started rotating counterclockwise and assuming a strike slightly oblique with respect to the new southeastward propagating STEP fault. The contemporary elastic rebound of the more western parts enhanced this rotation of the flexure line and produced at the same time the rotation of the southern Apennine and the formation of the Vavilov basin. The difference in the Euler poles of rotation of the southern Apennine and the Calabrian arc with respect to Europe determined strong extension between the two sectors and exhumation of most of the southern Apennine. The latter was previously part of the Calabrian arc accretionary wedge.

6. Conclusions

The tectonic reconstruction of the Tyrrhenian basin and the Apennine chain offered us the opportunity to propose a method that integrates plate kinematics tools with the methods of structural geology. The technique has proved to be adequate for performing quantitative deformation analyses at an intermediate scale such as that of the Mediterranean area.

Structural geology has made significant contributions for the application of plate kinematics methods to continental tectonics. Among these, we remind the identification of the components of the Apennine puzzle (micro-plates), the determination of the poles and angles of rotation, and the definition of the plate boundaries. With these fundamental parameters, we have formulated the hypothesis of a rotation model of the puzzle of micro-plates, built with the aid of the PCME software tool. The model provided a framework of kinematic constraints that allowed to hypothesize innovative solutions to the problems inherent the tectonic reconstruction of the Apennine chain. The kinematic framework is also useful for future research that can confirm, improve, or modify the model we propose.

The model can be viewed through an animation (Movie 1 in Supplementary Material) composed by a temporal sequence of images, generated by the software. The characteristic of this approach, well known in the global tectonics studies, is to provide and visualize, for each point along plate boundaries, the relative velocity vectors between the micro-plates. Finally, this important feature allowed us to predict the tectonic and structural implications along the plate boundaries. The expected implications of our rotation model have found, in our opinion, a very satisfactory confirmation for the tectonic structure we observe for the Apennine chain and the Tyrrhenian basin. We therefore believe that the initial hypotheses made on the recognition of micro-plates, on the determination of the rotation poles, on the activity times, and on the angles of rotation have been verified. The model also provides a precise balance of the areas, which is essential for paleogeographic reconstructions.

Due to these important characteristics, we believe that the rotation models constructed with this methodology, thanks to their ability to predict implications of tectonic and structural processes in a very broad framework, represent powerful tools suitable for responding to interpretative disputes for complex areas still debated. We are however fully aware that the transition from the rotation model to its tectonic implications requires a certain aptitude to visualize the third dimension, not represented in the rotation models.

For the Apennine and Tyrrhenian areas, the rotation model has responded well to the historical issues. Among these we recall the most significant. Starting from the North, they are: the transversal structures that interrupt the Umbria–Marche Arc and separate it from the Northern Apennine, while extending the Tyrrhenian, Tuscan–Lazio and Umbria areas in a direction transverse to the Apennine chain; the Ancona-Anzio transpressive lineament which, in the triple junction with the Sibillini and Gran Sasso front, form the Laga basin; the exhumation of the tectonic units of the Southern Apennine, which implies a great extension that pervades the entire chain; the formation of important basins, located near the foredeep of the southern chain (Santarcangelo, Sibari, Crotonese and Caltanissetta), the Paola basin, and Crati Valley located in an innermost position and the Catanzaro trough that crosses the entire Calabrian arc. All these basins exhibit extension along the direction of the chain that is not appreciable in the tectonic reconstructions of transects.

Supplementary Materials: The following are available online at https://www.mdpi.com/article/10.3390/geosciences11040177/s1, Plate 1: Kinematic Evolution of the Tyrrhenian–Apennine system; rotation model.rot; Movie 1: Kinematic Evolution of the Tyrrhenian–Apennine system.

Author Contributions: Conceptualization, methodology and writing—original draft preparation, E.T.; software, investigation and data curation, C.M., A.S., G.P.; funding acquisition, supervision and writing—review and editing, P.P.P. All authors have read and agreed to the published version of the manuscript.

Funding: This research was funded by the Italian Ministry of University and Scientific Research, PRIN Turco, prot. 2008YWPCWB and by the Università degli studi di Camerino (Far-Pierantoni), Grant number STI000030.

Institutional Review Board Statement: Not applicable.

Informed Consent Statement: Not applicable.

Data Availability Statement: The data presented in this study are available in Supplementary Material.

Acknowledgments: We thank the editor G. Wang and the guest editors D. Liotta, G. Molli and A. Cipriani for the opportunity to share our research in the Special Issue "The Apennines: Tectonics, Sedimentation, and Magmatism from the Palaeozoic to the Present". Special thanks to the anonymous reviewers for their excellent contribution in restructuring and improving the manuscript.

Conflicts of Interest: The authors declare no conflict of interest.

References

1. Sartori, R.; Torelli, L.; Zitellini, N.; Carrara, G.; Magaldi, M.; Mussoni, P. Crustal features along a W–E Tyrrhenian transect from Sardinia to Campania margins (Central Mediterranean). *Tectonophysics* **2004**, *383*, 171–192. [CrossRef]
2. Biju-Duval, B.; Dercourt, J.; Le Pichon, X. From the Tethys Ocean to the Mediterranean Seas: A plate tectonic model of the evolution of the Western Alpine system. In *Structural History of the Mediterranean Basins, Proceedings of the International Symposium on the Structural History of the Mediterranean Basins, Split, 1976, Paris, France, 25–29 October 1977*; Biju-Duval, B., Montadert, L., Eds.; Editions Technip: Paris, France, 1977; pp. 143–164.
3. Dercourt, J.; Zonenshain, L.P.; Ricou, L.E.; Kazmin, V.G.; Le Pichon, X.; Knipper, A.L.; Grandjacquet, C.; Sbortshikov, I.M.; Geyssant, J.; Lepvrier, C.; et al. Geological evolution of the Tethys belt from the Atlantic to the Pamirs since the Lias. *Tectonophysics* **1986**, *123*, 241–315. [CrossRef]
4. Malinverno, A.; Ryan, W.B.F. Extension in the Tyrrhenian Sea and shortening in the Apennines as result of arc migration driven by sinking of the lithosphere. *Tectonics* **1986**, *5*, 227–245. [CrossRef]
5. Kastens, K.A.; Mascle, J. Site 654: Upper sardinian margin. *Proc. ODP Init. Rep.* **1987**, *107*, 772.
6. Dewey, J.F.; Helman, M.L.; Turco, E.; Hutton, D.H.W.; Knott, S.D. Kinematics of the Western Mediterranean. In *Alpine Tectonics*; Coward, M.P., Dietrich, D., Park, R.G., Eds.; Geological Society: London, UK, 1989; Volume 45, pp. 265–283.
7. Boccaletti, M.; Ciranafi, N.; Cosentino, D.; Deiana, G.; Gelati, R.; Lentini, F.; Massari, F.; Moratti, G.; Pescatore, T.; Lucchi, F.R.; et al. Palinspastic restoration and paleogeographic reconstrution of the perithyrrenian area during the Neogene. *Palaeogeogr. Palaeoclimatol. Palaeoecol.* **1990**, *77*, 41–50. [CrossRef]
8. Carmignani, L.; Decandia, F.A.; Disperati, L.; Fantozzi, P.L.; Lazzarotto, A.; Liotta, D.; Oggiano, G. Relationships between the tertiary structural evolution of the Sardinia-Corsica-Provencal Domain and the Northern Apennines. *Terra Nova* **1995**, *7*, 128–137. [CrossRef]
9. Lavecchia, G.; Federico, C.; Stoppa, F.; Karner, G. La distensione tosco-tirrenica come possibile motore della compressione appenninica. *Studi Geol. Camerti* **1995**, 489–497. [CrossRef]
10. Faccenna, C.; Davy, P.; Brun, J.-P.; Funiciello, R.; Giardini, D.; Mattei, M.; Nalpas, T. The dynamics of back-arc extensions: An experimental approach to the opening of the Tyrrhenian Sea. *Geophys. J. Int.* **1996**, *126*, 781–795. [CrossRef]
11. Ferranti, L.; Oldow, J.S.; Sacchi, M. Pre-Quaternary orogen-parallel extension in the Southern Apennine Belt, Italy. *Tectonophysics* **1996**, *260*, 325–347. [CrossRef]
12. Turco, E.; Zuppetta, A. A kinematic model for the Plio-Quaternary evolution of the Tyrrhenian–Apenninic system; implications for rifting processes and volcanism. *J. Volcanol. Geoth. Res.* **1998**, *82*, 1–18. [CrossRef]
13. Jolivet, L.; Faccenna, C. Mediterranean extension and the Africa-Eurasia collision. *Tectonics* **2000**, *19*, 1095–1107. [CrossRef]
14. Faccenna, C.; Becker, T.W.; Lucente, F.P.; Jolivet, L.; Rossetti, F. History of subduction and back-arc extension in the Central Mediterranean. *Geophys. J. Int.* **2001**, *145*, 809–820. [CrossRef]
15. Rosenbaum, G.; Lister, G.S.; Duboz, C. Reconstruction of the tectonic evolution of the western Mediterranean since the Oligocene. *J. Virtual Explor.* **2002**, *8*, 107–126. [CrossRef]
16. Lavecchia, G.; Boncio, P.; Creati, N.; Brozzetti, F. Some aspects of the Italian geology not fitting with a subduction scenario. *J. Virtual Explor.* **2003**, *10*, 1–14. [CrossRef]
17. Peccerillo, A.; Turco, E. Petrological and geochemical variations of Plio-Quaternary volcanism in the Tyrhhenian Sea area: Regional distribution of magma types, petrogenesis and geodynamic implications. *Per. Miner.* **2004**, *73*, 231–251.
18. Turco, E.; Macchiavelli, C.; Mazzoli, S.; Schettino, A.; Pierantoni, P.P. Kinematic evolution of the Alpine Corsica in the framework of Mediterranean mountain belts. *Tectonophysics* **2012**, *579*, 193–206. [CrossRef]

19. Schettino, A.; Turco, E. Tectonic history of the western Tethys since the Late Triassic. *GSA Bull.* **2011**, *123*, 89–105. [CrossRef]
20. Johnston, S.T.; Mazzoli, S. The Calabrian Orocline: Buckling of a previously more linear orogen. *Geol. Soc. Lond. Mem.* **2009**, *327*, 113–125. [CrossRef]
21. Carminati, E.; Lustrino, M.; Doglioni, C. Geodynamic evolution of the central and western Mediterranean: Tectonics versus igneous petrology constraints. *Tectonophysics* **2012**, *579*, 173–192. [CrossRef]
22. Le Breton, E.; Handy, M.R.; Molli, G.; Ustaszewski, K. Post-20 Ma motion of the Adriatic Plate: New constraints from surrounding orogens and implications for crust-mantle decoupling. *Tectonics* **2017**, *36*, 3135–3154. [CrossRef]
23. Rosenbaum, G.; Lister, G.S. Neogene and Quaternary rollback evolution of the Tyrrhenian Sea, the Apennines, and the Sicilian Maghrebides. *Tectonics* **2004**, *23*, TC1013. [CrossRef]
24. Hosseinpour, M.; Williams, S.; Seton, M.; Barnett-Moore, N.; Müller, R.D. Tectonic evolution of Western Tethys from Jurassic to present day: Coupling geological and geophysical data with seismic tomography models. *Int. Geol. Rev.* **2016**, *58*, 1616–1645. [CrossRef]
25. Müller, R.D.; Zahirovic, S.; Williams, S.E.; Cannon, J.; Seton, M.; Bower, D.J.; Gurnis, M. A global plate model including lithospheric deformation along major rifts and orogens since the Triassic. *Tectonics* **2019**, *38*, 1884–1907. [CrossRef]
26. Van Hinsbergen, D.J.; Torsvik, T.H.; Schmid, S.M.; Maţenco, L.C.; Maffione, M.; Vissers, R.L.; Spakman, W. Orogenic architecture of the Mediterranean region and kinematic reconstruction of its tectonic evolution since the Triassic. *Gondwana Res.* **2020**, *81*, 79–229. [CrossRef]
27. Turco, E.; Schettino, A.; Macchiavelli, C.; Pierantoni, P.P. A plate kinematics approach to the tectonic analysis of the Tyrrhenian–Apennines System. *Geophys. Res. Abstr.* **2013**, *29*, 187–190.
28. Pierantoni, P.P.; Macchiavelli, C.; Penza, G.; Schettino, A.; Turco, E. Kinematics of the Tyrrhenian–Apennine system and implications for the origin of the Campanian magmatism. In *Vesuvius, Campi Flegrei, and Campanian Volcanism*, 1st ed.; De Vivo, B., Belkin, H.E., Rolandi, G., Eds.; Elsevier Inc.: Amsterdam, The Netherlands, 2019; p. 520.
29. Schettino, A. *Quantitative plate tectonics. Physics of the Earth–Plate Kinematics–Geodynamics*, 1st ed.; Springer International Publishing: New York, NY, USA, 2014; p. 14.
30. Knott, S.D. The Liguride complex of southern Italy—A Cretaceous to Paleogene accretionary wedge. *Tectonophysics* **1987**, *142*, 217–226. [CrossRef]
31. Ogniben, L. Schema introduttivo alla geologia del confine calabro-lucano. *Mem. Soc. Geol. Ital.* **1969**, *8*, 453–763.
32. Elter, P.; Perusati, P. Considerazioni sul limite Alpi-Appennino e sulle sue relazioni con l'arco delle Alpi occidentali. *Mem. Soc. Geol. Ital.* **1973**, *12*, 359–375.
33. Amodio-Morelli, L.; Bonardi, G.; Colonna, V.; Dietrich, D.; Giunta, G.; Ippolito, F.; LiguorI, V.; Lorenzoni, S.; Paglionico, A.; Perrone, V.; et al. L'arco calabro-peloritano nell'orogene appenninico-maghrebide. *Mem. Soc. Geol. Ital.* **1976**, *17*, 1–60.
34. Patacca, E.; Scandone, P. Post-Tortonian mountain building in the Apennines. The role of the passive sinking of a relic lithospheric slab. In *The Lithosphere in Italy: Advances in Earth Science Research*; Boriani, A., Ed.; Accademia Nazionale dei Lincei: Rome, Italy, 1989; Volume 80, pp. 157–176.
35. Elter, P.; Gilgia, G.; Tonigiorgi, M.; Trevisan, L. Tensional and compressional areas in the recent (Tortonian to present) evolution of the northern Apennines. *Boll. Di Geof. Teor. Appl.* **1975**, *17*, 3–18.
36. Barchi, M. The CROP 03 profile: A synthesis of results on deep structures of the Northern Apennines. *Mem. Soc. Geol. Ital.* **1998**, *52*, 383–400.
37. Carmignani, L.; Decandia, F.A.; Disperati, L.; Fantozzi, P.L.; Kligfield, R.; Lazzarotto, A.; Liotta, D.; Meccheri, M. Inner northern Apennines. In *Anatomy of an Orogen: The Apennines and Adjacent Mediterranean Basins;* Vai, G.B., Martini, I.P., Eds.; Kluwer Academic: Dordrecht, The Netherlands, 2001; pp. 197–213.
38. Centamore, E.; Rossi, D.; Tavernelli, E. Geometry and kinematics of Triassic to Recent structures in the Northern-Central Apennines: A rewiew and an original working hypothesis. *Ital. J. Geosci.* **2009**, *128*, 419–432.
39. Barchi, M. The Neogene-Quaternary evolution of the Northern Apennines: Crustal structure, style of deformation and seismicity. *J. Virtual Explor.* **2010**, *36*. [CrossRef]
40. Marroni, M.; Meneghini, F.; Pandolfi, L. A revised subduction inception model to explain the Late Cretaceous, double-vergent orogen in the precollisional western Tethys: Evidence from the Northern Apennines. *Tectonics* **2017**, *36*, 2227–2249. [CrossRef]
41. Parotto, M.; Praturlon, A. Geological summary of the Central Apennines. In Structural Model of Italy. *Quad. Ric. Sci. Ital.* **1975**, *90*, 257–311.
42. Deiana, G.; Pialli, G. The structural provinces of the Umbro-Marchean Apennines. *Mem. Soc. Geol. Ital.* **1994**, *48*, 473–484.
43. Calamita, F.; Centamore, E.; Deiana, G.; Ridolfi, M. Caratterizzazione geologico-strutturale dell'area marchigiano-abruzzese esterna (Appennino centrale). *Studi Geol. Camerti* **1995**, 171–182. [CrossRef]
44. Scarsella, F. Sulla zona di incontro dell'Umbria e dell'Abruzzo. *Boll. Serv. Geol. D'It.* **1951**, *71*, 155–165.
45. Cosentino, D.; Scoppola, C.; Scrocca, D.; Vecchia, P. Stile strutturale dei Monti Reatini e dei Monti Sabini settentrionali (Appennino centrale) a confronto. *Studi Geol. Camerti* **1991**, 55–61. [CrossRef]
46. Piana, F. Configurazione geometrica ed evoluzione cinematica della zona di convergenza strutturale tra l'arco umbro ed il dominio laziale-abruzzese (Appennino Centrale). *Studi Geol. Camerti* **1991**, 85–94. [CrossRef]
47. Centamore, E. *Carta Geologica dei Bacini della Laga e del Cellino e dei Rilievi Carbonatici Circostanti (Marche Meridionali, Lazio Nord-Orientale, Abruzzo Settentrionale;* Università degli studi di Camerino: Camerino, Italy, 1992.

48. Capotorti, F.; Fumanti, F.; Mariotti, G. Evoluzione tettonico-sedimentaria e strutturazione del settore di piattaforma carbonatica laziale-abruzzese nell'alta Valle del F. Velino. *Studi Geol. Camerti* **1995**, 101–111. [CrossRef]
49. Civitelli, G.; Brandano, M. Atlante delle litofacies e modello deposizionale dei Calcari a Briozoi e Litotamni nella Piattaforma carbonatica laziale-abruzzese. *Boll. Soc. Geol. Ital.* **2005**, *124*, 611.
50. Patacca, E.; Scandone, P. Geology of the Southern Apennines. *Boll. Soc. Geol. Ital. (Ital. J. Geosci.)* **2007**, *7*, 75–119.
51. Patacca, E.; Scandone, P. La zona di giunzione tra l'Arco Appenninico Settentrionale e l'Arco Appenninico Meridionale nell'Abruzzo e nel Molise. *Studi Geol. Camerti* **1991**, 417–441. [CrossRef]
52. Cello, G.; Mazzoli, S. Apennine tectonics in southern Italy: A review. *J. Geodyn.* **1998**, *27*, 191–211. [CrossRef]
53. Pescatore, T.; Renda, P.; Schiattarella, M.; Tramutoli, M. Stratigraphic and structural relationships between Meso-Cenozoic Lagonegro basin and coeval carbonate platforms in southern Apennines, Italy. *Tectonophysics* **1999**, *315*, 269–286. [CrossRef]
54. Mazzoli, S.; Barkham, S.; Cello, G.; Gambini, R.; Mattioni, L.; Shiner, P.; Tondi, E. Reconstruction of continental margin architecture deformed by the contraction of the Lagonegro Basin, southern Apennines, Italy. *J. Geol. Soc.* **2001**, *158*, 309–319. [CrossRef]
55. Van Dijk, J.P.; Bello, M.; Brancaleoni, G.P.; Cantarella, G.; Costa, V.; Frixa, A.; Zerilli, A. A regional structural model for the northern sector of the Calabrian Arc (southern Italy). *Tectonophysics* **2000**, *324*, 267–320. [CrossRef]
56. Bonardi, G.; Cavazza, W.; Perrone, V.; Rossi, S. Calabria-Peloritani terrane and northern Ionian Sea. In *Anatomy of an Orogen: The Apennines and Adjacent Mediterranean Basins*; Vai, G.B., Martini, I.P., Eds.; Springer: Dordrecht, The Netherlands, 2001. [CrossRef]
57. Finetti, I.R. *CROP Project: Deep Seismic Exploration of the Central Mediterranean and Italy*; Elsevier: Amsterdam, The Netherlands, 2005.
58. Selli, R. Thoughts on the geology of the Mediterranean region. In Proceedings of the Consiglio Nazionale delle Ricerche. International Conference, Rome, Italy, 22–23 May 1981; pp. 489–501.
59. Grad, M.; Tire, T.; ESC Working Group. The Moho depth map of the European Plate. *Geophys. J. Int.* **2009**, *176*, 279–292. [CrossRef]
60. Panza, G.F.; Calcagnile, G. The upper mantle structure in Balearic and Tyrrhe-nian bathyal plains and the Messinian salinity crisis. *Palaeogeogr. Palaeoclimatol. Palaeoecol.* **1979**, *29*, 3–14. [CrossRef]
61. Nicolich, R. Crustal structures in the Italian Peninsula and Surrounding Seas: A review of DDS data. In *Sedimentary Basins of the Mediter-ranean Margins. C.N.R. Italian Project of Oceanography*; Wezel, F.C., Ed.; Tectoprint: Bologna, Italy, 1981; pp. 489–501.
62. Panza, G.F. Structure of the lithosphere–asthenosphere system the Mediterranean region. *Ann. Geophys.* **1984**, *2*, 137–138.
63. Marani, M.P.; Gamberi, F. Structural framework of the Tyrrhenian Sea unveiled by seafloor morphology. *Mem. Descr. Carta Geol. Ital.* **2004**, *64*, 97–108.
64. Meloni, A.; Carmisciano, C.; Chiappini, M.; Faggioni, O.; Marson, I. Total Field Magnetic Anomaly Map of Italy and Surrounding Marine Areas at Sea Level. Recent Compilation and Scientific Significance. In Proceedings of the EGM 2007 International Workshop, Capri, Italy, 15–18 April 2007.
65. Spadini, G.; Cloething, S.; Bertotti, G. Thermo-mechanical modeling of the Tyrrhenian Sea: Lithosphere necking and kinematics of rifting. *Tectonics* **1995**, *14*, 629–644. [CrossRef]
66. Sartori, R. The main results of ODP leg 107 in the frame of neogene to recent geology of peri-tyrrhenian areas. *Proc. ODP Sci. Results* **1990**, *107*, 715–730.
67. Sartori, R. The Tyrrhenian back-arc basin and subduction of the Ionian lithosphere. *Episodes* **2003**, *26*, 217–221. [CrossRef] [PubMed]
68. Marani, M.P. Super-inflation of a spreading ridge through vertical accretion. *Mem. Descr. Carta Geol. D'It.* **2004**, *XLIV*, 185–194.
69. Milia, A.; Torrente, M.M.; Tesauro, M. From stretching to mantle exhumation in a triangular backarc basin (Vavilov basin, Tyrrhenian Sea, Western Mediterranean). *Tectonophysics* **2017**, *710–711*, 108–126. [CrossRef]
70. Schettino, A.; Turco, E. Plate kinematics of the Western Mediterranean region during the Oligocene and Early Miocene. *Geophysi. J. Int.* **2006**, *166*, 1398–1423. [CrossRef]
71. Speranza, F.; Cosentino, D.; Villa, I.M. Età della rotazione sardo-corsa: Nuovi dati paleomagnetici e geocronologici. *Riassunti Convegno Geoitalia* **1999**, *1*, 299–301.
72. Gattacceca, J.; Deino, A.; Rizzo, R.; Jones, D.S.; Henry, B.; Beaudoin, B.; Vadeboin, F. Miocene rotation of Sardinia: New paleomagnetic and geochronological constraints and geodynamic implications. *Earth Planet. Sci. Lett.* **2007**, *258*, 359–377. [CrossRef]
73. Milia, A.; Torrente, M.M. Early-stage rifting of the Southern Tyrrhenian region: The Calabria–Sardinia breakup. *J. Geodyn.* **2014**, *81*, 17–29. [CrossRef]
74. Greiner, B. Euler rotations in plate-tectonic reconstructions. *Comput. Geosci.* **1999**, *25*, 209–216. [CrossRef]
75. Turco, E.; Schettino, A.; Pierantoni, P.P.; Santarelli, G. The Pleistocene extension of the Campania Plain in the framework of the southern Tyrrhenian tectonic evolution: Morphotectonic analysis, kinematic model and implications for volcanism. In *Volcanism in the Campania Plain: Vesuvius, Campi Flegrei and Ignimbrites*; De Vivo, B., Ed.; Elsevier: Amsterdam, The Netherlands, 2006; Volume 9, pp. 27–51.
76. Kearey, P.; Vine, F.J. *Global Tectonics*; Wiley-Blackwell: Hoboken, NJ, USA, 1990.
77. Ross, M.I.; Scotese, C.R. A hierarchical tectonic model of the Gulf of Mexico and Caribbean region. *Tectonophysics* **1988**, *155*, 139–168. [CrossRef]
78. Schettino, A. Computer-aided paleogeographic reconstructions. *Comput. Geosci.* **1998**, *24*, 259–267. [CrossRef]
79. Müller, R.D.; Cannon, J.; Qin, X.; Watson, R.J.; Gurnis, M.; Williams, S.; Pfaffelmoser, T.; Seton, M.; Russell, S.H.J.; Zahirovic, S. GPlates: Building a virtual Earth through deep time. *Geochem. Geophys. Geosyst.* **2018**, *19*. [CrossRef]
80. Kastens, K.A.; Mascle, J. The geological evolution of the Tyrrhenian Sea: An introduction to the scientific results of ODP Leg 107. *Proc. ODP Sci. Results* **1990**, *107*, 26.

81. Pascucci, V.; Martini, I.P.; Sagri, M.; Sandrelli, F.; Nichols, G. Effects of transverse structural lineaments on the Neogene-Quaternary basins of Tuscany (inner Northern Apennines, Italy). *Sediment. Process. Environ. Basins* **2007**, *38*, 155–182.
82. Thinon, I.; Guennoc, P.; Serrano, O.; Maillard, A.; Lasseur, E.; Rehault, J.P. Seismic markers of the Messinian Salinity Crisis in an intermediate-depth basin: Data for understanding the Neogene evolution of the Corsica Basin (northern Tyrrhenian Sea). *Mar. Pet. Geol.* **2016**, *77*, 1274–1296. [CrossRef]
83. Pascucci, V. Tyrrhenian Sea extension north of the Elba Island between Corsica and western Tuscany (Italy). *Boll. Soc. Geol. Ital.* **2002**, *1*, 819–828.
84. Di Bucci, D.; Tozzi, M. La linea "Ortona-Roccamonfina": Revisione dei dati esistenti e nuovi contributi per il settore settentrionale (media valle del Sangro). *Studi Geol. Camerti* **1991**, 397–406. [CrossRef]
85. Caricchi, C.F.; Cifelli, L.; Sagnotti, F.; Sani, F.; Speranza, F.; Mattei, M. Paleomagnetic evidence for a post-Eocene 90° CCW rotation of internal Apennine units: A linkage with Corsica-Sardinia rotation? *Tectonics* **2014**, *33*, 374–392. [CrossRef]
86. Zuppetta, A.; Mazzoli, S. Analisi strutturale ed evoluzione paleotettonica dell'unità del Cilento nell'Appennino Campano. *Studi Geol. Camerti* **1995**, *13*, 103–114.
87. Cesarano, M.; Pierantoni, P.P.; Turco, E. Structural analysis of the Albidona Formation in the Alessandria del Carretto–Plataci area (Calabro-Lucanian Apennines, Southern Italy). *Ital. J. Geosci.* **2002**, *1*, 669–676.
88. Patacca, E.; Scandone, P. Late thrust propagation and sedimentary response in the thrust-belt foredeep system of the Southern Apennines (Pliocene-Pleistocene). In *Anatomy of an Orogen: The Apennines and Adjacent Mediterranean Basins*; Vai, G.B., Martini, I.P., Eds.; Kluwer Academic: Dordrecht, The Netherlands, 2001; pp. 401–440.
89. Capalbo, A.; Ascione, A.; Aucelli, P.P.; Mazzoli, S. Valutazione del tasso di erosione in Appennino Meridionale da dati geologico-geomorfologici. *Ital. J. Quat. Sci.* **2010**, *23*, 75–90.
90. Ghisetti, F.; Vezzani, L. Thin-skinned deformations of the western Sicily thrust belt and relationships with crustal shortening; mesostructural data on the Mt. Kumeta-Alcantara fault zone and related structures. *Boll. Soc. Geol. Ital.* **1984**, *103*, 129–157.
91. Roure, F.; Howell, D.G.; Müller, C.; Moretti, I. Late cenozoic subduction complex of Sicily. *J. Struct. Geol.* **1990**, *12*, 259–266. [CrossRef]
92. Catalano, R.; Valenti, V.; Albanese, C.; Accaino, F.; Sulli, A.; Tinivella, U.; Morticelli, G.M.; Zanolla, C.; Giustiniani, M. Sicily's fold-thrust belt and slab roll-back: The SI.RI.PRO. seismic crustal transect. *Geol. Soc. Lond. Mem.* **2013**, *170*, 451–464. [CrossRef]
93. Butler, R.W.H.; Maniscalco, R.; Pinter, P.R. Syn-kinematic sedimentary systems as constraints on the structural response of thrust belts: Re-examining the structural style of the Maghrebian thrust belt of Eastern Sicily. *Ital. J. Geosci.* **2019**, *138*, 371–389. [CrossRef]
94. Ghisetti, F.; Vezzani, L. Contribution of structural analysis to understanding the geodynamic evolution of the Calabrian arc (Southern Italy). *J. Struct. Geol.* **1981**, *3*, 371–381. [CrossRef]
95. Jongsma, D. The geometry and rates of microplate motions in the Eastern Mediterranean Sea-Quantitative constraints by using anoxic basins as piercing points. *Mar. Geol.* **1987**, *75*, 1–29. [CrossRef]
96. Catalano, S.; De Guidi, G.; Lanzafame, G.; Monaco, C.; Tortorici, L. Late Quaternary deformation on the island on Pantelleria: New constraints for the recent tectonic evolution of the Sicily Channel Rift (southern Italy). *J. Geodyn.* **2009**, *48*, 75–82. [CrossRef]
97. Guerrera, F.; Martín-Martín, M.; Raffaelli, G.; Tramontana, M. The Early Miocene "Bisciaro volcaniclastic event" (northern Apennines, Italy): A key study for the geodynamic evolution of the central-western Mediterranean. *Int. J. Earth Sci.* **2015**, *104*, 1083–1106. [CrossRef]
98. Mutti, E.; Lucchi, F.R. Le torbiditi dell'Appennino settentrionale introduzione all'analisi di facies. *Mem. Soc. Geol. Ital.* **1972**, *11*, 161–199.
99. Montuori, C.; Cimini, G.B.; Favali, P. Teleseismic tomography of the southern Tyrrhenian subduction zone: New results from seafloor and land recordings. *J. Geophys. Res.* **2007**, 112. [CrossRef]
100. Di Stefano, R.; Kissling, E.; Chiarabba, C.; Amato, A.; Giardini, D. Shallow subduction beneath Italy: Three-dimensional images of the Adriatic-European-Tyrrhenian lithosphere system based on high-quality P wave arrival times. *J. Geophys. Res.* **2009**, 114. [CrossRef]
101. Giacomuzzi, G.; Chiarabba, C.; De Gori, P. Linking the Alps and Apennines subduction systems: New constraints revealed by high-resolution teleseismic tomography. *Earth Planet. Sci. Lett.* **2011**, *301*, 531–543. [CrossRef]
102. Dallan Nardi, L.; Elter, P.; Nardi, R. Considerazioni sull'arco dell'Appennino settentrionale e sulla" linea" Ancona-Anzio. *Boll. Soc. Geol. Ital.* **1971**, *90*, 203–211.
103. Castellarin, A.; Colacicchi, R.; Praturlon, A.; Cantelli, C. The jurassic-lower pliocene history of the Ancona-Anzio Line (Central Italy). *Mem. Soc. Geol. Ital.* **1982**, *24*, 325–336.
104. Cipollari, P.; Cosentino, D. La linea Olevano-Antrodoco: Contributo della biostratigrafia alla sua caratterizzazione cinematica. *Studi Geol. Camerti* **1991**, 143–150. [CrossRef]
105. Cappelletti, F.; Panei, L.; Colucci, F.; Guandalini, R.; Moia, F.; Stella, G. ViDEPI Project Data Utilization for Sites Characterization in Range of CCS, Geothermal and Hydrocarbons Projects in Italy, 2009–2017. Available online: http://rse-web.it/documenti.page?RSE_originalURI=/documenti/documento/315786&RSE_manipulatePath=yes&country=ita (accessed on 12 April 2021).
106. Corradino, M.; Pepe, F.; Bertotti, G.; Picotti, V.; Monaco, C.; Nicolich, R. 3-D architecture and Plio-Quaternary evolution of the Paola Basin: Insights into the forearc of the Tyrrhenian-Ionian subduction system. *Tectonics* **2020**, *39*. [CrossRef]
107. Dewey, J.F. Plate tectonics. *Rev. Geophys.* **1975**, *13*, 326–332. [CrossRef]
108. Royden, L.; Patacca, E.; Scandone, P. Segmentation and configuration of subducted lithosphere in Italy: An important control on thrust-belt and foredeep-basin evolution. *Geology* **1987**, *15*, 714–717. [CrossRef]

109. Bigi, G.; Cosentino, D.; Parotto, M.; Sartori, R.; Scandone, P. *Structural Model of Italy Scale 1:500.000, Sheet 5. In C.N.R.*; Progetto Finalizzato Geodinamica. Structural Model of Italy; SELCA: Firenze, Italy, 1991.
110. Pondrelli, S.; Piromallo, C.; Serpelloni, E. Convergence vs. retreat in Southern Tyrrhenian Sea: Insights from kinematics. *Geophys. Res. Lett.* **2004**, *31*. [CrossRef]
111. Neri, G.; Barberi, G.; Oliva, G.; Orecchio, B. Spatial variations of seismogenic stress orientations in Sicily, south Italy. *Phys. Earth Planet. Inter.* **2005**, *148*, 175–191. [CrossRef]

geosciences

Article

Constraining the Passive to Active Margin Tectonics of the Internal Central Apennines: Insights from Biostratigraphy, Structural, and Seismic Analysis

Giovanni Luca Cardello [1,2,*], Giuseppe Vico [3], Lorenzo Consorti [4,5], Monia Sabbatino [6], Eugenio Carminati [1] and Carlo Doglioni [1,7]

1 Department of Earth Sciences, Sapienza University of Rome, 00185 Rome, Italy; eugenio.carminati@uniroma1.it (E.C.); carlo.doglioni@uniroma1.it (C.D.)
2 Department of Chemistry and Pharmacy, University of Sassari, 07100 Sassari, Italy
3 Department of Land, Environmental and Infrastructure Engineering (DIATI), Polytechnic of Turin, 10129 Turin, Italy; giuseppe.vico@polito.it
4 Department of Mathematics and Geosciences, University of Trieste, 34128 Trieste, Italy; lorenzo.consorti@isprambiente.it
5 Geological Survey of Italy (ISPRA), 00144 Rome, Italy
6 Department of Earth, Environmental and Resources Sciences (DISTAR), Federico II University of Naples, 80126 Naples, Italy; monia.sabbatino@unina.it
7 Istituto Nazionale di Geofisica e Vulcanologia, 00143 Rome, Italy
* Correspondence: glcardello@uniss.it

check for updates

Citation: Cardello, G.L.; Vico, G.; Consorti, L.; Sabbatino, M.; Carminati, E.; Doglioni, C. Constraining the Passive to Active Margin Tectonics of the Internal Central Apennines: Insights from Biostratigraphy, Structural, and Seismic Analysis. *Geosciences* **2021**, *11*, 160. https://doi.org/10.3390/geosciences11040160

Academic Editors: Domenico Liotta, Giancarlo Molli and Angelo Cipriani

Received: 1 February 2021
Accepted: 28 March 2021
Published: 1 April 2021

Publisher's Note: MDPI stays neutral with regard to jurisdictional claims in published maps and institutional affiliations.

Copyright: © 2021 by the authors. Licensee MDPI, Basel, Switzerland. This article is an open access article distributed under the terms and conditions of the Creative Commons Attribution (CC BY) license (https://creativecommons.org/licenses/by/4.0/).

Abstract: The polyphase structural evolution of a sector of the internal Central Apennines, where the significance of pelagic deposits atop neritic carbonate platform and active margin sediments has been long debated, is here documented. The results of a new geological survey in the Volsci Range, supported by new stratigraphic constraints from the syn-orogenic deposits, are integrated with the analysis of 2D seismic reflection lines and available wells in the adjacent Latin Valley. Late Cretaceous syn-sedimentary faults are documented and interpreted as steps linking a carbonate platform to the adjacent pelagic basin, located to the west. During Tortonian time, the pelagic deposits were squeezed off and juxtaposed as mélange units on top of the carbonate platform. Subsurface data highlighted stacked thrust sheets that were first involved into an initial in-sequence propagation with top-to-the-ENE, synchronous to late Tortonian foredeep to wedge-top sedimentation. We distinguish up to four groups of thrust faults that occurred during in-sequence shortening (thrusts 1–3; about 55–60 km) and backthrusting (thrust 4). During Pliocene to recent times, the area has been uplifted and subsequently extended by normal faults cross-cutting the accretionary wedge. Beside regional interest, our findings bear implications on the kinematic evolution of an orogenic wedge affected by far-traveled units.

Keywords: Central Apennines; passive margin inversion; mélange; pelagic deposits; thrust sheets; backthrust; cretaceous; Miocene; nannoplankton

1. Introduction

Carbonate platforms are a type of passive margin sedimentary succession that can be commonly involved in the thrust-sheet imbrication of an orogenic wedge [1–3]. During in-sequence ongoing deformation, the wedge propagates by incorporating new portions of the foreland, which is commonly made up of crystalline basement, clastic and/or carbonatic successions, and overriding foredeep/foreland clastics with variable thickness and composition [4–6]. The so formed fold-and-thrust belt, incorporating distinctive tectono-stratigraphic units, is the combined product of inherited syn-sedimentary structures and orogenic dynamics [7,8]. Thus, the wedge-related deformation style may strongly depend on the stratigraphic architecture and in particular on the presence and depth of décollement layers within the stratigraphic successions (i.e., salt [9]). In this sense, thick-skinned deformation (see, e.g., in [10]) can dominate when there is no suitable detachment horizon. On the contrary, when preferred slip-levels occur, thin-skinned tectonics develop, generating flat-ramp-flat geometries and disharmonic folding, which, for example, can occur within base-of-slope to pelagic successions [11,12]. At the transition between such structural domains, strain localization can occur, nucleating thrusts by inverting previous listric boundary extensional faults (see, e.g., in [13]).

During inversion of hyperextended passive margins, orogenesis forms far-traveled units that can reach a high-degree of internal deformation [14–16]. The chaotic structure of these so-formed mélange units is the result of the superposition of tectonic, sedimentary, and mud-diapiric processes [17], to which gravitative processes add, by incorporating both allochthonous and autochthonous blocks [18]. Despite the subsequent orogenic deformation overprint, occurring within far-traveled thrust-sheets, the structural heritage may be preserved and studied (see, e.g., in [19–22]).

The Apennines are a fold-and-thrust belt involving basinal and platform-derived thrust sheets and mélange units (Figure 1) that offer well-outcropping structures representative of inverted hyperextended passive margins. The present-day deep structure of the Apennines has been a long matter of debate, as the amount of thrust allochthony and the involvement of the crystalline basement are widely discussed (see, e.g., in [23–29]. In this frame, the recognition of inherited structures also bears implications on the reconstructions of the pre- to syn-orogenic evolution [30–33]. For the Central Apennines, timing of deformation and shortening rates through time were reconstructed by coupling kinematic reconstructions with dating of the deposits overlying the forebulge unconformity [34] or, more classically, by dating the siliciclastic syn-orogenic deposits of the foredeep and wedge-top basins by using biostratigraphy (see, e.g., in [35–38]). However, controversial age interpretations may be derived due to the occurrence of few index fossils or reworked specimens from cannibalized foredeep and wedge-top deposits (see, e.g., in [34–39]). Recently, thermo-chronological studies have provided absolute dating of calcite and fault-gouge that have supported the reconstruction of regional thrust evolution [40–43].

Considering that the central Apennines represent an orogen that involves large volumes of the Adriatic plate, identification and description of the most internal thrust sheets are fundamental to highlight the role of inherited structures in determining the dynamics of far-traveled thrusting. In particular, one of the most crucial problems is deciphering the degree of distance covered by the units after detachment within foreland, foredeep, and wedge-top basins during shortening. In this paper, we provide (i) a review of the existing literature of the Volsci Range (VR; Figure 1) and of the adjacent Latin Valley; (ii) a comprehensive stratigraphic and structural analysis based on new age determinations of the syn-orogenic deposits; and (iii) a reinterpretation of a composite dataset of public well data and seismic lines, integrated by unpublished data provided by Pentex Italia Limited. We recognize a polyphase structural evolution based on the documentation of the characteristic mélange structures in the Chaotic complex and the distinction of foreland-directed thrusts cross-cut by younger hinterland-directed reverse faults. As a brand-new outcome, the reconstruction of the pre-orogenic heritage and the syn-orogenic Miocene structures allows us to constrain a previously unpublished regional inversion tectonic process and its peculiar evolution of thrusting. In this frame, the internal Central Apennines represent an example of the kinematic evolution of platform and basin-derived thrust sheets. Our study can help unravel the evolution of similar belts worldwide, and more specifically contributes to the understanding of far-traveled thrust sheets.

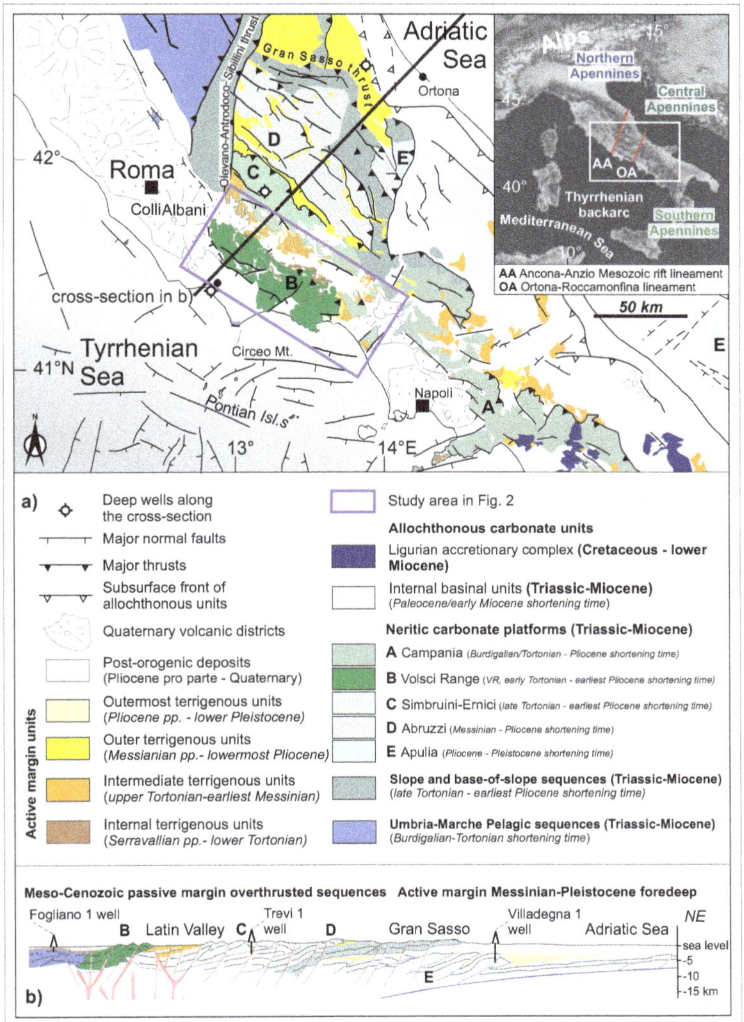

Figure 1. (**a**) Simplified Tectonic map of Central Italy (modified from the works in [30,38,44], showing the active margin units and the Meso-Cenozoic passive margin units. The shortening time is in italic. (**b**) Crustal cross-section (modified after the work in [45]). Deep well location is taken from in [23].

2. Geological Setting

2.1. The Central Apennines

The Apennines (Figure 1) are a ~1500 km long accretionary wedge made of different pre-orogenic and syn-orogenic units accreted together during the progressive E/NE-ward migration of leading-edge frontal thrusts and associated active margin units deposited within foredeep and wedge-top basins (see, e.g., in [46–50]. From Miocene time, the Apennine foreland became progressively involved in pre-thrusting bulging, uplift, and erosion resulting from the wedge migration [51–56]. Since Tortonian time (~11 Ma), the west-directed subduction of the Adriatic slab drove the development of the accretionary wedge now exposed in the central sector of the Apennine belt [49,54]. Subsequently, the fold-and-thrust belt underwent severe crustal stretching, related to back-arc extension that progressively migrated from the Sardinian margin to the axial part of the central

Apennines [49,57]. The chain is now uplifted and cross-cut by Quaternary normal faults and also affected by several volcanic centers along the Tyrrhenian margin [45,47,49,58,59].

The central Apennines constitute a mountain chain sector bounded by two major NNE-trending tectonic lines (Figure 1), comprised between two arcs with polyphase activity: the Ortona–Roccamonfina and the Olevano–Antrodoco–Sibillini lines [60,61]. The latter can be considered as the positive transpressive reactivation (see in [7] and the references therein) of a Mesozoic extensional fault system associated with continental rifting, the Ancona–Anzio line [62] (Figure 1).

The Mesozoic paleogeography was characterized by different domains defined by peculiar stratigraphic successions. West of the Olevano–Antrodoco–Sibillini line, Meso-Cenozoic pelagic sequences occur in the northern Apennines. East of the Ancona–Anzio line, the central Apennines are mainly formed by neritic carbonate platform units that are bounded by base-of-slope to basinal domains (e.g., Gran Sasso [30]). According to the works in [63,64], drowning of the Mesozoic carbonate platform of the VR occurred during the latest Cretaceous or Cenozoic times and is testified by basinal deposits lying on top of platform carbonates. More internal basinal/oceanic units, referred to the Sicilide and Ligurian Accretionary Complex, crop out both in the southern Apennines [44] and along the coast west of Rome (i.e., Tolfa region [65,66]; Figure 1). These units are traditionally recognized as allochthonous units that were involved into the wedge in Miocene time. The occurrence of similar internal allochthonous units in the central Apennines is still debated. A stratigraphic correlation between the deposits atop the neritic carbonates of the VR and the Ligurian-Sicilian basinal units of Sicily and southern Apennines was first made by [67]. A different interpretation was proposed by the authors of [65,66], who recognized the marly–terrigenous terrains atop the VR carbonates as the remobilization of the Cenozoic basinal succession.

The terrigenous units cropping out in the central Apennines mostly occur in NW-striking valleys (e.g., Latin Valley [68]; Figure 2). These units are representative of foreland basin deposits, whose formation nomenclature varies from region to region, i.e., the Frosinone Formation [64] shares similar timing and facies with the Termini and Pietraroja formations of the southern Apennines [69,70]. To harmonize their occurrence throughout the central and southern Apennines, we have grouped them in four different units, representative of progressively more external and younger stages of the wedge accretion towards the east (Figure 1a). To the south, as shown by well logs and outcrops in the Pontian islands and at Circeo Mt., Mesozoic basinal units overthrust Oligocene to early Miocene flysch units [42,71]. South of Naples (Figure 1), Serravallian to lower Tortonian flysch represent internal terrigenous foredeep units [44]. Serravallian syn-orogenic units, indicative of plate flexuration, were recognized as well in more internal positions within the Volsci Range [72]. Such flexural deposits rejuvenate towards the east suggesting a progressive shift of the wedge towards the outer portions of the arc. Intermediate terrigenous units of late Tortonian–earliest Messinian age occur in the Latin Valley and underneath the overthrusted platform carbonates of Campanian age.

North of the Latin Valley, the Simbruini-Ernici Mts are built up of NW-striking imbricate carbonate thrust sheets that overthrusted onto the outer terrigenous units of Messinian age (e.g., within the Latin valley, Figure 1 [73,74]). This is well evidenced by the Trevi well that shows the juxtaposition, at considerable depths (3000 m), of Triassic terrains onto Cretaceous and Miocene carbonates, testifying for the doubling of the Mesozoic succession [75]. A horizontal displacement in the order of 30 km and vertical offset of about 5 km has been proposed for this thrust [76,77], although field evidence from the Simbruini thrust front is at odds with this interpretation [49]. These ridges constitute the backbone of the internal sectors of the Central Apennines (internal Central Apennines), which first overthrust onto the outer active margin deposits and, during late Messinian time, were involved into renewed shortening [43]. Differently from the Internal Apennines, the axial and external parts of the chain, that occur more to the northeast, were involved into the wedge respectively during Messinian (Abruzzi) and Pliocene (Majella Mountain deformed Apulian terrains; Figure 1; see in [78]) times. During middle Pliocene time, the outermost terrigenous units experienced compression, while back-arc extension was affecting the internal part of the chain.

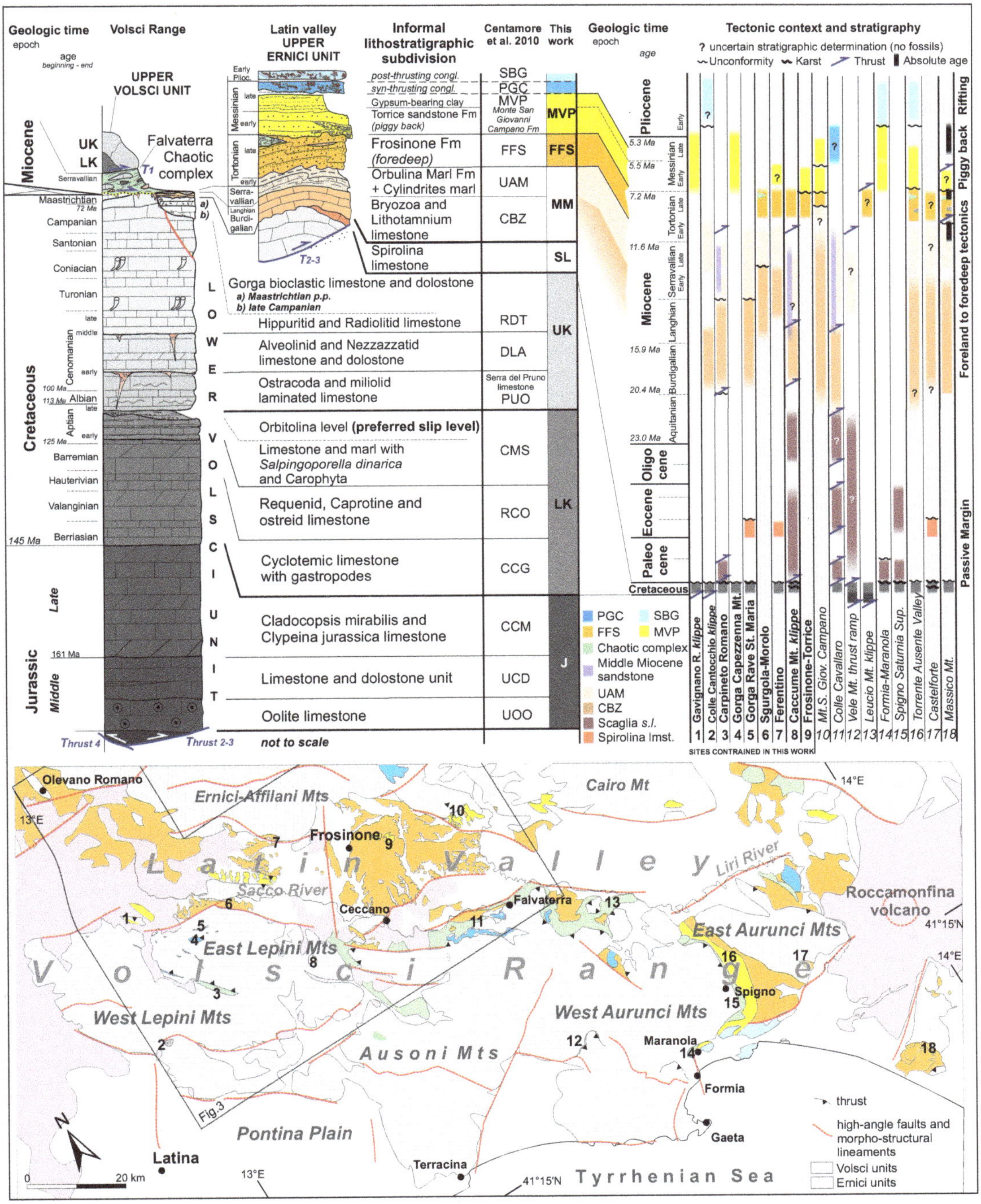

Figure 2. Simplified stratigraphic columns of the Volsci Range and the adjacent Latin Valley correlated, on the left, with the official cartography [64]. On the right, tectonic context and stratigraphy of basin deposits is reported from the literature (see Appendix A) and original data at representative localities. Localities from the Ernici unit are highlighted by vertical gray stripes. Below, the geological map of the study area with the studied locations and their respective numbers.

2.2. The Volsci Range and the Latin Valley

The VR is traditionally subdivided into major mountain groups, i.e., West Lepini, East Lepini, Ausoni, and West and East Aurunci Mts (Figure 2), that are separated by major valleys or mountain passes. More to the SW, the Mount Massico structural high occurs. These groups share a similar tectonic and stratigraphic evolution. The VR is mostly composed of passive margin Mesozoic neritic carbonates belonging to the Latium and Abruzzi platform or Apennine carbonate platform (see, e.g., in [79–81]). The Mesozoic dominant facies are representative of inner to rim carbonate platform environments (see, e.g., in [63,64,82,83]).

A compilation of the Mesozoic lithostratigraphic units cropping out in the Lepini sectors is presented in Figure 2. The Upper and Lower Volsci thrust sheets differ from the Upper Ernici unit on the basis of the Cenozoic stratigraphy. Of note, the VR succession generally bears a thin and incomplete succession of Paleocene to Miocene deposits [84] atop late Cretaceous formations of different ages, possibly due to progressive drowning of some sectors of the platform during Late Cretaceous time [63]. On the other hand, in the Latin Valley, the Ernici unit is thicker and also contains Eocene to early Tortonian foreland units and late Tortonian to earliest Pliocene active margin siliciclastic formations (see in [63] and the references therein).

Seismic interpretation studies in the Latin Valley, carried out by AGIP and other companies (www.videpi.com) (accessed on 20 January 2021), trace top-platform seismic horizons that allowed us to locally outline a fold-and-thrust structure [85]. According to the authors of [64,86], the VR front propagation affected the Latin Valley foredeep deposits that were doubled or even triplicated [45]. Upper and lower units in the Volsci Range and in the Ernici units of the Latin Valley were thus distinguished. As also shown in the cross sections in [64], thrusting involved the Cretaceous carbonates of the Ernici unit together with upper Tortonian foredeep sediments of the Frosinone Formation [63,64]. Finally, out-of-sequence thrusting during and after the Messinian salinity crisis was documented in [77,87], possibly related to backthrusting, like at Carpineto Romano [88]. The thrust front does not crop out, but according to the most recent reconstructions, it is offset by normal faults [45,86]. At least from Middle Pliocene time, the study area experienced regional uplift, accompanied by subaerial exposure and consequent diffuse erosional processes that generated erosional surfaces, now found at different elevations [63].

According to the authors of [89], just north of VR the uplift rate increased during the last 2.4 Myr. In the VR, no such detail was reached yet. However, early to late Pleistocene slope, river, and lacustrine paralic and continental deposits were mapped within depressions bounded by high-angle NW- and NE-striking normal faults that dissected the fold-and-thrust fabric. Further, E-striking transtensional faults contribute to generate middle Pleistocene wrench zones and basins between the Latin Valley and the Pontina Plain. Syn- to post-tectonic upper Pliocene–middle Pleistocene continental successions are preserved in the Middle Latin Valley, the Pontina Plain, and locally in the VR intermontane depressions [64]. Further, during late Pliocene to possibly Holocene times, the fold-and-thrust belt was progressively cross-cut by a system of conjugate synthetic and antithetic normal faults determining the formation of the coastal plain and intra-mountain depressions [64,90,91]. The VR hosts volcanic terrains of Pleistocene age from both nearby volcanic districts and local eruptive centers belonging to the Volsci Volcanic Field (VVF; Figure 1 [64,92]).

3. Materials and Methods

3.1. Stratigraphic Review and New Paleontological Determinations

The lithostratigraphic architecture of the Meso-Cenozoic carbonate platform succession has been reviewed, following the scheme in [45], and it has been integrated with a stratigraphic chart that compares eighteen different key localities representative of pre-orogenic passive margin to syn-orogenic foreland basin lithostratigraphic units throughout the study area (Figure 2). Erosive submarine and karst-related unconformities are reported to support the regional review of the syn-orogenic evolution, also constrained by the absolute ages provided in [43] for the Massico Mt ridge. The overall stratigraphic setting allowed us to correlate diachronous events among different structural units from the Volsci Range and Latin Valley. Lithologies not constrained by biomarkers are traced by a question mark, whereas lithologic and biostratigraphic information coming from the review of the existing literature is resumed in the table of Appendix A. We have harmonized the stratigraphic information published in the 1:100,000 maps (i.e., Latina, Frosinone, and Alatri; https://www.isprambiente.gov.it/) (accessed on 20 January 2021), and in the more recent and detailed 1:50,000 maps (i.e., Anagni, Ceccano, and Velletri; https://www.isprambiente.gov.it/) (accessed on 20 January 2021) as well as and in other papers (i.e., in [72,84,86–88,93,94] and, using the stratigraphic nomenclature after that in [64], then grouped the deposits into the broader informal lithostratigraphic subdivision of Figure 2.

New biostratigraphic information was acquired by studying Upper Cretaceous–Miocene to early Pliocene samples collected from fifteen localities at Colle Cantocchio, Gorga, Gavignano, Carpineto Romano, Caccume Mt., and Siserno Mt. (Figure 2). Further sampling through the Latin Valley at Morolo, Ferentino, and Frosinone localities was performed in order to determine facies and fossil content of syn-orogenic deposits. Hard rock samples have been prepared for thin sections analysis, which provided thirty-three new age determinations. Further, we collected seventeen samples for nannoplankton using samples prepared under smear slide technique, and following the procedures described in [95]. We observed the nannoplankton content through the polarized light microscope Zeiss Axioscop equipped with an $\times 100$ oil immersion objective lens. We performed a qualitative evaluation of the assemblages on all the samples, but only twelve of them proved to be fossiliferous, while five other ones are barren or poorly fossiliferous. Important time maker nannoplankton taxa were identified up to species level, as presented in Supplementary Material. We base our time determination on the micro-biostratigraphic frames in [82,96–98] for the shallow-water carbonate assemblage and the biostratigraphic scale in [99–101] for the nannoplankton.

3.2. Structural Analysis

A new structural-geological survey of the carbonate and siliciclastic succession integrates previous work of the Geological Survey of Italy (ISPRA) (i.e., in [64,102,103] and the references therein). The resulting new geological map is built also considering a specific review of the 1:50,000 geological sheets "Anagni" and "Ceccano" in order to avoid lithostratigraphic synonymy (see Appendix A) [64,103].

Bedding attitude was retrieved from existing map sheet tables at the scale 1:25,000 on a stripe of about one kilometer to each side of the main cross section (Figure 3). In order to constrain fault kinematics, field measurements of faults, fractures, and slicken-fibers were collected at key localities and plotted by means of TectonicsFP software [104] with lower-hemisphere projections and rose diagrams. In particular, at each locality eigen vectors are calculated from the bedding and are indicative of the orientation of the axes of deformation, where the gray circles are representative of the plane between the principal and minimal eigenvector. In general, an eigenvector is a vector which gets stretched, but not rotated, when operated on by the matrix. Considering that eigenvectors have corresponding eigenvalues, the amount of squeezing or stretching (the strain) is called the eigenvalue. Eigenvectors from key localities are reported in Table S1 (Supplementary material).

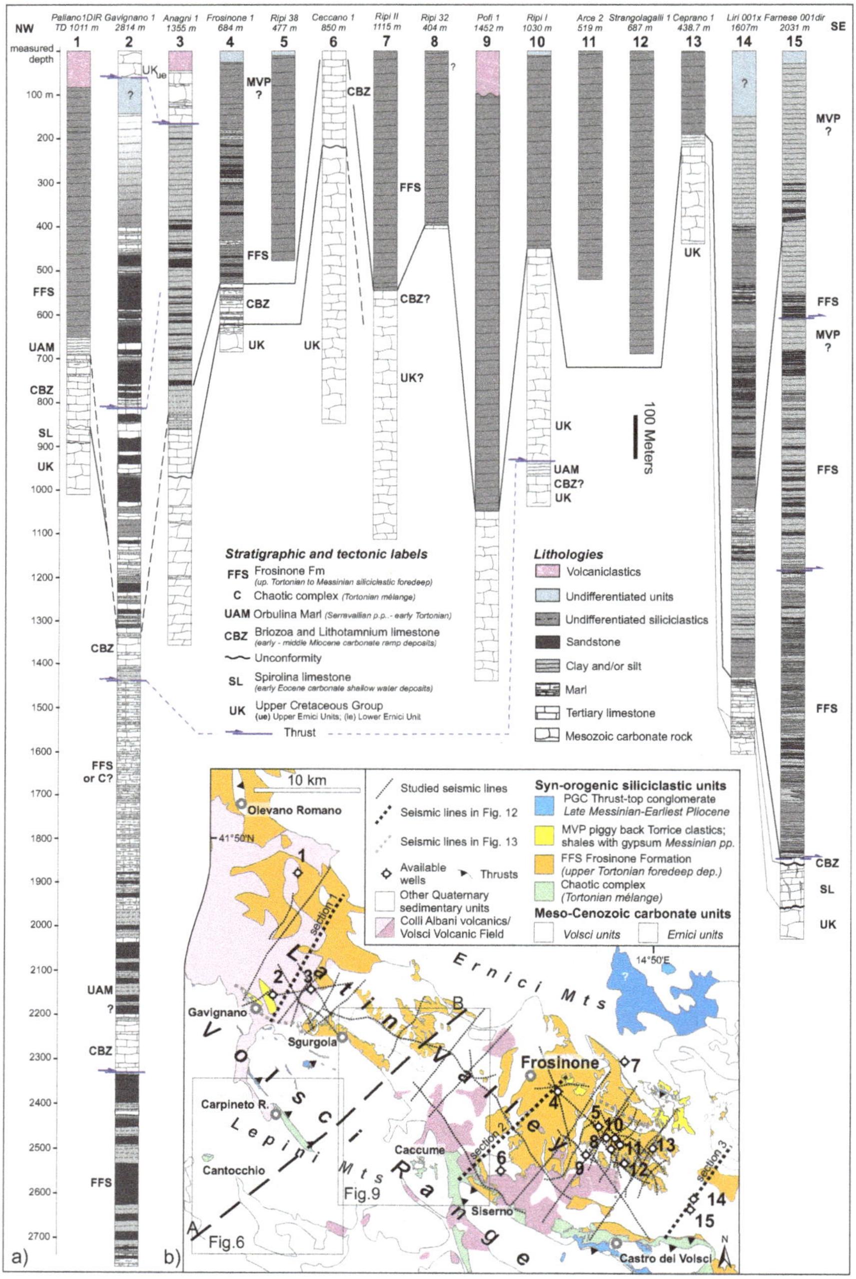

Figure 3. (**a**) Comparison among lithostratigraphic data from wells in the Latin Valley. (**b**) Sketched geological map with the location of the studied wells and seismic lines. Wells 1, 2, and 4 are from a public dataset (www.videpi.com) (accessed on 20 January 2021). Wells, 3, 5, 8, 11, 12, and 13 are provided by Pentex Limited Italia (see Acknowledgments). Wells 6, 7, 9, and 10 are reported in [105]. Full lines are stratigraphic correlations within the same structural unItal. Black dashed lines are uncertain stratigraphic and tectonic correlations (blue lines). Regional cross section AB (Figure 6) and the detailed structural maps of the Figures 6 and 9 are also shown.

3.3. Borehole Data from the Latin Valley

Composite well log data from the exploration and production of hydrocarbon activity were used to calibrate the seismic lines (Figure 3). Fifteen wells were drilled through the syn-orogenic lithologies, and they provide insights on late Miocene siliciclastic deposits. Four wells are from a public database (www.videpi.com) (accessed on 20 January 2021), the others were extrapolated from the literature [64,106,107] or confidential reports provided by Pentex Italia Ltd. The stratigraphic calibration of the seismic profiles was performed by using (i) the Frosinone 1 well, which is located within a relatively dense network of seismic lines and drilled at total depth of 684 m, reaching the Orbulina Marl Fm at 526 m and the CBZ at 551 m, while the Cretaceous carbonate platform top was encountered at 620 m, and (ii) the Anagni 1 well, which encountered mesozoic platform carbonates between 47 and 162 m and reached again the carbonate top at 862 m after having crossed a thick siliciclastic succession (Figures 2 and 3). Three wells were characterized by velocity data that allowed us to calibrate seismic data and/or calculate average and interval velocity for the identified macro-units. Where velocity logs were not available, an average interval velocity based on our calculations was applied to fit with the correspondent lithology and reflector detected on seismic profile. In few cases, velocity logs were available for a direct local time-depth chart; in the other cases, average velocity obtained by the analysis of the available logs and from literature was used. These two velocity laws were used to depth-convert the two-way-time interpretation on seismic dataset, in order to define thickness and depth of the main top interpreted horizons to set the geological cross section (Figure 14). Biostratigraphic data are available only for a few key wells (i.e., Paliano 1, Gavignano 1, Anagni 1, Frosinone 1, Liri 1, and Farnese 1) and have been anchored using the regional scale in [96].

3.4. Seismic Dataset

The structural setting of the Latin Valley presented in this study largely relies on thirty-eight 2D seismic reflection profiles irregularly arranged (map view Figure 3b). In the north, some seismic lines gather around the Gavignano 1 and the Anagni 1 wells, while in the south they occur together with different wells (Figure 3). The seismic sections originate from different acquisition campaigns carried out in the 1980s and 1990s for the exploration of hydrocarbons by AGIP and recently by Sovereign and Pentex. Most seismic lines are part of a public dataset (ViDEPI Project. Available online: https://www.videpi.com accessed on: 20 January 2021. This public network has been integrated by a few other seismic lines from different surveys, to better constrain the structural setting of the Latin Valley. The interpreted seismic dataset was a stack version. Public data were in raster format, so we produced segy files for each raster seismic line in order to be able to import all the dataset into the interpretation software (OpendTect). This was achieved using Kogeo© 2.7, a free and open software for 2D/3D seismic data analysis that allows to create a geo referenced seg-y file from a scanned seismic image (http://www.kogeo.de/index.htm) (accessed on 20 January 2021). Seismic quality is good to poor, probably due to a lack of reprocessing and therefore interpretation may be inaccurate in some points. In those cases, we have integrated the outcropping geological information to reconstruct a geological model along the seismic profile, identifying when possible the main reflectors.

The most evident reflectors are the unconformities at the top of the upper Cretaceous carbonates (Figure 4), and of the Orbulina Marl Fm. (UAM; Figure 2). To calibrate and detect the main reflectors/markers in the Latin Valley, a synthetic seismogram was created for the Anagni-1 well (Figure 4) by focusing on the following formation discontinuities (from the bottom to the top): at the top of the Cretaceous limestones (UK), at the top of the Bryozoa and Lithothamnium limestone (CBZ), and at the base of the Frosinone Formation (FFS). For the interpretation of the seismic profiles, we identified the top-CBZ as the key reflector with the strongest acoustic impedance contrast observed over the entire Latin Valley. This often corresponds to the UAM lithostratigraphic unit (Figure 2), which at the basin scale corresponds with one of the most used reflectors that tie wells with seismic lines [108–110]. Miocene and Cretaceous near-top reflectors are well recognizable because of the characteristic geometry and energy picks that are stronger than the adjacent reflectors. In particular, the marly layers reflect most of the down-going seismic energy, obscuring the siliciclastic sequence or the underthrusted carbonate units. Despite the limited thickness of UAM, this reflector was followed also on the poorer quality seismic lines.

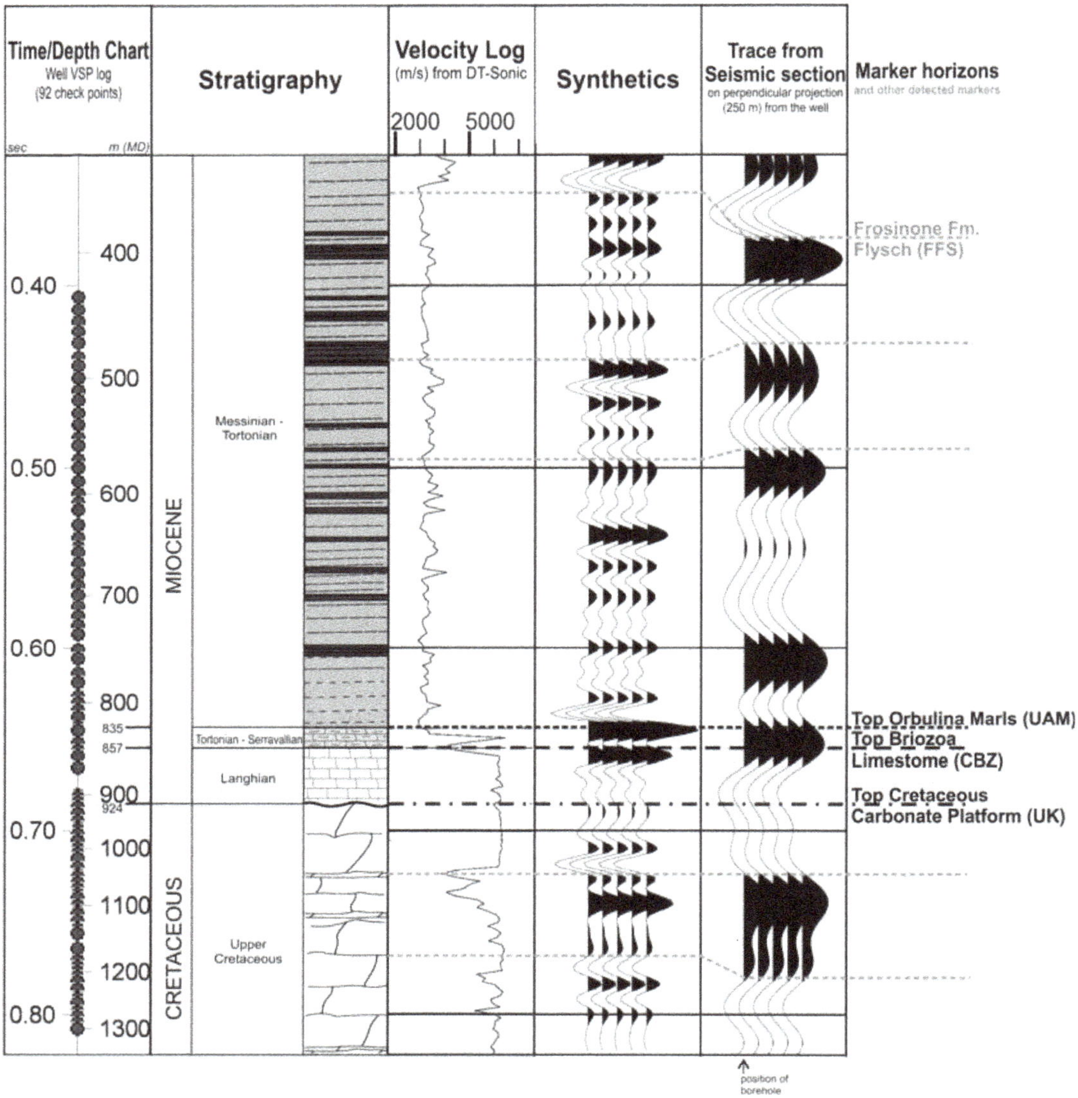

Figure 4. Simplified stratigraphic column of the Anagni 1 borehole with velocity log, synthetic seismogram, and an extract of a seismic section passing by the well (location in Figure 3). The seismic marker horizons and additional stratigraphic horizons interpreted in this study are also shown.

4. Results

4.1. Stratigraphic Constraints

4.1.1. Stratigraphic Review

The stratigraphy of the study area is schematically reported from the literature in Figure 2, where the lithostratigraphic units are anchored to the exposed sections at each of the eighteen localities presented in the map. The basics of the different tectonic units are exposed in Section 2.2. A new set of ages is proposed for the succession cropping out at the northern Volsci Range, as shown in the next section.

The Upper Campanian to Eocene carbonate platform succession that rest on the Hippuritid and Radiolitid limestone is generally missing [64], possibly due to a widespread depositional hiatus, although it locally crops out (e.g., at Gorga [103,111]). Note that the shallow-water Spirolina limestone (lower to mid-Eocene [112]), which crops out only in rare patches comprised between two unconformities—probably related to emersion events—was found at Gorga [112], Ferentino, and Castelforte (Figure 2; see also in [96,113], while it was recognized in well logs of Paliano 1 DIR and Farnese 001dir (Figure 3). In the Volsci Range, the Bryozoa and Lithothamnium limestone (CBZ) was dated as middle Miocene (see, e.g., in [64,87]). However, our data from the Volsci Range show that at least the CBZ base is early Miocene in age (see Section 4.1.2). Locally in the Volsci Range (e.g., Carpineto Romano, Figure 2), the CBZ lithotype is reported to occur within and beneath the allochthonous sub-Ligurian units [72], that can be compared with the Falvaterra Chaotic complex in [63].

Overall, the Falvaterra Chaotic complex is an ensemble of Paleocene to middle Miocene lithoclasts (from dm to decametric) wrapped within a matrix, whose best age constraints were provided mostly from the outcrops of Colle Cavallaro [114]. The basal contact of the Chaotic complex, although tectonically overprinted [63], is often marked by a ferruginous-limonitic veneer that occurs as a calcareous-detrital iron-oxide cruston. Differently from the classical carbonate hardgrounds, that are surfaces of synsedimentarily cemented carbonate layers that have been exposed on the seafloor under an extremely low sedimentation rate, the crustons of the Volsci Range could be either of karstic origin and/or the product of fluids involved into thrust faulting. Near Formia these crustons occur on top of peritidal limestones with benthic foraminifera (redetermined after the work in [63]) including *Spirolina* sp. [115], which can be possibly attributed to the early Eocene [111]. In particular, the foraminifera shown in [63] (their Figure 4) appear closer to some shallow-water discorbidae rather than planktonic forams. However, this need to be verified with new determinations. Our data constrain the top platform units providing new insights on the correlation, envisaged in [63], between these crustons and the Upper Cretaceous–lower Miocene succession preserved in the Chaotic complex (cf. Section 5.1 on the basis of the new stratigraphic constraints presented in Section 4.1.2.). Concerning the stratigraphic evidence from the Paleogene-early Aquitanian pelagic terms atop (Figure 2), they are mostly represented by Scaglia lithotypes (e.g., Formia and Spigno Saturnia, Figure 2). These lithotypes also crop out beneath the thrust south of Carpineto Romano, and beneath the Caccume Mt. and Colle Cavallaro klippen (Figure 2). Further, Scaglia *sensu latu* lithotypes were found as blocks of various dimensions wrapped in clayey matrix together with: early-middle Miocene lithoclasts (Figure 2; Appendix A), upper Serravallian cherty marl, and massive to laminated arcosic greywackes with mica [103]. The latter resulted sterile at the Caccume Mt. [84]. Lithologies of clasts involved into the Chaotic complex belong to a wide chronostratigraphic interval (i.e., Paleogene-Serravallian pro parte; Figure 2). More to the south, beneath the Vele Mt. thrust, siliciclastic marly deposits, mapped as Chaotic complex equivalent units, occur. Our data provide age constraints for the northern Volsci Range, see Section 4.1.2, and provide insights on the stratigraphic development of the sedimentary succession later deformed as Chaotic complex.

In the Latin Valley, the Frosinone Fm. was homogeneously attributed to late Tortonian time, while on the northeastern edge of the valley the base seems to be younger (i.e., uppermost Tortonian [87]). The upper part of the Frosinone Fm. unit bears olistoliths and olistostromes [115], from Mesozoic platform and Chaotic complex equivalent lithologies. They are reported at Sgurgola [35] and in the Torre Ausente Valley [64,116], although not as nicely cropping out as at the Massico Mt. [37].

Well data show a highly variable facies pattern of the siliciclastic units that include carbonate intercalations and thick marly successions with minor to rare sandstone horizons (Gavignano 1; Anagni 1; Frosinone 01; Farnese 001 wells; Figure 3). Due to tectonic juxtaposition, these successions may appear repeated at least twice and thus also reaching a total thickness of about 1.8 to 2.5 km at Gavignano and Liri and Farnese wells. Single thrust-bounded siliciclastic units are up to some 0.7 km thick.

In particular, the Gavignano 1 well hits four repeated siliciclastic-marly sequences bounded by thrust faults juxtaposing older terrains above younger ones. The uppermost unit is constituted of Upper Cretaceous (UK) limestones (cf. Anagni 1 well). The deeper fault-bounded units are about 600–900 m thick. Their siliciclastic sequence is defined by different lithofacies associations including alternations of sandstone, marl, and limestone. By correlating the wells providing detailed biostratigraphic information (e.g., Paliano, Gavignano, and Frosinone), we have correlated similar lithostratigraphic units, thus providing a formation identification. Biostratigraphic data from wells do not report Messinian taxa. Thus, we consider the Messinian Monte San Giovanni Campano unit (MVP) following the work in [63] and composed of wedge-top clastics [87], including other formally defined lithostratigraphic units (i.e., Torrice Sandstone Fm, Figure 2). Despite this lack of subsurface biostratigraphic information, its occurrence at depth cannot be excluded. Further, the correlation among conglomerates bearing exotic clasts of granitoids (SBG) is not clear as not supported by resolutive available stratigraphic information. However, their occurrence is of regional relevance as they could be representative of the transition from late orogenic [117] to backarc settings (i.e., Formia; Figure 2).

4.1.2. New Stratigraphic Constraints

New stratigraphic data from the northern Volsci Range and Latin Valley constrain the age of sedimentary units (Appendix B). The uppermost Cretaceous carbonate units were studied at different localities to reconstruct the tectono-stratigraphic setting of the top of the platform before thrusting. This information is provided by the variable thickness and facies distribution of the carbonate units between the Hippuritid and Radiolitid limestone and the ferruginous cruston on top, which usually marks the top of the platform. East of Gorga (Figure 2), the Hippuritid and Radiolitid limestone is overlain by some decameters of Maastrichtian bioclastic limestone and dolostone. This unit is truncated at the top by breccias, indicating an unconformity on the Upper Cretaceous succession. Those breccias are intercalated with a middle Burdigalian shallow-water marly level (Lep 12c, Appendix B) passing upward to typical CBZ limestone.

The Mesozoic platform top was found on top of the Lower Volsci Unit at the Caccume Mt., where it occurs as an encrusted breccia. At Carpineto Romano (Figure 5), atop of the platform succession of the Lower Volsci Unit, when preserved, discontinuous thin patches of proximal early Miocene CBZ limestone and middle Miocene Orbulina Marl formations occur (cf. Cosentino et al., 2002). At Colle Cantocchio (Figure 2; Appendix B), the early Miocene CBZ limestone was found disconformable on the Jurassic-Cretaceous limestone, which is marked by a hardground (structural details in Section 4.2.1).

Atop the Meso-Cenozoic carbonate units, the Chaotic complex occurs as a mélange that contains both native and exotic blocks, the latter being Cretaceous to Miocene basinal to distal ramp deposits that are coeval with the in situ formerly described proximal succession (Figure 5). Both block types are internally folded. South of Carpineto Romano (Figure 3), the deformed platform blocks involved within the Chaotic complex are stratigraphically comparable with the encrusted carbonates that are preserved at the top of the Lower Volsci Unit (cf. Figure 2). In particular, within the Chaotic complex, we have mapped several lenses of Cenomanian to early Campanian limestones covered by middle Campanian karstic breccias and ferruginous to limonitic cruston (Figure 5; structural details in Section 4.2.2).

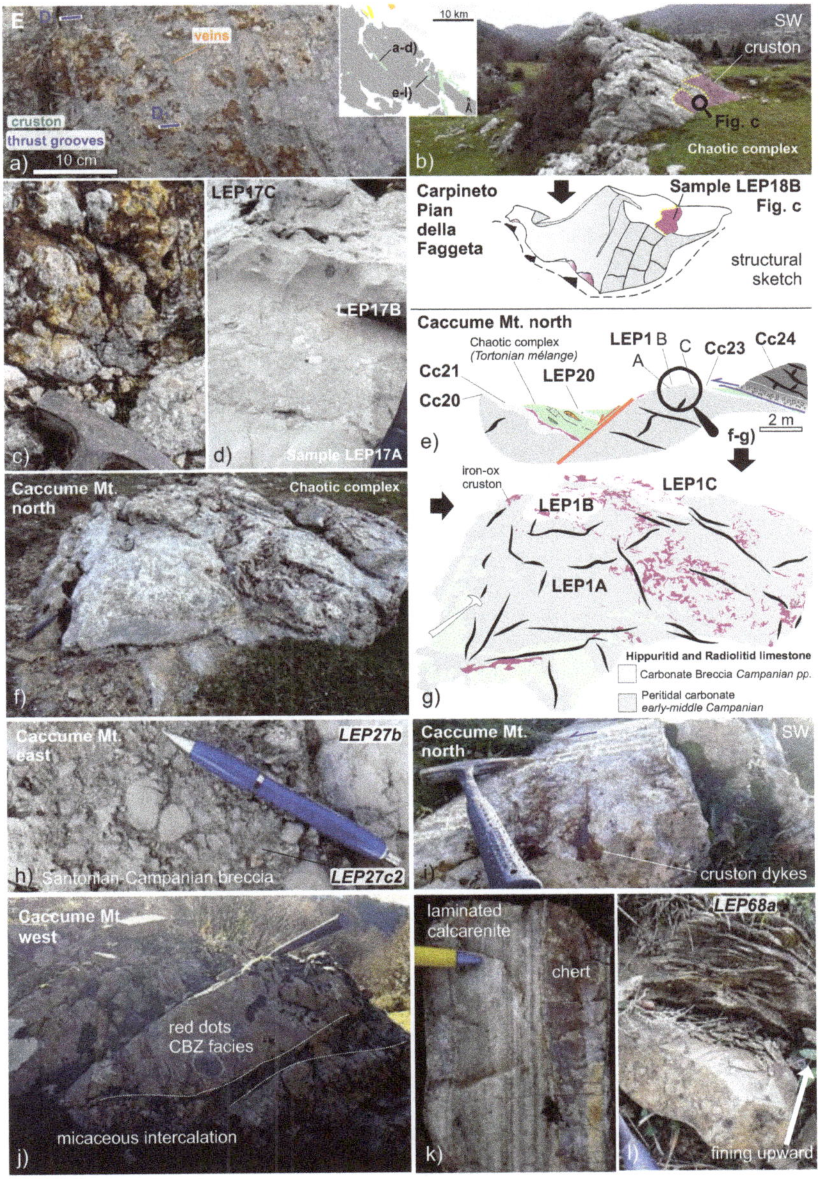

Figure 5. Sampled lithologies of the top of the platform and Chaotic complex. (**a**) Carpineto Romano (Pian della Faggeta; cf. Figures 6–8), encrusted top of platform crossed by E-trending thrust grooves and later veins having growth-fiber lineations plunging towards the NE (corresponding to plot 3 in Figure 8; 41°34′51″ N/13°6′30″ E); (**b–c**) encrusted native block within the Chaotic complex; (**d**) Campanian breccia beneath cruston; (**e**) Sampling site of the top of the Lower Volsci Unit north of Caccume Mt. and inherited paleo-topographic reconstruction; (**f–g**) outcrop detail of the cruston and underneath discordant units. (**h**) example of discordant Santonian-Campanian breccia beneath Chaotic complex. (**i**) Small-scale dykes of the grooved top platform cruston (41°34′32″ N/13°14′5″ E); (**j,k**) lower Miocene blocks 41°34′33″ N/13°13′20″ E; (**l**) Tortonian turbidites from Caccume Mt. north.

Differently from the native blocks, the Scaglia-type pelagic to hemipelagic limestones (with rare planktonic foraminifera and iron oxides) occur as exotic inclusions. In this category, at Carpineto Romano and Caccume Mt. (Figures 2 and 5; Appendix B), we have found CBZ blocks of early Miocene age represented by red dots (iron oxide spherules) glauconitic calcarenite associated with micaceous intercalations and chert (Figure 5). Minor lenses of hemipelagic middle Miocene marl and sandstone occur as well. Overall, the blocks are wrapped within a sandy-clayey matrix that is alternated with shales, foliated brownish marl, greenish arenaceous beds with exotic lithic, and coarse-grained micro-conglomerate with carbonatic and crystalline elements.

The matrix of the Chaotic complex at the base of the Caccume and Siserno mounts, includes Paleocene-Eocene, Oligocene-early Miocene, middle Miocene, and perhaps also late Tortonian-Messinian nannofossil assemblages (Appendix B). A similar wide span of ages was obtained from the shaly units of Colle Cantocchio (Figure 2), where Mesozoic to Tortonian nannoplankton reworked specimens were found beneath a major thrust (Appendix B; see also Section 4.2.1).

In the Latin Valley, the nannoplankton from the Frosinone Fm. can be referred, although rare or hardly diagnostic, to late Tortonian time. Wedge-top conglomerate deposits were studied at two key localities. At Gavignano (Figure 3), folded calcareous conglomerate occurs atop karstified Cenomanian limestones that according to the well data are juxtaposed on arenaceous deposits (cf. Figure 3). The clasts of mixed origin are from the Upper Cretaceous carbonates (i.e., Coniacian-Campanian and Albian-Cenomanian; see also Farinacci, 1965) and from the Tortonian Orbulina Marl Fm. The embedding matrix is made of abundant quartz grains along with reworked *Amphistegina* and *Elphidium* that make it possible to refer the whole Gavignano clastic deposit to the MVP unItal. In particular, the fining upward series with rare sandy matrix at the base (LEP10L) are dated to the latest Tortonian-earliest Zanclean and the clay marl at the top (LEP10M) to the Messinian. Thus, we consider this topmost constrain as indicative of the Messinian age of the MVP unit in the Latin Valley.

Within the eastern Lepini backbone, the conglomerates of Gorga are composed of pebbles and rounded blocks of reworked conglomerates whose clayey matrix and a bioturbated marly pebble were investigated. The age of these samples is late Tortonian for the marly pebble due to the presence of the coccolithophore *Discoaster surculus*, and top Tortonian–earliest Zanclean for the clay matrix bearing the marker *Amaurolithus primus*.

4.2. Structural Analysis of the Volsci Range

In this section, we document the field data used to reconstruct a geological cross section across the northern Volsci Range. The Western Lepini Mounts essentially consist of a 3 km thick Jurassic to Cretaceous carbonates dipping to (E)NE, whose local variations are shown in the stereoplots from 1 to 6 in Figure 6. The Neogene lithostratigraphic units atop are locally preserved beneath a few klippen structures that we document in detail in the next paragraphs. In the map and in the cross section of Figure 6, two areas are highlighted and described in detail as they preserve novel insights about pre-orogenic and syn-orogenic tectonics, which are presented from the oldest to the youngest event.

Near the western edge of the Western Lepini Mounts, a detailed survey performed at Colle Cantocchio allowed us to update the previous work by providing details on the stratigraphic contacts and fault kinematics (Figure 7). In particular, we integrate the data from in [93] by describing the pre-orogenic contacts and the low-angle fault juxtaposing Cretaceous rocks onto the Orbulina Marl Fm. As we can see from the panoramic view and cross section (Figure 7), lower Cretaceous calcareous dolostones (LK) are juxtaposed to a thick Jurassic-Cretaceous succession. The LK unit is downthrown towards the WSW and it overall consists of a striated proto-cataclasite of a normal fault (in orange). The fault has a cut off angle of about 40° with the footwall bedding. On top of this fault (paleofault, orange line in Figure 7), patches of lower Miocene CBZ occur sealing the contact (see Section 4.1.2). At the contact, an oxidized bluish rim of Mesozoic limestones marks the paleoescarpment (yellow dotted line in Figure 7), which is surrounded by altered shales (late Serravallian-Tortonian pp. Orbulina Marl Fm).

Such an inherited tectono-stratigraphic setting is preserved at the footwall of a thrust, whose hanging wall consists of a one-hundred-meter-thick pile of Upper Cretaceous (early-mid Campanian) limestone, and whose base constitutes the roof of a cave. The cave is defined by an iron oxide-rich striated principal slip surface. In the hanging wall, cataclastic bands are crosscut by minor mirror-like faults.

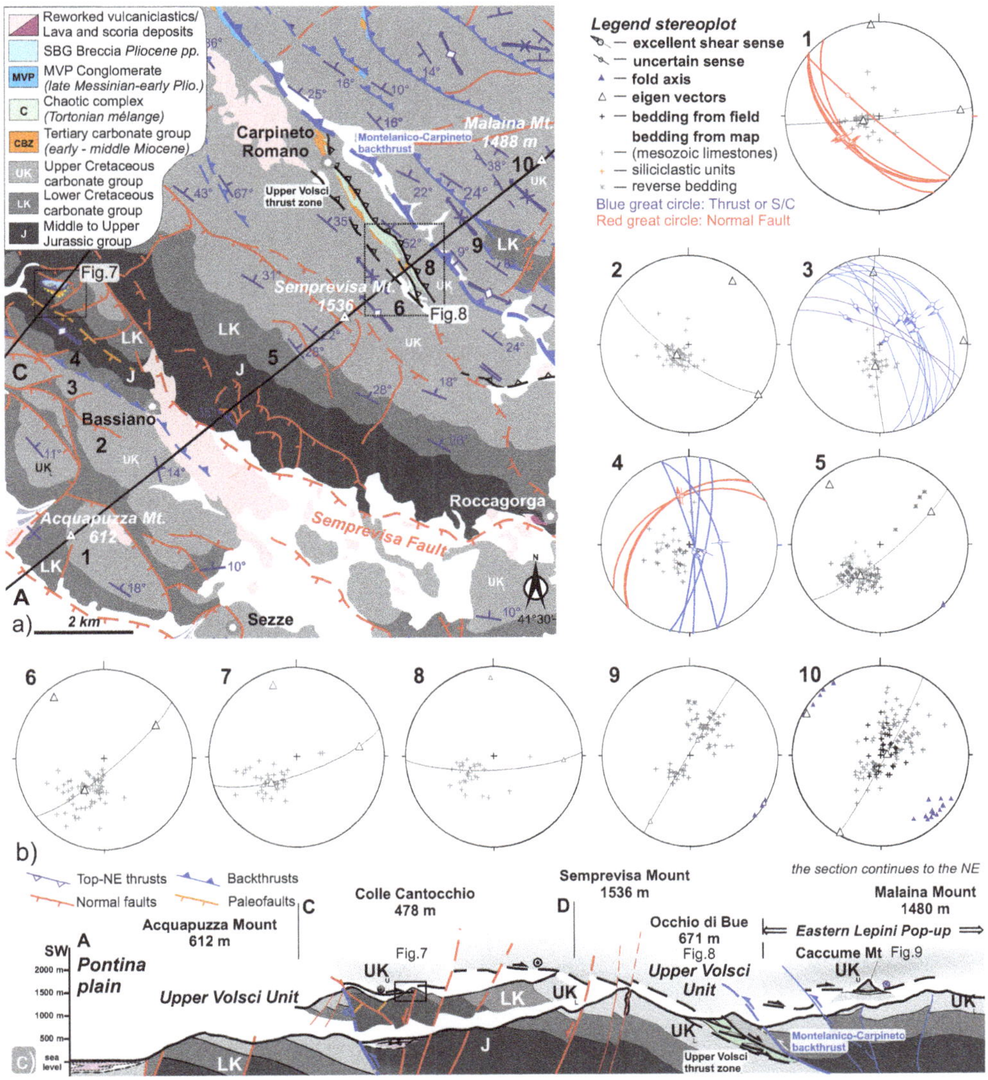

Figure 6. (**a**) Geological map of the western Lepini sector. (**b**) Stereoplots (lower hemisphere projection, equal area) summarizing orientation data for the structural elements representative of the subdivided areas in panel (**a**). Eigen vectors are indicative of the orientation of the axes of deformation calculated from the bedding, where the gray circles represent planes that contain the intermediate and maximum eigenvectors, as shown also by the data reported in the supplementary material. (**c**) Cross-section of the Volsci Range limited to the Malaina Mount to the northeast.

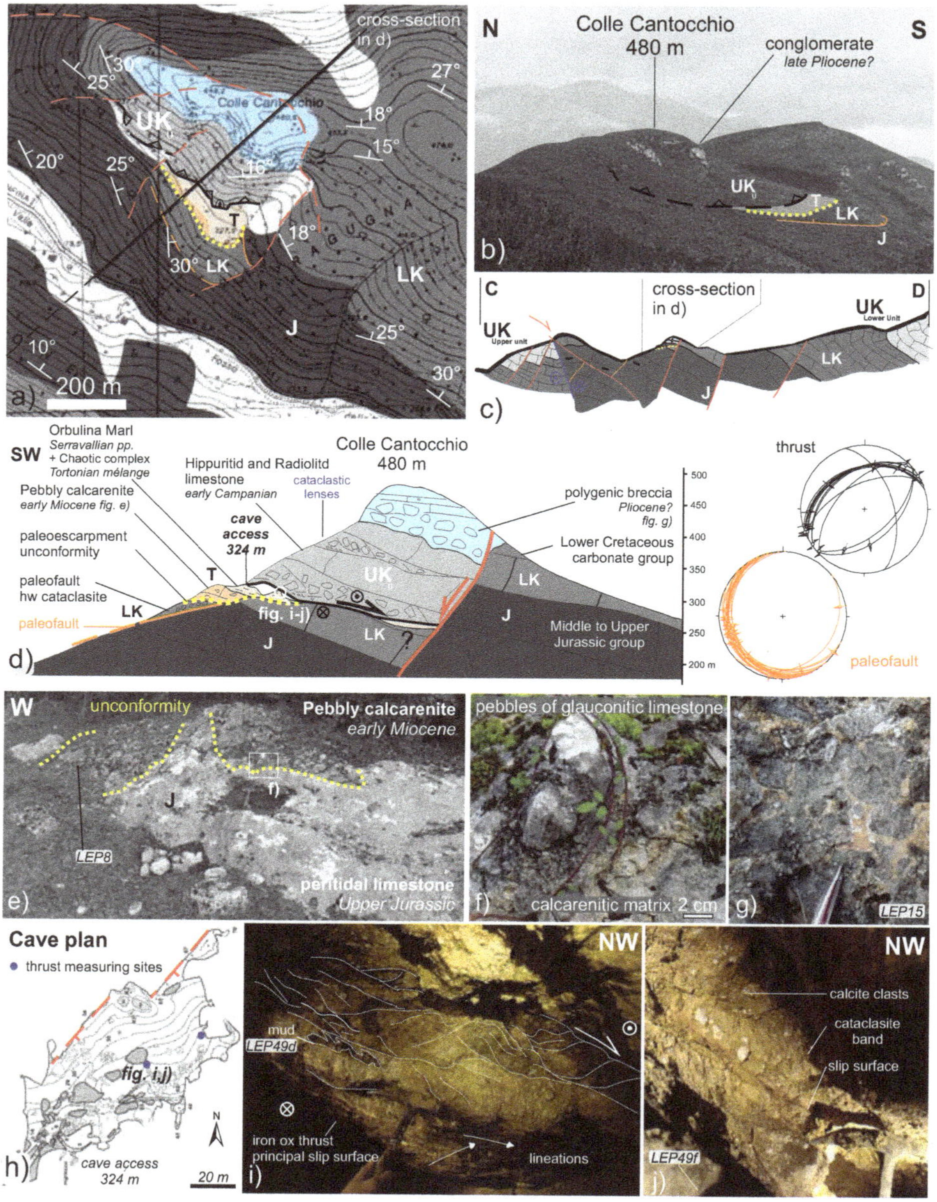

Figure 7. (**a**) Geological map of Colle Cantocchio modified after Cocozza and Praturlon (location in Figure 6 [93]). (**b**) Structural overview looking eastward. Blue line: thrusts and transpressive faults; yellow dotted line: paleoescarpment unconformity below Middle Miocene terrains (T); orange line (paleofault). (**c**) Larger geological cross section from Figure 6 and detailed (**d**) cross section (bold line traced in panel (**a**)) with stereoplots (lower hemisphere projection, equal area) of faults with slickenlines measured at the paleofault and in the roof of the cave. (**e**) Detail of the paleoescarpment contact of the pebbly calcarenite (**f**) over the hardground composed of oxidized Upper Jurassic peritidal limestones (41°34′29″ N/13°0′9″ E); (**g**) Polygenic breccia composed by Miocene and Cretaceous calcareous clasts with a reddish cement and calcareous matrix. (**h**) Cave details, grooved-base thrust fault zone constituted by foliated cataclasite bands (**i**,**j**). Sampling sites are referred to Appendix B.

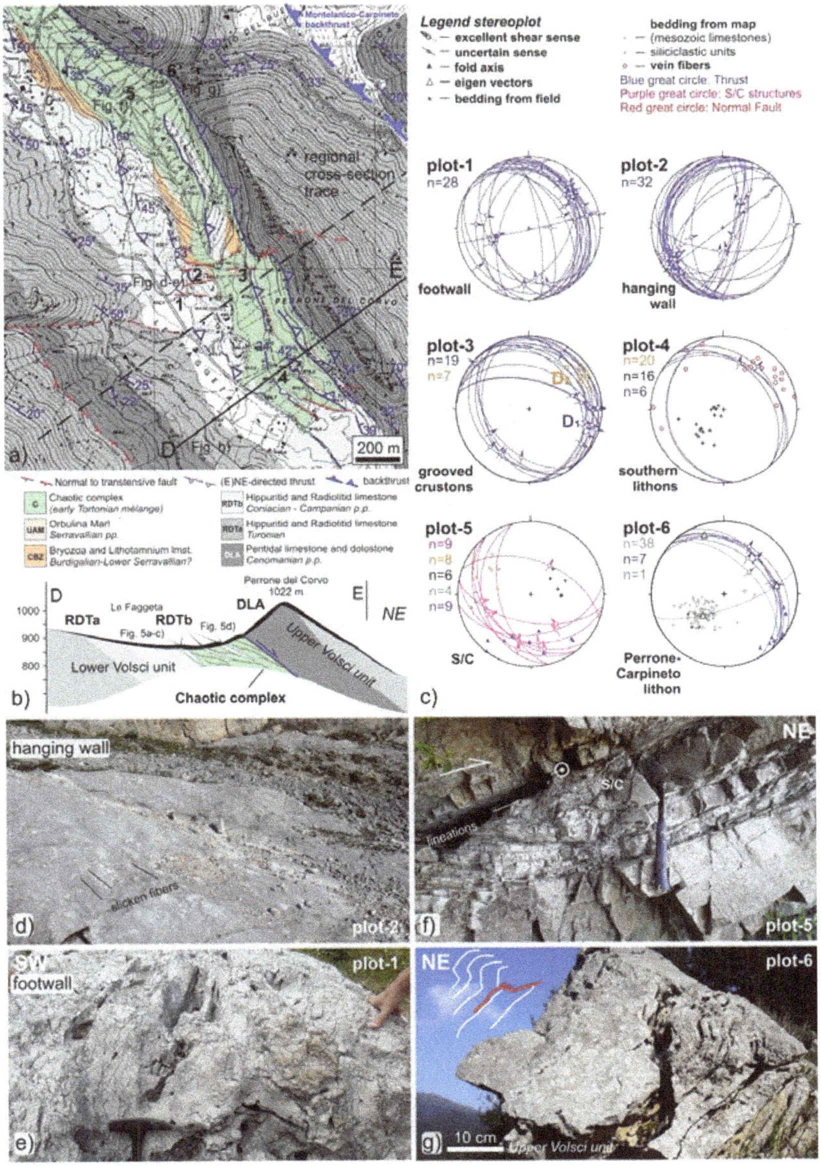

Figure 8. (**a**) Geological map of the Lower and Upper Volsci Unit deformation preserved between Pian della Faggeta and Occhio di bue localities (Figure 6) modified after [72] and the related geological cross section D-E in panel (**b**). (**c**) Stereoplots summarizing orientation data for the structural elements representative of the different key outcrops (from 1 to 6 in panel a). (**d**,**e**) Hanging wall and footwall of a (E)NE-directed thrust occurring in a cave near the top of the platform RTDb limestone (corresponding to plots 1–2; 41°34′48″ N/13°6′21″ E). (**f**) S/C top-to-the NE structures affecting lower–middle Miocene limestone and marl lithotypes (41°35′18.18″ N/13°6′16.40″ E). (**g**) Detail of the (E)NE verging fold (41°35′18.96″ N/13°6′19.28″ E) and striated bedding (41°36′11″ N/13°5′36″ E) of the Upper Volsci Unit (corresponding to plot 6). The sketch on top left shows the geometry of the outcrop that consists of a fault-propagation fold later tilted towards the foreland to the NE.

As constrained by nannoplankton analysis on samples from the fault core, both clasts and matrix (see Appendix B) are representative of different levels of a basinal sedimentary succession. The cataclasite also includes fragments of calcite mineralizations. The internal fabric is marked by the occurrence of slip surfaces associated with transpressive S/C structures indicating top-to-the-NE thrusting. Overall, the thrust seems to cut up-section although bounded and possibly tilted by later normal faults. The NW edge of the cave is bounded by a NE-striking normal fault with a displacement in the order of 20–40 of meters (red line in Figure 7h). At the top of the hill, the overall structure is topped by transgressive polygenic marine breccia composed by Miocene and Cretaceous calcareous clasts with a reddish cement and calcareous matrix, possibly crosscut by a SW-dipping normal fault with a displacement in the order of 150 m.

4.2.1. Thrusting at the top of Lower Volsci Unit

Figure 8 summarizes the kinematic indicators affecting the top of the Mesozoic platform and the Chaotic complex in six localities at the top of the Mesozoic succession of the Lower Volsci Unit in the Western Lepini Mounts.

Starting from the base of this deformed area, the Hippuritid and Radiolitid limestone (Campanian RDTb; Appendix B) of the Lower Volsci Unit is affected by bedding-parallel proto-cataclasite bands crossed at low-angle by striated curvy fault mirrors with dm^2 to m^2 dimensions (Figure 8). Across the most evident fault mirror (Figure 8), both footwall (plot-1) and hanging wall (plot-2) are characterized by top-to-the-NE slicken fibers, measured also on smaller fault mirrors. Crustons are disconformably topped by veined and laminated beige sandy calcarenites (plot-3). The thin carbonate blocks embedded in the Chaotic complex at Pian della Faggeta (plot-4) have variable thickness (up a few meters thick) and limited lateral extent (up to some dozens of meters). The native carbonate lithons are internally deformed and in places, display a sharp contact at their base with the siliciclastic units, and can be internally affected by top-to-the-(E)NE asymmetric folding. On the top of some of these slices, E-trending thrust grooves are cross-cut by NE-stretching mode-I veins. Beside the dominant NE-stretching, provided by the fiber direction of veins, more to the south (plot-4, Figure 8), veins crossing carbonate slices in similar structural positions also show NW-directed stretching.

At Occhio di bue locality (plot-5), a block of middle Miocene limestones and marls with chert topped by light green clay of late Serravallian age (c.f., Cosentino et al., 2003) is affected by S/C structures indicating top-to-the NE shear. Coherently, at the contact with the Cenomanian limestone on top, 1–2 m of foliated proto-cataclasite bands are topped by (E)NE verging folds (plot-6; Figure 8). In the same plot, top-to-the-NE striated bedding is reported as it crops out more to the north at the top of the same lithon. While bedding is folded around N- to NNW-striking axes (cf. stereoplots 7–8; Figure 6), northeast of a major backthrust it is folded around NW-striking axes of folds (stereoplots 9–10).

As the Chaotic complex is concerned, field data from the Eastern Lepini Mounts highlight the top-to-the-ENE juxtaposition of the Upper Volsci unit above the Chaotic complex (i.e., Caccume Mt., Siserno Mt.), which in the Volsci Range is preserved in a few klippen atop the Lower Volsci Unit, whereas in the Latin Valley it is found on top of the Frosinone Formation (Figures 9a and 10a). At the Caccume Mt., we report structural information from the juxtaposition of folded Cenomanian Lower Cretaceous limestone on the Chaotic complex. The regional folding affecting the Lower Volsci Unit defines a well-marked NW-striking open fold while the Upper Volsci unit of the Caccume Mt. displays rather dipping beds folded around an NNW-striking axis. The basal contact of the Chaotic complex is marked by thrust grooves and ferruginous faint slicken lines along the crustons, while at the top of the Chaotic complex, S/C and C′ structures display top-to-(E)NE shearing. Cross-cutting field relationships show that thrust grooves are further cross-cut by high-angle en-échelon shear zones and normal faults.

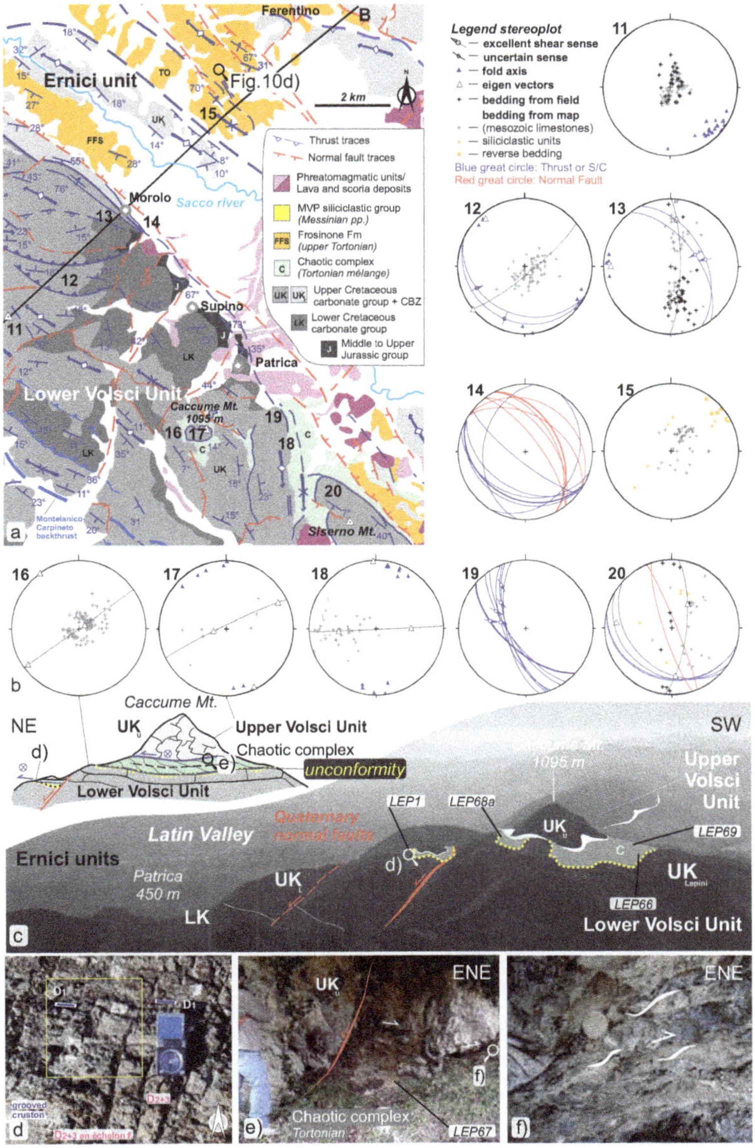

Figure 9. (**a**) Geological map of the Eastern Lepini sector and part of the Latin Valley. (**b**) Stereoplots (lower hemisphere projection, equal area for locations 11–20; numbering following after Figure 6) summarizing orientation data for the structural elements representative of the areas in panel a). Plot-13 shows E-striking folds interposed in the frontal thrust zone near Morolo, while plot-18 represents the N-S trending flank of a salient associated with transpressive S/C structures of Plot-19. (**c**) Sketched geological cross section and structural overview of the Volsci Range front (Caccume Mt. lower and upper unit, respectively, correspond to plots 16 and 17). Normal faults dip towards the NE, crosscut the Upper thrust. Sampling sites are reported in Appendix B. (**d**) Caccume Mt. front, detail of the encrusted top of the platform affected by E-trending D1 grooves and later crossed by oxides-rich (D_{2+3}) en-echelon fractures and later NW-striking oxides-free and cemented veins; 41°34′46″ N/13°13′60″ E). (**e**) Upper thrust juxtaposing the Cenomanian neritic limestone over the Chaotic complex (41°34′15.00″ N/13°13′55.13″ E), which, as shown as the sampling site of LEP67 on a lithotype that in panel (**f**), is affected by top-to-the-(E)NE S/C structures.

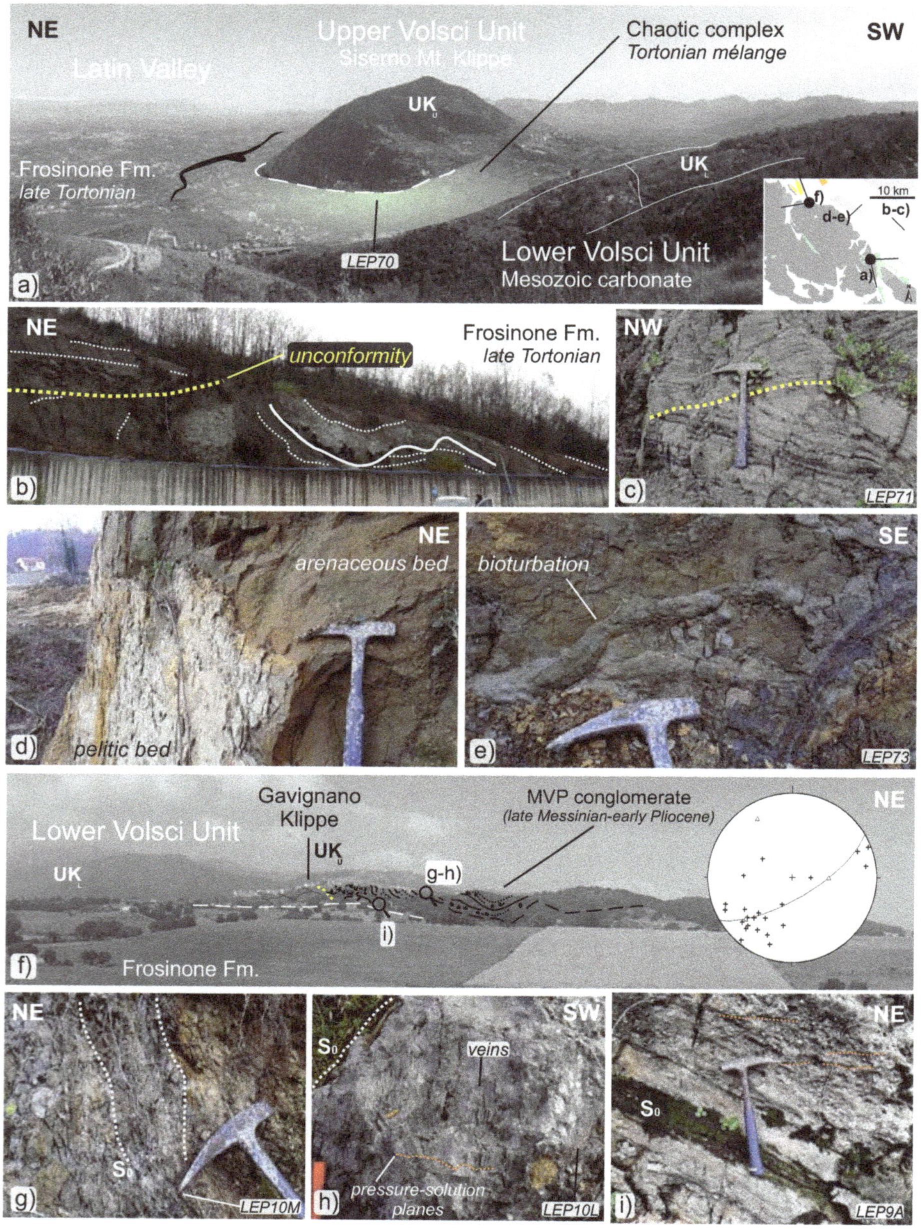

Figure 10. (**a**) Structural overview over two frontal klippen of the Latin Valley cropping out at Siserno Mt. where the Chaotic complex is juxtaposed to the Frosinone Fm. (**b**) Near Frosinone, an unconformity subdivides folded FFS units from the channelized facies on top. (**c**) Detail of the unconformity. (**d**) Vertical pelitic-arenaceous succession with (**e**) bioturbated levels. (**f**) Structural overview of the Gavignano area with stereoplot (lower hemisphere projection, equal area) of bedding and eigenvectors, that are indicative of the orientation of the axes of deformation related to the MVP thrust top conglomerates of Gavignano with (**g–i**) location of sampling localities. Conglomerates at the base are affected by pressure solutions and in the most calcareous beds also by veins. Sampling sites are referred to the Table in Appendix B.

4.2.2. The Volsci Range Thrust Front and the Latin Valley Structures

The geometries of the frontal part of the Volsci Range and Latin Valley are shown from the SW to the NE (stereoplots 11–15, Figure 9). The thrust front between the Ernici and Lower Volsci units occurs as a series of imbricates of overturned Cretaceous to CBZ layers (i.e., NW of Morolo; Figure 9). New data allowed us to recognize a salient at the front of the Eastern Lepini Mounts. This structure is accompanied by a change in the fold trend from NW to W (plots 12 and 13; Figure 9) and by transpressive top-to-the-NE kinematics. The frontal part is defined by a large-scale anticline in the west and a syncline in the east (Figure 9). The two folds are separated by a series of NNW-striking tear faults with inferred right-lateral kinematics (Figure 9). More to the east (plot-18), the N-S trending flank of the salient is associated with transpressive S/C structures in Cretaceous limestones (plot 19). Overall, the fold-and-thrust fabric is cross-cut by NE-dipping normal faults at the northeastern VR edge. As it is downfaulted, the thrust front does not outcrop further north. In the VR, a salient has been mapped between Morolo and Patrica (Figure 9), its most external point being characterized by the outcrop of Jurassic limestones. Upper Cretaceous units occur as klippe above the imbricated Chaotic complex juxtaposed to the foredeep deposits of the Frosinone Fm.

At the southern edge of the studied area of the Latin Valley (Figure 10a), the Chaotic complex was mapped as juxtaposed on the Frosinone Fm., and it reaches its maximum thickness west of the Siserno Mt. (about 250 m).

There (Figure 10a), we identify two thrusts: one juxtaposing the Upper Volsci Unit on the Chaotic complex (white dashed line) and the other juxtaposing the Chaotic complex onto the Frosinone Formation (black thrust). At Frosinone, a new road cut exposes a major intraformational unconformity within the Frosinone Fm. (yellow dotted line, Figure 10b, c) between folded layers beneath and sub-horizontal channelized deposits atop.

The channelized facies is made of arenaceous-pelitic associations with sets of thin pelitic-arenaceous and marly beds intercalated in thick massive arenaceous-pelitic layers. Southwest of Ferentino, paleocurrents are marked by a NW–SE direction, whereas the Frosinone formation is internally deformed and displays verticalized to overturned successions (Figures 9 and 10). There, the facies consists of an arenaceous association of amalgamated massive beds with arenaceous-pelitic and pelitic-arenaceous sets. As shown on the map (Figure 3), north of Sgurgola and north of the Siserno Mt., an anticline with upper Cretaceous and CBZ limestone belonging to the Ernici Unit emerges from the Latin Valley siliciclastics, which are locally bioturbated. In the syncline between this ridge and the Volsci Range, pelitic facies of the Frosinone Fm. occur.

At Gavignano (Figure 10f), the MVP Messinian calcareous conglomerate occurring on top of the Upper Volsci Unit overthrusting the Frosinone Formation is folded along an NNW-striking axis and is near vertical in places. In the most calcareous layers, pressure solution seams and veins crosscut the pebbles as typical of load-driven compaction.

4.2.3. Backthrusts and Normal Faults

Backthrusts best crop out in the northwestern part of the VR, where their presence is highlighted by some pockets of Messinian-earliest Pliocene heterogeneous conglomerate (Figure 11). Transpressive kinematics associated with a general top-to-the-(E)SE sense of thrusting was observed on the reverse faults along the Montelanico-Carpineto Backthrust. As typical of cannibalized wedge-top basins, blocks of conglomerates occur within a marly-conglomeratic matrix near Gorga (Figure 11).

In Figure 11, we sketch the structural setting related to the backthrusts, which cross-cut and preserve the top-to-the-(E)NE Chaotic complex at the footwall of the Montelanico-Carpineto Backthrust. This major backthrust (i.e., Montelanico-Carpineto Backthrust) bounds the East Lepini structure, a large-scale anticline with its culmination at the Malaina Mt. (Figures 6 and 11). The backthrust is accompanied by recumbent folds and minor high-angle reverse faults. In the southwestern sectors of the VR (Figure 11), normal faults cross-cut older contractional structures. More to the SW, another high-angle backthrust was mapped west of Bassiano (Figure 6). This structure allows the juxtaposition of the Jurassic and Early Cretaceous carbonate onto the upper Cretaceous and it is defined by transpressive kinematics (stereoplots in Figure 6).

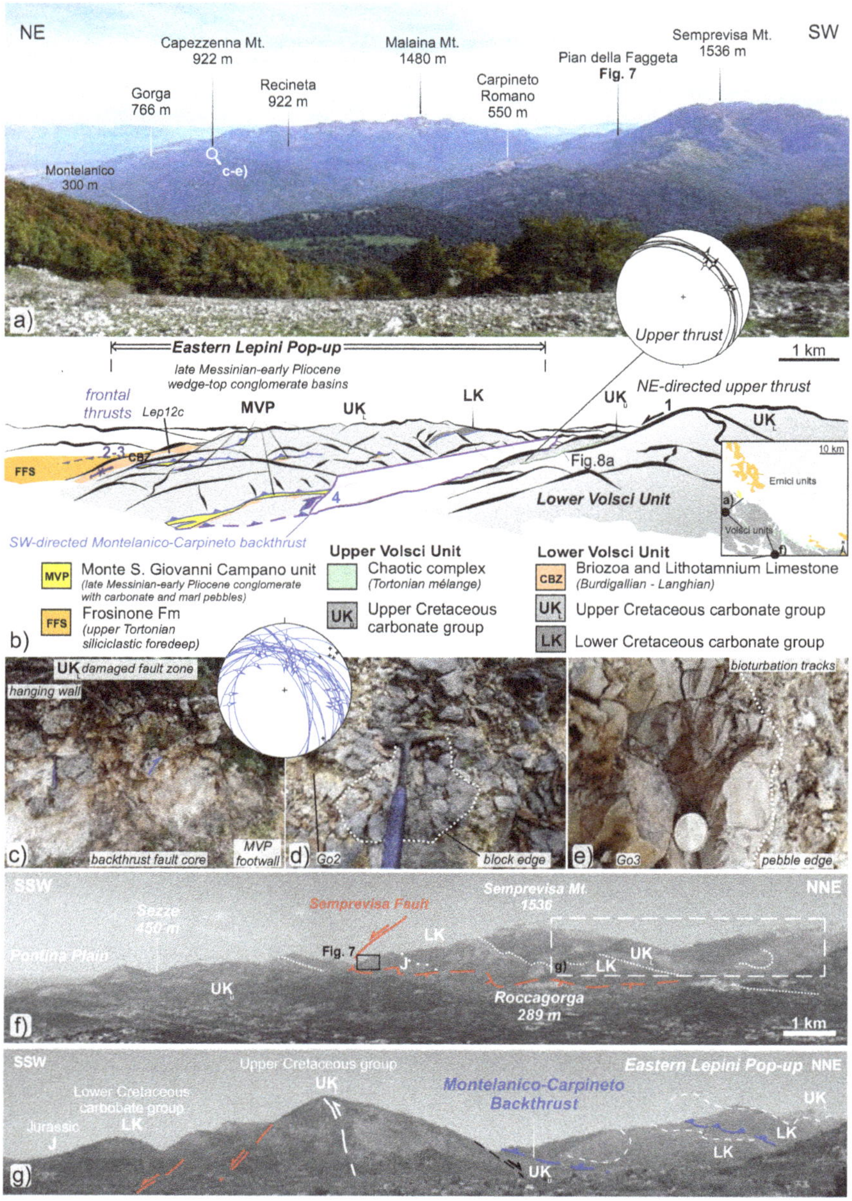

Figure 11. (**a**,**b**) Structural overview of the backthrusts in the northern Volsci Range with sampling sites (see Appendix B) and stereoplot (lower hemisphere projection, equal area) of bedding and backthrusts. (**c**) Thrust zone detail. (**d**) Block of conglomerate within conglomerate with clayish marly matrix. (e) Pebble of bioturbated marl with chondrites. (**f**,**g**) Structural overview of the Lepini sector and the Montelanico-Carpineto backthrust continuation towards the south beneath the Eastern Lepini Pop-up.

Along the southern slope of Semprevisa Mt. (i.e., the Semprevisa Fault), a major normal fault dissects the whole Jurassic-Upper Cretaceous succession, while along the northern slope, the top of the Mesozoic succession is overthrust by Upper Cretaceous units (documented in depth in the following sections). To the southwest, stepwise segments of normal faults bound the Pontina Plain (Figure 2). Further to the northeast, domino-like blocks are bounded by 2–3 km spaced faults, each with about 0.5 km downdip offset. More details on the Quaternary fault system are in [45].

4.3. Seismic Interpretation of the Latin Valley

By tracing the reflectors of the unconformable contact between the Meso-Cenozoic carbonates and the upper Miocene siliciclastic deposits on top (cf. Section 3.4), two major seismic units were recognized in the subsurface of the Latin Valley: (i) the Upper Ernici unit and (ii) the Lower Ernici unItal.

The Upper Ernici unit crops out at Ceccano (Figure 2), and northwest of Morolo (Figure 9), where it constitutes a carbonate ridge in the middle of the Latin Valley. Coupled seismic and field geological evidence shows that the ridge is represented by detached Upper Cretaceous carbonates topped by a thick CBZ succession sealed by UAM and FFS units. The Upper Ernici Unit was drilled by the Frosinone 1, Ripi I, Ripi II, Pofi 1, and Ceprano 1 wells (Figure 3). This thrust-bounded unit is composed of a stratigraphic succession that can be correlated with the upper units of the Gavignano-1 well.

The Lower Ernici unit, apart from the distinctive near-top reflections, displays a variable amplitude and frequency with a discontinuous and chaotic pattern of reflectors that generally is characterized by noisy seismic facies. We exclude that this reflector is a coherent noise (multiple) as it can be followed over the entire study area and it displaces geometries that roughly differ from the above reflectors. Due to the scarce penetration of the seismic signal, this unit can be considered as the acoustic substratum of the area. No boreholes reached this unItal. By comparison due to our reconstruction of the thrust geometry, the top of the Lower Ernici seismic unit is possibly represented by the Meso-Cenozoic carbonates that crop out northeast of the Latin Valley (Figure 2). Due to the above reported uncertainty, marks indicate the less-constrained portions of the interpreted cross sections.

Within the Latin Valley, minor thickness changes of the carbonate tectonic units occur. Due to the repetition of the top-CBZ reflector accompanied by underlying top-UK reflectors, we have recognized multiple repetitions of the Upper Ernici unit due the occurrence of several thrust faults. The Ripi I well [106]), although crossing a major thrust zone, shows no siliciclastic deposits under the Mesozoic carbonates, but rocks of the Orbulina Marl and CBZ formations.

To show the general structural trend of the research area, we present three representative seismic lines (Figure 12), constrained by field and borehole data, showing thrust sheets characterized by a general top-to-the-NE sense of shear. Major thrusts, although occurring in all of the seismic lines, are well evident but discontinuous in number and distribution from line to line. Four major groups of thrusts form before the occurrence of normal faulting (Figure 13). From the most internal to the outermost we describe them as (1) the first group (thrust-1) marks the juxtaposition of the Chaotic complex on top of the FFS units and it can be correlated with the Upper Volsci thrust. (2) Thrust-2 marks the translation and doubling of the Upper Ernici unit within the Frosinone foredeep domain. No clear indication of the front could be recognized in the study area, possibly due to subsequent erosion. This structure is also represented by a series of thrust splays that cross-cut the formerly formed the fold-and-thrust fabric. Carbonate thrust-sheet units as thick as 0.6–0.8 s intervals have undergone significant translation in the order of 20–25 km. Considering that no thrust ramp could be observed toward the SE, this is a minimum estimate calculated on the hanging wall flat. (3) Thrust-3 is a group of reverse faults with flat-ramp-flat geometries that involve both the Upper and Lower Ernici units. The thrust-3 records a minimum offset in the range of 5 to 8.5 km. (4) The latest reverse faults belonging to the thrust-4 include the backthrusts at the northern edge of the Latin Valley.

Such backthrusts cross-cut the previous 1–3 thrust faults and allow the formation of a triangle structure, during the deposition of the MVP deposits in the structural lows. In the southernmost section (Figure 12; Section 3), the cut-off relationships provided by the latest thrusts may have allowed the exposure of Thrust-2.

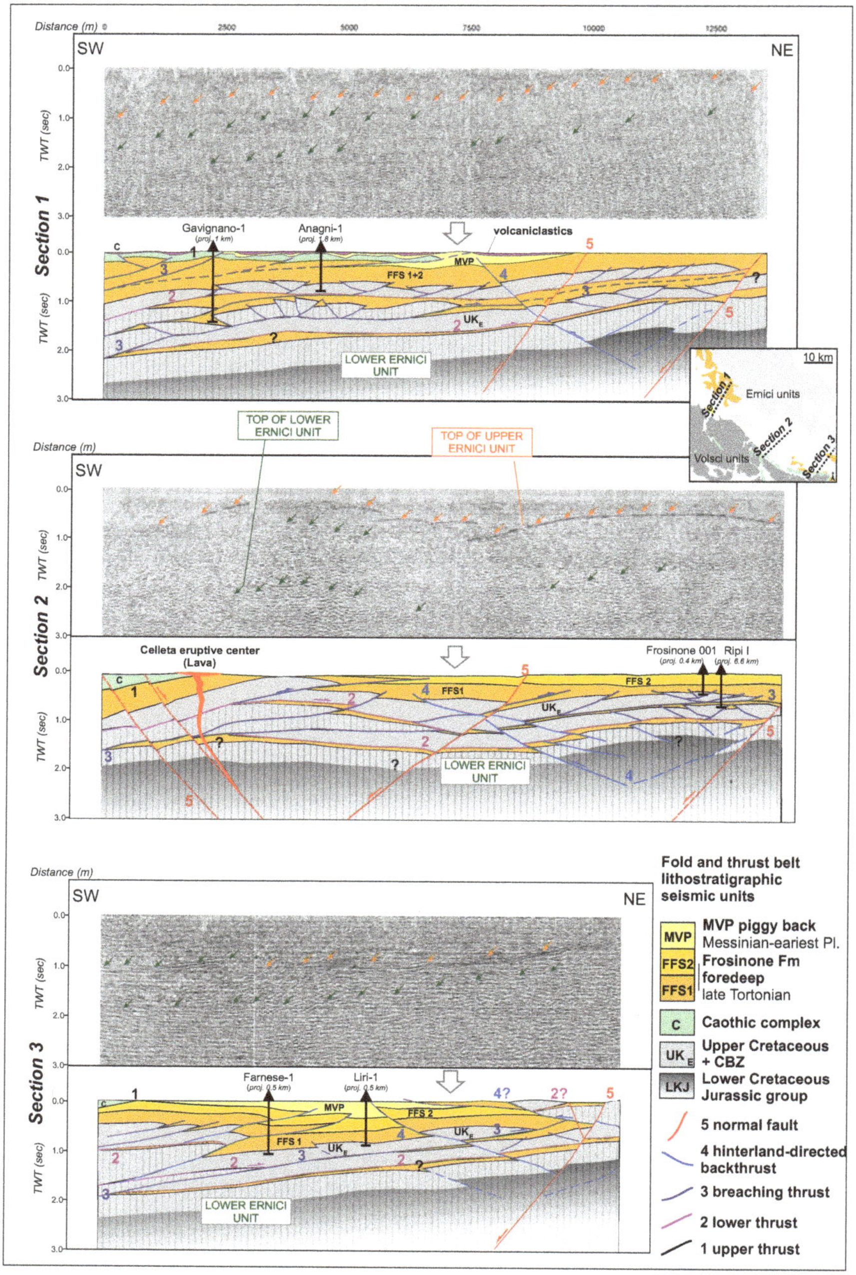

Figure 12. Two-times travel (TWT) seismic lines (from www.videpi.com (accessed on 20 January 2021); below, interpreted), also showing the projection of the wells. The location of both wells and seismic line traces of Section 1 (line label FR-309-80), Section 2 (FR-306-82), and Section 3 (FR-302-80) are in Figure 3. The vertical gray stripes highlight the Lower Ernici UnItal.

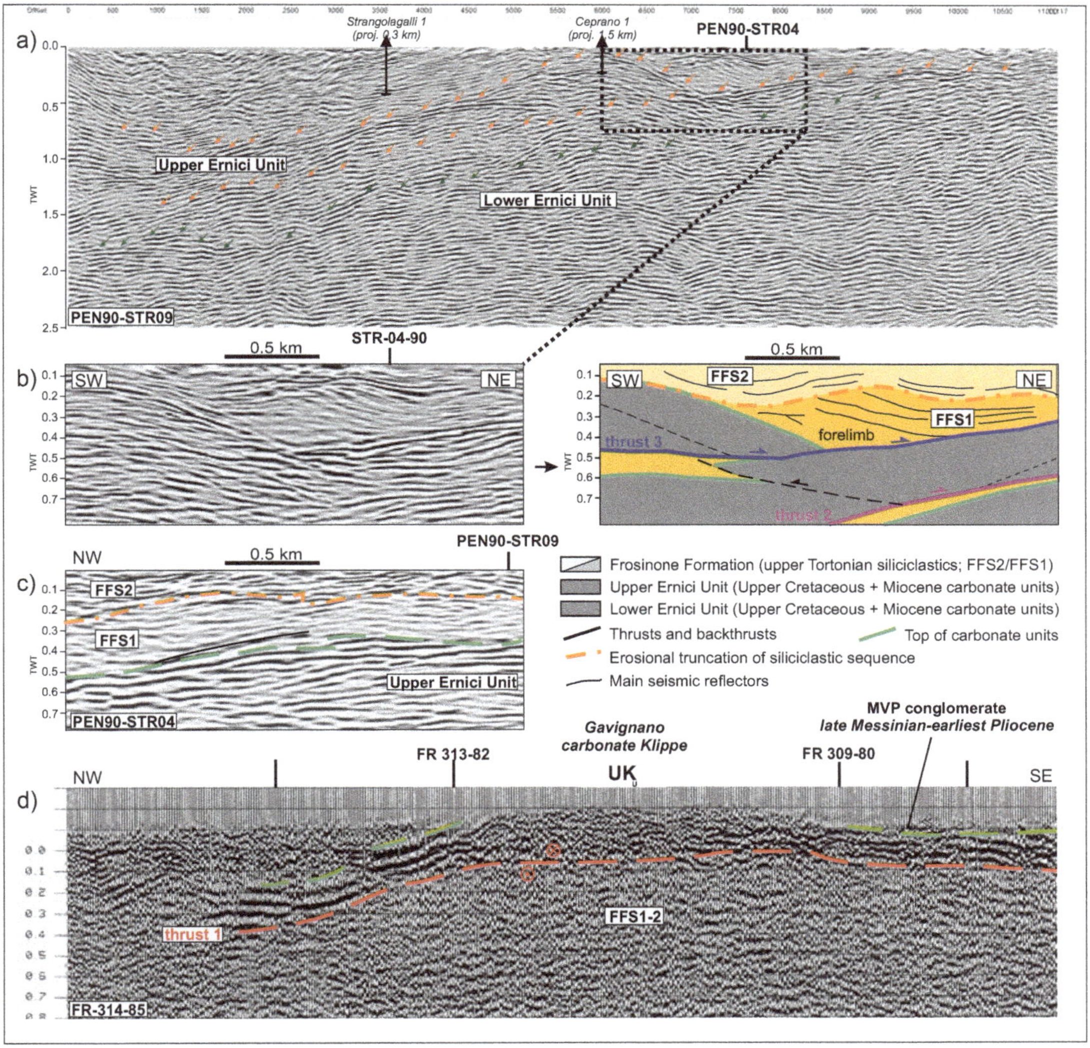

Figure 13. (**a**) Top-platform unconformities related to the upper (orange arrows) and lower Ernici units (green arrows) (Dip seismic line); (**b**) detail (right, interpreted) showing the angular unconformity between the Lower Frosinone seismic subunit (FFS1) and the Upper Frosinone seismic subunit (FFS2); (**c**) W-E view (Strike seismic line), showing the lateral variability of seismic facies FF1 and FFS2. (**d**) FR-314-82. Strike view of the Gavignano klippe, the purple dashed line marks the thrust onto the Frosinone Formation (transparent facies FFS), while the yellow dotted line highlights the top reflectors of the carbonates with the MVP conglomerate atop. Seismic line traces and well location in Figure 3.

The most prominent of this group of thrusts generates the outcrop of basal platform at the foothill of the VR Front. A few backthrusts were recognized at depth, with vertical displacement up to 1–2 km. In Figure 12, normal faults with appreciable offset were identified (labeled with number 5). NE-striking faults concentrate at the Latin Valley edges and do not clearly show in seismic lines. NW-striking faults bound Quaternary graben, where travertine, continental, and volcaniclastic deposits were cumulated. The normal fault trace in seismic lines was drawn when it is anchored to the outcrop evidence. In these cases, we have extended the minimum offset recognized at surface to the deeper structural levels.

The most distinctive unconformities occur at the top of the Mesozoic carbonate succession and above the Middle Miocene CBZ Fm., onlapped by late Serravallian-early Tortonian UAM horizons (Section 3 in Figures 12 and 13). At the borehole scale this contact may appear as a paraconformity but the discontinuous and variable thickness of both CBZ and UAM suggest that this is actually an unconformity with an irregular erosional surface. Three subunits, divided by two major unconformities, can be observed within the siliciclastics deposits and labeled as Lower Frosinone seismic subunit (FFS1), Upper Frosinone seismic subunit (FFS2), and Monte San Giovanni Campano seismic unit (MVP); the first two are made by the late Tortonian Frosinone Fm., while MVP is formed by the Messinian piggyback deposits (Monte San Giovanni Campano unit; see MVP in Figures 2 and 12).

The thickness of the syn-orogenic units varies depending on the fold-and-thrust belt structure, being the siliciclastic deposits thicker to the south and to the north (up to 0.600 sec) and thinner in the central part (usually limited to 0.180 sec). As shown in Figure 13, Subunit FFS1 is folded together with the underlying carbonates, showing a transparent seismic facies, while Subunit FFS2 is thicker in the syncline and thinner towards the anticline and it is possibly related to Thrust-2. In FFS2, minor internal unconformities, typical of syn-depositional antiforms in foredeep basins, are here expressed by lobate-type seismic facies. In detail, the antiformal-growth geometries are crestal erosional truncations and diverging/converging reflection patterns around the hinge of the anticlines. In the piggyback basins, the FFS2 is defined by well-reflecting horizons and is marked by an erosive unconformity that at Ceprano cross-cuts both FFS1 at anticline culminations (Figure 13). This anticline is sealed by FFS2 and is formed on top of Thrust-3. As shown by the strike section in Figure 13, the thrust-and-fold geometry changes laterally as also reported for the Gavignano klippe more to the north.

5. Discussion

The tectono-stratigraphic analysis of field and subsurface data enabled us to define different thrust units, providing insights for a time-deformation analysis of one of the innermost portions of the Central Apennines. Hereby, we present a geological cross section, interpretative of the deep structures produced after the integration of field and subsurface structures (Figure 14), that includes pre-orogenic passive margin deposits, mélange units, foredeep, and wedge-top deposits. In the following, we discuss the main novel features of the geologic history that led to the development of the geological setting of Figure 14. In the cross section, we correlate the Upper Volsci Unit remnants of the Colle Cantocchio, Carpineto Romano, and Caccume Mt klippen. Based on the mixed exotic-native composition of the blocks of the Chaotic complex, we recognize that they were overthrusted together with the Upper Volsci Unit on top of the Lower Volsci UnItal. As shown in the cross section, the Lower Volsci Unit of the Western Lepini Mounts is a monocline essentially composed of Jurassic to Cretaceous carbonates dipping to (E)NE, that together with the remnants of the upper units was further crossed by high-angle faults. In detail, the Montelanico-Carpineto backthrust, bounds the Eastern Lepini pop-up that is affected by small-scale folds and reverse faults, whose geometry suggests positive reactivation of pre-orogenic normal faults during shortening. The wedge-top pockets preserved by the backthrusts are infilled by MVP Messinian conglomerate that was deposited directly on the Lower Volsci Unit, when the Upper Volsci unit was already dismantled. Thrusts and folds are mostly evident in the Latin Valley (Figure 12), whose substrate has been reconstructed by applying a depth conversion on a structural model published in [45].

By studying the top of the Mesozoic carbonate platform both in the Lower Volsci Unit and in the blocks embedded in the Chaotic complex (Appendix B; Figure S1), we have reported the occurrence of an irregular surface at the top of the platform. Such a paleotopography was likely the result of Late Cretaceous syn-sedimentary tectonics. In such scenario, the most elevated structures might have been affected by karstism (possibly with the formation of ferruginous crustons) during the latest Cretaceous (see Section 4.1). The occurrence of a Late Cretaceous tectonics is supported by the lithostratigraphic unit we refer to the "Gorga bioclastic limestone and dolostone" upper Campanian to Maastrichtian in age, whose lateral change and abrupt facies shift points to syn-depositional tectonics (Figure 2). At Gorga (Figure 3), this unit is represented by about 250 m thick rock volume [112], that thins rapidly towards the west, whereas it lacks in the rest of the Volsci Range. In particular, as recognized at Caccume Mt. and near Carpineto Romano (Figure 5), the unconformity occurring at the top of the platform is marked by a very thin younger breccia partially overprinted by a dolomitic and ferrougeneous cruston (cf. Figures 5 and 9), whose age and origin need to be further constrained.

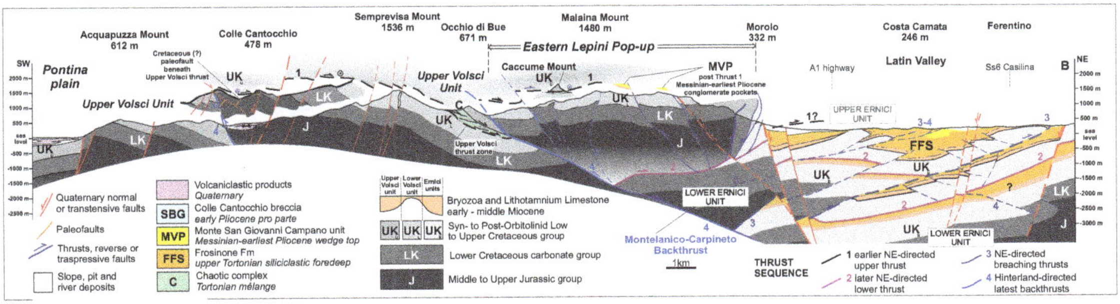

Figure 14. Geological cross section AB (trace in Figure 3) interpretative of both field and subsurface data converted to depth (see methods). Numbers related to the group of faults are disposed as in Figure 12. In the Volsci Range, the Upper Volsci Unit experiences about 25 km of thrusting (Thrust-1) towards the ENE. Thrust-2 accommodated the overthrust of the Volsci Range and Upper Ernici Unit on top of the Frosinone Formation. In Latin Valley, the Upper Ernici unit is doubled by the breaching of Thrust-3. Late reverse faults (Thrust-3 and -4) contribute to forming a triangle zone in the Latin Valley and backthrusts in the rear. Normal faults generate a graben in the Latin valley and SW dipping faults towards the Pontina Plain.

In the Apennine platform, the transition from the Upper Cretaceous carbonates to Paleocene–Eocene margin, slope, and Scaglia-type basin deposits was guided by a synchronous regional extension during Maastrichtian–Eocene time that affected both the Jurassic base-of-slope domains [30] and the demised sectors of the neritic platforms [118]. We recognize that the discordant stratigraphic contacts of Colle Cantocchio are due to the development of a submarine paleoescarpment, guided by normal faults down-stepping towards the WSW. The bluish hardground (highlighted by yellow dots Figure 15) can be interpreted as a submarine unconformity marking the onlap (escarpment contact) of the lower Miocene intraformational pebbly calcarenite on the Mesozoic carbonates. Similar facies have been reported elsewhere by the authors of [119] and are here interpreted as a diagenetic effect on the articulated inherited physiography of the previously unedited fault escarpment described in Figure 6. A simplified back-restoration of section C-D (Figure 7c) is attempted in Figure 15, where a fault step occurred to the south with an offset in the order of 700–1000 m due to the exposure of the Jurassic terrains and the downthrowing of the Cretaceous units in the hanging wall. The Semprevisa Fault can be still recognized laterally for over 10 km, although overprinted by later Pliocene-Quaternary tectonics, and possibly remarks at least part of this inherited structure. In our interpretation, as shown by the stratigraphic contacts, the Jurassic units of the southwestern slope of the Semprevisa Mt. were already exposed in early Miocene time (Figure 15). As suggested by the clasts within the Chaotic complex, coeval basinal sedimentation occurred more to the WSW [120]. In particular, the recognition of Cretaceous-Paleogene Scaglia lithotypes and of distal early Miocene CBZ limestones in the exotic blocks of the Chaotic complex (see Figures 7, 9 and 10) suggest that sedimentation occurred in a bypass slope setting during Paleogene-Neogene time. In particular, the Paleogene is recorded by a condensed to hemipelagic sedimentation, evolving during the Miocene to mixed calcareous-siliciclastic turbidites with chert. The Orbulina Marl Fm. (Serravallian pp.) sealed the pre-orogenic early Miocene topography.

The Colle Cantocchio pre-orogenic fault is a part of the normal fault system that produced the steps from the exposed Jurassic carbonates to the basin and is here proposed to be at least Eocene in age, although older ages cannot be excluded. Synthetizing, according to the new data, we propose a provenance of the Chaotic complex (i.e., including the exotic blocks) from a hemipelagic paleogeographic domain with slow depositional rates placed to the WSW of the present-day Volsci Range.

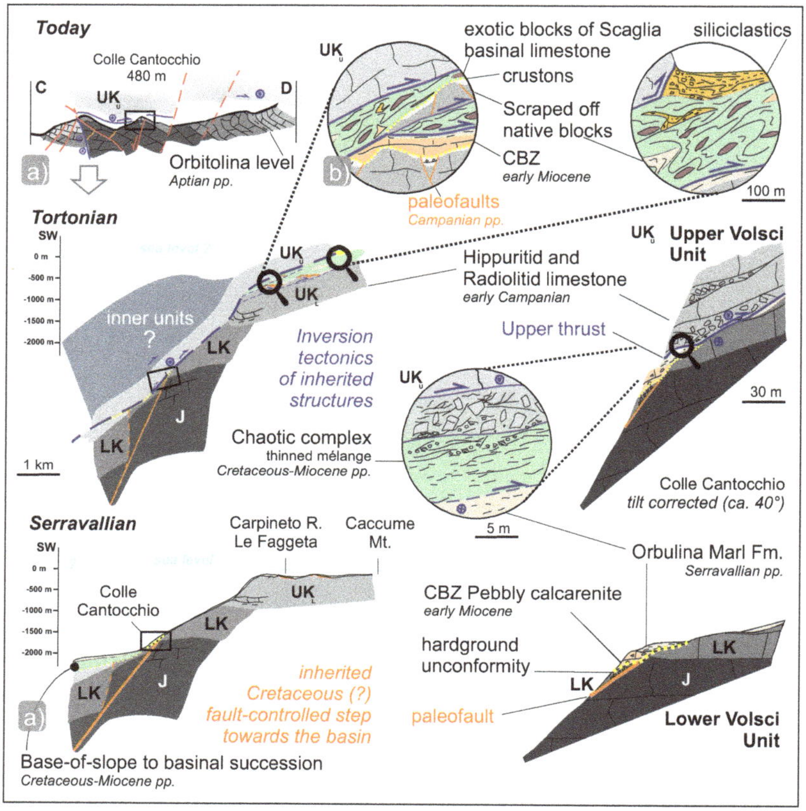

Figure 15. Reconstruction of the ramp-flat geometry of the upper thrust after restoration of section C-D by removing late backthrusts and normal faults. During Tortonian time, inversion tectonics of inherited structures occurred on a ramp by the overthrusting of the Upper Volsci UnItal. At the transition from ramp to flat, native blocks were scraped off from the Lower Volsci UnItal. On the right, peculiar settings inspired by field examples are contextualized to understand the mélange formation. During Serravallian time, the inherited structure was sealed by Orbulina Marl hemipelagic deposits. To the southwest, base-of-slope to basinal Cretaceous-Miocene deposits occurred on a fault-controlled step of the platform. On the right, a detail of the paleoscarpment setting prior to thrusting.

The ongoing research in the southern Volsci Range, is providing constraints for the determination of the age of the encrusted normal faults bounding the Formia plain and Spigno Saturnia areas, whose data from the literature are reinterpreted above (cf. Figure 2). A comparable syn-sedimentary setting, leading to the deposition of Scaglia deposits has been recorded nearby the VR [8,121] and documented at the western tip of the Volsci Range [122]. Of note, at Colle Cantocchio (Figure 5), the early Miocene transgression over the Jurassic-Lower Cretaceous rocks occurred on a step of the escarpment, where there was no record of Paleogene basinal sedimentation. In alternative, this sector could be associated with renewed normal faulting activity along a pre-existing Cretaceous-Paleogenic normal fault, which may have further exposed the Mesozoic rocks with its reactivation and allowed the CBZ-UAM units to settle on top prior to the Tortonian onset of thrusting.

5.1. Chaotic Complex Emplacement and Thrust Propagation

To define the overthrusting towards the (E)NE of the Upper Volsci Unit and to understand the evolution of the Chaotic complex, we correlated the carbonate klippen by documenting the stratigraphic and structural elements of the syn-orogenic deposits. This correlation was initially proposed by Accordi [71], but inherited structures, thrust kinematics, and age of the syn-orogenic deposits needed to be better constrained. With the degree of allochthony and origin of the Chaotic complex being long debated [45,64,67,86,123], in this section we discuss the Chaotic complex origin and the role of the thrust propagation towards the foreland into the late Miocene wedge growth.

Starting from the southwest, the Colle Cantocchio cataclasite and shale preserved underneath the Upper Volsci Thrust can be interpreted as a thin Chaotic complex unit juxtaposed on the paleo escarpment setting (c.f. Section 5.1). In this frame, the inherited topography produced a ramp in the upper thrust during shortening. A comparable setting occurs more to the south at the Vele Mt. (Figure 2), where the siliciclastic deposits underneath the thrust could be correlated with the Chaotic complex sliver of Colle Cantocchio (Figure 15). As commonly occurring in mélange complexes [124,125], the Chaotic complex formed at the expenses of the Lower Volsci Unit, whose inherited and articulated top was scraped off and grooved (see Figures 7–9). The Chaotic complex is a combination of (i) autochthonous "native" and (ii) allochthonous "exotic" blocks (Figure 15). The latter derive form a discontinuous series of Paleogene-Burdigalian pelagic deposits deposited more to the south and progressively mixed with lower Serravallian to upper Tortonian siliciclastic units bearing also crystalline clasts.

In particular, the matrix of the Chaotic complex shows the same composition of the embedded blocks, but it also shows the occurrence of late Tortonian-Messinian nannofossil assemblages, which may have deposited during the final stage of thrusting related to the Upper Volsci Thrust. Further, we are able to further narrow this time range to the late Tortonian, considering also the absence of *Amaurolithus* sp., typical marker of top Tortonian-Messinian. Provided that the overthrust of the pelagic elements of the Chaotic complex is due to the juxtaposition of the Upper Volsci Unit, which squeezed them out towards the foredeep, they must have originated from about the same distance reached by the Upper Volsci Thrust front (Thrust-1).

In this frame, the SE-ward termination of the Chaotic complex and the lens-like shape of the outcrop at Carpineto Romano (Figure 6) provide an example of interaction between inherited top-platform physiography and thrust geometry. In our interpretation, this structure is an inherited depression at the top of the platform that was later crosscut by the Upper Volsci Thrust. At its southern tip, as demonstrated by Accordi [71], this thrust still occurs as it doubles of the upper Cretaceous units although not involving anymore the Chaotic complex, whereas, as shown on the map (Figure 6), at the northern of the Upper Volsci Thrust, the younger Montelanico-Carpineto backthrust cross-cut it (Figure 11).

The Upper Volsci Unit is mainly composed by Upper Cretaceous neritic carbonates (e.g., Carpineto Romano, Figure 8), implying that this unit detached essentially above the uppermost Lower Cretaceous Orbitolina Marl level during shortening. However, although rare, older Mesozoic rocks can also be found. A second detachment level, highlighted by subsurface data, corresponds with the Orbulina Marl Fm, which allowed the doubling. The chronological relationship between Thrust-1 (marking the overthrust of the Upper Volsci Unit on to the Upper Ernici unit) and Thrust-2 (between the Ernici Units of the Latin Valley) is beneath the resolution of our data. However, provided their geometrical distribution, these thrusts are likely to represent a classical thrust propagation towards more external and lower structural levels through time (i.e., towards the foreland). The minimal shortening associated with Thrust-1 is of about 25–30 km, which corresponds with the approximated present-day distance between Colle Cantocchio and the frontal klippe along the ENE-directed Thrust-1; while Thrust-2 ranges about 20 to 25 km as shown by the thrust-2 structures in Figure 14. These amounts are comparable with the shortening estimated at the thrust fronts of the Gran Sasso Massif (>20 km [30]) and of the Apennine platform in the southern Apennines (>60 km [126]), while it is significatively lower than the translation that affected the Ligurian Accretionary Complex onto the foredeep units (> 100 km [127]). In this frame, the Tortonian southern Apennine platform thrusting [28] matches our thrust dynamics (Figure 15). As also typical of the far-traveled Sicilian platform units [128], the thrust geometry is characterized by long flats (10–15 km) and thin thrust-sheets, that in our case can be as thin as about 0.7 km near the front. This implies that the Orbitolina level and Orbulina Marl Fm preferred slip levels were very efficient in allowing far-traveled thrusting.

As shown by thickness and facies variations of the siliciclastic deposits of the Latin Valley, the Thrust-2 shortening stage was accompanied by syn-sedimentary folding of the deposits of the FFS2 seismic unit (Figure 13). In our interpretation, while the unconformable FFS1 contact with the CBZ limestone marks the flexuration of the foredeep, the unconformable contact associated with wedge shape and channelized FFS2 facies marks the growth of pop-up anticlines, thus being representative of wedge-top settings initially developed during Thrust-2.

The channelized facies may be, respectively, representative of syn-tectonic fringe and lobe deposits and of inner channelized sand bodies, while pelitic facies are rather typical of outer fans [129]. In particular, the observed syn-sedimentary folded channelized structures (Figure 10b), show that, the deposition of the Frosinone Fm. thus encompassed an increasing input (mostly during the FFS1 stage), later followed by a progressive channelization of turbidity flows onto the synclines during the FFS2 stage. As already suggested in [130] for the Latin Valley on the channelization of the foredeep to wedge-top sediments, the active margin possibly followed a comparable evolution similar to what elsewhere envisaged in the southern Apennines by Casciano et al. [131].

At the front of our study area, a transition between the mélange and the flysch units occurs. Based on published maps [64], wells, and seismic lines on the southwestern edge of the Latin Valley, we also confirm that the Chaotic complex is juxtaposed to the Frosinone Fm. of the upper Ernici unit (cf. Gavignano; Figures 10 and 13). For this feature, the authors of [132] proposed an olistostrome origin, while Centamore et al. (2007) proposed gravitational sliding of the Chaotic complex off the Volsci Upper UnItal. Further, this level can correlate with the mélange levels of the Massico Mt. [43,133].

To explain the abrupt thickening of the Chaotic complex east of the Caccume Mt. (Figure 10), we suggest that a growth structure was forming during the initial uplift of the Volsci Range front as testified by fault-propagation fold (Figure 14) at the hanging wall of thicker FFS units with syn-sedimentary folds (Figures 10 and 12). This generated the glide of the Chaotic complex on top of the FFS units. Similar contexts were reconstructed for other mélange units at thrust fronts, where the remobilization of the formerly emplaced thrust sheets, allows the incorporation of the extrabasinal (exotic) lithologies within the foredeep [18,134]. An alternative possible explanation to allow the juxtaposition of the Upper Volsci unit onto the FFS units, would envisage thrusting to occur during the uppermost Tortonian-earliest Messinian.

5.2. The Late Stages of Shortening

As observed in seismic lines (Figures 12 and 13), thrust-3 produced the doubling of the flat of the far-traveled Thrust-2, by involving deeper carbonates in the thrust ramps. We have also shown that in the area break back thrusting occurred [135] (Figure 16). As shown near Ceprano well (Figure 13), MVP wedge-top deposits that include calcareous pebbles from the CBZ unit [87] were directly deposited on Mesozoic carbonates deformed by an anticline. This contact is representative of a wedge with regional subsidence slower than local antiformal growth [136]. Nannoplankton determination finally allowed constraining the age of the folded conglomerates and atop marls of Gavignano, thus allowing a correlation with the MVP stratigraphic unit (Figure 10). This unit represents a folded Messinian thrust top deposit and this constraint attributes this late folding stage to late Messinian-earliest Pliocene time. As supported by subsurface data (Figures 3 and 13) the Gavignano klippe was involved into the renewed deformation of the VR front, which would correspond with the latest stage of thrusting and veining dated in [43] at the late Messinian on the Massico Mt. (cf. Figure 2). Those absolute constraints can be used to review the regional thrust kinematics. In this sense, the ages determined along the thrusts in areas more to the south can be compared to what provided in [114]. These authors have attributed a late Miocene-Pliocene age to the clayey matrix beneath the thrust at the front of the Siserno Mt. Similar to what reported for the Chaotic complex in this work (Appendix B), they have also reported that the exotic clasts are representative of a wide range of ages, from Late Cretaceous (including Scaglia Rossa pelagic limestone) to early-middle Miocene. The degree of fragmentation of microfauna embedded within the Chaotic complex [114] suggests active deposition during the late Miocene-Pliocene as well. Therefore, we can envisage a late involvement of Pliocene deposits into the reactivated thrust zones at the VR front. In this interpretation, the Chaotic complex was already exhumed likely after the strong erosion related to the Messinian salinity crisis [137–140], which also affected the Ernici Mts [77], implying reactivation in the rear [49].

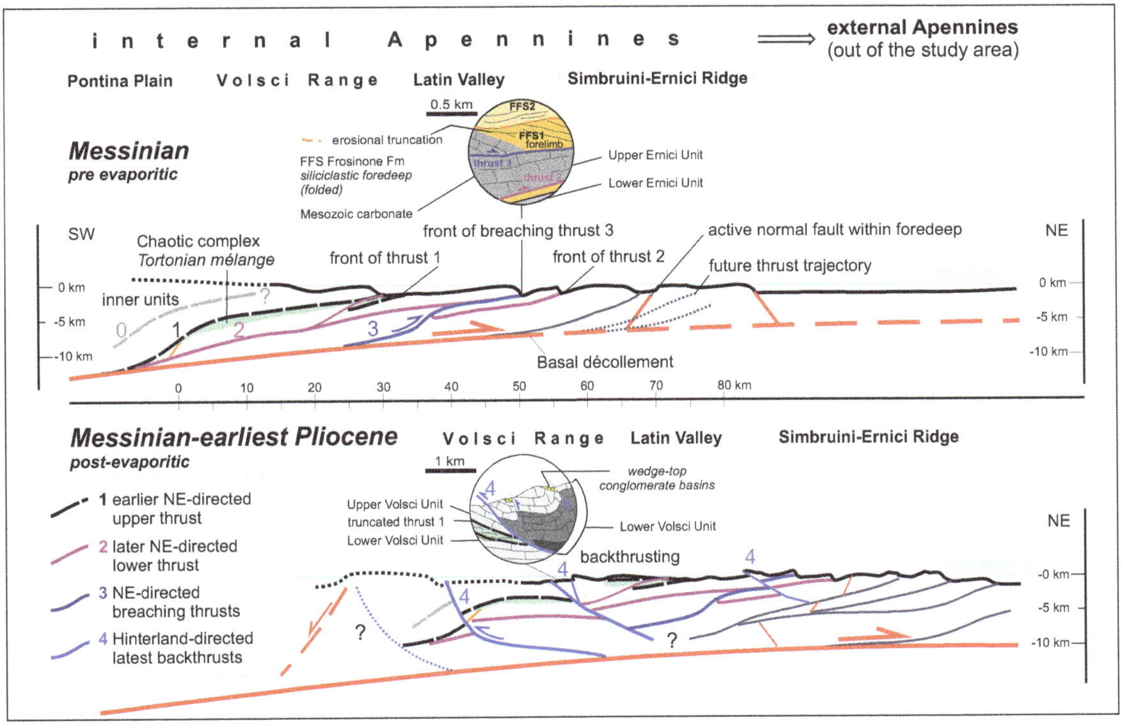

Figure 16. Sketch of relative timing and geometries of fore- and back-thrust involving different generations of thrusts (1–4) within the Apennine wedge through time. Backthrusts generate at progressively lower depths, moving towards the hinterland (to the left), due to the dip of the basal detachment.

In this context, the late Messinian shortening event could be correlated with the late orogenic structures in the northern VR that are crossed by a series of SW-directed backthrusts (Figure 11). In our interpretation, the SW-directed Montelanico-Carpineto backthrust cross-cuts the top-to-the (E)NE older Upper Volsci Thrust. Despite the lack of valuable data from the main lineament, minor thrusts show that top to the SW-backthrusting, could be accounted as partially reactivating the older fabric. Further, the fault strike of the backthrusts diverges about 20° from the trend of the upper Volsci Thrust that is underthrusted beneath the Eastern Lepini pop-up (Figures 5 and 11).

So far, scarce constraints of top-to-the-SW shear were found, although backthrusting is possibly localized more to the NE of the studied area of Figure 11. Our stratigraphic constraints (Appendix B) from the MVP conglomerate near Gorga, document Messinain Lago-Mare conglomerates that are produced after iterative cannibalization of older wedge-top deposits. The further occurrence of upper Messinian deposits in the Pian della Faggeta area (Figure 5), is a possible clue indicating depositional activity on top of the Volsci Range during the Messinian salinity crisis (5.96–5.33 Ma). During that time, the area was exposed to linear erosion followed by the deposition of sandy gravels that Centamore et al. (2010) dated at the early Pliocene (south of Castro dei Volsci; Figure 3). This implies that the major valleys were already formed before the latest orogenic compressional events affected both the VR and Latin Valley [117]. Field evidence in the rear (Figure 5), suggests the presence of a major backthrust with transpressive kinematics further south, possibly implying that a deeper backthrust affected the southwestern slope of the VR during the early Pliocene. At that time, the Apennines experienced renewed shortening with frontal thrusting accompanied by backthrusting and tilting toward the foreland to the northeast (Figure 16).

During late orogenic deformation, thrust front migrated towards the outermost active margin units (Figure 1), and the inherited fold-and-thrust belt of the external Apennines was folded together with lower Pliocene syn-orogenic conglomerates (i.e., Rigopiano conglomerate [30,78]). Meanwhile, the previous in-sequence structure of the internal Apennines was truncated by triangle zones (Figure 12) and by more internal backthrusts (e.g., in the Volsci Range, Figures 11 and 14).

In our interpretation (Figure 16), the backthrusting roots at deeper levels, by following the dip of the basal detachment towards the backarc. In this sense, moving to the inner parts of the wedge, the inner wedge is remobilized, affecting a larger volume with respect to the external part. In the case of late orogenic deformation affecting only the sedimentary cover, shortening localizes within the weakest stratigraphic levels, possibly by reactivating the décollement of the older fore-thrusts [136,141–145], while in the rear faulting tends to broaden and possibly involve also deeper structural levels.

Finally, Pleistocene to Holocene NW- and NE-trending normal faults deeply affected the fold-and-thrust belt structure. In particular, the almost constant NE-dip shown by the bedding planes of the studied carbonates might be interpreted as the result of the activity of the major NW-striking and SW-dipping listric normal faults bordering the Pontina Plain, which were also documented at depth [146].

6. Conclusions

This study contributes to constraining the timing of initiation and progressive development of platform-derived thrust sheets, mélange units, foreland, foredeep, and wedge-top sediments of the internal Central Apennines. The main phases of the evolution of the belt are as follows:

1. Late Cretaceous extensional tectonics. The dismembering of the carbonate platform into shallower and deeper domains is constrained by the finding of crustons that may testify moments of subaerial exposure, characterizing the top of the Lower Volsci UnItal. Cave exploration and field mapping allowed us to recognize a previously unreported fault-controlled paleo-escarpment constituted by Cretaceous and Jurassic carbonates sealed by early Miocene deposits that were previously dated as middle Miocene. These units seal a hardground settling on a platform edge facing to the west, where basinal to bypass slow-rate sedimentation occurred till Burdigalian time.
2. Tortonian Chaotic complex emplacement (Thrust-1) and foreland-directed (in-sequence) thrust propagation (Thrust-2). During the overthrusting of the Upper Volsci Unit, Paleogene to Neogene basinal deposits were squeezed off towards the Foredeep and

juxtaposed as a mélange unit on top of the carbonate platform together with early to middle Miocene calcareous-cherty-siliciclastics. The Chaotic complex also bears highly deformed basinal exotic and native blocks of neritic carbonates, the latter being scrapped off by the overthrust of the embedding Chaotic complex, whose Paleogene-Miocene matrix includes up to Tortonian nannoplankton. Seismic analysis supported by well logs at the regional scale highlighted repeated carbonate thrust sheets that have first been involved into an initial in-sequence propagation towards the foreland to the ENE occurred during foredeep to wedge-top sedimentation.

3. Intra Messinian thrusting (Thrust-3) breached the thrust front by doubling the flat of previous thrust fronts. Subsurface data show that during alternated phases of wedge-top deposition and erosion, the Upper Ernici unit was shortened approximately 5–8.5 km in the Latin Valley.
4. Messinian to early Pliocene backthrusting (Thrust-4). New biostratigraphic data constrain the thrust top deposits in the Volsci Range and in the Latin Valley, where SE-directed backthrusts contributed to the tilt and cross-cut of previous Thrust-2 and -3 structures.
5. Late Pliocene to Holocene normal faulting. Post-shortening extension has determined NE- and NW-striking orthogonal normal faults or WNW–ESE-trending right-lateral transtensional faults. These faults may have locally intercepted pre-existing normal faults that had been passively transported within the thrust sheets.

Finally, our findings bear implications on platform derived thrust sheets associated with active margin successions and mélange units. The far-traveled thrust sheets, hereby documented both in the field and in the subsurface, constitute a key aspect for the development of the internal Apennines, whose degree of allochthony and role of inherited structures was long debated. Furthermore, at the light of our new interpretation, the deeper platform units could be a new focus for hydrocarbon accumulation and may provide targets for geothermal and/or hydrocarbon research in the area. Beside the regional geological aspects, this work bears implications on the modes of involvement of mélange units at the transition from passive margin to foreland basin systems.

Supplementary Materials: The following are available online at https://www.mdpi.com/article/10.3390/geosciences11040160/s1.

Author Contributions: Conceptualization, G.L.C., G.V., L.C., M.S., and E.C.; methodology, G.L.C., and G.V.; software, G.V.; validation, E.C. and C.D.; formal analysis, G.L.C., G.V., L.C., and M.S.; investigation, G.L.C., G.V., L.C., M.S., and E.C.; resources, E.C., L.C., and C.D.; data curation, G.L.C., G.V., L.C., and M.S.; Writing—Original draft preparation, G.L.C., G.V., and L.C., Writing—Review and editing, G.L.C., G.V., L.C., M.S., and E.C.; visualization, G.L.C., G.V., L.C., and M.S.; supervision, E.C. and C.D.; project administration, G.L.C. and E.C.; funding acquisition, E.C., L.C., and C.D. All authors have read and agreed to the published version of the manuscript.

Funding: This research was funded by Progetti di Ateneo 2016 (C. Doglioni), 2017 and 2019 (C. Doglioni and E. Carminati) and by the Spanish Ministry of 'Economía y Competitividad' (projects CGL2012-33160 and CGL2015-69805-P).

Data Availability Statement: Data used for seismic interpretation and model reconstruction can mainly be found in the public VIDEPI database (www.videpi.com) (accessed on 20 January 2021) and in the Pentex Ltd. database. Data available at the ENI data room were also viewed.

Acknowledgments: We gratefully acknowledge Pentex Limited Italia and Luigi Albanesi for the permission to analyze their subsurface data and publish the seismic lines of the Strangolagalli Oil Concession area. We acknowledge ENI for the permission to participate at the data room as requested in San Donato Milanese, Milan, and in particular to analyze some seismic lines on old Permit Areas. We thank the G. Wang and the D. Liotta, G. Molli and A. Cipriani for the opportunity to share our research in the Special Issue "The Apennines: Tectonics, Sedimentation, and Magmatism from the Palaeozoic to the Present". Enrico Tavarnelli, Andrea Artoni and an anonymous reviewer are acknowledged for insightful comments and suggestions. We are grateful to "Gruppo Grotte Castelli

Romani", "Federazione Speleologica del Lazio", Andrea Cesaretti, Piero Ciccaglione, Pio Di Manna, Simone Fabbi, Luca Forti, Angelo Giuliani, Domenico Mannetta and Anne Mérienne for their support.

Conflicts of Interest: The authors declare no conflict of interest. The funders had no role in the design of the study; in the collection, analyses, or interpretation of data; in the writing of the manuscript; or in the decision to publish the results.

Appendix A

In the following, we report the Biostratigraphic and lithostratigraphic data of outcrops and stratigraphic units available from the literature related to syn-orogenic deposits shown in the representative stratigraphic logs of Figure 2 in the main manuscript. The formation labels are also related to Figure 2.

Site n°	Group of Localities	Latitude	Longitude	Tectonic Unit	Formation	Lithology	Biomarkers	Age Range	Comments	Author
1	Gavignano R. *klippe*	41°42′9.16″ N	13°20′38.15″ E	Upper Volsci unit	DLA	limestone	*Cisalveolina fraasi*	Cenomanian	in situ	[147]
		41°42′9.16″ N	13°20′38.15″ E	Upper Volsci unit	MVP	calcareous conglomerate	*Globorotalia apertura, G. involuta, G. concinna, Globigerina falconensis*	uppermost Tortonian-lowermost Zanclean	reworked	[103], this work
		41°42′9.16″ N	13°20′38.15″ E	Upper Volsci unit	PGC	polygenic conglomerate	no data	upper Messinian (?)		[103]
2	Colle Cantocchio *klippe*	41°34′29.48″ N	13°0′1.49″ E	lower Volsci unit	RDT	limestone	*Dicyclina schlumbergeri, Accordiella conica, Orbitoides*	Campanian	in situ	[93]
		41°34′29.48″ N	13°0′1.49″ E	lower Volsci unit	CBZ	calcarenite	echinid, *Ditrupa, Elphidium*, bryozoa, *Miogypsina, Amphistegina, Operculina, Heterostegina, Lepidocyclina*	Burdigalian	in situ	[93]; this work
		41°34′29.48″ N	13°0′1.49″ E	lower Volsci unit	UAM	gray-yellowish clay	*Orbulina universa, O. suturalis, O. bilobata, Globorotalia* aff. *Menardii, Globorotalia opima, Globorotalia scitula ventriosa, Globigerinoides trilobus, Globigerina eggeri, Globigerina* cf. *Bulloides, Globigerina concinna, Globoquadrina dehiscens, Globoquadrina altispira, Bolivinoides miocenicus, Valvulina pennatula italica.*	upper Serravallian—Tortonian p.p.	reworked	[93]
		41°34′29.48″ N	13°0′1.49″ E	Upper Volsci unit	SBG?	polygenic breccia	no data	Pliocene-Pleistocene (?)	reworked	[93]
3	Carpineto Romano	41°35′17.12″ N	13°06′15.66″ E	lower Volsci unit	RDT	limestone	rudist, *Dicyclina shlumbergeri, Rotalispira*	Coniacian-Campanian	in situ	[103]
		41°35′17.12″ N	13°06′15.66″ E	lower Volsci unit	CBZ	limestone	*Amphistegina, Heterostegina, briozoa, Operculina, Miogyspina globulina*	Burdigalian—Langhian	in situ?	[103]
		41°35′17.12″ N	13°06′15.66″ E	Upper Volsci unit	RDT	limestone	*Rotalispira maxima*	lower Campanian	native block within C	this work—sample LEP 17C

Site n°	Group of Localities	Latitude	Longitude	Tectonic Unit	Formation	Lithology	Biomarkers	Age Range	Comments	Author
		41°35′17.12″ N	13°06′15.66″ E	Upper Volsci unit	RDT	limestone with iron crust (ancient karstification ?)	*Decastronema, ostracodae, discorbidae*	Campanian	native block within C	this work—sample LEP 18B
		41°35′17.12″ N	13°06′15.66″ E	Upper Volsci unit	Scaglia s.l.	limestone	Heteroelicidae, Hantkeninidae, *Schackoina, Guembelina, Clavihedbergella, Globorotalia*	Albian?; Maastrichtian-early Eocene	exotic block within C	[71,148]
		41°35′17.12″ N	13°06′15.66″ E	Upper Volsci unit	C	glauconitic calcarenite	*Orbulina, Globigerinoides sacculiferus, Globoquadrina altispira, Globigerina parabulloides, Bigenerina nodosaria*	upper Serravallian-Tortonian p.p.	reworked	This work; [103]
		41°35′17.12″ N	13°06′15.66″ E	Upper Volsci unit	UAM	marl with cylindrites and calcarenite	*Sphenolithus heteromorphus, Cyclicargolithus floridanus, Reticulofenestra pseudoumbilicus, Coccolithus miopelagicus, Helicosphaera walbersdorfensis, Calcidiscus premacintyrei, Neogloboquadrina continuosa, Neogloboquadrina acostaensis.*	upper Serravallian	exotic block?	[72]
4	Gorga, Capezzenna Mt.	41°37′38.45″ N	13°08′32.37″ E	lower Volsci unit	RDT	limestone	*Rotalispira maxima, Dicyclina schlumbergeri*	Santonian-Campanian	in situ	[103]
		41°37′38.45″ N	13°08′32.37″ E	lower Volsci unit	MVP	marl pebble in conglomerate	*Amaurolithus primus*	uppermost Tortonian—basal Pliocene	in situ	this work—GO2
		41°37′38.45″ N	13°08′32.37″ E	lower Volsci unit	MVP	bioturbated marl pebble in conglomerate	*Discoaster surculus, Helicosphera wallichii. Calcidiscus leptoporus, C. macintyrei, Coccolithus pelagicus, Discoaster multiradiatus, Helicosphaera carteri, Reticulofenestra minuta, R. pseudombilicus, Sphenolithus moriformis, S. radians, Zygrhablithus bijugatus*	uppermost Tortonian—basal Pliocene	in situ	this work—GO3
5	Gorga, Rave St. Marie	41°39′36.01″ N	13°07′9.51″ E	lower Volsci unit	RDT	limestone	*Orbitoides medius, Sivasella monolateralis*	Maastrichtian	in situ	[111]
		41°39′36.01″ N	13°07′9.51″ E	lower Volsci unit	Spirolina lmst.	limestone	*Spirolina*, carofita	lower Eocene	in situ	[111]
		41°39′36.01″ N	13°07′9.51″ E	lower Volsci unit	CBZ	limestone and marl	*Cyclicargolithus floridanus, Sphenolithus conicus, Miogypsina cf. globulina*	lower Miocene (not younger than middle Burdigalian)	in situ	This work—LEP12C
		41°39′36.01″ N	13°07′9.51″ E	lower Volsci unit	UAM	marl with cylindrites	*Globorotalia menardii; Globorotalia ventriosa, Globigerina nepenthes*	lower Tortonian	in situ	[103]

Site n°	Group of Localities	Latitude	Longitude	Tectonic Unit	Formation	Lithology	Biomarkers	Age Range	Comments	Author
6	Sgurgola	41°40′38.97″ N	13°9′34.03″ E	upper Ernici unit	CBZ	limestone	*Amphistegina, Heterostegina*, briozoa, *Operculina, Miogyspina globulina, Cycloclypeus, Globorotalia scitula, Globigerinoides trilobus, G. sacculifer, G. bisphaericus, Orbulina universa, Orbulina suturalis, Globoquadrina dehiscens, Globorotalia mayeri*	Langhian—lower Serraval-lian	in situ	[103]
		41°40′38.97″ N	13°9′34.03″ E	upper Ernici unit	UAM	marl with cylindrites	*Globorotalia menardii, Globigerina nepenthes, Globorotalia ventriosa, Globorotalia acostaensis, G. bulloides, G. parabulloides, G. pseudopachyderma, G. apertura, Globigerinoides obliquus, Globoquadrina globosa, Orbulina universa*	upper Serravallian—Tortonian p.p.	in situ	[103]
		41°40′38.97″ N	13°9′34.03″ E	upper Ernici unit	FFS	arenaceous-argillous turbidite	*Globorotalia menardii, Globigerina nepenthes, Globorotalia ventriosa*	upper Tortonian	in situ	[103]
7	Ferentino	41°41′22.95″ N	13°14′41.78″ E	upper Ernici unit	RDT	limestone	*Rotalispira scarsellai, Accordiella conica, Cuvillierinella salentina*	middle Campa-nian	in situ	[103]
		41°41′22.95″ N	13°14′41.78″ E	upper Ernici unit	Spirolina lmst.	limestone	*Spirolina*, carofita, discorbidae	lower Eocene	in situ	[103]
		41°41′22.95″ N	13°14′41.78″ E	upper Ernici unit	CBZ	limestone	*Amphistegina, Heterostegina, briozoa, Operculina, Miogyspina globulina, Cycloclypeus, Globorotalia scitula, Globigerinoides trilobus, G. sacculifer, G. bisphaericus, Orbulina universa, Orbulina suturalis, Globoquadrina dehiscens, Globorotalia mayeri*	upper Langhian—upper Serraval-lian	in situ	[103]
		41°41′22.95″ N	13°14′41.78″ E	upper Ernici unit	UAM	marl with cylindrites	*Globorotalia menardii, Globigerina nepenthes, Globorotalia ventriosa, Globorotalia acostaensis, G. bulloides, G. parabulloides, G. pseudopachyderma, G. apertura, Globigerinoides obliquus, Globoquadrina globosa, Orbulina universa*	upper Serravallian—Tortonian p.p.	in situ	[103]
		41°41′22.95″ N	13°14′41.78″ E	upper Ernici unit	FFS	arenaceous-pelitic turbidite	no data	upper Tortornian		[103]
8	Caccume Mt. *klippe*	41°34′46.13″ N	13°14′0.62″ E	lower Volsci unit	RDT	limestone	*Rotalispira scarsellai, Accordiella conica, Thaumatoporella, Nezzazatinella*	Santonian—Campanian	in situ	this work—LEP1A-C; LEP27
		41°34′46.13″ N	13°14′0.62″ E	lower Volsci unit	RDT	limestone with iron cruston	*Rotalispira scarsellai, Accordiella conica, Thaumatoporella, Nezzazatinella*	Santonian—Campanian	in situ	this work—LEP1A-C; LEP28
		41°34′46.13″ N	13°14′0.62″ E	upper Volsci unit	RDT	rudstone	*Rotalispira maxima, Accordiella conica*	Campanian	native block within C	this work LEP20

Site n°	Group of Localities	Latitude	Longitude	Tectonic Unit	Formation	Lithology	Biomarkers	Age Range	Comments	Author
		41°34′46.13″ N	13°14′0.62″ E	upper Volsci unit	C	glauconitic calcarenite with bryozoa; white -mica-bearingsandstone, brownish folded calcareous sandstone, green sandstone, veined and fracture calcareous marls; pinkish marl, Orbulina Marl lenses		(Paleocene-Serravallian) Tortonian p.p.	exotic block within C	[45], this work
9	Torrice	41°38′1.75″ N	13°24′27.41″ E	upper Ernici unit	FFS	pelitic-arenaceous; arenaceous -pelitic facies	*Glogerinoides extremus, Globigerinoides obliquus, Neogloboquadrina acostaensis, Globorotalia humerosa, Orbulina suturalis, Orbulina universa*	upper Tortonian—basal Messinian	in situ	[64,129]
		41°38′1.75″ N	13°24′27.41″ E	upper Ernici unit	MVP	arenaceous-pelitic facies	NN 11 la (CN 9a) subzone. *Discoaster* cf. *quinqueramus* and *D.* cf. *berggrenii*	late Tortonian	in situ	[35], this work
10	Monte San Giovanni Campano	41°37′59.37″ N	13°30′31.89″ E	upper Ernici unit	UAM	marly limestone and gray marl	NN 11 la (CN 9a) subzone. *Discoaster* cf. *quinqueramus* and *D.* cf. *berggrenii*	late Tortonian	in situ	[87]
		41°37′59.37″ N	13°30′31.89″ E	upper Ernici unit	FFS	arenitic (sandstone) and pelitic facies	NN 11 la (CN 9a) subzone. *Discoaster* cf. *quinqueramus* and *D.* cf. *berggrenii*	late Tortonian	in situ	[87]
		41°37′59.37″ N	13°30′31.89″ E	upper Ernici unit	MVP	clay with gypsum, sandstone and conglomerate	*Turborotalia multiloba, Aurila albicans, Discoaster variabilis, Discoaster intercalaris*	lower Messinian	in situ	[87]
11	Colle Cavallaro	41°30′53.47″ N	13°26′25.33″ E	Volsci Thrust Front	C	Briozoa-bearing detrital limestone	Molluschi e di Echinoderms, *Elphidium* sp., Lagenidae, Rotaliidae, Melobesie.	middle Miocene	reworked	[114]
		41°30′53.47″ N	13°26′25.33″ E	Volsci Thrust Front	C	Marly gray-greenish limestone	*Heterohelix* sp., Globigerinidae, *Globorotalia* spp.	earliest Paleocene	reworked	[114]
		41°30′53.47″ N	13°26′25.33″ E	Volsci Thrust Front	C	white limestone	*Ticinella* sp., *Gavelinella*, sp.	Aptian-Albian	reworked	[114]
		41°30′53.47″ N	13°26′25.33″ E	Volsci Thrust Front	C	Detrital-organogen limestone	*Heterohelix*, sp., *Globigerinella* sp., arenacous foraminifera	early Paleocene	reworked	[114]
		41°30′53.47″N	13°26′25.33″E	Volsci Thrust Front	C	Oxided detrital-organogen limestone	Briozoa, *Globorotalia* sp. e Lagenidae	Eocene	reworked	[114]
		41°30′53.47″ N	13°26′25.33″ E	Volsci Thrust Front	C	cherty limestone ox-bearing	Radiolarians, Lagenidae, Sponge spiculae	(?)Oligocene—early Aquitanian (?)	reworked	[114]
		41°30′53.47″ N	13°26′25.33″ E	Volsci Thrust Front	C	marl, sandstone, greenish clay	Globigerinidae, *Ammodiscus, Haplophragmoides*	late Miocene-earliest Pliocene	reworked	[114]
		41°30′53.47″ N	13°26′25.33″ E	Volsci Thrust Front	C	marl, sandstone, greenish clay (matrix)	*Rotalipora* cfr. *appenninica, Globotruncana lapparenti lapparenti,*	Cenomanian	reworked	[114]

Site n°	Group of Localities	Latitude	Longitude	Tectonic Unit	Formation	Lithology	Biomarkers	Age Range	Comments	Author
		41°30′53.47″ N	13°26′25.33″ E	Volsci Thrust Front	C	marl, sandstone, greenish clay (matrix)	*Racemiguembelina fructicosa*	Maastrichtian	reworked	[114]
		41°30′53.47″ N	13°26′25.33″ E	Volsci Thrust Front	C	marl, sandstone, greenish clay (matrix)	*Globigeraspis* sp., *Globigerina* cfr. *dissimilis*, *Globorotalia aequa*, *Globorotalia quetra*	middle -upper Eocene	reworked	[114]
		41°30′53.47″ N	13°26′25.33″ E	Volsci Thrust Front	C	marl, sandstone, greenish clay (matrix)	*Cassidulina subglobosa horizontalis*, *Globoquadrina dehiscens*, *Globoquadrina* cfr. *quadraria*, *Globigerinoides bisphaericus*	early Miocene	reworked	[114]
		41°30′53.47″ N	13°26′25.33″ E	Volsci Thrust Front	C	marl, sandstone, greenish clay (matrix)	*Haplophragmoides sp.*, *Eggerella brady*, *Nodosaria ovicula*, *Elphidium complanatum*, *Elphidium macellum*, *Nonion boueanum*, *Nonion umbilicatum*, *Pullenia bulloides*, *Plectofrondicularia diversicostata*, *Plectofrondicularia semicosta*, *Orthomorphina cfr. proxima*, *Robertina bradyi*, *Bulimina aculeata*, *Bulimina costata*, *Bulimina fusiformis*, *Bulimina inflata*, *Bolivina arta*, *Bolivina cistina*, *Bolivina punctata*, *Bolivinoides miocenicus*, *Uvigerina canariensis*, *Uvigerina laviculata*, *Uvigerina peregrina*, *Uvigerina rutila*, *Angulogenerina angulosa*, *Valvulineria bradyana*, *Valvulineria complanata*, *Gyroidina longispira*, *Gyroidina longispira miocenica*, *Gyroidina soldanii*, *Gyroidina soldanii altiformis*, *Eponides haidingeri*, *Eponides umbonatus stellatus*, *Rotalia beccarii inflata*, *Siphonina reticulata*, *Cassidulina laevigata carinata*, *Cassidulina oblonga*, *Cassidulina subglobosa*, *Sphaeroidina bulloides*, *Globigerina bulloides*, *Globigerina concinna*, *Globigerina eggeri*, *Sphaeroidinella cfr. dehiscens*, *Globigerinoides trilobus*, *Globigerinoides rubra*, *Orbulina suturalis*, *Catapsidrax unicavus*, *Globigerinita naparimaensis*, *Globorotalia cfr. bononiensis*, *Globorotalia scitula*, *Globorotalia aff. scitula*, *Globorotalia mayeri*	Pliocene?	Not reworked	[114]
12	Vele Mt. Thrust ramp	41°21′12.97″ N	13°31′7.49″ E	Upper Volsci unit	LK	limestone	*Cladocoropsis mirabilis*, *Salpingoporella dinarica*, *Orbitolina lenticularis*, *Cuneolina laurentii*, *C. camposauri*, *Salpingoporella annulata*	uppermost Jurassic—lower Cretaceous	in situ	[102]

Site n°	Group of Localities	Latitude	Longitude	Tectonic Unit	Formation	Lithology	Biomarkers	Age Range	Comments	Author
		41°21′12.97″ N	13°31′7.49″ E	Upper Volsci unit	UK	limestone	*Accordiella conica, Dicyclina schlumbergeri, Sellialveolina viallii*	Cenomanian-Santonian	in situ	[102]
		41°21′12.97″N	13°31′7.49″E	Upper Volsci unit	C	clay mélange	no data		reworked	[102]
13	Leucio Mt. *klippe*	41°28′0.12″ N	13°37′5.79″ E	Upper Volsci unit	LK	limestone	*Cladocoropsis mirabilis, Salpingoporella dinarica, Orbitolina lenticularis, Cuneolina laurentii, C. camposauri, Salpingoporella annulata*	Uppermost Jurassic—lower Cretaceous	in situ	[102]
		41°28′0.12″ N	13°37′5.79″ E	Upper Volsci unit	C	clay mélange	no data	upper Tortonian—lower Messinian		[102]
14	Formia-Maranola	41°17′23.93″ N	13°36′35.81″ E	lower Volsci unit	UK	limestone	*Accordiella conica, Rotalispira scarsellai, Dicyclina schlumbergeri, Moncharmontia apenninica*	Santonian—Campanian	in situ	[115]
		41°17′23.93″ N	13°36′35.81″ E	lower Volsci unit	UK	limestone with iron crust	*Scandonea; Ticinella sp., Hedbergella sp.*	Campanian?-early Eocene?		[63,115]
		41°17′23.93″ N	13°36′35.81″ E	lower Volsci unit	C	sitly clays, marls and sandstone; Pietra paesina; Scaglia-type limestone; marly limestones and Mg-bearing sandstones	radiolarians, heterohelicidae, *Globotruncana, Hedbergella, Globigerinoides* sp. lagenidae, globigerinidae		exotic blocks	[115]
		41°17′23.93″ N	13°36′35.81″ E	lower Volsci unit	MVP	mica-rich silty clays and argillous sands with gypsum	*Glorotalia*, globorotaloidea, *Globorotalia incompta, G. mayeri, G. obesa, G. pseudopachyderma, G. scitula, Globigerinoides* spp., *Globigerina quinqueloba, Orbulina* sp.	middle to upper Messinian	in situ	[115]
		41°17′23.93″ N	13°36′35.81″ E		SBG	polygenic breccia	*Bolivina leonardi, Cibicides italicus, Elphidium complanatuma, lenticulina clerici, Marginulina costata, Nodosaria pentecostata, Glorotalia puncticulata, G. bononiensis*	uppermost Messinian—lower Pliocene	in situ	[115]
15	Spigno Saturnia	41°18′45.84″ N	13°41′59.17″ E	lower Volsci unit	RDT	limestone	*Rotorbinella scarsellai, Accordiella conica*	Aptian-Turonian	in situ	[96]
		41°18′45.84″ N	13°41′59.17″ E	lower Volsci unit	RDT	limestone with iron crust	no data			[115]
		41°18′45.84″ N	13°41′59.17″ E	lower Volsci unit	C	sitly clay, marl and sandstone; Pietra paesina; Scaglia-type limestone; marly limestones and Mg-bearing sandstones	radiolarians, heterohelicidae, *Globotruncana, Hedbergella, Globigerinoides* sp., lagenidae, globigerinidae		exotic blocks	[115]

Site n°	Group of Localities	Latitude	Longitude	Tectonic Unit	Formation	Lithology	Biomarkers	Age Range	Comments	Author
16	Torrente ausente Valley	41°21′53.62″ N	13°43′55.16″ E	upper Ernici unit	CBZ	limestone	*Amphistegina, Elphidium, Heterostegina, Gypsina*	lower Aquitanian—lower Serraval-lian (?)	in situ	[149]
		41°21′53.62″ N	13°43′55.16″ E	upper Ernici unit	UAM	marl with cylindrites	*Orbulina universa, O. suturalis, Globorotalia menardii, Globorotalia scitula ventriosa, Globigerinoides trilobus, Globigerina* cf. *bulloides, Globigerina concinna, Globoquadrina dehiscens*	Serravallian p.p.—Tortonian p.p.	in situ	[115]
		41°21′53.62″ N	13°43′55.16″ E	upper Ernici unit	FFS	sandstone with olistolites and olistostromes	*Globorotalia menardii, Globigerina nepenthes, Globorotalia ventriosa, G. parabulloides, Globigerinoides obliquus, Globoquadrina globosa*	upper Tortonian	in situ with native blocks	[115]
		41°21′53.62″ N	13°43′55.16″ E	upper Ernici unit	MVP	calcarenite and conglomerate with quartz grains	*Amphistegina, Elphidium,* Textularidae, Miliolidae, Globigerinidae, *Globotruncane, Nummulites*	lower Messinian	reworked	[115,116, 150]
		41°21′53.62″ N	13°43′55.16″ E	upper Ernici unit	MVP	subordianate marl and gypsum towards the top	*Globorotalia acostaensis, Globigerina bulloides, Globigerina vanazuelana, Orbulina bilobata, Orbulina suturalis, Orbulina universa*	lower Messinian	in situ	[115]
17	Castelforte	41°17′55.49″ N	13°49′54.89″ E	upper Ernici unit	RDT	limestone	*Accordiella conica, Rotalispira scarsellai, Dicyclina schlumbergeri, Moncharmontia apenninica, Laffitteina*	Santonian-Maas-trichtian	in situ	[113]
		41°17′55.49″ N	13°49′54.89″ E	upper Ernici unit	Spirolina *lmst.*	limestone	*Spirolina, Coskinolina liburnica, Alveolina ellipsoidalis*	upper Paleocene—lower Eocene	in situ	[113]
		41°17′55.49″ N	13°49′54.89″ E	upper Ernici unit	CBZ	limestone	*Amphistegina, Cibicides, Operculina, Eponides*	Langhian (?)—Serravallian p.p.	in situ	[115]
		41°17′55.49″ N	13°49′54.89″ E	upper Ernici unit	UAM	marl and sandstone	*Orbulina universa, O. suturalis, Globorotalia menardii, Globorotalia scitula ventriosa, Globigerinoides trilobus, Globigerina cf. bulloides, Globigerina concinna, Globoquadrina dehiscens*	Serravallian p.p.—Tortonian p.p.	in situ	[115]
		41°17′55.49″ N	13°49′54.89″ E	upper Ernici unit	FFS	arenaceous -siltous clay turbidite	no data	Tortonian		[115]
18	Massico Mt.	41°9′39.27″ N	13°54′16.24″ E	upper Ernici unit ?	RDT	limestone	*Dicyclina schlumbergeri, Accordiella conica, rudist*	Campanian	in situ	[37]
		41°9′39.27″ N	13°54′16.24″ E	upper Ernici unit ?	CBZ	limestone	*Amphistegina, bryozoa, Ditrupa, ostreidae*	Burdigalian—Langhian	in situ	[37]
		41°9′39.27″ N	13°54′16.24″ E	upper Ernici unit ?	UAM	marls and sandstone	*Orbulina, Globorotalia menardii*	Serravallian—lower Tortonian p.p.	in situ	[37,115]
		41°9′39.27″ N	13°54′16.24″ E	upper Ernici unit ?	FFS	clay and sandstone with olistolithes	*Globorotalia mayeri, G. scitula, Globigerinoides trilobus, Bolivina sp., Discoaster brouweri, D. variabilis, D. surculus, Helicosphaera wallichii, Sphenolithus abies*	upper Tortonian—lower Messinian(?)	in situ	[37,115]

Appendix B

In the following, we report the new Stratigraphic constraints and age determination of the samples collected from twenty-five different localities in the study area representative stratigraphic logs of Figure 2 in the main manuscript.

Sampling Locality	Latitude	Longitude	Sample	Fm	Lithology	Texture and Components	Biomarker	Age Range
Gavignano, Contrada Fornarelli	41°41′54″ N	13°3′30″ E	LEP9A	MVP	Calcarenitic/arenitic matrix with aboundat quartz grains	*Elphidium and Amphistegina*; clasts of Cretaceous age (with *Rotalispira*) and clasts with *Orbulina*		Late Miocene
	41°41′54″ N	13°3′30″ E	LEP9A2	MVP	Calcarenitic/arenitic matrix with aboundat quartz grains	reworked *Elphidium* e *Amphistegina* and clasts with planctonic forams		Late Miocene
Gavignano, promenade	41°42′10″ N	13°2′39″ E	LEP10B	MVP	Lithoclast of Rudist (radiolitids) limestone with benthic foraminifera	Peloidal packstone with *Rotalispira*	*Moncharmontia apenninica, Rotalispira*	Coniacian-Campanian
	41°42′10″ N	13°2′39″ E	LEP10A	MVP	Calcarenitic/arenitic matrix with aboundat quartz grains	reworked *Elphidium* e *Amphistegina* and clasts with planctonic forams		Late Miocene
	41°42′10″ N	13°2′39″ E	LEP10	MVP	Lithoclast wackestone with planktonic foraminifera	*Orbulina*	*Orbulina*	Serravallian-Tortonian
	41°42′8.67″ N	13°2′38.40″ E	LEP10L	MVP	Calcarenitic/arenitic matrix		*Coccolithus pelagicus, Discoaster surculus, Helicosphaera wallichii, Reticulofenestra bisecta, Reticulofenestra minuta*	uppermost Tortonian—lowermost Zanclean
	41°42′11.18″ N	13°2′42.66″ E	LEP10M	MVP	Marl and clay		*Amaurolithus primus, Coccolithus pelagicus, Discoaster surculus, Discoaster variabilis, Nicklithus amplificus, Sphenolithus abies*	Messinian
Bassiano, Colle Cantocchio	41°34′28.33″ N	13°0′02.26″ E	LEP8A	CBZ	Conglomerate of calcarenitic pebbles with glauconite	Pebbly grainstone with echinid, ditrupae, *Elphidium, bryozoa, Miogypsina, Amphistegina, Operculina, Heterostegina, Lepidocyclina*	*Miogypsina, Elphidium*, bryozoa	early Miocene
	41°34′28.33″ N	13°0′02.26″ E	LEP8C	CBZ	Conglomerate of calcarenitic pebbles with glauconite	matrix made up with echinoderm and ostreid fragments with reworked Cretaceous clasts with *Thaumatoporella* orpeloidal facies	Ostreid and echinoderms	early Miocene
	41°34′32″ N	13°0′9″ E	LEP16	CBZ	Conglomerate with Cretaceous clasts	Echinid, *Elphidium*, bryozoa, *Amphistegina, Heterostegina, Lepidocyclina*	*Elphidium, Amphistegina*	early Miocene
Bassiano, Colle Cantocchio Bat Cave	41°34′31.92″ N	13°0′2.08″ E	LEP49d	UAM	Clay within thrust		*Braarudosphaera bigelowii, Catinaster cf. coalitus, C. cf. glenos, Coccolithus cf. eopelagicus, C. cf. miopelagicus, C. pelagicus, Cyclicargolithus abisectus, Cy. floridanus, Helicosphaera carteri, H. walberdosfensis, Ortorhadus cf. rugosus, O. serratus, Pontosphaera multipora, Reticulofenestra bisecta, R. cf. dictyoda, R. minuta, R. cf. pseudoumbilicus, Sphenolithus moriformis, S. radians, Triquetrorhabdulus carinatus, T. challegeri, Watznaueria barnesiae*	Mesozoic, Paleocene-Eocene; Oligocene-middle Miocene; Serravallian-Tortonian?

Sampling Locality	Latitude	Longitude	Sample	Fm	Lithology	Texture and Components	Biomarker	Age Range
	41°34′31.92″ N	13°0′2.08″ E	LEP49f	UAM	Clay beneath thrust		*Braarudosphaera bigelowii, Calcidiscus leptoporus, Catinaster cf. coalitus, C. glenos, Chiasmolithus sp., Coccolithus cf. eopelagicus, C. cf. miopelagicus, C. pelagicus, C. tenuiforatus, Cruciplacolithus sp., Cyclicargolithus abisectus, Cy. floridanus, Discoaster cf. deflandrei, D. multiradiatus gr., D. sp., Helicosphaera carteri, H. walberdosfensis, Nannotetrina fulgens, Ortorhadus serratus, Pontosphaera multipora, Reticulofenestra bisecta, R. cf. dictyoda, R. minuta, R. cf. pseudoumbilicus, Sphenolithus moriformis, S. radians, Triquetrorhabdulus carinatus, Watznaueria barnesiae, Zygrhablithus bijugatus*	Mesozoic, Paleocene-Eocene; Oligocene-middle Miocene; Serravallian-Tortonian?
Bassiano, Colle Cantocchio near top	41°34′29″ N	13°0′9″ E	LEP15	SBG	Breccia with Mesozoic clasts	Reddish matrix with echinoderm fragments	echinoderms	
Gorga, Rave Santa Maria	41°39′18″ N	13°7′18″ E	LEP 14B	UK	Grainstone/ Packstone with rudists	*Rotalispira* and ostracods	*Rotalispira*	Campanian
	41°39′35″ N	13°7′8″ E	LEP12A	UK	Calcarenite with resedimented rudist	bioclastic detritus (echinoderms) with *Orbitoides* and *Murciella*	*Orbitoides* and *Murciella*	Upper Campanian-lower Maastrichtian
	41°39′35″ N	13°7′8″ E	LEP12B	CBZ	Breccia of encrusted K pebbles, calcarenitic matrix	Bivalve, echinoid fragments and reworked Cretaceous clasts		early Miocene
	41°39′35″ N	13°7′8″ E	LEP12C	CBZ	Marl level with tiny limestone clasts		*Cyclicargolithus floridanus, Sphenolithus conicus*	not younger than middle Burdigalian
Gorga, Capezzenna Mt.	41°39′6.86″ N	13°4′22.07″ E	GO2	MVP	Marly lens within conglomerate		*Amaurolithus primus*	uppermost Tortonian—lowermost Zanclean
	41°39′6.86″ N	13°4′22.07″ E	GO3	MVP	Marly clast within conglomerate	Clast with *Chondrites* bioturbations	*main markers: Discoaster surculus, Helicosphera wallichii. Other components: Calcidiscus leptoporus, C. macintyrei, Coccolithus pelagicus, Discoaster multiradiatus, Helicosphaera carteri, Reticulofenestra minuta, R. pseudombilicus, Sphenolithus moriformis, S. radians, Zygrhablithus bijugatus*	upper Tortonian
Marroni, Morolo	41°39′38.68″ N	13°10′27.54″ E	LEP72	FFS	Marl and clay		*Discoaster variabilis, Helicosphaera carteri, Reticulofenestra bisecta, R. minuta, Sphenolithus procerus, Zygrhablithus bijugatus*	not older than upper Tortonian
Colle Fatuccio, Ferentino	41°40′21.60″ N	13°13′45.14″ E	LEP73	FFS	Marl and clay	Crustaceans bioturbations	sterile	
Carpineto, Pian della Faggeta	41°34′34″ N	15°06′53″ E	LEP17C	UK	Carbonatic breccia at the top of a lithon	Packstone with Rudist and foraminifera	*Rotalispira*	Santonian-Campanian
	41°34′34″ N	16°06′53″ E	LEP17D	C	Fine-grained calcarenite lens	Grainstone-packstone with *Amphistegina*, Miogypsinids and echinoderm fragments	Miogypsinids	early Miocene
	41°34′34″ N	15°06′53″ E	LEP17E	C	Microconglomerate lens with carbonatic and crystalline pebbles		sterile	

Sampling Locality	Latitude	Longitude	Sample	Fm	Lithology	Texture and Components	Biomarker	Age Range
	41°34′34″ N	15°06′53″ E	LEP17F	C	Sandstone with carbonatic e and crystalline clasts		sterile	
	41°34′35″ N	18°06′41″ E	LEP18B	UK	sStriated carbonate cruston at the top of a lithon	Wackestone with *Decastronema*, ostracoda, discorbidae	*Decastronema*	Upper Creteceous
	41°34′35″ N	18°06′41″ E	LEP18D	C	Microconglomerate lens with crystalline clasts		Coccolithus pelagicus, Cyclicargolithus abisectus, Cy. floridanus, Reticulofenestra cf. pseudoumbilicus	early Miocene?
	41°34′33″ N	19°06′39″ E	LEP19A	UK	Rudist rudstone/floatstone at the top of a lithon		rudists	Upper Cretaceous
	41°34′33″ N	19°06′39″ E	LEP19B	UK	Carbonate breccia	Wackestone with *Thaumatoporella*, and benthic foraminifera	*Nezzazatinella, Rotalispira*	Coniacian-Campanian
Patrica, Caccume Mt. north	41°34′46.50″ N	13°13′58.92″ E	CC20	UK	Limestone	Wackestone with *Thaumatoporella*, and benthic foraminifera	*Accordiella conica, Rotalispira maxima*	Santonian Campanian
	41°34′46.50″ N	13°13′58.92″ E	CC21	UK	Limestone breccia on top	Wackestone with *Thaumatoporella*, and benthic foraminifera	*Nezzazatinella, Rotalispira maxima*	Santonian Campanian
	41°34′46.50″ N	13°13′58.92″ E	CC23	UK	Limestone breccia on top	Wackestone with *Thaumatoporella*, and benthic foraminifera	*Rotalispira maxima*	Santonian Campanian
	41°34′39.30″ N	13°14′3.01″ E	CC24	UK	Cataclastic limestone			Cenomanian?
	41°34′47″ N	13°13′59″ E	LEP1A	UK	Limestone below unconformity	Benthic foraminifera and small debris of rudist shells fragments	*Accordiella conica and Rotalispira maxima*	Campanian
	41°34′47″ N	13°13′59″ E	LEP1B	UK	Dolomitic limestone above unconformity			Upper Cretaceous?
	41°34′47″ N	13°13′59″ E	LEP1C	UK	Encrusted carbonatic breccia on top	Wackestone with benthic foraminifera, few intraclasts	*Rotalispira scarsellai*	Santonian-Campanian
	41°34′47″ N	13°13′59″ E	LEP20	UK	Limestone breccia within Chaotic complex	Wackestone with benthic foraminifera, Cretaceous intraclasts	*Rotalispira maxima and Accordiella conica*	Campanian
	41°34′32″	13°14′5″	LEP68a	C	Clay		*Coccolithus pelagicus, C. miopelagicus, Cyclicargolithus abisectus, Cy. floridanus, Discoaster sp., D. berggrenii, D. brouweri, D. deflandrei, D. formosus, D. multiradiatus, D. pentaradiatus, D. quinqueramus, D. variabilis, Helicosphaera recta, H. stalis, H. walberdosfensis, Orthorhabdus rugosus, Reticulofenestra bisecta, R. minuta, R. pseudoumbilicus, Reticulofenestra sp., Sphaenolithus abies, S. ciperoensis, S. disbelemnos, S. heteromorphus, S. moriformis, Zygrhablithus bijugatus*	Paleocene-Eocene; Oligocene-early Miocene; middle Miocene; upper Tortonian—Messinian?
Giuliano di Roma, Caccume Mt. east	41°34′19″ N	13°14′55″ E	LEP27B	UK	Limestone below unconformity	Packstone with benthic foraminifera	*Rotalispira scarsellai*	Upper Turonian-Campanian
	41°34′19″ N	13°14′55″ E	LEP27C2	UK	Breccia above unconformity			Santonian-Campanian
	41°34′19″ N	13°14′55″ E	LEP27D	UK	Limestone	Packstone with algae and benthic foraminifera	*Thaumatoporella, Rotalispira scarsellai, Rotalispira maxima, Moncharmontia apenninica*	Santonian-Campanian
Giuliano di Roma, Caccume Mt. east	41°34′19″ N	13°14′55″ E	LEP27F	UK	Limestone	Wackestone with benthic foraminifera	*Nezzazatinella, Rotalispira, Pseudocyclammina sphaeroidea*	Turonian-Santonian
Giuliano di Roma, Caccume Mt. Scorciapane	41°34′ 14.93″ N	13°13′ 54.82″ E	LEP67	C	Clay beneath thrust		sterile	

Sampling Locality	Latitude	Longitude	Sample	Fm	Lithology	Texture and Components	Biomarker	Age Range
Giuliano di Roma, Caccume Mt. west	41°34′40″N	13°13′23″E	LEP69a	C	Clay		*Calcidiscus leptoporus, Clarolithus ellipticus, Coccolithus pelagicus, C. miopelagicus, D. berggrenii, D. brouwerii, D. decorus, D. deflandrei, D. multiradiatus, D. pentaradiatus, D. quinqueramus, D. variabilis, Orthorhabdus rugosus, O. striatus, Pontosphaera discopora, P. multipora, Reticulofenestra bisecta, R. minuta, R. pseudoumbilicus, Sphaenolithus abies, S. moriformis, Zygrhablithus bijugatus*	Paleocene-Eocene; Oligocene-early Miocene; middle Miocene; upper Tortonian—Messinian?
Giuliano di Roma, Siserno Mt.	41°32′17″ N	13°18′02″ E	LEP 30	UK	Limestone	Wackestone with Ostracods	ostracods	Campanian?
	41°32′17″ N	13°18′02″ E	LEP 31	UK	Limestone	Mudstone with Dolomitized intraclasts		Upper Cretaceous
	41°32′17″ N	13°18′02″ E	LEP 32	UK	Limestone	Wackestone with miliolidae, ostracoda and dolomite crystals	ostracods and miliolids	Campanian
	41°32′17″ N	13°18′02″ E	LEP 32B	UK	Limestone	Wackestone with miliolidae, ostracoda, discorbidae and porcelaneous foraminifera	ostracods and miliolids	Campanian
Patrica, il Patricano	41°33′4.78″ N	13°16′2.89″ E	LEP36	UK	Limestone	Wackestone with iscorbidae	discorbidae	Campanian
	41°34′17″ N	13°15′54″ E	LEP39A	UK	Limestone	Wackestone with rudist fragments, miolidae, ostracoda and *Thaumatoporella*	ostracods and discorbide	Campanian
	41°34′17″ N	13°15′54″ E	LEP39B	UK	Limestone	Wackestone with ostracods and miliolidae	*Rotalispira scarsellai, Accordiella conica,*	Campanian
	41°34′20″ N	13°15′55″ E	LEP 40A	UK	Limestone	Wackestone/Packstone with ostracods and discorbidae	*Thaumatoporella, Nezzazata, Moncharmontia apenninica*	Campanian
Giuliano di Roma west	41°33′7″	13°16′10″	LEP70	C	Marl and clay		*Coccolithus pelagicus, Reticulofenestra bisecta, R. minuta, Sphenolithus moriformis, Zygrhablithus bijugatus*	Paleocene-Tortonian
Frosinone, Le Fornaci cinema	41°37′4.88″ N	13°20′26.32″ E	LEP71	FFS	Marl and pelite with coal		*Coccolithus pelagicus, Cyclicargolithus abisectus, Reticulofenestra bisecta, R. minuta, Pontosphaera sp.*	Oligocene-early Miocene; Tortonian?

References

1. Read, J.F. Carbonate platform facies models. *Aapg Bull.* **1985**, *69*, 1–21. [CrossRef]
2. Ford, M.; Stahel, U. The geometry of a deformed carbonate slope-basin transition: The Ventoux-Lure fault zone, SE France. *Tectonics* **1995**, *14*, 1393–1410. [CrossRef]
3. Carminati, E.; Doglioni, C. Alps vs. Apennines: The paradigm of a tectonically asymmetric Earth. *Earth-Sci. Rev.* **2012**, *112*, 67–96. [CrossRef]
4. Davis, D.; Suppe, J.; Dahlen, F.A. Mechanics of fold-and-thrust belts and accretionary wedges. *J. Geophys. Res. Solid Earth* **1983**, *88*, 1153–1172. [CrossRef]
5. Roure, F. Foreland and hinterland basins: What controls their evolution? *Swiss J. Geosci.* **2008**, *101*, 5–29. [CrossRef]
6. Grool, A.R.; Ford, M.; Vergés, J.; Huismans, R.S.; Christophoul, F.; Dielforder, A. Insights into the crustal-scale dynamics of a doubly vergent orogen from a quantitative analysis of its forelands: A case study of the Eastern Pyrenees. *Tectonics* **2018**, *37*, 450–476. [CrossRef]
7. Butler, R.W.; Tavarnelli, E.; Grasso, M. Structural inheritance in mountain belts: An Alpine–Apennine perspective. *J. Struct. Geol.* **2006**, *28*, 1893–1908. [CrossRef]
8. Tavani, S.; Storti, F.; Lacombe, O.; Corradetti, A.; Muñoz, J.A.; Mazzoli, S. A review of deformation pattern templates in foreland basin systems and fold-and-thrust belts: Implications for the state of stress in the frontal regions of thrust wedges. *Earth-Sci. Rev.* **2015**, *14*, 82–104. [CrossRef]
9. Santantonio, M.; Carminati, E. Jurassic rifting evolution of the Apennines and Southern Alps (Italy): Parallels and differences. *Gsa Bull.* **2011**, *123*, 468–484. [CrossRef]

10. Shiner, P.; Beccaccini, A.; Mazzoli, S. Thin-skinned versus thick-skinned structural models for Apulian carbonate reservoirs: Constraints from the Val d'Agri Fields, S Apennines, Italy. *Mar. Pet. Geol.* **2004**, *21*, 805–827. [CrossRef]
11. Livani, M.; Scrocca, D.; Arecco, P.; Doglioni, C. Structural and stratigraphic control on salient and recess development along a thrust belt front: The Northern Apennines (Po Plain, Italy). *J. Geophys. Res. Solid Earth* **2018**, *123*, 4360–4387. [CrossRef]
12. Cardello, G.L.; Di Vincenzo, G.; Giorgetti, G.; Zwingmann, H.; Mancktelow, N. Initiation and development of the Pennine Basal Thrust (Swiss Alps): A structural and geochronological study of an exhumed megathrust. *J. Struct. Geol.* **2019**, *126*, 338–356. [CrossRef]
13. Herwegh, M.; Pfiffner, O.-A. Tectono-metamorphic evolution of a nappe stack: A case study of the Swiss Alps. *Tectonophysics* **2005**, *404*, 55–76. [CrossRef]
14. Festa, A.; Pini, G.A.; Dilek, Y.; Codegone, G. Mélanges and mélange-forming processes: A historical overview and new concepts. *Int. Geol. Rev.* **2010**, *52*, 1040–1105. [CrossRef]
15. Festa, A.; Dilek, Y.; Pini, G.A.; Codegone, G.; Ogata, K. Mechanisms and processes of stratal disruption and mixing in the development of mélanges and broken formations: Redefining and classifying mélanges. *Tectonophysics* **2012**, *568*, 7–24. [CrossRef]
16. Ogawa, Y.; Anma, R.; Dilek, Y. *Accretionary Prisms and Convergent Margin Tectonics in the Northwest Pacific Basin*; Springer Science & Business Media: Berlin, Germany, 2011; Volume 8. [CrossRef]
17. Codegone, G.; Festa, A.; Dilek, Y. Formation of Taconic mélanges and broken formations in the Hamburg Klippe, central Appalachian Orogenic Belt, eastern Pennsylvania. *Tectonophysics* **2012**, *568*, 215–229. [CrossRef]
18. Ogata, K.; Festa, A.; Pini, G.A.; Pogačnik, Ž.; Lucente, C.C. Substrate deformation and incorporation in sedimentary mélanges (olistostromes): Examples from the northern Apennines (Italy) and northwestern Dinarides (Slovenia). *Gondwana Res.* **2019**, *74*, 101–125. [CrossRef]
19. Mohn, G.; Manatschal, G.; Masini, E.; Müntener, O. Rift-related inheritance in orogens: A case study from the Austroalpine nappes in Central Alps (SE-Switzerland and N-Italy). *Int. J. Earth Sci.* **2011**, *100*, 937–961. [CrossRef]
20. Bertok, C.; Martire, L.; Perotti, E.; d'Atri, A.; Piana, F. Kilometre-scale palaeoescarpments as evidence for Cretaceous synsedimentary tectonics in the External Briançonnais Domain (Ligurian Alps, Italy). *Sediment. Geol.* **2012**, *251*, 58–75. [CrossRef]
21. Cardello, G.L.; Mancktelow, N.S. Cretaceous syn-sedimentary faulting in the Wildhorn Nappe (SW Switzerland). *Swiss J. Geosci.* **2014**, *107*, 223–250. [CrossRef]
22. Balestro, G.; Festa, A.; Tartarotti, P. Tectonic significance of different block-in-matrix structures in exhumed convergent plate margins: Examples from oceanic and continental HP rocks in Inner Western Alps (northwest Italy). *Int. Geol. Rev.* **2015**, *57*, 581–605. [CrossRef]
23. Tozer, R.S.J.; Butler, R.W.H.; Corrado, S. Comparing thin-and thick-skinned thrust tectonic models of the Central Apennines, Italy. *Egu Stephan Mueller Spec. Publ. Ser.* **2002**, *1*, 181–194. [CrossRef]
24. Calabrò, R.A.; Corrado, S.; Di Bucci, D.; Robustini, P.; Tornaghi, M. Thin-skinned vs. thick-skinned tectonics in the Matese Massif, Central–Southern Apennines (Italy). *Tectonophysics* **2003**, *377*, 269–297. [CrossRef]
25. Tavarnelli, E.; Butler, R.W.H.; Decandia, F.A.; Calamita, F.; Grasso, M.; Alvarez, W.; Renda, P. Implications of fault reactivation and structural inheritance in the Cenozoic tectonic evolution of Italy. *Geol. Italy Spec.* **2004**, *1*, 209–222.
26. Scrocca, D.; Carminati, E.; Doglioni, C. Deep structure of the southern Apennines, Italy: Thin-skinned or thick-skinned? *Tectonics* **2005**, *24*. [CrossRef]
27. Billi, A.; Barberi, G.; Faccenna, C.; Neri, G.; Pepe, F.; Sulli, A. Tectonics and seismicity of the Tindari Fault System, southern Italy: Crustal deformations at the transition between ongoing contractional and extensional domains located above the edge of a subducting slab. *Tectonics* **2006**, *25*. [CrossRef]
28. Mazzoli, S.; D'errico, M.; Aldega, L.; Corrado, S.; Invernizzi, C.; Shiner, P.; Zattin, M. Tectonic burial and "young"(<10 Ma) exhumation in the southern Apennines fold-and-thrust belt (Italy). *Geology* **2008**, *36*, 243–246. [CrossRef]
29. Molli, G.; Menegon, L.; Malasoma, A. Switching deformation mode and mechanisms during subduction of continental crust: A case study from Alpine Corsica. *Solid Earth* **2017**, *8*, 767–788. [CrossRef]
30. Cardello, G.L.; Doglioni, C. From mesozoic rifting to Apennine orogeny: The gran Sasso range (Italy). *Gondwana Res.* **2015**, *27*, 1307–1334. [CrossRef]
31. Vitale, S.; Tramparulo, F.D.A.; Ciarcia, S.; Amore, F.O.; Prinzi, E.P.; Laiena, F. The northward tectonic transport in the southern Apennines: Examples from the Capri Island and western Sorrento Peninsula (Italy). *Int. J. Earth Sci.* **2017**, *106*, 97–113. [CrossRef]
32. Storti, F.; Balsamo, F.; Mozafari, M.; Koopman, A.; Swennen, R.; Taberner, C. Syn-Contractional Overprinting Between Extension and Shortening Along the Montagna Dei Fiori Fault During Plio-Pleistocene Antiformal Stacking at the Central Apennines Thrust Wedge Toe. *Tectonics* **2018**, *37*, 3690–3720. [CrossRef]
33. Cipriani, A.; Bottini, C. Early Cretaceous tectonic rejuvenation of an Early Jurassic margin in the Central Apennines: The "Mt. Cosce Breccia". *Sediment. Geol.* **2019**, *387*, 57–74. [CrossRef]
34. Sabbatino, M.; Vitale, S.; Tavani, S.; Consorti, L.; Corradetti, A.; Cipriani, A.; Arienzo, I.; Parente, M. Constraining the onset of flexural subsidence and peripheral bulge extension in the Miocene foreland of the southern Apennines (Italy) by Sr-isotope stratigraphy. *Sediment. Geol.* **2020**, *401*. [CrossRef]
35. Cipollari, P.; Cosentino, D. Miocene unconformities in the Central Apennines: Geodynamic significance and sedimentary basin evolution. *Tectonophysics* **1995**, *252*, 375–389. [CrossRef]

36. Bigi, S.; Milli, S.; Corrado, S.; Casero, P.; Aldega, L.; Botti, F.; Moscatelli, M.; Stanzione, O.; Falcini, F.; Marini, M.; et al. Stratigraphy, structural setting and burial history of the Messinian Laga basin in the context of Apennine foreland basin system. *J. Mediterr. Earth Sci.* **2009**, *1*, 61–84.
37. Vitale, S.; Prinzi, E.P.; Ciarcia, S.; Sabbatino, M.; Tramparulo, F.D.A.; Verazzo, G. Polyphase out-of-sequence thrusting and occurrence of marble detritus within the wedge-top basin deposits in the Mt. Massico (southern Apennines): Insights into the late Miocene tectonic evolution of the central Mediterranean. *Int. J. Earth Sci.* **2019**, *108*, 501–519. [CrossRef]
38. Vitale, S.; Prinzi, E.P.; Tramparulo, F.D.A.; De Paola, C.; Di Maio, R.; Piegari, E.; Sabbatino, M.; Natale, J.; Notaro, P.; Ciarcia, S. Late Miocene-Early Pliocene Out-of-Sequence Thrusting in the Southern Apennines (Italy). *Geosciences* **2020**, *10*, 301. [CrossRef]
39. De Capoa, P.; Di Staso, A.; Guerrera, F.; Perrone, V.; Tramontana, M. The extension of the Maghrebian Flysch basin in the Apenninic chain: Paleogeographic and paleotectonic implications. Atti Congr. "Etat des connaissances géologiques des régions nord du Maroc: La Chaîne rifaine dans son cadre Méditerranéen occidental, Rabat (Maroc). *Trav. Inst. Sci. RabatGéogr. Phys.* **2003**, *21*, 77–92.
40. Carboni, F.; Viola, G.; Aldega, L.; van der Lelij, R.; Brozzetti, F.; Barchi, M.R. K-Ar fault gouge dating of Neogene thrusting: The case of the siliciclastic deposits of the Trasimeno Tectonic Wedge (Northern Apennines, Italy). *Ital. J. Geosci.* **2020**, *139*, 300–308. [CrossRef]
41. Curzi, M.; Aldega, L.; Bernasconi, S.M.; Berra, F.; Billi, A.; Boschi, C.; Carminati, E. Architecture and evolution of an extensionally-inverted thrust (Mt. Tancia Thrust, Central Apennines): Geological, structural, geochemical, and K–Ar geochronological constraints. *J. Struct. Geol.* **2020**, 104059. [CrossRef]
42. Curzi, M.; Billi, A.; Carminati, E.; Rossetti, F.; Albert, R.; Aldega, L.; Cardello, G.L.; Conti, A.; Gerdes, A.; Smeraglia, L.; et al. Disproving the Presence of Paleozoic-Triassic Metamorphic Rocks on the Island of Zannone (Central Italy): Implications for the Early Stages of the Tyrrhenian-Apennines Tectonic Evolution. *Tectonics* **2020**, *39*, e2020TC006296. [CrossRef]
43. Smeraglia, L.; Aldega, L.; Billi, A.; Carminati, E.; Di Fiore, F.; Gerdes, A.; Richard, A.; Rossetti, F.; Vignaroli, G. Development of an intrawedge tectonic mélange by out-of-sequence thrusting, buttressing, and intraformational rheological contrast, Mt. Massico ridge, Apennines, Italy. *Tectonics* **2019**, *38*, 1223–1249. [CrossRef]
44. Vitale, S.; Ciarcia, S. Tectono-stratigraphic and kinematic evolution of the southern Apennines/Calabria–Peloritani Terrane system (Italy). *Tectonophysics* **2013**, *583*, 164–182. [CrossRef]
45. Cardello, G.L.; Consorti, L.; Palladino, D.M.; Carminati, E.; Carlini, M.; Doglioni, C. Tectonically controlled carbonate-seated maar-diatreme volcanoes: The case of the Volsci Volcanic Field, central Italy. *J. Geodyn.* **2020**, *139*, 101763. [CrossRef]
46. Roure, F.; Casero, P.; Vially, R. Growth processes and melange formation in the southern Apennines accretionary wedge. *Earth Planet. Sci. Lett.* **1991**, *102*, 395–412. [CrossRef]
47. Molli, G. Northern Apennine–Corsica orogenic system: An updated overview. *Geol. Soc. Lond. Spec. Publ.* **2008**, *298*, 413–442. [CrossRef]
48. Rosenbaum, G.; Gasparon, M.; Lucente, F.P.; Peccerillo, A.; Miller, M.S. Kinematics of slab tear faults during subduction segmentation and implications for Italian magmatism. *Tectonics* **2008**, *27*. [CrossRef]
49. Carminati, E.; Fabbi, S.; Santantonio, M. Slab bending, syn-subduction normal faulting, and out-of-sequence thrusting in the Central Apennines. *Tectonics* **2014**, *33*, 530–551. [CrossRef]
50. Faccenna, C.; Becker, T.W.; Miller, M.S.; Serpelloni, E.; Willett, S.D. Isostasy, dynamic topography, and the elevation of the Apennines of Italy. *Earth Planet. Sci. Lett.* **2014**, *407*, 163–174. [CrossRef]
51. Boccaletti, M.; Guazzone, G. Remnant arcs and marginal basins in the Cainozoic development of the Mediterranean. *Nature* **1974**, *252*, 18–21. [CrossRef]
52. Malinverno, A.; Ryan, W.B. Extension in the Tyrrhenian Sea and shortening in the Apennines as result of arc migration driven by sinking of the lithosphere. *Tectonics* **1986**, *5*, 227–245. [CrossRef]
53. Royden, L.; Patacca, E.; Scandone, P. Segmentation and configuration of subducted lithosphere in Italy: An important control on thrust-belt and foredeep-basin evolution. *Geology* **1987**, *15*, 714–717. [CrossRef]
54. Doglioni, C. A proposal for the kinematic modelling of W-dipping subductions-possible applications to the Tyrrhenian-Apennines system. *Terra Nova* **1991**, *3*, 423–434. [CrossRef]
55. Jolivet, L.; Faccenna, C. Mediterranean extension and the Africa-Eurasia collision. *Tectonics* **2000**, *19*, 1095–1106. [CrossRef]
56. Van Hinsbergen, D.J.; Torsvik, T.H.; Schmid, S.M.; Maţenco, L.C.; Maffione, M.; Vissers, R.L.; Gürer, D.; Spakman, W. Orogenic architecture of the Mediterranean region and kinematic reconstruction of its tectonic evolution since the Triassic. *Gondwana Res.* **2020**, *81*, 79–229. [CrossRef]
57. Sartori, R.; Torelli, L.; Zitellini, N.; Carrara, G.; Magaldi, M.; Mussoni, P. Crustal features along a W–E Tyrrhenian transect from Sardinia to Campania margins (Central Mediterranean). *Tectonophysics* **2004**, *383*, 171–192. [CrossRef]
58. Acocella, V.; Funiciello, R. Transverse systems along the extensional Tyrrhenian margin of central Italy and their influence on volcanism. *Tectonics* **2006**, *25*. [CrossRef]
59. Beaudoin, A.; Augier, R.; Jolivet, L.; Jourdon, A.; Raimbourg, H.; Scaillet, S.; Cardello, G.L. Deformation behavior of continental crust during subduction and exhumation: Strain distribution over the Tenda massif (Alpine Corsica, France). *Tectonophysics* **2017**, *705*, 12–32. [CrossRef]
60. Locardi, E. The origin of the Apenninic arcs. *Tectonophysics* **1988**, *146*, 105–123. [CrossRef]
61. Pizzi, A.; Galadini, F. Pre-existing cross-structures and active fault segmentation in the northern-central Apennines (Italy). *Tectonophysics* **2009**, *476*, 304–319. [CrossRef]

62. Castellarin, A.; Colacicchi, R.; Praturlon, A. Fasi distensive, trascorrenze e sovrascorrimenti lungo la <<Linea Ancona-Anzio>>, dal Lias Medio al Pliocene. *Geol. Romana* **1978**, *17*, 161–189.
63. Centamore, E.; Di Manna, P.; Rossi, D. Kinematic evolution of the Volsci Range: A new overview. *Ital. J. Geosci.* **2007**, *126*, 159–172.
64. Centamore, E.; Dramis, F.; Di Manna, P.; Fumanti, F.; Milli, S.; Rossi, D.; Palombo, M.R.; Palladino, D.M.; Trigila, R.; Zanon, V.; et al. Note illustrative del Foglio 402 Ceccano. In *Carta Geol. D'italia 1:50.000. Serv. Geol. D'italia (ISPRA)*; Instituto Superiore per la Protezione e la Ricerca Ambientale: Rome, Italy, 2010.
65. Fazzini, P.; Gelmini, R.; Mantovani, P.; Pellegrini, M. Geologia dei Monti della Tolfa (Lazio settentrionale; prov. di Viterbo e Roma). *Mem. Soc. Geol. Ital.* **1972**, *11*, 65–144.
66. Parotto, M.; De Rita, D.; Giordano, G.; Cecili, A.; Chiocci, F.L.; La Monica, G.B. Note illustrative del Foglio 187 Albano Laziale. In *Carta Geol. D'italia 1:50.000. Serv. Geol. D'italia (ISPRA)*; Instituto Superiore per la Protezione e la Ricerca Ambientale: Rome, Italy, 2009.
67. Ogniben, L. Schema Introduttivo alla Geologia del confine Calabro-Lucano. *Mem. Soc. Geol. Ital.* **1969**, *8*, 453–763.
68. Acocella, V.; Faccenna, C.; Funiciello, R. Elementi strutturali della media Valle Latina. *Boll. Soc. Geol. Ital.* **1996**, *115*, 501–518.
69. Selli, R. Sulla trasgressione del Miocene nell'Italia meridionale. *Museo Geologico Giovanni Capellini* **1957**, *2*, 1–54.
70. De Blasio, I.; Lima, A.; Perrone, V.; Russo, M. Nuove vedute sui depositi miocenici della Penisola Sorrentina. *Boll. Soc. Geol. Ital.* **1981**, *100*, 55–70.
71. Accordi, B. La componente traslativa nella tettonica dell'Appennino laziale-abruzzese. *Geol. Romana* **1966**, *5*, 355–406.
72. Cosentino, D.; Cipollari, P.; Di Donato, V.; Sgrosso, I.; Sgrosso, M. The Volsci Range in the kinematic evolution of the northern and southern Apennine orogenic system. *Boll. Della Soc. Geol. Ital.* **2002**, *121*, 209–218.
73. Devoto, G. Note geologiche sul settore centrale dei Monti Simbruini ed Ernici (Lazio nord-orientale). *Boll. Soc. Nat. Napoli* **1967**, *76*, 1–112.
74. Fabbi, S.; Santantonio, M. First report of a Messinian coralgal facies in a terrigenous setting of Central Apennines (Italy) and its palaeogeographic significance. *Geol. J.* **2019**, *54*, 1756–1768. [CrossRef]
75. Dondi, L.; Papetti, I.; Tedeschi, D. Stratigrafia del pozzo Trevi 1 (Lazio). *Geol. Romana* **1966**, *5*, 249–262.
76. Cavinato, G.; Cerisola, R.; Sirna, M. Strutture compressive pellicolari e tettonica distensiva nei Monti Ernici sudoccidentali (Appennino centrale). *Mem. Sgi* **1990**, *45*, 539–553.
77. Cavinato, G.P.; Parotto, M.; Sirna, M. Geological summary of the Central Apennines. Four decades later. *RendOnline Soc. Geol. Ital.* **2012**, *23*, 31–44.
78. Brozzetti, F.; Cerritelli, F.; Cirillo, D.; Agostini, S.; Lavecchia, G. The Roccacaramanico Conglomerate (Maiella Tectonic Unit) in the frame of the Abruzzo early Pliocene Foreland Basin System: Stratigraphic and structural implications. *Ital. J. Geosci.* **2020**, *139*, 266–286. [CrossRef]
79. Accordi, G.; Carbone, F. Sequenze carbonatiche meso-cenozoiche. In *Note Illustrative della Carta delle Litofacies del Lazio-Abruzzo ed Aree Limitrofe*; Consiglio Nazionale delle Ricerche: Rome, Italy, 1988; pp. 11–92.
80. Civitelli, G.; Brandano, M. Atlante delle litofacies e modello deposizionale dei Calcari a Briozoi e Litotamni nella Piattaforma carbonatica laziale-abruzzese. *Boll. Della Soc. Geol. Ital.* **2005**, *124*, 611.
81. Cosentino, D.; Cipollari, P.; Marsili, P.; Scrocca, D. Geology of the central Apennines: A regional review. *J. Virtual Explor.* **2010**, *36*, 1–37. [CrossRef]
82. Consorti, L.; Frijia, G.; Caus, E. Rotaloidean foraminifera from the Late Cretaceous carbonates of Central and Southern Italy. *Cretac. Res.* **2017**, *70*, 226–243. [CrossRef]
83. Romano, M.; Manni, R.; Venditti, E.; Nicosia, U.; Cipriani, A. First occurrence of a Tylosaurinae mosasaur from the Turonian of the Central Apennines, Italy. *Cretac. Res.* **2019**, *96*, 196–209. [CrossRef]
84. Angelucci, A.; Devoto, G. Geologia del Monte Caccume. *Geol. Romana* **1966**, *5*, 177–196.
85. Casero, P. Structural setting of petroleum exploration plays in Italy. In *Special Volume of the Italian Geological Society for the IGC 32 Florence-2004*; Crescenti, U., d'Offizi, S., Merlino, S., Sacchi, L., Eds.; Italian Geological Society: Rome, Italy, 2004; pp. 189–199.
86. Sani, F.; Del Ventisette, C.; Montanari, D.; Coli, M.; Nafissi, P.; Piazzini, A. Tectonic evolution of the internal sector of the Central Apennines, Italy. *Mar. Pet. Geol.* **2004**, *21*, 1235–1254. [CrossRef]
87. Pasquali, V.; Castorina, F.; Cipollari, P.; Cosentino, D.; Lo Mastro, S. I depositi tardo-orogenici della Valle Latina meridionale: Stratigrafia e implicazioni cinematiche per l'evoluzione dell'Appennino centrale. *Boll. Soc. Geol. Ital.* **2007**, *126*, 101–118.
88. Parotto, M.; Tallini, M. Geometry and kinematics of the Montelanico-Carpineto Backthrust (Lepini Mts., Latium) in the hanging wall of the early Messinian thrust front of the central Apennines: Implications for the Apennine chain building. *Ital. J. Geosci.* **2013**, *132*, 274–289. [CrossRef]
89. Delchiaro, M.; Fioramonti, V.; Della Seta, M.; Cavinato, G.P.; Mattei, M. Fluvial inverse modelling for inferring the timing of Quaternary uplift in the Simbruini range (Central Apennines, Italy). In *Proceedings of the Geomorphometry 2020 Conference*; Alvioli, M., Marchesini, I., Melelli, L., Guth, P., Eds.; Consiglio Nazionale delle Ricerche: Rome, Italy, 2020. [CrossRef]
90. Boni, C.; Bono, P.; Calderoni, G.; Lombardi, S.; Turi, B. Indagine idrogeologica e geochimica sui rapporti tra ciclo carsico e circuito idrotermale nella Pianura Pontina. *Geol. Appl. E Idrogeol.* **1980**, *15*, 204–247.
91. Marra, F.; Bahain, J.-J.; Jicha, B.R.; Nomade, S.; Palladino, D.M.; Pereira, A.; Tolomei, C.; Voinchet, P.; Anzidei, M.; Aureli, D.; et al. Reconstruction of the MIS 5.5, 5.3 and 5.1 coastal terraces in Latium (central Italy): A re-evaluation of the sea-level history in the Mediterranean Sea during the last interglacial. *Quat. Int.* **2019**, *525*, 54–77. [CrossRef]

92. Marra, F.; Cardello, G.L.; Gaeta, M.; Jicha, B.; Montone, P.; Niespolo, E.M.; Nomade, S.; Palladino, D.M.; Pereira, A.; De Luca, G.; et al. The Volsci Volcanic Field (central Italy): An open window on continental subduction processes. *Int. J. Earth Sci.* **2021**. [CrossRef]
93. Cocozza, T.; Praturlon, A. Note geologiche sul colle Cantocchio (Lepini sud-occidentali, Lazio). *Geol. Rom* **1966**, *5*, 323–334.
94. Cosentino, D.; Cipollari, P.; Pipponzi, G. Il sistema orogenico dell'Appennino centrale: Vincoli stratigrafici e cronologia della migrazione. Evoluzione cinematica del sistema orogenico dell'Appennino centro-meridionale: Caratterizzazione stratigrafico-strutturale dei bacini sintettonici. In *Studi Geologici Camerti - Convegno-escursione COFIN '99*; Numero Speciale; Cipollari, P., Cosentino, D., Eds.; Università degli Studi di Camerino: Camerino, Italy, 2003; pp. 85–99.
95. Bown, P.R.; Young, J.R. Techniques. In *Calcareous Nannofossil Biostratigraphy*; Bown, P.R., Ed.; Springer: Dordrecht, The Netherlands, 1998; pp. 16–28.
96. Chiocchini, M.; Mancinelli, A. Microbiostratigrafia del Mesozoico in facies di piattaforma carbontica dei Monti Aurunci (Lazio Meridionale). *Studi Geol. Camerti* **1977**, *3*, 109–152.
97. Brandano, M.; Giannini, E.; Schiavinotto, F.; Varrubbi, V. Miogypsina globulina (Michelotti) from Lower Miocene Villa, S. Lucia section (M. te Cairo, Central Apennines). *Geol. Romana* **2007**, *40*, 119–127.
98. Chiocchini, M.; Pampaloni, M.L.; Pichezzi, R.M. Microfacies and microfossils of the Mesozoic carbonate successions of Latium and Abruzzi (Central Italy). In *Memorie per Servire alla Descrizione della Carta Geologica D'Italia (ISPRA) Dipartimento Difesa del Suolo*; Instituto Superiore per la Protezione e la Ricerca Ambientale: Rome, Italy, 2012; Volume 17, p. 269.
99. Martini, E. Standard Tertiary and Quaternary calcareous nannoplankton zonation, in: Proceedings of the Second Planktonic Conference, Roma 1970. *Tecnoscienza* **1971**, *2*, 739–785.
100. Okada, H.; Bukry, D. Supplementary modification and introduction of code numbers to the low-latitude coccolith biostratigraphic zonation (Bukry, 1973; 1975). *Marine Micropaleontology* **1980**, *5*, 321–325. [CrossRef]
101. Backman, J.; Raffi, I.; Rio, D.; Fornaciari, E. Biozonation and biochronology of Miocene through Pleistocene calcareous nannofossils from low and middle latitudes. *Newsl. Stratigr.* **2012**, *47*, 131–181. [CrossRef]
102. Accordi, B.; Angelucci, A.; Sirna, G. Note illustrative della Carta Geologica d'Italia alla scala 1:100.000. Foglio 159 (Frosinone) e 160 (Cassino). In *Servizio Geologico d'Italia (ISPRA), Roma*; Instituto Superiore per la Protezione e la Ricerca Ambientale: Rome, Italy, 1967.
103. Alberti, A.; Bergomi, C.; Catenacci, V.; Centamore, E.; Cestar, G.; Chiocchini, M.; Chiocchini, U.; Manganelli, V.; Molinari-Paganelli, V.; Panseril-Crescenzi, C.; et al. Note illustrative del Foglio 389 Anagni. In *Carta Geol. D'italia 1:50.000. Serv. Geol. D'italia (ISPRA)*; Instituto Superiore per la Protezione e la Ricerca Ambientale: Rome, Italy, 1975.
104. Ortner, H.; Retier, F.; Acs, P. Easy handling tectonic data: The programs VB for Mac and Tectonics FP for windows. *Comput. Geosci.* **2002**. [CrossRef]
105. Novarese, V. Il Miocene della Valle Latina. *Boll. Del Reg. Uff. Geol. D'italia. Col. Roma* **1943**, *68*, 29–48.
106. Novarese, V. I terreni petroliferi della Valle Latina. *Boll. Soc. Geol. Ital.* **1923**, *42*, 347–367.
107. Carboni, S.; Lombardi, L. Su alcuni affioramenti in facies di flysch della Valle Latina. *Boli. Soc. Geol. Ital.* **1956**, *75*, 109–113.
108. Bally, A.W.; Burbi, L.; Cooper, C.; Ghelardoni, R. Balanced sections and seismic reflection profiles across the Central Apennines. *Mem. Soc. Geol. Ital.* **1986**, *35*, 257–310.
109. Cassinis, R.; Scarascia, S.; Lozej, A. The deep crustal structure of Italy and surrounding areas from seismic refraction data; a new synthesis. *Boll. Soc. Geol. Ital.* **2003**, *122*, 365–376.
110. Mirabella, F.; Barchi, M.R.; Lupattelli, A. Seismic reflection data in the Umbria Marche Region: Limits and capabilities to unravel the subsurface structure in a seismically active area. *Ann. Geophys.* **2008**, *51*. [CrossRef]
111. Chiocchini, M.; Mancinelli, A. Sivasella monolateralis Sirel and Gunduz, 1978 (Foraminiferida) in the Maastrichtian of Latium (Italy). *Rev. Micropaleontol.* **2001**, *44*, 267–277. [CrossRef]
112. Vecchio, E.; Barattolo, F.; Hottinger, L. Alveolina horizons in the Trentinara Formation (Southern Apennines, Italy): Stratigraphic and paleogeographic implication. *Riv. Ital. Paleontol. E Stratigr.* **2007**, *113*, 21–42. [CrossRef]
113. Romano, A.; Urgera, A. Geologia del Paleogene dei Monti Aurunci orientali (Lazio meridionale). *Studi Geol. Camerti* **1995**, *13*, 29–38.
114. Angelucci, A.; Devoto, G.; Farinacci, A. Le "argille caotiche" di Colle Cavallaro a est di Castro dei Volsci (Frosinone). *Geol. Romana* **1963**, *2*, 305–329.
115. Bergomi, C.; Catenacci, V.; Cestari, G.; Manfredini, M.; Manganelli, V. Note illustrative della Carta Geologica d'Italia, F° 171. In *Gaeta. Servizio Geologico d'Italia, Napoli*; Instituto Superiore per la Protezione e la Ricerca Ambientale (ISPRA): Rome, Italy, 1969.
116. Putignano, M.L.; Ungaro, A. Considerazioni sulla provenienza delle intercalazioni carbonatiche nella successione terrigena Messiniana nella Valle dell'Ausente (Appennino centro-meridionale). *Mem. Soc. Geol. Ital.* **1996**, *51*, 351–362.
117. Angelucci, A. Tectonic marks on pebbles of Middle Latina Valley (Central Italy). *Geol. Romana* **1966**, *5*, 313–322.
118. Vitale, S.; Amore, O.F.; Ciarcia, S.; Fedele, L.; Grifa, C.; Prinzi, E.P.; Tavani, S.; Tramparulo, F.D.A. Structural, stratigraphic, and petrological clues for a Cretaceous–Paleogene abortive rift in the southern Adria domain (southern Apennines, Italy). *Geol. J.* **2017**, *53*, 660–681. [CrossRef]
119. Brandano, M. Tropical/subtropical inner ramp facies in Lower Miocene «Calcari a Briozoi e Litotamni» of the Monte Lungo Area (Cassino Plain, Central Apennines, Italy). *Boll. Della Soc. Geol. Ital.* **2003**, *122*, 85–98.
120. Lombardi, L. Il pozzo Fogliano nei pressi di Latina e la paleogeografia dell'area. *Boll. Soc. Geol. Ital.* **1968**, *87*, 13–18.
121. Corda, L.; Madonna, S.; Mariotti, G. Late Cretaceous to early Miocene evolution of the Southern Prenestini Mountains (Central Apennines): From fault-block platforms to carbonate ramp. *J. Mediterr. Earth Sci.* **2020**, *12*, 15–31. [CrossRef]

122. Funiciello, R.; Giordano, G.; Capelli, G.; De Benedetti, A.; Del Monaco, F.; Mazza, R.; Tallini, M. Note illustrative del Foglio 388 Velletri. Carta Geologica 1:50.000. In *Serv. Geol. D'Italia*; Instituto Superiore per la Protezione e la Ricerca Ambientale (ISPRA): Rome, Italy, 2018.
123. Beneo, E. Dalla Valle Latina a Carpineto Romano. *Boll. Soc. Geol. Ital.* **1950**, *69*, 600–601.
124. Fagereng, Å.; Sibson, R.H. Melange rheology and seismic style. *Geology* **2010**, *38*, 751–754. [CrossRef]
125. Festa, A.; Ogata, K.; Pini, G.A.; Dilek, Y.; Alonso, J.L. Origin and significance of olistostromes in the evolution of orogenic belts: A global synthesis. *Gondwana Res.* **2016**, *39*, 180–203. [CrossRef]
126. Corrado, S.; Aldega, L.; Di Leo, P.; Giampaolo, C.; Invernizzi, C.; Mazzoli, S.; Zattin, M. Thermal maturity of the axial zone of the southern Apennines fold-and-thrust belt (Italy) from multiple organic and inorganic indicators. *Terra Nova* **2005**, *17*, 56–65. [CrossRef]
127. Carlini, M.; Artoni, A.; Aldega, L.; Balestrieri, M.L.; Corrado, S.; Vescovi, P.; Bernini, M.; Torelli, L. Exhumation and reshaping of far-travelled/allochthonous tectonic units in mountain belts. New insights for the relationships between shortening and coeval extension in the western Northern Apennines (Italy). *Tectonophysics* **2013**, *608*, 267–287. [CrossRef]
128. Catalano, R.; Valenti, V.; Albanese, C.; Accaino, F.; Sulli, A.; Tinivella, U.; Morticelli, M.G.; Zanolla, C.; Giustiniani, M. Sicily's fold–thrust belt and slab roll-back: The SI. RI. PRO. seismic crustal transect. *J. Geol. Soc.* **2013**, *170*, 451–464. [CrossRef]
129. Angelucci, A.; Bellotti, P.; Valeri, P. Analisi di facies dei sedimenti terrigeni tortoniani nella zona di Frosinone. *Geol. Romana* **1979**, *18*, 127–135.
130. Critelli, S.; Le Pera, E.; Galluzzo, F.; Milli, S.; Moscatelli, M.; Perrotta, S.; Santantonio, M. Interpreting siliciclastic-carbonate detrital modes in foreland basin systems: An example from Upper Miocene arenites of the central Apennines, Italy. *Spec. Pap. Geol. Soc. Am.* **2007**, *420*, 107. [CrossRef]
131. Casciano, C.I.; Patacci, M.; Longhitano, S.G.; Tropeano, M.; Mccaffrey, W.D.; Di Celma, C. Multi-scale analysis of a migrating submarine channel system in a tectonically-confined basin: The Miocene Gorgoglione Flysch Formation, southern Italy. *Sedimentology* **2019**, *66*, 205–240. [CrossRef]
132. Cipollari, P.; Cosentino, D.; Pipponzi, G. Il sistema orogenico dell'Appennino centrale: Vincoli stratigrafici e cronologia della migrazione. *Studi Geol. Camerti* **2003**, 87–101. [CrossRef]
133. Smeraglia, L.; Aldega, L.; Bernasconi, S.M.; Billi, A.; Boschi, C.; Caracausi, A.; Carminati, E.; Franchini, S.; Rizzo, A.L.; Rossetti, F.; et al. The role of trapped fluids during the development and deformation of a carbonate/shale intra-wedge tectonic mélange (Mt. Massico, Southern Apennines, Italy). *J. Struct. Geol.* **2020**, *138*. [CrossRef]
134. Artoni, A.; Bernini, M.; Papani, G.; Rizzini, F.; Barbacini, G.; Rossi, M.; Rogledi, S.; Ghielmi, M. Mass-transport deposits in confined wedge-top basins: Surficial processes shaping the messinian orogenic wedge of Northern Apennine of Italy. *Ital. J. Geosci.* **2010**, *129*, 101–118. [CrossRef]
135. Butler, R.W.H. Thrust sequences. *J. Geol. Soc.* **1987**, *144*, 619–634. [CrossRef]
136. Doglioni, C.; Harabaglia, P.; Merlini, S.; Mongelli, F.; Peccerillo, A.T.; Piromallo, C. Orogens and slabs vs. their direction of subduction. *Earth-Sci. Rev.* **1999**, *45*, 167–208. [CrossRef]
137. Roveri, M.; Bassetti, M.A.; Lucchi, F.R. The Mediterranean Messinian salinity crisis: An Apennine foredeep perspective. *Sediment. Geol.* **2001**, *140*, 201–214. [CrossRef]
138. Jolivet, L.; Augier, R.; Robin, C.; Suc, J.P.; Rouchy, J.M. Lithospheric-scale geodynamic context of the Messinian salinity crisis. *Sediment. Geol.* **2006**, *188*, 9–33. [CrossRef]
139. Roveri, M.; Flecker, R.; Krijgsman, W.; Lofi, J.; Lugli, S.; Manzi, V.; Sierro, F.J.; Bertini, A.; Camerlenghi, A.; De Lange, G.; et al. The Messinian salinity crisis: Past and future of a great challenge for marine sciences. *Mar. Geol.* **2014**, *349*, 113–125. [CrossRef]
140. Andreetto, F.; Aloisi, G.; Raad, F.; Heida, H.; Flecker, R.; Agiadi, K.; Lofi, J.; Blondel, S.; Bulian, F.; Camerlenghi, A.; et al. Freshening of the Mediterranean Salt Giant: Controversies and certainties around the terminal (Upper Gypsum and Lago-Mare) phases of the Messinian Salinity Crisis. *Earth-Sci. Rev.* **2021**, *216*, 103577. [CrossRef]
141. Tavarnelli, E. The effects of pre-existing normal faults on thrust ramp development: An example from the northern Apennines, Italy. *Geol. Rundsch.* **1996**, *85*, 363–371. [CrossRef]
142. Patacca, E.; Scandone, P.; Di Luzio, E.; Cavinato, G.P.; Parotto, M. Structural architecture of the central Apennines: Interpretation of the CROP 11 seismic profile from the Adriatic coast to the orographic divide. *Tectonics* **2008**, *27*. [CrossRef]
143. Scisciani, V. Styles of positive inversion tectonics in the Central Apennines and in the Adriatic foreland: Implications for the evolution of the Apennine chain (Italy). *J. Struct. Geol.* **2009**, *31*, 1276–1294. [CrossRef]
144. Turienzo, M.; Sánchez, N.; Dimieri, L.; Lebinson, F.; Araujo, V. Tectonic repetitions of the Early Cretaceous Agrio Formation in the Chos Malal fold-and-thrust belt, Neuquén basin, Argentina: Geometry, kinematics and structural implications for Andean building. *J. South Am. Earth Sci.* **2014**, *53*, 1–19. [CrossRef]
145. Schori, M.; Mosar, J.; Schreurs, G. Multiple detachments during thin-skinned deformation of the Swiss Central Jura: A kinematic model across the Chasseral. *Swiss J. Geosci.* **2015**, *108*, 327–343. [CrossRef]
146. Milia, A.; Torrente, M.M. Tectono-stratigraphic signature of a rapid multistage subsiding rift basin in the Tyrrhenian-Apennine hinge zone (Italy): A possible interaction of upper plate with subducting slab. *J. Geodyn.* **2015**, *86*, 42–60. [CrossRef]
147. Farinacci, A. Breccias and laminated dolomites of the Gavignano exposure. *Geol. Romana* **1965**, *4*, 129–144.
148. Civitelli, G.; Funiciello, R.; Lombardi, S. Alcune considerazioni sulla genesi della «Pietra Paesina». *Geol. Romana* **1970**, *9*, 195–204.

149. Angrisani, A.C.; Calcaterra, D.; Cappelletti, P.; Colella, A.; Parente, M.; Přikryl, R.; de'Gennaro, M. Geological features, technological characterization and weathering phenomena of the Miocene Bryozoan and Lithothamnion limestones (central-southern Italy). *Ital. J. Geosci.* **2011**, *130*, 75–92.
150. Cipollari, P.; Cosentino, D. Considerazioni sulla strutturazione della Catena dei Monti Aurunci: Vincoli stratigrafici. *Studi Geol. Camerti* **1991**, *2*, 151–156.

geosciences

Review

The Alps-Apennines Interference Zone: A Perspective from the Maritime and Western Ligurian Alps

Fabrizio Piana [1,*], Luca Barale [1], Carlo Bertok [2], Anna d'Atri [1,2], Andrea Irace [1] and Pietro Mosca [1]

1 Italian National Research Council, Institute of Geosciences and Earth Resources, 10125 Torino, Italy; luca.barale@cnr.it (L.B.); anna.datri@unito.it (A.d.); andrea.irace@cnr.it (A.I.); pietro.mosca@cnr.it (P.M.)
2 Department of Earth Sciences, University of Torino, 10124 Torino, Italy; carlo.bertok@unito.it
* Correspondence: fabrizio.piana@cnr.it

Abstract: In SW Piemonte the Western Alps arc ends off in a narrow, E-W trending zone, where some geological domains of the Alps converged. Based on a critical review of available data, integrated with new field data, it is concluded that the southern termination of Western Alps recorded the Oligocene-Miocene activity of a regional transfer zone (southwestern Alps Transfer, SWAT) already postulated in the literature, which should have allowed, since early Oligocene, the westward indentation of Adria, while the regional shortening of SW Alps and tectonic transport toward the SSW (Dauphinois foreland) was continuing. This transfer zone corresponds to a system of deformation units and km-scale shear zones (Gardetta-Viozene Zone, GVZ). The GVZ/SWAT developed externally to the Penninic Front (PF), here corresponding to the Internal Briançonnais Front (IBF), which separates the Internal Briançonnais domain, affected by major tectono-metamorphic transformations, from the External Briançonnais, subjected only to anchizonal metamorphic conditions. The postcollisional evolution of the SW Alps axial belt units was recorded by the Oligocene to Miocene inner syn-orogenic basin (Tertiary Piemonte Basin, TPB), which rests also on the Ligurian units stacked within the adjoining Apennines belt in southern Piemonte. The TPB successions were controlled by transpressive faults propagating (to E and NE) from the previously formed Alpine belt, as well as by the Apennine thrusts that were progressively stacking the Ligurian units, resting on the subducting Adriatic continental margin, with the TPB units themselves. This allows correlation between Alps and Apennines kinematics, in terms of age of the main geologic events, interference between the main structural systems and tectonic control exerted by both tectonic belts on the same syn-orogenic basin.

Keywords: tectonics; sedimentation; exhumation; Western Alps; Apennines

check for updates

Citation: Piana, F.; Barale, L.; Bertok, C.; d'Atri, A.; Irace, A.; Mosca, P. The Alps-Apennines Interference Zone: A Perspective from the Maritime and Western Ligurian Alps. *Geosciences* **2021**, *11*, 185. https://doi.org/10.3390/geosciences11050185

Academic Editors: Jesus Martinez-Frias, Giancarlo Molli, Domenico Liotta and Angelo Cipriani

Received: 9 February 2021
Accepted: 19 April 2021
Published: 25 April 2021

Publisher's Note: MDPI stays neutral with regard to jurisdictional claims in published maps and institutional affiliations.

Copyright: © 2021 by the authors. Licensee MDPI, Basel, Switzerland. This article is an open access article distributed under the terms and conditions of the Creative Commons Attribution (CC BY) license (https://creativecommons.org/licenses/by/4.0/).

1. Introduction

In the southern part of the Piemonte region (NW Italy) the Western Alps arc ends in a narrow, E-W trending zone (here named "southern termination of Western Alps"), where some of the main geologic domains of the Alps are now strictly juxtaposed. The Alps and the Apennines presently join in southern Piemonte where they have been intergrowing since the Paleogene: see [1–10] with references therein.

The southern termination of the Western Alps consists of three main geomorphologic sectors: a northwestern sector comprised between the Maddalena Pass and the Stura di Demonte valley, a central sector comprehending the Gesso Valley and extended eastward to the Tenda Pass and Vermenagna valley, and an eastern part, the western Ligurian Alps, here considered as the mountain range comprised between the Tenda Pass and the Tanaro valley (Figure 1). This region is very well suited for studying the relations between the Alps and the Apennines orogenic systems in terms of both the age of formation and the way in which the two main tectonic belts developed. This is mainly because: (i) the Maritime and Ligurian Alps formed later than other sectors of Western Alps [11–14]; (ii) they preserve, on top of their polymetamorphic basements, extensive Mesozoic to Oligocene sedimentary successions, which provide chronological constraints to the reconstruction of the regional tectono-sedimentary evolution, and (iii) the adjoining syn-orogenic basins recorded the tectonic history of both the two belts.

At the crustal scale, the SW Alps are interpreted as a composite tectonic belt detached at a depth of about 15 km in its north-eastern part and 5 km in the southwestern part, on the European crust. A high-density body (southern prolongation of the Ivrea body Auct.) occurs at a depth of about 40 km below the eastern margin of the SW Alps in southern Piemonte (Figure 2). The Ivrea body could have played an important role in the postcollisional Alpine tectonic evolution [15–17].

The first part of this paper concerns the analysis of the concepts (e.g., the Penninic Front and the Briançonnais Front) used in the literature to subdivide the SW Alps orogenic belts into domains that have different paleogeographic pertinence and/or show different geodynamic evolution, so that their effective relations with the other sectors of the Western Alps can be clearly defined. Once the main Alpine subdivisions are traced within the study area, a further analysis is carried out in the second part of the paper investigating how, when and at which extent the SW Alps were involved in the Apennine dynamics, which started during the late Oligocene when the westward-directed subduction of the Adriatic plate began [11–14,18].

This paper is thus based on the assumption that the evolution of the north-westernmost part of the Apennines can be studied referring to Alpine geodynamics because, although the Alps and the Apennines are two distinct geomorphologic and geophysical entities at the scale of the Western Mediterranean area [14], they share consistent kinematic evolution and common synorogenic basins in their junction zone of NW Italy. The steps of the Alps-Apennines evolution have been clearly recorded by a set of regional scale Oligocene to Pleistocene unconformities that can be continuously traced at surface in the southern part of the Piemonte region and in the subsurface of the western Po plain [19].

2. Geological Setting of the Southern Termination of the Western Alps

The southern termination of the Western Alps comprehends several tectonic units juxtaposed by NW-SE striking, mainly steeply dipping Alpine tectonic contacts. These units constitute the southern part of a double vergent structure developed at the regional scale [20,21] that involves the Briançonnais Domain in the internal northeastern side, and the Dauphinois-Provençal Domain in the external southwestern side [22,23]. The Briançonnais Domain, referred to as the distal part of the European continental palaeomargin of the Alpine Tethys [24], is subdivided by the Internal Briançonnais Front into an internal sector (Internal Briançonnais, mostly cropping out in the Cottian and Ligurian Alps) affected by HP-LT metamorphism [21], and an external sector affected by very low-grade metamorphism [25,26]. The Dauphinois-Provençal Domain, representing the proximal part of the European continental palaeomargin, was affected only by anchizone metamorphism, and is bounded along its inner side by the External Briançonnais Front. It may be subdivided into a basinal area where a several km-thick and clay-rich Mesozoic succession was deposited (Dauphinois succession), and a shallow water area, which is characterized by a reduced succession with carbonate platform facies (Provençal succession).

The more eastern part of the Western Alps southern termination, i.e., the western Ligurian Alps, shows a tectono-metamorphic and geometric setting [23,25–29] quite similar to that of the southern Cottian and Maritime Alps, although the fan-like, double-vergent structure is less pronounced than in the Maritime and southern Cottian Alps.

Finally, to the south of the External Briançonnais Front, in the investigated area the Western Ligurian Flysch units are present. These units, made up of Helminthoides Flysch-type successions [22,30–32] (also known as "Embrunais–Ubaye nappes" north of the Argentera Massif [33,34] and San Remo-M.Saccarello Unit to the SE of it [29,35,36]) are a stack of tectonic units composed of Lower Cretaceous–Lower Paleocene deep-water sediments referred to as the proximal Ligurian Domain and detached from their original substrate (i.e., the European continental margin). In the study area, the Western Ligurian Flysch, which were detached and emplaced in the early stages of the alpine tectonics [37], were later thrust over the Alpine Foreland Basin and/or the Dauphinois-Provençal succession, and are in turn involved in the Dauphinois-Briançonnais fold and thrust belt [23].

Tectono-Stratigraphic Evolution of the Southern Termination of the Western Alps

In the southern termination of the Western Alps, the Briançonnais domain represents a part of the more internal, uplifted sector, of the European distal margin, close to the Mesozoic Tethyan ocean (residual H-block [38,39]). The polymetamorphic Briançonnais basement crops out discontinuously in the Acceglio zone where it consists of micaschists, metabasites and granite. The overlying succession starts with Permian volcanic and volcaniclastic deposits and Lower Triassic fluvial to littoral conglomerate, quartzite sandstone and lagoonal pelite, followed by a Middle Triassic peritidal carbonate succession. The top of the Triassic succession is truncated by an unconformity due to a regional uplift and related subaerial exposure during the Tethyan syn-rift stage [40–42]. The succession continues with Middle Jurassic platform carbonates and Upper Jurassic pelagic plateau limestone, followed by mineralised Lower Cretaceous hard ground and Upper Cretaceous hemipelagic sediments.

The Dauphinois-Provençal domain represents the proximal margin of the Mesozoic Tethyan ocean [43,44] developed above the continental crust (i.e., the Argentera Massif in the study area). The succession starts with Carboniferous–Permian continental sediments and Lower Triassic coastal and lagoonal deposits, followed by Middle Triassic peritidal carbonates and Upper Triassic evaporites and lagoonal pelites. Starting from the Late Triassic–Early Jurassic, the Dauphinois-Provençal domain was affected by intracontinental rifting, and partitioned into fault-bounded rift-basins [41]. From the Early Jurassic to Early Cretaceous, the rift basins (Dauphinois domain) progressively subsided, and thick successions of deep-water marl, limestone and shale with interbedded resedimented calcirudite and calcarenite layers were deposited. Toward the south the Dauphinois domain passed laterally to a structural high (Provençal domain) that remained during the Middle Jurassic-earliest Cretaceous in shallow water conditions with the development of carbonate platforms [45]. In the Valanginian, the carbonate platform drowned and was covered by a few metres to a few tens of metres of condensed, open marine deposits locally rich in authigenic minerals (Hauterivian-Albian; [35,46]). The Dauphinois-Provençal succession ends with Upper Cretaceous hemipelagic marly limestone, locally rich in resedimented, mainly siliciclastic, layers [27,47,48].

The present geological setting of the Briançonnais and Dauphinois-Provençal domains of the southern termination of the Western Alps is mainly due to the progressive involvement of the European continental palaeomargin, which these domains belonged to in the Alpine tectonic belt since the middle Eocene (e.g., [34]). The first stages of this tectonic evolution caused the development, on top of the Mesozoic succession, of a regional unconformity corresponding to a hiatus spanning the latest Cretaceous–middle Eocene, overlain by the Alpine Foreland Basin succession [49], middle Eocene discontinuous continental to lagoonal deposits (Microcodium Formation), middle Eocene mixed carbonate–siliciclastic ramp deposits (Nummulitic Limestone), upper Eocene hemipelagic sediments (Globigerina Marl) and upper Eocene–lower Oligocene turbidite succession (Grès d'Annot, [50]).

Both the continental margin and the foreland basin successions have experienced, since the latest Eocene, a multistage tectonic evolution characterized first by southwestward brittle–ductile thrusting and superposed foldings, then by northeastward back-vergent folding and, lastly, by southward brittle thrusting and flexural folding [23]. The regional structural setting resulted from a transpressional regime [23,25,51], documented by a post-early Oligocene NW-SE transcurrent shear zone (Limone Viozene Zone) extending for some tens of kilometres through the study area. This shear zone is probably superimposed on a long-lived shearing corridor, active since the Jurassic–Cretaceous and reactivated during the Cenozoic [23,52].

In the southern termination of the Western Alps, the Alpine Foreland Basin succession is overthrust by the Helminthoides Flysch units (Western Ligurian Flysch units in [23]). The Helminthoides Flysch units [22,30–32] are composed of Lower Cretaceous–lower Paleocene deep-water sediments detached from their original substrate and referred to as the Ligurian Domain. These units consist of carbonate-poor, thin-bedded varicoloured pelites (the basal complex) interpreted as basin plain deposits, thick-bedded, coarse-grained sandstones deposited in internal deep-sea fans and thick-bedded, mainly carbonate turbidite successions deposited in external deep-sea fans [31].

3. The SW Alps Transfer

On the basis of the paleogeographic reconstructions proposed in the literature (see [53] with references therein), the geological evolution of the southern termination of the Western Alps arc (Figures 1 and 2) was controlled, since at least the Early Cretaceous, by major transcurrent fault zones.

In the latest Eocene (35 Ma ago), the onset of the Europe-Adria continental collision induced the westward indentation of the Adria continental block, together with its high-density roots (Ivrea geophysical body [17]) and its counterclockwise rotation with respect to Europe, marking a dramatic change in convergence and thrusting direction ("Oligocene revolution" *sensu* [54]).

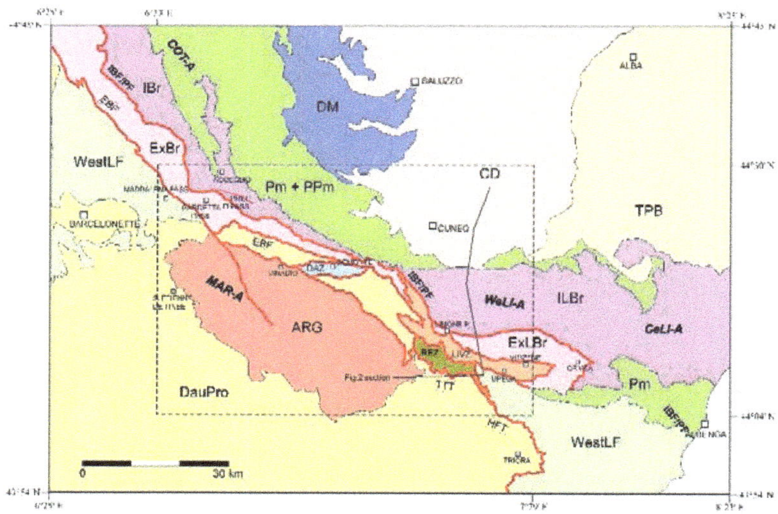

Figure 1. Geological sketch of the southern termination of the Western Alps. The main tectonic domains are represented with different colours. The black dashed rectangle corresponds to the study area of Figure 3. Legend: ARG: Argentera Massif; EBr: External Briançonnais Domain; IBr: Internal Briançonnais Domain; Pm: Piemonte Domain; PPm: Pre-Piemonte Domain; DM: Dora-Maira Domain; DauPro: Dauphinois—Provençal Domain; ExLBr: External Ligurian Briançonnais Domain; ILBr: Internal Ligurian Briançonnais Domain; TPB: Pliocene sediments and Tertiary Piemonte Basin; WestLF: Western Ligurian Flysch Domain; EBF: External Briançonnais Front; IBF/PF: Internal Briançonnais Front/Penninic Front; LiVZ: Limone-Viozene Zone; DAZ: Demonte-Aisone Zone; REZ: Refrey Zone; HFT: Helminthoides Flysch Basal Thrust; TTT: Tenda-tunnel Thrust; COT-A: Cottian Alps; MAR-A: Maritime Alps; WeLI-A: Western Ligurian Alps; CeLi-A: Central Ligurian Alps. The black line within the dashed inset refers to trace of the section of Figure 2. Adapted with permission from ref. [23]. Copyright 2016 Springer-Verlag Berlin Heidelberg.

At the same time, south of the Ligurian Alps, a switch in the subduction polarity occurred from a SE-dipping polarity to a W-dipping polarity, and the eastward retreat of the Apennine slab began, accompanied by onset of back-arc rifting in the Liguro-Provençal area [13,14,18,55]. Both these processes were quite absent to the north east of the Ligurian Alps, suggesting the presence of a major transfer zone between these two realms, which should have been oriented NNW-SSE in the first (early Oligocene) stages and WNW-ESE in the last (early-middle Miocene) ones ([9,23]. The activity of this major transfer zone, with active steep crustal tectonic features, extended at depth and allowed the west-ward indentation of Adria [34] that in the northern part of the Western Alps was accommodated by dextral shearing along the Periadriatic line [56]. Then, since the Oligocene, the transfer zone should have acted as a strike-slip fault zone along the southern margin of the Adria indenter [54]. This major transfer zone placed at the southern termination of the Western Alps may have developed partly on inherited Mesozoic extensional and transtensional faults active before the onset of the Alpine collisional regime (as discussed in [57]), which may have compensated for the different seafloor spreading rate of the Ligurian ocean to the south and Piemonte ocean to the north [34,53,54,58] Consequently, the geological units presently placed at the southern termination of the Western Alps arc should bear clear and widespread evidence of such distinctive geological evolution in terms of sedimentary evolution, deformational history and related hydrothermal activity.

A large amount of recently published data [23,25–28,46,52,57,59–63] document that since Early Cretaceous up to at least the early Miocene, the geological units presently placed at the southern termination of the Western Alps experienced a tectono-sedimentary evolution mainly related to large scale strike-slip and transpressive faulting, with significative migration of syntectonic fluids.

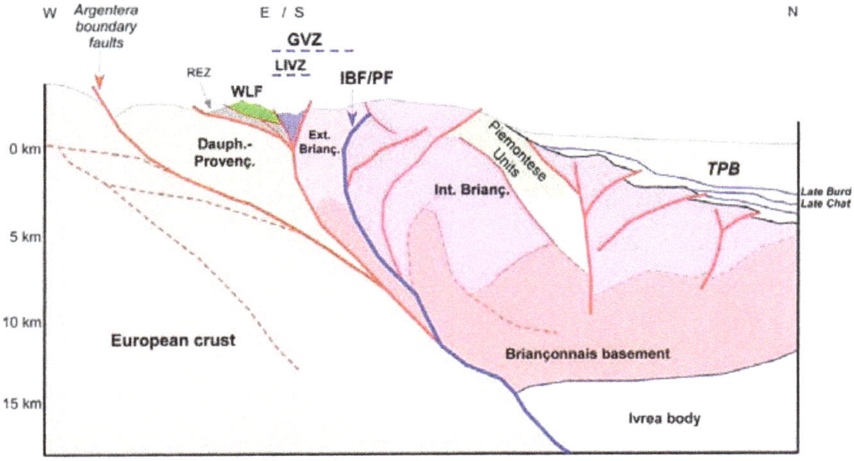

Figure 2. Schematic geological section (location in Figure 1) across the Argentera massif, the western Ligurian Alps and the TPB syn-orogenic basin. Reference data from [17,23,25,26,64,65]. Legend: TPB: Tertiary Piemonte Basin; GVZ: Gardetta-Viozene Zone; LIVZ: Limone-Viozene Zone; REZ: Refrey Zone; WLF: Western Ligurian Flysch (Helminthoides Flysch); IBF/PF: Internal Briançonnais Front/Penninic Front; Late Burd, Late Chat: seismic imaged unconformities of inferred late Burdigalian and late Chattian age.

The field evidence of such an important, often invoked, transfer zone has been reported by [23] who described the southern termination of Western Alps arc as an assemblage of juxtaposed tectonic units, mainly belonging to the Briançonnais and Dauphinois-Provençal sedimentary domains of the palaeo-European margin and to the overlying Alpine Foreland basin, elongated on average ESE-WNW direction [20,47] within a km-scale transpressive deformation zone (SW Alps Transfer, SWAT). A progressive transition from high pressure metamorphic rocks in the internal (NE) part of the transfer zone, to very low grade and nonmetamorphic rocks in the external ones (SW) can be presently observed [21,23,26,66]

(Figures 1 and 3). The SWAT is probably rooted within the underlying European continental basement [16,17] (Figure 3) made up of mono and polymetamorphic rocks. These units crop out in the Argentera Massif, an ellipsoid-shaped body exhumed in Miocene times [67] through the Briançonnais, Provençal and Dauphinois sedimentary successions, from which it is presently separated on both sides by Miocene-Pliocene boundary fault systems [68] (Sanchez et al., 2011a) (Figures 1–3).

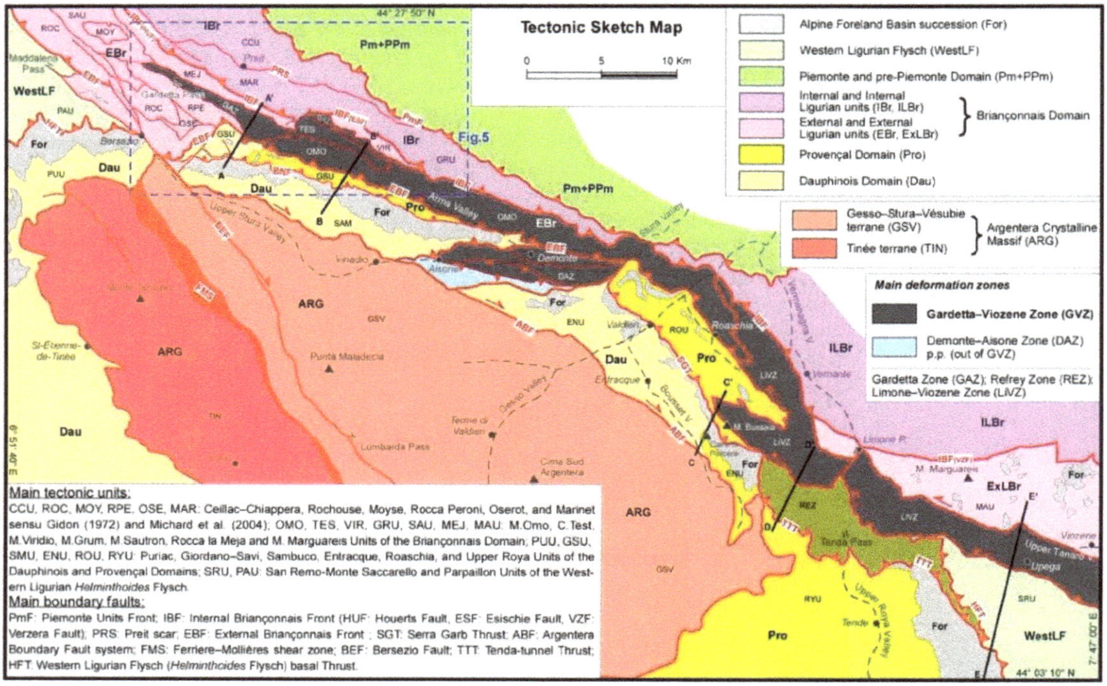

Figure 3. Geological sketch map of the southern termination of the Western Alps adapted from [23]. The Gardetta-Viozene Zone is highlighted (in grey). The main tectonic units are represented referring to their different paleogeographic pertinence (different colors) and geometrical position with respect to the main tectonic boundaries and deformation zones. The location of geological cross-sections of Figure 4 (sections A-A′, B-B′, C-C′, D-D′, E-E′) is reported. The Internal Briançonnais Front (IBF) here corresponds to the external boundary of the green-schist to eclogite facies metamorphic axial belt of the Alps (*sensu* [69]) and can be thus considered as the surface expression of the Penninic Front.

4. The Penninic Front at the Southern Termination of Western Alps

To discuss the evolution of the southern termination of Western Alps in the frame of the collisional and postcollisional kinematics of the orogenic belt, it is necessary to unambiguously define the intended meaning of the "Penninic Front concept", in order to understand what are the relations of the SWAT with the ideal main tectonic boundary of the Western Alps belt, i.e., the so-called Penninic Front. Before starting to debate on this matter, a brief examination of the historical intended meaning(s) of the "Penninic Front concept" is given in the following.

The Penninic nappe stack can be seen, in a very simplified view, as the suture zone between the European (including the Briançonnais domain) and Adriatic plates, comprising remnants of the former subducted oceanic crust or exhumed mantle, as well as the extended European continental crust [70]. A slightly different, definition of the Penninic zone is given by Dal Piaz (ets al.) [71], as a stack of generally metamorphic nappes scraped off the subducting oceanic lithosphere and European passive continental margin (distal part), mainly accreted during the Paleogene, whose outer boundary is the Penninic frontal thrust.

In many other papers [16,34,72,73] the Penninic domain has been conceived differently from the above cited papers because the Penninic frontal thrust (Penninic Front) has been placed in the Southwestern Alps at the boundary between the Briançonnais and Dauphinois-Provençal domains, irrespective of the distribution of the metamorphism. Further south, the Penninic Front has been placed along the external boundary of the nonmetamorphic Western Ligurian Flysch Unit (Helminthoides Flysch Units) of the SW Alps, which is detached over the Dauphinois-Provençal domain, and labelled as Penninic Basal Contact or "AlpineThrust Front" [29,37].

In this paper, the Penninic Front is intended (in the sense of [69]) as the frontal thrust which bounds the Alpine Axial Belt, which comprises continental units derived from the Adriatic margin ("Austroalpine units"), as well as parts of the European continental margin together with oceanic units originated from the Mesozoic Piemonte–Liguria Ocean, labelled together as "Penninic Units" (Figure 1). All these units of the Alpine Axial Belt underwent alpine metamorphism ranging from greenschist facies to ultra-high-pressure eclogite facies conditions. The Penninic Front bounds the penninic units from the less metamorphosed external parts of the paleoEuropean margin, which can pertain to the External Briançonnais, Helvetic, and Dauphinois-Provençal domains.

5. Structure of the Southern Termination of the Western Alps: Increasing Deformation, Metamorphism and Relations with the Adjoining Alps-Apennines Syn-Orogenic Basins

A review of the tectonic setting and evolution of the southern termination of the Western Alps, as well as the dating of its main stages, is necessary to provide the elements to correlate it with the evolution of the northernmost part of the Apennines belt. We specify that the following description of the metamorphic conditions of the tectonic units refers to the "metamorphic degree", defined on the basis of mineralogical indexes (such as the 'crystallinity' degree of phyllosilicate minerals [26]), even for the very poorly transformed rocks of the external Briançonnais and Dauphinois-Provençal domains, which were subjected to anchizonal or very low-grade metamorphic conditions [26]. Conversely, if textural criteria are used to define the metamorphic features in the study area [19,74], the rocks of the studied external Briançonnais and Dauphinois-Provençal domains would be classified as "nonmetamorphic", since their primary features are mostly preserved from the effect of secondary petrogenetic processes and do not show any major internal reorganization or recrystallization.

The overall aspect of the southern termination of the Western Alps is presently a fan-like setting that evokes a flower structure, well known in the Briançonnais domain of the Guillestre, Briançon and Moûtiers transects [75–79], as well as in the Maritime Alps (Stura valley and Tenda Pass area [27]) and western Ligurian Alps [23,25,26,29,80].

This double-vergent tectonic system [20,21] involves the Briançonnais Domain in the internal, northeastern side and the Dauphinois-Provençal Domain in the external, southwestern side [22,23]. This system consists of NE-vergent thrusts and transpressive fault systems developed to the north of the External Briançonnais Front, and SW-vergent ones to the south of it. In map view (Figure 1), the external and internal Briançonnais Fronts, as well as the frontal thrust of the Piemonte Zone, get closer to each other from west to east, while changing their directions from NW–SE to west–east, almost merging together in the Stura valley, east of Aisone (Figures 1 and 2). Due to the deflection of tectonic fronts and the consequent strong reduction in thickness of the main structural domains, the transition from the HP-LT metamorphic units of the internal Brianconnais to the very low-grade units of the external Briançonnais and Dauphinois-Provençal domains occurs over a relatively short distance (about ten kilometers) along a NE-SW oriented geological section (Figures 1–3). The eastward prolongation of the tectonic stack made up of the Piemonte Zone and the Internal Briançonnais tectonic domains, is concealed by the sediments of TPB and overlying Pliocene and Quaternary successions: see [19] with references therein. At the southern termination of the Western Alps, the transition between the internal Alpine metamorphic tectonic prism and the external Briançonnais-Dauphinois-Provençal fold-and-thrust belt occurs.

This geological transect is suitable for investigation of how the postsubduction Alpine tectonic evolution has been recorded since the early Oligocene by the adjoining syn-orogenic basins of the Piemonte region, whose tectono-sedimentary setting was controlled, at least since the late Oligocene, by the beginning of the Apennines subduction and related surface tectonics. In this way, a correlation between the Alpine and Apennine main tectonic events can be attempted, as discussed in the following sections. As reported in Section 3, in the southern termination of the Western Alps there is evidence of a major transfer zone (the SWAT) represented by an assemblage of juxtaposed tectonic units, mainly belonging to the Briançonnais, Dauphinois-Provençal and Alpine Foreland Basin sedimentary domains, elongated on average in an ESE-WNW direction [23]. The SWAT represents an ideal macroscale tectonic feature, whose existence can be inferred on the base of kinematic modeling at the regional scale [15,34,81]. A number of effective deformation units (*sensu* GeosciML), and/or strongly deformed tectonic units, exist in the southern termination of the Western Alps, which are defined in Figures 2, 4 and 5 and listed as follows, starting from the more internal units in the footwall of the IBF: (a) the steepened Triassic dolostone succession of the Rocca la Meja unit; (b) the Gardetta Deformation Zone, consisting of several deformation units made up of Carboniferous (?)-Permian (?) volcanics and schists, Triassic quartzarenite, quartzite and km-scale slices of gypsum, intensively folded and steepened along the WNW-ESE faults that bound and dissect the deformation zone; (c) the steepened and strongly sheared external Briançonnais M.Omo Unit and Provençal Giordano-Savi Unit; and (d) the Demonte- Aisone deformation unit of [23]; (e) the tectonic slice system of the Roaschia Unit [27], which passes laterally to the transpressive Limone-Viozene Zone [23,25]. All these units share a succession of geologic events and related mesoscale regional foliations, as well as consistent kinematic features, that formed in response to a succession of deformation phases as described in the following. The assemblage of the above-described tectonic units is here labelled as the Gardetta-Viozene Zone (GVZ), which can be interpreted as the effective representation of the SWAT.

5.1. Rock Deformation and Metamorphism across the IBF in Southern Cottian Alps

The transition from the green-schist facies metamorphic rocks of the internal Briançonnais to the anchizonal external Briançonnais, which occurs across the IBF, can be observed through the watershed between the Grana and Arma valleys (San Magno-Fauniera Pass area) and along the Preit-Gardetta transect in the right side of the Maira valley (Figure 5). Near San Magno, the Permian (?)-Early Triassic conglomerate and quarzarenite [47] of the internal Briançonnais (site GRA4 in Figure 5) are transformed into greenschist gneisses, while in the Fauniera Pass area (External Briançonnais) they crop out as a very poorly stretched conglomerate with pink quartz and rhyolite pebbles (sample PRV in Figures 5 and 6a,c). Similarly, in the Preit valley, the Permian (?)-Early Triassic quartzite have a marked gneissic structure (sample SRV in Figures 5 and 6b,d), while the quartzarenite and dolostone cropping out SW of the IBF, in the Rocca la Meja tectonic slice do not show evidence of major metamorphic transformations, as also suggested by the presence of preserved microfacies in the Triassic dolostone (sample PRE6, Figure 6e) and very well-preserved Archosauriform footprints recently discovered [82] a few hundred meters from the IBF. The IBF cuts across the Middle-Upper Triassic evaporites and the Lower Triassic quartzite/quartzarenite layers, which, conversely, in the external (SW) part of the GVZ and in the fold and thrust belt comprised between the IBF and the Argentera northern boundary faults, are not displaced by high angle faults and seem to have played a role of regional detachment horizons.

5.2. Field Constraints to the Age of Tectonic Events and Kinematic Interpretation

5.2.1. Tectonic Phases

The southern termination of the Western Alps shows structural and metamorphic characters acquired through a succession of three "phases of deformation" (D1–D3) with prevalent folding and thrusting, followed by a transpressive, mostly disjunctive phase (D4) and a late (D5) extensional phase.

- D1: this phase is associated with an early phase of décollement tectonics responsible for the piling up of the Briançonnais duplexes by mostly bedding-parallel thrusts, mainly vergent to the SW (present geographic coordinates), as shown by [76] in the Briançon area, [21] in the Ubaye-Maira transect and by [23,25,83] in the Maritime and Western Ligurian Alps. The D1 phase is assumed to be roughly consistent with a regional E-W shortening direction [84].

The D1 was probably characterized by transpressional kinematics, since the sedimentary successions were shortened by duplexes and steepened (Figure 7a) after being detached from the Triassic gypsum and quartzite/quartzarenite levels, while the S1 foliation developed mostly parallel to the steepened bedding [21,25]. With the D1 related strain, a regional scale transpressional deformation unit formed, developed mainly at the expense of the external Briançonnais and inner Dauphinois-Provençal domains, which can be still recognised from the high Stura valley to the Tanaro valley (Figure 2) [23] within the GVZ. Gypsum masses of Triassic age [47] are involved in the GVZ (Figures 7b and 8e,f), probably dragged from the basal detachment level placed at the base of the external Briançonnais and Dauphinois-Provençal succession where these rocks largely occur [23,85,86].

The D1 phase generated the oldest composite tectonic foliation (S1, [87], locally axial-planar to F1, SW verging recumbent folds and syn-D1 reverse/sinistral shear zones ("charriages" Auctorum; [20,88]) Gidon, 1972; Lefèvre, 1983), within which the S1 foliation (Figure 7c,d,f) formed mostly parallel to the boundary shear planes [23]. Evidence of the D1 phase in Dauphinois-Provençal Domain is poorly documented at the mesoscale, except for the Giordano-Savi unit close to the EBF in the left side of the Stura valley, where D1 folds are preserved (Figure 8b) but rotated and displaced by syn-D2/D3, SW-vergingSW vergent thrusts, with related drag folds (M. Bodoira, Figure 4).

At the end of the D1 phase, local transtension should have occurred in some parts of the GVZ, as inferred by the activity of relatively hot fluids, which circulated along NW-SE fault systems, that led to the formation of hydrothermal marbles at the expense of the Jurassic-Cretaceous Dauphinois succession (Valdieri marble, [63]) and to intense recrystallization of some parts of the Alpine Foreland Basin succession (Aisone Flysch, [23,63,89]).

- D2/D3: these two cogenetic tectonic phases generated double-vergent fold systems (with predominant top to NE and subordinate top to SW vergence sense of shearing, D2) evolving to a transpressi shearing onal regime (D3), with high-medium angle reverse and strike-slip faults (Figure 8a) and related minor folds, roughly consistent with E-W regional shortening directions (D2), shifting to NE-SW directions (D3) [9,23,90]. Diffuse fault block rotations on vertical axis and reactivation of faults occurred during the D2, as in the case of the sinistral Preit fault (Figure 2). This major strike-slip fault displaces the boundary of the Internal/External Briançonnais units (i.e., the Internal Briançonnais Front, IBF), attesting that the D2 phase mostly post-dates the metamorphic evolution of the units involved in the southern termination of Western Alps.

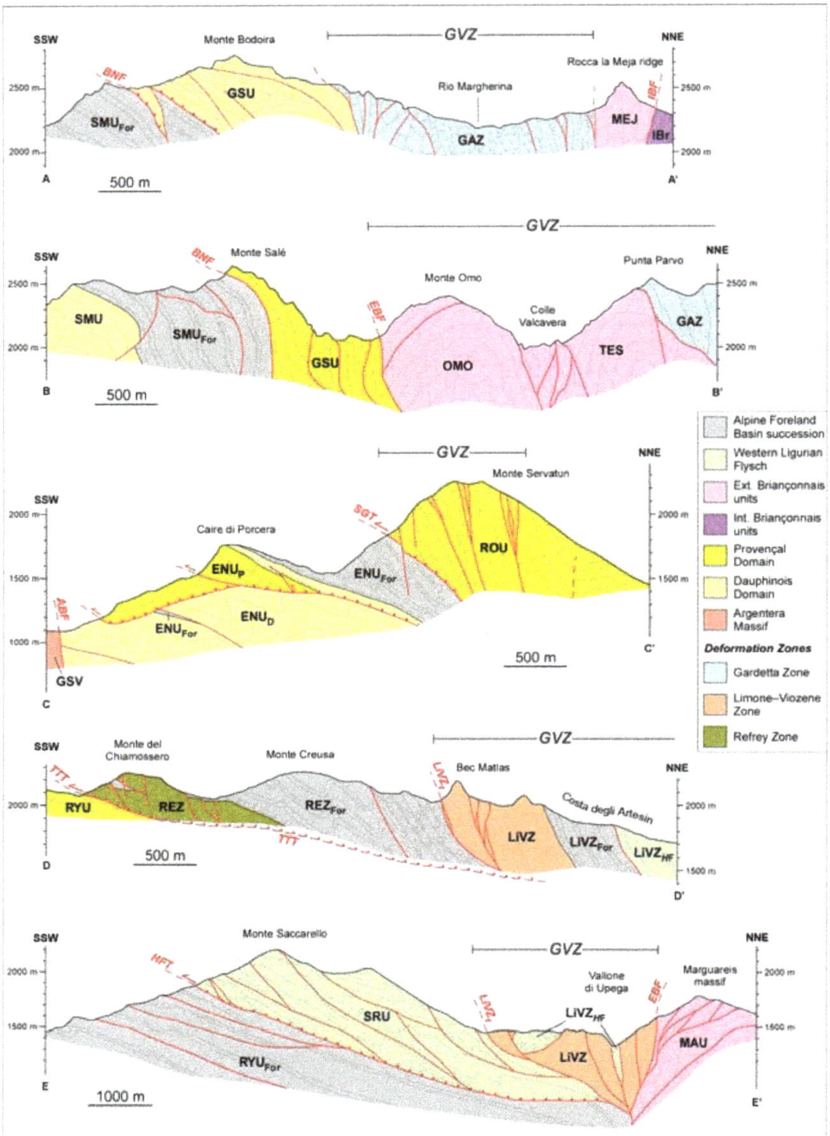

Figure 4. Geological cross-sections across the Gardetta-Viozene Zone (GVZ; location in Figure 3); modified from [20] (sections AA', BB'); [27] (sections CC', DD'); [26] (section EE'). Tectonic units: ENUD: Dauphinois succession of Entracque Unit; ENU, For: Foreland Basin succession of Entracque Unit; ENUP: Provençal succession of Entracque Unit; GAZ: Gardetta Zone; GSU: Giordano–Savi Unit; GSV: Gesso–Stura–Vésubie terrane; IBr: internal Briançonnais units; LiVZ: Limone–Viozene Zone; LiVZFor: tectonic slices of Foreland Basin succession involved in the Limone–Viozene Zone; LiVZHF: tectonic slices of Western Ligurian Flysch Unit involved in the Limone–Viozene Zone; MAU: Monte Marguareis Unit; MEJ: Rocca la Meja Unit; OMO: Monte Omo Unit; SMU: Sambuco Unit; SMUFor: Foreland Basin succession of Sambuco Unit; TES: Cima Test Unit; REZ: Refrey Zone; REZFor: Foreland Basin succession of the Refrey Zone; ROU: Roaschia Unit; RYU: Upper Roya Unit; RYUFor: Foreland Basin succession of Upper Roya Unit; SRU: San Remo–Monte Saccarello Unit. Main boundary faults: ABF: Argentera boundary fault system; BNF: Bersaio–Nebius Fault; EBF: External Briançonnais Front; HFT: Helminthoides Flysch basal Thrust; IBF: Internal Briançonnais Front; LiVZf: external boundary faults; SGT: Serra Garb Thrust; TTT: Tenda-tunnel Thrust.

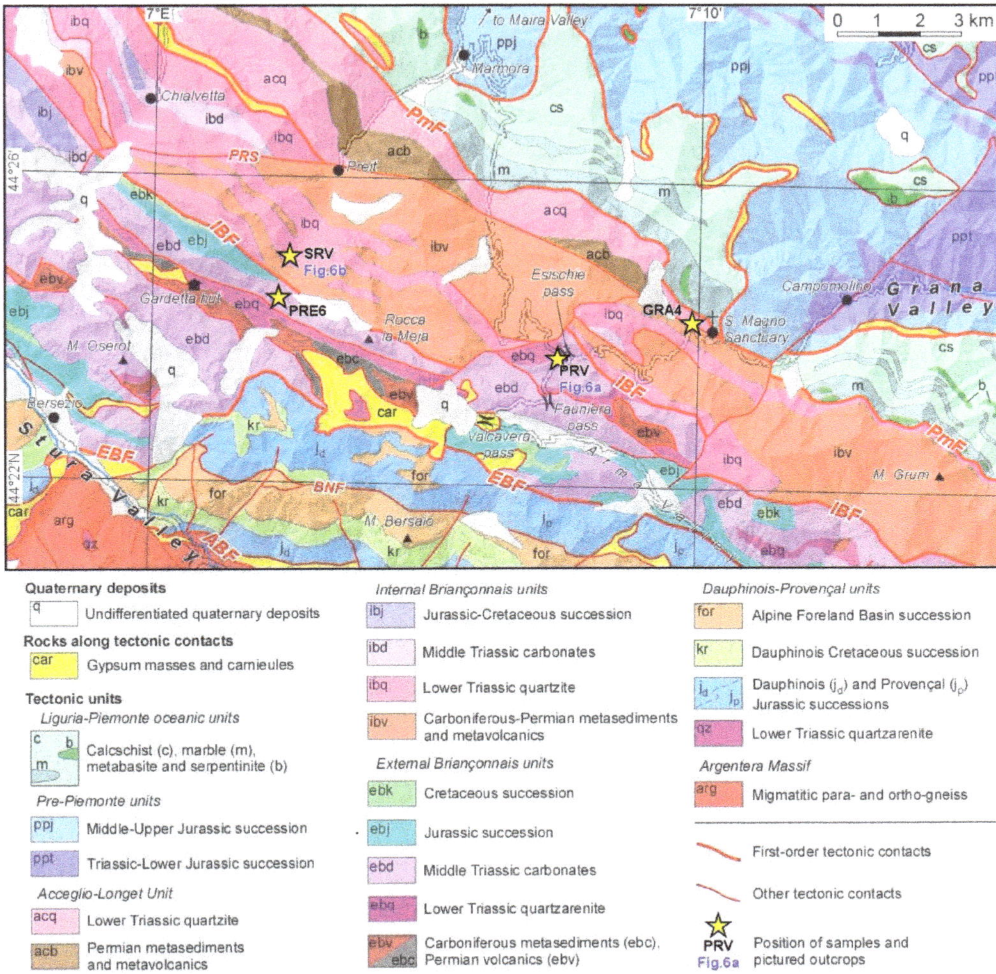

Figure 5. Geological map of the Gardetta highland area (modified from [19]), showing the sampling points of samples PRE6, PRV, and SRV and the position of outcrops pictured in Figure 6a,b. ABF: Argentera boundary fault system; BNF: Bersaio-Nebius Fault; EBF: External Briançonnais Front; IBF: Internal Briançonnais Front; PmF: Liguria-Piemonte units Front; PRS: Preit fault ("Preit Scar").

The folding phase D2 folded the D1 duplexes and reactivated the D1 steep transpressive shear zones (Figure 8c,d). A well-developed spaced crenulation cleavage (S2) is associated with F2 folds (Figure 7c–f). S2 planes are in several places reactivated and reoriented by shear deformation (S2-shear, generated by the D3, Figure 7e,f) kinematically consistent with the D2 phase that often induced the displacement or partial transposition of F2 folds hinge zones [25]. In the major transpressive shear zones (LiVZ), the S1 and S2 surfaces are almost subparallel and often form a composite foliation [25] (Figure 6c).

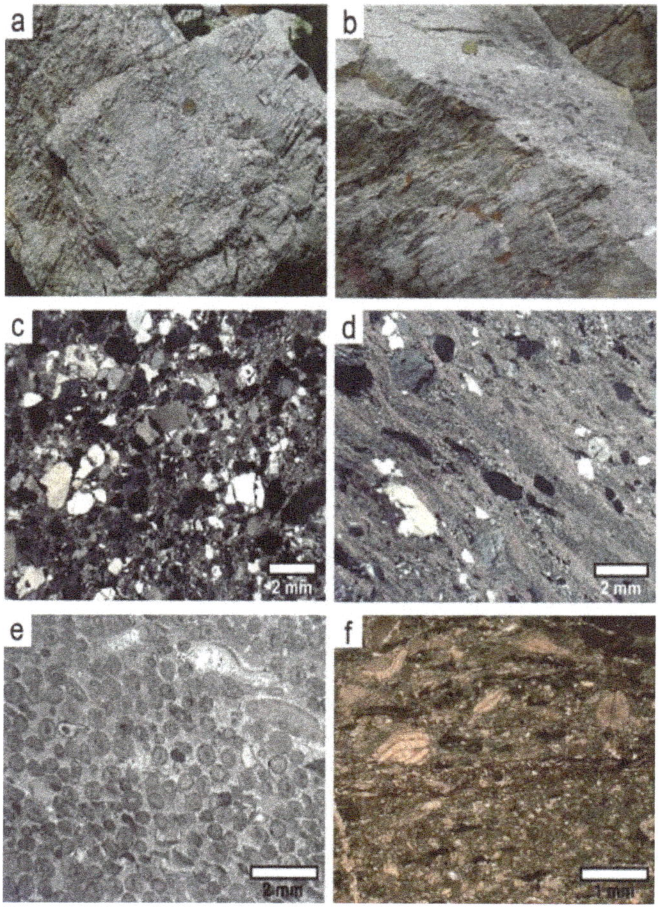

Figure 6. (**a**) Permian (?)-Early Triassic interbedded conglomerates and sandstones made up mainly of quartz and volcanics clasts. External Briançonnais Domain (Viridio Unit), Fauniera Pass area. (**b**) Permian (?)-Early Triassic quartzitic-rhyolitic metaconglomerate and metasandstone, Internal Briançonnais Domain (near Preit village), showing isooriented and stretched clasts that mark a gneissic structure. (**c**) Transmitted-light, crossed polars photomicrograph of sample PRV (external Briançonnais Domain, Fauniera Pass area; location in Figure 5), a coarse lithic sandstone corresponding to the finer levels of Figure 7a. A poorly defined lamination is evidenced by grain size variations. (**d**) Transmitted-light, crossed polars photomicrograph of sample SRV (internal Briançonnais Domain, Preit Valley; location in Figure 5), a metalithic sandstone corresponding to the finer levels of Figure 7b. A well-defined foliation is evidenced by the iso-orientation of the arenitic grains and by the occurrence of sub-millimetre thick levels of neoblastic iso-oriented white mica. (**e**) Transmitted-light photomicrograph of sample PRE 6 (Rocca la Meja tectonic slice, near the GVZ inner boundary, location in Figure 5), a dolomitized grainstone with oolites, echinoderm fragments and other bioclasts. Note the well-preserved morphology and internal structure of the oolites. (**f**) Transmitted-light photomicrograph of bioclastic arenaceous limestone (Eocene Nummulitic Limestone, Alpine Foreland Basin succession) from a tectonic slice within the LIVZ, Lago dei Signori pass, Marguareis area. Although a mm-scale foliation is present, evidenced by pressure dissolution seams and cataclastic levels, no volumetric dissolution or recrystallization occurred, as documented by the very well-preserved macroforaminifera and bryozoan.

The S1/S2 foliation is clearly reoriented and reactivated by the D3 shearing event in all the sectors of the GVZ, from the Gardetta Pass to the Marguareis area. The S2 cleavage is well exposed in the Ligurian Briançonnais units (Marguareis area [25,83]) due to the presence of the Upper Cretaceous marly limestone (Upega Formation) that recorded it with clear evidence. Conversely, this important regional foliation is poorly represented in the external Briançonnais units cropping out to the North of the Stura Valley, where, although present, it can be less frequently observed (Figure 7d).

- D4: faulting events related to the final uplifting stages of the Argentera Massif, inducing reactivation of D3 reverse and strike-slip fault systems and a minor rearrangement of the D1/D2 structural setting. During this stage, the compression direction is assumed to have rotated from NE–SW and then to N–S, inducing the reactivation of SW-verging reverse faults into dextral transpressive faults [15,17,34,68,90].

Among the main evidence for the D4 phase are the boundary faults of the Argentera Massif, with the related huge mass of gypsum-bearing brecciated fault rocks (Carnieules Auct. [47]) aligned all along the boundary faults of the massif.

- D5: the late stage of the tectonic evolution of the southern termination of the Western Alps was achieved in a general extensional-transtensional regime [67,68] as it represents a transitional zone between the regions affected by extension, located in the inner part of chain, and those affected by strike-slip and contractional tectonics, located in the outer parts of the alpine belt. A general frictional reactivation of the previously existing NW-SE to E-W fault systems occurred at this stage [90].

5.2.2. Age of Tectonic Phases and Metamorphism

The southern termination of Western Alps shows a metamorphic evolution ranging from anchizonal and very low-grade facies in the Dauphinois-Provençal and external Briançonnais domains to low grade, high-pressure greenschist facies and carpholite-quartz (blueschist) facies in the internal Briançonnais domain: see [21] with references therein) and [26]. In the more internal Briançonnais and prePiemontese successions, as well as the Acceglio zone, adjoining to the study area the metamorphic degree reached the eclogite facies; see [91] and [21] with references therein.

The metamorphic conditions and the ages related to the above-described tectonic phases are described in the following.

- D1 phase. The HP metamorphic transformations recorded by the Internal Briançonnais units were acquired before the D1 phase, as evidenced by the relations between the HP minerals and the tectonic foliations (D0 in this paper, D1 [21]). The metamorphism occurred in a time span between the age of the phengite in the Triassic quartzites (37 Ma [92]) of the adjoining, more internal, Pelvo d'Elva unit, and that of the very low grade S1 foliation (see below). In the external sectors, the D1 phase described in this paper, here intended as the older deformation phase that gave origin to a penetrative foliation, occurred later and at a temperature lower than 300 °C [21,26]. Its age should be younger than 33–34 Ma, i.e., the age of the upper part of Grès d'Annot successions [93] involved in the external Briançonnais tectonic units [20,23,47]. Furthermore, the metamorphic HP-LT transformation along the Penninic Front fault rocks in the Pelvoux area, during its later reactivation, has been dated at 34–30 Ma [77,84], thus ascribing it to the D1 phase.

Figure 7. (**a**) The sub-vertical Triassic dolostone succession of Rocca la Meja (Meja unit, GVZ, Gardetta highland). In the foreground (right lower corner) an outcrop of steepened Lower Triassic quartzarenite of the Gardetta Deformation Zone is visible. (**b**) Intensively folded and steepened gypsum within the Gardetta Deformation Zone. (**c**) Upper Cretaceous marly limestones in the external part of the GVZ (Dauphinois-Provençal succession) close to the Tenda pass showing a well-developed spaced crenulation cleavage (S2, black lines), crosscutting the older S1 foliation (black dashed lines). Note that in this area the S2 is NE-dipping. (**d**) Upper Cretaceous marly limestone in the External Briançonnais M. Omo Unit, close to Valcavera pass showing a well-developed spaced crenulation cleavage (S2, black lines), crosscutting the S1 foliation (black dashed lines). (**e**) Strain localization along the lithological contact between Upper Cretaceous marly limestone (Upega Fm.) and Upper Jurassic grey limestone (M. Marguareis, Ligurian Briançonnais domain). The S2 spaced cleavage (black lines) is dragged by S2-shear planes (S2-sh, white dashed lines) developed during the late stages of D2 and probably also during the D3 phase. (**f**) Upper Cretaceous marly limestones of the External Briançonnais succession in the Limone-Viozene Zone, Marguareis area. A well-developed spaced crenulation cleavage (S2, black lines) was generated as axial planar surfaces of F2 folds that folded both bedding and older foliation S1 (black dashed lines). Locally, S2 surfaces evolved from an axial plane foliation to a slip cleavage (S2sh, white dashed lines), giving origin, in some places, to dm-thick shear zones, which reoriented all the pre-existing foliations.

The radiometric age of the late D1 hydrothermal fluids responsible for the formation of the Valdieri marble (see above), defined at 30–31.6 Ma by U-Pb analyses on recrystallised and vein carbonate and neoblastic silicate minerals, provides a useful constraint for dating not only the D1, but also the D2 phase, which clearly postdates the marble [63]. It is remarked that a major gap in the distribution of metamorphism is not recorded across an ideal internal Briançonnais-Dauphinois cross section, i.e., across the front of the alpine tectonic prism, as occurring in other orogens (e.g., across the Himalayan main thrust front [94]), but the decrease of the metamorphic grade from the inner to the outer tectonic units seems to occur gradually (see above, Section 5.1), although a relatively major boundary can be established along the internal Briançonnais Front, corresponding to the "Penninic Front" in the sense of the assumption defined in Section 4, following [69].

- D2/D3 phase: since the metamorphic HP-LT transformation along the Penninic Front fault rocks in the Pelvoux area, during its syn-D1 reactivation, has been dated at 34–30 Ma [77,84], the D2 phase, which is consistent mainly with very low-grade metamorphism and pressure-dissolution processes of carbonate rocks [25,26], should be younger than 30 Ma. Zircon fission track data ([9] with reference therein) indicate that the western Ligurian Briançonnais basement was at about 265–215 °C between 32 and 29 Ma, suggesting that the onset of the very low-grade to anchizonal D2 phase should have occurred after that time span, i.e., at least since the late Rupelian. U-Pb radiometric age on recrystallized carbonate matrix (24.7 $\pm$ 6,9 Ma) and calcite veins (26 $\pm$ 11 Ma) in Lower Cretaceous deposits of the Entracque Unit [95], which could be referred to as the last event of diffuse recrystallization and pressure dissolution (ascribable to the D2/D3 phase), thus occurred around 25 Ma, confirming what suggested above.

Other constraints for dating the D2/D3 phases are provided by the faults of the northern part of the Argentera Massif [96]. The kinematics of these faults have been related to the shifting of the regional shortening to N-S directions, which is here thought as the reason for the shearing evolution of the D2 folding phase, i.e., the D3 phase. This occurred probably between 26 Ma and 20 Ma when the D3 phase induced the main uplifting of the Argentera Massif, controlled by dextral transpression along NW-SE fault systems [68,97]. In this time span, the internal sectors of the uplifting SW Alps crossed the apatite fission track closure temperature of 120 °C [9], while in the external sectors (Argentera and Dauphinois-Provençal) this occurred at about 14–12 Ma [68]. These data suggest that after the D3 phase, the tectonic evolution (D4 and D5 phases) should have been characterized by development of purely brittle discrete fault systems and/or individual faults.

The D2 NE-vergent tectonics recorded in the external Ligurian Briançonnais have been ascribed to the late Oligocene also by [29].

During the D2/D3 time span the uplifting was recorded by the sedimentary evolution of the adjoining internal syn-orogenic basin, the so-called Tertiary Piemonte Basin, which in the sector close to the Ligurian Alps is characterized by localized fault bounded basins whose tectono-sedimentary evolution has been interpreted as controlled by transtensional and strike-slip tectonics since the late Rupelian [9,10,65].

- D4 phase: N-S regional shortening continued to be active from the early Miocene until the Tortonian-Messinian from the Ligurian Alps to the Apennines (Padane) thrust front, and is recorded in the southern termination of the Western Alps mainly as dextral transpression along the NW-SE fault systems subparallel to the main Briançonnais and Dauphinois tectonic fronts and the boundary faults of the Argentera Massif [68,98].
- D5 phase: the late extensional and transtensional regime occurred during the late Miocene-Pliocene [67,68], which was coeval to the strike slip and contractional tectonic regime of the outer sectors of the SW Alpsto the south of the Argentera Massif and in the northern margin of the present Ligurian basin [64,90].

Figure 8. (**a**) Slickensided strike-slip fault surface bounding the GDZ on its southern side along the EBF (Dogger limestones of M. Bodoira); (**b**) steepened folds affecting the Jurassic-Cretaceous successions close to the EBF in the M. Bodoira tectonic unit; (**c**) subvertical, intensively sheared beds in the hundred-meters scale "Servagno transcurrent zone". This zone is located on the GDZ southern boundary, along the EBF and it affects the Jurassic-Cretaceous succession of M.Bodoira near Servagno pass. (**d**) Detail of Figure 10c, drag folds showing vertical axes and vertical axial planes developed along the individual strike-slip faults of the Servagno transcurrent zone; (**e**) tight to isoclinal folds in the gypsum masses of the GDZ to the west of Valcavera pass; (**f**) major gypsum tectonic slices in the central part of the GDZ, bounded by steeply dipping faults belonging to the EBF fault system (in the background the vertical tectonic slices of Monte Salè, reported in the cross-section of Figure 4).

5.3. *Subsurface Stratigraphic Constraints to the Uplifting Stages*

The eastern extension of the SWAT/GVZ system (as described in previous sections) can be traced in the Cuneo-Mondovì area. In this area, outcrop and subsurface data constrain the tectono-depositional evolution of Oligocene to Miocene synorogenic basins developed on the western Alpine basements: see [7] and references therein. The line drawing in Figure 9 (SL1, [65]) refers to a N-S seismic line running from the Ligurian Alps to the adjacent plain. It shows strike-slip faults with flower geometries, deep-seated in the western Ligurian Alps basement. During the Oligocene and early Miocene, these faults (roughly E-W striking in map view) controlled evolution of basins (presently buried below younger successions) filled by continental to marginal marine/slope successions [7] and laterally equivalent to the cropping out eastern TPB successions (Molare and San Paolo formations [99,100]). Unconformities and onlap terminations point out the progressive Oligocene to Miocene syntectonic uplift of this part of the Ligurian Alps.

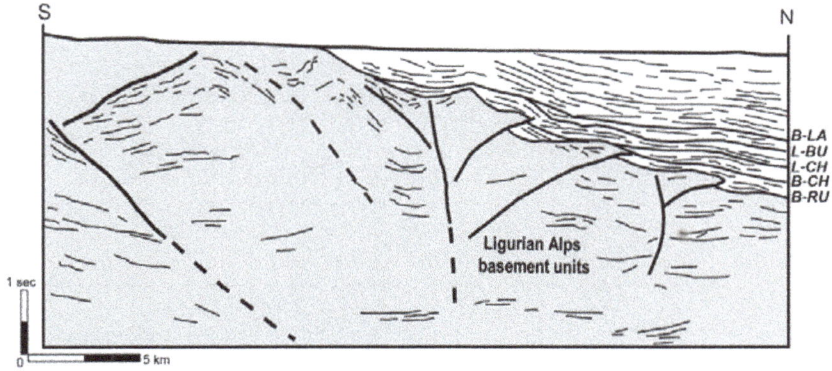

Figure 9. Seismic line drawing crossing the TPB sediments in the subsurface of the Cuneo-Mondovì area. The unconformities are base Rupelian (B-RU), base Chattian (B-CH), late Chattian (L-CH), late Burdigalian (L-BU) and base Langhian (B-LA). From [7,65].

6. Discussion: The Late Eocene to Miocene Evolution of the Internal SW Alps in the Frame of the Adria Indenter Kinematics and First Stages of the Apennines Orogenesis

Some points for discussion on the relations between the kinematic evolution of the SW Alps and the Adria indenter in the frame of the ongoing Apennines orogenesis, are here proposed.

(a) The first point concerns the dynamic context in which the exhumation of the alpine units occurred. As reviewed in the above sections, it can be suggested that the exhumation of the tectonic units derived from the Briançonnais, Dauphinois-Provençal and Alpine Foreland basin domains occurred in a transpressional regime. It was stated by [21] that whereas the exhumation of the adjoining Alpine units, such as the Dora-Maira UHP-HP eclogite rocks and the Monviso meta-ophiolites, mostly occurred through extrusion in the subduction channel and then late extensional tectonics, "the Briançonnais nappes were exhumed mostly through transpressional deformation at the bottom of a collapsing and eroded orogenic wedge". This interpretation is consistent with the ideas sustained in this paper, and is supported by: (i) the overall structural setting of the Briançonnais-Dauphinois-Provençal transect in the SW Alps, corresponding to a mega-macroscale fan-like geometry evoking a flower structure; (ii) the transpressional characters of the structural associations developed in the D1–D3 phases, consisting of several juxtaposed tectonic domains dominated, in turn, by steep reverse and strike-slip faults, low-angle thrust surfaces and folded domains with alternating low-angle axial surface folds and vertical axes folds; (iii) the progressive and gradual decrease of the metamorphic grade from the internal Briançonnais to the external Dauphinois-Provençal units, indicating that no major vertical offset, and no related metamorphic gap, occurred across the tectonic unit boundaries; (iv) the tectono-sedimentary setting of the adjoining syn-orogenic basins placed in the internal side of the Western Alps, in southern Piemonte, that recorded the kinematics and uplifting stages of the Alps tectonic belt.

The subpoint (ii) demands further considerations on the geometrical setting of the GVZ, which has been described in detail only for its eastern part (LIVZ, [23,25]). The western part of the GVZ, consists of a number of tectonic slivers made up of different rock types arranged in a km-scale deformation zone (Gardetta Deformation Zone, Figure 10) developed between the internal (IBF) and the external (EBF) Briançonnais fronts. The tectonic slices are made up of both the rocks of the internal Briançonnais domain (metavolcanics and volcanoclastite, phyllite, quartzite and meta quartzarenite) and of the external Briançonnais (volcanoclastite, quartzarenite and parts of the Triassic-Cretaceous carbonate succession), as well as gypsum masses derived probably from the base of the Briançonnais or the Dauphinois successions. The slices are separated by steeply dipping tectonic contacts, oblique to the GDZ boundaries, and consisting of individual faults showing strike-slip slickensided surfaces (Figure 8a), or transcurrent shear zones. The internal setting of the slices can be represented by contractional structural associations (thrusts with related ramp folds, symmetrical folds in the core of the GDZ, Figure 8b) and/or by transcurrent strained domains showing drag folds with subvertical axes and axial planes (Figure 8c,d), and subhorizontal extension lineations. The huge gypsum masses involved in the GDZ show isoclinal to tight folds with axial planes subparallel to the bedding (Figure 8e) but are always bounded by steeply dipping faults (Figure 8f). Although the detailed description of the GDZ structural setting is beyond the scope of the paper, we believe that a qualitative interpretation of the GDZ as a strain-partitioned transpressive zone should be accountable on the basis of the above reported observations, as well as the overall geometric setting on map view (Figure 10). The macro and mesoscale features of the GDZ seem to fit with the diagnostic features described for the transpressive zones in the basic works of [101–103]. Furthermore, the abundant presence within the GDZ of lithotypes that rarely occur in the external Briançonnais domain (e.g., the basic volcanics of the Becco Nero slices, Figure 10) suggest that the GDZ could have originated on some prealpine lithological inhomogeneity.

(b) The second discussion point refers to the effective relations of the transpressive tectonics, discussed at point (a), with the inferred presence of the regional transfer zone, here named SWAT (see Section 3), that should have contributed to the west-ward indentation of Adria and its counterclockwise rotation with respect to Europe [34]. This transfer kinematics, coeval with the continuing shortening due to the Adria-Europe indentation, seems to be effectively recorded by the structural setting of the southern termination of Western Alps, namely by the Gardetta-Viozene Zone (GVZ), consisting of an assemblage of transpressive deformation units [20,23], which can be followed quite continuously from the NW in the Cottian Alps to the Tanaro valley in the western Ligurian Alps (Figure 3).

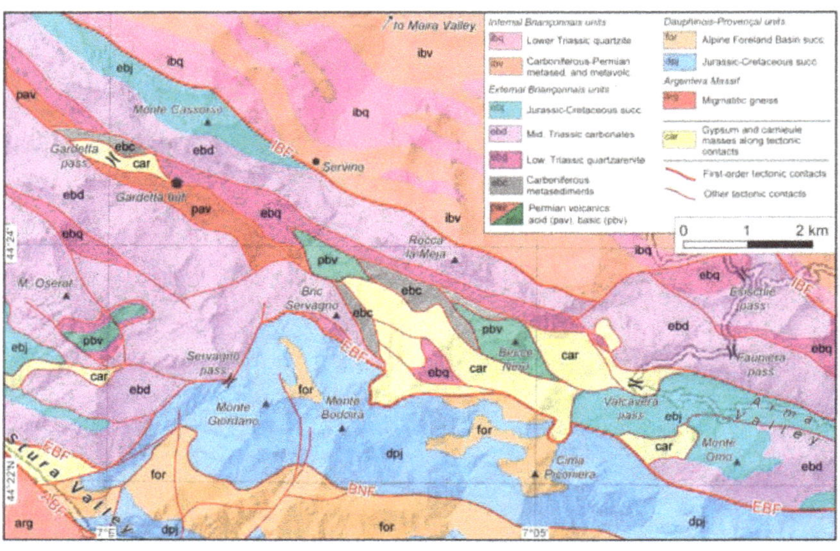

Figure 10. Geotectonic map of the Gardetta Deformation Zone, developed between the internal Briançonnais front (IBF) and the external Briançonnais front (EBF) (modified from [19]).

The structural setting of the SWAT/GVZ corresponds to a double-vergent tectonic system, mainly developed during the D1/D2/D3 phases, where NW-trending major folds, and both NE dipping and SW dipping reverse and strike slip faults developed. In the NW sectors [75–77], the NE-vergent (back-vergent) branch of the transect is indeed developed mostly inwards (NE) of the GVZ, in the internal Briançonnais units [21] (Figure 4). Conversely, in the SE sectors the NE-vergent fold systems are intensively developed within the GVZ itself [25,80,83], while the SW-vergent folds are less abundant, except near the tip zone of the SW-vergent thrusts (Figure 4). The presence of steepened slices of anhydrite and gypsum in the northern branch of the GVZ (Valcavera-Gardetta highland at the altitude of more than 2000 m), which have been found more to the East, at the northern termination of the LIVZ (not far from the IBF, at an altitude of less than 700 m, in the "Buzzi tunnel" between Roaschia and Robilante [104]) and, more extensively, along the Tenda Tunnel thrust at about 1000–1200 m a.s.l [23,86], suggests that the transpressive GVZ probably merged into the basal detachment of the external Briançonnais-Dauphinois-Provençal thrust belt from where it could have dragged, up to higher geometric positions, the above cited gypsum and anhydrite-bearing tectonic slices. The presence, within the fault core zone (namely in the restraining bends), of duplexes and slices extruded from deeper levels, is indirect evidence of transpressional fault setting [101,105,106]. The development of the GVZ unlikely occurred as the result 6tof an homogeneous transpression, as it shows a marked internal strain partitioning. Further analysis is required to ascertain if the strain partitioning was achieved mostly during the first deformation stage (D1) or continuously during the D1 to D3 stages.

(c) As the SWAT activity was recorded by the tectonic evolution of the basement and covers (Mesozoic succession and Eocene-Oligocene Foreland Basin succession) at the southern termination of Western Alps, a third crucial point, concerning the sedimentary recordings of this tectonics in the syn-orogenic basins is to be discussed. The data reported in the above sections indicate that the exhumation/uplifting of the southern Cottian-Maritime-western Ligurian system occurred during the formation of the Alps-Apennines syn-orogenic basin known as TPB (Tertiary Piemonte Basin [9,19,74,107] with references therein). The TPB succession (Figure 1) was deposited starting from the early Oligocene on the exhuming metamorphic complexes of the Western Alps, as well as on the top of the overthrusting Ligurian units involved in the northwestern Apennines. The TPB successions recorded the Alps-Apennines tectonics through some regional scale unconformities related to main Geologic Events (Figure 11) that divided the succession into a number of unconformity-bounded stratigraphic units (Synthems [108]) continuous at the regional scale [19]. The analysis of a seismic line (Figure 9) available for the TPB sectors adjoining the Western Ligurian Alps [65], evidenced that the uplifting of the basement occurred during the depositions of the Oligocene-early Miocene succession, as evidence by late Chattian and Burdigalian reflectors that onlap distinct tectonic units, sealing the progressively younger activity of the faults branching from the main steep fault systems, interpreted by the authors as roughly E-W strike-slip faults deep-seated in the alpine basement. We suggest that this fault system could be consistent (as close to it and showing similar geometric and kinematic features) with the inferred eastward prolongation of the GVZ system (Figure 2), whose evidence in this sector of the western Ligurian Alps (i.e., the Limone-Viozene Zone [25,28]) has been confirmed by surface data. The activity of the GVZ/SWAT system thus occurred first during the early Oligocene sedimentation stages, when the basement of the TPB underwent a stretching, consistent with the sinistral transcurrent tectonics of the D1/D2 phase, which controlled the deposition of the lower Oligocene continental and coarse-grained marine sediment (Molare Fm. [99]), and induced an intense vertical mobility leading to a main regional denudation episode [109] and the onset of differentially subsiding sub-basins, bounded by high angle transtensional faults ([6,10] with references therein) and flanked by fastly uplifting areas. This stage was concomitant with the rifting phase of the Balearic basin [110] that in the internal part of the Ligurian Alps was less pronounced and rapidly decreased toward the North, maybe due to the hindering effect of the southern prolongation of the Ivrea high-density body [9]. The gradual decrease of the rifting was probably partitioned by the transpressive faults of the GVZ/SWAT system, as suggested by [10], during the D1/D2 phase.

All the early Oligocene sub-basins of the southern TPB then underwent, in late Rupelian and early Chattian times, a general drowning and subsidence, with the deposition of outer shelf and slope marly sediments (Rocchetta and Rigoroso Fms. [107]). This event occurred in a transtensional regime concomitant with crustal thinning and opening of back-arc basins [55] and was coeval with the initial stages of the Apennines dynamics, i.e., with the beginning of the Adriatic (Apennines) subduction [111].

Then, a marked inversion of southern TPB structures occurred during the Aquitanian-Early Burdigalian in response to an important geologic event induced by the change in the direction of motion of the Adriatic indenter with respect to Europe from NW-ward to WNW-ward at about 20 My [53,112,113]. This resulted in conditions of oblique convergence and increased collisional tectonics, whose effects are recorded in a large part of the western Mediterranean area and caused the switch from transtensional to contractional and transpressional regime in the southern TPB, inducing inversion of a great number of the formerly active structures [5,7,9,10,114–116]. In this period (early to middle Miocene) the TPB underwent a counterclockwise rotation of ca. 50° with respect to Africa [117].

In the southern western Alps, the Aquitanian-Burdigalian tectonic stage induced a marked regional uplift, with subsequent high denudation rates. This caused the eastward migration of fan-delta systems, prograding from the western margin of the TPB (the "Saluzzo-Monregalese belt" of [7]), accompanied by significant change in sediment composition [118]. In the western Ligurian Alps, the uplifting stage can be referred to the D3 phase, coeval with the main uplifting of the Argentera Massif [68,97] induced by dextral transpression along the northern boundary faults of the massif and D2/D3 back-thrusting. These thrusts are clearly sealed by the late Burdigalian reflectors reported in the line drawing of Figure 9.

The D2/D3 phase developed a penetrative regional foliation through large sectors of the Maritime and western Ligurian Alps: this foliation postdates the early Oligocene (see also [63]) and could be ascribed to compressional stages coeval and dynamically consistent with those recorded in the adjoining TPB successions, i.e., the main Apennines related tectonics that occurred at the Aquitanian-Burdigalian boundary, active while the Maritime and western Ligurian Alps continued their uplifting. The propagation of the N- and NE-vergent thrust front in the westernmost part of the Apennines was thus hindered by the uplifting Alpine basement and related covers. The E and NE-vergent transpressive faults propagating from the previously formed alpine belt also involved and stacked the Ligurian units resting on the Adria crust, as well as the same syn-orogenic sedimentary successions that were flanking the alpine units while they were uplifting [5,51,65]. It becomes clear now, in our view, that the NE-vergent contractional tectonic systems affecting the TPB can be defined alpine or Apennines-related depending on the substrate they displaced (alpine metamorphic units vs. Ligurian non-metamorphic units), but they developed indeed within the same geodynamic context since the late Oligocene, making it appropriate, for the westernmost Alpine-Padane realm, to refer to a single "Alps-Apennines orogenic system", as in [19,74].

Geologic Events Table

Geologic Event/ Tectonic Phase	Age	Structural associations	Tectonic Domains of reference for this paper	Sedimentary record in TPB	Geodynamics
D0 *(D1 of Michard et al., 2004)*	Late Eocene (about 42-37 Ma)	SW- vergent thrusts with metamorphic HP-LT foliation	Internal Briançonnais	— this event was sealed by the epimesoAlpine basin (*sensu* Mutti et al., 1995)	Late stages of Alpine subduction
D1 *T < 280°C in Ext. Briançonnais*	Early Oligocene (about 34 - 30 Ma)	SW vergent, bedding parallel thrusting, local transpression, metamorphic foliations in Internal. Briançonnais	Internal Brianc., External Brianç., Dauph.-Provenç. with Alpine foreland basin succession (AFB)	U1 unconformity, transgression on uplifting Alpine units, fault-bounded basins	Collisional décollement tectonics, exhumation of the Brianç.-Dauph-Provenc. stacking
D2 *T < 230-280°C in Ext. Briançonnais*	Early- Late Oligocene (about 30 - 24 Ma)	Double vergent (NE- and SW) fold systems and E W to NW SE transpressive sinistral faults	Internal Brianc., External Brianç., Dauph.-Provenç., AFB, Tertiary Piemonte Basin (TPB)	Minor unconformities between U1 and U2. Platform drawning	Uplifting in SW internal Alps. Transtensional tectonics in TPB. SWAT/GVZ activation
D3 *T < 200°C in Ext. Briançonnais*	Latest Oligocene Early Miocene (about 24 - 18 Ma)	Shearing of D2 folds, dextral reactivation of NW-SE transpressive faults, double vergent (NE-SW) thrusting	External Brianç., Dauph. Provenç., AFB, TPB, Argentera Massif	U2 and U3 unconformities	Beginning of Adria "Apennine" subduction, first TPB basin inversion
D4 *T < 120°C in Ext. Briançonnais*	Middle Miocene Late Miocene (about 18 - 8 Ma)	Dextral E W and NW SE transpressive faults, NNE SSW transtensive faulting	External Brianç., Dauph. Provenç., AFB, TPB, Argentera Massif	U4 and U5 unconformities, major TPB subsidence rates	NE-propagation of Apennines tectonic fronts, onset and N-ward shifting of main TPB depocenters
D5 *T < 70°C in Argentera Massif*	Latest Miocene Pliocene	Extensional - transtensional faulting on differently oriented fault systems	TPB, Argentera Massif	U6 and U7 unconformities, Messinian erosion, Pliocene transgression	Transtensional regime at Western Alps arc south termination, Messinian crisis, Pliocene general subsidence in Western Po plain

Figure 11. List of the Alpine Geologic events (D1–D5) recorded in the southern termination of Western Alps since Early Oligocene. The U1, U2 ... D7 unconformities correspond to the D1, D2 ... D7 regional unconformities of the TPB, as defined in [19,74]. The column "geodynamics" describes the relations with the main stages of the tectono-sedimentary evolution of the southern termination of the Western Alps and of north-western Apennines.

7. Conclusions

Based on a critical review of surface and subsurface geological data, integrated with new data and interpretations, it is concluded that the southern termination of the Western Alps arc recorded the Oligocene-Miocene activity of a regional transfer zone (the southwestern Alps Transfer, SWAT) whose existence has been often postulated in literature [15,34,81] and that should have allowed, since early Oligocene, the westward indentation of Adria and its counterclockwise rotation with respect to Europe. This "virtual" transfer zone, inferred on the basis of geodynamic constraints and reconstructions, could be partially seen, at shallow crustal level, in an effective system of deformation units and km-scale shear zones, here defined as the Gardetta-Viozene Zone (GVZ). The GVZ is developed externally to the internal Briançonnais Front (IBF), involving the external Briançonnais and Dauphinois-Provençal domains and the overlying Eocene-Oligocene sediments of the Alpine Foreland basin. The IBF, which represents the inner boundary of the SWAT, is thought to correspond to the Penninic Front, here intended as the frontal thrust which bounds the Alpine Axial Belt, i.e., the metamorphic orogenic prism (in the sense of [69]). Thus, in the southern termination of western Alps, the Penninic Front divides the external from the internal Briançonnais domains. Consequently, it can be argued that, in this area, the Briançonnais domain did not experience subduction and exhumation as a whole. The internal Briançonnais underwent major tectono-metamorphic transformations, while the external Briançonnais was subjected only to anchizonal P-T conditions. The relatively gradual transition (although stepwise across distinct tectonic fronts) from HP-LT metamorphism and very low-grade to anchizone metamorphism through the Briançonnais-Dauphinois transect of SW Alps, suggests a low entity of the thrust vertical offsets, as expected in an overall transpressive or strike-slip regional context.

The southwestern Alps Transfer acted outwardly of the IBF, in a foreland fold and thrust belt, consisting of the external Briançonnais and the Dauphinois-Provençal domains with related Alpine Foreland Basin successions, which was detached above quartzites and anhydrite-gypsum levels of inferred Triassic age, now locally involved in the core of the SWAT shear zones. Conversely, the IBF cut across the Triassic evaporite and quartzite level, bounding the external domains affected by "cover tectonics" from the internal levels, where the stacking involved the Permian metavolcanics and some levels of the underlying polymetamorphic basement [119,120]. The Oligocene to Miocene kinematic evolution of the above-described Alpine units was well recorded by the tectono-sedimentary evolution of the inner syn-orogenic basins, i.e., the so-called Tertiary Piemonte Basin, as evidenced by stratigraphic, sedimentological and geophysical data. This allows correlation with the Apennines kinematics and dynamics, in terms of the age of the main geologic events, the interference between the main structural systems and the tectonic control exerted by both the tectonic belts on the same syn-orogenic basin.

Author Contributions: Conceptualization, F.P.; investigation: A.d., L.B. and C.B.; data curation, F.P., A.d., L.B., C.B., A.I. and P.M.; writing—original draft preparation, F.P.; writing—review and editing, F.P., A.d., L.B., C.B., A.I. and P.M. All authors have read and agreed to the published version of the manuscript.

Funding: This research received no external funding.

Data Availability Statement: The study did not report any organised data sets.

Acknowledgments: Two anonymous reviewers are kindly thanked for their useful suggestions.

Conflicts of Interest: The authors declare no conflict of interest.

References

1. Gelati, R.; Gnaccolini, M. Evoluzione tettonico-sedimentaria della zona limite tra Alpi ed Appennini tra l'inizio dell'Oligocene ed il Miocene medio. *Mem. Soc. Geol. Ital.* **1982**, *2*, 183–191.
2. Mutti, E.; Papani, L.; di Biase, D.; Davoli, G.; Mora, S.; Segadelli, S.; Tinterri, R. Il Bacino Terziario Epimesoalpino e le sue implicazioni sui rapporti tra Alpi ed Appennino. *Mem. Sci. Geol.* **1995**, *48*, 233–291.
3. Piana, F.; Polino, R. Tertiary structural relationships between Alps and Apennines: The critical Torino Hill and Monferrato area. Northwestern Italy. *Terra Nova* **1995**, *7*, 138–143. [CrossRef]
4. Biella, G.C.; Polino, R.; de Franco, R.; Rossi, P.M.; Clari, P.; Corsi, A.; Gelati, R. The crustal structure of the western Po plain: Reconstruction from integrated geological and seismic data. *Terra Nova* **1997**, *9*, 28–31. [CrossRef]
5. Piana, F. Structural setting of western Monferrato (Alps-Apennines junction zone, NW Italy). *Tectonics* **2000**, *19*, 943–960. [CrossRef]
6. Mutti, E.; di Biase, D.; Fava, L.; Mavilla, N.; Sgavetti, M.; Tinterri, R. The Tertiary Piemonte Basin. In *Revisiting Turbidites of the Marnoso-Arenacea Formation and their Basin-Margin Equivalents: Problems with Classic Models*; Mutti, E., Ricci Lucchi, F., Roveri, M., Eds.; Excursion Notes-Part II. Excursion Guidebook of "Turbidite Workshop"; Dipartimento di Scienze della Terra, Università di Parma, ENI Spa: Parma, Italy, 2002; p. 25.
7. Rossi, M.; Mosca, P.; Polino, R.; Rogledi, S.; Biffi, U. New outcrop and subsurface data in the Tertiary Piedmont Basin (NW Italy): Unconformity-bounded stratigraphic units and their relationships with basin modification phases. *Riv. Ital. Paleontol. Stratigr.* **2009**, *115*, 305–335.
8. Ghielmi, M.; Minervini, M.; Nini, C.; Rogledi, S.; Rossi, M. Late Miocene-Middle Pleistocene sequences in the Po Plain—Northern Adriatic Sea (Italy): The stratigraphic record of modification phases affecting a complex foreland basin. *Marine Petrol. Geol.* **2013**, *42*, 50–81. [CrossRef]
9. Maino, M.; Decarlis, A.; Felletti, F.; Seno, S. Tectono-sedimentary evolution of the Tertiary Piedmont Basin (NW Italy) within the Oligo-Miocene central Mediterranean geodynamics. *Tectonics* **2013**, *32*, 593–619. [CrossRef]
10. Ghibaudo, G.; Massari, F.; Chiambretti, I.; d'Atri, A.; Fornaciari, E. Birth and tectono-sedimentary evolution of the Tertiary Piedmont Basin (NW Italy). *J. Medit. Earth Sci.* **2019**, *11*, 5–112. [CrossRef]
11. Doglioni, C.; Mongelli, F.; Pialli, G.P. Boudinage of the Alpine belt in the Apenninic back-arc. *Mem. Soc. Geol. It.* **1998**, *52*, 457–468.
12. Doglioni, C.; Gueguen, E.; Harabaglia, P.; Mongelli, F.C. On the origin of west-directed subduction zones and applications to the western Mediterranean. *Geol. Soc. Lond. Spec. Publ.* **1999**, *156*, 541–561. [CrossRef]
13. Brunet, C.; Monié, P.; Jolivet, L.; Cadet, J.P. Migration of compression and extension in the Tyrrhenian Sea, insights from 40Ar/39Ar ages on micas along a transect from Corsica to Tuscany. *Tectonophysics* **2000**, *321*, 127–155. [CrossRef]

14. Carminati, E.; Doglioni, C. Europe—Mediterranean tectonics. In *Encyclopedia of Geology*; Selley, R., Cocks, R., Plimer, J., Eds.; Elsevier: Oxford, UK, 2004; pp. 135–146.
15. Ford, M.; Duchene, S.; Gasquet, D.; Vanderhaeghe, O. Two-phase orogenic convergence in the external and internal SW Alps. *J. Geol. Soc.* **2006**, *163*, 815–826. [CrossRef]
16. Lardeaux, J.M.; Schwartz, S.; Tricart, P.; Paul, A.; Guillot, S.; Béthoux, N.; Masson, F. A crustal-scale cross-section of the south-western Alps combining geophysical and geological imagery. *Terra Nova* **2006**, *18*, 412–422. [CrossRef]
17. Schreiber, D.; Lardeaux, J.-M.; Martelet, G.; Courrioux, G.; Guillen, A. 3-D modelling of Alpine Mohos in Southwestern Alps. *Geoph. J. Intern.* **2010**, *180*, 961–975. [CrossRef]
18. Daniel, J.M.; Jolivet, L.; Goffe, B.; Poinssot, C. Crustal-scale strain partitioning: Footwall deformation below the Alpine Oligo-Miocene detachment of Corsica. *J. Struct. Geol.* **1996**, *18*, 41–59. [CrossRef]
19. Piana, F.; Fioraso, G.; Irace, A.; Mosca, P.; d'Atri, A.; Barale, L.; Falletti, P.; Monegato, G.; Morelli, M.; Tallone, S.; et al. Geology of Piemonte region (NW Italy, Alps–Apennines interference zone). *J. Maps* **2017**, *13*, 395–405. [CrossRef]
20. Gidon, M. Les chaînons briançonnais et subbriançonnais de la rive gauche de la Stura entre Bersezio et le Val de l'Arma (province de Cuneo Italie). *Géol. Alpine* **1972**, *48*, 87–120.
21. Michard, A.; Avigad, D.; Goffé, B.; Chopin, C. The high-pressure metamorphic front of the south Western Alps (Ubaye-Maira transect, France, Italy). *Schweiz. Min. Petr. Mitt.* **2004**, *84*, 215–235.
22. Malaroda, R.; Carraro, F.; Dal Piaz, G.; Franceschetti, B.; Sturani, C.; Zanella, E. Note illustrative della Carta Geologica del Massiccio dell'Argentera alla scala 1: 50.000. *Mem. Soc. Geol. Ital.* **1970**, *9*, 557–663.
23. D'Atri, A.; Piana, F.; Barale, L.; Bertok, C.; Martire, L. Geological setting of the southern termination of Western Alps. *Int. J. Earth Sci.* **2016**, *105*, 1831–1858. [CrossRef]
24. Mohn, G.; Manatschal, G.; Beltrando, M.; Masini, E.; Kusznir, N. Necking of continental crust in magma-poor rifted margins: Evidence from the fossil Alpine Tethys margins. *Tectonics* **2012**, *31*, TC1012. [CrossRef]
25. Piana, F.; Musso, A.; Bertok, C.; d'Atri, A.; Martire, L.; Perotti, E.; Varrone, D.; Martinotti, G. New data on post-Eocene tectonic evolution of the External Ligurian Brianconnais (Western Ligurian Alps). *Boll. Soc. Geol. Ital.* **2009**, *128*, 353–366.
26. Piana, F.; Battaglia, S.; Bertok, C.; d'Atri, A.; Ellero, A.; Leoni, L.; Martire, L.; Perotti, E. Illite (KI) and chlorite (AI) "crystallinity" indices as a constraint for the evolution of the External Briançonnais Front in Western Ligurian Alps (NW Italy). *Ital. J. Geosci.* **2014**, *133*, 445–454. [CrossRef]
27. Barale, L.; Bertok, C.; d'Atri, A.; Martire, L.; Piana, F.; Domini, G. Geology of the Entracque–Colle di Tenda area (Maritime Alps, NW Italy). *J. Maps* **2016**, *12*, 359–370. [CrossRef]
28. Bertok, C.; Musso, A.; D'Atri, A.; Martire, L.; Piana, F. Geology of the Colle di Tenda—Monte Marguareis area (Ligurian Alps, NW Italy). *J. Maps* **2018**, *14*, 542–551. [CrossRef]
29. Mueller, P.; Maino, M.; Seno, S. Progressive Deformation Patterns from an Accretionary Prism (Helminthoid Flysch, Ligurian Alps, Italy). *Geosciences* **2020**, *10*, 26. [CrossRef]
30. Sagri, M. Le Arenarie di Bordighera: Una conoide sottomarina nel bacino di sedimentazione del Flysch a Elmintoidi di San Remo (Cretaceo superiore, Liguria Occidentale). *Boll. Soc. Geol. Ital.* **1980**, *98*, 205–226.
31. Sagri, M. Litologia, stratimetria e sedimentologia delle torbiditi di piana di bacino del flysch di San Remo (Cretaceo superiore, Liguria occidentale). *Mem. Soc. Geol. Ital.* **1984**, *28*, 577–586.
32. Vanossi, M. Les unités géologiques des Alpes Maritimes entre l'Ellero et la Mèr Ligure: Un aperçu schématique. *Mem. Ist. Geol. Mineral. Univ. Padova* **1980**, *3*, 101–142.
33. Kerckhove, C. La "zone du flysch" dans les nappes de l'Embrunais-Ubaye (Alpes occidentales). *Géol. Alpine* **1969**, *45*, 5–204.
34. Dumont, T.; Schwartz, S.; Guillot, S.; Simon-Labric, T.; Tricart, P.; Jourdan, S. Structural and sedimentary records of the Oligocene revolution in the Western Alpine arc. *J. Geodyn.* **2012**, *56*, 18–38. [CrossRef]
35. Lanteaume, M. *Contribution à l'étude Géologique des Alpes Maritimes Franco-Italiennes*; Carte Géologique de France, Imprimerie Nationale: Paris, France, 1968; pp. 1–405.
36. Di Giulio, A. The evolution of the Western Ligurian Flysch Units and the role of mud diapirism in ancient accretionary prisms (Maritime Alps, Northwestern Italy). *Int. J. Earth Sci.* **1992**, *81*, 655–668. [CrossRef]
37. Seno, S.; Dallagiovanna, G.; Vanossi, M. A kinematic evolutionary model for the Penninic sector of the central Ligurian Alps. *Int. J. Earth Sci.* **2005**, *94*, 114–129. [CrossRef]
38. Lavier, L.; Manatschal, G. A mechanism to thin the continental lithosphere at magma-poor margins. *Nature* **2006**, *440*, 324–328. [CrossRef]
39. Decarlis, A.; Manatschal, G.; Haupert, I.; Masini, E. The tectono-stratigraphic evolution of distal, hyper-extended magma-poor conjugate rifted margins: Examples from the Alpine Tethys and Newfoundland-Iberia. *Mar. Petr. Geol.* **2015**, *68*, 54–72. [CrossRef]
40. Decarlis, A.; Lualdi, A. Late Triassic-early Jurassic paleokarst from the Ligurian Alps and its geological significance (Siderolitico Auct., Ligurian Briançonnais domain). *Swiss J. Geosci.* **2008**, *101*, 579–593. [CrossRef]
41. Lemoine, M.; Bas, T.; Arnaud-Vanneau, A.; Arnaud, H.; Dumont, T.; Gidon, M.; Bourbon, M.; de Graciansky, P.-C.; Rudkiewicz, J.L.; Megard-Galli, J.; et al. The continental margin of the Mesozoic Tethys in the Western Alps. *Mar. Petrol. Geol.* **1986**, *3*, 179–199. [CrossRef]
42. Claudel, M.E.; Dumont, T. A record of multistage continental break-up on the Briançonnais marginal plateau (Western Alps): Early and middle–late Jurassic rifting. *Eclogae Geol. Hel.* **1999**, *92*, 45–61.

43. Debelmas, J.; Kerckhove, C. II: Les Alpes franco-italiennes. In: France: Introduction à la géologie du Sud-Est. *Géol. Alpine* **1980**, *56*, 282.
44. Mohn, G.; Manatschal, G.; Müntener, O.; Beltrando, M.; Masini, E. Unravelling the interaction between tectonic and sedimentary processes during lithospheric thinning in the Alpine Tethys margins. *Int. J. Earth Sci.* **2010**, *99*, 75–101. [CrossRef]
45. Faure-Mauret, A. *Études Géologiques sur le Massif de l'Argentera-Mercantour et ses Enveloppes Sédimentaires*; Mémoires. Carte Géologique: Paris, France, 1955; p. 336.
46. Barale, L.; Bertok, C.; d'Atri, A.; Martire, L.; Piana, F. Stratigraphy, sedimentology and syndepositional tectonics of the Jurassic–Cretaceous succession at the transition between Provençal and Dauphinois domains (Maritime Alps, NW Italy). *Riv. Ital. Paleont. Strat.* **2017**, *123*, 355–378.
47. Malaroda, R. Carta Geologica del Massiccio dell'Argentera alla scala 1: 50.000. *Mem. Soc. Geol. Ital.* **1970**, *9*, 557–663.
48. Bersezio, R.; Barbieri, P.; Mozzi, R. Redeposited limestones in the Upper Cretaceous succession of the Helvetic Argentera Massif at the Italy–France border. *Eclogae Geol. Helv.* **2002**, *95*, 15–30.
49. Sinclair, H.D. Tectono–stratigraphic model for underfilled peripheral foreland basins: An Alpine perspective. *GSA Bull.* **1997**, *109*, 324–346. [CrossRef]
50. Apps, G.; Peel, F.; Elliott, T. The structural setting and palaeogeographical evolution of the Grès d'Annot Basin. *Geol. Soc. London Spec. Publ.* **2004**, *221*, 65–96. [CrossRef]
51. Molli, G.; Crispini, L.; Malusà, M.G.; Mosca, P.; Piana, F.; Federico, L. Geology of the Western Alps-Northern Apennine junction area: A regional review. *J. Virt. Expl.* **2010**, *36*, 1–48. [CrossRef]
52. Bertok, C.; Martire, L.; Perotti, E.; d'Atri, A.; Piana, F. Kilometre-scale palaeoescarpments as evidence for Cretaceous synsedimentary tectonics in the External Briançonnais Domain (Ligurian Alps, Italy). *Sedim. Geol.* **2012**, *251*, 58–75. [CrossRef]
53. Handy, M.R.; Schmid, S.M.; Bousquet, R.; Kissling, E.; Bernoulli, D. Reconciling plate-tectonic reconstructions of Alpine Tethys with the geological–geophysical record of spreading and subduction in the Alps. *Earth Sci. Rev.* **2010**, *102*, 121–158. [CrossRef]
54. Dumont, T.; Simon-Labric, T.; Authemayou, C.; Heymes, T. Lateral termination of the north-directed Alpine orogeny and onset of westward escape in the Western Alpine arc: Structural and sedimentary evidence from the external zone. *Tectonics* **2011**, *30*, TC5006. [CrossRef]
55. Vignaroli, G.; Rossetti, F.; Rubatto, D.; Theye, T.; Lisker, F.; Phillips, D. Pressure-temperature- deformation-time (P-T-d-t) exhumation history of the Voltri Massif HP complex, Ligurian Alps, Italy. *Tectonics* **2010**, *29*, TC6009. [CrossRef]
56. Laubscher, H. The arc of the Western Alps today. *Eclogae Geol. Helv.* **1991**, *84*, 631–659.
57. Tavani, S.; Bertok, C.; Granado, P.; Piana, F.; Salas, R.; Vigna, B.; Muñoz, J.A. The Iberia-Eurasia plate boundary east of the Pyrenees. *Earth Sci. Rev.* **2018**, *187*, 314–337. [CrossRef]
58. Vignaroli, G.; Faccenna, C.; Jolivet, L.; Piromallo, C.; Rossetti, F. Subduction polarity reversal at the junction between the Western Alps and the Northern Apennines, Italy. *Tectonophysics* **2008**, *450*, 34–50. [CrossRef]
59. Bertok, C.; Martire, L.; Perotti, E.; d'Atri, A.; Piana, F. Middle-Late Jurassic syndepositional tectonics recorded in the Ligurian Briançonnais succession (Marguareis–Mongioie area, Ligurian Alps, NW Italy). *Swiss J. Geosci.* **2011**, *104*, 237–255. [CrossRef]
60. Perotti, E.; Bertok, C.; d'Atri, A.; Martire, L.; Piana, F.; Catanzariti, R. A tectonically-induced Eocene sedimentary mélange in the West Ligurian Alps, Italy. *Tectonophysics* **2012**, *568*, 200–214. [CrossRef]
61. Barale, L.; Bertok, C.; d'Atri, A.; Domini, G.; Martire, L.; Piana, F. Hydrothermal dolomitization of the carbonate Jurassic succession in the Provençal and Subbriançonnais Domains (Maritime Alps, North-Western Italy). *Compt. Rend. Geosci.* **2013**, *345*, 47–53. [CrossRef]
62. Barale, L.; Bertok, C.; Salih Talabani, N.; d'Atri, A.; Martire, L.; Piana, F.; Préat, A. Very hot, very shallow hydrothermal dolomitization: An example from the Maritime Alps (north-west Italy-south-east France). *Sedimentology* **2016**, *63*, 2037–2065. [CrossRef]
63. Bertok, C.; Barale, L.; d'Atri, A.; Martire, L.; Piana, F.; Rossetti, P.; Gerdes, A. Unusual marbles in a non-metamorphic succession of the SW Alps (Valdieri, Italy) due to early Oligocene hydrothermal flow. *Int. J. Earth Sci.* **2019**, *108*, 693–712. [CrossRef]
64. Larroque, C.; Delouis, B.; Godel, B.; Nocquet, J.-M. Active deformation at the southwestern Alps–Ligurian basin junction (France–Italy boundary): Evidence for recent change from compression to extension in the Argentera massif. *Tectonophysics* **2009**, *467*, 22–34. [CrossRef]
65. Mosca, P.; Polino, R.; Rogledi, S.; Rossi, M. New data for the kinematic interpretation of the Alps–Apennines junction (Northwestern Italy). *Int. J. Earth Sci.* **2009**, *99*, 833–849. [CrossRef]
66. Goffe, B.; Schwartz, S.; Lardeaux, J.-M.; Bousquet, R. Explanatory notes to the map: Metamorphic structure of the Alps western and Ligurian Alps. *Mitt. Österr. Miner. Ges.* **2004**, *149*, 125–144.
67. Bigot-Cormier, F.; Sosson, M.; Poupeau, G.; Stéphan, J.-F.; Labrin, E. The denudation history of the Argentera Alpine External Crystalline Massif (Western Alps, France-Italy): An overview from the analysis of fission tracks in apatites and zircons. *Geodin. Acta* **2006**, *19*, 455–473. [CrossRef]
68. Sanchez, G.; Rolland, Y.; Jolivet, M.; Brichau, S. Exhumation controlled by transcurrent tectonics: The Argentera–Mercantour massif (SW Alps). *Terra Nova* **2011**, *23*, 116–126. [CrossRef]
69. Beltrando, M.; Compagnoni, R.; Lombardo, B. (Ultra-) High-pressure metamorphism and orogenesis: An Alpine perspective. *Gond. Res.* **2010**, *18*, 147–166. [CrossRef]

70. Herwegh, M.; Berger, A.; Baumberger, R.; Wehrens, P.; Kissling, E. Large-Scale Crustal-Block-Extrusion during Late Alpine Collision. *Sci. Rep.* **2017**, *7*, 413. [CrossRef]
71. Dal Piaz, G.V.; Bistacchi, A.; Massironi, M. Geological outline of the Alps. *Episodes* **2003**, *26*, 175–180. [CrossRef]
72. Ford, M.; Lickorish, W.; Kusznir, N. Tertiary foreland sedimentation in the Southern Subalpine Chains, SE France: A geodynamic appraisal. *Basin Res.* **1999**, *11*, 315–336. [CrossRef]
73. Tricart, P. From extension to transpression during the final exhumation of the Pelvoux and Argentera massifs, Western Alps. *Eclogae Geol. Helv.* **2004**, *97*, 429–439. [CrossRef]
74. Piana, F.; Barale, L.; Compagnoni, R.; d'Atri, A.; Fioraso, G.; Irace, A.; Mosca, P.; Tallone, S.; Morelli, M. Geological map of Piemonte region at 1:250,000 scale. Explanatory Notes. *Acc. Sc. Torino Mem. Sc. Fis.* **2017**, *41*, 1–143.
75. Debelmas, J.; Lemoine, M. *Carte Géologique de France au 1/50 000, Feuille Guillestre. Notice Explicative*; Bureau. Recherchés. Géologique. Miniéres: Orléans, France, 1965; p. 20.
76. Tricart, P. Les rétrocharriages dans les Alpes franco-italiennes. Évolution des structures sur la transversale Embrunais-Queyras (Hautes-Alpes). *Sci. Géol. Bull. Strasbourg* **1975**, *28*, 239–259. [CrossRef]
77. Tricart, P. From passive margin to continental collision: A tectonic scenario for the Western Alps. *Am. J. Sci.* **1984**, *284*, 97–120. [CrossRef]
78. Caby, R. Low-angle extrusion of high-pressure rocks and the balance between outward and inward displacements of Middle Penninic units in the Western Alps. *Eclogae Geol. Helv.* **1996**, *89*, 229–267.
79. Bucher, S.; Schmid, S.M.; Bousquet, R.; Fügenschuh, B. Late-stage deformation in a collisional orogen (Western Alps): Nappe refolding, back-thrusting or normal faulting? *Terra Nova* **2003**, *15*, 109–117. [CrossRef]
80. Vanossi, M.; Cortesogno, L.; Galbiati, B.; Messiga, B.; Piccardo, G.; Vanucci, R. Geologia delle Alpi Liguri: Dati, problemi, ipotesi. *Mem. Soc. Geol. Ital.* **1984**, *28*, 5–75.
81. Giglia, G.; Capponi, G.; Crispini, L.; Piazza, M. Dynamics and seismotectonics of the West-Alpine arc. *Tectonophysics* **1996**, *267*, 143–146. [CrossRef]
82. Petti, F.M.; Furrer, H.; Collo, E.; Martinetto, E.; Bernardi, M.; Delfino, M.; Romano, M.; Piazza, M. Archosauriform footprints in the Lower Triassic of Western Alps and their role in understanding the effects of the Permian-Triassic hyperthermal. *PeerJ* **2020**, *8*, e10522. [CrossRef]
83. Carminati, E.; Gosso, G. Structural map of a Ligurian Briançon- nais cover nappe (Conca delle Carsene, Monte Marguareis, Ligurian Alps, Italy) and explanatory notes. *Mem. Sci. Geol.* **2000**, *52*, 93–99.
84. Simon-Labric, T.; Rolland, Y.; Dumont, T.; Heymes, T.; Authemayou, C.; Corsini, M.; Fornari, M. 40Ar/39Ar dating of Penninic Front tectonic displacement (W Alps) during the Lower Oligocene (31–34 Ma). *Terra Nova* **2009**, *21*, 127–136. [CrossRef]
85. Baldacci, L.; Franchi, S. Studio geologico della galleria del Colle di Tenda: (linea Cuneo–Ventimiglia). *Boll. R. Com. Geol. Ital.* **1900**, *31*, 33–87.
86. Cavinato, G.P.; Di Luzio, E.; Moscatelli, M.; Vallone, R.; Averardi, M.; Valente, A.; Papale, S. The new Col di Tenda tunnel between Italy and France: Integrated geological investigations and geophysical prospections for preliminary studies on the Italian side. *Engen. Geol.* **2006**, *88*, 90–109. [CrossRef]
87. Brizio, F.D.; Deregibus, C.; Eusebio, A.; Gallo, M.; Gosso, G.; Rattalino, E.; Rossi, F.; Tosetto, S.; Oxilia, M. Guida all'escursione: I rapporti tra la zona Brianzonese Ligure e il Flysch a Elmintoidi, Massiccio del Marguareis (Limone Piemonte-Certosa di Pesio, Cuneo, 14/15 Settembre 1983). *Mem. Soc. Geol. Ital.* **1983**, *26*, 579–595.
88. Lefèvre, R. La cicatrice de Preit: Une discontinuité structurale majeure au sein de la zone briançonnaise entre Acceglio et l'Argentera (Alpes Cottiennes méridionales). *C. R. Acad. Sci. Paris* **1983**, *296*, 1551–1554.
89. Malaroda, R. Il Flysch di Demonte-Aisone e il sondaggio di Aisone. *Acc. Sc. Torino Atti Sc. Fis.* **2000**, *24*, 1–37.
90. Beucher, R.; Sue, C.; Tricart, P. Orogen-parallel brittle extension as a major tectonic imprint in the Neogene evolution of the south-western Alpine arc. *Int. J. Earth Sci.* **2017**, *106*, 2973–2990. [CrossRef]
91. Lefèvre, R.; Michard, A. Les nappes briançonnaises internes et ultra-briançonnaises de la bande d'Acceglio (Alpes franco-italiennes); une étude structurale et pétrographique dans le faciès des schistes bleus à jadéite. *Sci. Géol. Bull.* **1976**, *29*, 183–222. [CrossRef]
92. Monié, P. Preservation of Hercynian 40Ar/39Ar ages through high-presssure low-temperature Alpine metamorphism in the Western Alps. *Eur. J. Mineral.* **1990**, *2*, 343–361. [CrossRef]
93. Du Fornel, E.; Joseph, P.; Desaubliaux, G.; Eschard, R.; Guillocheau, F.; Lerat, O.; Muller, C.; Ravenne, C.; Sztràkos, K. The southern Grès d'Annot outcrops (French Alps): An attempt at regional correlation. *Geol. Soc. Lond. Spec. Publ.* **2004**, *221*, 137–160. [CrossRef]
94. Yin, A. Cenozoic tectonic evolution of the Himalayan orogen as constrained by along-strike variation of structural geometry, exhumation history, and foreland sedimentation. *Earth Sci. Rev.* **2006**, *76*, 1–131. [CrossRef]
95. Barale, L.; Bertok, C.; d'Atri, A.; Martire, L.; Piana, F.; Bernasconi, S.M.; Birgel, D.; Czuppon, G.; Gerdes, A. U/Pb dating and geochemical constraints to hydrothermal dolomitization in the Provençal Domain (Maritime Alps): Implications for Early Cretaceous tectonics. *Ofioliti* **2021**. accepted for publication.
96. Baietto, A.; Cadoppi, P.; Martinotti, G.; Perello, P.; Perrochet, P.; Vuataz, F.D. Assessment of thermal circulations in strike-slip fault systems: The Terme di Valdieri case (Italian Western Alps). *Geol. Soc. Lond. Spec. Publ.* **2008**, *299*, 317–339. [CrossRef]

97. Sanchez, G.; Rolland, Y.; Schneider, J.; Corsini, M.; Oliot, E.; Goncalves, P.; Verati, C.; Lardeaux, J.-M.; Marquer, D. Dating low-temperature deformation by 40Ar/39Ar on white mica, insights from the Argentera-Mercantour Massif (SW Alps). *Lithos* **2011**, *125*, 521–536. [CrossRef]
98. Bauve, V.; Plateaux, R.; Rolland, Y.; Sanchez, G.; Bethoux, N.; Delouis, B.; Darnault, R. Long-lasting transcurrent tectonics in SW Alps evidenced by Neogene to present-day stress fields. *Tectonophysics* **2014**, *621*, 85–100. [CrossRef]
99. Lorenz, C. Evolution stratigraphique et structurale des Alpes Ligures depuis l'Eocène supérieur. *Boll. Soc. Geol. Ital.* **1984**, *28*, 211–228.
100. Casnedi, R.; Mosna, S. Segnalazione di una serie miocenica inferiore nel Monregalese (Bacino Terziario Piemontese). *Acc. Naz. Lincei Cl. Sc. Fis. Mat. Nat.* **1970**, *48*, 146–155.
101. Dewey, J.F.; Holdsworth, R.E.; Strachan, R.A. Transpression and transtension zones. *Geol. Soc. Lond. Spec. Publ.* **1998**, *135*, 1–14. [CrossRef]
102. Jones, R.; Tunner, P.V.G. Strain partitioning in transpression zones. *J. Struct. Geol.* **1995**, *17*, 793–802. [CrossRef]
103. Jones, R.; Holdsworth, R.; Clegg, P.; McCaffrey, K.; Tavarnelli, E. Inclined transpression. *J. Struct. Geol.* **2004**, *26*, 1531–1548. [CrossRef]
104. Minoja, M. *Bene Comune e Comportamenti Responsabili. Storie di Imprese e di Istituzioni*; EGEA: Milano, Italy, 2015; p. 410, ISBN 9788823814035.
105. Woodcock, N.H.; Rickards, B. Transpressive duplex and flower structure: Dent Fault System, NW England. *J. Struct. Geol.* **2003**, *25*, 1981–1992. [CrossRef]
106. Cunningham, W.D.; Mann, P. Tectonics of strike-slip restraining and releasing bends. *Geol. Soc. Lond. Spec. Publ.* **2007**, *290*, 1–12. [CrossRef]
107. Gelati, R.; Gnaccolini, M. Synsedimentary tectonic and sedimentation in the Tertiary Piedmont Basin, northwestern Italy. *Riv. Ital. Paleont. Strat.* **1998**, *104*, 193–214.
108. Chang, K.H. Unconformity bounded stratigraphic units. *Geol. Soc. Am. Bull.* **1975**, *86*, 1544–1552. [CrossRef]
109. Bertotti, G.; Mosca, P.; Juez, J.; Polino, R.; Dunai, T. Oligocene to Present kilometres scale subsidence and exhumation of the Ligurian Alps and the Tertiary Piedmont Basin (NW Italy) revealed by apatite (U-Th)/He thermochronology: Correlation with regional tectonics. *Terra Nova* **2006**, *18*, 18–25. [CrossRef]
110. Réhault, J.P.; Boillot, G.; Mauffret, A. The western Mediterranean basin geological evolution. *Mar. Geol.* **1984**, *55*, 447.
111. Jolivet, L.; Faccenna, C. Mediterranean extension and the Africa-Eurasia collision. *Tectonics* **2000**, *19*, 1095–1106. [CrossRef]
112. Edel, J.B.; Dubois, D.; Marchant, R.; Hernandez, J.; Cosca, M. La rotation miocène inférieur du bloc corso-sarde; nouvelles contraintes paléomagnétiques sur la fin du mouvement. *Bull. Soc. Géol. France* **2001**, *172*, 275–283. [CrossRef]
113. Schmid, S.M.; Kissling, E. The arc of the Western Alps in the light of geophysical data on deep crustal structure. *Tectonics* **2000**, *19*, 62–85. [CrossRef]
114. Falletti, P.; Gelati, R.; Rogledi, S. Oligo-Miocene evolution of Monferrato and Langhe, related to deep structures. In *Atti del Convegno "Rapporti Alpi-Appennino" e Guide alle Escursioni*; Peveragno (CN) 31 Maggio–1 Giugno 1994; Polino, R., Sacchi, R., Eds.; Accademia Nazionale delleScienze: Verona, Italy, 1995; Volume 14, pp. 1–19.
115. Festa, A.; Piana, F.; Dela Pierre, F.; Malusà, M.G.; Mosca, P.; Polino, R. Oligocene-Neogene kinematic constraints in the retroforeland basin of the northwestern Alps. *Rend. Soc. Geol. Ital.* **2005**, *1*, 107–108.
116. Piana, F.; Tallone, S.; Cavagna, S.; Conti, A. Thrusting and faulting in metamorphic and sedimentary units of Ligurian Alps: An example of integrated field work and geochemical analyses. *Int. J. Earth Sci.* **2006**, *95*, 413–430. [CrossRef]
117. Maffione, M.; Speranza, F.; Faccenna, C.; Cascella, A.; Vignaroli, G.; Sagnotti, L. A synchronous Alpine and Corsica-Sardinia rotation. *J. Geophys. Res.* **2008**, *113*, 10–25. [CrossRef]
118. Carrapa, B.; Bertotti, G.; Krijgsman, W. Subsidence, stress regime and rotation (s) of a tectonically active sedimentary basin within the western Alpine Orogen: The Tertiary Piedmont Basin (Alpine domain, NW Italy). *Geol. Soc. Lond. Spec. Publ.* **2003**, *208*, 205–227. [CrossRef]
119. Lefèvre, R. La structure et le style tectonique de la bande d'Acceglio en Val Maira (Alpes cottiennes italiennes). *Géol. Alpine* **1968**, *44*, 27–36.
120. Schwartz, S.; Lardeaux, J.M.; Tricart, P. La zone d'Acceglio (Alpes Cottiennes): Un nouvel exemple de croûte continentale éclogitisée dans les Alpes occidentales. *C. R. Acad. Sci.* **2000**, *330*, 859–866. [CrossRef]

geosciences

MDPI

Article

Reconsidering the Variscan Basement of Southern Tuscany (Inner Northern Apennines)

Enrico Capezzuoli [1,*], Amalia Spina [2], Andrea Brogi [3,4], Domenico Liotta [3,4], Gabriella Bagnoli [5], Martina Zucchi [3], Giancarlo Molli [5] and Renzo Regoli [6]

1 Dipartimento di Scienze della Terra, Università di Firenze, 50121 Firenze, Italy
2 Dipartimento di Scienze fisiche e della Terra, Università di Perugia, 06123 Perugia, Italy; amalia.spina@unipg.it
3 Dipartimento di Scienze della Terra e Geoambientali, Università di Bari, 70121 Bari, Italy; andrea.brogi@uniba.it (A.B.); domenico.liotta@uniba.it (D.L.); martina.zucchi@uniba.it (M.Z.)
4 CNR-IGG, 56124 Pisa, Italy
5 Dipartimento di Scienze della Terra, Università di Pisa, 56126 Pisa, Italy; gabriella.bagnoli@unipi.it (G.B.); giancarlo.molli@unipi.it (G.M.)
6 Associazione Mineralogica Paleontologica Senese, 53100 Siena, Italy; renzoregoli@gmail.com
* Correspondence: enrico.capezzuoli@unifi.it; Tel.: +39-0552757530

Abstract: The Pre-Mesozoic units exposed in the inner Northern Apennines mostly consist of Pennsylvanian-Permian successions unconformably deposited on a continental crust consolidated at the end of the Variscan orogenic cycle (Silurian-Carboniferous). In the inner Northern Apennines, exposures of this continental crust, Cambrian?-Devonian in age, have been described in Northern Tuscany, Elba Island (Tuscan Archipelago) and, partly, in scattered and isolated outcrops of southern Tuscany. This paper reappraises the most significant succession (i.e., Risanguigno Formation) exposed in southern Tuscany and considered by most authors as part of the Variscan Basement. New stratigraphic and structural studies, coupled with analyses of the organic matter content, allow us to refine the age of the Risanguigno Fm and its geological setting and evolution. Based on the low diversification of palynoflora, the content of sporomorphs, the structural setting and the new field study, this formation is dated as late Tournaisian to Visean (Middle Mississippian) and is not affected by pre-Alpine deformation. This conclusion, together with the already existing data, clearly indicate that no exposures of rocks involved in the Variscan orogenesis occur in southern Tuscany.

Keywords: northern Apennines; Risanguigno Formation; Carboniferous; southern Tuscany; Monticiano-Roccastrada Unit; Tuscan Palaeozoic; palynology

check for updates

Citation: Capezzuoli, E.; Spina, A.; Brogi, A.; Liotta, D.; Bagnoli, G.; Zucchi, M.; Molli, G.; Regoli, R. Reconsidering the Variscan Basement of Southern Tuscany (Inner Northern Apennines). *Geosciences* **2021**, *11*, 84. https://doi.org/10.3390/geosciences11020084

Academic Editors: Jesus Martinez-Frias and Rodolfo Carosi

Received: 5 January 2021
Accepted: 7 February 2021
Published: 12 February 2021

Publisher's Note: MDPI stays neutral with regard to jurisdictional claims in published maps and institutional affiliations.

Copyright: © 2021 by the authors. Licensee MDPI, Basel, Switzerland. This article is an open access article distributed under the terms and conditions of the Creative Commons Attribution (CC BY) license (https://creativecommons.org/licenses/by/4.0/).

1. Introduction

Stratigraphic reconstructions of the deep successions involved in orogens later affected by post-collisional extensional tectonics are always tempting, since these are normally metamorphosed and involved in polyphase deformation, are laterally segmented and, consequently, are exposed in scattered outcrops. This is even crucial for the metamorphosed, deep successions of the Northern Apennines [1], which experienced the Variscan sedimentary and tectonic evolution (Devonian-Carboniferous), then the Alpine cycle (Triassic-Oligocene), and ultimately the extensional process leading to the opening of the Tyrrhenian Basin (Miocene-Quaternary). Nowadays, the so-called Tuscan Crystalline Basement (Cambrian?-Devonian [2]) is discontinuously exposed, and the scarcity of fossils remains inhibits precise age determination [3–5]. Thus, in absence of fossil records, the Tuscan Basement is traditionally related to the well-known and better exposed Palaeozoic succession of southeastern Sardinia, where the Alpine deformation is relatively minor [6–9].

To strengthen this approach, several studies of the pre-Alpine metamorphic rocks of the Tuscan Archipelago and Apuan Alps have incorporated palaeontological, stratigraphic and tectonic data [2,9–14]. On the other hand, a few studies have focused on southern Tuscany, where contrasting interpretations on datings and palaeogeographic reconstructions have been proposed [15–20]. More recently, radiometric (Ar/Ar, U/Th [21–23]) and palynological studies [24,25] have served to motivate a review of the entire Tuscan Palaeozoic successions based on bio/chronological markers. These studies contribute more precise age datings and provide new evolutionary scenarios in the context of two distinct Palaeozoic cycles (Mississippian-early Permian and middle-late Permian) [5,26]. Accordingly, we re-consider the Risanguigno Formation, which is regarded as part of the Tuscan Crystalline Basement and constitutes the oldest outcrops in southern Tuscany. In this view, this isolated and scarcely studied formation strongly influenced the reconstruction of the entire Pre-Alpine Apenninic succession. Therefore, the aim of this paper is to document and illustrate a newly discovered palynofloral content and, consequently, to provide the precise age of the Risanguigno Fm, together with its structural setting. This will lead to determination of Variscan formations in this key sector of Northern Apennines from a stratigraphic and palaeogeographic perspective.

2. Geological Outline of the Palaeozoic Units of Tuscany

The inner Northern Apennines (Figure 1) resulted from the convergence (Cretaceous-Eocene) and collision (Oligocene-early Miocene) between the European Corsica-Sardinia massif and the Adria microplate of the Africa pertinence. This process produced the stacking of tectonic units deriving from oceanic and continental palaeogeographic domains [27,28].

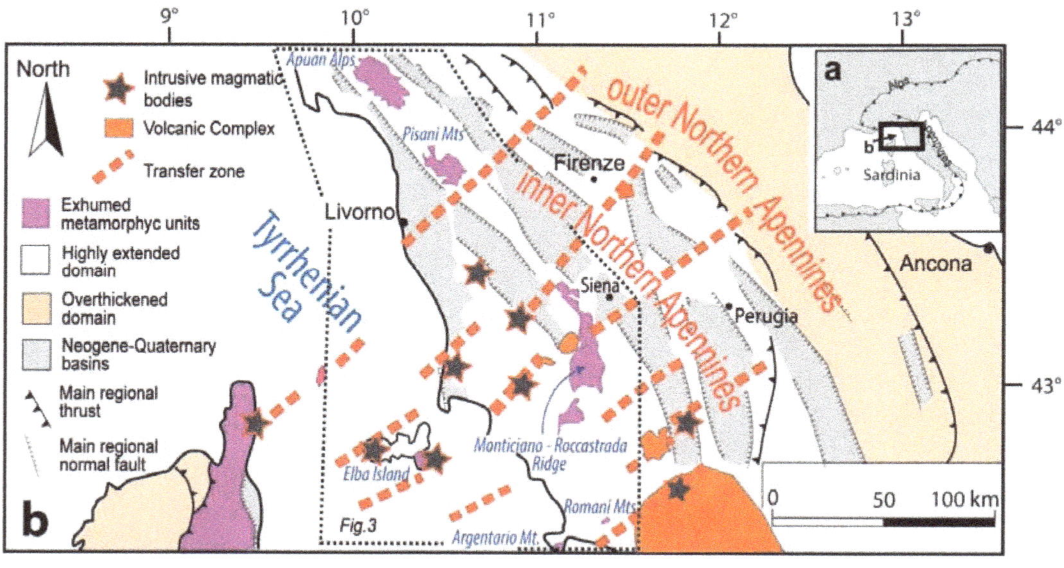

Figure 1. Structural sketch map of (**a**) the Northern Apennines and (**b**) Northern Tyrrhenian Sea.

In southern Tuscany these are, from top to bottom (Figure 2): (a) the Ligurian and Sub-Ligurian Units, consisting of remnants of Jurassic oceanic and transitional crust and their related Cretaceous–Oligocene sedimentary cover; (b) the Tuscan Units including the Triassic-early Miocene sedimentary (Tuscan Nappe) and Palaeozoic-Triassic metamorphic succession. According to [29,30], this metamorphic succession can be broadly subdivided in (i) a late Cambrian?-Mississippian basement (affected by deformation during the Variscan orogenesis) and (ii) a Late Pennsylvanian to Triassic sedimentary cover, deposited during the Variscan post-collisional events.

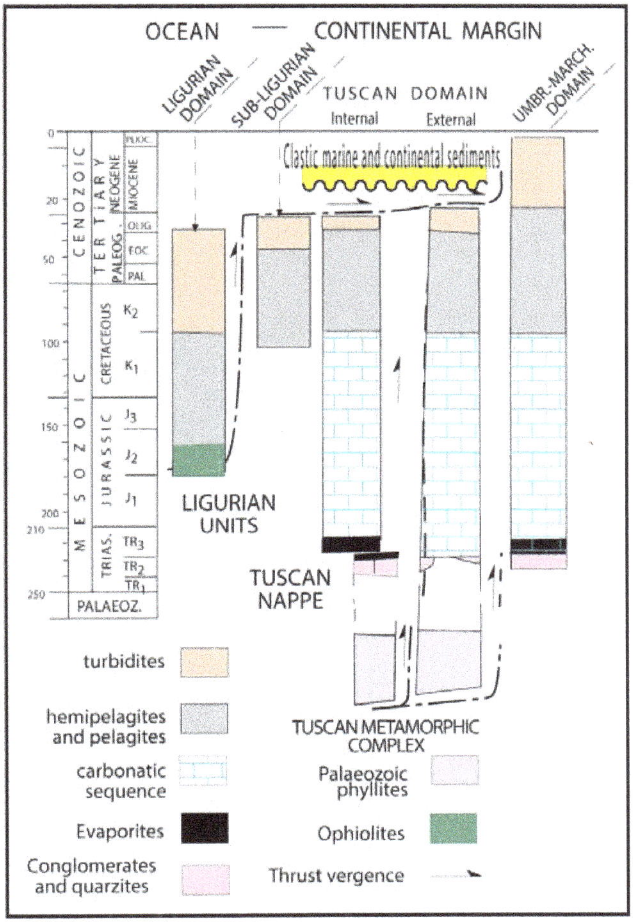

Figure 2. Tectono-stratigraphic columns illustrating the main features of the paleogeographic domains of the inner Northern Apennines (redrawn from [27,31]).

The Palaeozoic basement is made up of quartzites and phyllite with acidic to intermediate metavolcanic rock (porphyroid). At their top, black shale, radiolarite (lydian stone) and metacarbonate deposits (dolostone, calcschist) have been detected [4,10,32–34]. This succession, attributed to the Cambrian?-Devonian, on the basis of scattered fossils [4] and U/Th radiometric dating [21–23], is classically related to the central-southern Sardinia succession [35] and is considered to be involved in the Variscan orogeny during the early Carboniferous [12]. In the Inner Northern Apennines, Palaeozoic rocks involved in the Variscan deformation extensively crop out in the La Spezia-Apuan Alps-Mt. Pisani area, while smaller exposures are located in the Tuscan Archipelago (Elba island) and southern Tuscany (Figure 3). It is noteworthy that some deep wells in northern Tuscany (Pontremoli) and in the geothermal area of southern Tuscany are believed to have intersected deformed Variscan rocks [2,8,9,12,14,36–40].

The "post-Variscan" Palaeozoic-Triassic sedimentary succession (referable to the Phyllite-Quartzitic Group of [41]) is mostly exposed in the Monticiano-Roccastrada Unit (Figures 3 and 4), along the Middle Tuscan Ridge, in three different main tectonic units, as defined by [19]: Iano Sub-Unit 1; Monte Quoio-Montagnola Senese Sub-Unit 2; Monte Leoni-Farma Sub-Unit 3—Figure 4. Only minor outcrops are present elsewhere, and locally drilled by boreholes [1,9,10,37,42–46].

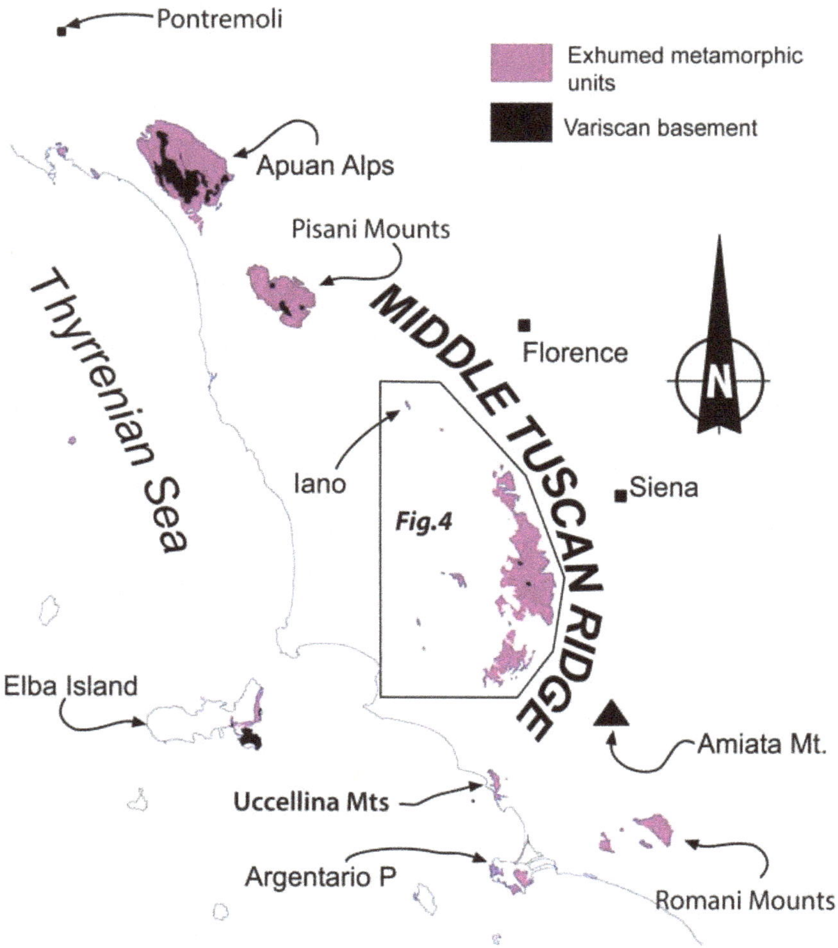

Figure 3. Distribution of metamorphic units in Tuscany, including Variscan deposits (in black).

This succession is formed by phyllite, metasandstone, metaconglomerate with local carbonate levels attributed to Mississippian-late Permian on the basis of the fossil [16,18,44,47–49], palynoflora content [24,25] and radiometric dating [23]. Its evolution is related to rifting [1], transcurrent/transtensive pull-apart basins [5,50] or to late Variscan compressional events [9,19].

The uppermost part of the succession is represented by the typical Triassic continental quartz-dominated clastic sedimentation belonging to the Verrucano Group [51–53].

During the Apennines collisional stages, the above-mentioned Palaeozoic-Triassic successions were involved in duplex structures, up to HP-LT conditions ($P \geq 1.1$ GPa and $T \sim 350$–400 °C) and retrograde green schist metamorphic conditions [54–61]. Their exhumation was favoured by the development of Miocene extensional detachments [26,62], which produced extensional horses (i.e., megaboudins [63]) and the lateral segmentation of the previously stacked tectonic units.

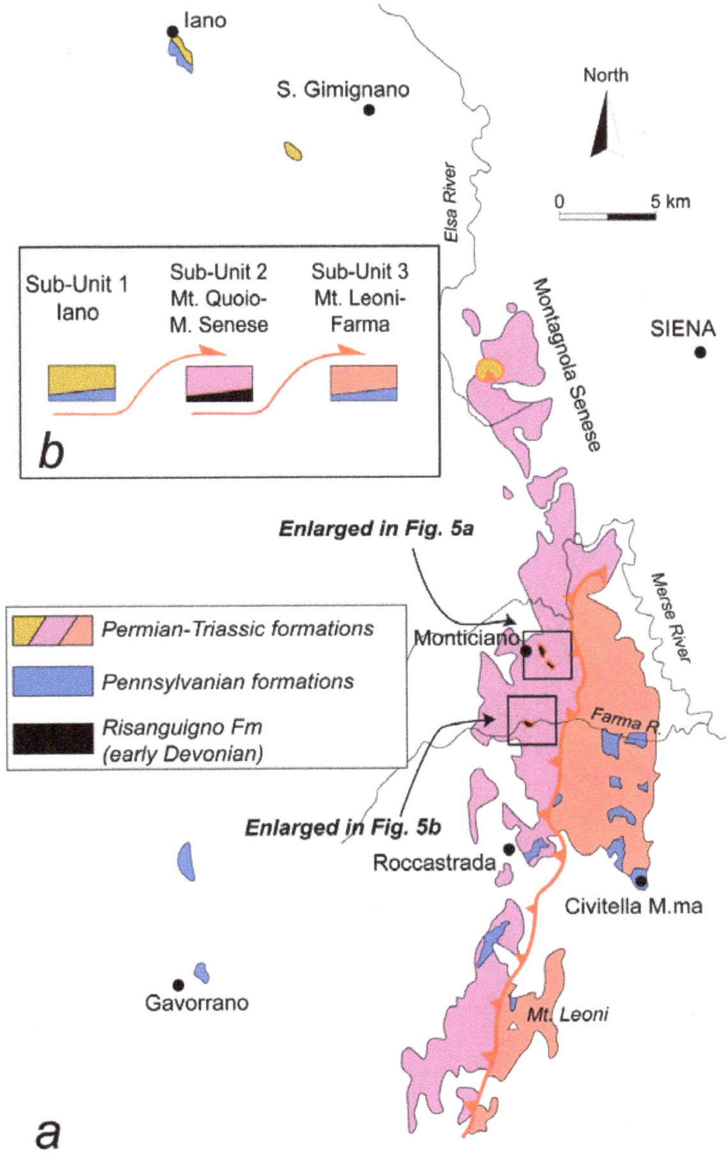

Figure 4. (**a**) Schematic sketch of the Middle Tuscan Ridge with geographic distribution of the three structural Sub-Units; (**b**) Simplified stratigraphy and tectonic relation among the three Sub-Units (redrawn from [64]).

Variscan Basement in Southern Tuscany: The Risanguigno Formation

In this framework, the Risanguigno Fm represents the only cropping out unit in southern Tuscany assigned to the Palaeozoic basement. It is part of Sub-Unit 2 (Monte Quoio-Montagnola Senese) of the Monticiano-Roccastrada Unit (Figure 4).

Such a formation was initially defined in the type locality of the Risanguigno Creek by [4]. These main exposures were previously described by [47,65], although interpreted as part of another formation (Boccheggiano Fm). Subsequently, [20,66,67] related other outcrops exposed in the surroundings (Farma River, Figures 4 and 5) to the Risanguigno Fm, furthermore recognized in a few boreholes [68].

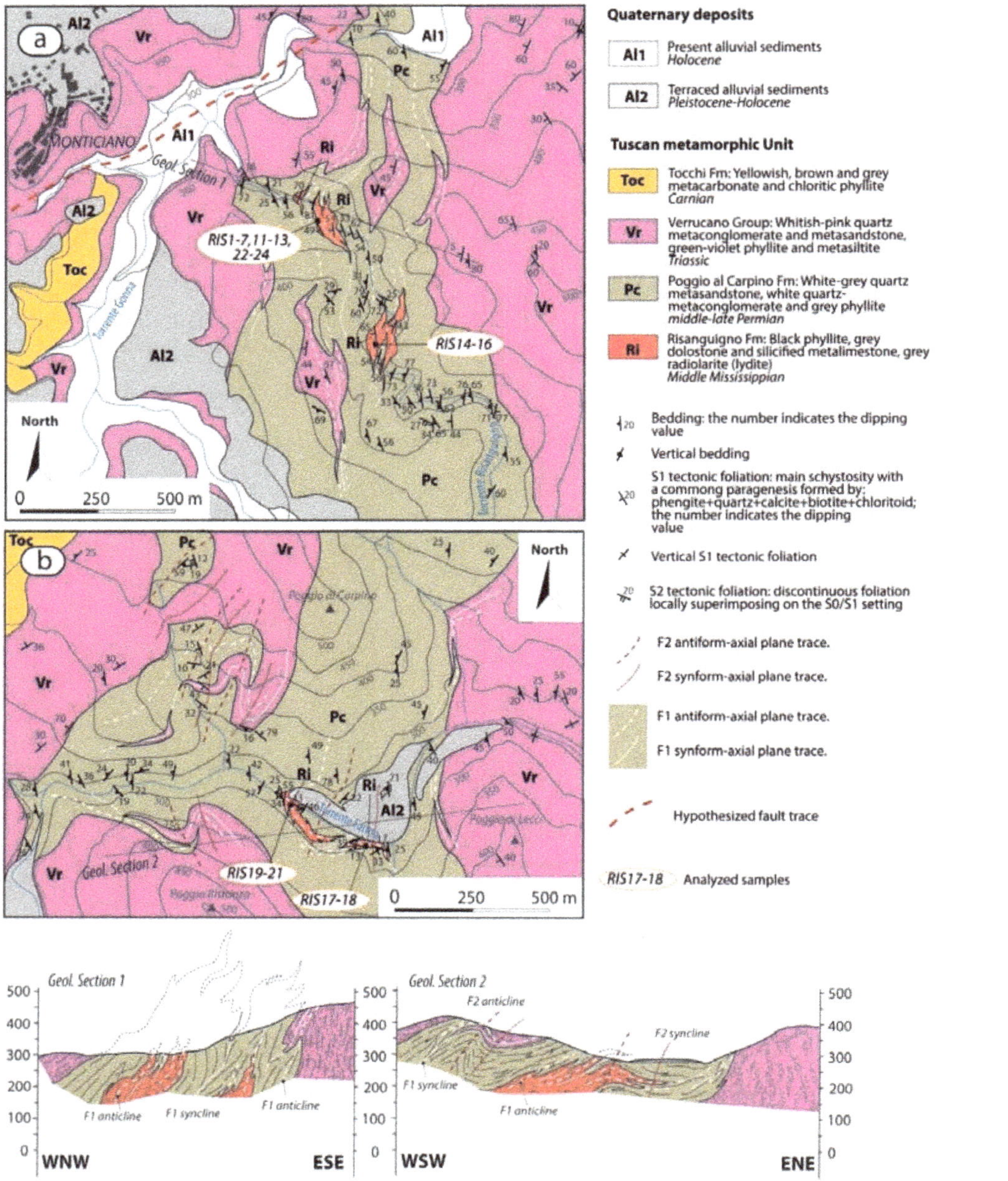

Figure 5. New geological maps of the study areas and related geological cross sections: (**a**) Risanguigno Creek; (**b**) Farma River (previous maps from [20,64,65]). See Figure 4 for their location.

The base of the formation is never exposed, although [20] postulated the presence of a basal stratigraphic unconformity separating the Risanguigno Fm from the underlying Variscan deformed units.

At the top, the Risanguigno Fm is in contact with the Poggio al Carpino Fm [64], a middle-late Permian [19,25,69] clast- to matrix-supported polymictic conglomerate, often alternated with grey quartzose sandstone and subordinate dark grey phyllite [70]. The contact between the Risanguigno and Poggio al Carpino formations is described as an angular unconformity by [18], in contrast with [20], indicating that the Poggio al Carpino Fm stratigraphically overlies the lower unit.

From a sedimentary point of view [2,18,19], the Risanguigno Fm is composed by black-grey graphitic to bituminous phyllite intercalated with cm- to dm-thick alternations of: (i) grey-greenish to black quartzose, granolepidoblastic metasandstone and siltstone with iron-rich carbonate matrix and detritic mica, (ii) cm-thick microcrystalline, granoblastic dolostone rich in detritic quartz and white mica, (iii) silicified grey metalimestone, and (iv) thinly bedded, grey-greenish to black chert and radiolarian lydite. A chert sub-sequence, up to 4.5 m thick and intercalated with fine-grained clastics, was also recognized in close outcrops by [66], and later correlated with the small chert sequence present also in the Risanguigno type locality [20]. Anhydrite in the silicified limestone is reported by [17,65], while this is not described by [68]. Local post-tectonic chloritoid needles are reported in the metasandstones and metasiltstones by [2].

Rocks are strongly deformed, making the stratigraphic reconstruction difficult. By this, and due to the fact that the basal contact is not exposed, the thickness of the Risanguigno Fm is unknown and only inferred in 40 m, at least [20].

Regarding the fossiliferous content, [4,71] reported a conodont fauna, characterized by *Ozarkodina denckmanni*, *Panderodus unicostatus* and *Icriodus* sp. This fauna was recovered from the dolostone levels in the type locality at the altitude of 304 m along the Risanguigno Creek.

Regarding the chert-subsequence, [4,68] accounted for the presence of recrystallized radiolaria, often well preserved although flattened during deformation.

The formation, originally attributed to a generic Carboniferous by [47,65], was ascribed to the Early Devonian on the basis of the conodont fauna [4]. Alternatively, [66] suggested a Tournaisian-Visean age based on the radiolaria observed in the chert sub-sequence, while [17,20] related these siliceous portions to late Devonian-early Carboniferous (late Emsian to Visean?) on the basis of the lithological correlation with similar deposits in the circum-Mediterranean area.

Similarly, the interpretation of the depositional environment is matter of debate. Ref. [4] proposed a shallow marine origin, while [17] favoured a moderately deep water basin origin, owing to the presence of the siliceous portions. In contrast, [71] suggested an epicontinental shelf characterized by recurrent anoxic conditions, while [20] accounted for a highly condensed sequence deposited in a starved, low energetic, distal and relatively deep marine environment.

From a tectonic point of view, the Risanguigno Fm is described as intensely deformed and marked by a metamorphic grade higher than the one affecting the overlying formation, i.e., the Poggio al Carpino Fm [4]. In this view, according to [9,37], the Risanguigno Fm evidences relics of a pre-Alpine deformation, interpreted as a Variscan syn-metamorphic tectonic foliation relatable to the Sudetic event.

3. Materials and Methods

A detailed field survey was carried out in key areas where the Risanguigno Fm is exposed. The fieldwork was dedicated to field mapping and data collection for describing the deformation affecting the Risanguigno Fm and the overlying units. During the survey, 21 samples of black phyllite, metasiltstone and metacarbonate (16 samples from the Risanguigno Creek and 5 samples from the Farma River (Table 1) were collected for petrographic, microfacies and biostratigraphic studies. Palynological samples (c. 20 g each for phyllite and metasiltstone lithologies and 100 g each for metacarbonate samples) were treated by standard palynological acid maceration (with 37% HCl, 50% HF, boiling HCl 10%), density separation of the organic matter (using a $ZnCl_2$ solution) and filtration of the organic-rich residue at 10 µm. As a result of the high degree of thermal alteration, the organic residue was treated with Schultz solution and filtered with 10 µm sieve.

Table 1. Analysed samples, with related geographical coordinates, lithology and content (quotes in meter above sea level—m a.s.l.).

Sample	Locality	Quote	Latitude	Longitude	Lithology	Analysis	Content
RIS 1	Risanguigno Creek	304 m a.s.l.	43°8′3.85″ N	11°11′38.13″ E	Black phyllite	Palynology	productive
RIS 2	Risanguigno Creek	304 m a.s.l.	43°8′4.03″ N	11°11′37.84″ E	Fine metasand-stone	Palynology	barren
RIS 3	Risanguigno Creek	304 m a.s.l.	43°8′4.34″ N	11°11′33.84″ E	Black phyllite	Palynology	productive
RIS 4	Risanguigno Creek	304 m a.s.l.	43°8′4.43″ N	11°11′33.74″ E	Black phyllite	Palynology	productive
RIS 5	Risanguigno Creek	304 m a.s.l.	43°8′8.01″ N	11°11′32.15″ E	Black phyllite and lidyte	Palynology	productive
RIS 6	Risanguigno Creek	304 m a.s.l.	43°8′7.95″ N	11°11′32.12″ E	Black phyllite and lidyte	Palynology	productive
RIS 7	Risanguigno Creek	304 m a.s.l.	43°8′7.98″ N	11°11′32.11″ E	Black phyllite	Palynology	productive
RIS 11	Risanguigno Creek	304 m a.s.l.	43°8′3.77″ N	11°11′40.46″ E	Black phyllite	Palynology	barren
RIS 12	Risanguigno Creek	304 m a.s.l.	43°8′4.72″ N	11°11′34.02″ E	Dolostone	Palynology	barren
RIS 13	Risanguigno Creek	304 m a.s.l.	43°8′7.76″ N	11°11′31.90″ E	Dolostone	Palynology	barren
RIS 14	Risanguigno Creek	324 m a.s.l.	43°7′46.76″ N	11°11′43.98″ E	Black phyllite	Palynology	barren
RIS 15	Risanguigno Creek	324 m a.s.l.	43°7′47.91″ N	11°11′43.53″ E	Black phyllite	Palynology	productive
RIS 16	Risanguigno Creek	324 m a.s.l.	43°7′47.95″ N	11°11′43.44″ E	Black phyllite	Palynology	productive
RIS 17	Farma River	265 m a.s.l.	43°5′15.41″ N	11°11′23.57″ E	Metasiltstone	Palynology	productive
RIS 18	Farma River	265 m a.s.l.	43°5′15.32″ N	11°11′24.59″ E	Black phyllite	Palynology	productive
RIS 19	Farma River	270 m a.s.l.	43°5′20.19″ N	11°11′13.15″ E	Black phyllite	Palynology	barren
RIS 20	Farma River	270 m a.s.l.	43°5′20.13″ N	11°11′13.22″ E	Metacarbonate	Palynology	barren
RIS 21	Farma River	270 m a.s.l.	43°5′20.10″ N	11°11′13.28″ E	Metacarbonate	Palynology/ Conodonts	barren
RIS 22	Risanguigno Creek	304 m a.s.l.	43°8′4.22″ N	11°11′33.97″ E	Dolostone	Palynology/ Conodonts	barren
RIS 23	Risanguigno Creek	304 m a.s.l.	43°8′4.28″ N	11°11′33.79″ E	Dolostone	Palynology/ Conodonts	barren
RIS 24	Risanguigno Creek	304 m a.s.l.	43°8′4.18″ N	11°11′35.98″ E	Dolostone	Palynology/ Conodonts	barren

Light microscope observations were performed on palynological slides using a Leica DM1000 microscope (Leica, Wetzlar, Germany) using the differential interference contrast technique in transmitted light. Images were captured using the camera on the digital microscope and successively corrected for contrast and brightness using the open-source Gimp software. The palynological slides are stored at the Sedimentary Organic Matter Laboratory of the Department of Physics and Geology, University of Perugia, Italy. Metacarbonate samples were collected from dolostone levels for the analyses of conodont content and processed by standard procedures using 10% acetic acid. The residue was washed through a 71 μm sieve.

4. Results

The results are summarized in different sections, according to the main issues of lithology, fossil content, and deformation.

4.1. Lithological Characteristics

The outcrops exposed in the Risanguigno Creek and Farma River were revisited (Figure 5).

In the Risanguigno Creek, the formation crops out in two small windows (quote 304 m and quote 324 m a.s.l.) in correspondence of the riverbed (Figure 5a). A small supplementary outcrop, never described before, was discovered along the riverbed at quote 300 m. The Risanguigno Fm is mostly dominated by black to grey phyllite, locally intercalated by cm-thick level and lenses of metasandstone and metasiltstone (Figure 6a,b). Only in the outcrop of quote 304 m is phyllite intercalated with cm-thick beds and lenses of microcrystalline dolostone and silicified grey metacarbonate (Figure 6c). These are geometrically positioned below a small succession (max 2 m thick) displaying alternation of phyllite and chert beds/lydite in thinly bedded laminae (Figure 6d). The transition from the Risanguigno Fm to the overlying grey quartzose metasandstone and metaconglomerate formation (Poggio al Carpino Fm) is marked by a sharp angular unconformity (Figure 6e).

Along the Farma River, outcrops are located close at the Ferriera locality, on the right bank of the riverbed at quote 270 m and 265 m a.s.l. (Figure 5b). Black bituminous phyllite, locally smelly and rich in millimetric-sized crystals of pyrite, is the dominant lithotype. Conversely, metasandstone and metasiltstone, as also dolostone and metacarbonate, are less diffuse. The 4.5-m-thick chert sequence evidenced by [66] constitutes the main lithological variation and morphological prominence (Figure 6f,g). Similarly to the Risanguigno Creek, here, also, the chert beds are positioned at the top of the succession, immediately below the Poggio al Carpino Fm.

In both valleys, the complete stratigraphic reconstruction is prevented by an intense folding (see the next paragraph).

4.2. Fossiliferous Content

Twenty-one samples were obtained from almost all the analysed outcrops. In the Risanguigno Creek, three samples were from quote 324 (RIS14, 15, 16) and thirteen from quote 304 (RIS1, 2, 3, 4, 5, 6, 7, 11, 12, 13, 22, 23, 24). In the Farma River, two samples were from quote 265 (RIS17,18) and three from quote 270 (RIS19, 20, 21). All of them were analysed for their palynological content. Four samples were analysed for conodont content (Table 1).

4.2.1. Palynological Content

Ten samples out of 21 were productive, even yielding strongly degraded palynofloral assemblages (Table 1). Degradation was mainly due to intense in situ pyritization affecting the exine of miospores. A low metamorphic grade with a temperature of about 350–400 °C [56] was recognized. Nonetheless, different organic microfossils were recognized, adding new data for the age determination of the Risanguigno Fm. The palynological assemblage mainly consists of ornamented forms as *Auroraspora balteola, Claytonispora distincta, Retialetes radforthii, Vallatisporites? hystricosus, Perotrilites magnus, Spelaeotriletes balteatus, S. pretiosus* and *Grandispora* sp. Different tetrads of indeterminate apiculate spores also occur (Figure 7).

Figure 6. Lithological characteristics of the Risanguigno Fm: (**a**) view of the dominant black phyllite; (**b**) thin intercalations of metasiltstone lenses; (**c**) examples of metacarbonate beds; (**d**) example of thin laminae and beds of lydite; (**e**) upper contact of the Risanguigno Fm with the Poggio al Carpino Fm with evidence of the angular unconformity; (**f**) chert sequence cropping out along the Farma River; (**g**) detail of thinly laminated chert sequence.

4.2.2. Conodont Content

All processed samples were barren in terms of conodont content.

4.3. Deformation

The Risanguigno Fm, together with the overlying units (Permian Poggio al Carpino Fm and Triassic Verrucano Group) exposed in the Risanguigno Creek and Farma River (Figure 5), are commonly involved in polyphase folding characterized by superposed F_1 and F_2 folds, with NS and NS-NNE axial trends, respectively.

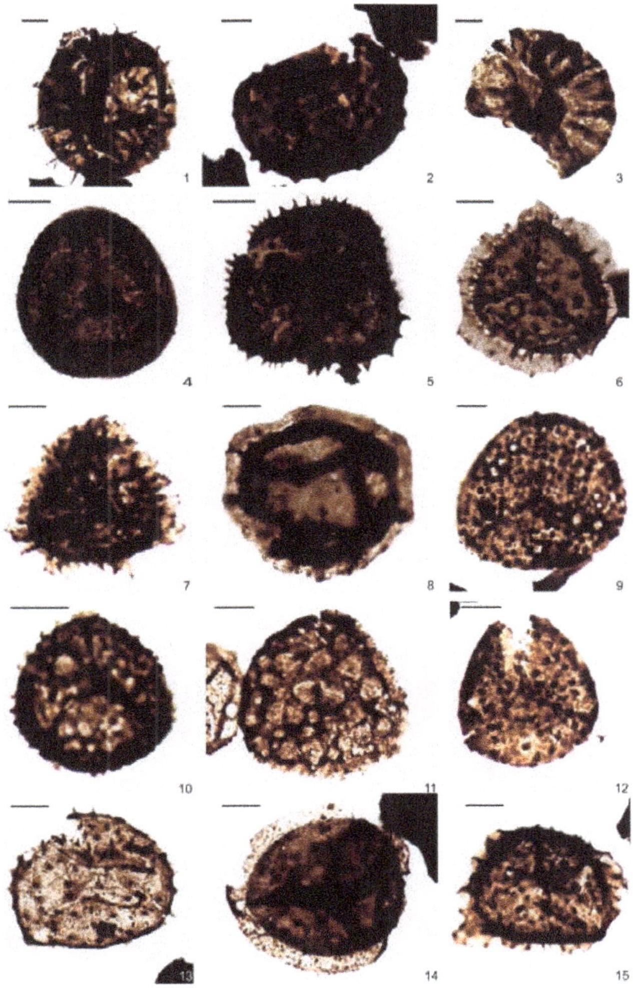

Figure 7. Miospores from Risanguigno Formation. Scale bar indicates 10 μm. (**1**) *Claytonispora distincta* Playford and Melo 2012 (slide: RIS 3). (**2**) *Pustulatisporites* sp. (slide: RIS 3). (**3**) *Retialetes radforthii* Staplin 1960 (slide: RIS 15). (**4,15**) Indeterminate miospore (slide: RIS 15). (**5**) Tetrad of indeterminate apiculate spores (slide: RIS 3). (**6,7**). *Vallatisporites? hystricosus* (Winslow) Byvscheva 1985 (slide: RIS 15). (**8**) *Auroraspora balteola* Sullivan 1964 (slide: RIS 3); (**9**) *Spelaeotriletes balteatus* (Playford) Higgs 1975 (slide: RIS 15). (**10,11**) Indeterminate spore with a heavily pyritized exine. (slide: RIS 15). (**12**) *Spelaeotriletes pretiosus* (Playford) Neves and Belt 1970 (slide: RIS 18). (**13**) *Grandispora* sp. (slide: RIS 15). (**14**) *Perotrilites magnus* Hughes and Playford 1961 (slide: RIS 15).

Both folding events are referrable to the Alpine evolution. F1 folds are the prominent structures and involve the entire succession, defining the main shape and geometries of the exposures (Figures 8 and 9). F1 folds range from map-scale to outcrop-scale and have hectometre- to decimetre sizes (Figure 8). These consist of tight and isoclinal recumbent folds, with axial planes steeply dipping toward west. F1 hinge lines mostly dip gently toward S-SE (Figure 8), although in some rare cases they dip toward N-NW (Figure 9).

Figure 8. N-S trending F_1 folds affecting the metasandstone and metapelite succession of the Poggio al Carpino Fm exposed in the Risanguigno Creek. (**a**) F1 sub-isoclinal folds and related S1 axial planar tectonic foliation and related stereographic diagram (lower hemisphere, equiareal diagram); (**b**,**c**) enlarged sector of the F_1 hinge zones indicated in (**a**) and showing the S_0/S_1 angular relationships also shown in the stereographic diagram (lower hemisphere, equiareal diagram); (**d**,**e**) hinge zone of F_1 fold with a pervasive axial planar S_1 tectonic foliation developed within the metapelite; relationships between S_0 and S_1 are indicated in the stereographic diagram (lower hemisphere, equiareal diagram).

In the quartz-metasandstone and metaconglomerate, S_1 is a rough cleavage, as highlighted by the differentiated domains when alternating quartzitic and micaceous layers are present. The L_1 object lineation occurs in the phyllite and metasilstone, defined by elongated quartz and mica lenses tracking the x axis of the finite strain ellipse.

Figure 9. Macroscopic scale deformation pattern of the phyllite and metacarbonate belonging to the Risanguigno Fm. (**a**) detail of the contact separating the Risanguigno Fm from the Poggio al Carpino Fm and geometrical relationships between S_0/S_1 and S_2 foliations (see the text for more details) also indicated in the stereographic diagram (lower hemisphere, equiareal diagram). (**b**) Penetrative S1 foliation crossing the metaconglomerate level belonging to the basal part of the Poggio al Carpino Fm. (**c**) Centimetre-scale isoclinal F_1 folds and the S_1 axial planar tectonic foliation crossed by the S_2 tectonic foliation. (**d**,**e**) F_2 open folds affecting the S_0/S_1 foliations affecting phyllite and metacarbonate levels.

At the microscopic scale, S_1 relates to a continuous foliation, mainly defined by elongate quartz layers, formed by flattened and dynamically recrystallized grains, alternated with mica-rich domains (Figure 10a,b). Mica domains are mainly composed of fine-grained white mica and biotite (Figure 10a–d) with locally developed chloritoid crystals, grown both along the main foliation and crossing it (Figure 10d).

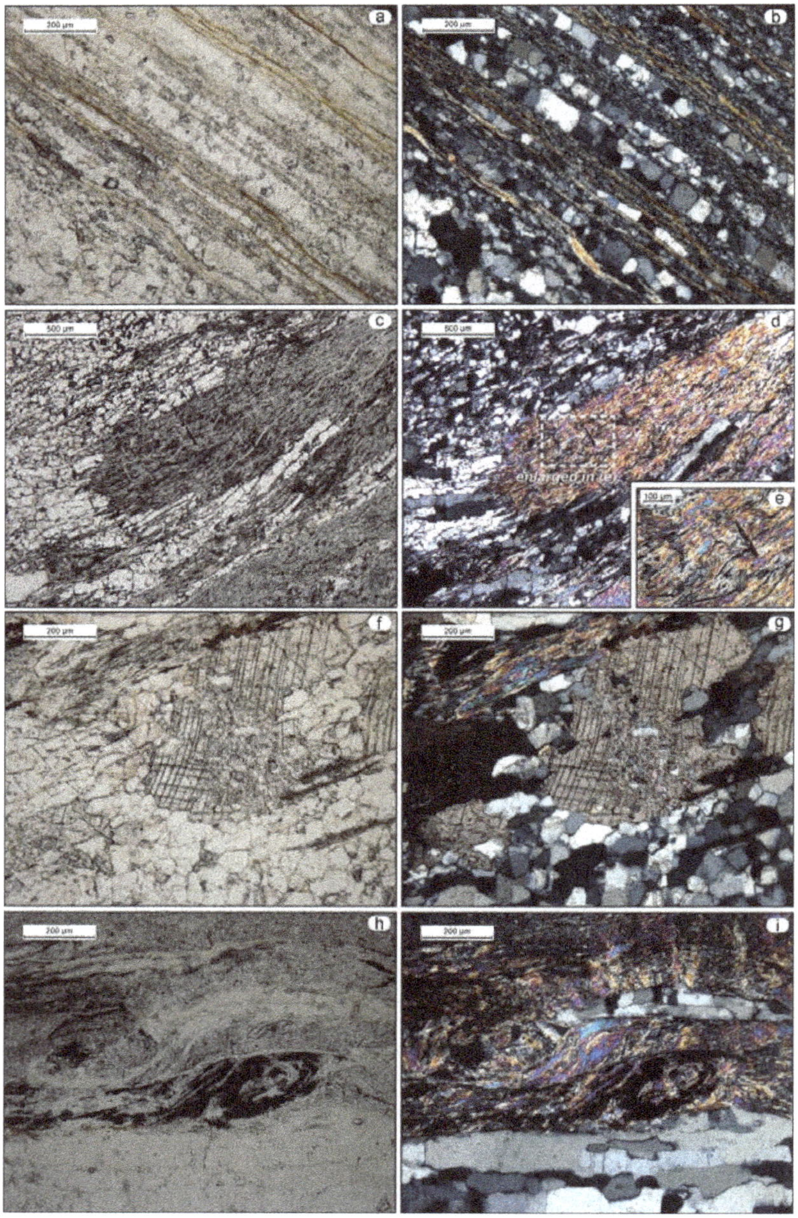

Figure 10. (**a**,**b**) Phyllitic quartzite (Risanguigno Fm) showing the S_1 foliation consisting of metamorphic layering made up of quartz and white mica + biotite levels ((**a**) plane polarized light; (**b**) crossed polars). (**c**,**d**) Microscale F_1 fold with associated S_1 foliation mainly formed by quartz + white mica + biotite + chloritoid. This latter is also represented by post-kinematic bigger crystals (see (**e**)), suggesting syn- and post-S1 development ((**c**) plane polarized light; (**d**) crossed polars). (**f**,**g**) Mineralogical association of the S1 foliation developed within carbonate rich levels, mainly composed by qtz + cc + white mica + biotite + chloritoid ((**f**) plane polarized light; (**g**) crossed polars). (**h**,**i**) s-c shear zone (top-to-the right) affecting the organic matter-bearing phyllite and developed coevally with the S_1 foliation; the latter is affected by a later crenulation cleavage possible related to the F_2 folding event ((**h**) plane polarized light; (**i**) crossed polars).

Syn-kinematic calcite locally developed within the polycrystalline quartz-rich layers (Figure 10e,f). Localized mylonitic layers with mica fish structures (Figure 10 g,h) developed mainly at the boundary between quartz- and mica-dominated domains. F_2 folds are also recognisable both at the map and outcrop scale and display hectometre to decimetre sizes (Figures 8 and 9). F_2 folds deformed F_1 isoclinal folds and their related S_1 axial planar foliation. F_1/F_2 fold interferences have been reconstructed in the Farma Creek area (Figure 8), where F_1 folds affecting the Triassic and Palaeozoic succession have been deformed by top-to-the-East verging F_2 folds. F_2 folds consist of gentle to close folds, in some cases overturned. Axial planes are gently to moderately dipping toward West. F_2 hinge lines are sub-horizontal or deep gently toward SSW (Figure 9).

An axial-planar foliation (S_2) is associated with F_2 folds, well developed only in the metapelite levels (Figure 9). It ranges from spaced disjunctive to a crenulation cleavage. At the microscopic scale, the S_2 consists of a spaced foliation often producing zonal crenulation cleavage defined by symmetric or asymmetric microfolds.

5. Discussion

The newly obtained data, especially from the bio-chronological perspective, allow us to frame the Risanguigno Fm in a new scenario with fallouts in the Palaeozoic palaeogeography of Gondwana. We describe this in the following sections.

5.1. New Bio-Chronological Framework of the Risanguigno Fm

The palynological assemblage shows similar compositional characteristics to those documented in the Mississippian successions of Western Europe, northern Gondwana and other areas.

Auroraspora balteola was documented in the mid-Visean within the *Knoxisporites triradiatus-Knoxisporites stephanephorus* (TS) Zone of Kammquartzite Formation in the Rhenohercynian Zone (Germany; [72]) and in the late Visean of England [73] in assemblage with *Spelaeotriletes pretiosus* in the Tournaisian of eastern Scotland [74]. This last taxon marks the base of the *Spelaeotriletes pretiosus-Raistrickia clavata* (PC) Zone attributed to the late Tournaisian and first described from SW Britain [75] and from Ireland [76]. Later, in the latter country, [77] also documented the PC biozone, characterized by the occurrence of *S. balteatus* and *Claytonispora distincta* within a stratigraphic interval attributed to middle-late Tournaisian on the basis of conodont fauna. In Belgium, the base of the PC biozone occurred within the upper *Siphonodella crenulata* conodont Zone (late Tournaisian, [78]). *Spelaeotriletes pretiosus* was also reported in assemblage with *S. balteatus* from other Mississippian sequences of Western Europe [79–82], North America [83–86] and China [87]. The species was also documented from similar-aged rocks in some regions of Northern Gondwana. In particular, in North Africa, [88,89] considered microfloristic assemblage marked by the occurrence of *S. pretiosus*, *S. balteatus* and *Vallatisporites vallatus* of late Tournaisian age, without excluding a younger early Visean age. In Algeria, *S. pretiosus* occurred in the middle Tournasian-lower Visean palynozones [90,91]. In Libya, a similar microflora was found in the late Tournaisian-Visean time interval (palynozones XI and XII [92]; palynozones 13 and 14 [93,94]). Analogous palynoflora also occurs in the Tournaisian-early Visean of Saudi Arabia [95,96] and the Central Iranian Basin [97]. In Southeastern Turkey, [98] tentatively correlated the *Spelaeotriletes pretiosus-Aratrisporites saharensis* assemblage, where *Vallatisporites hystricosus* also occurs, with the PC biozone of Western Europe. In Western Gondwana regions, a similar assemblage also characterizes the late middle to early late Tournaisian *Spelaeotriletes pretiosus-Colatisporites decorus* Biozone documented from Brazil [99–105]. On the other hand, in the northern Gondwana regions, *S. balteatus* was also documented in slightly younger time-intervals (e.g., Visean of Libya [89]; Visean of Morocco [106]; Visean-Bashkirian of Saudi Arabia [95,107]).

Regarding the conodont content previously reported by [4], the new investigation carried out in the same levels was not productive. This negative evidence, coupled with the contemporaneous presence of a younger-aged rich microflora, suggests that the previously reported conodonts were reasonably reworked fossils, deriving from older deposits.

Therefore, based on the stratigraphic range of the recorded microflora, we can confirm the age of Risanguigno Fm as being late Tournaisian-Visean, as already suggested by [66] on the basis of radiolarian fossil content.

5.2. Paleoenvironmental Insights

The Risanguigno Fm depositional environment was highly debated in previous studies and alternatively attributed to shallow [4], moderate [7,64] or relatively deep marine environments [20]. The presence of Middle Mississippian metacarbonate/dolostone and siliceous portions (lydite beds) seems consistent with carbonate-to-radiolarite platform environment, also recognized in several lower Carboniferous tectofacies (eastern Southern Alps, Karawanken Mountains, external Dinarides, southern margin of the Pannonian Basin, Aegean islands, Calabria and southern Sardinia [7,107]) of the central Mediterranean area. Accordingly, the lydite deposits do not necessarily require a deep-water environment since these can develop in different depositional areas [108,109], especially if associated with a local silica-enrichment related to volcanic activity in nearby zones [4]. In this view, it is worth remembering that the Variscan evolution was associated with a widespread magmatism during the late Carboniferous [110–112], as well as during the Mississippian [112–115]. On the other hand, the organic-rich property of the phyllite supports the deposition in a starved, oxygen-deficient environment. In fact, the finding of spores characterized by pseudosculpture induced by deposition of pyrite crystals in the wall (exine) interstices is indicative of syn-depositional pyrite, suggesting that the water/sediment interface was in a strongly reducing state [116–119]. Regarding the bathymetric definition of this anoxic environment, the interpretation remains difficult. Nonetheless, the type and morphology of the recovered microflora are indicative of a shallow-marine-to-epicontinental depositional environment: the presence of ornamented spores and tetrads suggest a proximal depositional environment since the spores were selected according to their hydrodynamic equivalence, and the tetrads did not maintain their integrity along the distal direction [120].

Consequently, we interpreted the Risanguigno Fm as being deposited during the Middle Mississippian in a shallow-marine-to-epicontinental setting, characterized by starved, anoxic condition in its lower portion and progressively evolving to carbonate-radiolarite platform. Some authors [121] have evidenced that chert sedimentation dominated during the late Devonian and Mississippian in the tropical Palaeotethys strait, and associated their development with sea-level rise.

The organic-rich deposit could also be related to oceanic anoxic events known for the late Frasnian to Late Mississippian age and influenced by global climatic and oceanographic changes. One of these corresponds to the mid-Tournaisian carbon isotope excursion (TICE) [122,123], as indicated by the largest positive δ13C excursion in the Phanerozoic. This is related to the climatic transition between the Devonian greenhouse and the late Paleozoic ice age [124]. Such TICE was interpreted as being the result of either Oxygen Carbon sequestration in foreland basin deposits (tectonic-sedimentation driver [122,125]) or oxygen minimum zone expansion (marine anoxia driver [126–129]).

5.3. Stratigraphic Setting

The new palynological evidence frames the Risanguigno Fm in the Mississippian, thus implying a reconsideration of the southern Tuscany Palaeozoic setting.

The Risanguigno Fm represented an issue in the lateral juxtaposition with the other southern Tuscany Palaeozoic deposits belonging to the three sub-units of the Monticiano Roccastrada Unit (i.e., Sub-Unit 1: Scisti di Iano Fm—[130,131]; Sub-Unit 3: Calcari di S.Antonio-Scisti a Spirifer formations—[16,18]—Figure 4b). The new attribution of the Risanguigno Fm to the Middle Mississippian implies a stratigraphic correlation with all these Carboniferous deposits as representing different portions of a same marine depositional environment, evolving through time.

In this view, the Risanguigno Fm is interpreted as the older cropping out deposits of the basin. This shallow marine-to-epicontinental setting was progressively evolving, in its upper part, to a Moscovian shale-carbonate deposition (Calcare di S.Antonio Fm—Scisti a Spirifer Fm; [16,18]—Figure 11) and open marine environment during Upper Pennsylvanian (Scisti di Iano Fm. [131]). A similar age (up to lower Permian) is also testified for the continental succession (Scisti di San Lorenzo Fm; [132]) exposed in the northernmost area of Tuscany (Figure 11).

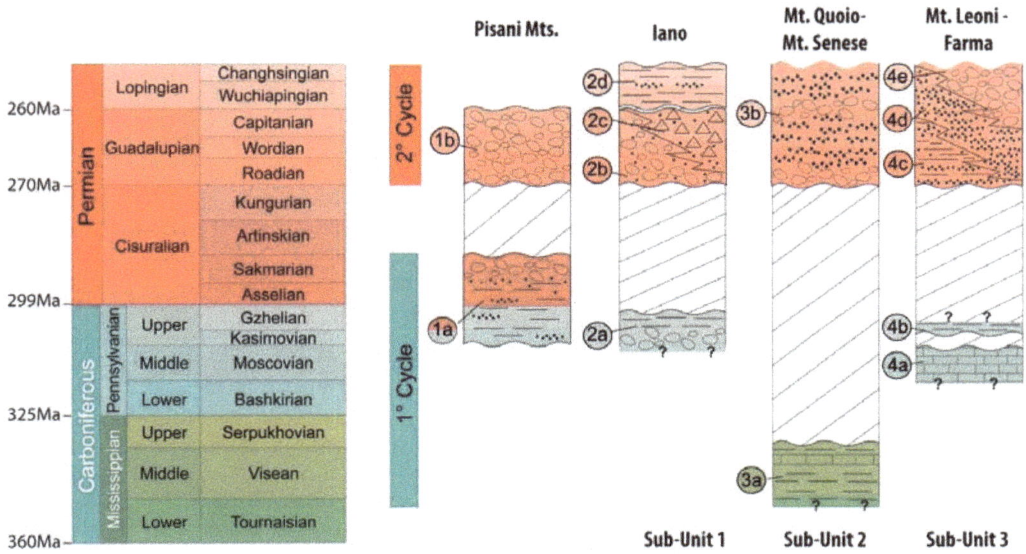

Figure 11. Stratigraphic chart relating the different successions of the Monticiano-Roccastrada Unit located in the Middle Tuscan Ridge (for location, see Figure 4a): Pisani Mts: 1a—Scisti di San Lorenzo Fm; 1b—Breccia di Asciano Fm; Iano (Sub-Unit 1): 2a—Scisti di Iano Fm; 2b—Breccia e Conglomerati di Torri Fm; 2c—Scisti Porfirici Fm; 2d—Fosso del Fregione Fm; Mt. Quoio-Mt. Senese (Sub-Unit 2): 3a—Risanguigno Fm; 3b—Poggio al Carpino Fm; Mt. Leoni- Farma River (Sub-Unit 3): 4a—Calcare di Sant'Antonio Fm; 4b—Scisti a Spirifer Fm; 4c—Farma Fm—Falsacqua Fm; 4d—Carpineta Fm—Quarziti di Poggio alle Pigne Fm; 4e—Le Cetine Fm.-Conglomerato di Fosso Pianacce Fm. See the main text for references.

This Carboniferous-lower Permian deposition was succeeded by a second middle-late Permian sedimentary cycle (see, e.g., [5] for a review) where the older deposits were partially dismantled and accumulated in the new one. This is also testified in the Monticiano-Roccastrada area by the occurrence of numerous clastic fragments relatable to the Risanguigno Fm [4], or by the presence of middle Carboniferous (late Visean-early Namurian: [16,47–49,133,134]) clasts, bioclasts and olistoliths embedded within the middle-late Permian Farma and Carpineta formations [70].

5.4. Deformation Insights

The deformation evidenced by the structural survey indicates that the Risanguigno Fm shared its tectono-metamorphic evolution with the overlying middle-late Permian Poggio al Carpino Fm and Triassic Verrucano Group, therefore highlighting their involvement in the Alpine deformational history. Noteworthy, neither outcrop-scale nor microscopic-scale evidence suggests the involvement of the Risanguigno Fm in a pre-Alpine deformation. This implies that the depositional environment of the Risanguigno Fm remained reasonably external to the orogenesis of Variscan chain, even during the formation of foreland and/or piggy-back basins. In this view, the presence of a Variscan tectonic phase in explaining the angular unconformity separating the Risanguigno Fm with the overlying Poggio al Carpino Fm and attributed to the Sudetic [15,43] or Bretonian phase [9,18,38] is denied. Therefore, such an angular unconformity is to be considered as having developed during the Carboniferous-Permian post-collisional tectonic regime [32,45,53], giving rise to short-lived, possibly pull-apart basins, dominated by continental to shallow-marine conditions [5].

5.5. Paleogeographical Implications

According to several reconstructions [1,5,135–141], during the Variscan evolution the Mississippian foredeep and piggy-back basin facies are always represented by coarse-dominated deposits (Culm facies—[141–144]) rapidly involved in the orogenesis and then progressively dismantled during exhumation and uplift. Accordingly, this foredeep basin was considered as possibly having been affected by late Variscan deformation [9,19], thus determining basins and rises, bringing to highly diverse depositional settings [20]. Coupling this latter interpretation with the results of the new structural survey (which rules out the Variscan deformation), we conclude that the Risanguigno Fm is not related to the Culm deposits. Thus, we propose the Risanguigno Fm as the oldest deposits of this sedimentary succession promoted in the "stable" Gondwana foreland that developed within fairly narrow continental or epicontinental domains. These depositional features could have favoured the low-energy, anoxic environments.

These settings evolved during the Late Pennsylvanian-Permian [32,45,64], originating graben/semigraben [1] or transcurrent/transtensive pull-apart basins [5,32,145] dominated by continental (Scisti di San Lorenzo Fm [131]) to shallow-marine conditions (Scisti di Iano Fm [130]), or local development to carbonate platform (Calcare di S. Antonio Fm [7]).

6. Conclusions

The new palynological-fossiliferous data for the Risanguigno Fm, coupled with its sedimentary and deformational setting, make it possible to assign it to the Middle Mississippian (late Tournasian-Visean) and to exclude its encompassment in the Variscan basement.

For this reason, it is possible now to exclude in southern Tuscany the outcrops of successions deformed during Variscan Orogenesis. Consequently, the Tuscan Crystalline Basement (Cambrian?-Devonian) is only exposed in the northern Tuscany (Apuan Alps, Pisani Mts and La Spezia area) and Tuscan Archipelago (Elba Island).

Sedimentation of Risanguigno Fm occurred in a shallow-marine-to-epicontinental setting, characterized by starved, anoxic conditions. This setting, localized in the Variscan foreland, evolved to open marine during the Pennsylvanian-Permian without any involvement in the Variscan Orogenesis.

On these bases, the Tuscan Palaeozoic-Triassic sedimentary succession (Phyllite-Quartzitic Group of [40]), classically considered as "post-Variscan" and now comprising the Middle Mississippian Risanguigno Fm, is no more to be related to the Variscan Orogenesis.

Author Contributions: Conceptualization, E.C., D.L., A.B. and A.S.; methodology, A.S., A.B. and E.C.; software, E.C., A.B., A.S. and M.Z.; formal analysis, A.S., G.B., A.B. and E.C.; investigation, E.C, A.B., A.S. and R.R.; writing—original draft preparation, E.C.; writing—review and editing, A.B., D.L., A.S., G.B., M.Z., G.M. and R.R.; project administration, E.C., A.S.; funding acquisition, E.C. and A.S. All authors have read and agreed to the published version of the manuscript.

Funding: The Research Project of the Department of Physics and Geology (University of Perugia) 'RED-AB—REDiscovering the Apennine Basement: a multidisciplinary correlation among the Northern Apennine Upper Paleozoic–Triassic successions (CAPEZBASE2017) is acknowledged by E.C. and A.S. E.C. thanks to the Research Project 2021 of the Department of Earth Sciences (University of Firenze) "Circolazione di fluidi nel sottosuolo e travertini: esempi in aree chiave del Settore Tirrenico dell'Appennino Settentrionale (Italia) e dell'Anatolia (Turchia)". A.S. thanks also PRIN 2017RX9XXXY.

Institutional Review Board Statement: Not applicable.

Informed Consent Statement: Not applicable.

Data Availability Statement: The presented data are available on request from the corresponding author.

Acknowledgments: Enrico Capezzuoli and AmaliaSpina warmly thanks Mauro Aldinucci (Eni) and Geoff Clayton (University of Sheffield, UK) for the discussions on the treated subjects. The authors thanks to Ausonio Ronchi, and an anonymous reviewer for their constructive comments.

Conflicts of Interest: The authors declare no conflict of interest. The funders had no role in the design of the study; in the collection, analyses, or interpretation of data; in the writing of the manuscript, or in the decision to publish the results.

References

1. Vai, G.B. Basement and early (pre-Alpine) history. In *Anatomy of an Orogen: The Apennines and Adjacent Mediterranean Basins*; Vai, G.B., Martini, I.P., Eds.; Kluwer Academic Publishers: Dordrecht, The Netherlands, 2001; pp. 121–150.
2. Bagnoli, G.; Giannelli, G.; Puxeddu, M.; Rau, A.; Tongiorgi, M. A tentative stratigraphic reconstruction of the Tuscan Palaeozoic basement. *Mem. Soc. Geol. Ital.* **1979**, *20*, 99–116.
3. Vai, G.B. Evidence of Silurian in the Apuane Alps (Tuscan, Italy). *G. Geol.* **1972**, *38*, 349–379.
4. Bagnoli, G.; Tongiorgi, M. New Fossiliferous Silurian (Mt. Corchia) and Devonian (Monticiano) Layers in the Tuscan Paleozoic. *Mem. Soc. Geol. Ital.* **1979**, *20*, 301–313.
5. Molli, G.; Brogi, A.; Caggianelli, A.; Capezzuoli, E.; Liotta, D.; Spina, A.; Zibra, I. Late Palaeozoic tectonics in Central Mediterranean: A reappraisal. *Swiss J. Geosci.* **2020**, *113*, 1–32. [CrossRef]
6. Carmignani, L.; Carosi, R.; Di Pisa, A.; Gattiglio, M.; Musumeci, G.; Oggiano, G.; Pertusati, P.C. The Hercynian chian in Sardinia. *Geodin. Acta* **1994**, *7*, 31–47. [CrossRef]
7. Carmignani, L.; Cocozza, T.; Ghezzo, C.; Pertusati, P.C.; Ricci, C.A. *Guide-Book to the Excursion on the Paleozoic Basement of Sardinia*; IGCP Project n. 5, Newsletter, Special Issue; Pacini Editore: Pisa, Italy, 1986; p. 102.
8. Conti, P.; Di Pisa, A.; Gattiglio, M.; Meccheri, M. The Pre-Alpine Basement in the Alpi Apuane (Northern Apennines, Italy). In *Pre-Mesozoic Geology in the Alps*; Von Raumer, J.F., Neubauer, F., Eds.; Springer: Berlin/Heidelberg, Germany, 1993; pp. 609–621.
9. Pandeli, E.; Gianelli, G.; Puxeddu, M.; Elter, F.M. The Paleozoic basement of the northern Apennines: Stratigraphy, tectono-metamorphic evolution and Alpine hydrothermal processes. *Mem. Soc. Geol. Ital.* **1994**, *48*, 627–654.
10. Gattiglio, M.; Meccheri, M.; Tongiorgi, M. Stratigraphic correlation forms of the Tuscan Palaeozoic basement. *Rend. Soc. Geol. Ital.* **1989**, *12*, 247–257.
11. Elter, F.M.; Pandeli, E. Structural features of the metamorphic Paleozoic-Triassic sequences in deep geothermal drillings of the Monte Amiata area (SE Tuscany, Italy). *Boll. Soc. Geol. Ital.* **1991**, *110*, 511–522.
12. Conti, P.; Gattiglio, M.; Meccheri, M. The overprint of the Alpine tectonometamorphic evolution on the Hercynian orogen: An example from the Apuane Alps (Northern Apennines, Italy). *Tectonophysics* **1991**, *191*, 335–346. [CrossRef]
13. Conti, P.; Carmignani, L.; Massa, G.; Meccheri, M.; Patacca, E.; Scandone, P.; Pieruccioni, D. *Note Illustrative della Carta Geologica d'Italia alla Scala 1:50.000—"Foglio 249 Massa Carrara"*; Servzio Geologico d'Italia: Roma, Italy, 2019; p. 290.
14. Rau, A. Il Paleozoico Toscano: Da segmento della catena varisica sud- Europea a margine passivo alpidico peri-adriatico. *Mem. Soc. Geol. Ital.* **1994**, *49*, 325–334.
15. Bagnoli, G.; Giannelli, G.; Puxeddu, M.; Rau, A.; Squarci, P.; Tongiorgi, M. Segnalazione di una potente successione clastica di età probabilmente carbonifera nel basamento della Toscana meridionale. *Mem. Soc. Geol. Ital.* **1980**, *21*, 127–136.
16. Pasini, M. I Fusulinidi della valle del Torrente Farma (Toscana meridionale). *Mem. Soc. Geol. Ital.* **1980**, *20*, 323–342.
17. Cocozza, T.; Decandia, F.A.; Lazzarotto, A.; Pasini, M.; Vai, G.B. The marine Carboniferous sequence in Southern Tuscany: Its bearing for Hercinynian palaeogeography and tectofacies. In *Pre-Variscan and Variscan Events in the Alpine-Mediterranean Mountain Belts*; Flugel, H.V., Sassi, F.P., Grecula, P., Eds.; Mineralia Slovaca-Monography: Bratislava, Slovakia, 1987; pp. 135–144.
18. Costantini, A.; Decandia, F.A.; Lazzarotto, A.; Sandrelli, F. L'Unità di Monticiano-Roccastrada fra la Montagnola Senese e il Monte Leoni (Toscana meridionale). *Atti Tic. Sci. Terra* **1988**, *31*, 382–420.
19. Lazzarotto, A.; Aldinucci, M.; Cirilli, S.; Costantini, A.; Decandia, F.A.; Pandeli, E.; Sandrelli, F.; Spina, A. Stratigraphic correlation of the Upper Palaeozoic-Triassic successions in Tuscany, Italy: A review. *Boll. Soc. Geol. Ital.* **2003**, *1*, 25–35.
20. Engelbrecht, H. Carboniferous continental margin deposits in southern Tuscany, Italy: Results from geological mapping of the geotopes Farma Valley and San Antonio Mine Area. *Geol. J.* **2008**, *43*, 279–305. [CrossRef]
21. Musumeci, G.; Mazzarini, F.; Tiepolo, M.; Di Vincenzo, G. U–Pb and 40Ar–39Ar geochronology of Palaeozoic units in the northern Apennines: Determining protolith age and alpine evolution using the Calamita Schist and Ortano Porphyroids. *Geol. J.* **2011**, *46*, 288–310. [CrossRef]
22. Sirevaag, H.; Jacobs, J.; Ksienzyk, A.K.; Rocchi, S.; Paoli, G.; Jørgensen, H.; Košler, J. From Gondwana to Europe: The journey of Elba Island (Italy) as recorded by U–Pb detrital zircon ages of Paleozoic metasedimentary rocks. *Gondwana Res.* **2016**, *38*, 273–288. [CrossRef]
23. Paoli, G.; Stokke, H.H.; Rocchi, S.; Sirevaag, H.; Ksienzyk, A.K.; Jacobs, J.; Košler, J. Basement provenance revealed by U-Pb detrital zircon ages: A tale of African and European heritage in Tuscany, Italy. *Lithos* **2017**, *277*, 376–387. [CrossRef]

24. Aldinucci, M.; Brogi, A.; Spina, A. Middle-Late Permian sporomorphs from the Farma Formation (Monticiano-Roccastrada Ridge, southern Tuscany): New constraints for the tectono-sedimentary history of the Tuscan Domain. *Boll. Soc. Geol. Ital.* **2008**, *127*, 581–597.
25. Spina, A.; Capezzuoli, E.; Brogi, A.; Cirilli, S.; Liotta, D. Middle-late Permian microfloristic evidences in the metamorphic successions of Northern Apennines: Insights for age constraining and palaeogeographic correlations. *J. Geol. Soc.* **2019**, *176*, 1262–1272. [CrossRef]
26. Cassinis, G.; Perotti, C.; Ronchi, A. Permian continental basins in the Southern Alps (Italy) and peri-mediterranean correlations. *Int. J. Earth Sci. (Geol. Rundsch.)* **2012**, *101*, 129–157. [CrossRef]
27. Carmignani, L.; Decandia, F.A.; Disperati, L.; Fantozzi, P.L.; Lazzarotto, A.; Liotta, D.; Meccheri, M. Tertiary extensional tectonics in Tuscany (northern Apennines, Italy). *Tectonophysics* **1994**, *238*, 295–315. [CrossRef]
28. Molli, G. Northern Apennine-Corsica orogenic system: An updated overview. *Geol. Soc. Lond. Spec. Publ.* **2008**, *298*, 413–442. [CrossRef]
29. Vai, G.B. Structure and Stratigraphy: An overview. In *Anatomy of an Orogen: The Apennines and Adjacent Mediterranean Basins*; Vai, G.B., Martini, I.P., Eds.; Kluwer Academic Publishers: Dordrecht, The Netherlands, 2001; pp. 15–31.
30. Conti, P.; Cornamusini, G.; Carmignani, L. An outline of the geology of the Northern Apennines (Italy), with geological map at 1:250,000 scale. *Ital. J. Geosci.* **2020**, *139*, 149–194. [CrossRef]
31. Brogi, A.; Capezzuoli, E.; Liotta, D.; Meccheri, M. The Tuscan Nappe structures in the Monte Amiata geothermal area (central Italy): A review. *Ital. J. Geosci.* **2015**, *134*, 219–236. [CrossRef]
32. Conti, P.; Costantini, A.; Decandia, F.A.; Elter, F.M.; Gattiglio, M.; Lazzarotto, A.; Meccheri, M.; Pandeli, E.; Rau, A.; Sandrelli, F.; et al. Structural frame of Tuscan Paleozoic: A review. *Boll. Soc. Geol. Ital.* **1991**, *110*, 523–541.
33. Bortolotti, V.; Fazzuoli, M.; Pandeli, F.; Principi, G.; Babbini, A.; Corti, S. Geology of central and eastern Elba Island, Italy. *Ofioliti* **2001**, *26*, 97–150.
34. Carmignani, L.; Decandia, F.A.; Disperati, L.; Fantozzi, P.L.; Kligfield, R.; Lazzarotto, A.; Liotta, D.; Meccheri, M. Inner Northern Apennines. In *Anatomy of an Orogen: The Apennines and Adjacent Mediterranean Basins*; Vai, G.B., Martini, I.P., Eds.; Kluwer Academic Publishers: Dordrecht, The Netherlands, 2001; pp. 197–214.
35. Pandeli, E.; Puxeddu, M. Paleozoic age for the Tuscan upper metamorphic sequences of Elba and its implications for the geology of the Northern Apennines (Italy). *Eclogae Geol. Helv.* **1990**, *83*, 123–142.
36. Pandeli, E.; Decandia, F.A.; Tongiorgi, M. The Paleozoic basement through the 500Ma history of the Northern Apennines. In Proceedings of the 32nd International Geological Congress from PR01 to B15, Pre-Congress B05, Florence, Italy, 20–28 August 2004; Volume 1, pp. 1–36.
37. Vai, G.B. Tentative correlation of Palaeozoic Rocks, Italian Peninsula and Islands. *Oest. Ak. Wiss. Schr. Erd. Kom.* **1978**, *3*, 313–329.
38. Pandeli, E.; Puxeddu, M.; Gianelli, G.; Bertini, G.T.; Castellucci, P. Paleozoic sequences crossed by deep drillings in the Monte Amiata geothermal region (Italy). *Boll. Soc. Geol. Ital.* **1988**, *107*, 593–606.
39. Pandeli, E.; Gianelli, G.; Morelli, M. The crystalline units of the middle-upper crust of the Larderello geothermal region (southern Tuscany, Italy): New data for their classification and tectono-metamorphic evolution. *Boll. Soc. Geol. Ital.* **2005**, *3*, 139–155.
40. Franceschelli, M.; Gianelli, G.; Pandeli, E.; Puxeddu, M. Variscan and Apine metamorphic events in the Northern Apennines, Italy: A review. *Period. Mineral. Spec. Issue* **2004**, *2*, 43–56.
41. Batini, F.; Brogi, A.; Lazzarotto, A.; Liotta, D.; Pandeli, E. Geological features of the Larderello–Travale and Mt. Amiata geothermal areas (southern Tuscany, Italy). *Episodes* **2003**, *26*, 239–244. [CrossRef] [PubMed]
42. Tongiorgi, M.; Bagnoli, G. Stratigraphie du socle paleozoic de la bordure continentale de l'Apennin septentrional (Italie centrale). *Bull. Soc. Geol. Fr.* **1981**, *23*, 319–323. [CrossRef]
43. Vai, G.B.; Cocozza, T. Tentative schematic zonation of the Hercynian chain in Italy. *Bull. Soc. Géol. Fr.* **1986**, *8*, 95–114. [CrossRef]
44. Pandeli, E.; Bertini, G.; Castellucci, P. The tectonic wedges complex of the Larderello area (southern Tuscany). *Boll. Soc. Geol. Ital.* **1991**, *110*, 621–629.
45. Pandeli, E. Sedimentary-Tectonic evolution of the Tuscan area (Northern Apennines, Italy) from Late «Autunian» to Carnian. *Boll. Soc. Geol. Ital.* **2002**, *1*, 251–262.
46. Brogi, A. The structure of the Monte Amiata volcano-geothermal area (Northern Apennines, Italy): Neogene-Quaternary compression versus extension. *Int. J. Earth Sci.* **2008**, *97*, 677–703. [CrossRef]
47. Cocozza, T. Il Carbonifero nel Gruppo Monticiano-Roccastrada (Toscana). *Ric. Sci. Rend.* **1965**, *8*, 1–38.
48. Pasini, M. Further paleontological records from the Upper Paleozoic outcrops of the Farma Valley (southern Tuscany). In *Report on the Tuscan Paleozoic Basement*; Tongiorgi, M., Ed.; CNR Internal Report of the Progetto finalizzato Energetica—Sottoprogetto Energia Geotermica: Pisa, Italy, 1978; pp. 71–75.
49. Pasini, M. Gli Archaeodiscidi dell'affioramento di Poggio alle pigne nella Valle del Farma (Toscana meridionale). *Mem. Soc. Geol. Ital.* **1980**, *20*, 343–346.
50. Rau, A. Evolution of the Tuscan Domain between the Upper Carboniferous and Middle Triassic: A new hypothesis. *Boll. Soc. Geol. Ital.* **1990**, *109*, 231–238.
51. Rau, A.; Tongiorgi, M. Geologia dei Monti Pisani a Sud-Est della Valle del Guappero. *Mem. Soc. Geol. Ital.* **1974**, *13*, 216–280.
52. Cassinis, G.; Elter, G.; Rau, A.; Tongiorgi, M. Verrucano: A tectofacies of the Alpine-Mediterranean Southern Europe. *Mem. Soc. Geol. Ital.* **1980**, *20*, 135–149.

53. Aldinucci, M.; Gandin, A.; Sandrelli, F. The Mesozoic continental rifting in the Mediterranean area: Insights from the Verrucano tectofacies of southern Tuscany (Northern Apennines, Italy). *Int. J. Earth Sci. (Geol. Rundsch.)* **2008**, *97*, 1247–1269. [CrossRef]
54. Franceschelli, M.; Leoni, L.; Memmi, I.; Puxeddu, M. Regional distribution of Al-silicates and metamorphic zonation in the low—Grade Verrucano metasediments from the Northern Apennines, Italy. *J. Metam. Geol.* **1986**, *4*, 309–321. [CrossRef]
55. Theye, T.; Reinhardt, J.; Goffé, B.; Jolivet, L.; Brunet, C. Ferro and magnesiocarpholite from Mt. Argentario (Italy): First evidence for high-pressure metamorphism of the metasedimentary Verrucano sequence, and its significance for the P-T path reconstruction. *Eur. J. Mineral.* **1997**, *9*, 859–873. [CrossRef]
56. Giorgetti, G.; Goffé, B.; Memmi, I.; Nieto, F. Metamorphic evolution of Verrucano metasediments in northern Apennines: New petrological constraints. *Eur. J. Mineral.* **1998**, *10*, 1295–1308. [CrossRef]
57. Brunet, C.; Monié, P.; Jolivet, L.; Cadet, J.P. Migration of compression and extension in the Tyrrhenian Sea, insights from 40Ar/39Ar ages on micas along a transect from Corsica to Tuscany. *Tectonophysics* **2000**, *321*, 127–155. [CrossRef]
58. Rossetti, F.; Faccenna, C.; Jolivet, L.; Goffé, B.; Funiciello, R. Structural signature and exhumation P–T–t paths of the blueschist units exposed in the interior of the Northern Apennine chain, tectonic implication. *Boll. Soc. Geol. Ital.* **2002**, *1*, 829–842.
59. Liotta, D. D2 asymmetric folds and their vergence meaning in the Montagnola Senese metamorphic rocks (inner northern Apennines, central Italy). *J. Struct. Geol.* **2002**, *24*, 1479–1490. [CrossRef]
60. Montomoli, C.; Carosi, R.; Pertusati, P.C. Tectonic history of the Monti dell'Uccellina range, Southern Tuscany, Italy. *Boll. Soc. Geol. Ital.* **2009**, *128*, 515–526.
61. Brogi, A.; Giorgetti, G. Tectono-metamorphic evolution of the siliciclastic units in the Middle Tuscan Range (inner Northern Apennines): Mg–carpholite bearing quartz veins related to syn-metamorphic syn-orogenic foliation. *Tectonophysics* **2012**, *526–529*, 167–184. [CrossRef]
62. Liotta, D.; Cernobori, L.; Nicolich, R. Restricted rifting and its coexistence with compressional structures: Results from the Crop03 traverse (Northern Apennines, Italy). *Terra Nova* **1998**, *10*, 16–20. [CrossRef]
63. Brogi, A. Kinematics and geometry of Miocene low-angle detachments and exhumation of the metamorphic units in the hinterland of the Northern Apennines (Italy). *J. Struct. Geol.* **2008**, *30*, 2–20. [CrossRef]
64. Aldinucci, M.; Pandeli, E.; Sandrelli, F. Tectono-sedimentary evolution of the Late Palaeozoic-Early Mesozoic metasediments of the Monticiano-Roccastrada Ridge (southern Tuscany, Northern Apennines, Italy). *Boll. Soc. Geol. Ital. (Ital. J. Geosci.)* **2008**, *127*, 567–579.
65. Cocozza, T.; Costantini, A.; Lazzarotto, A.; Sandrelli, F. Continental Permian in Southern Tuscany (Italy). In *Report on the Tuscan Paleozoic*; Tongiorgi, M., Ed.; Consiglio Nazionale delle Ricerche, Rapporto interno del Sottoprogetto Energia Geotermica, Progetto Finalizzato Energetica: Pisa, Italy, 1978; pp. 35–49.
66. Pasini, M.; Vai, G.B. Review and updating of the Moscovian to Artinskian marine rocks in peninsula Italy. In *Peri-Tethys: Stratigraphic Correlations*; Crasquin-Soleau, S., De Wever, P., Eds.; Geodiversitas: Paris, France, 1997; Volume 19, pp. 187–191.
67. Lazzarotto, A.; Decandia, F.A.; Aqué, R.; Bellagotti, E.; Costantini, A.; Sandrelli, F. Carta Geologica Toscana scala 1:10.000, Foglio 307120. 2005. Available online: http://web.rete.toscana.it/pta/servlet/Scheda?foglio=307 (accessed on 5 December 2020).
68. Puxeddu, M.; Raggi, G.; Tongiorgi, M. Descrizione di alcuni sondaggi e osservazioni geologiche nel Paleozoico della zona di Monticiano (Siena). *Mem. Soc. Geol. Ital.* **1979**, *20*, 233–242.
69. Spina, A.; Cirilli, S.; Decandia, F.A.; Lazzarotto, A. Palynological data from the Poggio al Carpino sandstones Fm. (Southern Tuscany, Italy). In Proceedings of the International Congress on «Stratigraphic and Structural Evolution of the Late Carboniferous to Triassic Continental and Marine Successions in Tuscany (Italy). Regional Reports and General Correlation», Siena, Italy, 30 April–7 May 2001; pp. 65–66.
70. Engelbrecht, H. Deposition of Tempestites in the Eastern Rheic Strait: Evidence from the Upper Palaeozoic of Southern Tuscany (Italy). *Facies* **2000**, *43*, 103–122. [CrossRef]
71. Bagnoli, G. Segnalazione di Conodonti devoniani nel Paleozoico della Maremma senese (Nota preliminare). *Atti Soc. Tosc. Sci. Nat. Mem. A* **1979**, *86*, 23–26.
72. Jäger, H. Palynology of the lower carboniferous (Mississippian) kammquartzite formation in the rhenohercynian zone, Germany. *Senckenbergiana Lethaea* **2002**, *82*, 609–637. [CrossRef]
73. Sullivan, H. *Miospores from the Lower Lime- Stone Shales (Tournaisian) of the Forest of Dean Basin, Gloucestershire. Compte Rendu 5-eme Congrés International de Stratigraphie et de Géologie du Carbonifère, Paris (1963)*; Académie des Sciences: Paris, France, 1964; pp. 1249–1259.
74. Clayton, G. A Lower Carboniferous miospore assemblage from the Calciferous Sandstone Measures of the Cockburnspath region of eastern Scotland. *Pollen Spores* **1971**, *12*, 577–600.
75. Dolby, G. Spore assemblage from the Devonian-Carboniferous transition measures in South-West Britain and Southern Eire. In *Colloque sur la Stratigraphie de Carbonifere*; Congres et Colloques, Univ. Liège: Liége, France, 1971; Volume 55, pp. 267–274.
76. Higgs, K. Upper Devonian and Lower Carboniferous miospore assemblages from Hook Head, County Wexford, Ireland. *Micropaleontology* **1975**, *21*, 393–419. [CrossRef]
77. Clayton, G.; Johnston, I.S.; Smith, D.G. Micropalaeontology of a Courceyan (Carboniferous) borehole section from Ballyvergin, County Clare, Ireland. *J. Earth Sci.* **1980**, *3*, 81–100.
78. Higgs, K.T.; Dreesen, R.; Dusar, M.; Streel, M. Palynostratigraphy of the Tournaisian (Hastarian) rocks in the Namur Synclinorium, West Flanders, Belgium. *Rev. Palaeobot. Palynol.* **1992**, *72*, 149–158. [CrossRef]

79. Clayton, G.; Coquel, R.; Doubinger, J.; Gueinn, K.J.; Loboziak, S.; Owens, B.; Streel, M. Carboniferous miospores of Western Europe: Illustration and zonation. *Meded. Rijks Geol. Dienst.* **1977**, *29*, 1–71.
80. Clayton, G.; Higgs, K.; Keegan, J.B.; Sevastopulo, G.D. Correlation of the palynological zonation of the Dinantian of the British Isles. *Palinología* **1978**, *1*, 137–147.
81. Clayton, G.; Turnau, E. Correlation of the Tournaisian miospore zonation of Poland and the British Isles. *Ann. Soc. Geol. Pol.* **1990**, *60*, 45–58.
82. Higgs, K.T. Taxonomic and systematic study of some Tournaisian (Hastarian) spores from Belgium. *Rev. Palaeobot. Palynol.* **1996**, *93*, 269–297. [CrossRef]
83. Playford, G. Miospores from the Mississippian Horton Group, eastern Canada. *Geol. Surv. Can.* **1964**, *107*, 1–47.
84. Utting, J. Palynology of the Lower Carboniferous Windsor Group and Windsor-Canso boundary beds of Nova Scotia, and their equivalents in Quebec, New Brunswick and Newfoundland. *Geol. Surv. Can.* **1987**, *374*, 1–93.
85. Utting, J. Palynostratigraphic investigation of the Albert Formation (Lower Carboniferous) of New Brunswick, Canada. *Palynology* **1987**, *11*, 73–96. [CrossRef]
86. Utting, J.; Keppie, J.D.; Giles, P.S. Palynology and stratigraphy of the Lower Carboniferous Horton Group, Nova Scotia. Contributions to Canadian Paleontology. *Geol. Surv. Can.* **1989**, *396*, 117–143.
87. Gao Lianda 1984. Carboniferous spore assemblages in China. IX Congrès International de Stratigraphie et de Géologie du Carbonifère (Washington and Champaign-Urbana 1979). *Compte Rendu* **1984**, *2*, 103–108.
88. Clayton, G.; Loboziak, S. Early Carboniferous (Early Viséan-Serpukhovian) Palynomorphs. *J. Micropalaeontol.* **1985**, 4–5. [CrossRef]
89. Loboziak, S.; Clayton, G. The early Carboniferous palynology and stratigraphy of Northeast Libya. In *Subsurface Palynostratigraphy of Northeast Libya*; El-Arnauti, A., Owens, B., Thusu, B., Eds.; Garyounis University: Benghazi, Libya, 1989; pp. 129–149.
90. Attar, A.; Fournier, J.; Candilier, A.M.; Coquel, R. Etude palynologique du Dévonien terminal et du Carbonifère inférieur du Bassin d'Illizi (Fort Polignac) Algérie. *Rev. l'Institut Français Pétrole* **1980**, *35*, 585–619.
91. Coquel, R.; Abdesselam-Rouighi, F. Palynostratigraphical revision of the uppermost Devonian-Lower Carboniferous in the Western Great Erg (Bechar Basin) Algerian Sahara. *Rev. Micropaléontologie* **2000**, *43*, 353–364. [CrossRef]
92. Massa, D.; Coquel, R.; Loboziak, S.; Taugourdeau-Lantz, J. Essai de synthèse stratigraphique et palynologique du Carbonifère en Libye Occidentale. *Ann. Soc. Géologique Nord.* **1980**, *99*, 429–442.
93. Grignani, D.L.; Anzoni, E.; Elatrash, H. Palaeozoic and Mesozoic subsurface palynostratigraphy in Al Kufrah basin, Libya. In *The Geology of Libya. Third Symposium on the Geology of Libya*; Salem, M.J., Hammuda, O.S., Eliagoubi, B.A., Eds.; Elsevier: Amsterdam, The Netherlands, 1991; pp. 1159–1227.
94. Tawadros, E.; Rasul, S.M.; Elzaroug, R. Petrography and palynology of quartzites in the Sirte Basin, central Libya. *J. Afr. Earth Sci.* **2001**, *32*, 373–390. [CrossRef]
95. Clayton, G. Carboniferous miospore and pollen assemblages from the Kingdom of Saudi Arabia. *Rev. Palaeobot. Palynol.* **1995**, *89*, 115–123. [CrossRef]
96. Clayton, G.; Owens, B.; Al-Hajri, S.; Filatoff, J. Latest Devonian and Early Carboniferous miospore assemblages from Saudi Arabia. In *Stratigraphic Palynology of the Palaeozoic Of Saudi Arabia*; Al-Hajri, S., Owens, B., Eds.; Geoarabia Special Publication; Gulf PetroLink: Manama, Bahrain, 2000.
97. Aria-Nasab, M.; Spina, A.; Cirilli, S.; Daneshian, J. The palynostratigraphy of the Lower Carboniferous (middle Tournaisian–upper Viséan) Shishtu Formation from the Howz-e-Dorah section, southeast Tabas, central Iranian Basin. *Palynology* **2016**, *40*, 247–263. [CrossRef]
98. Higgs, K.T.; Finucane, D.; Tunbridge, I.P. Late Devonian and early Carboniferous microfloras from the Hakkarı Province of southeastern Turkey. *Rev. Palaeobot. Palynol.* **2002**, *118*, 141–156. [CrossRef]
99. Loboziak, S.; Streel, M.; Caputo, M.V.; Melo, J.H.G. Middle Devonian to Lower Carboniferous miospore stratigraphy in the central Parnaíba Basin (Brazil). *Ann. Société Géologique Belg.* **1992**, *115*, 215–226.
100. Loboziak, S.; Streel, M.; Caputo, M.V.; Melo, J.H.G. Middle Devonian to Lower Carboniferous miospores from selected boreholes in Amazonas and Parnaíba Basins (Brazil): Additional data, synthesis, and correlation. *Doc. Lab. Géologie Lyon* **1993**, *125*, 277–289.
101. Loboziak, S.; Melo, J.H.G.; Streel, M. Reassessment of Viséan miospore biostratigraphy in the Amazon Basin, northern Brazil. *Rev. Palaeobot. Palynol.* **1998**, *104*, 143–155. [CrossRef]
102. Melo, J.H.G.; Loboziak, S.; Streel, M. Latest Devonian to early Late Carboniferous biostratigraphy of northern Brazil: An update. *Bull. Cent. Rech. Elf Explor.-Prod.* **1999**, *22*, 13–33.
103. Playford, G.; Dino, R.; Marques-Toigo, M. The upper Paleozoic miospore genus Spelaeotriletes Neves and Owens, 1966, and constituent Gondwanan species. *J. South Am. Earth Sci.* **2001**, *14*, 593–608. [CrossRef]
104. Melo, J.H.G.; Loboziak, S. Devonian-Early Carboniferous miospore biostratigraphy of the Amazon Basin, Northern Brazil. *Rev. Palaeobot. Palynol.* **2003**, *124*, 131–202. [CrossRef]
105. Loboziak, S.; Vachard, D.; Fadli, D.; Streel, M. Datation par miospores et Foraminifères du Tournaisien et du Viséen de l'Oued Zemrine (Massif des Mdakra, Maroc). *J. Afr. Earth Sci.* **1990**, *11*, 113–118. [CrossRef]
106. Owens, B.; Filatoff, J.; Clayton, G.; Al-Hajri, S. Evidence of mid-Carboniferous miospore assemblages from central Saudi Arabia. Stratigraphic Palynology of the Palaeozoic of Saudi Arabia. *GeoArabia Spec. Publ.* **2000**, *1*, 154–167.
107. Bouillin, J.P.; Majesté-Menjoulas, C.; Baudelot, S.; Cygan, C.; Fournier-Vinas, C. Les formations paléozoiques de l'Arc Calabro-Péloritain dans leur cadre structural. *Boll. Soc. Geol. Ital.* **1987**, *106*, 683–698.

108. Randon, C. Siliceous deposits within a carbonate sequence: A significant event. In *Carboniferous Conference Cologne. From Platform to Basin. Sept 4–10 2006*; Aretz, M., Herbig, H.-G., Eds.; Program and Abstracts; Kölner Forum für Geologie und Paläontologie: Köln, Germany, 2006; Volume 15, p. 106.
109. Randon, C.; Caridroit, M. Age and origin of Mississippian lydites: Examples from the Pyrénées, southern France. *Geol. J.* **2008**, *43*, 261–278. [CrossRef]
110. Renna, M.R.; Tribuzio, R.; Tiepolo, M. Interaction between basic and acid magmas during the latest stages of the post-collisional Variscan evolution: Clues from the gabbro-granite association of Ota (Corsica-Sardinia batholith). *Lithos* **2006**, *90*, 92–110. [CrossRef]
111. Edel, J.B.; Casini, L.; Oggiano, G.; Rossi, P.; Schulmann, K. Early Permian 90 clockwise rotation of the Maures–Esterel–Corsica–Sardinia block confirmed by new palaeomagnetic data and followed by a Triassic 60 clockwise rotation. *Geol. Soc. Lond. Spec. Publ.* **2014**, *405*, 333–361. [CrossRef]
112. Casini, L.; Cuccuru, S.; Puccini, A.; Oggiano, G.; Rossi, P. Evolution of the Corsica–Sardinia Batholith and late-orogenic shearing of the Variscides. *Tectonophysics* **2015**, *646*, 65–78. [CrossRef]
113. Paquette, J.L.; Ménot, R.-P.; Pin, C.; Orsini, J.B. Episodic and short-lived granitic pulses in a post-collisional setting: Evidence from precise U–Pb zircon dating through a crustal cross-section in Corsica. *Chem. Geol.* **2003**, *198*, 1–20. [CrossRef]
114. Li, X.-H.; Faure, M.; Lin, W. From crustal anatexis to mantle melting in the Variscan orogen of Corsica (France): SIMS U-Pb zircon age constraints. *Tectonophysics* **2014**, *634*, 219–230. [CrossRef]
115. Rossi, P.; Cocherie, A.; Fanninng, C.M. Evidence in Variscan Cosica of a brief and voluminous Late Carboniferous to Early Permian volcanic–plutonic event contemporaneous with a high-temperature/low-pressure metamorphic peak in the lower crust. *Bull. Soc. Géol. Fr.* **2015**, *186*, 171–192. [CrossRef]
116. Neves, R.; Sullivan, H.J. Modification of fossil spore exines associated with the presence of pyrite crystals. *Micropaleontology* **1964**, *10*, 443–452. [CrossRef]
117. Ouyang, S.; Utting, J. Palynology of Upper Permian and Lower Triassic rocks, Meishan, Changxing County, Zhejiang Province, China. *Rev. Palaeobot. Palynol.* **1990**, *66*, 65–103. [CrossRef]
118. Turnau, E. Note on the preservational nature of ornamentation in sphaeromorphs assignable to Tapajonites Sommer & van Boekel, 1963 (Prasinophyta?). *J. Micropalaeontology* **2000**, *19*, 159–162. [CrossRef]
119. Traverse, A. *Paleopalynology*, 2nd ed.; Springer: Dordrecht, The Netherlands, 2008; p. 813. [CrossRef]
120. Tyson, R.V. *Sedimentary Organic Matter; Organic Facies and Palynofacies*, 1st ed.; Chapman and Hall: Davon, UK, 1995; p. 615.
121. Gursky, H.-J. Pelagic facies of the Early Carboniferous—From Laurasia's margin to Iberia. In *Carboniferous Conference Cologne. From Platform to Basin, Sept 4–10 2006*; Aretz, M., Herbig, H.-G., Eds.; Program and Abstracts; Kolner Forum fur Geologie und Palaontologie: Köln, Germany, 2006; Volume 15, p. 35.
122. Saltzman, M.R.; Groessens, E.; Zhuravlev, A. Carbon cycle models based on extreme changes in δ13C: An example from the Lower Mississippian. *Palaeogeogr. Palaeoclimatol. Palaeoecol.* **2004**, *213*, 359–377. [CrossRef]
123. Cheng, K.; Elrick, M.; Romaniello, S.J. Early Mississipian ocean anoxia triggered organic carbon burial and late Paleozoic cooling: Evidence from uranium isotopes recorded in marina limestone. *Geology* **2020**, *48*, 363–367. [CrossRef]
124. Montañez, I.P.; Poulsen, C.J. The Late Paleozoic Ice Age: An evolving paradigm. *Annu. Rev. Earth Planet. Sci.* **2013**, *41*, 629–656. [CrossRef]
125. Saltzman, M.R.; González, L.A.; Lohmann, K.C. Earliest Carboniferous cooling step triggered by the Antler orogeny? *Geology* **2000**, *28*, 347–350. [CrossRef]
126. Saltzman, M.R. Late Paleozoic ice age: Oceanic gateway or pCO2? *Geology* **2003**, *31*, 151–154. [CrossRef]
127. Buggisch, W.; Joachimski, M.M.; Sevastopulo, G.; Morrow, J.R. Mississippian δ13C carb and conodont apatite δ18O recordsTheir relation to the late Palaeozoic glaciation. *Palaeogeogr. Palaeoclimatol. Palaeoecol.* **2008**, *268*, 273–292. [CrossRef]
128. Liu, J.; Algeo, T.J.; Qie, W.; Saltzman, M.R. Intensified oceanic circulation during early Carboniferous cooling events: Evidence from carbon and nitrogen isotopes. *Palaeogeogr. Palaeoclimatol. Palaeoecol.* **2018**, *531*, 108962. [CrossRef]
129. Maharjan, D.; Jiang, G.; Peng, Y.; Henry, R.A. Paired carbonate–organic carbon and nitrogen isotope variations in Lower Mississippian strata of the southern Great Basin, western United States. *Palaeogeogr. Palaeoclimatol. Palaeoecol.* **2018**, *490*, 462–472. [CrossRef]
130. Pandeli, E. Permo-Triassic siliciclastic sedimentation in the Northern Apennines: New data from the Iano metamorphic inlier (Florence). *Mem. Soc. Geol. Ital.* **1998**, *53*, 185–206.
131. Costantini, A.; Elter, F.M.; Pandeli, E.; Sandrelli, F. Geologia dell'area di Iano (Toscana meridionale, Italia). *Boll. Soc. Geol. Ital.* **1998**, *117*, 187–218.
132. Marini, F.; Pandeli, E.; Tongiorgi, M.; Pecchioni, E.; Orti, L. The Carboniferous-mid Permian successions of the Northern Apennines: New data from the Pisani Mts. inlier (Tuscany, Italy). *Ital. J. Geosci.* **2020**, *139*, 212–232. [CrossRef]
133. Pasini, M. Edmondia scalaris (M'Coy, 1844) e Phestia cf. attenuata (Fleming 1828) nella Formazione Carpineta (Paleozoico superiore della Toscana meridionale). *Riv. Ital. Paleont. Stratigr.* **1978**, *84*, 849–864.
134. Ferraresi, E.; Pasini, M. Some fossil algae from the Carboniferous of the Farma Gorge (southeastern Tuscany, Italy). *Palaeopelagos* **1996**, *6*, 31–43.
135. Von Raumer, J.F.; Stampfli, G.M.; Borel, G.D.; Bussy, F. The organisation of pre-Variscan basement areas at the north-Gondwanan margin. *Int. J. Earth Sci.* **2002**, *91*, 35–52. [CrossRef]

136. Von Raumer, J.F.; Stampfli, G.M.; Bussy, F. Gondwana-derived microcontinents-the constituents of the Variscan and Alpine collisional orogens. *Tectonophysics* **2003**, *365*, 7–22. [CrossRef]
137. Von Raumer, J.F.; Bussy, R.; Schaltegger, U.; Schulz, B.; Stampfli, G.M. Pre-Mesozoic Alpine basements-Their place in the European Paleozoic framework. *Geol. Soc. Am. Bull.* **2013**, *125*, 89–108. [CrossRef]
138. Spiess, R.; Cesare, B.; Mazzoli, C.; Sassi, R.; Sassi, F.P. The cristalline basement of the Adria microplate in the eastern Alps: A review of the paleostructural evolution from Neoproterozoic to Tertiary. *Rend. Lincei* **2010**, *12*, 31–50. [CrossRef]
139. Cassinis, G.; Perotti, C.; Santi, G. Post-Variscan Verrucano-like deposits in Italy, and the onset of the alpine tectono-sedimentary cycle. *Earth-Sci. Rev.* **2018**, *185*, 476–497. [CrossRef]
140. Ballevre, M.; Manzotti, P.; Dal Piaz, G.V. Pre-Alpine (Variscan) Inheritance: A Key for the Location of the Future Valaisan Basin (Western Alps). *Tectonics* **2018**, *737*, 86–817. [CrossRef]
141. Barca, S.; Farci, G.; Forci, A. I depositi sinorogenici ercinici del Sulcis (Sardegna sud-occidentale). *Boll. Soc. Geol. Ital.* **1998**, *117*, 407–419.
142. Carmignani, L.; Conti, P.; Barca, S.; Cerbai, N.; Eltrudis, A.; Funedda, A.; Oggiano, G.; Patta, E.D. *Carta Geologica d'Italia alla Scala 1:50.000 "Foglio 549—Muravera"*; Servizio Geologico d'Italia: Roma, Italy, 2001.
143. Carmignani, L.; Oggiano, G.; Barca, S.; Conti, P.; Salvadori, I.; Eltrudis, A.; Funedda, A.; Pasci, S. *Geologia della Sardegna: Note Illustrative della Carta Geologica della Sardegna in Scala 1:200.000, Memorie Descrittive della Carta Geologica d'Italia*; Servizio Geologico d'Italia: Roma, Italy, 2001; Volume 60, p. 283.
144. Funedda, A. Foreland- and hinterland-verging structures in fold-and- thrust belt: An example from the Variscan foreland of Sardinia. *Int. J. Earth Sci.* **2009**, *98*, 1625–1642. [CrossRef]
145. Rau, A. Lineamenti profondi nel basamento pre-Triassico della Toscana continentale (Italia). *Studi Geol. Camerti* **1991**, *1*, 141–148.

geosciences

MDPI

Article

Timing of Contractional Tectonics in the Miocene Foreland Basin System of the Umbria Pre-Apennines (Italy): An Updated Overview

Francesco Brozzetti [1,2], Daniele Cirillo [1,2,*] and Lucina Luchetti [3]

[1] Dipartimento di Scienze Psicologiche, della Salute e del Territorio, Università G. d'Annunzio, via dei Vestini 31, 66100 Chieti, Italy; francesco.brozzetti@unich.it
[2] Centro inteRUniversitario per l'analisi SismoTettonica tridimensionale (CRUST), 66100 Chieti Scalo, Italy
[3] ARTA-Abruzzo, District of Chieti, Via D. Spezioli 52, 66100 Chieti, Italy; l.luchetti@artaabruzzo.it
* Correspondence: d.cirillo@unich.it; Tel.: +39-0871-355-6389

Abstract: A large dataset of lithostratigraphic and biostratigraphic data, concerning the Early-Late Miocene turbidite succession of the Umbria pre-Apennines, is presented and analyzed. The data come from the study of 24 sections that are representative of all the main tectonic units cropping out between the front of the Tuscan allochthon and the Umbria-Marche calcareous chain. The sections have been dated using quantitative calcareous nannofossil biostratigraphy and, wherever possible, they were correlated through key-beds recognition. Such a multidisciplinary approach allowed us to reconstruct the evolution of the Umbria foredeep over time and to unveil the chronology of compressive deformations by defining: (i) the onset of the foredeep stage in each structural unit, (ii) the age of depocenter-shifting from a unit to the adjacent one, (iii) the progressive deactivation of the western sector of the foredeep due to the emplacement of allochthon units, and (iv) the internal subdivisions of the basin due to the presence of foreland ramp faults or thrust-related structures. A further original outcome of our study is having brought to light the Late Burdigalian "out-of-sequence" reactivation of the Tuscan allochthon which bounded westward the foredeep, and the subsequent protracted period of tectonic stasis that preceded the deformations of the Umbrian parautochthon.

Keywords: Umbria pre-Apennines; foreland basin systems evolution; timing of contractional tectonics; biostratigraphic constraints to foredeep deposition

check for updates

Citation: Brozzetti, F.; Cirillo, D.; Luchetti, L. Timing of Contractional Tectonics in the Miocene Foreland Basin System of the Umbria Pre-Apennines (Italy): An Updated Overview. *Geosciences* **2021**, *11*, 97. https://doi.org/10.3390/geosciences11020097

Academic Editors: Domenico Liotta, Giancarlo Molli, Angelo Cipriani and Jesus Martinez-Frias

Received: 5 January 2021
Accepted: 8 February 2021
Published: 19 February 2021

Publisher's Note: MDPI stays neutral with regard to jurisdictional claims in published maps and institutional affiliations.

Copyright: © 2021 by the authors. Licensee MDPI, Basel, Switzerland. This article is an open access article distributed under the terms and conditions of the Creative Commons Attribution (CC BY) license (https://creativecommons.org/licenses/by/4.0/).

1. Introduction

It is widely recognized that, in a foreland basin system, the spatial-temporal distribution of sedimentary processes reflects the tectonic deformations which drove its structuring [1,2]. Such a close link between the evolution of the chain and the depositional events that occurred in the foredeep and the associated satellite basins is well documented in any stage of the Apennines orogenic system [3–7].

During their eastward migration, the foreland basin systems of the Apennines included the following tectono-sedimentary zones [1,2,8]: (i) piggyback or thrust-top basins, characterized by continental and shallow marine deposition occurred unconformably above older foredeep successions, (ii) foredeep basin, a high-subsidence basin hosting deep-water turbidites lying in slight unconformity on pre-orogenic successions, and (iii) forebulge sector, where sedimentation of hemipelagic ramp-mud [9] occurred, in substantial conformity, on the foreland monocline.

We tried to reconstruct the nature and timing of the tectonic events which affected the Umbria pre-Apennines based on a detailed litho- and bio-stratigraphic analysis of the sin-tectonic successions outcropping in the sections shown in Figure 1, which have been selected taking into account also their present structural setting.

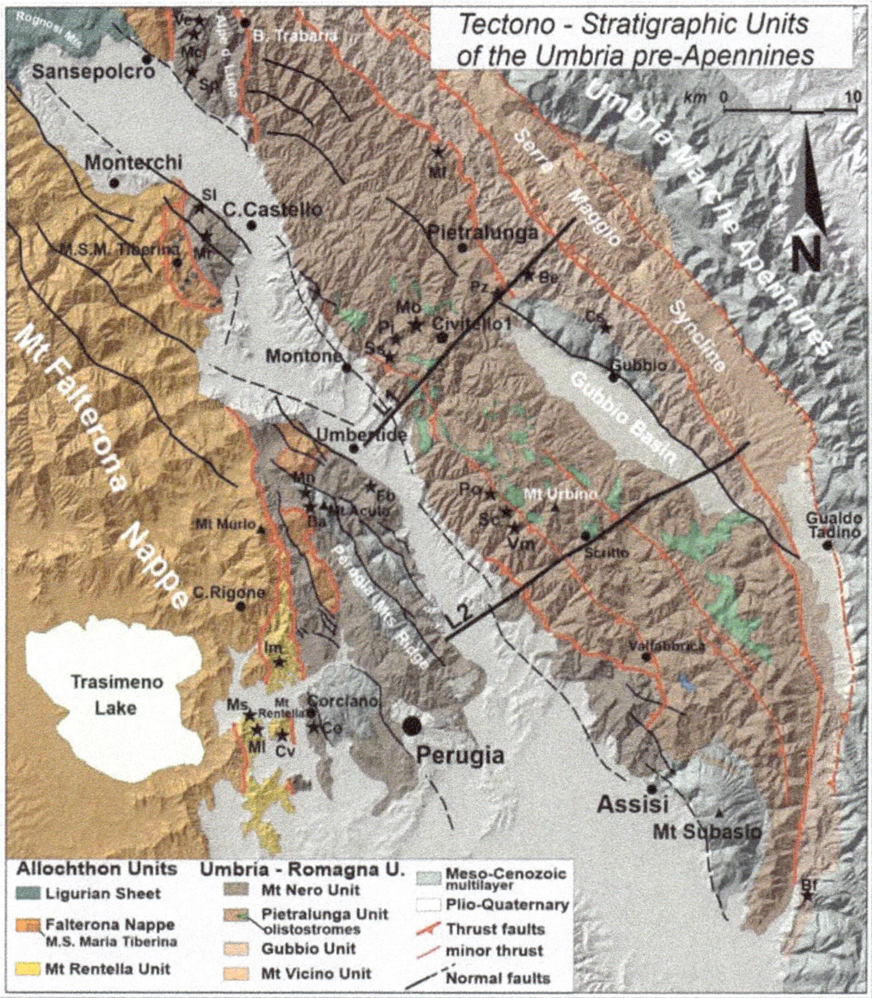

Figure 1. Structural Geological scheme of the study area showing the main tectono-stratigraphic units in which the Early-Middle Miocene Marnoso Arenacea basin is presently split according to the data reported in [10–16]. Asterisks and acronyms locate the studied sections as explained in the following: REN Unit: Ml = I Molini, Ms = Monte Sperello, Cv = Castelvieto, Im = Il Molino; Mt Nero Unit: Ba = Balconcelli, Co = Corciano, Mn = Monestevole, Fb = Fosso della Badia, Sp = Case Spertaglia, Mc = Monte Casale, Ve = Vesina; M. S. Maria Tiberina area: Sl = San Lorenzo, Mr = Monte Cedrone; Pietralunga Unit: Pz = Piazze, Sc = Santa Cecilia, Po = Portole, Vm = Valmarcolone, Ss = Sassarone, Pt = Pietralata, Mo = Moravola; Gubbio Unit: Cs = Contessa, Be = Bavelle; Mf = Madonna dei 5 Faggi, Bf = Belfiore.

The Umbria pre-Apennine is a ~40 km-wide belt delimited to the west by the front of the Tuscan Falterona nappe and to the east by the inner border of the Umbria-Marche calcareous chain (Figure 1) [10–14].

Its shallow structure consists mainly of an east-verging imbricate fan formed by the stacking of Early to Late Miocene successions of turbidites and hemipelagites. Within the pre-Apennine, the Meso-Cenozoic Umbria-Marche multilayer crops out only in correspondence of narrow culminations located in the Perugia Mts ridge (west of the Tiber river) and in the Gubbio and Mt Subasio doubly plunging anticlines (eastern Umbria, Figure 1).

The major thrust-sheets correspond to regional tectono-stratigraphic units referred to, in the literature, as Mt Rentella, Mt Nero, Pietralunga, Gubbio, and Mt Vicino units, from west to east (Figure 1) [11–15]. The turbidite succession characterizing each one of these units differs from the neighboring ones as regards the age of the top of the basal ramp-mud, the time interval during which the turbidite sedimentation took place, the feeding areas of the gravity flows, and their dispersal pattern [15–20].

A clear eastward-younging trend is documented by the ages through which each unit evolved, from the foreland to the foredeep stage, and was progressively incorporated into the chain.

In fact, as highlighted by previous biostratigraphic data, thrusting rejuvenated from the Late Burdigalian, in the Mt Rentella Unit, to the Early-Late Tortonian, in the easterner Gubbio ad Mt Vicino units [15–17,21–23].

The various tectonic stages which involved the basin were marked by major changes in local and regional stratigraphy, in the sedimentary rates, in turbidite facies, and arenite composition.

An additional factor that had a significant influence on foredeep sedimentation and its duration over time was the emplacement, on the western border of the basin, of allochthon terrains of Tuscan and Ligurian pertinence. The latter became, for a long period, the source areas of transversal sedimentary inputs and of regional-scale olistostromic bodies that were embedded at various levels within the Marnoso Arenacea Fm [17,19,24,25].

In such a general picture, an overall and updated synthesis of the depositional architecture and the structural setting of the pre-Apennine foreland basin is still lacking, and the timing of deformation has been defined only in broad terms, except for restricted sectors.

The first objective of our work is therefore to revise the stratigraphy of the Miocene successions at the scale of the entire basin, defining the vertical and lateral relationships among the different units.

Once the stratigraphic framework has been reconstructed, we highlight those depositional events that were driven by tectonics and are useful to define a high-resolution timing of the deformational events that affected the foredeep, from its onset to the complete accretion into the chain.

In particular, an issue that was not previously explored in-depth, and deserves to be addressed, is the influence exerted on sedimentation by the concurrent activity of contractional structures and foreland ramp faults, both displacing, during the Middle-Late Miocene, the Meso-Cenozoic carbonate substratum [7,21,22].

We have pursued the aforementioned goals by analyzing, reviewing, and correlating a total of 24 stratigraphic sections representative of the entire foreland basin system of the Umbria pre-Apennine, most of which unpublished. The sections, which have been studied through quantitative nannofossil biostratigraphy, are located in the six following key-areas; from west to east, these are (Figure 1): (1) Mt Rentella, (2) Perugia Mts ridge, (3) Alpe della Luna, (4) M. S. Maria Tiberina, (5) Pietralunga-Mt Urbino ridge and (6) Gubbio-Serra Maggio.

This paper aims to provide an overview of all turbidite successions referred, in the literature, to the Umbrian domain, giving useful chronological constraints to future research on the tectonics and geodynamics of the Tyrrhenian-Apennines system, focusing on the Miocene deformational stages.

We are also confident that our work can be explanatory of a methodological approach, based on the intimate integration of stratigraphy and tectonics, that we consider highly effective in the study of foreland basin systems.

2. Materials and Methods

2.1. Physical Stratigraphy

We started from an accurate revision of a huge bulk of unpublished lithostratigraphic data, acquired during original research surveys and the CARG project 289-Città di Castello Sheet of the Carta Geologica d'Italia [20].

These data, which had already been calibrated by nannofossils biostratigraphy [26], were subsequently homogenized with the datasets reported in previous works dealing with the tectono-stratigraphic setting of the Mt Rentella [15], M. S. Maria Tiberina [7,21,27], and Mt Nero [18] units. Further stratigraphic sections have been recently studied in sectors of the pre-Apennine where, in the course of this study, the need for further integrative data emerged.

An accurate physical-stratigraphic analysis was carried out on all the sections, to distinguish the stratigraphic units of various ranks within the Miocene successions, such as formations, members, facies associations, and key beds.

Logging was carried out at a 1:50 scale factor, which allowed us to collect and describe all the major sedimentary features characterizing the different types of strata and their composition and provenance, and ultimately formulate hypotheses on the depositional mechanisms and the types of causative gravity flows.

An expeditious compositional determination was also performed by comparing hand specimens with the petrofacies described in several type-sections of the high Tiber Valley by [28–31]. These determinations, integrated with facies analysis (based on the classification scheme defined by [32]) and paleocurrent measurements, allowed us to distinguish the arenites of alpine provenance, generally characterized by fine grain size, distal facies, and low thickness, from the deposits originated by transversal supply [25], which were fed from piggyback basins lying above the Apennines front.

2.2. Biostratigraphy

During stratigraphic logging, the sections were sampled for micropaleontological purposes.

The marly samples collected for the biostratigraphic analysis were prepared as smear slides according to the procedures indicated by [33–35] and subsequently studied under a polarizing optical microscope at 1000× with both parallel and crossed nicols.

The microscopic analyses of nannofossil assemblages were referred to the biozonations schemes proposed by [35–37] for the Late Oligocene and Early-Middle Miocene of the Mediterranean domain (Figure 2).

The adopted schemes still show a better resolution either with respect to the "standard" ones [38,39] or to the recent upgrades of the tertiary nannofossil biostratigraphy [40,41], which have been mainly defined based on ocean bio-events recognized on a global scale. Other updated schemes of the Miocene Mediterranean nannofossils biostratigraphy are available but, having reviewed only some certain time-intervals (the Burdigalian stage in [42] and the Langhian stage in [43]), their use would lead to some problems of chronostratigraphic correlations with the biozonation proposed by [35–37].

The latter works, including the entire Miocene, are therefore still to be preferred for the study of stratigraphic successions that cover a large part of this epoch.

In each sample, the identification of marker species has been integrated by quantitative analysis, carried out following the methodologies developed by several authors [35,44].

It consists of two kinds of counting: the first counts the species percentages within a population of 300 specimens of the whole assemblage; the second is carried out only for the most significant species whose percentage is determined in comparison with a fixed number of coccoliths pertaining to the same genus (f.i. counting of *S. heteromorphus* within 100 sphenoliths).

The application of quantitative methodologies has also allowed the aforesaid authors to establish new bio-horizons based on variations in the abundance of some marker species, such as the First Common Occurrence (FCO), the Beginning of the Paracme (PB), and the End of the Paracme (PE) of *Sphenolithus heteromorphus.*

Finally, in the case of the individuals belonging to the genus *Reticulofenestra*—which, although having been studied in depth [34,45–49], still present classification problems—a biometric approach has been applied. The latter led to referring to *Reticulofenestra pseudoumbilicus* individuals with size > 7 [35,37,44,50] and to distinguish them from those characterized by a smaller size (which have been, in turn, shared in two classes, respectively characterized by size <6 µm and >6 µm).

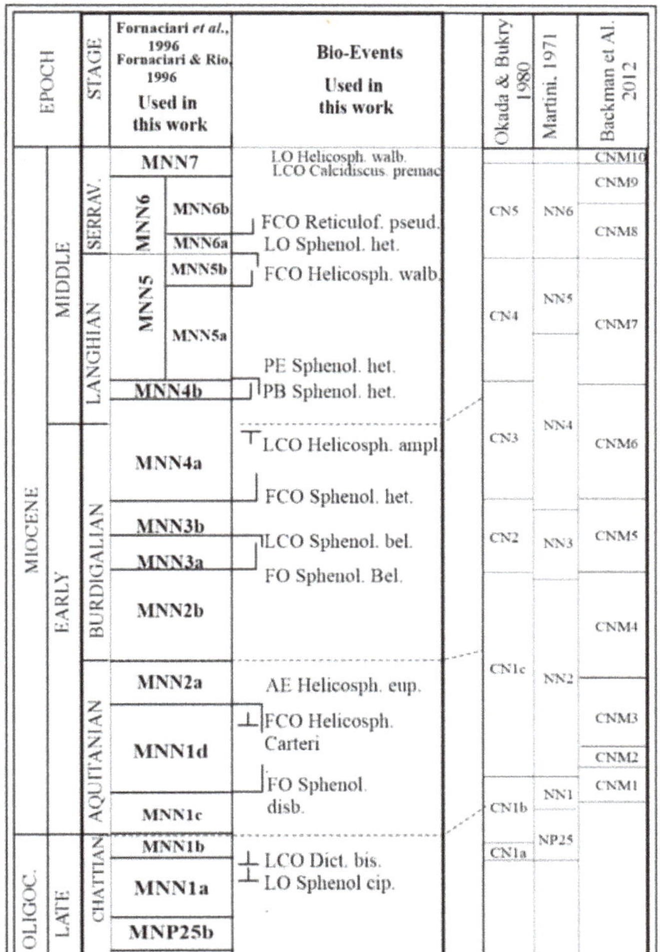

Figure 2. Biostratigraphic scheme for the Upper Oligocene-Middle Miocene interval. The scheme correlates the nannofossils biozonations proposed by [35,37], used in this work, with the Standard ones by [38,39], recently updated by [41]. The scheme is also integrated (bold rectangle) with the main bio-event characterizing the Mediterranean area.

The variation in abundance of the marker species, within each succession, was highlighted by plotting the relative distribution diagrams alongside the corresponding lithostratigraphic Log.

2.3. Field Geology

A large part of the fieldwork was carried out in the past years using the traditional geological mapping techniques. These former surveys, however, were recently revised and integrated with digital surveys and new data collection, aided by an Apple iPad-Pro in which a dedicated mapping software, whose details are given in the following section, was installed.

Due to the considerable extension of the study area, most of the structural survey was performed in selected key outcrops to characterize the major tectonic contacts and the spatial relationships between the Tectono-Stratigraphic units, synthesized in Figure 1.

Two commercial seismic lines provided by [51], hereinafter referred to as L1 and L2 (traces in Figure 1), have been interpreted to assess the present thickness of Miocene foredeep deposits and detect the geometry of the contractional deformations. The seismic interpretation through Move software was performed on both the aforesaid lines.

3. An Overview on the Tectono-Stratigraphic Units of the Umbria Pre-Apennines

3.1. Mt Rentella Unit

The Mt Rentella unit (REN) [15] crops out in a narrow ~NS striking belt, tectonically sandwiched between the Tuscan Falterona Nappe, to the west, and the Marnoso Arenacea Fm of the Mt Nero Unit, to the east. Its peculiar stratigraphy justifies its classification as an independent tectono-stratigraphic unit derived from a paleogeographic domain located in an intermediate position between the two aforementioned units.

In the type locality (sections Im, Ms, Ml, and Cv, Figure 1), the REN includes, from the bottom, the polychromic marls of Mt Rentella, the cherty marls of Mt Sperello, and the Castelvieto turbiditic sandstones (also referred to as La Montagnaccia Fm by [13,52–55]). The nannofossil assemblages [15] show that the polychromic marls are referable to the MNP25b-MNN1d biozones (Late Oligocene-Aquitanian), the cherty-marly interval to the MNN1d-MNN2a biozones (approximating the Aquitanian-Burdigalian boundary), and the turbidite succession deposited during the MNN2a-MNN2b biozones (Early Burdigalian) [15,53,54]. A bio-chronostratigraphic correlation table, showing the overall stratigraphy, reconstructed "mending" the portions of the REN succession logged in four reference sections (I Molini = Ml, Monte Sperello = Ms, Castelvieto = Cv, and Il Molino = Im, location in Figure 1) is provided in Figure S1 (Supplementary Materials). Based on these results, the Mt Rentella sequence cannot be biostratigraphically correlated either to the topmost part of the Falterona Nappe, which is older [52–57], or to the outcropping turbiditic sequence of the Umbria domain, which is younger [7,17,18].

As regards the compositional data concerning the fine-grained lithic fragments, the arenites characterizing the turbidite succession of the REN differ sensibly from the Macigno sandstones, whereas they are comparable with some Alpine-fed beds of the Marnoso Arenacea Fm [31–51,58,59].

The reported litho- and bio-stratigraphic constraints suggest that the REN represents a tectonic slice originated by the innermost marginal sector of the Umbria domain which, during the Chattian and most of the Aquitanian, was located in the peripheral bulge of the Tuscan foredeep, which will never be reached later on, by the deposition of the Macigno turbidites. Such a hypothesis would account for the remarkable facies affinity between the polychromic marls of Mt Rentella and the "Scisti Policromi" Auctt. occurring at the top of the Tuscan Scaglia, the former being characterized by a larger content of calcium carbonate.

In the study area, the succession of the REN is rootless, being detached on the polychromic marls and tectonically superimposed to the Marnoso Arenacea Fm of the Mt Nero unit. For this reason, its stratigraphic passage to the original carbonate substrate was nowhere observed.

According to [15], this substrate would correspond to the succession cropping out in the Mt Peglia-Mt Piatto ridge (south-western Umbria), characterized by an "Umbria type" facies but showing some affinities with the Tuscan terms, during the Oligocene-Early Miocene [60].

3.2. Units Derived from the Marnoso Arenacea Foredeep

The Marnoso Arenacea foredeep basin [24,61,62] developed, since Early-Middle Burdigalian times, in front of a tectonic pile made of the REN unit and the overlying Falterona Nappe. The turbidite deposition lasted, within this basin, until the Late Miocene, when it was tectonized and completely incorporated within the Umbria pre-Apennines.

After the seminal work of [10], several studies contributed to identify, within the deformed Marnoso Arenacea of the Umbria area (MAR hereinafter), a number NW-SE trending sectors characterized by significant stratigraphic differences and diachronic tectonic evolutions [7,11–18,21,28,29,61–66].

These sectors, delimited by regional-scale thrust surfaces, can be considered as tectono-stratigraphic units, namely thrust sheets whose internal stratigraphy is substantially homogeneous but showing significant differences compared to the adjacent ones (Figure 3). In the present work, for reasons of conciseness, the term "Unit" is used with the meaning above.

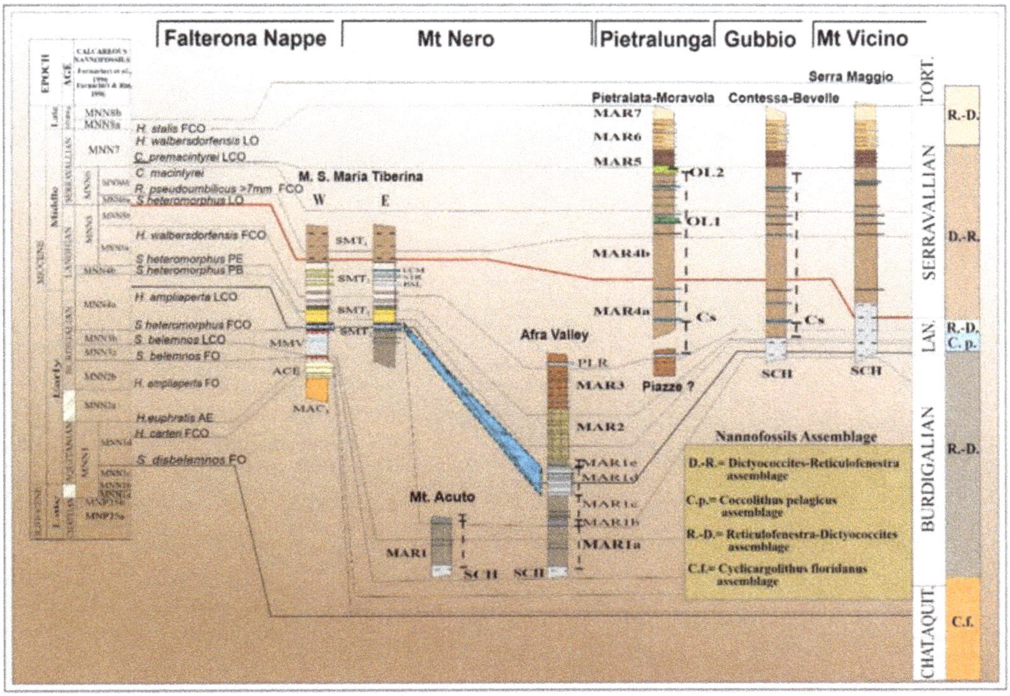

Figure 3. Comprehensive stratigraphic scheme of the Umbria pre-Apennine Marnoso Arenacea succession cropping out East of the Falterona Nappe and Mt Rentella frontal thrusts to the inner side of the carbonate chain. The columns represent, to all effects, composite Logs obtained by correlating the studied sections, located in Figure 1.

Within each Unit, sedimentation was mainly controlled by the following tectonic processes: (a) flexural retreat of the regional monocline, which caused the system depocenter-ramp-peripheral bulge of the foredeep to shift eastward [9], (b) progressive emplacement, along the western border of the foredeep, of allochthon units of Tuscan and Ligurian pertinence [7,15], (c) activity of foreland ramp synsedimentary normal faults, and (d) nucleation and progressive growth of compressional structures in the carbonatic substratum of the basin.

Ultimately, the MAR Units derive from pre-contractional ~ chain-parallel isopic zones which differed from each other in what concerns the age of onset of turbiditic sedimentation, its duration over time, and, partly, the sedimentological and compositional characteristics of the beds.

Based on the aforesaid criteria, in the Umbria pre-Apennines, the Mt Nero, Pietralunga, Gubbio, and Mt Vicino Units have been recognized, from west to east (Figures 1, 3 and 4). Their sedimentary and tectonic evolution is synthetically described below.

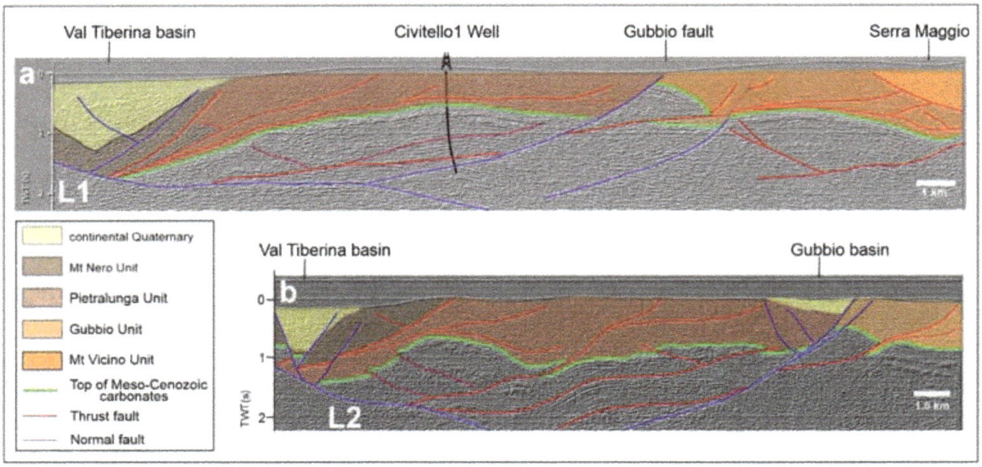

Figure 4. Interpreted seismic reflection profiles crossing the deformed Marnoso Arenacea, cropping out between the Tiber valley Quaternary graben and the Apennine calcareous chain; (**a**) = line L1, (**b**) = line L2 whose traces are shown in Figure 1. Both the "clean" sections are taken from [51] and here re-interpreted in light of the different aims of our work.

3.2.1. Mt Nero-Unit

The Mt Nero Unit includes the more internal successions of the MAR, which originated from the Burdigalian-Langhian depocenter of the Umbria foredeep.

From a tectonic point of view, it belongs to the so-called "internal Umbria Romagna parautochthon" [16], which overrides, along a NW-SE trending thrust fault (Bocca Trabaria thrust), the Pietralunga Unit (Figure 1). Clear exposures of the Mt Nero succession can be observed in the Perugia Mts ridge and on the east sides of the northern Valtiberina, where these successions are also referred to as "Alpe della Luna sequence" (Figure 1).

In previous studies, the lithostratigraphy of the Mt Nero succession had been pointed out for both these areas [17,28,64,65,67], but an updated biostratigraphic calibration was defined only during the recent survey of the 289-Città di Castello Sheet of the Carta Geologica d'Italia [20].

In the Alpe della Luna (Figure 1), the stratigraphic sections are quite continuous and, despite some minor gaps, allow for an investigation of the entire succession lying over the Schlier Fm (Figure 3 and Figure S2). In this area, the MAR, from the bottom, is composed by:

(1) a pelitic-arenitic member (mr hereinafter), named as MAR1 or Case Spertaglia mr, extending from the MNN3a to the MNN4b nannofossil biozones; it includes (section Spertaglia = Sp) a lower facies characterized by prevalent thin-bedded alpine-supplied turbidites (fine-grained and mica-rich arenites with NW-SE provenances) and an upper facies dominated by thick-bedded transversally-supplied calcarenites and hybrid arenites, containing coarse-grained lithoclasts of Ligurian origin;
(2) an intermediate massive sandstone member, consisting of arkosic transversal-supplied turbidites referred to as MAR2 or Mt Casale mr, developing across the MNN4b to MNN5a biozones (section Monte Casale = Mc);
(3) an upper pelitic-arenaceous member referred to as MAR3 or Vesina mr, straddling the MN5a-MN5b biozones, mainly consisting of alpine-supplied turbidites showing sub-arkosic composition and laminated tractive intervals (Section Vesina = Ve); in the uppermost part of this member, the Poggio La Rocca Mega-bed, a coarse-grained calcarenite, which in the past was incorrectly correlated with the Contessa key-bed [28], occurs.

In the Perugia Mts ridge, where only the two lower mrs of the MAR crop out, the exposures are discontinuous (Figure 1) and separated by a thrust fault which splits the Mt Nero Unit in two distinct thrust sheets, the western or "Mt Acuto" and the eastern or "Mt Corona" tectonic elements. Consequently, the stratigraphic continuity between the two members cannot be directly observed.

The Mt Acuto element consists only of the pelitic-arenaceous lower facies of the MAR1 mr, referable to the MNN3a-MNN4a biozones (sections Balconcelli = Ba, Corciano = Co and Monestevole = Mn, location in Figure 1, stratigraphy in Figure 3 and Figure S2) [18].

The Mt Corona element is instead characterized, at the surface, by the higher interval of the MAR1 mr and part of the MAR2 mr (Section Fosso della Badia = Fb in Figure 1), which, consistently with the section of Mt Casale, have been dated to the Middle-Late Langhian, MNN4b-MNN5a zones [18].

Synthesizing, within the entire Mt Nero Unit, the beginning of the MAR sedimentation (i.e., the onset of the foredeep stage s.s.) can be referred to the upper part of the Early Burdigalian, being the lowest passage to the Schlier Fm identifiable in the MNN3a biozone.

Conversely, the top of the Fm is markedly diachronic because it corresponds, in the western sector (Mt Acuto) to the intermediate portion of the MAR1 (MNN4a subzone, Late Burdigalian) and in the eastern sector (Alpe della Luna) to the uppermost MAR3 (MNN5b, Late Langhian; Figure S2).

The Mt Acuto succession does not include the higher part of the MAR1 because, in the internal sector of the foredeep, the MAR sedimentation was interrupted, during the MNN4a biozone, by the emplacement of the Falterona Nappe.

East of the leading edge of the Falterona Nappe, in the easternmost portion of the Mt Nero Unit, the topmost MAR1 mr and the following MAR2 (Late Burdigalian-Early Langhian) and MAR3 (Early-Late Langhian) members continued regularly to deposit, at least up to the top of the MNN5a biozone.

3.2.2. The Monte Santa Maria Tiberina Succession

The M. S. Maria Tiberina area has long been considered a crucial area to investigate the connection between the stratigraphic evolution of the Marnoso Arenacea Fm and the progressive emplacement of the Falterona Nappe along the western boundary of the foredeep.

Conflicting interpretations were formulated in the past on the outcropping Miocene succession, which has recently been re-named as M. S. Maria Tiberina Fm (SMT) in the 239-Città di Castello Sheet of the Carta Geologica d'Italia [30].

The stratigraphic reconstructions formulated by [7,21,30], based on robust biostratigraphic constraints, solved the debate between "autochthonist" [28,68] and "allochtonist" [10,11] interpretations, showing the SMT to have sedimented upon two different substrates, that are the Falterona Nappe and its semi-allochthon cover, in the internal sector, and the lower part of the Marnoso Arenacea Fm (MAR1 mr), in the external one (Figure S3).

More precisely: (i) to the West, the SMT Fm deposited unconformably over the Vicchio Marls of the Late Burdigalian age (MMN4a nannofossil subzone) and, after that, the Falterona Nappe was thrusted upon the inner side of the Umbria foredeep; (ii) in the intermediate zone (Gioiello syncline), the SMT covered the Nappe front and sealed it; and (iii) in the external sector (Mt Cedrone-Uppiano, Section Mr, in Figure 1), the SMT Fm deposited conformably on the Marnoso Arenacea MAR1 mr, whose topmost beds also show a Late Burdigalian age (MNN4a subzone) (Figure S3).

The integrated stratigraphic analysis of the SMT Fm (Section Sl in Figure 1) provides additional information about the timing of the former contractional phases affecting the succession of the Umbria domain.

In fact, the distribution of benthic foraminiferal taxa in the intermediate-to-upper part of this Fm clearly points to a shallowing upward trend of the paleodepth [27] during the Early Serravallian (the boundary between the MNN5b-MNN6a biozones).

This marked change, preluding the end of SMT sedimentation and also the emplacement of a minor olistostromic body, topping the formation [30], can be related to the growth of incipient anticlines involving the underlying carbonate multilayer of the Mt Nero Unit.

According to this interpretation, the contractional tectonics affected the western Umbria domain from the Middle-Late Serravallian, leading to the thrusting and folding of the MAR and the re-folding of the overlying Falterona Nappe front.

3.2.3. Pietralunga Unit

The Pietralunga Unit consists of a succession of turbidites and hemipelagites of Middle Langhian-Late Serravallian age [12,16] deformed by east-verging folds and minor thrusts (Figures 1 and 3).

Its structural arrangement and the thickness of its succession can be well appreciated in two SW-NE striking commercial seismic lines, published by [51], that cross nearly the entire Umbria-pre-Apennines, from the internal contact with the Mt Nero Unit as far as the Gubbio Normal Fault to the East (Figure 4, traces in Figure 1).

Although these profiles have been acquired for deeper targets, deformations affecting the Pietralunga Marnoso Arenacea MAR with the adjacent Mt Nero (to the west) and Gubbio (to the East) Units.

The proposed interpretations allow us to refer the Pietralunga Unit to an imbricate fan, detached on the top of the Meso-Cenozoic carbonates, in which each splay thrust branches upward, in-sequence, from the basal sole thrust.

Nevertheless, some faults seem to penetrate the carbonates, in correspondence with the major structures of Perugia Mts and Gubbio, refolding the overhanging shallow thrust.

In the intermediate part of the L2 line, in a sector unaffected by significant thrust and tectonic doubling, the top of the Meso-Cenozoic carbonates is located at a pseudo-depth of ~1 s TWT, which, assuming a v = 4.0 km/s seismic velocity [51], provides, for the Miocene succession including the Bisciaro, Schlier, and MAR Fms, a maximum thickness of ~2000 m.

Within the Pietralunga Unit, the age of the basal passage of the turbidites to the Schlier Fm is undetermined, as well as the latero-vertical stratigraphic relationships with the Mt Nero succession, because the lower part of the Unit is buried, and the occurrence, below it, of the MAR1-MAR3 members is uncertain.

The presence of Early Langhian terms, possibly with reduced thickness compared to the western Umbria succession, is suggested by some limited outcrops of a pelitic-arenaceous succession containing a 7 m-thick calcarenite, sampled in the Piazza area (5 km SE of Pietralunga). This section (Pz in Figure 1) has been tentatively related, in age and composition, with the topmost MAR3 of the Mt Nero Unit including the Poggio La Rocca marker bed (Figures 1 and 3).

The Pietralunga succession shows an outcropping thickness > 1000 m and, based on the facies analysis, can be divided into four members, hereinafter referred to as MAR4–MAR7.

The ~750 m-thick MAR 4 mr spans in age from the highest part of the MNN5b to the topmost MNN6b nannofossil biozones (Late Langhian-Middle Serravallian) according to [26] and our unpublished data (Figure S4), whereas all the remaining MAR5-7 mrs fall within the MNN7 biozone of the Late Serravallian age.

The overall stratigraphic reconstruction of this member was carried out by correlating eight sections (Vm, Sc, Po, Pz, Ss, Pi, Mo, and Mf, full names and location in Figure 1), that have been logged and analyzed from the litho- and bio-stratigraphic points of view.

The MAR4 mr is mainly characterized by the typical association of marls, arenite layers, and scattered media to thick-bedded calcarenites and hybrid arenites, with a predominant A/P ratio varying from <1 to <<1.

Despite an apparent monotonous facies, this member embodies several outstanding layers, including the well-known Contessa mega-bed, the "Colombine" calcarenites (occurring above the Contessa bed [25,66]) and some very thick and laterally continuous arenite and hybrid arenite beds, useful as key-layers for both geological mapping and basin analysis purposes (Figures S4 and S5 and detailed stratigraphy in [26]).

As aforesaid, the upper part of the Pietralunga succession is characterized by three further members, all of them falling in the MNN7 biozone, Late Serravallian in age.

The MAR5 mr corresponds to a quite thin (nearly 150 m) pelitic-arenaceous interval, characterized by thin-bedded fine-grained alpine-supplied arenites including some slump episodes.

The passage to the MAR6 mr is marked by a sharp increase in bed thickness (thick to very thick beds with an arenitic portion up to 8 m) and by the complete absence of calcarenite layers.

Finally, the MAR7 mr consists mainly of a pelitic-arenaceous facies association, characterized by a thinning-upward trend and a gradually decreasing A/P ratio, until the complete disappearance of the arenite beds.

The upper part of the MAR4 mr contains at least two large episodes of submarine sliding [12,24,25,66], the lower of which, referred to as "Lame-Castiglione olistostrome", occurs nearly 600 m above the Contessa bed [66], whereas the uppermost one, or "San Faustino-Scritto olistostrome", is located at least another 400 m above the former (Figure 3 and Figure S4).

The two olistostromes consist of wide lenses of chaotic materials within which disrupted rock fragments of different origin and size (up to decametric strata-fragments) are scattered in a polychromic sandy-clayey, and locally scaly, matrix. These lithotypes are derived from the Tuscan succession (Scaglia Toscana, jaspers, Triassic gypsum, and anhydrites) and from the Liguride units (Alberese and Palombini shales).

Generally, a thick interval, with a few meters, characterized by slumped intraformational materials, occurs at the base of both the olistostromes.

In the southern part of the study area the upper olistostrome consists of a quite continuous and undamaged stack of strata ("Scritto Limestones", according to [17,24]) which should be correlated with the sub-Ligurian Canetolo succession and the Ligurian units [17]. This attribution implies that, during the Middle Serravallian, the Apennine allochthon had reached the frontal part of the Falterona Nappe.

The detailed lithostratigraphy of the sections studied within the Pietralunga Unit exceeds the objectives of this work.

Anyhow, a composite Log, synthesizing the reconstructed stratigraphy of the entire Unit and showing the occurrence and position of the significant key layers, used for horizontal correlations, is provided in Figure S4.

For a more detailed description of all the sections mapped in Figure 1, the reader is sent back to the original work by [26].

3.2.4. Gubbio Unit

The Marnoso Arenacea of the Gubbio Unit overlies the Meso-Cenozoic multilayer of the homonym anticline in eastern Umbria [12,42,69].

In the reference sections of the Contessa Road and Bevelle (Cs and Be, in Figure 1), the Late Langhian—Serravallian succession, from the Contessa mega-bed (MNN5b biozone) upward [26], is well correlated with the Pietralunga one, in what concerns the dominant facies and the occurrence of the major key-beds (compare Figures S4 and S5).

Conversely, in the aforesaid sections, the age of the basal stratigraphic passage to the Schlier Fm differs from that found in all the western outcrops, being localized in the Early Langhian, highest MNN5a subzone.

The absence of the MAR1-MAR3 members (Figure 2) implies that, during the entire Burdigalian-Early Langhian time span, the Marnoso Arenacea Fm did not deposit in this outer sector of the foredeep.

Actually, in the "Contessa" section (Figure 1), the homonym key-layer is placed nearly 105 m above the base of the MAR (Figure 2 and Figure S5), whereas it was drilled in the Mt Civitello well (location in Figure 1, Log available at: https://www.videpi.com/videpi/pozzi/dettaglio.asp?cod=3896; accessed on 10 February 2021) at least ~700 m above.

The significant reduction of thickness of the pre-Contessa turbidite succession passing from the Pietralunga to the Gubbio Units occurs quite abruptly, straddling the Gubbio normal fault, and can be explained by its synsedimentary activity during Burdigalian-Langhian times.

The interpretation of seismic reflection data by [70] confirms this hypothesis, showing that the Langhian terms of the Pietralunga Unit display an eastward-thickening wedge-shape geometry that is compatible with a growth on the Gubbio fault.

The hypothesis of an early structuring of the Gubbio anticline suggested in [12] could also explain the absence of the lower Langhian Marnoso Arenacea in this area. However, even if it cannot be ruled out a priori, we note that it does not explain the eastward-thickening of the succession highlighted by [70], west of the Gubbio fault.

A further significant feature, which differentiates the Gubbio succession from the more internal ones, is the absence of olistostromic bodies at all stratigraphic levels.

The lateral continuity of the Serravallian terms from the Pietralunga to the Gubbio units (including the related marker beds) leads us to exclude that the lack of olistostromes was due, in that time, to a subdivision of the foredeep into several smaller basins. More reasonably, it was caused by the emplacement mechanism of the olistostromes that, moving by sliding processes on a tectonically unstable slope, could not reach the outermost areas of the basin.

Finally, our biostratigraphic data do not confirm the occurrence of the Tortonian in the highest part of the succession, as already suggested in [12,69].

3.3. Mt Vicino Unit

The Mt Vicino Unit, which is the easternmost unit of the Umbria pre-Apennine, is located just at the back of the present carbonate chain. In this area the upper boundary of the Schlier Fm was dated to the Early Serravallian and the MAR Fm to the Early Serravallian-Early Tortonian (section Bf, Figure 1) [26,71].

During the uppermost part of Early Tortonian and the lower part of the Middle Tortonian, the turbidite flows continued to affect only a residual furrow east of the Gubbio structure, leading to the deposition of the Mt Vicino Fm [72,73].

The M. Vicino Fm consists of a 600 to 1400 m thick turbidite succession in which, from the bottom, the following facies associations have been recognized: (i) thin-bedded pelite alternating with subordinate thin to medium bedded arenites, (ii) middle to thick-bedded arenaceous-pelitic turbidites, (iii) medium to thick-bedded arenites and bioclastic hybrid arenites.

The Mt Vicino Fm reaches its maximum thickness, of a few hundred meters, in the axial zone of the narrow depression which also corresponds to the core of the Serra Maggio syncline, suggesting that the latter was already in an advanced stage of structuring during sedimentation.

In other words, as proposed by [69], the Mt Vicino Fm deposited when the innermost folds of the Apennines were already growing, marking the late evolution of the MAR foredeep into the wedge-top basin stage.

Such a suggestion is confirmed by the occurrence, at intermediate stratigraphic levels within the succession, of slumped layers and coarse-grained poorly cemented turbiditic sandstones, showing Apennine provenance and both NW and SE paleocurrents [72,73].

4. Reconstructed Timing of Deformations

In the following, we reconstruct the timing of the main tectono-sedimentary stages and events which can be inferred based on the sedimentary evolution of the Marnoso Arenacea basin, described in the previous section.

Such stages, graphically schematized in the steps of Figures 5 and 6, document the setting, evolution, and progressive tectonization of the Umbria foreland basin system during the Miocene. Their temporal constraint is provided by the high-resolution biostratigraphic analysis performed on all the 24 studied sections, scattered throughout the entire northern Umbria pre-Apennines.

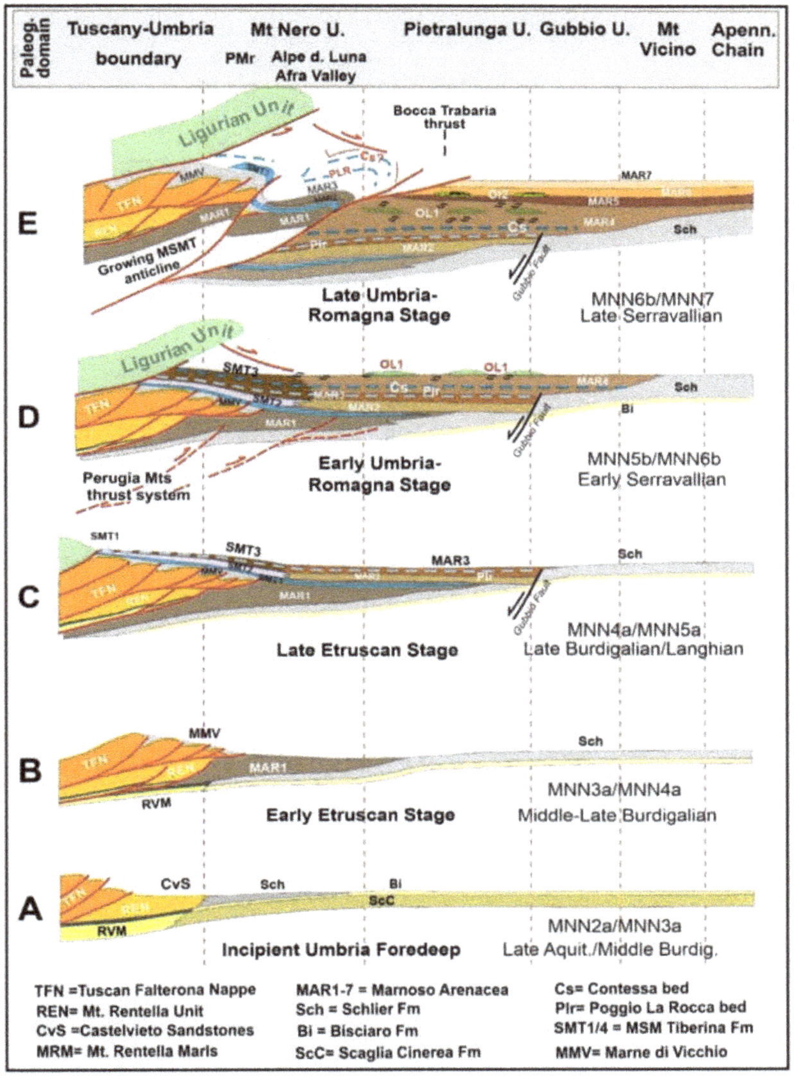

Figure 5. Exemplified tectono-sedimentary evolution of the Umbria pre-Apennines foreland basin system, during the Early and Middle Miocene; the stages, referred to the biozonation scheme proposed by [35–37] synthesized in Figure 2, are: (**A**) Incipient foredeep, (**B**) early Etruscan; (**C**) late Etruscan; (**D**) early Umbria-Romagna, and (**E**) late Umria Romagna stages (detail in the text; n.b.: sections are schematic and not in scale).

4.1. Incipient Umbria Foredeep (Aquitanian-Burdigalian Boundary; MNN2a-MNN3a Nannofossil Biozones)

The beginning of the regional subsidence related to the embryonic stage of the Umbria foredeep can be identified in the passage from hemipelagic (cherty marls of Mt Sperello) to turbidite deposition (Castelvieto-Montagnaccia Sandstones), within the REN Unit (Figures 5A and 6). At this stage the Apennine tectonic pile bounding the basin to the west included the Falterona Nappe, passively carrying the Liguride l.s. units and some wedge-top basins in which piggyback deposition took place (Celle Sandstones and Vicchio Marls, [7,74].

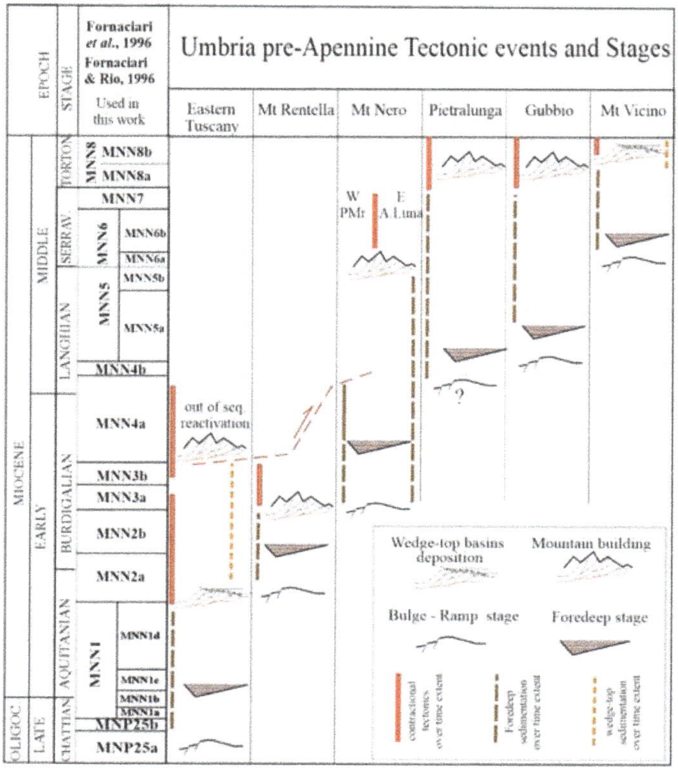

Figure 6. High-resolution timing of sedimentary and tectonic events (referred to the bio-chronostratigraphic scheme of Figure 2) recognized in each of the Units of the Umbria-pre-Apennines (detailed explanations in the text). The scheme is inspired by that proposed for the central Apennines by [6].

During the uppermost MNN2b-MNN3a zones (Middle Burdigalian), the REN Unit was affected by contractional deformations and progressively accreted into the orogen.

4.2. Early Etruscan Stage (Middle-Late Burdigalian; MNN3a-MNN4a Nannofossil Biozones)

The eastward migration of the orogenic wedge led to the setup of the Etruscan depocenter of the foredeep, in which the Mt Nero succession settled down (Figure 5B).

Up to the earlier Langhian, turbidite sedimentation affected exclusively this westernmost sector of the Umbria domain.

This inference is suggested by the stratigraphy of the Civitello1 borehole (Figures 1 and 4a, https://www.videpi.com/videpi/pozzi/dettaglio.asp?cod=3896; accessed on 10 February 2021), in which the base of the MAR of the Pietralunga Unit was attributed to an unspecified "Langhian"; moreover, the lithological descriptions of these lowermost turbidites show no facies affinity with the MAR1 and MAR2 members of Mt Nero Unit.

During the MNN3b-MNN4a (p.p.) interval, the Falterona Nappe underwent an out-of-sequence reactivation (Figures 5B and 6) and, after overtaking the deformed REN unit, arrived at the west side of the foredeep and covered its westernmost sector, interrupting the sedimentation.

At the beginning of Late Burdigalian (lower part of the MNN4a zone), its leading-edge can reasonably be thought to have reached its present location. This latter assertion is corroborated by the reciprocal consistency of the following points: (i) the topmost beds of MAR1, sampled just below the Falterona thrust in the Perugia Mts ridge (Monestevole = Mn section in Figure 1 and Figure S2, [18]), provided a MNN4a age, (ii) this same age was determined in the lowermost strata of the M. S. Maria Tiberina Fm, which has been proven to seal the front of the Tuscan allochthon in the Mt Cedrone area [21,27] (Figure S3), and (iii) the lacking evidence, in the younger and more eastern MAR of the Mt Nero Unit, of any signal of subsequent reactivation of the allochthon front.

4.3. Late Etruscan Stage (Late Burdigalian-Late Langhian; MNN4a-MNN5a/b Nannofossil Biozones)

This stage corresponds to the time which followed the emplacement of the Tuscan allochthon and pre-dates the nucleation of the major contractional structures within the western Umbria carbonates. During such an interval, most of the turbidite flows sedimented in the outer sector of the basin, giving rise to the upper MAR1 and the following MAR2 and MAR3 mrs.

Further west, in the wedge-top structural position, and also above the slope connecting the chain front with the foredeep depocenter (Figure 5C) Apennine-supplied low- and high-density turbidites, slumped bodies and pelite mud drapes continued to deposit (SMT3 and SMT4 mrs of the M.S. Maria Tiberina Fm).

No reliable data allows us to constrain the transversal extent of the foredeep in this stage beeing the lateral continuity of the Langhian succession not strictly constrained.

Nevertheless, the aforementioned stratigraphy of the Mt Civitello1 well indicates that, during the Langhian, the depocenter of the basin included the whole Mt Nero and part of the Pietralunga Units.

We can exclude with certainty that, during this stage, the MAR deposited east of the area where the Gubbio and Mt Subasio anticlines are presently located. This area was placed, up to the Early Langhian, in the peripheral bulge of the MAR foredeep, and the sedimentation of the Schlier Fm was still going on (Figures 5C and 6).

4.4. Early Umbria-Romagna Stage (Late Langhian-Early Serravallian; MNN5b-MNN6b Nannofossil Biozones)

We refer to this stage as "Umbria-Romagna" because our stratigraphic data, compared with the literature data collected over the past decades [3,9–13], undoubtedly demonstrates that a single and undivided foredeep developed in front of the Early Miocene northern paleo-Apennines.

The main argument supporting this inference is the lateral continuity of several key-layers over the entire MAR of the Umbria and Romagna area. These layers include the Contessa mega-bed, tens of "Colombina type" calcarenites, and some other noticeable arenite beds (Figure 3, Figures S4 and S5) [3,9,25,66].

During this stage, which was characterized by quite stationary sedimentary conditions, the depocenter of the foredeep was localized in correspondence to the present Pietralunga Unit, whereas its eastern side extended to the Gubbio zone (Figures 5D and 6).

In other words, at least since the MNN5b zone, the MAR Fm was settling, with fairly uniform thickness, into a single wide basin.

The peripheral bulge was shifted to the Mt Vicino and the inner chain areas, where up to the MNN6a-(6b?) zone, the deposition of the Schlier ramp-muds persisted (Figures 3, 5D, 6 and S6).

Orogenic contraction began to affect the older and more internal succession deposited during the Etruscan stage.

The onset of thrusting and, in particular, the growth of the Mt Cedrone anticline (see geological section of Figure S3) has been referred by [21] to the Early Serravallian (MNN6a subzone).

Actually, the sudden fining-upward evolution of the M. S. Maria Tiberina Fm and the concurrent shallowing of the seafloor [27] may be interpreted as the first sign of ongoing compression at shallow crustal levels.

Uplift processes appear to have involved the entire western sector of the basin since the very Early Serravallian, as no turbidites younger than Late Langhian (MNN5b zone) have ever been detected at the top of the Mt Nero Unit.

As better explained in the following Section 5, we hypothesize that this regional-scale uplift is the shallow effect of the activation of the regional thrust fault displacing the Permo-Triassic basement in the subsurface of the Perugia Mts ridge [75–77].

4.5. Late Umbria-Romagna Stage (Middle-Late Serravallian; MNN6b-MNN7)

We place the beginning of this stage, in correspondence with the earliest involvement of the Serravallian MAR, in the orogenic deformations.

In particular, we interpret the emplacement of the Lame-Castiglione olistostrome (topmost MNN6b biozone) as evidence that the front of the chain, which had now extended to the completely detached Mt Nero Unit, was thrusted over the inner Pietralunga succession (Figure 5E). The inner part of this latter Unit was subsequently detached from the underlying Meso-Cenozoic carbonates and, at least since the earlier MNN7 biozone, was progressively shortened "in-sequence". Due to such progressive deformation, during the upper MNN7 biozone the width of the foredeep underwent a severe reduction and, as a consequence, the higher stratigraphic members of the succession (MAR6–MAR 7 mrs) deposited only in the central-east part of the basin (Figure 5E).

The end of this stage can be located straddling the Serravallian-Tortonian boundary (top of MNN7 biozone), as we have not found Upper Miocene deposits in the Pietralunga and Gubbio successions (Figure 6 and Figure S6).

4.6. Wedge-Top and Accretion Stage (Early-Middle Tortonian; MNN8-MNN9)

At the beginning of the Late Miocene, the Langhian-Serravallian successions of both the Pietralunga and Gubbio Units were diffusely affected by contractional tectonics, giving rise to structures of variable scale: (i) regional macro-folds affecting the underlying Meso-Cenozoic carbonates (Figure 4), (ii) low-wavelength folds (some hundred meters to 1–2 km wide), detached at the Schlier Fm (Figure 4), and (iii) mesoscopic folds nucleated above local intra-formational decollements.

In the easternmost sector of the pre-Apennine, the growth of the Gubbio and Mt Subasio anticlines further reduced the active part of the foredeep, which, during the earlier Tortonian, was restricted to the Mt Vicino-Serra Maggio trough (Figure 1). This area was the only one located west of the chain, which hosted the MAR sedimentation during the Tortonian.

Above the MAR, the Mt Vicino sandstone (upper part of the Early Tortonian) deposited unconformably.

Reasonably, at the end of Early Tortonian, the further eastward progression of the contractional front caused the Mt Vicino sheet to override the innermost Umbria-Marche anticlines (Scheggia-Mt Maggio anticlines), which, according to [78] were in their initial stages of growth.

5. Discussion and Final Remarks

The systematic litho- and bio-stratigraphic study of the Umbria turbidite successions allowed for the recognition of the main evolutionary stages of the Umbria pre-Apennines' foredeep and the definition of the tectonic events that have involved the Miocene foreland basin system in this sector of the orogen.

Based on such high-resolution tectono-stratigraphic history, schematized in the steps of Figure 5, we have elaborated a kinematic model which describes the distribution of the structural paleo-domain over time (Figure 7).

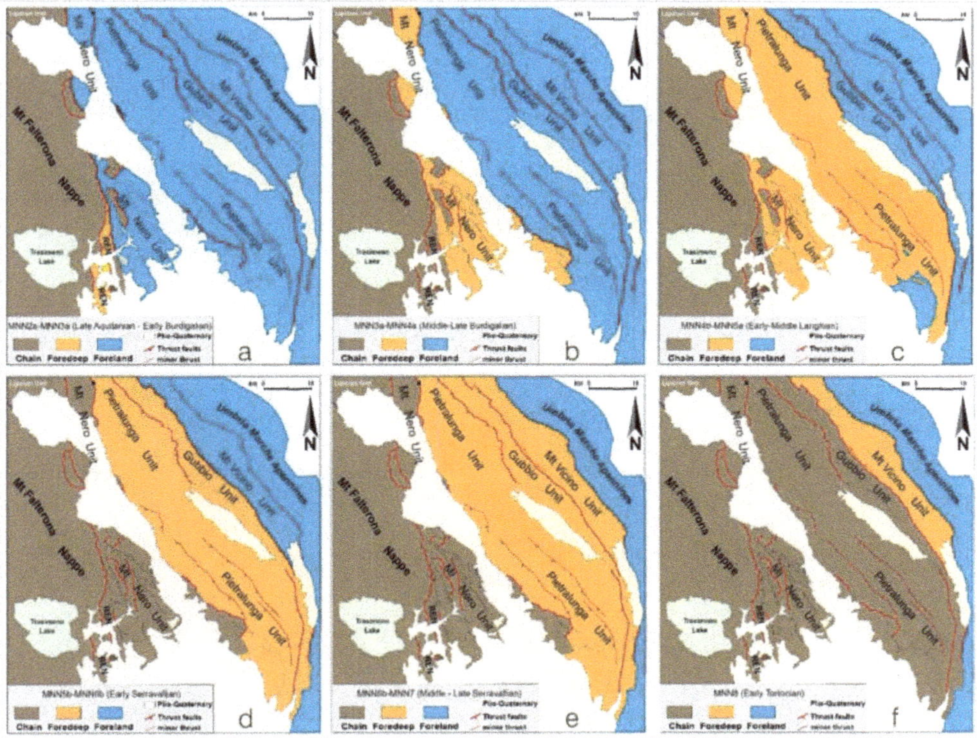

Figure 7. Distribution of the pre-Apennines structural paleo-domain, as inferred by the high-resolution tectono-stratigraphic timing summarized in Figure 6, during the: (**a**) Late Aquitanian-Early Burdigalian, (**b**) Middle-Late Burdigalian, (**c**) Early-Middle Langhian, (**d**) Early Serravallian, (**e**) Late Serravallian; (**f**) Early Tortonian. Note that the migrating paleo-domains are shown by the brown, orange, and blue colors and that the thrust faults refer to the present configuration, taken from Figure 1.

A first fundamental achievement of our reconstruction was the precise definition of the age of emplacement of the thrust-stack bounding west of the foredeep, which, at the end of the Aquitanian, included the Tuscan Falterona Nappe and the overlying allochthon units of Ligurian (l.s.) pertinence.

The over-thrusting of such tectonic pile above the Umbria domain is bracketed between the very Early Burdigalian and the Late Burdigalian and was articulated in two distinct steps: (a) "in-sequence" thrusting above the REN, during the MNN2b-MNN3a zones [15], and (b) "out-of-sequence" thrusting above the MAR, after overtaking the REN, which was, in turn, tectonically superimposed on the MAR of the Mt Nero Unit (Figure 6—Eastern Tuscany column- and Figure 7).

The out-of-sequence phase of the Falterona Nappe had not been previously recognized in Umbria and deserves to be verified in other sectors of the chain, e.g., in the Tuscan-Emilian Apennines, to understand if it must be considered a significant event on the geodynamic scale.

The eastward advancement of the Nappe, in Middle-Late Burdigalian times, progressively narrowed the Umbria foredeep of the early Etruscan stage and interrupted the sedimentation in its western sector [7,18]. The reduction of the basin-width, resulting in a lower availability of the accommodation space for the incoming gravity flows, might explain the increase in the average thickness of turbidite beds, observed at the passage from the lower member (MAR1-MNN3a-MNN4a biozones) to the intermediate one (MAR2 mr-MNN4b-MNN5a biozones) of the Mt Nero Unit (Figure S2). Such an increase is consistent with a sharp change in the foredeep physiography that, during the late Etruscan stage, evolved from an open submarine plain to a strongly subsiding and transversely confined basin. In this regard, we must also consider that, during the aforementioned stages, the foredeep basin did not reach its maximum W-E extent, being delimited eastward by a set of west-dipping sinsedimentary normal faults that offset the foreland ramp.

Although the influence of foreland extensional faults on the MAR sedimentation is poorly investigated, the Gubbio normal fault seems to have played a significant role [70], producing a sharp uplift of the peripheral bulge that was located, in Early-Middle Langhian times, in its footwall block (Figure 7c).

After the emplacement of the allochthon units, the Umbria domain appears to have experienced an interval with absent or scarce deformation, at least from the Middle Burdigalian to the Early Serravallian (MNN4b-MNN6a biozones)—that is, for most of the Early Umbria-Romagna stage (Figure 7b,c). Such a ~3.5 ma-long standstill can be bracketed between the time at which the front of the Falterona Nappe was sealed by the sedimentation of the M. S. Maria Tiberina Fm (lower MNN4a zone) and the first clear evidence of compression and uplift affecting the western Umbria carbonate multilayer (uppermost MNN6a zone).

We point out that the corresponding time interval, spanning from ~17 to ~13.5 Ma, matches the transition between the collisional and post-collisional stages of the Apennine orogenesis [79–83]. The latter, which is still in progress, is associated with the Tyrrhenian rifting and is characterized by the eastward migration of two coupled-sub-parallel and synchronous tectonic belts causing contraction at the front and extension at the rear of the chain.

The evidence of a break in the shift of compressive deformation toward the foreland indicates that the passage between these two stages did not occur with continuity, at least in the more peripheral zone of the orogen.

In such a zone, the change in the geodynamic regime seems not to have produced significant tectonic manifestations during the time required for the new stress-field to propagate outside the collisional suture.

We are aware that such a hypothesis is preliminary and highly speculative. Anyhow, we want to stress that having highlighted such a protracted stasis in the orogenic deformations is an unexpected result, worthy of being further investigated, to verify its consistency also in other areas of the Middle Miocene Apennine front.

We refer the compression of the western Umbria domain (Figure 7d) to the Early Serravallian (MNN6a-MNN6b biozones), as suggested by the shallowing-upward trend of the topmost M. S. Maria Tiberina Fm and by the subsequent interruption of its sedimentation [21,27].

During this time interval, a major thrust fault nucleated within the Permian-Triassic basement [51,75–77]. It displaced the Meso-Cenozoic Multilayer of the Perugia Mts ridge and, branching at shallower structural levels, gave rise to an imbricate fan and associated folds affecting the MAR.

We identify the main surface expression of this regional thrust in the frontal thrust of the Mt Nero Unit, which is well-exposed in the Alpe della Luna (Afra Valley) and has been recognized to discontinuously crop out also even further south, in the hydrographic left of the Tiber (Figures 1 and 4).

Considering the amount of the associated displacement and the considerable continuity of this structure, all along the entire central Umbria (Figure 1), we hypothesize its possible correlation with the "Mandriacce Line", described in the Marnoso Arenacea Romagnola [84], which causes the systematic superposition of Langhian terms above Serravallian ones.

According to our reconstruction, thrusting in western Umbria went on until the Late Serravallian, as shown by the timing of emplacement of the upper olistostromic slices (earlier MNN7a zones) within the MAR of the Pietralunga Unit (Figures 5D,E and 7e).

The few public subsurface data, available in the area comprised between the Tiber Valley and the Umbria-Marche chain, highlights that the contractional deformations affecting the Pietralunga MAR are arranged in a shallow imbricate whose basal decollement is localized at the top of the Meso-Cenozoic Carbonate multilayer (Figure 4a,b).

The geometry of such a thrust system, joined to the observation that the uppermost members of the succession (MAR5-7) occur only in the easternmost sector of the Pietralunga Unit, supports the inference of an "in-sequence" thrusting, within this unit, during the very Late Serravallian-Early Tortonian (uppermost MNN7b-MNN8 biozones, Figures 5 and 6).

However, it should be noted that such argument is not conclusive, as it cannot be excluded that the complete absence of such members, in the innermost areas, might also be due to widespread erosion, rather than their non-deposition.

Just after the beginning of the Early Tortonian, the Umbria foredeep was entirely undergoing compression, whereas the origin of the Serra Maggio syncline can be confidently placed in the Late Tortonian, taking into account the piggyback sedimentation of the Mt Vicino sandstones.

Subsequently, the so-called "intra-Messinian" phase [22,72] split the previously unitary foredeep in several minor basins interposed to the rising Umbria–Marche anticlines. Inside them, a terrigenous-evaporitic sedimentation occurred up to the earlier Pliocene.

At the end of this latter period, after the uplift and accretion into the chain of such residual elongated furrows, the tectono-sedimentary history of the Umbria foreland basin system was over.

Supplementary Materials: The following are available online at https://www.mdpi.com/2076-3263/11/2/97/s1, Figures S1–S6.

Author Contributions: F.B. and L.L. conceived the work. F.B. and D.C. performed the fieldwork. L.L. carried out the nannofossil analyses and the biostratigraphic determinations. F.B. wrote the manuscript. D.C. and F.B. interpreted the seismic lines and prepared the figures. D.C. geo-referenced the maps and carried out the final review of the manuscript. All authors have read and agreed to the published version of the manuscript.

Funding: This work was funded by DiSPUTer Departmental Research grants 2020 (Resp. F. Brozzetti).

Acknowledgments: We thank two anonymous referees for the constructive comments, that have been useful for improving the manuscript.

Software Used in This Work: Field survey was partly performed through Fieldmove software (Move™, produced by Midland Valley Exploration Ltd. 2018 Glasgow, and Petroleum Experts Edinburgh, Scotland UK—https://www.petex.com/products/move-suite/digital-field-mapping/, accessed on 10 February 2021) installed on an Apple iPad-Pro. The newly acquired data were managed in a GIS database elaborated through ArcGIS v.10.8. Figure 1 was created using ArcGIS v.10.8 (http://desktop.arcgis.com; accessed on 10 February 2021). Figures 2–6 and all the supplementary figures were drawn using CorelDRAW graphics suite 2020 (https://www.coreldraw.com; accessed on 10 February 2021). The interpretation of seismic lines and their conversion to depth (Figure 4) were carried out with the Move suite software 2019.1.8 (https://www.petex.com/products/move-suite/; accessed on 10 February 2021), in particular using the 2D depth conversion tool. Microsoft Excel was used to elaborate the distribution diagrams for the marker species of nannofossils, placed alongside the stratigraphic columns of Figures S2, S4 and S5.

Conflicts of Interest: The authors declare no conflict of interests.

References

1. DeCelles, P.G.; Giles, K.A. Foreland basin systems. *Basin Res.* **1996**, *8*, 105–123. [CrossRef]
2. Ori, G.G.; Friend, P.F. Sedimentary basins formed and carried piggyback on active thrust sheets. *Geology* **1984**, *12*, 475–478. [CrossRef]
3. Argnani, A.; Ricci Lucchi, F.R. Tertiary Silicoclastic Turbidite Systems of the Northern Apennines. In *Anatomy of an Orogen: The Apennines and Adjacent Mediterranean Basins*; Vai, G.B., Martini, I.P., Eds.; Kluver Academic Publishers: Amsterdam, The Netherlands, 2001; pp. 327–350. [CrossRef]

4. Patacca, E.; Scandone, P.; Bellatalla, M.; Perilli, N.; Santini, U. La zona di giunzione tra l'arco appenninico settentrionale e l'arco appenninico meridionale nell'Abruzzo e nel Molise. *St. Geol. Camerti* **1991**, 417–441. [CrossRef]
5. Cipollari, P.; Cosentino, D.; Esu, D.; Girotti, O.; Gliozzi, E.; Praturlon, A. Thrust-top lacustrine-lagoonal basin development in accretionary wedges: Late Messinian (Lago-Mare) episode in the central Apennines (Italy). *Palaeogeogr. Palaeocl.* **1999**, *151*, 149–166. [CrossRef]
6. Cosentino, D.; Cipollari, P.; Marsili, P.; Scrocca, D. Geology of the central Apennines: A regional review. *J. Virtual Explor.* **2010**, *36*, 1–37. [CrossRef]
7. Brozzetti, F.; Boncio, P.; Pialli, G. Early-Middle Miocene evolution of the Tuscan Nappe-Western Umbria foredeep system: Insights from stratigraphy and structural analysis. *Boll. Soc. Geol. It. Vol. Spec.* **2002**, *1*, 319–331.
8. Mutti, E.; Tinterri, R.; Benevelli, G.; di Biase, D.; Cavanna, G. Deltaic, mixed and turbidite sedimentation of ancient foreland basins. *Mar. Petrol. Geol.* **2003**, *20*, 733–755. [CrossRef]
9. Ricci Lucchi, F. The Oligocene to Recent foreland basins of the northern Apennines. *Spec. Publ. Int. Ass. Sediment.* **1986**, *8*, 105–139.
10. Ten Haaf, E.; Van Wamel, W.A. Nappes of the Alta Romagna. In *Fixism, Mobilism or Relativism: Van Bemmelen's Search for Harmony*; Van Der Linden, W.J.M., Ed.; Royal Geological and Mining Society of The Netherlands: Utrecht, The Netherlands, 1979; Volume 58, pp. 145–152.
11. De Feyter, A.J. The structure of the Northern Umbria Apennines, Italy. *Geol. Mijnbouw* **1982**, *16*, 183–189.
12. Menichetti, M.; Pialli, G. Geologia strutturale del Preappennino umbro tra i Monti di Gubbio e la catena del M. Petrano-M. Cucco (Appennino Umbro-Marchigiano). *Mem. Soc. Geol. Ital.* **1986**, *35*, 371–388.
13. Conti, P.; Cornamusini, G.; Carmignani, L.; Pirro, A.; Pizziolo, M.; Daniele, G.; Lavorini, G.; Motti, A.; Natali, N.; Bettucci, C. *Carta Geologica delle Regioni Emilia-Romagna, Marche, Toscana e Umbria. Scala 1:250000*. 2019. Available online: https://www.geological-map.it/pdf.html (accessed on 10 February 2021).
14. Conti, P.; Cornamusini, G.; Carmignani, L. Note Illustrative della Carta Geologica delle Regioni Emilia-Romagna, Marche, Toscana e Umbria. Scala 1:250000. 2019. Available online: https://www.geological-map.it/expl.html (accessed on 10 February 2021).
15. Brozzetti, F.; Luchetti, L.; Pialli, G. La successione di M. Rentella (Umbria occidentale): Biostratigrafia a nannofossili calcarei ed ipotesi per un inquadramento tettonico regionale. *Boll. Soc. Geol. It.* **2000**, *119*, 407–422.
16. Menichetti, M.; De Feyter, A.; Corsi, M. CROP03—Il tratto Val Tiberina—Mare Adriatico: Sezione Geologica e caratterizzazione tettonico-sedimentaria delle avanfosse della zona Umbro-Marchigiano-Romagnola. *Studi Geol. Cam.* **1991**, *1*, 279–293.
17. Damiani, A.V.; Pannuzzi, L.; Pialli, G. Osservazioni geologiche nelle aree comprese fra i massicci perugini ed i rilievi di Gubbio. *Giorn. Geol.* **1983**, *45*, 127–150.
18. Luchetti, L. Biostratigrafia a nannofossili calcarei delle successioni torbiditiche mioceniche dell'area di M. Acuto e di M. Corona (Umbria nord-occidentale, Italia centrale). *Boll. Soc. Geol. It.* **1998**, *117*, 333–335.
19. Zattin, M.; Zuffa, G.G. Unravelling the source rocks of Late Eocene-Miocene orogenic wedge and foredeep arenites of the northern Apennines and southern Alps. *Boll. Soc. Geol. It.* **2004**, *123*, 67–76.
20. Pialli, G.; Plesi, G.; Damiani, V.A.; Brozzetti, F. Note Illustrative della Carta Geologica d'Italia alla Scala 1:50.000; Foglio 289—Città di Castello. 2009. Available online: https://www.isprambiente.gov.it/Media/carg/note_illustrative/289_Citta_di_Castello.pdf (accessed on 10 February 2021).
21. Brozzetti, F. The Umbria Preapennines in the Monte Santa Maria Tiberina area: A new geological map with stratigraphic and structural notes. *Boll. Soc. Geol. It.* **2007**, *126*, 511–529.
22. Barchi, M.R. The Neogene-Quaternary evolution of the Northern Apennines: Crustal structure, style of deformation and seismicity. *J. Virtual Explor.* **2010**, *36*, 11. [CrossRef]
23. Guerrera, F.; Tramontana, M.; Donatelli, U.; Serrano, F. Space/time tectono-sedimentary evolution of the Umbria-Romagna-Marche Miocene Basin (Northern Apennines, Italy): A foredeep model. *Swiss J. Geosci.* **2012**, *105*, 325–341. [CrossRef]
24. Pialli, G. Osservazioni geologiche sulle formazioni flyschoidi di Castiglione Aldobrandi (foglio 123 IV NW). *Mem. Soc. Geol. It.* **1966**, *5*, 365–386.
25. Ricci Lucchi, F.; Pialli, G. Apporti secondari nella marnoso arenacea: 1. torbiditi di conoide e di pianura sottomarina ad est-nordest di Perugia. *Boll. Soc. Geol. It.* **1973**, *92*, 669–712.
26. Luchetti, L. Stratigrafia a Nannofossili Calcarei della Marnoso Arenacea dell'Umbria Settentrionale Fra il Fronte Dell'unità Falterona e la Dorsale Eugubina. Ph.D. Thesis, Università di Perugia, Perugia, Italy, 1997; 118p.
27. Luchetti, L.; Brozzetti, F.; Nini, C.; Nocchi, M.; Rettori, R. Lithostratigraphy, integrated biostratigraphy and paleoenvironmental analysis of the Monte Santa Maria Tiberina Miocenic succession (Umbria—central Italy). *Boll. Soc. Geol. It.* **2002**, *121*, 589–602.
28. Centamore, E.; Chiocchini, U. Le unità torbiditiche della Marnoso Arenacea nell'alta Valle del Tevere. *Stud. Geol. Camerti* **1985**, *10*, 37–59.
29. Chiocchini, U.; Chiocchini, M.; Cipriani, N.; Torricini, F. Petrografia delle unità torbiditiche della Marnoso-arenacea nell'alta Valle Tiberina. *Mem. Soc. Geol. It.* **1986**, *35*, 57–73.
30. Servizio Geologico d'Italia. Carta Geologica d'Italia alla Scala 1:50.000, Foglio 289 Città di Castello. 2010. Available online: http://www.isprambiente.gov.it/Media/carg/289_CITTA_CASTELLO/Foglio.html (accessed on 10 February 2021).
31. Gandolfi, G.; Paganelli, L.; Zuffa, G.G. Petrology and dispersal pattern in the Marnoso Arenacea formation (Miocene, northern Apennines). *J. Sediment. Res.* **1983**, *53*, 493–507.
32. Mutti, E. *Turbidite sandstones*; San Donato Milanese; Agip-Istituto di Geologia, Università di Parma: Parma, Italy, 1992; 275p.

33. Thierstein, H.R.; Geitzenauer, K.R.; Molfino, B. Global synchroneity of late Quaternary coccolith datum levels: Validation by oxygen isotopes. *Geol. Soc. Am. Bull.* **1977**, *5*, 400–404. [CrossRef]
34. Backman, J. Miocene-Pliocene nannofossils and sedimentation rates in the Hatton-Rockall Basin, NE Atlantic Ocean. *Stockh. Contrib. Geol.* **1980**, *36*, 1–91.
35. Rio, D.; Fornaciari, E.; Raffi, I. Late Oligocene through Early Pleistocene calcareous nannofossils distribution patterns in the western Mediterranean. *Proc. Ocean. Drill. Program. Sci. Res.* **1990**, *107*, 175–235.
36. Fornaciari, E.; Rio, D. Latest Oligocene to Early Miocene quantitative calcareous nannofossils biostratigraphy in the mediterranean region. *Mycropaleontology* **1996**, *42*, 1–36. [CrossRef]
37. Fornaciari, E.; Di Stefano, A.; Rio, D.; Negri, A. Middle Miocene quantitative calcareous nannofossils biostratigraphy in the mediterranean region. *Micropoaleontology* **1996**, *42*, 37–63. [CrossRef]
38. Martini, E. Standard Tertiary and Quaternary calcareous nannoplankton zonation. In Proceedings of the Second International Conference on Planktonic Microfossils, Rome, Italy, 1971; Farinacci, A., Ed.; Volume 2, pp. 739–785. Available online: https://onlinelibrary.wiley.com/doi/epdf/10.1002/iroh.19720570511 (accessed on 19 February 2021).
39. Okada, H.; Bukry, D. Supplementary modification and introduction of code numbers to the low-latitude coccolith biostratigraphic zonation (Bukry 1973, 1975). *Mar. Micropaleontol.* **1980**, *5*, 321–325. [CrossRef]
40. Backman, J.; Raffi, I.; Rio, D.; Fornaciari, E.; Pälike, H. Biozonation and biochronology of Miocene through Pleistocene calcareous nannofossils from low and middle latitudes. *Newslett. Stratigr.* **2012**, *45*, 221–244. [CrossRef]
41. Raffi, I.; Agnini, C.; Backman, J.; Catanzariti, R.; Pälike, H. A Cenozoic calcareous nannofossil biozonation from low and middle latitudes: A synthesis. *J. Nannoplankton Res.* **2016**, *36*, 121–132.
42. Fabbrini, A.; Baldassini, N.; Caricchi, C.; Foresi, L.M.; Sagnotti, L.; Dinarès-Turell, J.; Di Stefano, A.; Lirer, F.; Menichetti, M.; Winkler, A.; et al. In search of the Burdigalian GSSP: New evidence from the Contessa Section (Italy). *Ital. J. Geosci.* **2019**, *138*, 274–295. [CrossRef]
43. Di Stefano, A.; Foresi, L.M.; Iaccarino, S.M.; Turco, E.; Amore, F.O.; Mazzei, R.; Morabito, S.; Salvatorini, G.; Aziz, H.A. Calcareous nannoplancton high-resolution bio-magnetostratigraphy for the Langhian of the Mediterranean area. *Riv. Ital. Paleont. Stratigr.* **2008**, *114*, 51–76.
44. Backman, J.; Shackleton, N.J. Quantitative bio chronology of Pliocene and early Pleistocene calcareous nannoplankton from the Atlantic, Indian and Pacific Oceans. *Mar. Micropaleontol.* **1983**, *8*, 141–170. [CrossRef]
45. Pujos, A. Late Eocene to Pleistocene medium-sized and small-sized "Reticulofenestrids". *Abh. Geologischen Bundesanst.* **1987**, *39*, 239–277.
46. Flores, J.A.; Sierro, F.J. Calcareous nannoflora and planktonic foraminifera in the Tortonian-Messinian boundary interval of East Atlantic DSDP sites and their relation to Spanish and Moroccan sections. In *Nannofossils and Their Applications: Chichester*; Crux, J.A., van Heck, S.E., Eds.; Ellis Horwood: Chichester, UK, 1989; pp. 249–266.
47. Young, J.R. Neogene. In *Calcareous Nannofossil Biostratigraphy*; Bown, P.R., Ed.; Kluwer Academic: London, UK, 1998; pp. 225–265.
48. Takayama, T. Notes on Neogene calcareous nannofossil biostratigraphy of the Ontong Java Plateau and size variations of Reticulofenestra coccoliths. In Proceedings of the ODP, Scientific Results 130 (Ocean Drilling Program), College Station, TX, USA, 1993; Berger, W.H., Kroenke, L.W., Mayer, L.A., Eds.; pp. 179–229. Available online: http://www-odp.tamu.edu/publications/130_SR/VOLUME/CHAPTERS/sr130_11.pdf (accessed on 19 February 2021). [CrossRef]
49. Kameo, K.; Sato, T.; Takayama, T. Late Pliocene calcareous nannofossil datums and biohorozons. In *5th INA Conference in Salamanca Proceedings*; Flores, J.A., Sierro, F.J., Eds.; Universidad de Salamanca Press: Salamanca, Spain, 1995; pp. 87–98.
50. Raffi, I.; Rio, D. Calcareous nannofossil biostratigraphy of DSDP Site 132, Leg 13 (Tyrrhenian Sea-Western Mediterranean). *Riv. It. Paleontol. Stratigr.* **1979**, *85*, 127–172.
51. Mirabella, F.; Brozzetti, F.; Lupattelli, A.; Barchi, M. Tectonic evolution of a low-angle extensional fault system from restored cross-sections in the Northern Apennines (Italy). *Tectonics* **2011**, *30*, TC6002. [CrossRef]
52. Bucefalo Palliani, R.; Luchetti, L.; Nini, C.; Nocchi, M.; Rettori, R. Age and palaeoecological inferences of the upper Monte Falterona Sandstones Formation (Lonnano Member, Early Miocene), Northern Apennines. *Giorn. Geologia* **1997**, *59*, 143–168.
53. Servizio Geologico d'Italia. Carta Geologica d'Italia alla Scala 1:50.000, Foglio 310 Passignano sul Trasimeno. 2014. Available online: https://www.isprambiente.gov.it/Media/carg/310_PASSIGNANO/Foglio.html (accessed on 10 February 2021).
54. Barchi, M.; Marroni, M. Note Illustrative della Carta Geologica d'Italia alla Scala 1:50.000; Foglio 310 Passignano sul Trasimeno. 2014. Available online: https://www.isprambiente.gov.it/Media/carg/note_illustrative/310_Passignano_sul_Trasimeno.pdf (accessed on 10 February 2021).
55. Barsella, M.; Boscherini, A.; Botti, F.; Marroni, M.; Meneghini, F.; Motti, A.; Palandri, S.; Pandolfi, L. Oligocene-Miocene foredeep deposits in the Lake Trasimeno area (Central Italy): Insights into the evolution of the Northern Apennines. *Boll. Soc. Geol. It.* **2009**, *128*, 341–352. [CrossRef]
56. Aruta, G.; Bruni, P.; Cipriani, N.; Pandeli, E. The siliciclastic turbidite sequences of the Tuscan Domain in the Val di Chiana-Val Tiberina area (Eastern Tuscany and North-Western Umbria). *Mem. Soc. Geol. It.* **1998**, *52*, 579–593.
57. Plesi, G.; Damiani, A.V.; Boscherini, A.; Martini, E.; Luchetti, L.; Palandri, S.; Rettori, R.; Tuscano, F.; Botti, F.; Daniele, G.; et al. Note Illustrative della Carta Geologica d'Italia alla Scala 1:50.000, F. 299 Umbertide. ISPRA-Rome. 2010. Available online: https://www.isprambiente.gov.it/Media/carg/note_illustrative/299_Umbertide.pdf (accessed on 10 February 2021).

58. Costa, E.; Di Giulio, A.; Negri, A.; Plesi, G. CROP 03. Settore compreso tra Castiglion Fiorentino e Bocca Trabaria: Nuovi dati stratigrafici, petrografici e strutturali. *Stud. Geol. Cam.* **1991**, *1*, 217–234.
59. Andreozzi, M.; Di Giulio, A. Stratigraphy and petrography of the Mt. Cervarola Sandstone in the type area, Modena Province. *Mem. Soc. Geol. It.* **1994**, *48*, 351–360.
60. Servizio Geologico Nazionale. Carta Geologica d'Italia alla Scala 1:100.000; Foglio 130—Orvieto. Available online: http://sgi.isprambiente.it/geologia100k/mostra_foglio.aspx?numero_foglio=130 (accessed on 10 February 2021).
61. Signorini, R. Struttura dell'Appennino tra la Val Tiberina e l'Urbinate. *Giorn. Geol.* **1941**, *15*, 17–29.
62. D'Offizi, S.; Minelli, G.; Pialli, G. Foredeeps and thrust systems in the Northern Apennines. *Boll. Geof. Teor. Appl.* **1994**, *36*, 91–102.
63. Van Wamel, W.A.; Zwart, P.E. The structural geology and basin development of the Romagna-Umbrian zone (Upper Savio and Upper Bidente Valleys, N.Italy). *Geol. Mijnb.* **1990**, *69*, 53–68.
64. Delle Rose, M.; Guerrera, F.; Moretti, E.; Rusciadelli, G. Evoluzione del segmento interno dell'Avanfossa Appenninica durante il Miocene medio (spartiacque tosco-romagnolo). *Giorn. Geol.* **1990**, *52*, 135–158.
65. Delle Rose, M.; Guerrera, F.; Moretti, E.; Ruscadelli, G.; Corona, F. The Marnoso-Arenacea Fm. betweeen Bocca Trabaria and Campigna (Northern Apennines): Lithostratigraphy from Schlier to the Verghereto Marls Fms., in a new structural context. *Giorn. Geol.* **1991**, *53*, 131–145.
66. Ridolfi, A.; Luchetti, L.; Menichetti, M.; Negri, A.; Pialli, G. Use of key-bed in stratigraphical and structural analysis of the Marnoso Arenacea of the Assino Valley area (Northern Umbria pre-Apennines). *Giorn. Geol.* **1995**, *57*, 113–129.
67. Boscherini, A.; Nocchi, M.; Pialli, G. Geologia della Riva Etrusca del Tevere tra le confluenze del T. Niccone e del T. *Nese. Rend. Acc. Sc. Fis. Nat. Soc. Naz. Napoli* **1982**, *48*, 409–444.
68. Jacobacci, A.; Bergomi, C.; Centamore, E.; Malatesta, A.; Malferrari, N.; Martelli, G.; Pannuzzi, L.; Zattini, N. *Note illustrative della Carta Geologica d'Italia alla scala 1: 100.000—Fogli 115 «Città di Castello»—122 «Perugia»—130 «Orvieto»*; Generico: Salerno, Italy, 1970; p. 151.
69. De Feyter, A.J.; Molenaar, N.; Pialli, G.; Menichetti, M.; Veneri, F. Paleotectonic Significance of Gravity Displacement Structures in the Miocene Turbidite Series of the M Pollo Syncline (Umbro-Marchean Apennines, Italy). *Geol. Mijnb.* **1990**, *69*, 69–86.
70. Mirabella, F.; Ciaccio, M.; Barchi, M.; Merlini, S. The Gubbio normal fault (central Italy): Geometry, displacement distribution and tectonic evolution. *J. Struct. Geol.* **2004**, *26*, 2233–2249. [CrossRef]
71. Cantalamessa, G.; Centamore, E.; Chiocchini, U.; Micarelli, A.; Potetti, M.; Di Lorito, L. II Miocene delle Marche. In *La Geologia delle Marche*; Centamore, E., Deiana, G., Eds.; Università di Camerino, Dipartimento di Scienze della Terra: Camerino, Italy, 1986; pp. 35–55.
72. Centamore, E.; Chiocchini, U.; Cipriani, N.; Deiana, G.; Micarelli, A. Analisi dell'evoluzione tettonico-sedimentaria dei "bacini minori" torbiditici del Miocene medio-superiore nell'Appennino umbro-marchigiano e laziale-abruzzese: (5) risultati degli studi in corso. *Mem. Soc. Geol. It.* **1978**, *18*, 135–170.
73. Centamore, E.; Chiocchini, U.; Micarelli, A. Analisi dell'evoluzione sedimentaria dei "bacini minori" torbiditici del Miocene medio-superiore nell'Appennino umbro-marchigiano e laziale-abruzzese: (3) Le Arenarie di M. Vicino, un modello di conoide sottomarina affogata (Marche set tentrionali). *Stud. Geol. Camerti* **1977**, *3*, 7–56.
74. Plesi, G.; Luchetti, L.; Boscherini, A.; Botti, F.; Brozzetti, F.; Bucefalo Palliani, R.; Daniele, G.; Motti, A.; Nocchi, M.; Rettori, R. The Tuscan succession of high Tiber valley (F.289 Cittá di Castello): Biostratigraphic, petrographic and structural features, regional correlations. *Boll. Soc. Geol. Ital.* **2002**, *121*, 425–436.
75. Minelli, G.; Motti, A.; Pialli, G. Evoluzione tettonica dei Massicci Perugini; area di Monte Torrazzo. *Mem. Soc. Geol. It.* **1986**, *35*, 389–398.
76. Barchi, M.R.; De Feyter, A.; Magnani, M.B.; Minelli, G.; Pialli, G.; Sotera, B.M. The structural style of the Umbria-Marche fold and thrust belt. *Mem. Soc. Geol. It.* **1998**, *52*, 557–578.
77. Minelli, G.; Menichetti, M. Tectonic evolution of the Perugia massifs area (central Italy). *Boll. Soc. Geol. Ital.* **1990**, *109*, 445–453.
78. De Feyter, A.; Koopman, A.; Molenaar, N.; Van den Ende, C. Detachment tectonics and sedimentation, Umbro-Marchean Apennines, Italy. *Boll. Soc. Geol. It.* **1986**, *105*, 65–85.
79. Carmignani, L.; Decandia, F.A.; Disperati, L.; Fantozzi, P.L.; Lazzarotto, A.; Liotta, D.; Meccheri, M. Tertiary extensional tectonics in Tuscany (Northern Apenines, Italy). *Tectonophysics* **1994**, *238*, 295–315. [CrossRef]
80. Liotta, D.; Cernobori, L.; Nicolich, R. Restricted rifting and its coexistence with compressional structures: Results from the Crop03 traverse (Northern Apennines, Italy). *Terra Nova* **1998**, *10*, 16–20. [CrossRef]
81. Mattei, M.; Cipollari, P.; Cosentino, D.; Argentieri, A.; Rossetti, F.; Speranza, F.; Di Bella, L. The Miocene tectonosedimentary evolution of the Southern Tyrrhenian Sea: Stratigraphy, structural and paleomagnetic data from the onshore Amantea Basin (Calabrian Arc Italy). *Basin Res.* **2002**, *14*, 147–168. [CrossRef]
82. Milia, A.; Torrente, M.M. Early-stage rifting of the southern Tyrrhenian region: The Calabria-Sardinia breakup. *J. Geodyn.* **2014**, *81*, 17–29. [CrossRef]
83. Milia, A.; Valente, A.; Cavuoto, G.; Torrente, M.M. Miocene progressive forearc extension in the Central Mediterranean. *Tectonophysics* **2017**, *710–711*, 232–248. [CrossRef]
84. Farabegoli, E.; Benini, A.; Martelli, L.; Onorevoli, G.; Severi, P. Geologia dell'Appennino Romagnolo da Campigna a Cesenatico. *Mem. Descr. Carta Geol. It.* **1991**, *46*, 165–184.

geosciences

MDPI

Article

Miocene Seep-Carbonates of the Northern Apennines (Emilia to Umbria, Italy): An Overview

Stefano Conti [1], Claudio Argentino [2], Chiara Fioroni [1], Aura Cecilia Salocchi [1] and Daniela Fontana [1,*]

[1] Department of Chemical and Geological Sciences, University of Modena and Reggio Emilia, 41125 Modena, Italy; stefano.conti@unimore.it (S.C.); chiara.fioroni@unimore.it (C.F.); aura.salocchi@gmail.com (A.C.S.)

[2] CAGE—Centre for Arctic Gas Hydrate, Environment and Climate, Department of Geosciences, UiT the Arctic University of Norway, 9037 Tromsø, Norway; claudio.argentino@uit.no

* Correspondence: daniela.fontana@unimore.it

Abstract: The natural emission of methane-rich fluids from the seafloor, known as cold seepage, is a widespread process at modern continental margins. The studies on present-day cold seepages provide high-resolution datasets regarding the fluid plumbing system, biogeochemical processes in the sediment, seafloor seepage distribution and ecosystems. However, the long-term (hundreds of thousands to millions of years) evolution of cold seepage remains elusive. The identification and study of outcrop analogous now exposed on land represent a valuable method for better understanding the effects of geological processes and climate forcing on the development of cold seepage systems. Here, we provide an overview on Miocene seep-carbonate deposits of the northern Apennines (from Emilia to the Umbria-Marchean sector, Italy), based on decades of field research integrated with detailed sedimentological and geochemical investigations. We report a total of 13 seep-carbonate outcrops, which formed in three different structural settings of the paleo-accretionary wedge corresponding to wedge-top basins, outer slope and intrabasinal highs at the deformational front. We discuss the recurring lithostratigraphic occurrence of seep deposits and the main compositional features (carbonate facies, carbon and oxygen stable isotopes) in order to interpret the seepage dynamics, duration and infer the contribution of methane-rich fluids released by paleo-gas hydrates. The datasets presented in this study represent a valuable complete record of cold seepage spanning ~12 Myr, that can be used to better understand factors controlling the regional-scale spatial and temporal evolution of cold seepage systems at modern active continental margins.

Keywords: seep-carbonates; cold seepage; Miocene; northern Apennines; accretionary wedge

check for updates

Citation: Conti, S.; Argentino, C.; Fioroni, C.; Salocchi, A.C.; Fontana, D. Miocene Seep-Carbonates of the Northern Apennines (Emilia to Umbria, Italy): An Overview. *Geosciences* **2021**, *11*, 53. https://doi.org/10.3390/geosciences11020053

Academic Editors: Domenico Liotta and Jesús Martínez-Frías

Received: 23 December 2020
Accepted: 25 January 2021
Published: 28 January 2021

Publisher's Note: MDPI stays neutral with regard to jurisdictional claims in published maps and institutional affiliations.

Copyright: © 2021 by the authors. Licensee MDPI, Basel, Switzerland. This article is an open access article distributed under the terms and conditions of the Creative Commons Attribution (CC BY) license (https://creativecommons.org/licenses/by/4.0/).

1. Introduction

The seepage of methane-rich fluids at the seafloor, also known as cold seeps, has been frequently observed in accretionary wedges, where active tectonics generate pore-fluid overpressures and induce fluid migration through the sediments [1–5]. Cold seepage along the slope of continental margins is often associated with a large variety of sedimentary processes (e.g., landslides, mud volcanism, diapirism) and fluid escape structures (e.g., pockmarks, carbonate mounds) [6–10].

Due to the fluxes of reduced carbon and sulfur compounds reaching the seafloor, these environments are inhabited by peculiar microbial consortia and chemosymbiotic macrofaunal assemblages [11–16] and marked by specific geochemical imprinting [17,18]. Fossil analogous to modern systems have been recognized in exposed sedimentary successions on all continents (except Antarctica) and have allowed the investigation of the long-term evolution of hydrocarbon seepage in relation to tectonic processes and climate change [12]. Spectacular seep-carbonate examples have been reported from Miocene outer shelf and upper slope deposits at Hikurangi Margin, New Zealand [10,19,20] and in Cretaceous deposits linked to cold seepage in forebulge setting (Tepee Buttes carbonate mounds) cropping out in Colorado, USA [21]. It is worth mentioning the well-exposed and widely-studied Mesozoic seep deposits in the shelf and fore-arc successions of Japan

Ref [22], and seep carbonates in Oligocene flysch related to foreland basins of the Outer Carpathians, Poland [23]. The global sedimentary record of cold seepage provided statistical evidence for the fact that sea-level forcing and rates of organic carbon burial in ocean basins have been the main factors controlling overall seepage activity since the early Cretaceous [24], resulting in cycles with periodicity in the order of tens of Myr.

Fossil seeps have been widely reported from the Italian Apennine chain, in the form of seep-carbonate outcrops, and mostly hosted in Miocene successions [25]. These deposits are historically known under the informal lithostratigraphic name of Calcari a *Lucina*, as they include densely-packed large lucinid bivalves [26]. These authigenic carbonates are characterized by very negative $\delta^{13}C$ values (<-30‰ VPDB), peculiar chemosynthetic fauna (vesicomyid and lucinid clams, bathymodiolid mussels) and distinctive carbonate facies related to fluid expulsion processes [27,28], and testify a long history of cold seepage during the Neogene building phase of the Apennine chain. Seep-carbonate precipitation and fluid expulsion processes occurred in different tectonic settings of the Apennine foreland, from wedge-top basins through the outer slope of the accretionary prism, and at the leading edge of the deformational front in the inner foredeep, in correspondence of fault-related anticlines. Outcrop distribution highlights a causal relationship between tectonics and seepage occurrence. Recent studies also showed that some seep deposits could have originated during sea-level low-stands [29–31] and from gas hydrate destabilization events [32,33].

In this paper we provide an overview of the main geochemical, sedimentological and stratigraphical features of the northern Apennine seep-carbonates located in the Emilia to Umbria-Marchean area (Figure 1), included in Burdigalian to Messinian successions. The extraordinary dataset reported here, covers ~10,000 km^2 area and the examined carbonate outcrops are representative of different structural settings along the paleo-Apennine wedge, thus serving as a record of the regional-scale spatial and temporal evolution of cold seepage systems which can be used to better understand analogous systems in other modern convergent geo-settings.

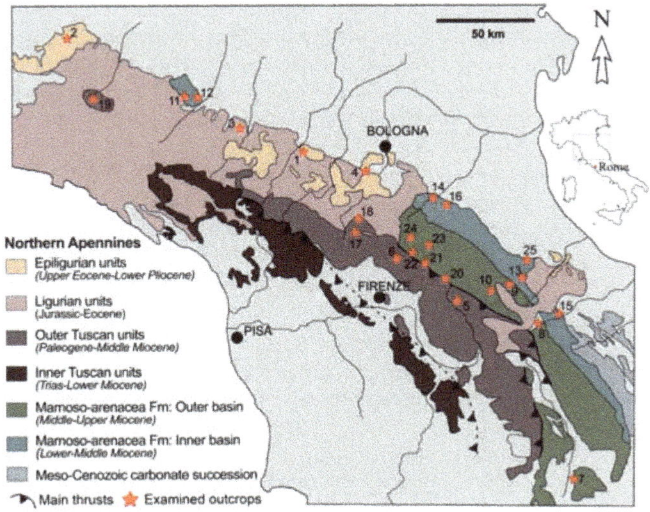

Figure 1. Simplified geological map of the studied area in the northern Apennines. Numbers refer to the examined and correlatable outcrops. 1. Montebaranzone; 2. Cappella Moma; 3. Traversetolo; 4. Lavino Valley; 5. Moggiona; 6. Fosso Riconi; 7. Deruta; 8. Poggio Campane; 9. S. Sofia-S. Vernicio; 10. S. Sofia-Case Buscarelle; 11. Salsomaggiore-Case Gallo; 12. Salsomaggiore-Case Cagnotti-Zappini; 13. Montepetra; 14. L Lame-Pietralunga; 15. Caresto; 16. Brisighella; 17. Brasimone–Suviana; 18. Poggio Michelino; 19. Telecchio; 20. Castagno d'Andrea-Corella; 21. Acquadalto (Podere Filetta-Monte Citerna); 22. Bibbiana-Le Fogare; 23. Colline-Mondera; 24. Prati Piani; 25. Case Bandirola.

2. Geological Setting

The northern Apennine chain results from the convergence and collision between the European and African plates, with the interposition of the Adria and Corsica-Sardinia microplates. During the early stages of the collision (late Oligocene), the internal oceanic units (Ligurides) were placed over the adjacent thinned margin of the continental Adria microplate, represented by Subligurian units [34]. From Miocene to Recent, the thrust system migrated towards the foreland, involving the continental Tuscan and Umbria-Marchean units deposited on the Adria microplate. This collisional stage involved the subduction of the Adria under the Corsica-Sardinia lithosphere coupled with the flexuring of the foreland and the formation of foredeep basins, progressively migrating towards the northeast [35,36].

The progressive migration of the foredeep produced a segmentation of its inner part [37] by the growing of anticlines on top of synsedimentary blind thrust faults, creating intrabasinal highs. Sedimentation on top of thrust-related anticlines consisted of hemipelagites and diluted turbidites (drape mudstones) forming up to a hundred meter thick fine-grained intervals. The structural evolution of these thrust-related anticlines created favorable conditions for gas accumulation and gas hydrate formation at the ridge crest, promoting the development of cold seepage systems and inducing sediment remobilization along the ridge flanks [28]. The progressive closure of the foredeep, with its involvement in the accretionary wedge (closure phase), is marked by the deposition of slope marls characterized by sediment instability due to the strong tectonic activity of the substrate [38]. Slope sedimentation is stopped by the overriding of the Ligurian units. During the overriding of the Ligurian units, from the Burdigalian to the early Tortonian, sedimentation also occurred in wedge-top basins (Epiligurian succession) at the top of the accretionary prism [39], and marked by intense seepage processes. Authigenic seep-carbonates in Upper Miocene sediments of the Tertiary Piedmont Basin are reported in [40,41].

The age of the Apennine successions becomes younger moving toward northeast, following the direction of wedge advancement and foredeep migration. In the Burdigalian, the closure of the Tuscan foredeep (filled by turbidites of the Falterona-Cervarola Fm) is followed by the deposition of fine-grained hemipelagites of the Vicchio Fm. During the Langhian, the newly formed foredeep is filled by the thick turbidite succession of the Marnoso-arenacea Fm (Umbro-Marchean units) partially sealed by the Verghereto Marls (late Serravallian-early Tortonian) and by the Ghioli di letto mudstones (late Tortonian-Messinian).

3. Structural Distribution of Apennine Seep-Carbonates

Miocene seep-carbonates occur in specific positions of the Apennine wedge-foredeep system, reported below from the inner to the outer sectors (Figure 2; Table 1):

(1) wedge-top basins: seep-carbonates mainly occurring in the Epiligurian units (Termina Fm), associated with diapiric processes;
(2) slope of the accretionary wedge, during two different tectonic phases of the foredeep closure: early phase. Seep-carbonates are hosted in fine-grained sediments draping buried ridges constituted by the older accreted turbiditic units [28]: final phase. Seep-carbonates are hosted in slope hemipelagites preceding the overriding of Ligurian units.
(3) inner foredeep, at the leading edge of the deformational front, in thrust-related anticlines (intrabasinal highs) standing above the seafloor. Seep-carbonates are hosted in fine-grained intervals deposited above these structures.

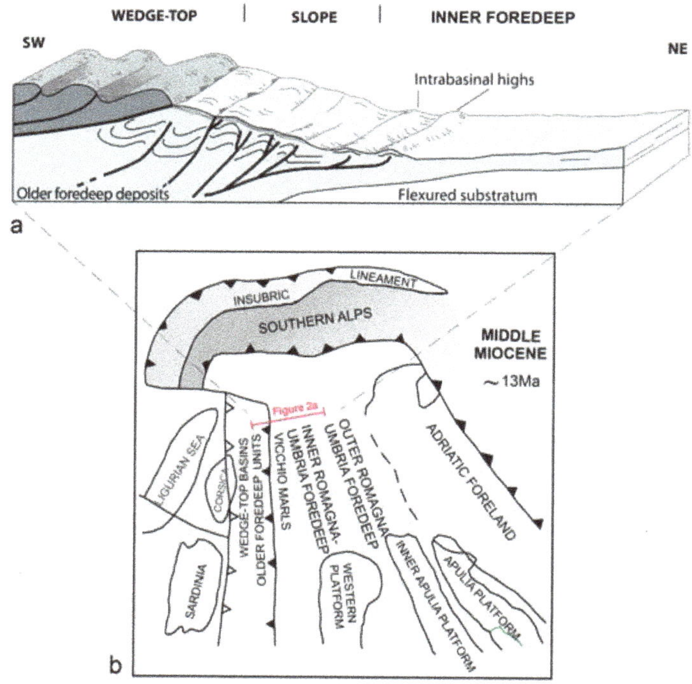

Figure 2. (**a**) Block diagram showing the northern Apennine wedge-foredeep system during the Miocene. Seep carbonates occurred in wedge-top basins, on the slope and in the inner foredeep basin. (**b**) Simplified paleogeographic sketch of the Apennine domains during the middle Miocene (modified from [26]).

Table 1. Structural distribution and paleogeographic domain of the examined outcrops (described in the text).

GEO-SETTING	APENNINIC DOMAIN		OUTCROPS
WEDGE-TOP			Montebaranzone
SLOPE	Tuscan	early phase final phase	Moggiona Fosso Riconi
	Umbria-Marchean inner basin	early phase final phase	Deruta, Poggio Campane Santa Sofia
	Umbria Marchean outer basin	early phase final phase	Montepetra Brisighella, Minor basins
INNER FOREDEEP	Tuscan Umbria Marchean		Brasimone-Suviana Castagno, Corella, Acquadalto

It is worth noting that the temporal/structural relationship between the slope and the foredeep is complicated by the migration of the accretionary wedge. This leads to the progressive incorporation of the foredeep units in the slope, and the creation of a new foredeep in a more external position.

In this work we report the location (Figure 1), morphology (Figure 3), facies (Figures 4 and 5), geochemistry ($\delta^{13}C$, $\delta^{18}O$; Figure 6) and stratigraphic distribution (Figure 7) of 13 seep-carbonate outcrops representative of the different geological settings (Table 1).

Figure 3. Seep-carbonate outcrops from the three examined geological settings. (**a**) Seep-carbonates hosted in the wedge-top basin (Termina Fm) in the Montebaranzone area (outcrop 1 in Figure 1), dotted line marks the geometry of the seep-carbonate bodies. (**b**) Castagno d'Andrea seep-carbonates enclosed in pelitic interval of the inner foredeep (Marnoso-arenacea Fm), (outcrop 20 in Figure 1). (**c**) Montepetra seep-carbonates hosted in slope fine-grained sediments (Ghioli di letto Fm), (outcrop 13 in Figure 1).

3.1. Wedge-Top Epiligurian Basins

Montebaranzone

Location. Emilia Apennines. Hundreds of seep-carbonate bodies crop out extensively between Parma and Bologna Apennines (Figure 1), enclosed within marly sediments of the Termina Fm (late Serravallian—early Messinian), made up of slope deposits up to 500 m thick. In the upper portion, the Termina Fm is represented by the Montardone mélange, a chaotic body made up of polygenic and heterometric blocks dispersed in a fine-grained matrix. Seep-carbonate deposits reach the maximum concentration in the Montebaranzone syncline (Modena area), associated with the Montardone mélange (for geological details see [42]).

Carbonate size and distribution. Seep-carbonates consist of stratiform bodies and large pinnacles with lateral extent from a few meters to 100 m and a maximum thickness of 25–30 m, mainly concentrated within a 50 m thick stratigraphic interval of slope marlstones. The lithologies are micritic limestones and calcareous marls, with wide portions of carbonate breccias. The contacts with the host sediment vary from sharp to transitional. Conversely, seep-carbonates occurring on the western side of the mélange consist of small and laterally isolated marly and marly-calcareous bodies, which exhibit a lenticular, domed, columnar to irregular shape. Their dimensions vary from a few tens of centimeters to 4–5 m; the lateral contact to enclosing sediment is gradual.

Seep-related facies. Polygenic breccias occur at the base of seep-carbonate bodies close to the contact with the Montardone mélange and form units of variable thickness from some centimeters to a few meters, often interdigitated with fine-grained carbonate cemented sediments. Polygenic breccias contain clasts of different dimensions and provenance, carbonate, arenitic and pelitic, chaotically floating in the micritic matrix. Extraformational clasts largely derive from the basal complex of the Ligurian units. Clasts are heterometric (from some millimeters to ~50 cm), generally angular (Figure 5a). Intraformational clasts are sourced from previously precipitated seep-carbonates and from the Termina marls. Disarticulated and isolated lucinid shells have been observed, as well as fragments of shells forming packstones or grainstones. Pseudo-fluidal textures and soft sediment deformations are observed. Dense localized semi-infaunal chemosynthetic fauna often in living position (mainly lucinid and vesicomyid bivalves) are present in the middle upper portions of the outcrops (Figure 4a). In thin section, authigenic minerals consist of micro- to cryptocrystalline micrite and sparry calcite with minor dolomite. Micrite is the dominant authigenic phase and includes abundant shell fragments, associated with planktonic foraminifera and terrigenous particles in variable amounts.

Carbon and oxygen isotopic composition. Carbon isotopic composition differs for various carbonate bodies and inside a single mass (δ^{13}C from −39.1 to −18.2‰ VPDB), with the most negative δ^{13}C values obtained in sparry calcite cement in the brecciated portions. Samples are significantly enriched in δ^{18}O (δ^{18}O between +0.3 and +5.5‰ VPDB); the δ^{18}O composition reaches the maximum values in the brecciated portions, close to the contact with the Montardone mélange.

Biostratigraphy of host sediments. The interval hosting seep-carbonates has been ascribed to the Serravallian to early Tortonian based on planktonic foraminiferal assemblages (MMi7-MMi9 Biozone). The oldest ages (Serravallian) are recorded in eastern bodies closest to the vertical contact with the mélange.

Correlatable outcrops. Cappella Moma (δ^{13}C from −19.0 to −16.0‰ VPDB; δ^{18}O from +2.1 to +2.9‰) Traversetolo, Lavino Valley (Figure 1).

3.2. Outer Slope of the Accretionary Prism

In slope mudstones of the Tuscan and Umbria-Marchean domains (Tuscan-Romagna-Marchean-Umbria Apennines), two different phases of seepage are recognized: the first during the initial part of the foredeep closure stage; the second marks the final part of the closure stage when slope mudstones were topped by Ligurian overthrust (Table 1).

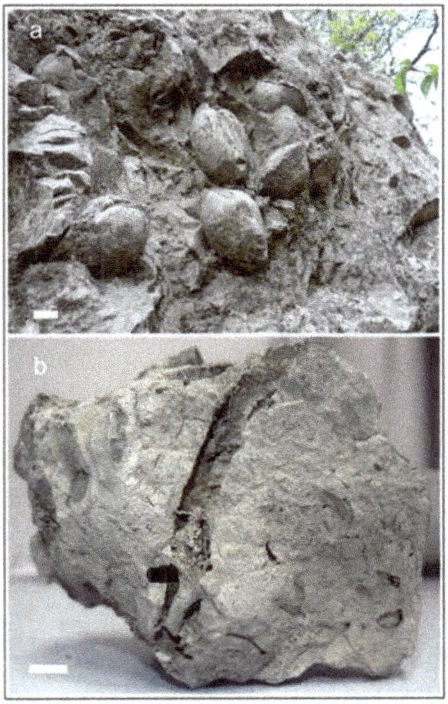

Figure 4. Representative seep-carbonate facies. (**a**) Densely packed articulated bivalves from the Sasso delle Streghe carbonate pinnacle (Montebaranzone area, outcrop 1 in Figure 1). (**b**) Large conduit and a complex network of vuggy fabric and doughnut structures from Poggio Campane outcrop (8 in Figure 1). Bar length = 2 cm.

3.2.1. Tuscan Domain

Moggiona

Location. Seep-carbonates are present in the basal member of Vicchio Fm made up of marls and silty marls with centimeter thick layers of fine-grained arenites (early phase of Table 1). Seep-carbonates are located in the footwall syncline at the base of the thrust within a stratigraphic interval 20–30 m thick, with a lateral extent of ~200 m. Carbonate bodies have a vertical attitude and are concordant to host sediments (geological detail in [25,28]).

Carbonate size and distribution. Seep-carbonate bodies are 5–100 m wide and up to 8 m thick; two smaller meter-sized blocks are also present. Morphologies are mostly stratiform with an irregular profile and strong lateral thickness variations. The contacts between carbonates and host sediments vary from sharp to gradual. Pinch-out lateral terminations, bifurcations, multiple interdigitation of carbonates with enclosing marls, lateral repetitions of rounded concretions and nodules are observed.

Seep-related facies. The basal portion of large carbonate bodies is characterized by a dense arrays of conduits, vertical to subhorizontal, with crosscutting relationships. Conduits are generally a few centimeters in diameter, with circular sections, and a few tens of centimeters long; conduit infilling consists of silty particles typically associated with shell debris. The top of the carbonate bodies is commonly characterized by assemblages of chemosynthetic fauna, either articulated and in life position or dismembered shells oriented parallel to bedding.

Carbon and oxygen isotopic composition. $\delta^{13}C$ ranges from -40.2 to $-13.6‰$; $\delta^{18}O$ from-9.9 to $+ 0.7‰$.

Biostratigraphy of host sediments. Based on nannofossil assemblages of the enclosing marls, the age of the seep carbonates is ascribed to the Burdigalian MNN3b biozone.

Correlatable outcrops. Poggio Corniolo.

Fosso Riconi

Location. The outcrop (final phase of the closure) is situated in the Tuscan Units (Vicchio Fm) cropping out in the Mugello area. The outcrop includes one of the most extensive exposures of a fossil seep system in the Apennines. About 80 lenses of authigenic carbonates are hosted in the topmost 30 m of marls with a lateral extent of about 500 m; the attitude is conformable to bedding of the enclosing marls (geological detail in [32,43]).

Carbonate size and distribution. Seep-carbonates have various geometries from elongated bed-like to lenticular bodies and pinnacles. The thickness of each body ranges from 1.5 m to 6 m and the lateral extent is from 1.5 to 10 m. The transition to host sediments vary from sharp to gradual. Nodular structures (2–3 cm in diameter), cylindrical to encircling concretions (4–5 cm in diameter) are present in the marginal portion of the seep-carbonates and arranged along stratification. Fossils are irregularly concentrated and consist of densely packed articulated and disarticulated bivalves up to 25 cm long. The most common lithotype is marly limestone.

Seep-related facies. Common facies are mottled carbonates, with irregular patches of micrite, pervaded of sinuous pipes, conduits and tubules, (Figure 4b), varying in diameter from 2–3 mm to 1–2 cm. Conduit sections are circular to elliptical, with the central hole filled by authigenic micrite, sparry calcite, coarser sandy sediment and shell fragments. In thin section, calcite is the dominant authigenic phase, associated with minor ankerite and dolomite. A very subordinate detrital fraction is made of illite-muscovite, chlorite, quartz and albite.

Carbon and oxygen isotopic composition. $\delta^{13}C$ values range from −39 to −4.7‰ VPDB and $\delta^{18}O$ values from −2 to +4‰ VPDB.

Biostratigraphy of host sediments. Based on planktonic foraminifera and nannofossil, the seep-carbonate precipitation approximates the Langhian/Serravallian boundary (MNN6a Biozone).

3.2.2. Umbria-Marchean Domain (Serravallian to Early Tortonian)

Seep outcrops are included in the Verghereto Marls and other coeval slope deposits. As previously described for the Tuscan slope, two different seepage phases mark respectively the early phase of the closure stage, and the final part of the closure stage when slope mudstones were topped by Ligurian overthrust.

Deruta

Location. Seep-carbonates in Deruta (early phase of the closure) are included in the Marnoso-arenacea Fm at different stratigraphic levels associated with coarse-grained pebbly sandstones and conglomerates (delta-slope and large-scale mass-wasting deposits) in the wedge-top area, prograding into the inner Marnoso-arenacea foredeep. The release of abundant methane-rich fluids through thrust faults pervaded coarse-grained sediments, causing the precipitation of authigenic seep-carbonates both along the delta-slope and the adjacent foredeep (geological details in [44]).

Carbonate size and distribution. Authigenic seep-carbonates occur as large (50 m wide, 10 m thick) fossiliferous lenses, as concretions and cements in previously reworked coarse-grained deposits, and as reworked blocks (cobble and boulder) in slide/slump horizons.

Seep-related facies. Common facies in large lenses are mottled micrite and biomicrite with densely packed seep-bivalves (lucinids, vesicomyids) mainly articulated. Disarticulated shells are scattered in brecciated portions. Vuggy structures are present, with void infilling made up of carbonate cements and/or coarser-grained sediments (Figure 5c). Monogenic breccias consist of angular clasts (a few millimeters to 5–10 cm) made up of previously precipitated micrite. In fine- to coarse-grained calcarenitic limestones, concentric and radial patterns of carbonate veins and micritic patches are frequent; fossils are absent. In delta-front conglomerates and sandstones authigenic micrite precipitated as intergranular cement, dense irregular networks of carbonate-filled veins and extensional fractures are frequent. Veins contain abundant black iron sulfides. The fossil content is scarce, with scattered disarticulated clams or articulated lucinids. Stratified marlstones cemented by authigenic micrite barren of fossils represent a transitional facies to normal marine conditions.

Carbon and oxygen isotopic composition. $\delta^{13}C$ largely differs in the different lithofacies, ranging from −46.0 to −11.0‰ PDB, and $\delta^{18}O$ ranges from −4.7 to +2.4‰.

Biostratigraphy of host sediments. Nannofossil analyses of the enclosing marls indicate the MNN6b subzone, Serravallian in age.

Figure 5. Recurrent seep-carbonate facies. (**a**) Polygenic breccias with clasts sourced from Ligurian units (Montebaranzone area, outcrop 1 in Figure 1). (**b**) Polygenic breccias with disarticulated bivalves; clasts sourced from previously formed seep-carbonates and from the underlying Marnoso-arenacea Fm (Colline, outcrop 23 in Figure 1). (**c**) Complex network of veins and conduits (Deruta, outcrop 7 in Figure 1). (**d**) Layered structures associated with disarticulated bivalves and mottled micrite (Montepetra, outcrop 13 in Figure 1). Bar length = 5 cm.

Poggio Campane

Location. Seep carbonates are included in Verghereto marls capping the inner Marnoso-arenacea unit coinciding with a main phase of the Apennine overriding (Figure 1).

Carbonate size and distribution. Seep-carbonates consist of three large stratiform bodies and several large pinnacles ranging in extent from a few meters to 20 m and with a maximum thickness of 8–10 m. Several minor irregular blocks and lenses (30 cm to 1 m in extent) are scattered around the main bodies. The lithologies are calcarenites and calcareous marls, with wide portions of carbonate breccias.

Seep-related facies. Marly limestones, calcareous marls, fine to very fine calcarenitic limestones, grey to pale brown in color, with abundant fossil content. Fossils are densely-packed articulated and disarticulated lucinid clams, with maximum diameter up to 25 cm. Other facies include mottled micrites with veins and shell debris, vuggy marly limestones with fractures and cavities, complex vein networks and doughnut structures (Figure 4b). Monogenic breccias are common.

Carbon and oxygen isotopic composition. $\delta^{13}C = -32.2‰$; $\delta^{18}O = +2.2$.

Biostratigraphy of host sediments. Based on nannofossil assemblages, the host sediment indicates the MNN7 biozone.

Santa Sofia

Location. Seep-carbonate blocks are hosted in slope sediments of the Verghereto Marls consisting of laminated, fine grained sandstones and marly-muddy beds, deposited by low density turbulent flows. Seep-carbonates have a wide extent and mark the closure stage of the foredeep before the overriding of the Ligurian units.

Carbonate size and distribution. Several tens of bodies of various dimensions and shapes from large stratiform (up to 50 m in lateral extent and 10 m thick) to irregular metric blocks and lenses. Carbonates consist of lightly colored, micritic limestones rich in mussels and clams [45]. The passage to enclosing marls is gradual.

Seep-related facies. Common facies are: polygenic and monogenic breccias with isolated articulated or disarticulated clams, centimetric conduits and doughnut fabric, network of conduits filled by calcite cements, laminated micritic limestones with alternance of whitish and brownish laminae.

Carbon and oxygen isotopic composition. $\delta^{13}C$ from −36.4 to −27.2‰; $\delta^{18}O$ from −0.3 to +3.6‰.

Biostratigraphy of host sediments. The nannofossil biostratigraphy indicates the MNN8 biozone (San Vernicio outcrop) and MNN8–9 (Case Buscarelle outcrop).

Correlatable outcrops. Salsomaggiore: Case Gallo, Case Cagnotti and Zappini ($\delta^{13}C$ from −41.4 to −8.7‰; $\delta^{18}O$ from −2.4 to +2.8‰) (Figure 1) [46].

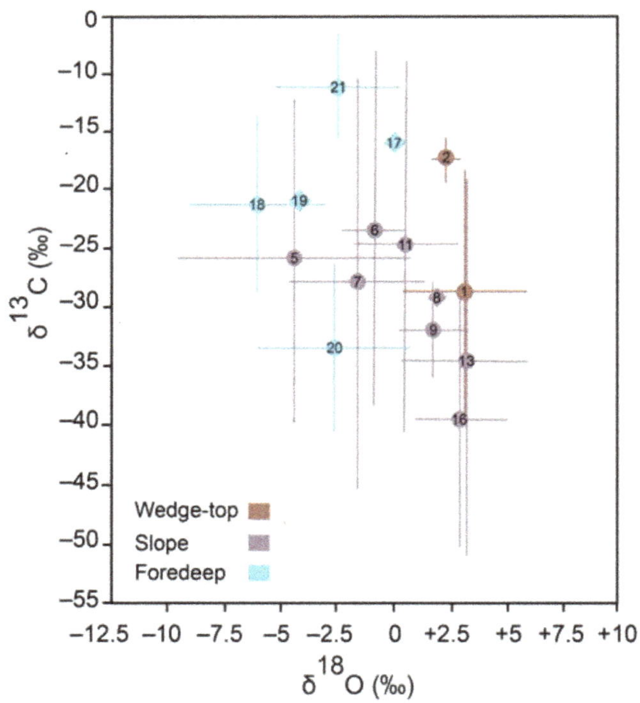

Figure 6. Carbon and oxygen isotopic values measured in seep-carbonates from the examined outcrops. The bar length represents the range of variability of the isotopic values; circles indicate the average of the endmember values and numbers specify the outcrop as listed in the caption of Figure 1. Rhombic features represent single measurement values.

3.2.3. Umbria-Romagna Domain (Tortonian to Messinian)

Seep carbonates are hosted in the Ghioli di letto mudstones and coeval slope deposits: during the early phase of the closure stage (Montepetra outcrops) and at the end of the closure stage below the contact with the Gessoso-solfifera Fm. Seep carbonates are also hosted in minor basin formed during the final stages of the Umbria-Marchean foredeep.

Montepetra

Location. The Montepetra outcrop formed along the outer slope of the accretionary prism, close to the front of the orogenic wedge. Seep-carbonates are hosted in fine-grained sediments (Ghioli di letto Fm, late Tortonian-early Messinian) draping thrust-bounded folds and buried ridges, constituted by the older accreted turbiditic units. The Montepetra outcrop is located in the south-eastern edge of a regional anticline, extending for more than twenty kilometers in a NW–SE direction. Seep-carbonates crop out in the hinge zone of the anticline and at the top of the mass transport deposits (geological details in [47]).

Carbonate size and distribution. Seep-carbonates consist of irregular metric lenses (up to 25 m) and blocks with different morphologies: lenticular-amygdaloid, mound-like irregular bodies, pinnacles, concretions of variable thickness (Figure 3c).

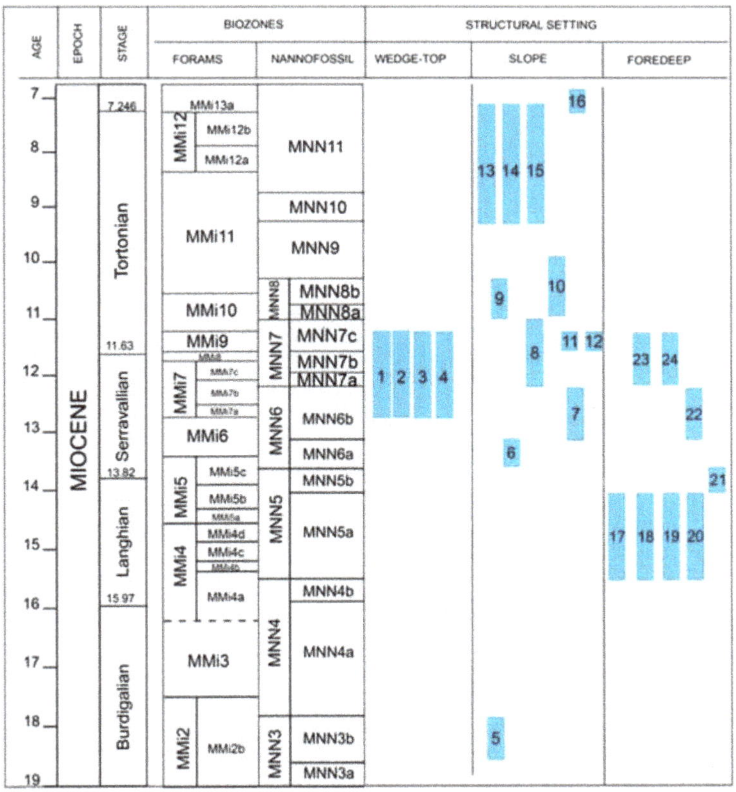

Figure 7. Nannofossil and foraminifera biostratigraphic framework indicating the distribution of the examined outcrops (numbers refer to Figure 1).

Seep-related facies. Monogenic and polygenic breccias, with intraformational and extraformational clasts (from Ligurian units), networks of conduits and veins, with scarce disarticulated and reworked fossils prevail in the basal portion of seep-carbonate bodies, indicating repeated phases of carbonated precipitation followed by fracturing. Moderate to strong fluid seepages are suggested by laminated and mottled marly limestones with vuggy fabrics chaotically mixed with polygenic microbreccias rich in articulated and/or disarticulated lucinid clams (Figure 5d). Stratified micritic limestones and fine-grained calcarenites with plane-parallel laminations and articulated lucinid-like bivalves occur in the upper portion of the seep bodies during phases of low and diffuse fluid circulation.

Carbon and oxygen isotopic composition. Seep-carbonates yielded depleted $\delta^{13}C$ values with a large dispersion from −52.7 to −19.1‰, and positive or slightly negative oxygen ($\delta^{18}O$ from −0.7 to +6.0‰). Brecciated levels show the most depleted carbon isotope values ($\delta^{13}C$ from −52.7 to −36.0‰), associated with heavy $\delta^{18}O$ values (from +2 to +6.0‰).

Biostratigraphy of host sediments. Based on planktonic foraminifera, the age spans from the late Tortonian to the early Messinian MMi12-13 Zones (Figure 7).

Correlatable outcrops. Le Lame, Pietralunga [28,48], Caresto (geological description in [49].

Brisighella

Location. Numerous seep-carbonate bodies are located at the top of the euxinic marls of the Ghioli di letto Fm, and mark the contact with the overlying Gessoso-solfifera Fm. Seep-carbonates and evaporitic levels are involved in several back-thrusts striking parallel to the main NW−SE Apennine structures; the detachment level is located at the base of the euxinic marls bearing carbonate bodies. The attitude of seep-carbonate bodies is concordant with the Gessoso-solfifera Fm.

Carbonate size and distribution. Seep-carbonates vary from stratiform to pinnacular (from 1 to 30 m in extent and up 10 m thick), to minor irregular lenses and blocks.

Seep-related facies. Monogenic and polygenic breccias are a common facies, with isolated articulated or disarticulated clams and conduits (diameter of a few cm). Other facies are marly limestones rich in densely packed bivalves (giant lucinids and rarely mussels), massive to laminated micritic limestones resembling *Beggiatoa* beds, stratified marly limestones with small lucinids and gastropods. Spongy fabric with a complex network of cavities and veins and doughnut fabric, are common.

Carbon and oxygen isotopic composition. $\delta^{13}C$ from −51.7 to −27.4‰; $\delta^{18}O$ from −1.6 to +5.0‰ [50].

Biostratigraphy of host sediments. Early Messinian.

Correlatable outcrops. See outcrops reported in [50,51].

Minor Basins

Location. Several small seep-carbonates in the Borello and Savio valley of the Romagna Apennines (Case Bandirola outcrop of Figure 1) are hosted in late Tortonian proximal turbidites belonging to minor basins of the Romagna-Marchean foredeep, known in literature under various names (Molasse grossolane, Fontanelice Sandstones, Urbania Sandstones),

Carbonate size and distribution. Carbonates are present as small blocks (thickness ranging from tens of centimeters to meters) in arenaceous coarse-grained turbidites, as crusts or as arenite cement.

Seep-related facies. Common facies are mottled micrite with small articulated bivalves, monogenic breccias with disarticulated bivalves and spongy fabric.

Carbon and oxygen isotopic composition. $\delta^{13}C$ values are reported in [50].

Biostratigraphy of host sediments. Late Tortonian.

3.3. *Intrabasinal Highs of the Inner Foredeep*

3.3.1. Tuscan Foredeep (Cervarola Fm)

Brasimone-Suviana

Location. The pelitic interval enclosing seep-carbonates in the M. Cervarola Fm crops out in the northern limb of the Granaglione-Montepiano overturned anticline [52]. The pelitic interval consists of fine-grained marly turbidites up to 40–50 m thick and with a lateral extent of 20 km. Carbonate bodies are located in proximity of the tectonic contact separating the M. Cervarola Fm from the Sestola-Vidiciatico Unit. Geological details in [52].

Carbonate size and distribution. Numerous carbonate lenses (a few decimeters to several meters wide, 20 cm to 4–5 m thick) are made up of micritic and calcarenitic lenses, strongly brecciated and rich in densely packed lucinid bivalves. The micritic groundmass is commonly associated with pyrite, abundant bioclastic debris (planktonic foraminifera and fragments of shells) and locally fine-grained sand grains, made of quartz, feldspars and low-grade metamorphic rock fragments (phyllite, serpentinite, chlorite-schist) similar to those of the M. Cervarola arenites.

Seep-related facies. Common facies are mottled micrite, strongly bioturbated, monogenic and polygenic breccias frequently barren of fossils, usually disarticulated, micrite with dense networks of calcite veins and extensive vuggy fabrics. Micritic doughnuts, nodular and cylindrical concretions and pipe-like structures are interpreted as fluid-flow conduits. Chaotic structures and soft sediment deformation are common and consist of small slumps involving marly and carbonate deposits.

Carbon and oxygen isotopic composition. $\delta^{13}C = -15.9‰$; $\delta^{18}O = +1.4$; Poggio Michelino: $\delta^{13}C$ from -29.5 to $-14.7‰$; $\delta^{18}O$ from -8.7 to $-2.6‰$; Telecchio: $\delta^{13}C = -21.0$, $\delta^{18}O = -4.0$.

Biostratigraphy of host sediments. The nannofossil biostratigraphy indicates the MNN5a Biozone.

Correlatable outcrops. Telecchio, Poggio Michelino (Figure 1).

3.3.2. Umbria-Marchean Foredeep (Marnoso-Arenacea Fm)

Thousands of seep-carbonates crop out in pelitic intervals draping intrabasinal highs in the Marnoso-arenacea foredeep. We recognized several pelitic intervals, each containing numerous seep-carbonate outcrops (Figure 1): Castagno-Corella (geological details in [25,28]; Acquadalto (Podere Filetta, Monte Citerna, Capanne di Favaglie and correlate outcrop of Poggio Cavalmagra); Susinello-Romiceto-Casaglia-M.Colonna-Nasseto described in [53]; Bedetta-Archetta (Colline, Mondera), Visignano (Prati Piani, case Termine). Here we describe the most representative outcrops. The more ancient pelitic interval (Castagno d'Andrea-Corella) contains numerous (up to 40) seep carbonate bodies scattered at various stratigraphic intervals; here we describe the two main outcrops, at the base (Castagno d'Andrea) and at the top (Corella) of the Castagno-Corella pelitic interval.

Castagno D'Andrea

Location. Four large seep-carbonate bodies are vertically stacked and distributed on three stratigraphic horizons concordant to the enclosing sediments and to the main structural trends. Small-scale slumps occur locally in host sediments above the carbonate bodies (geological details in [25,28]).

Carbonate size and distribution. Seep-carbonates vary from pinnacle-like to stratiform, 12 to 30 m wide and 5 to 10 m thick. Stratiform bodies are connected by pinnacular structures (Figure 3b). The basal contact is highly irregular marked by <0.5 m sized micritic concretions. Lateral transitions to enclosing sediments are sharp to gradual, with lateral repetitions of small concretions.

Seep-related facies. Stratiform bodies are characterized by an irregular framework of fractures and drusy-like cavities, associated with branched veins; in the upper portions, dense arrangements of lucinid-like clams are observed. In pinnacle-like bodies polygenic breccias prevail, with mixing of clasts from older underlying successions and from previously precipitated carbonate crusts and disarticulated clams. Monogenic breccias generated by autoclastic processes are common.

Biostratigraphy of host sediments. Calcareous nannofossils of the enclosing marls indicate the Langhian MNN5a subzone [54].

Corella

Location. The outcrop is located within the same pelitic interval as Castagno d'Andrea, approximately 10 km northwest, in a higher stratigraphic position. Six large carbonate bodies and several minor meter-sized blocks are concentrated in two horizons (geological details in [25,28]).

Carbonate size and distribution. Carbonate bodies are stratiform to lenticular, with a lateral extent from 50 m to 230 m and thickness up to 30 m. Basal and upper surfaces are flat. Lateral contacts with the host sediment are usually sharp, with pinch-out terminations. Larger carbonate bodies are vertically connected by irregular minor meter-sized bodies, or by highly cemented sediment.

Seep-related facies. The basal portion of the bodies is characterized by an irregular framework of fractures, conduits, drusy-like cavities and polygenic breccias (mm to cm sized) associated with disarticulated bivalves (Figure 5b). Monogenic breccias occur at various levels. A dense concentration of articulated bivalves is observed at the top of the bodies.

Carbon and oxygen isotopic composition. $\delta^{13}C$ ranges from −42.3 to −26.6‰ VPDB. $\delta^{18}O$ ranges from −5.7 to +1.2‰ VPDB.

Biostratigraphy of host sediments. Calcareous nannofossils of the enclosing marls indicate the Langhian MNN5a subzone.

Acquadalto

Location. The outcrops (Podere Filetta-Monte Citerna) are located within the pelitic interval of Acquadalto (geological detail in [54]). The interval has a thickness from 40 to 75 m and is cut by normal faults with Apenninic direction. Seep-carbonates are numerous at different levels commonly aligned along bedding. Some bodies show evidence of moderate reworking within the slopes of the intrabasinal high.

Carbonate size and distribution. Seep-carbonates consist of numerous marly-calcareous lenses, or irregular column-like and stratiform bodies, ranging in size from some decimeters to 3–4 m, and with a thickness from 20–30 cm to 3 m. The lateral contact with host sediments is gradual and interfingering; the transition is marked by carbonate-rich marly nodules.

Seep-related facies. The basal portion of bodies is characterized by an irregular framework of fractures, large conduits, drusy-like cavities and polygenic breccias (mm to cm sized) associated with disarticulated bivalves (Figure 5b). Monogenic breccias occur at various levels. Densely packed articulated lucinid-like bivalves occur at the top of the bodies. Mottled micrites pervaded by doughnut-like structures are common. Many marly nodules are connected to seep-carbonates by irregular and intertwined conduits.

Carbon and oxygen isotopic composition. $\delta^{13}C$ ranges from −15.8 to −3.6‰ VPDB. $\delta^{18}O$ ranges from −4.5 to −0.4‰ VPDB.

Biostratigraphy of host sediments. Calcareous nannofossils of the enclosing marls indicate the Langhian MNN5b subzone.

Similar features characterize the outcrops of Colline Mondera (Bedetta pelitic interval) (more details in [28]), and Prati Piani (Visignano pelitic interval) [55], all referable to the nannofossil MNN7 biozone.

The Prati Piani interval has a lateral extent of about 15 km and a thickness from 60 to 120 m, mainly constituted by hemipelagic marls and thin bedded fine-grained turbidites rich in ichnofossils. Extraformational bodies are present mainly in the basal portion, whereas the top is marked by glauconitic arenites. Seep-carbonates occur both at the base and in the upper portion of the interval.

Correlatable outcrops. Le Caselle-Pontevecchio [56].

4. Discussion

We report the spatial and stratigraphic distribution of seep-carbonate outcrops over an area of ~10,000 km^2 in the Emilia and Umbria-Marchean sector of the Apennine chain, northern Italy. The examined 13 outcrops are representative of three different structural positions along the Miocene Apenninic accretionary system, and document a clear causal relationship between active tectonics and their origin and distribution. The outcrops formed within wedge-top basin, along the outer continental slope close to the orogenic front, and in a more external position in the inner foredeep, in correspondence with fault-related anticlines. The widely diverse outcrop-scale spatial patterns, facies and morphologies of seep-carbonates reported in our study indicate that several factors influenced seepage activity. From the examination of this dataset it follows that:

- there is a clear link between the onset and evolution of seepage and specific structural positions with respect to main structural elements of the migrating Apenninic thrust-wedge. Anticlinal structures represent the most recurring setting for methane-rich fluid emissions and seep-carbonate formation. Folded structures along the outer slope were constituted by thrust-bounded folds composed of the older imbricated units. In the inner foredeep, at the leading edge of the deformation front, gentle anticlines were generated by blind faults connected to the basal detachment. Upward migrating fluids were conveyed toward the incipient anticline promoting seepage at the forelimb, likely in correspondence of the propagation of the thrust fault to the seafloor as observed in modern systems worldwide [28]. With the proceeding of the deformation, the fault-related folding caused the progressive growth of the ridge. The seepage shifted toward the hinge zone of the anticline, as the extensional stresses created an effective system that provided pathways for migrating fluids at the crest. In some cases, the predominance of diffusive methane seepage led to the precipitation of tens of meter-sized and spatially-scattered carbonate bodies.
- Seep-carbonates provide important constraints to the duration of seepage activity. The majority of studied outcrops are constituted by several meters thick carbonate deposits spanning hundreds of thousands of years of precipitation [28,47]. Often, the seep deposits are represented by numerous vertically-stacked carbonate bodies testifying an intermittent activity of seepage and/or complex fluid flow patterns in the sedimentary column (e.g., Fosso Riconi outcrop, [43]).
- The presence of gas hydrates was an additional controlling factor for seepage distribution on the paleo-wedge of the Apennine and other mountain chains [23,57]. The modelling of the paleo-gas hydrate stability recently reported by [33] for the examined outcrops, indicated that pure methane hydrates were stable at a water depth of 1000 m, within the uppermost few tens of meters to 400 m of sediment (assuming Miocene bottom water temperatures from 4 °C to 10 °C). The growth and the uplift of the thrust-related anticlines created favorable conditions for gas hydrate destabilization reducing the hydrostatic pressure and bringing gas hydrates out of their stability zone (e.g., [58]). The focusing of fluids underneath the crest of the anticline was also favored by the upward shift of the base of gas hydrate stability zone in proximity of the main fault, due to the advection of warm fluids [5,59].
- In the examined wedge-top basin, seep-carbonate precipitation appears strictly related with diapiric processes of the Montardone mélange. Diapirism involved old Epiligurian olistostromes and favored the ascending methane-rich fluids along the faulted flank of the diapir, with the subsequent precipitation of seep-carbonates characterized by chemosynthetic fauna. Diapirism was promoted by thrust loading of underconsolidated mud-breccias. Polygenic breccias sourced from the ascending mélange formed during fast, explosive seepage phases. In the last stage, the mélange reached the sea-bottom and slope failure processes (debris flows, olistostromes) occurred along the flanks of the mud-volcano, involving also blocks of authigenic carbonates.
- Finally, we identify a potential relationship between the lithostratigraphic distribution of seep-carbonates and eustatic/climatic implication as a whole [29,30,60]. Examined seep-carbonates are roughly concentrated in three main intervals of the Miocene: in the Langhian (MNN5a), in the late Serravallian—early Tortonian (MNN6b-MNN7) and the late Tortonian—early Messinian (MNN10-MNN11) (Figure 7). We are aware that the biostratigraphic resolution is conditioned by the duration of the biozone

that could be longer than the seepage activity. It is also evident that the tectonic processes played a primary role in favoring the development of seepage systems, by creating the structural pathways for fluid advection (thrust faults) as well as trapping mechanisms for the accumulation of fluids (fold structures). Nevertheless when comparing seep distributions with third-order eustatic curves [61], they seem matching phases of sea-level low-stands. In particular the detailed study of one of the best exposed Apennine outcrops (Fosso Riconi [43]) indicates that the onset of the seepage approximates to the Mi3b cooling event (13.82 Ma). These results also match with results of previous authors that proposed the correspondence of seep-carbonates with cold climate and sea-level lowering. In this proposed scenario, the reduction in hydrostatic pressure acting on the plumbing system, and related to sea-level falls, would shift the bottom of the gas hydrate stability zone to shallower depths, inducing gas hydrate destabilization.

5. Summary

The Miocene seep-carbonate outcrops of the northern Apennines (Italy) reflect a long history of methane-rich fluid emissions along the paleo-accretionary wedge. In the last decades, studies on the lithostratigraphic distribution of seep-carbonate deposits in the Emilia to Umbria-Marchean sector of the Apennine chain highlighted the fact that cold seeps developed in three main tectonic settings, corresponding to wedge-top basins, outer slope and intrabasinal highs located at the deformational front. Structural and biostratigraphic analyses conducted over the years provided solid evidence for the causal relationship between tectonic phases related to the building of the Apennine wedge and the evolution of fluid plumbing systems, and critical estimates of the duration of methane emissions on the paleo-seafloor, that in the case of the largest deposits can reach several hundreds of thousands of years of carbonate precipitation. In some cases, a climate forcing has been proposed as a contributing factor to the inception of seepage. Detailed sedimentological (facies, microstructures) and geochemical ($\delta^{13}C$, $\delta^{18}O$) investigations of seep-carbonates revealed the involvement of paleo-gas hydrates, but further studies are required in order to support the hypothesis of a regional-scale hydrate destabilization.

This study provided an overview on seep-carbonate deposits of the northern Apennines (Emilia to Umbria-Marchean sector) and factors controlling the regional-scale spatial and temporal evolution of Miocene cold seepage systems, representing a remarkably complete record that can be used to better understand fluid plumbing systems at modern convergent margins.

Author Contributions: Methodology—writing—review & editing, S.C., C.A., C.F., A.C.S. and D.F. All authors have read and agreed to the published version of the manuscript.

Funding: This research was founded by University of Modena and Reggio Emilia (FAR 2018-2020).

Institutional Review Board Statement: Not applicable.

Informed Consent Statement: Not applicable.

Data Availability Statement: Not applicable.

Acknowledgments: Our thanks to Luca Martire and anonymous reviewer for constructive criticism that led us to improve the paper.

Conflicts of Interest: The authors declare no conflict of interest.

References

1. Bohrmann, G.; Heeschen, K.; Jung, C.; Weinrebe, W.; Baranov, B.; Cailleau, B.; Heath, R.; Huhnerbach, V.; Hort, M.; Masson, D.; et al. Widespread fluid expulsion along the seafloor of the Costa Rica convergent margin. *Terra Nova* **2002**, *14*, 69–79. [CrossRef]
2. Ding, F.; Spiess, V.; Fekete, N.; Murton, B.; Brüning, M.; Bohrmann, G. Interaction between accretionary thrust faulting and slope sedimentation at the frontal Makran accretionary prism and its implications for hydrocarbon fluid seepage. *J. Geophys. Res. Solid Earth* **2010**, 115. [CrossRef]

3. Han, X.; Suess, E.; Liebetrau, V.; Eisenhauer, A.; Huang, Y. Past methane release events and environmental conditions at the upper continental slope of the South China Sea: Constraints by seep carbonates. *Int. J. Earth Sci.* **2014**, *103*, 1873–1887. [CrossRef]
4. Suess, E. Marine cold seeps and their manifestations: Geological control, biogeochemical criteria and environmental conditions. *Int. J. Earth Sci.* **2014**, *103*, 1889–1916. [CrossRef]
5. Klaucke, I.; Berndt, C.; Crutchley, G.; Chi, W.C.; Lin, S.; Muff, S. Fluid venting and seepage at accretionary ridges: The four way closure ridge offshore SW Taiwan. *Geo-Mar. Lett.* **2015**, *36*, 165–174. [CrossRef]
6. Judd, A.; Hovland, M. *Seabed Fluid Flow: The Impact on Geology, Biology and the Marine Environment*; Cambridge University Press: Cambridge, UK, 2009.
7. Leifer, I.; Kamerling, M.J.; Luyendyk, B.P.; Wilson, D.S. Geologic control of natural marine hydrocarbon seep emissions, Coal Oil Point seep field, California. *Geo-Mar. Lett.* **2010**, *30*, 331–338. [CrossRef]
8. Talukder, A.R. Review of submarine cold seep plumbing systems: Leakage to seepage and venting. *Terra Nova* **2012**, *24*, 255–272. [CrossRef]
9. Morley, C.K.; Warren, J.; Tingay, M.; Boonyasaknanon, P.; Julapour, A. Reprint of: Comparison of modern fluid distribution, pressure and flow in sediments associated with anticlines growing in deepwater (Brunei) and continental environments (Iran). *Mar. Petrol. Geol.* **2014**, *55*, 230–249. [CrossRef]
10. Nelson, C.S.; Nyman, S.L.; Campbell, K.A.; Rowland, J.R. Influence of faulting on the distribution and development of cold seep-related dolomitic conduit concretions at East Cape, New Zealand. *N. Z. J. Geol. Geophys.* **2017**, *60*, 487–496. [CrossRef]
11. Levin, L.A. Ecology of Cold Seep Sediments: Interactions of Fauna with Flow, Chemistry and Microbes. *Oceanogr. Mar. Biol. Annu. Rev.* **2005**, *43*, 1–46.
12. Campbell, K.A. Hydrocarbon seep and hydrothermal vent paleoenvironments and paleontology: Past developments and future research directions. *Palaeogeogr. Palaeoclimatol. Palaeoecol.* **2006**, *232*, 362–407. [CrossRef]
13. Kiel, S.; Tyler, P.A. Chemosynthetically-Driven Ecosystems in the Deep Sea. In *The Vent and Seep Biota*; Kiel, S., Ed.; Springer: Dordrecht, The Netherlands, 2010; pp. 1–14.
14. Pierre, C.; Blanc-Valleron, M.; Demange, J.; Boudouma, O.; Foucher, J.P.; Pape, T.; Himnler, T.; Fekete, N.; Spiess, P. Authigenic carbonates from active methane seeps offshore southwest Africa. *Geo-Mar. Lett.* **2012**, *32*, 501–513. [CrossRef]
15. Levin, L.A.; Baco, A.R.; Bowden, D.A.; Colaco, A.; Cordes, E.E.; Cunha, M.R.; Demopoulos, A.W.; Gobin, J.; Grupe, B.M.; Le, J.; et al. Hydrothermal vents and methane seeps: Rethinking the sphere of influence. *Front. Mar. Sci.* **2016**, *3*, 1–23. [CrossRef]
16. Kiel, S.; Taviani, M. Chemosymbiotic bivalves from Miocene methane-seep carbonates in Italy. *J. Paleontol.* **2017**, *91*, 444–466. [CrossRef]
17. Hu, Y.; Chen, L.; Feng, D.; Liang, Q.; Xia, Z.; Chen, D. Geochemical record of methane seepage in authigenic carbonates and surrounding host sediments: A case study from the South China Sea. *J. Asian Earth Sci.* **2017**, *138*, 51–61. [CrossRef]
18. Bojanowski, M.J.; Bagiński, B.; Guillermier, C.; Franchi, I.A. Carbon and oxygen isotope analysis of hydrate-associated Oligocene authigenic carbonates using NanoSIMS and IRMS. *Chem. Geol.* **2015**, *416*, 51–64. [CrossRef]
19. Nyman, S.L.; Nelson, C.S.; Campbell, K.A. Miocene tubular concretions in East Coast Basin, New Zealand: Analogue for the subsurface plumbing of cold seeps. *Mar. Geol.* **2010**, *272*, 319–336. [CrossRef]
20. Campbell, K.A.; Francis, D.A.; Collins, M.; Gregory, M.R.; Nelson, C.S.; Greinert, J.; Aharon, P. Hydrocarbon seep-carbonates of a Miocene forearc (East Coast Basin), North Island, New Zealand. *Sediment. Geol.* **2008**, *204*, 83–105. [CrossRef]
21. Metz, C.L. Tectonic Controls on the Genesis and Distribution of Late Cretaceous, Western Interior Basin Hydrocarbon-SeepMounds (Tepee Buttes) of North America. *J. Geol.* **2010**, *118*, 201–213. [CrossRef]
22. Majima, R.; Nobuhara, T.; Kitazaki, T. Review of fossil chemosynthetic assemblages in Japan. *Palaeogeogr. Palaeoclimatol. Palaeoecol.* **2005**, *227*, 86–123. [CrossRef]
23. Bojanowski, M.U. Oligocene cold-seep carbonates from the Carpathyans and their inferred relation to gas hydrates. *Facies* **2007**, *53*, 347–360. [CrossRef]
24. Oppo, D.; De Siena, L.; Kemp, D.B. A record of seafloor methane seepage across the last 150 million years. *Sci. Rep.* **2020**, *10*, 2562. [CrossRef]
25. Conti, S.; Fioroni, C.; Fontana, D. Correlating shelf carbonate evolutive phases with fluid expulsion episodes in the foredeep (Miocene, northern Apennines, Italy). *Mar. Pet. Geol.* **2017**, *79*, 351–359. [CrossRef]
26. Ricci Lucchi, F.; Vai, G.B. A stratigraphic and tectonofacies framework of the "calcari a *Lucina*" in the Apennine Chain, Italy. *Geo-Mar. Lett.* **1994**, *14*, 210–218. [CrossRef]
27. Taviani, M. Fluid venting and associated processes. In *Anatomy of an Orogen: The Apennines and Adjacent Mediterrenean Basins*; Vai, G.B., Martini, I.P., Eds.; Springer Sciences & Business Media: Berlin, Germany, 2001; pp. 351–366.
28. Argentino, C.; Conti, S.; Crutchley, G.J.; Fioroni, C.; Fontana, D.; Johnson, J.E. Methane-derived authigenic carbonates on accretionary ridges: Miocene case studies in the northern Apennines (Italy) compared with modern submarine counterparts. *Mar. Pet. Geol.* **2019**, *102*, 860–872. [CrossRef]
29. Kiel, S. Global hydrocarbon seep-carbonate precipitation correlates with deep-water temperatures and eustatic sea-level fluctuations since the Late Jurassic. *Terra Nova* **2009**, *21*, 279–284. [CrossRef]
30. Teichert, B.M.A.; Eisenhauer, A.; Bohrmann, G.; Haase-Schramm, A.; Bock, B.; Linke, P. U/Th Systematics and ages of authigenic carbonates from Hydrate Ridge, Cascadia Margin: Recorders of fluid flow variations. *Geochim. Cosmochim. Acta* **2003**, *67*, 3845–3857. [CrossRef]

31. Watanabe, Y.; Nakai, S.; Hiruta, A.; Matsumoto, R.; Yoshida, K. U-Th dating of carbonate nodules from methane seeps off Joetsu. Eastern Margin of Japan Sea. *Earth Planet. Sci. Lett.* **2008**, *272*, 89–96. [CrossRef]
32. Fontana, D.; Conti, S.; Grillenzoni, C.; Mecozzi, S.; Petrucci, F.; Turco, E. Evidence of climatic control on hydrocarbon seepage in the Miocene of the northern Apennines: The case study of the Vicchio Marls. *Mar. Pet. Geol.* **2013**, *48*, 90–99. [CrossRef]
33. Argentino, C.; Conti, S.; Fioroni, C.; Fontana, D. Evidences for Paleo-Gas Hydrate Occurrence: What We Can Infer for the Miocene of the Northern Apennines (Italy). *Geosciences* **2019**, *9*, 134. [CrossRef]
34. Malusà, M.G.; Danišík, M.; Kuhlemann, J. Tracking the Adriatic-slab travel beneath the Tethyan margin of Corsica–Sardinia by low-temperature thermochronometry. *Gondwana Res.* **2016**, *31*, 135–149. [CrossRef]
35. Faccenna, C.; Becker, T.W.; Lucente, F.P.; Jolivet, L.; Rossetti, F. History of subduction and back arc extension in the central Mediterranean. *Geophys. J. Int.* **2001**, *145*, 809–820. [CrossRef]
36. Barchi, M.; Landuzzi, A.; Minelli, G.; Pialli, G. Outer Northern Apennines. In *Anatomy of an Orogen: The Apennines and Adjacent Mediterranean Basins*; Vai, G.B., Martini, I.P., Eds.; Springer Sciences & Business Media: Berlin, Germany, 2001; pp. 215–254.
37. Tinterri, R.; Magalhaes, P.M. Synsedimentary structural control on foredeep turbidites: An example from Miocene Marnoso-arenacea Formation, Northern Apennines, Italy. *Mar. Pet. Geol.* **2011**, *28*, 629–657. [CrossRef]
38. Ricci Lucchi, F. The Oligocene to Recent Foreland Basins of the Northern Apennines. In *Foreland Basins*; Allen, P.A., Homewood, P., Eds.; International Association of Sedimentologists: Oxford, UK, 1986; pp. 105–139.
39. Conti, S.; Fioroni, C.; Fontana, D.; Grillenzoni, C. Depositional history of the Epiligurian wedge-top basin in the Val Marecchia area (northern Apennines, Italy): A revision of the Burdigalian-Tortonian succession. *Ital. J. Geosci.* **2016**, *135*, 324–335. [CrossRef]
40. De La Pierre, F.; Martire, L.; Natalicchio, M.; Clari, P.; Petrea, C. Authigenic carbonates in Upper Miocene sediments of the Tertiary Piedmont Basin (NW Italy): Vestiges of an ancient gas hydrate stability zone? *GSA Bull.* **2010**, *122*, 994–1010. [CrossRef]
41. Martire, L.; Natalicchio, M.; Petrea, C.C.; Cavagna, S.; Clari, P.; Pierre, F.D. Petrographic evidence of the past occurrence of gas hydrates in the Tertiary Piedmont Basin (NW Italy). *Geo-Mar. Lett.* **2010**, *30*, 461–476. [CrossRef]
42. Conti, S.; Fontana, D.; Lucente, C.C.; Pini, G.A. Relationships between seep-carbonates, mud volcanism and basin geometry in the Late Miocene of the northern Apennines of Italy: The Montardone mélange. *Int. J. Earth Sci.* **2014**, *103*, 281–295. [CrossRef]
43. Grillenzoni, C.; Monegatti, P.; Turco, E.; Conti, S.; Fioroni, C.; Fontana, D.; Salocchi, A.C. Paleoenvironmental evolution in a high-stressed cold-seep system (Vicchio Marls, Miocene, northern Apennines, Italy). *Palaeogeogr. Palaeoclimatol. Palaeoecol.* **2017**, *487*, 37–50. [CrossRef]
44. Conti, S.; Fontana, D.; Lucente, C.C. Sedimentary filling of a wedge-top basin and relationship with the foredeep (Middle Miocene Marnoso-arenacea Formation, northern Apennines). *Facies* **2008**, *54*, 479–498. [CrossRef]
45. Taviani, M. The calcari a *Lucina* macrofauna reconsidered: Deep- sea faunal oases from Miocene-age cold vents in the Romagna Apennine, Italy. *Geo-Mar. Lett.* **1994**, *14*, 185–191. [CrossRef]
46. Artoni, A.; Conti, S.; Turco, E.; Iaccarino, S. Tectonic and climatic control on deposition of seep-carbonates: The case of middle-late Miocene Salsomaggiore Ridge (Northern Apennines, Italy). *Riv. Ital. Paleontol. Stratigr.* **2014**, *120*, 317–335.
47. Conti, S.; Fontana, D.; Mecozzi, S.; Panieri, G.; Pini, G.A. Late Miocene seep-carbonates and fluid migration on top of the Montepetra intrabasinal high (Northern Apennines, Italy): Relations with synsedimentary folding. *Sediment. Geol.* **2010**, *231*, 41–54. [CrossRef]
48. Peckmann, J.; Thiel, V.; Reitner, J.; Taviani, M.; Aharon, P.; Michaelis, W. A microbial mat of a large sulfur bacterium preserved in a Miocene methane-seep limestone. *Geomicrobiol. J.* **2004**, *21*, 247–255. [CrossRef]
49. De Feyter, A.J. Gravity tectonics and sedimentation of the Montefeltro, Italy. *Geol. Ultraiectina* **1991**, *35*, 1–168.
50. Terzi, C.; Lucchi, F.R.; Vai, G.B.; Aharon, P. Petrography and stable isotope aspects of cold-vent activity imprinted on Miocene-age "calcari a *Lucina*" from Tuscan and Romagna Apennines, Italy. *Geo-Mar. Lett.* **1994**, *14*, 177–184. [CrossRef]
51. Iadanza, A.; Sanpalmieri, G.; Cipollari, P.; Mola, M.; Cosentino, D. The "Brecciated Limestones" of Maiella, Italy: Rheological implications of hydrocarbon-charged fluid migration in the Messinian Mediterranean Basin. *Palaeogeogr. Palaeoclimatol. Palaeoecol.* **2013**, *390*, 130–147. [CrossRef]
52. Conti, S.; Fontana, D. A Pelitic Interval Enclosing Primary Chemoherms in the M. Cervarola Formation (Northern Apennines): Evidence for Synsedimentary Tectonics during the Miocene. *Boll. Soc. Geol. Ital.* **2002**, *1*, 499–508.
53. Lucente, C.C.; Taviani, M. Chemosynthetic communities as fingerprints of submarine sliding-linked hydrocarbon seepage, Miocene deep-sea strata of the Tuscan–Romagna Apennines, Italy. *Palaeogeogr. Palaeoclimatol. Palaeoecol.* **2005**, *227*, 176–190. [CrossRef]
54. Conti, S.; Fontana, D.; Gubertini, A.; Sighinolfi, G.; Tateo, F.; Fioroni, C.; Fregni, P. A multidisciplinary study of middle Miocene seep-carbonates from the northern Apennine foredeep (Italy). *Sediment. Geol.* **2004**, *169*, 1–19. [CrossRef]
55. Argentino, C.; Johnson, J.E.; Conti, S.; Fioroni, C.; Fontana, D. Preservation of 34 S-enriched sulfides in fossil sulfate-methane transition zones: New evidence from Miocene outcrops of the northern Apennines (Italy). *Geo-Mar. Lett.* **2020**, *40*, 379–390. [CrossRef]
56. Berti, M.; Cuzzani, M.G.; Landuzzi, A.; Taviani, M.; Aharon, P.; Vai, G.B. Hydrocarbon-derived imprints in olistostromes of the Early Serravallian Marnoso-arenacea Formation, Romagna Apennines (Northern Italy). *Geo-Mar. Lett.* **1994**, *14*, 192–200. [CrossRef]
57. Greinert, J.; Bohrmann, G.; Suess, E. Gas hydrate associated carbonates and methane venting at Hydrate Ridge: Classification, distribution, and origin of authigenic lithologies. In *Natural Gas Hydrates: Occurrence, Distribution, and Detection*; Paull, C.K., Dillon, W.P., Eds.; American Geophysical Union Geophysical Monograph Series: Washington, DC, USA, 2001; Volume 124, pp. 99–113.

58. Paull, C.K.; Buelow, W.J.; Ussler, W., III; Borowski, W.S. Increased continental-margin slumping frequency during sea-level lowstands above gas hydrate-bearing sediments. *Geology* **1996**, *24*, 143–146. [CrossRef]
59. Laird, A.P.; Morley, C.K. Development of gas hydrates in a deep-water anticline based on attribute analysis from three-dimensional seismic data. *Geosphere* **2011**, *7*, 240–259. [CrossRef]
60. Peckmann, J.; Thiel, V. Carbon cycling at ancient methane-seeps. *Chem. Geol.* **2004**, *205*, 443–467. [CrossRef]
61. Haq, B.U.; Hardenbol, J.A.N.; Vail, P.R. Chronology of fluctuating sea levels since the Triassic. *Science* **1987**, *235*, 1156–1167. [CrossRef]

geosciences

MDPI

Article

Plio–Quaternary Structural Evolution of the Outer Sector of the Marche Apennines South of the Conero Promontory, Italy

Mario Costa [1], Jessica Chicco [2], Chiara Invernizzi [3,*], Simone Teloni [3] and Pietro Paolo Pierantoni [3]

1 Via Selvelli, 6, 61032 Fano, Italy; ioset@libero.it
2 Department of Earth Sciences, University of Turin, via Valperga Caluso, 35, 10125 Torino, Italy; jessica.chicco@unito.it
3 School of Science and Technology, Geology Division, University of Camerino, via Gentile III da Varano, 62032 Camerino, Italy; simone.teloni@unicam.it (S.T.); pietropaolo.pierantoni@unicam.it (P.P.P.)
* Correspondence: chiara.invernizzi@unicam.it; Tel.: +39-3288604234

Abstract: Some new results and preliminary remarks about the Plio–Quaternary structural and evolutionary characteristics of the outer Marche Apennines south in the Conero promontory are presented in this study. The present analysis is based on several subsurface seismic reflection profiles and well data, kindly provided by ENI S.p.A. and available on the VIDEPI list, together with surface geologic–stratigraphic knowledge of Plio–Quaternary evolution from the literature. Examples of negative vs. positive reactivation of inherited structures in fold and thrust belts are highlighted. Here, we present an example from the external domain of the Marche Apennines, which displays interesting reactivation examples from the subsurface geology explored. The study area shows significant evolutionary differences with respect to the northern sector of the Marche region previously investigated by the same research group. The areal distribution of the main structures changes north and south of the ENE–WSW oriented discontinuity close to the Conero promontory. Based on the old tripartite classification of the Pliocene, the results of this work suggest a strong differential subsidence with extension occurring during the Early Pliocene and principal compressive deformation starting from the Middle Pliocene and decreasing or ceasing during the Quaternary. The main structure in this area is the NNW–SSE Coastal Structure, which is composed of E-vergent shallow thrusts and high-angle deep-seated normal faults underneath. An important right-lateral strike–slip component along this feature is also suggested, which is compatible with the principal NNE–SSW shortening direction. As mentioned, the area is largely characterized by tectonic inversion. Starting from Middle Pliocene, most of the Early Pliocene normal faults became E-vergent thrusts.

Keywords: Plio–Quaternary evolution; outer Marche Apennines; seismic reflection profiles; tectonic inversion; Coastal Structure; extensional and contractional deformation

check for updates

Citation: Costa, M.; Chicco, J.; Invernizzi, C.; Teloni, S.; Pierantoni, P.P. Plio–Quaternary Structural Evolution of the Outer Sector of the Marche Apennines South of the Conero Promontory, Italy. *Geosciences* **2021**, *11*, 184. https://doi.org/10.3390/geosciences11050184

Academic Editors: Jesus Martinez-Frias, Domenico Liotta, Giancarlo Molli and Angelo Cipriani

Received: 30 January 2021
Accepted: 20 April 2021
Published: 24 April 2021

Publisher's Note: MDPI stays neutral with regard to jurisdictional claims in published maps and institutional affiliations.

Copyright: © 2021 by the authors. Licensee MDPI, Basel, Switzerland. This article is an open access article distributed under the terms and conditions of the Creative Commons Attribution (CC BY) license (https://creativecommons.org/licenses/by/4.0/).

1. Introduction

In the Apennines of Italy, and especially the Adriatic foreland domain, it is possible to infer the foreland deformation process and explore the impacts of inherited faults and basins on the subsequent evolution thanks to the milder deformation in the area and the good geological and geophysical record documenting an interaction between normal, thrust, and strike–slip faults.

Foreland domains are often affected by inherited rift-related or flexure-related syn-sedimentary normal faults becoming involved in the advancing fold-and-thrust belt. This introduces an element of further complications into the evolution of the foredeep systems subsequently involved in the mountain belts, as evidenced by numerous studies in different contexts, such as the Northern Apennines, Po Plain, and South-Eastern Pyrenean foreland basins ([1–4], among others).

The tectonic and structural features of the Umbria-Marche Apennines (Figure 1) are widely described in the literature, and several models have been proposed. The most important model is found in [5], which proposes a thin-skinned imbricate belt detached above the crystalline basement (see also [6]). This model indicates strong shortening (in the order of hundreds of kilometres) and important repetitions of the sedimentary cover. Further studies on the geometries and evolution of the outer Marche sector, as well as their extent, style, and age of deformation, are thoroughly reported in many works. Among others, [7–13] mainly focus on stratigraphic record, geological setting, and sedimentary evolution; [14] on the anatomy of the Apennine orogen; [15–20] on the structural and deformation style; and [14,21] on the role of inherited structures and tectonic inversion.

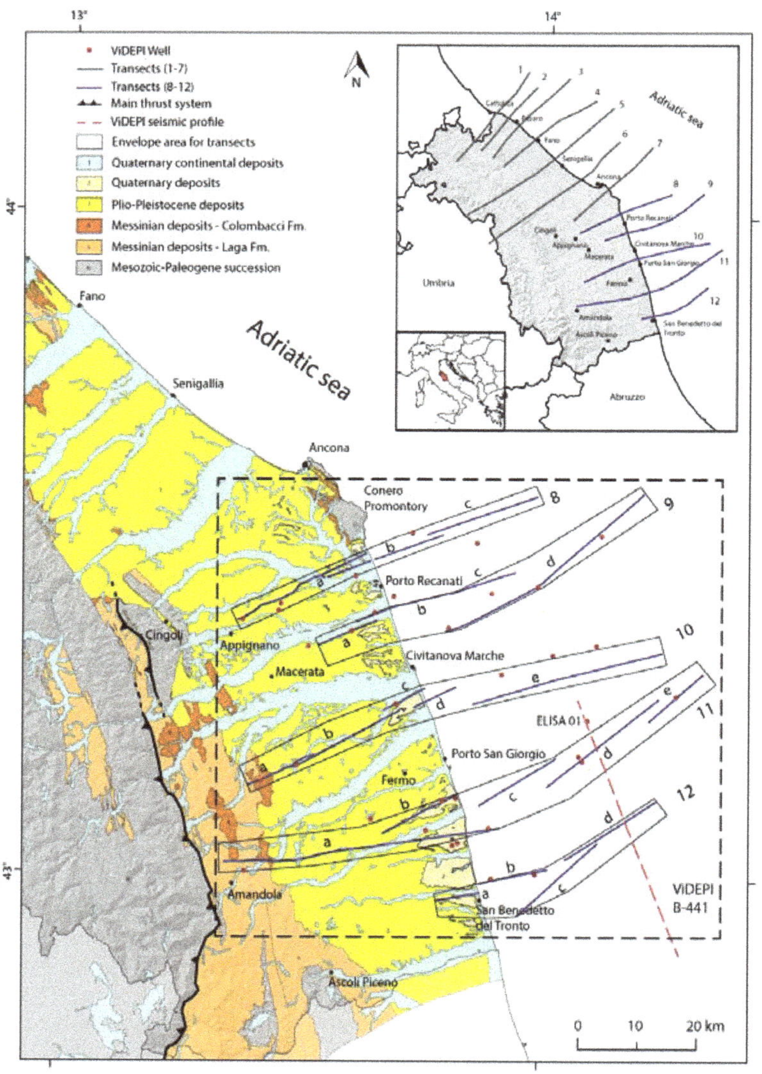

Figure 1. On shore schematic geological map of the Marche region (modified from [22]). The work in [23] was considered for the thrust location. Dashed square: study area; dashed red line: ViDEPI seismic profile B-441 with Elisa 1 well (Figure 3). Inset: geographic location of the study and location of the Transects. Transects: results for 1 to 7 are published in [24,25]. Transects 8 to 12: this work.

The acquisition of new data (such as the CROP project, well stratigraphy and seismic reflection profiles, and sedimentological and paleo-thermometric data) have shed new light on the evolution of the area and introduced models that indicate the crystalline basement's involvement (thick-skinned model) and the reactivation of inherited faults (inversion tectonic model). As a main outcome, the amount of shortening affecting this area was progressively reduced from hundreds of kilometres to tens of kilometres [26].

New observations about onshore and offshore outcropping and buried Neogene–Quaternary structures, as well as their possible implications for deep geothermal fluid circulation, were recently integrated into the tectonic framework of the northern outer Marche Apennines [24,25]. These studies highlighted new findings mainly characterized by the presence of positive flower structures to be considered as common features along the whole outer sector of the Northern Apennine chain [24]. This suggests the more relevant influence of strike–slip kinematics in recent times, with implications for seismic assessment and deep fluid circulation [25].

The southern sector of the outer Marche Apennines has been long investigated by authors who addressed specific features in this area as related to a complex foreland–foredeep geometry. In particular, several works explore the influence of thrust-system propagation on the distribution of sedimentary sequences, their 3D geometric organization, and the burial and exhumation history of these units [27–31]. These features were identified as the link between the inner, uplifted, and Early Miocene Apennine fold-and-thrust belt and the outer and younger belt to the east [31]. The interpretations of integrated structural and stratigraphic studies indicate this to be the result of turbidite deposition in a complex foredeep, strongly affected by tectonic activity and Messinian–Pliocene climate changes ([29,32] and references therein).

This paper represents a continuation of the above-mentioned studies [24,25] and aims at highlighting the significant structural and depositional differences between the northern and southern outer Marche Apennine, as well as discussing the timing and style of deformation in the outermost sector of the belt toward the Adriatic foreland, where milder deformation and mainly buried structures are present.

To this end, a detailed study along the sector south of the Conero promontory to S. Benedetto del Tronto was conducted (Figure 1) using seismic reflection profile interpretations and well data for hydrocarbon purposes, kindly provided by ENI S.p.A. Available online data: https://www.videpi.com/videpi/sismica/sismica.asp (accessed on 30 March 2021), and published studies further contributed to acquiring complete information and enriching the results in our previous works.

2. Geological Setting

At the continental scale, the Alps and Apennines orogens are located in the hanging wall of two opposite subduction zones. The Alps resulted from the Cretaceous to present via the European plate being subducted beneath the Adriatic plate to the east, whereas the Apennines resulted from the Eocene to the present via subduction of the Adriatic plate to the west ([33] and references therein). The Adriatic plate itself is also subducted below the Dinarides in its easternmost part [33,34].

The arcuate-shaped, NE-verging Umbria–Marche Apennines form the external part of the Northern Apennines foreland fold-and-thrust belt (see [35] and references therein). This belt resulted from the convergence between a mosaic of minor blocks of the Africa–Eurasia plates, such as the European Corsica–Sardinia plate to the West and the African–Adria plate to the east ([36–39] and references therein).

In the Umbria–Marche area, starting from the Miocene, the previously rifted and telescoped African-bearing continental margin was involved in the compressive phase. Here, different styles and degrees of the positive inversion of pre-orogenic faults controlled the location, geometry, and evolution of compressive structures in several cases [16,28,40–44]. In addition, the inner portion of the chain was involved in the Late Miocene to present day extension [14,23,45,46], with episodes of negative inversion [43,47–49].

The study area belongs to the outer Umbria–Marche Apennines chain. The general tectonic–sedimentary evolution of the Umbria–Marche sequence can be framed in three main stages: pre-, syn-, and post-orogenic sedimentation [23]. The pre-orogenic sedimentation is characterized by basin carbonates and marly lithostratigraphic units (Late Triassic to Paleogene in age; [23,35] and references therein, Figure 1). Both syn- and post-orogenic sedimentation is characterized by prevalent terrigenous deposits from Neogene to Quaternary in age and hosted in a wide foredeep basin (Periadriatic Foredeep; [50]). This basin was generated by the flexure of the Adria plate under the Apennine Chain [51], migrating eastward. The foredeep filling includes siliciclastic turbidites (e.g., the Messinian Laga Basin), Plio–Pleistocene marine deposits [51–53], and wedge-top basin sediments [31]. The foredeep itself was gradually involved into the fold-and-thrust belt during the Late Miocene to present.

In the present study, we investigate an area lying in the outer portion of the southern Marche Apennines between the Conero promontory and S. Benedetto del Tronto (Figure 1). In particular, the double effect of the Sibillini thrust to the west and the Gran Sasso thrust to the south (the Abruzzo area in Figure 1) influences the Messinian foredeep geometry and depth. The foredeep hosts thick, internally deformed, turbiditic fan complexes (the Laga Formation; [30,54]) and some positive structures (Acquasanta, Montagna dei Fiori and Coastal Structure) described in the literature (see [2,4,7,30]). The outcropping succession consists of Messinian turbiditic deposits (Figure 2), including a thick, arenaceous basal member whose source is partially provided by the Eocene–Oligocene westernmost chain [7,54] and shallow water facies (S. Donato and Argille a Colombacci Formations). The Argille a Colombacci Formation is always above S. Donato Formation, while the latter may rest discordantly above different members of the Laga Formation (see [22] and references therein). Messinian deposits are followed by the Pliocene succession, whose base marks the marine transgression that occurred after the "lago-mare" phase (sedimentation breck-off; [55]) and the subsequent filling of the Central Adriatic foredeep [56].

The Plio–Pleistocene foredeep basin is associated with deep marine to alluvial sedimentation that shows progressive infill of the basin and a final vertical regressive trend [41]. The infill mainly consists of hemipelagic mudstones deeply incised by coarse-grained canyon-fill deposits [57,58] indicating slope degradation and sediment supply from the uplifted Apennines [32]. Many authors associate these deposits with the outermost part of the orogenic wedge [28,32]; with the formation of thrust fronts and folded structures in the Early Pliocene [11,52]; followed by intense deep water clayey sedimentation in the deepest areas until the Pleistocene; and a new compressive phase right after, likely linked to the reactivation of Late Pliocene thrusts [10]. Deformation of the foredeep via thrusting likely yielded open piggy-back basins and structural highs filled up by shallow-water deposits, likely due to the tightness and blockage of the system against a stable platform, as hypothesized in [11]. The sedimentation within the basin was also partially controlled by the Pliocene–Pleistocene obliquity/precession cycles of the Earth's orbit driving climatic changes, as suggested in [28].

In its outermost portion, the belt shows compressive to transpressive flower structures, which are NW–SE oriented and generally covered by Plio–Pleistocene deposits or partial outcropping on the seafloor. These structures were identified and described in [25] and are located further north of the study area as well as some NE-SW trending faults, which affects the continuity of the previous structures.

In the considered area, the main structural element is the NNW–SSE trending Coastal Structure ("Struttura Costiera" in [11]) which is located near the coastline. This structure continues southwards in the Abruzzo area with similar characteristics [50].

Two main deformation events in the area were recognized by previous authors: an extensional Messinian–Early Pliocene event due to the Adria plate flexure [50] followed by a compressive phase ascribable to the late Early Pliocene [55] or to the Middle Pliocene [50].

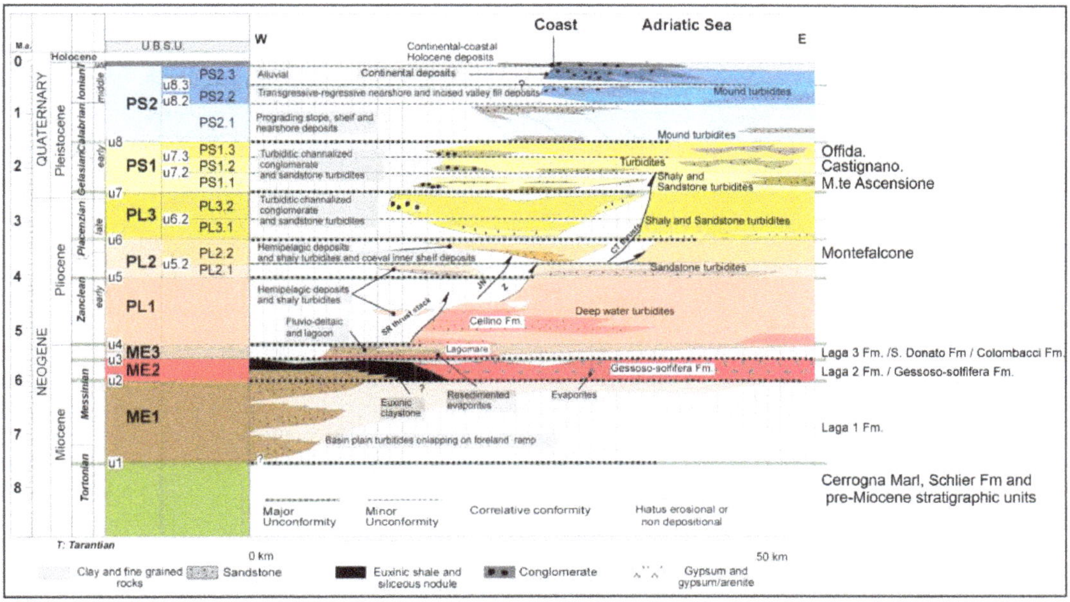

Figure 2. Synthetic stratigraphic scheme of the Messinian–Pleistocene of the Central Periadriatic Basin (CPB; slightly modified from [32]). This scheme includes the stratigraphic schemes of previous studies [11,13,27,50,59], but in the right column we list only the units and members of our study area.

3. Dataset and Working Methods

In the onshore and offshore areas between the Conero promontory and San Benedetto del Tronto locality, numerous ENI S.P.A. seismic profiles have been interpreted in addition to those available from VIDEPI (https://www.videpi.com/videpi/sismica/sismica.asp; accessed on 30 March 2021). The ENI seismic profiles were migrated, while the VIDEPI ones were stacked and already interpreted. Some ENI profiles were of a MERGE type and good quality, resulting from advanced reprocessing. All the wells available in the area corresponding to the interpreted seismic profiles were used for the interpretation. However, the materials supplied by ENI are confidential, and we are thus not able to represent them on the seismic profiles. Only a general well location was reported. For the seismic velocities of the sedimentary sequences, we referred to [20,60–62]. We then compared the seismic stratigraphy of the seismic profile VIDEPI B-441 001 (Figure 1) with the log data of the Elisa 1 well placed on it (Figure 3). This comparison indicates that velocity, Vp, for the Plio–Pleistocene sequence is 2000 m/s in agreement with the literature in the same area [50,63].

Seismic profiles were then homogenized and scaled to 1:100.000 horizontally and 1 s TWT = 2 cm vertically. In this way, the horizontal and vertical scales were harmonized for the Plio–Pleistocene sequence of the seismic profile, and the geometries of the tectonic and seismic–stratigraphic elements were preserved. As velocity increased at depths below the lower Pliocene deposits, the dip angles of these structures became higher.

To determine the Plio–Quaternary's tectono–stratigraphic evolution, specific seismo–stratigraphic horizons were considered, as follows:

- Top of the Messinian/Pre-Pliocene;
- Near the top of the Early Pliocene;
- Near the base of the Quaternary;
- Unconformities.

Within the interpreted profiles, the seismo–stratigraphic horizons are highlighted with different colours (see Figures 3 and 4 and Plate 1 in Supplementary materials). Unconformities are shown in green dots (see Figures 4, 5, and 7). Some additional reflectors are also highlighted (light blue) because these reflectors allow the main structures to be better marked and identified. The boundary between the Pliocene and Quaternary deposits has been always defined based on the available well stratigraphy, where Calabrian is considered to be the base of the Quaternary, while the new bio–stratigraphic scale from https://stratigraphy.org/ (accessed on 30 March 2021) includes the Gelasian (2.58–1.8 Ma) to Quaternary. This scale could introduce some differences compared to recent cartography [22] but is consistent with [7,62].

Some of the best seismic profiles were selected and organized in 5 almost-parallel Transects with a SW–NE direction within the above-mentioned area. Each Transect is composed of several seismic profiles that are aligned or partially overlapping and aim at realizing a single element.

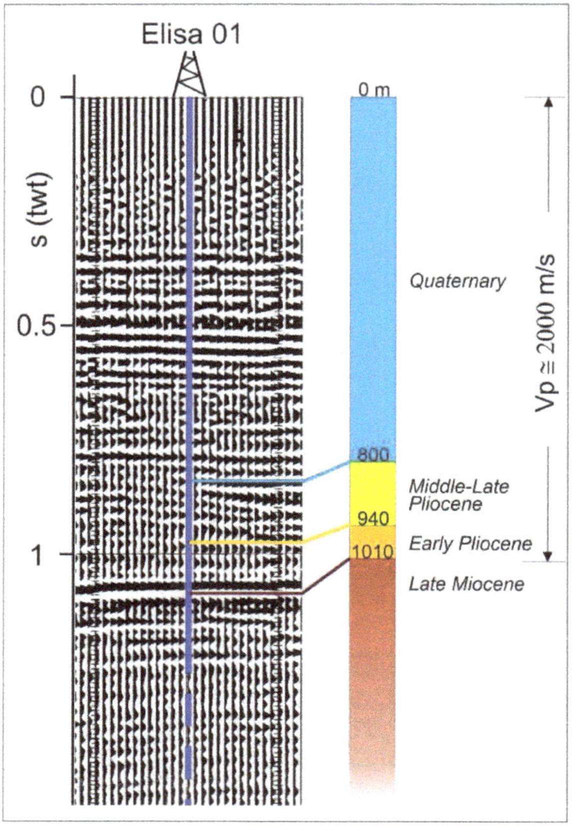

Figure 3. Stratigraphic correlation between the ELISA 1 well and a segment of the ViDEPI B-441 seismic profile where the well is placed (Figure 1).

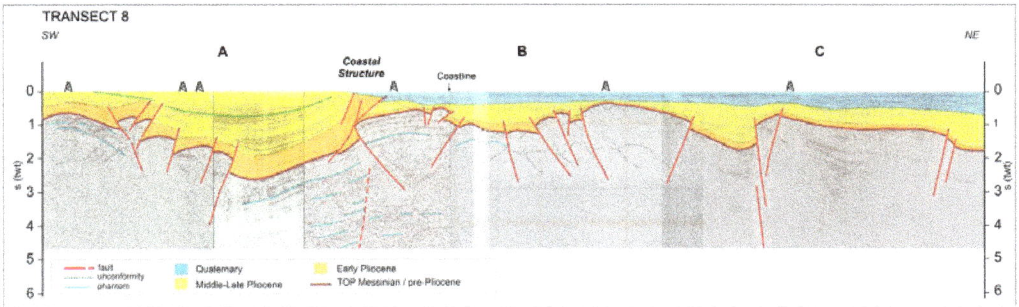

Figure 4. Transect 8. The Transect is composed of several seismic profiles labelled with letters *A*, *B*, *C*, see Figure 1). Quaternary deposits are highlighted in blue, Middle–Upper Pliocene deposits are in yellow, Lower Pliocene deposits are in orange, and top Messinian/Pre-Pliocene deposits are in brown. No colour is used for the pre-Pliocene sequence. Unconformities are shown in green dots. Light blue: undefined seismic reflectors. This legend is valid for all Transects (Figures 5–8).

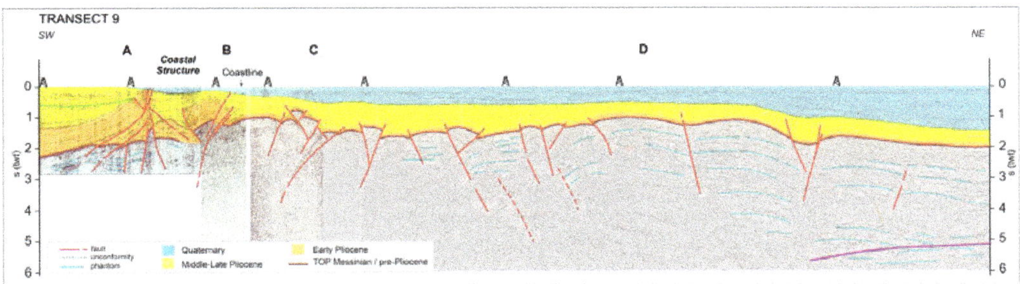

Figure 5. Transect 9. The purple line shows the hypothetical basement.

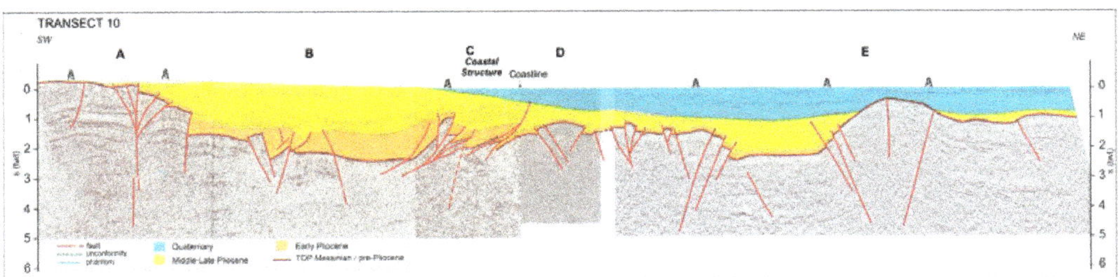

Figure 6. Transect 10.

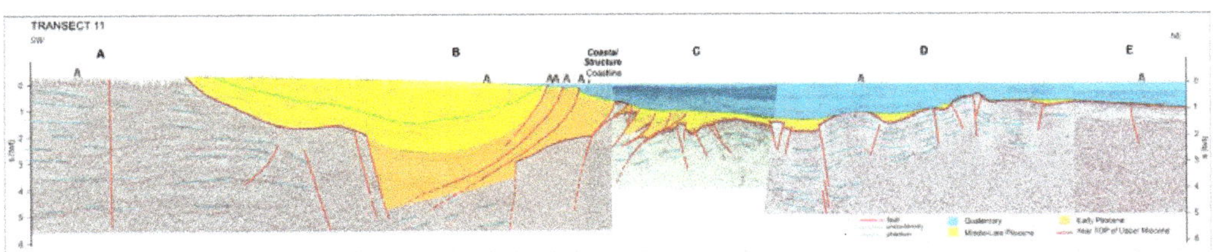

Figure 7. Transect 11.

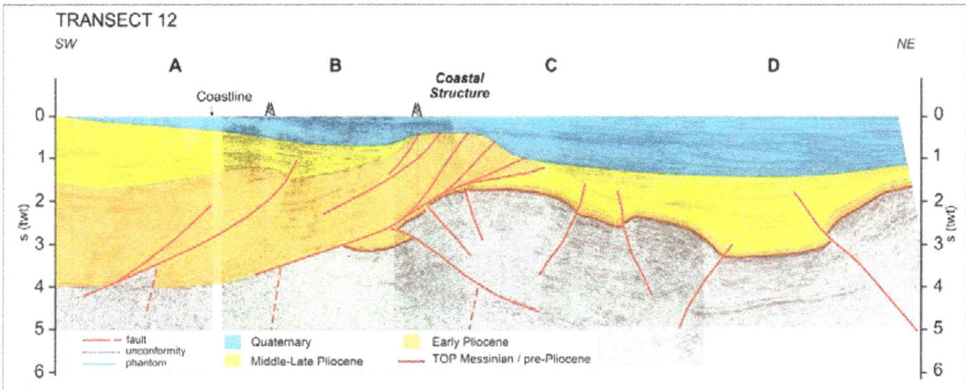

Figure 8. Transect 12.

Figure 1 presents traces of the Transects and each seismic profile within them. These traces complete our previous analyses of the outer Apennine Marche sector north of the Conero promontory, where seven Transects have already been observed [24,25]. For this reason, the new Transects are numbered from 8 to 12.

These Transects are represented individually in Figures 4–8 and are reported using a high-resolution plate in the supplementary material (Plate 1).

4. Results by Wells and Seismic Profiles Interpretation

4.1. Transects

The northernmost Transect (number 8 in this work; Figure 1) developed in the onshore and offshore areas just south of the Conero promontory and includes the seismic reflection profiles *A*, *B*, and *C* (Figure 4). Overall, the quality of these seismic reflection profiles is good; among these profiles, profile *B*, which was already interpreted in the VIDEPI catalogue, was further interpreted here using different colours.

In the onshore area, a WSW–ENE seismic profile (*A*) extends from the east of the Appignano locality to the coastline (ENE). Along this profile, some wells (Figure 1) allow good calibration of the top of the Messinian/Pre-Pliocene and near the top of the Early Pliocene seismo–stratigraphic horizons. An unconformity is also present within the Middle–Upper Pliocene deposits. In the offshore area, two seismic reflection profiles are present (*B* and *C*) and are aligned along a WSW–ENE direction. In particular, the *B* profile partly overlaps the *A* profile, and two hydrocarbon wells occur nearby.

Transect 8 is characterized by three structurally well-defined areas. On the western area a wide syncline is present, affecting a large thickness of about 3000 m Pliocene deposits. Lower Pliocene deposits cover the Messinian deposits in transgression. These deposits have an almost constant thickness, while those above the Middle–Upper Pliocene feature have variable thickness (ranging between 1800 m in the syncline core and about 1000 m along the limbs). Quaternary deposits have a thickness of a few hundred meters.

The western limb of the syncline is affected by a high-angle, W-dipping reverse fault system. This system deforms the whole Lower to Middle Pliocene sequence without involving that of the Upper Pliocene. However, in the westernmost part, some faults deform the overlying unconformity placed within the uppermost part of the Middle–Upper Pliocene sequence.

In the central area, a very complex compressive and uplifted structure ("Coastal Structure") is present. This structure is characterized by shallow East-verging thrusts affecting the Lower–Middle Pliocene sequence and, marginally, the Upper Miocene sequence. Below this structure, an E-dipping reverse fault and a slightly W-dipping sub-vertical fault reaching the relevant depths (>4 s TWT) are present. Quaternary deposits were likely involved in the deformation of the upper and frontal sectors of this Coastal Structure. As highlighted in the *A* seismic profile, Quaternary deposits outcropping on the western side of the Coastal Structure show reduced thickness compared to those on the eastern side. The Lower Pliocene deposits are indeed less than 100 m in thickness in front of the Coastal Structure and about 1000 m within the syncline behind.

On the eastern area (seismic profiles *B* and *C*), numerous reverse high angle W- and E-dipping faults are present. Overall, the vertical displacement of these faults is moderate, rarely exceeding 500 m. Lower Pliocene deposits have an average thickness of about 100 m or can be absent in the proximity of some structural highs (see *C* in Figure 4 and the wells presented here). The Middle–Upper Pliocene deposits show more variable thickness, from about one hundred meters on the structural highs to more than 1000 m in the proximity of faults and in structurally deeper areas. This extreme variability together with the characteristics of unevenness and chaos of the seismic horizons suggest a syn–tectonic origin of these deposits. The middle lower part of this sequence is certainly affected by reverse faults, while the upper part does not appear to be involved in deformation (seismic profiles *B* and *C* in Figure 4). Indeed, in this area the Quaternary deposits show a regular trend—increasing their thickness toward the east—and are not involved in deformation.

Transect 9 (Figure 5), which includes several wells, shows similar structural and stratigraphic characteristics to those of Transect 8. These differences relate to the greater thickness of the Lower Pliocene and Quaternary deposits facing the Coastal Structure and the high angle faults that are more evident below this structure. A Middle-Upper Pliocene unconformity is also clear in this area and was displaced by frontal thrusts. In this Transect, seismic reflection profile *A* overlaps profile *B* moving eastward toward the coastline. This profile exhibits a shallow compressive structure characterized by east- and west-verging thrusts.

In this structure, the Lower Pliocene deposits are concordant with the Messinian ones, featuring a thickness of about 800 m and more than 1000 m. Eastward, the thickness is notably reduced (about 150 m). Moreover, an unconformity present in the Middle-Upper Pliocene deposits separates the upper portion of the sequence, which is characterized by onlap geometry, from the lower one featuring pinch-out geometry. Furthermore, below the surface thrusts of the Coastal Structure, seismic profiles *A* and *B* from Figures 4 and 5 show very evident high angle W- and E-dipping faults. Offshore, seismic reflection profiles *C* and *D* show pre-Pliocene bedrock widely deformed by high angle west- and east-dipping compressive faults forming gentle pop-up structures with reduced vertical displacement. The thickness of the Lower Pliocene deposits is always very low (<100 m, as also reported in well stratigraphy), while the Middle–Upper Pliocene deposits are syn–tectonic with high variable thickness (from a few to several hundreds of meters) close to compressive structures. Quaternary deposits have a relatively constant thickness (about 600–800 m) and are not affected by deformation. All the other Transects (10–12, see Figures 6–8) show similar characteristics to those described above. As already mentioned, due to the different resolutions of seismic profiles and/or local factors, certain features are clearer than others.

In the westernmost sector of Transects 10 and 11, a deeply rooted sub-vertical structure is highlighted. Transect 10 (Figure 6) shows a branched flower structure that separates the Laga Formation units by the Colombacci Formation at the surface (Figure 1; [25]). This is a branched structure with a possible strike–slip component. In these two Transects, the compressive, W-dipping structures observed in Transect 8 are absent. Furthermore, along Transect 11 (profile *A* in Figure 7), an important normal E-dipping fault (more than 3000 m of vertical displacement) defines the Lower Pliocene basin to the west and is covered by transgressive deposits of the Middle–Upper Pliocene. In the same Transect, the above-mentioned unconformity within the Middle–Upper Pliocene sequence is clearly visible within the syncline. Transect 11 shows that during the Middle/Upper Pliocene, there was simultaneous subsidence (with transgression) in the current onshore to the west together with compression and uplift to the east (Coastal Structure, profiles *A-B* in Figure 7). In both

Transect 11 and 12 (Figure 8), compressive E-dipping faults under the Coastal Structure thrusts are clearly present, as in Transects 8 and 9.

Some thrusts of the Coastal Structure, as shown in Transect 11, affect the Quaternary deposits, such as in Transect 8. In Transect 12, only the shallowest Quaternary deposits are transgressive and are not involved in the deformation. Instead, in Transect 10 the thrusts affect only the Middle–Upper Pliocene sequence. In all Transects, the Quaternary succession covering the offshore flower structure is undeformed. Furthermore, evidence of fore-set Quaternary sedimentation is present in Transects 10 and 12 (Figures 6 and 8).

4.2. Characteristics and Distribution of the Plio–Quaternary Deformation

4.2.1. Early Pliocene

Based on well data logs and interpretations of both VIDEPI and ENI seismic profiles, we achieved a reconstruction of the thicknesses and distribution of the Lower Pliocene stratigraphic sequence (Figure 9).

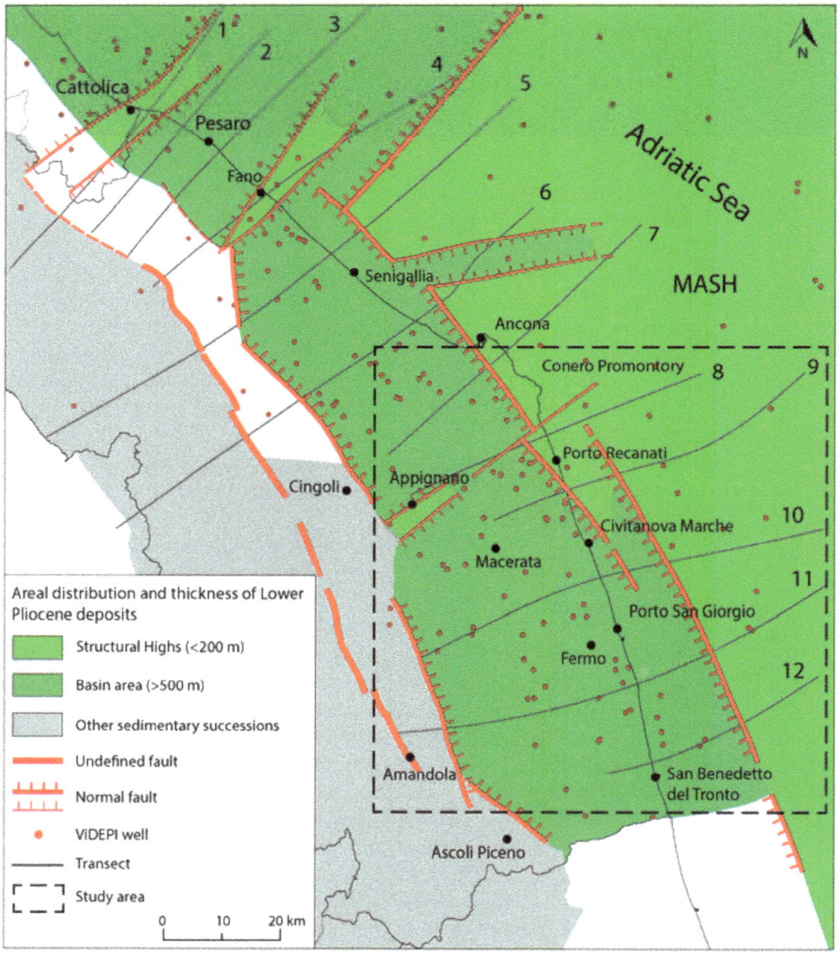

Figure 9. Schematic map of the distribution and thickness of Lower Pliocene deposits within the Marche region and the adjacent Adriatic Sea. Marche Adriatic Structural High (MASH) and smaller structural highs and basins are highlighted with the same colour. The study area is located in the dashed square.

This sequence has significant and sudden variations in thickness, and we were able to distinguish between the true sedimentary thicknesses and local tectonic repetitions or highly inclined bedding in proximity of compressive structures. This does not allow us to reconstruct a reliable isopaches map. Afterward, for an immediate view of the Lower Pliocene deposits distribution, we identified two distinct thickness classes (less than 200 m and greater than 500 m). This simple representation makes it possible to easily locate the position, geometry, and kinematics of the faults affecting this sequence. Finally, in Figure 9 the whole outer Marche Apennine sector has been reproduced to show the distribution of this succession.

The thickness distribution of the Lower Pliocene deposits shows evidence of a wide semi-submerged Marche Adriatic Structural High (MASH) area mainly located on the current Adriatic offshore (light green in Figure 9), as well as evidence of a wide basin area located in the northern portion of the same offshore area and in the south onshore area (dark green in Figure 9). This area also includes the northern part of the Marche territory, which is only marginally examined in this work.

Within the MASH area, the thickness of these deposits is very modest (a few tens of meters and, locally, not more than 200 m). In the basin area, the thickness rapidly increases, ranging from more than 500 to 3330 m. The limit between the plateau and basin areas features an NNW–SSE trend south of the Conero promontory lies slightly eastward of the current coastline, which shows instead an NE–SW trend in proximity to the Fano offshore area.

This spatial distribution is likely due to a normal or transtensive fault system that separates the wide and stable MASH area, which appears to be slightly subsident and located in the central–southern Adriatic offshore, from a basin area that is strongly subsident towards its western and northern portions.

Furthermore, the western side of the basin is marked by a normal fault system (Transects 10*A*, 11*A*; Figures 6 and 7). This transtensive fault system was active soon before the onset of the compressive phase highlighted within the Transect.

As underlined in the previous section, this normal fault system consists of syn–sedimentary high angle W- and E-dipping faults characterized by remarkable vertical displacement reaching thousands of meters, which is clearly detectable in the interpreted seismic profiles.

The main faults were likely placed in proximity of the NNW–SSE and NE–SW boundaries of uplifted and subsident areas. Other minor faults further disarticulated both the basin and the MASH areas, defining local thickness variations in the sequence.

4.2.2. Middle-Late Pliocene-Quaternary

Based on our investigation, three structurally well-defined areas along a W–E direction are identified (Figure 10).

The western area is characterized by a wide syncline. In the northern part of this area, the syncline is locally intersected by W-dipping high-angle reverse faults (Figure 4); in the southern portion, Lower Pliocene deposits end against a high-angle E-dipping normal fault to the West, covered by transgressive Middle–Upper Pliocene deposits. The syncline axis is about N–S oriented. W of the syncline, a sub-vertical fault system deeply rooted with a N–S trend can be observed.

The central area is marked by a compressive structure (Coastal Structure). This structure consists of a series of E-verging thrusts within the shallower sequence, mainly affecting the Lower–Middle Pliocene deposits and only marginally affecting the Messinian ones. Thrust displacements are rapidly reduced within the Messinian and Lower Pliocene deposits. Just below this horizon, E-dipping reverse faults and deeper high-angle W-dipping faults are present. The Coastal Structure shows an NNW–SSE, almost continuous, trend, and sometimes crops out close to the coastline. This structure was formed starting during the Middle Pliocene, and its deformation continued until the Upper Pliocene, in some parts up to the Quaternary.

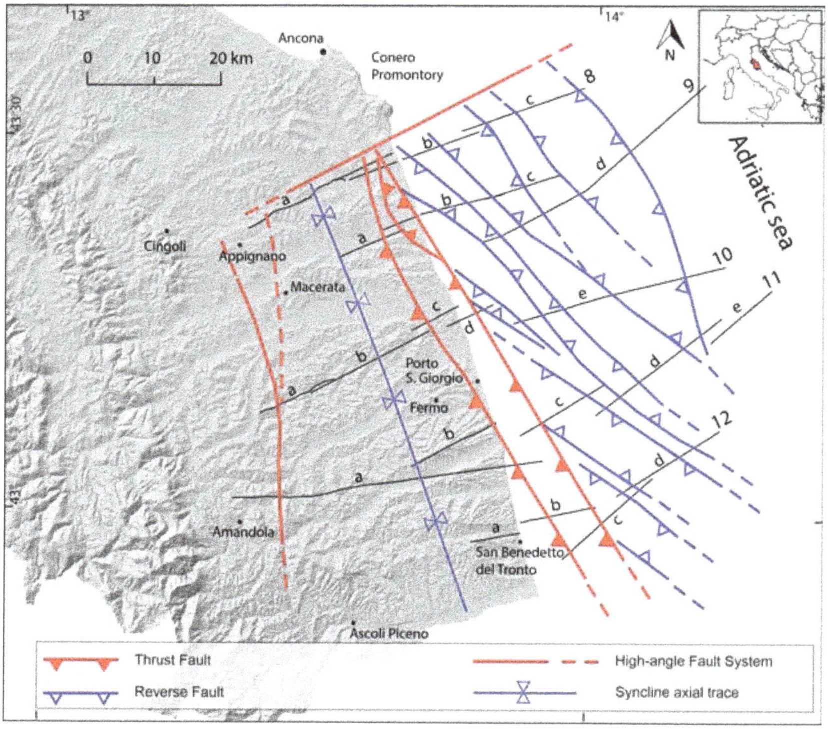

Figure 10. Structural sketch map of the outer Marche area south of Conero promontory. The main Plio–Quaternary structures are highlighted.

The eastern area is characterized by W- and E-verging high-angle reverse faults, giving rise to gentle flower structures with an NW–SE trend. These structures were formed from the Middle Pliocene and developed structural highs, some of which were still emerging during the Upper Pliocene/Pleistocene (Transects 10, 11; Figures 6 and 7). Compressive deformation stopped during the Upper Pliocene, and Quaternary deposits were not affected.

5. Discussion

During the Messinian, this part of the Marche Apennines outer sector emerged or was close to emersion (the "lago-mare" succession in Figure 2), with sedimentary break-off [55]. The top Messinian/pre-Pliocene seismic horizon was always clearly evident in the examined seismic profiles, with frequent characteristics of an erosive surface (Figures 6 and 7). The Lower Pliocene deposits, however, are often transgressive or discordant over the underlying Messinian or pre-Pliocene ones. Furthermore, no important evidence of Messinian active tectonics was found in this area. This part of the sector started to deform during the Early Pliocene when normal or transtensive faults with an NNW–SSE trend were enucleated. These faults separated heavily subsident basin areas from almost-stable structural highs (Figure 9).

The basin area was mainly located along the current onshore, while the Marche Adriatic Structural High (MASH) was located in the current offshore. This feature continued southward in the Abruzzo region with quite similar characteristics, as described in [28]. According to these authors, in Abruzzo, the basin formed due to horst and graben structures starting in the Messinian–Pliocene transition due to flexural extension of the under-thrusting Adria Plate. In our study area, this extensional phase started in the Early

Pliocene, as indicated by the erosive top Messinian and the transgressive and discordant Pliocene deposits above it.

North of the Conero promontory, the area's slightly more complex setting was also due to an important NE–SW trending fault that segmented the MASH, yielding a basin area to the NW (Figure 9). This structure, already identified in [24], continues from the Fano offshore to the SW along the river valley (Figure 1). Other local features with an NE–SW trend segmented both the basin and the MASH, forming lower-ranking depressions and structural highs (Figure 9). The northernmost transverse structures correspond to the Cattolica seismogenic system [24].

Starting from the Middle Pliocene, a compressive regime was established in the sector south of the Conero promontory, growing the structures underlined in the Transects and in Figure 10.

In more detail, in the study area we highlighted a wide syncline with an almost N–S trend to the west, the Coastal Structure with an NNW–SSE trend in the central portion, and the NW–SE-trending gentle-flower structure system to the east. The syncline was thus formed in correspondence with the Lower Pliocene basin, and the Coastal Structure formed in correspondence with the normal faults bordering the same basin to the east (Figures 9 and 10). The Middle Pliocene deposits are continuous with those of the Lower Pliocene at the syncline core while resting on the same deposits with a pinch-out feature and reduced thickness in proximity to the growing Coastal Structure western flank (Transects 9, 11, and 12; Figures 5, 7 and 8). Variable thickness, with greater thickness close to the faults, attests to the syn–tectonic origins of these deposits in the offshore area.

Further to the west of the syncline, the N–S Amandola-positive flower structure (Figure 6) separates different Messinian units [7]. This structure is high-angle and deeply rooted (Transects 10 and 11; Figures 6 and 7), likely extending farther than the representation in Figure 10. The push-up in the western part of Transect 8a (Figure 4) is likely a continuation of the Amandola structure or one of its branches. All these structural elements are slightly divergent from each other and are interrupted along a complex transverse structure approximately ENE–WSW oriented and located immediately south of the Conero Promontory (Figure 10).

The compressive deformation phase ended in the western and eastern areas during the Late Pliocene-Early Pleistocene. The unconformity within the Middle–Upper Pliocene deposits (Transects 9 and 11; Figures 5 and 7) indicates that the syncline has not deepened since the Late Pliocene. Upper Pliocene deposits rest in an on-lap over the underlying ones above the unconformity and reduce their thickness in proximity of the western flank of the Coastal Structure. These features indicate that, within the syncline, the lower parts of the Middle–Upper Pliocene deposits are syn–tectonic, while those of the upper part are post-tectonic.

The flower structures of the Adriatic offshore are sealed by the Quaternary deposits. In the central area, the Coastal Structure continued its activity even during the Quaternary, as shown in several areas (Transects 8,9, and 11 in Figures 4, 5 and 7). Therefore, all these structures were formed during the Middle Pliocene. Most of these were deactivated at the end of the Late Pliocene, while a few others were probably still active during the Early Pleistocene (Transects 10–12 in Figures 6–8).

The Coastal Structure is characterized by low-angle faults close to the surface and high angle faults at depth. Low-angle faults are mainly involved in the Lower Pliocene deposits, making their repetition clearly visible in all Transects. The underlying Messinian deposits were, instead, not significantly involved, likely due to detachment between the two sequences. In [11], however, Messinian deposits were considered to be strongly involved in deformation. At the western edge of the syncline, and underneath the highly deformed close-to-the-surface succession (Transect 11), parts of the original faults bordering the Lower Pliocene basin are still recognizable. Looking at the Lower Pliocene deposits distribution map (Figure 9), it can be seen that the Coastal Structure is nucleated in the same position as the faults bordering the previous Lower Pliocene basin to the east and perfectly follows the trend of the latter. Therefore, this structure was formed by partially inverting or deforming (Figure 11) the previous high-angle normal/transtensive faults (see also Figures 5–8). These faults may have acted initially as a barrier, forcing the involved sequences to climb upwards; in some cases (Figures 5, 7 and 8), the innermost thrusts show a higher angle than the external ones. Subsequently, as the shortening increased, the upper parts of the Early Pliocene faults were decapitated (see [43]) and included within the superficial low-angle, E-verging thrust sheets, which mainly affect the Lower Pliocene succession that is partially detached from the underlying one (Figure 11).

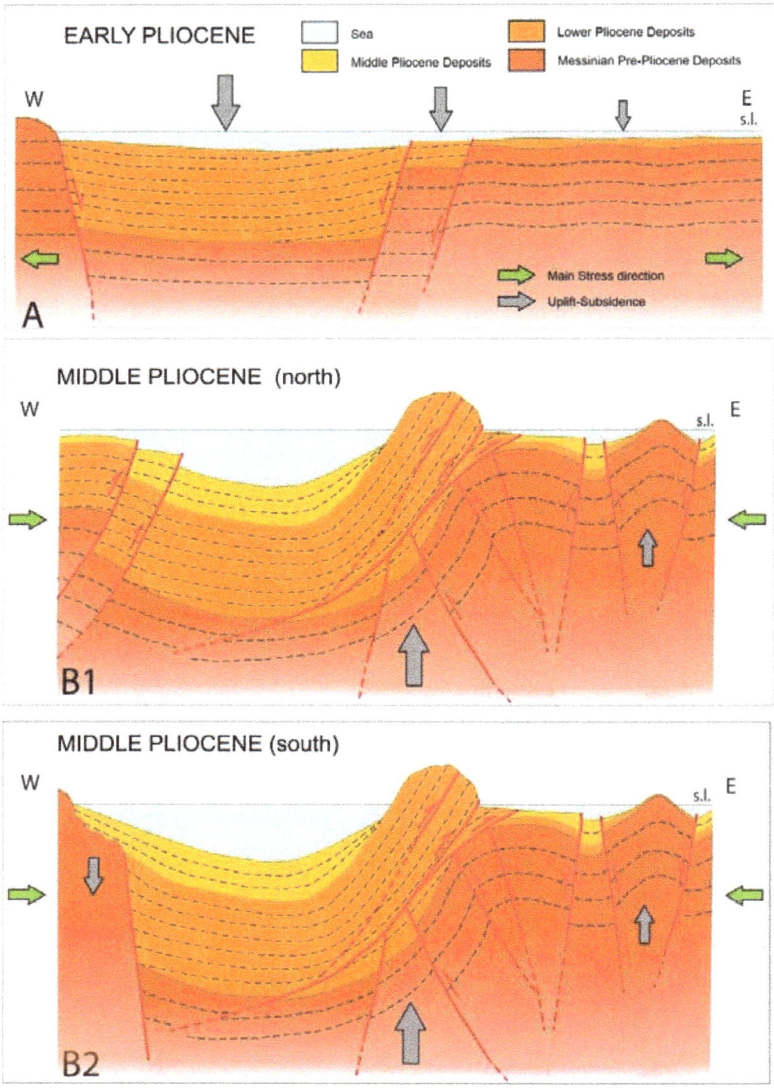

Figure 11. Sketch diagram showing the evolution of the Coastal Structure from the Early to Middle Pliocene across two representative cross-sections (not to scale). The size of the grey arrows is proportional to the intensity of vertical movement.

Thus, in the later stage (Late Pliocene–Quaternary), some of the thrust sheets partially covered the westernmost flower structures of the eastern Adriatic area (Transects 8, 10, and 12 in Figures 4, 6 and 8). The compressive Coastal Structure formed due to the inversion of previous extensional features following the "interaction of extensional and contractional deformation" model proposed by [43,64] and in [18,65] for the nearby Montagna dei Fiori and Cingoli structures (Figure 11).

The Coastal Structure continues southward, in the Abruzzo region, with quite similar litho–structural characteristics and ages of deformation [50,65]. Additionally, in that area, E-vergent thrusts are mainly found in the Lower Pliocene deposits, which, in this case, are completely detached from the underlying Messinian ones. However, unlike the process proposed for our study area, these authors suggest that the previous Early Pliocene normal faults bordering the basin were already enucleated during the Messinian. Furthermore, these normal faults were not involved in compressive deformation but were simply covered by the thrust sheets. According to these authors, compressive deformation began in the Early Pliocene.

As a result of the compression that determined the Coastal Structure's development, tilting of the block between this structure and the Amandola structure to the W likely also have occurred. During the Middle Pliocene, there was simultaneous uplift of the eastern front (enucleation and uplift of the Coastal Structure) and subsidence of the western side (transgression of the Middle Pliocene deposits on those previously raised during the pre-Pliocene time; Transect 11). The horizontal rotational axis may correspond to the syncline axis. This mechanism is similar to that described in [66] for the Po Valley. During the Late Pliocene–Quaternary this rotation ceased, and the deposits of the same age became horizontal.

The Amandola structure, the syncline, and the Coastal Structure show a straight and regular trend. As previously mentioned, the trend of these main onshore structures is somewhat divergent from the offshore one, even though they all formed during the same time interval. This can be attributed to the influence of pre-existing features inherited by previous deformation phases such as the faults shown in Figure 9. These structures are compatible with the main local shortening oriented in an NNE–SSW direction during the Middle–Late Pliocene (compression with the P axis about NNE–SSW; Figure 12), which emerged in the northern sector of the Marche region [24,67,68] and, more generally, in the overall Central Adriatic area [69,70].

In this context, right-lateral transpression developed along the Coastal Structure and likely enhanced the gentle flower systems of the Adriatic offshore (Figure 12)

The Coastal Structure schematically represented as continuous and regular in Figure 10 is most likely composed of several structures, some of which were still active during the Quaternary, as shown by fairly significant earthquake sequences (M_W = 5, Porto San Giorgio sequence, [71,72]) that occurred recently (Figure 12).

As previously mentioned, the described structures were somehow interrupted to the north along a transverse ENE–WSW-oriented structure. The existence of transverse faults has been highlighted in literature by various authors, particularly in the Marche-Abruzzo onshore (see [7,24,73] and reference therein). In our study area, several structures underwent sudden changes in characteristics (age of deformation, geometry, and direction) that are observable when compared to those mapped in [24] in the areas west, east, and north of the Conero promontory. Furthermore, the structures present immediately northward of this transverse element and of our study area, e.g., the Early Pliocene transpressive structure of Strada-Moie and S. Andrea di Suasa (see Figure 7 in [23]) are no longer present in the south. Indeed, in this southern area, extension was still occurring during the Early Pliocene.

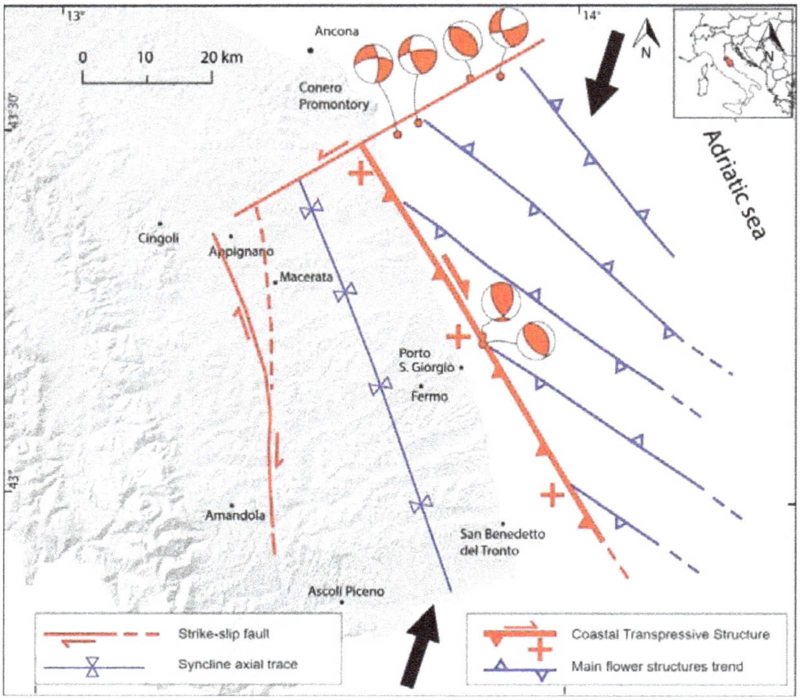

Figure 12. Kinematic sketch map. Red lines indicate fault systems still active during the Quaternary. The black arrow represents the main shortening direction, red arrows describe the right lateral strike–slip component, and the plus symbol (+) is the narrow strongly uplifted area. Focal mechanisms (beach balls) of the main earthquakes of 1987 Porto S. Giorgio and 2013 south of Conero seismic sequences are shown.

The transverse structure south of the Conero promontory, already partially present in [7], interrupts structures with Quaternary activity, i.e., the Coastal Structure to the south and the Conero compressional structure [25] to the north. Therefore, this tectonic element must be Quaternary itself, as also attested by recent earthquakes and seismic sequences in the offshore along the element (Figure 12). These focal mechanisms are predominantly strike–slip, with P-axes oriented around the ENE–WSW and sub-vertical planes [67,68]. Furthermore, the epicentres of the seismic sequences described by these authors are aligned ENE–WSW.

6. Conclusions

Seismic profile interpretations and well stratigraphic data allowed us to describe the Plio–Quaternary evolution of the outer Marche Apennines south of the Conero promontory. The main results can be summarized as follows:

- During the Early Pliocene, the area was affected by extensional or transtensive tectonics, resulting in the formation of a strongly subsident basin and a more stable structural high. More than 3000 m of sediment accumulated in the basin zone, while the structural high (MASH) hosts less than 200 m of Lower Pliocene deposits.
- The basins and structural highs are separated by an approximately NNW–SSE normal and transtensive fault system located close to the current coastline. Other normal faults with an NNW–SSE trend developed in the current onshore area and border the basin to the W. The structural high is instead located in the current offshore area.
- Starting from the Middle Pliocene, the entire area underwent compression, with the P axis oriented about NNE–SSW leading the formation, from W to E, of the NNW-SSE

dextral strike-slip Amandola structure, the NNW–SSE dextral transpressive Coastal Structure, and an NW–SE-striking system of gentle flower structures (offshore).

- The Coastal Structure is the most complex and important structure in the study area. It consists of an E-vergent thrust system at surface and high-angle E and W-vergent faults at depth. Shallow thrusts mainly affected the Pliocene deposits and, locally, the Quaternary ones. The mainly Messinian underlying deposits were marginally involved in deformation. Deeper faults affect Mio-Pliocene and older deposits. As a result, in the shallower part of the Coastal Structure, pre-existing normal faults were inverted or crosscut and incorporated into the ongoing thrusts, while at depth, they were not deformed.
- The trends of the Coastal Structure and the flower structures within the offshore are slightly divergent despite being contemporaneous because the former was strongly influenced by inherited structures.
- The compressive phase was finished during the Late Pliocene in the syncline, as well as along the flower structures. The Coastal Structure was still active during the Quaternary. This is also testified by recently recorded seismic activity.
- A complex transverse structure with a general ENE–WSW trend (at least partially active and seismogenic) traces the boundary between the outer areas north and south of the Conero promontory, where the styles, geometries and times of deformation of the Plio–Quaternary structures are significatively different.

Based on our results, we conclude that during the Plio–Quaternary times, this portion of the outer Apennine sector is mainly affected by a right-lateral transpressive deformation, and by widespread kinematic inversion of pre-existing structures. Former studies proposed a simple E-vergent compressive deformation for the same area.

Supplementary Materials: The following are available online at https://www.mdpi.com/article/10.3390/geosciences11050184/s1.

Author Contributions: Investigation, M.C.; resources, J.C., M.C., S.T, and P.P.P.; data curation, S.T., P.P.P., and M.C.; writing—original draft preparation, C.I., M.C., and J.C.; writing—review and editing, M.C., C.I., and J.C.; funding acquisition, P.P.P. and C.I. All authors have read and agreed to the published version of the manuscript.

Funding: This research was supported by grants from P.P. Pierantoni (FAR—Fondo di Ricerca Ateneo, Università di Camerino) and C. Invernizzi (Interreg Coastenergy project ID: 10045844; FAR—Fondo di Ricerca Ateneo, Università di Camerino).

Institutional Review Board Statement: Not applicable.

Informed Consent Statement: Not applicable.

Acknowledgments: ENI S.p.A. is thanked for supplying seismic profiles, for their facilities, and for permission to publish the data. Stefano Mazzoli and Claudio Di Celma are kindly thanked for their useful discussions. The authors also thank the reviewers for several insightful comments that consistently improved the manuscript and Giancarlo Molli, Angelo Cipriani, and Domenico Liotta for their editorial work.

Conflicts of Interest: The authors declare no conflict of interest.

References

1. Boccaletti, M.; Calamita, F.; Deiana, G.; Gelati, R.; Massari, F.; Moratti, G.; Ricci-Lucchi, F. Migrating Foredeep-Thrust Belt Systems in the Northern Apennines and Southern Alps *Palaeogeogr. Palaeoclimatol. Palaeoecol.* **1990**, *77*, 3–14. [CrossRef]
2. Ghielmi, M.; Minervini, M.; Nini, C. Sedimentary and Tectonic Evolution in the Eastern Po-Plain and Northern Adriatic Sea Area from Messinian. *Rend. Fis. Acc. Lincei.* **2010**, *21*, 131–166. [CrossRef]
3. Turrini, C.; Toscani, G.; Lacombe, O.; Roure, F. Influence of Structural Inheritance on Foreland-Foredeep System Evolution: An Example from the Po Valley Region (Northern Italy). *Mar. Pet. Geol.* **2016**, *77*, 376–398. [CrossRef]
4. Odlum, M.L.; Stockli, D.; Capaldi, T.N.; Thomson, K.D.; Clarck, J.; Puigdefàbregas, C.; Fildani, A. Tectonic and Sediment Provenance Evolution of the South Eastern Pyrenean Foreland Basins During Rift Margin Inversion and Orogenic Uplift. *Tectonophysics* **2019**, *765*, 226–248. [CrossRef]

5. Bally, A.W.; Burbi, L.; Cooper, C.; Ghelardoni, R. Balanced Sections and Seismic Reflecion Profiles across the Central Apennines. *Mem. Della Soc. Geol. Ital.* **1986**, *35*, 257–310.
6. Calamita, F.; Deiana, G. Evoluzione Strutturale Neogenico-Quaternaria Dell'appennino Umbro-Marchigiano. *Stud. Geol. Camerti Vol. Spec. "La Geol. delle Marche"* **1986**, 91–98. [CrossRef]
7. Centamore, E.; Pambianchi, G.; Minetti, A. *Ambiente Fisico Delle Marche: Geologia Geomorfologia Idrogeologia Regione Marche, Giunta Regionale*; Firenze: S.E.L.C.A.: Florence, Italy, 1991.
8. Pierantoni, P.P.; Deiana, G.; Romano, A.; Paltrinieri, W.; Borraccini, F.; Mazzoli, S. Geometrie strutturali lungo la thrust zone del fronte montuoso umbro-marchigiano-sabino. Bollettino della Società geologica italiana. *Bollettino della Società geologica italiana* **2005**, *124*, 395–411.
9. Amadori, C.; Toscani, G.; Di Giulio, A.; Maesano, F.E.; D'Ambrogi, C.; Ghielmi, M.; Fantoni, R. From Cylindrical to Non-Cylindrical Foreland Basin: Pliocene–Pleistocene Evolution of the Po Plain–Northern Adriatic Basin (Italy). *Basin. Res.* **2019**, *31*, 991–1015. [CrossRef]
10. Cantalamessa, G.; Centamore, E.; Cristallini, C.; Invernizzi, C.; Matteucci, R.; Micarelli, A.; Piccini, M.; Pontoni, F.; Potetti, M. Nuovi Dati Sulla Geologia Dell'area di Porto San Giorgio (Ascoli P iceno, Marche). *Geol. Rom.* **1987**, *26*, 359–369.
11. Ori, G.G.; Serafini, G.; Ricci Lucchi, F.; Casnedi, R.; Colalongo, M.L.; Mosna, S. The Plio-Pleistocene Adriatic Foredeep (Marche and Abruzzo, Italy): An Integrated Approach to Surface and Subsurface Geology. In Proceedings of the 3rd E.A.P.G. Conference, Adriatic Foredeep, Florence, Italy, 26–30 May 1991; pp. 1–86.
12. Invernizzi, C. Relazioni tra Costruzione Della Catena e Migrazione Dell'avanfossa Nell'area Marchigiana Esterna: Alcune Osservazioni. *Stud. Geol. Camerti* **1992**, *12*, 71–78. [CrossRef]
13. Ghielmi, M.; Minervini, M.; Nini, C.; Rogledi, S.; Rossi, M. Late Miocene-Middle Pleistocene Sequences in the Po Plain Northern Adriatic Sea (Italy): The stratigraphic Record of Modification Phases Affecting a Complex Foreland Basin. *Mar. Pet. Geol.* **2013**, *42*, 50–81. [CrossRef]
14. Barchi, M.R.; Alvarez, W.; Shimabukuro, D.H. The Umbria-Marche Apennines as a Double Orogen: Observations and Hypotheses. *Ital. J. Geosci.* **2012**, *131*, 258–271. [CrossRef]
15. De Donatis, M.; Invernizzi, C.; Landuzzi, A.; Mazzoli, S.; Potetti, M. CROP03: Structure of the Montecalvo in Foglia-Adriatic Sea Segment. *Mem. della Soc. Geol. Ital.* **1998**, *52*, 617–630.
16. Mazzoli, S.; Pierantoni, P.P.; Borraccini, F.; Paltrinieri, W.; Deiana, G. Geometry, Segmentation Pattern and Displacement Variations Along a Major Apennine Thrust Zone, Central Italy. *J. Struct. Geol.* **2005**, *27*, 1940–1953. [CrossRef]
17. Scisciani, V.; Calamita, F.; Tavarnelli, E.; Rusciadelli, G.; Ori, G.G.; Paltrinieri, W. Foreland-Dipping Normal Faults in the Inner Edges of Syn-Orogenic Basins: A Case from the Central Apennines, Italy. *Tectonophysics* **2001**, *330*, 211–224. [CrossRef]
18. Scisciani, V.; Tavarnelli, E.; Calamita, F. The Interaction of Extensional and Contractional Deformations in the Outer Zones of the Central Apennines, Italy. *J. Struct. Geol.* **2002**, *24*, 1647–1658. [CrossRef]
19. Scisciani, V.; Montefalcone, R. Coexistence of Thin- and Thick-Skinned Tectonics: An Example from the Central Apennines Italy. *Spec. Pap. -Geol. Soc. Am.* **2005**, *2414*, 33–54. [CrossRef]
20. Mancinelli, P.; Scisciani, V. Seismic Velocity-Depth Relation in a Siliciclastic Turbiditic Foreland Basin: A Case Study from the Central Adriatic Sea. *Mar. Pet. Geol.* **2020**, *120*, 104554. [CrossRef]
21. Tavarnelli, E.; Scisciani, V.; Patruno, S.; Calamita, F.; Pace, P.; Iacopini, D. The Role of Structural Inheritance in the Evolution of Fold-and-Thrust Belts: Insights from the Umbria-Marche Apennines, Italy. *GSA Spec. Pap.* **2019**, *542*. [CrossRef]
22. Conti, P.; Cornamusini, G.; Carmignani, L. An Outline of the Geology of the Northern Apennines (Italy), with Geological Map at 1:250,000 Scale *Ital. J. Geosci.* **2020**, *139*, 149–194. [CrossRef]
23. Pierantoni, P.P.; Deiana, G.; Galdenzi, S. Stratigraphic and Structural Features of the Sibillini Mountains (Umbria- Marche Apennines, Italy). *Ital. J. Geosci.* **2013**, *132*, 497–520. [CrossRef]
24. Pierantoni, P.P.; Chicco, J.; Costa, M.; Invernizzi, C. Plio-Quaternary Transpressive Tectonics: A Key Factor in the Structural Evolution of the Outer Apennine–Adriatic System, Italy. *J. Geol. Soc. Lond.* **2019**, *176*, 1273–1283. [CrossRef]
25. Chicco, J.M.; Pierantoni, P.P.; Costa, M.; Invernizzi, C. Plio-Quaternary Tectonics and Possible Implications for Geothermal Fluids in the Marche Region (Italy). *Tectonophysics* **2019**, *755*, 21–34. [CrossRef]
26. Scisciani, V.; Agostini, S.; Calamita, F.; Paolo, P.; Cilli, A.; Giori, I.; Paltrinieri, W. Positive Inversion Tectonics in Foreland Fold-and-Thrust Belts: A Reappraisal of the Umbria-Marche Northern Apennines (Central Italy) by Integrating Geological and Geophysical Data. *Tectoniphysics* **2014**, *637*, 218–337. [CrossRef]
27. Bigi, S.; Calamita, F.; Cello, G.; Centamore, E.; Deiana, G.; Paltrinieri, W.; Pierantoni, P.P.; Ridolfi, M. Tectonics and Sedimentation Within a Messinian Foredeep in the Central Apennines, Italy. *J. Pet. Geol.* **1999**, *22*, 5–18. [CrossRef]
28. Artoni, A. The Pliocene-Pleistocene Stratigraphic and Tectonic Evolution of the Central Sector of the Western Periadriatic Basin of Italy. *Mar. Pet. Geol.* **2013**, *42*, 82–106. [CrossRef]
29. Milli, S.; Moscatelli, M.; Stanzione, O.; Falcini, F. Sedimentology and Physical Stratigraphy of the Messinian Turbidite Deposits of the Laga Basin (Central Apennines, Italy). *Boll. della Soc. Geol. Ital.* **2007**, *126*, 255–281.
30. Bigi, S.; Milli, S.; Corrado, S.; Casero, P.; Aldega, L.; Botti, F.; Moscatelli, M.; Stanzione, O.; Falcini, F.; Marini, M.; et al. Stratigraphy, Structural Setting and Burial History of the Messinian Laga Basin in the Context of Apennine Foreland Basin System. *J. Mediterr. Earth Sci.* **2009**, *1*, 61–84. [CrossRef]

31. Bigi, S.; Casero, P.; Ciotoli, G. Seismic Interpretation of the Laga Basin; Constraints on the Structural Setting and Kinematics of the Central Apennines. *J. Geol. Soc. Lond.*. **2011**, *168*, 179–189. [CrossRef]
32. Ghielmi, M.; Serafini, G.; Artoni, A.; di Celma, C.; Pitts, A. From Messinian to Pleistocene: Tectonic Evolution and Stratigraphic Architecture of the Central Adriatic Foredeep (Abruzzo and Marche, Central Italy). In Proceedings of the 34th IAS International Meeting of Sedimentology, Rome, Italy, 10–13 September 2019.
33. Carminati, E.; Doglioni, C. Alps vs. Apennines: The paradigm of a Tectonically Asymmetric Earth. *Earth-Sci. Rev.* **2012**, *112*, 67–96. [CrossRef]
34. Cuffaro, M.; Riguzzi, F.; Scrocca, D.; Antonioli, F.; Carminati, E.; Livani, M.; Doglioni, C. On the Geodynamics of the Northern Adriatic Plate *Rend. Lincei* **2010**, *21*, 253–279. [CrossRef]
35. Deiana, G.; Pialli, G. The Structural Provinces of the Umbria-Marchean Apennines. *Mem. Soc. Geol. It* **1994**, *48*, 473–484.
36. Dewey, J.F.; Helman, M.L.; Knott, S.D.M.L.; Turco, E.; Hutton, D.H.W. Kinematics of the Western Mediterranean. *Geol. Soc. London Spec. Publ.* **1989**, *45*, 265–283. [CrossRef]
37. Schettino, A.; Turco, E. Plate Kinematics of the Western Mediterranean Region During the Oligocene and Early Miocene. *Geophys. J. Int.* **2006**, *166*, 1398–1423. [CrossRef]
38. Turco, E.; Macchiavelli, C.; Mazzoli, S.; Schettino, A.; Pierantoni, P.P. Kinematic Evolution of Alpine Corsica in the Framework of Mediterranean Mountain Belts. *Tectonophysics* **2012**, *579*, 193–206. [CrossRef]
39. Pierantoni, P.P.; Penza, G.; Macchiavelli, C.; Schettino, A.; Turco, E. Kinematics of the Tyrrhenian-Apennine System and Implications for the Origin of the Campanian Magmatism. In *Vesuvius, Campi Flegrei, and Campanian Volcanism*; De Vivo, B., Belkin, H., Rolandi, G., Eds.; Elsevier: Amsterdam, The Netherlands, 2020. [CrossRef]
40. Calamita, F.; Scisciani, V.; Montefalcone, R.; Paltrinieri, W.; Pizzi, A. L'ereditarietà del Paleomargine Dell'adria Nella Geometria Del Sistema Orogenico Centro-Appenninico: L'area Abruzzese Esterna. *Mem. Soc. Geol. Ital.* **2002**, *55*, 355–368.
41. Tavarnelli, E.; Butler, R.W.H.; Decandia, F.A.; Calamita, F.; Grasso, M.; Alvarez, W.; Renda, P. Implications of Fault Reactivation and Structural Inheritance in the Cenozoic Evolution of Italy. *Spec. Vol. Ital. Geol. Soc. IGC 32 Florence-2004* **2004**, 209–222.
42. Centamore, E.; Rossi, D. Neogene-Quaternary Tectonics and Sedimentation in the Central Apennines. *Boll. della Soc. Geol. Ital.* **2009**, *128*, 73–88.
43. Scisciani, V. Styles of Positive Inversion Tectonics in the Central Apennines and in the Adriatic Foreland: Implications for the Evolution of the Apennine Chain (Italy). *J. Struct. Geol.* **2009**, *31*, 1276–1294. [CrossRef]
44. Scisciani, V.; Patruno, S.; Tavarnelli, E.; Calamita, F.; Pace, P.; Iacopini, D. Multi-Phase Reactivations and Inversions of Paleozoic-Mesozoic Extensional Basins During the Wilson Cycle: Case Studies from the North Sea (UK) and Northern Apennines (Italy). *Geol. Soc. Lond., Spec. Publ.* **2019**, *470*, 205–243. [CrossRef]
45. Bonini, L.; Basili, R.; Burrato, P.; Cannelli, V.; Fracassi, U.; Maesano, F.E.; Melini, D.; Tarabussi, G.; Tiberti, M.M.; Vannoli, P.; et al. Testing Different Tectonic Models for the Source of the Mw 6.5, 30 October 2016, Norcia Earthquake (Central Italy): A Youthful Normal Fault, or Negative Inversion of an Old Thrust? *Tectonics* **2019**, *38*, 990–1017. [CrossRef]
46. Mazzoli, S.; Deiana, G.; Galdenzi, S.; Cello, G. Miocene Fault-Controlled Sedimentation and Thrust Propagation in the Previously Faulted External Zones of the Umbria-Marche Apennines. *Stephan Mueller Spec. Publ. Ser.* **2002**, *1*, 195–209. [CrossRef]
47. Calamita, F.; Coltorti, M.; Piccinini, D.; Pierantoni, P.P.; Pizzi, A.; Ripepe, M.; Scisciani, V.; Turco, E. Quaternary Faults and Seismicity in the Umbro-Marchean Apennines (Central Italy): Evidence from the 1997 Colfiorito Earthquake. *J. Geodyn.* **2000**, *29*, 245–264. [CrossRef]
48. Cello, G.; Deiana, G.; Ferelli, L.; Marchegiani, L.; Maschio, L.; Mazzoli, S.; Michetti, A.; Serva, L.; Tondi, E.; Vittori, T. Geological Constraints for Earthquake Faulting Studies in the Colfiorito Area (Central Italy). *J. Seismol.* **2000**, *4*, 357–364. [CrossRef]
49. Tondi, E.; Jablonská, D.; Volatili, T.; Michele, M.; Mazzoli, S.; Pierantoni, P.P. The Campotosto Linkage Fault Zone Between the 2009 and 2016 Seismic Sequences of Central Italy: Implications for Seismic Hazard Analysis. *GSA Bull.* **2020**, 1–16. [CrossRef]
50. Carruba, S.; Casnedi, R.; Perotti, C.R.; Tornaghi, M.; Bolis, G. Tectonic and Sedimentary Evolution of the Lower Pliocene Periadriatic Foredeep in Central Italy. *Int. J. Earth Sci.* **2006**, *95*, 665–683. [CrossRef]
51. Lucchi, F.R. The Oligocene to Recent Foreland Basins of the Northern Apennines. *Spec. Publs. Ass. Sediment.* **1986**, *8*, 105–139. [CrossRef]
52. Cantalamessa, G.; Cantemore, E.; Chiocchini, U.; Colalongo, M.L.; Micarelli, A.; Nanni, T.; Pasini, G.; Potetti, M.; Ricci Lucchi, F.; Cristallini, C.; et al. Il Plio-Pleistocene Marchigiano-Abruzzese. Available online: http://193.204.8.201:8080/jspui/bitstream/1336/199/1/Vol.%20Geologia%20Marche%20Capitolo%205.pdf (accessed on 30 March 2021).
53. Marini, M.; Milli, S.; Ravnås, R.; Moscatelli, M. A Comparative Study of Confined vs. Semi-Confined Turbidite Lobes from the Lower Messinian Laga Basin (Central Apennines, Italy): Implications for Assessment of Reservoir Architecture. *Mar. Pet. Geol.* **2015**, *63*, 142–165. [CrossRef]
54. Artoni, A. Messinian Events Within the Tectono-Stratigraphic Evolution of the Southern Laga Basin (Central Apennines, Italy). *Boll. della Soc. Geol. Ital.* **2003**, *122*, 447–465.
55. De Alteriis, G. Different Foreland Basins in Italy: Examples from the Central and Southern Adriatic Sea. *Tectonophysics* **1986**, *252*, 349–373. [CrossRef]
56. Cantalamessa, G.; Centamore, E.; Micarelli, A.; Potetti, M. Il Miocene nelle Marche. Available online: http://193.204.8.201:8080/jspui/bitstream/1336/196/1/Vol.%20Geologia%20Marche%20Capitolo%203.pdf (accessed on 30 March 2021).

57. Di Celma, C. Sedimentology, Architecture, and Depositional Evolution of a Coarse-Grained Submarine Canyon fill from the Gelasian (Early Pleistocene) of the Peri-Adriatic Basin, Offida, Central Italy. *Sediment. Geol.* **2011**, *238*, 233–253. [CrossRef]
58. Di Celma, C.; Cantalamessa, G.; Didaskalou, P. Stratigraphic Organization and Predictability of Mixed Coarse-Grained and Fine-Grained Successions in an Upper Slope Pleistocene Turbidite System of the Peri-Adriatic Basin. *Sedimentology* **2013**, *60*, 763–799. [CrossRef]
59. Roveri, M.; Manzi, V.; Ricci Lucchi, F.; Rogledi, S. Sedimentary and Tectonic Evolution of the Vena del Gesso Basin (Northern Apennines, Italy): Implications for the Onset of the Messinian Salinity Crisis. *Bull. Geol. Soc. Am.* **2003**, *115*, 387–405. [CrossRef]
60. Bigi, S.; Conti, A.; Casero, P.; Ruggiero, L.; Recanati, R.; Lipparini, L. Geological Model of the Central Periadriatic Basin (Apennines, Italy). *Mar. Pet. Geol.* **2013**, *42*, 107–121. [CrossRef]
61. Porreca, M.; Minelli, G.; Ercoli, M.; Brobia, A.; Mancinelli, P.; Cruciani, F.; Giorgetti, C.; Carboni, F.; Mirabella, F.; Cavinato, G.; et al. Seismic Reflection Profiles and Subsurface Geology of the Area Interested by the 2016–2017 Earthquake Sequence (Central Italy). *Tectonics* **2018**, *37*, 1116–1137. [CrossRef]
62. Centamore, E.; Cantalamessa, G.; Micarelli, A. Stratigrafia e Analisi di Facies dei Depositi del Miocene e del Pliocene Inferiore Dell'avanfossa Marchigiano-Abruzzese e Delle Zone Limitrofe. *Stud. Geol. Camerti* **1991**, *11*, 125–131.
63. Pace, P.; Scisciani, V.; Calamita, F.; Butler, R.W.H.; Iacopini, D.; Esestisme, P.; Hodgson, N. Inversion Structures in a Foreland Domain: Seismic Examples from the Italian Adriatic Sea. *Interpretation* **2015**, *3*, 161–176. [CrossRef]
64. Williams, G.D.; Powell, C.M.; Cooper, M. Geometry and Kinematics of Inversion Tectonics. *Geol. Soc. Lond. Spec. Publ.* **1989**, *44*, 3–15. [CrossRef]
65. Bolis, G.; Carruba, S.; Casnedi, R.; Perotti, C.R.; Ravaglia, A.; Tornaghi, M.; Tavarnelli, E.; Calamita, F.; Paltrinieri, W. Compressional Tectonics Overprinting Extensional Structures in the Abruzzo Periadriatic Foredeep (Central Italy) during Pliocene Times. *Ital. J. Geosci.* **2003**, *122*, 251–266.
66. Costa, M. The Buried, Apenninic Arcs of the Po Plain and Northern Adriatic Sea (Italy): A New Model. *Italian Journal of Geosciences.* **2003**, *122*, 3–23.
67. Mazzoli, S.; Santini, S.; Macchiavelli, C.; Ascione, A. Active tectonics of the outer northern Apennines Adriatic vs. Po Plain seismicity and stress fields. *J. Geodyn.* **2015**, *84*, 62–76. [CrossRef]
68. Mazzoli, S.; Macchiavelli, C.; Ascione, A. The 2013 Marche Offshore Earthquakes: New Insights into the Active Tectonic Setting of the Outer Northern Apennines. *J. Geol. Soc. Lond.* **2014**, *171*, 457–460. [CrossRef]
69. Mantovani, E.; Viti, M.; Babbucci, D.; Tamburelli, C.; Cenni, N. Possible Location of the Next Major Earthquakes in the Northern Apennines: Present Key Role of the Romagna-Marche-Umbria Wedge. *Int. J. Geosci.* **2017**, *8*, 1301–1314. [CrossRef]
70. Mantovani, E.; Viti, M.; Babbucci, D.; Tamburelli, C.; Cenni, N. Geodynamics of the Central-Western Mediterranean Region: Plausible and Non-Plausible Driving Forces. *Mar. Pet. Geol.* **2020**, *113*, 104121. [CrossRef]
71. Riguzzi, F.; Tertulliani, A.; Gasparini, C. Study of the Seismic Sequence of Porto San Giorgio (Marche)-3 July 1987. *Nuovo Cim. C.* **1989**, *12*, 453–466. [CrossRef]
72. Battimelli, E.; Adinolfi, G.M.; Amoroso, O.; Capuano, P. Un Nuovo Studio Della Sequenza Sismica del 1987 di Porto San Giorgio. In Proceedings of the 37th National Conference GNGTS, Bologna, Italy 19–21 November 2018; pp. 13–15.
73. Pierantoni, P.P.; Centamore, E.; Costa, M. Geological and Seismological data Review of the 2009 L'Aquila Seismic Sequence (Central Apennines, Italy): Deep Seated Seismogenic Structures and Seismic Hazard. *Ital. J. Eng. Geol. Environ.* **2017**, *17*, 5–40.

geosciences

MDPI

Article

Burial-Deformation History of Folded Rocks Unraveled by Fracture Analysis, Stylolite Paleopiezometry and Vein Cement Geochemistry: A Case Study in the Cingoli Anticline (Umbria-Marche, Northern Apennines)

Aurélie Labeur [1], Nicolas E. Beaudoin [1], Olivier Lacombe [2,*], Laurent Emmanuel [2], Lorenzo Petracchini [3], Mathieu Daëron [4], Sebastian Klimowicz [2] and Jean-Paul Callot [1]

[1] Laboratoire des Fluides Complexes et Leurs Réservoirs, Universite de Pau et des Pays de l'Adour, E2S UPPA, CNRS, TOTAL, LFCR 64000 Pau, France; alabeur@univ-pau.fr (A.L.); nicolas.beaudoin@univ-pau.fr (N.E.B.); jean-paul.callot@univ-pau.fr (J.-P.C.)

[2] Institut des Sciences de la Terres de Paris–ISTeP, CNRS-INSU, Sorbonne Université, 75252 Paris, France; laurent.emmanuel@sorbonne-universite.fr (L.E.); sebastian.klimowicz.sk@gmail.com (S.K.)

[3] Consiglio Nazionale delle Ricerche, IGAG, 00015 Roma, Italy; lorenzo.petracchini@igag.cnr.it

[4] Laboratoire des Sciences du Climat et de l'Environnement, UMR82212, LCE/IPSL, CEA-CNRS-UVSQ, Université Paris-Saclay, Gif-sur-Yvette, 91190 Saclay, France; daeron@lsce.ipsl.fr

* Correspondence: olivier.lacombe@sorbonne-universite.fr

check for updates

Citation: Labeur, A.; Beaudoin, N.E.; Lacombe, O.; Emmanuel, L.; Petracchini, L.; Daëron, M.; Klimowicz, S.; Callot, J.-P. Burial-Deformation History of Folded Rocks Unraveled by Fracture Analysis, Stylolite Paleopiezometry and Vein Cement Geochemistry: A Case Study in the Cingoli Anticline (Umbria-Marche, Northern Apennines). *Geosciences* **2021**, *11*, 135. https://doi.org/10.3390/geosciences11030135

Academic Editors: Domenico Liotta, Giancarlo Molli, Angelo Cipriani and Jesus Martinez-Frias

Received: 30 January 2021
Accepted: 1 March 2021
Published: 13 March 2021

Publisher's Note: MDPI stays neutral with regard to jurisdictional claims in published maps and institutional affiliations.

Copyright: © 2021 by the authors. Licensee MDPI, Basel, Switzerland. This article is an open access article distributed under the terms and conditions of the Creative Commons Attribution (CC BY) license (https://creativecommons.org/licenses/by/4.0/).

Abstract: Unravelling the burial-deformation history of sedimentary rocks is prerequisite information to understand the regional tectonic, sedimentary, thermal, and fluid-flow evolution of foreland basins. We use a combination of microstructural analysis, stylolites paleopiezometry, and paleofluid geochemistry to reconstruct the burial-deformation history of the Meso-Cenozoic carbonate sequence of the Cingoli Anticline (Northern Apennines, central Italy). Four major sets of mesostructures were linked to the regional deformation sequence: (i) pre-folding foreland flexure/forebulge; (ii) fold-scale layer-parallel shortening under a N045 σ_1; (iii) syn-folding curvature of which the variable trend between the north and the south of the anticline is consistent with the arcuate shape of the anticline; (iv) the late stage of fold tightening. The maximum depth experienced by the strata prior to contraction, up to 1850 m, was quantified by sedimentary stylolite paleopiezometry and projected on the reconstructed burial curve to assess the timing of the contraction. As isotope geochemistry points towards fluid precipitation at thermal equilibrium, the carbonate clumped isotope thermometry (Δ_{47}) considered for each fracture set yields the absolute timing of the development and exhumation of the Cingoli Anticline: layer-parallel shortening occurred from ~6.3 to 5.8 Ma, followed by fold growth that lasted from ~5.8 to 3.9 Ma.

Keywords: Apennines; fold-and-thrust belt; burial and tectonic history; fractures; stylolites; fluid flow; clumped isotope thermometry; paleopiezometry

1. Introduction

The reconstruction of the burial-deformation history of sedimentary rocks is a complex issue but an essential exercise to understand the tectonic and sedimentary history in fold-and-thrust belts and foreland basins, with numerous implications spanning from the evolution of the fluid-flow system and associated resources to the understanding of the long-term behavior of the upper crust [1–7]. Mesostructures observed in fold-and-thrust belts and related forelands, such as faults, veins, and stylolites provide essential information for understanding the deformation pattern [8]. Numerous studies have indeed linked the development of fractures to the large-scale long-term folding history and geometry, either through a descriptive field-based approach [9–11], paleostrain and paleostress reconstructions [12–17], mechanical simulation (e.g., [18,19]), or through geochemical approaches (e.g., [20]). Besides, studies of distributed subseismic fractures demonstrated that mesostructures are markers of the local deformation sequences and provide access to the evolution of the associated paleostresses [8–10,15,21–36]. When it comes to the burial history, however, the depth-time paths are reconstructed from temperature-dependent proxies, such as organic matter thermal maturity, temperature-dependent clay minerals (e.g., [37–39]), fluid inclusion microthermometry, or low temperature thermochronology (apatite fission tracks, U-Th/He on apatite crystals [40–46]). All these methods rely on the occurrence of specific features (organic matter, fluid inclusions, specific minerals like zircons, or Ar-rich clays) and return only partial information (e.g., maximum temperature, timing of reaching closure temperature …) while requiring assumptions about the past geothermal gradient. The latter is often argued over in fold-and-thrust belts and foreland basins where uplift and erosion are common events, thus it remains strongly challenging to reconstruct the burial at which the deformation occurred [7].

We propose to build on recent methodological advances to better constrain the timing of deformation in fold-and-thrust belts and foreland basins. The development of U-Pb dating on calcite cement enables absolute dating on the tectonic vein filling and synkinematic clay minerals (e.g., [47–49]), clumped isotopes can be reliably used to reconstruct the exact temperature of precipitation under 90–100 °C [50], and its combination to fluid inclusion microthermometry allows to reconstruct the temperature-pressure-time path in sedimentary sequence [51]. Moreover, the understanding of how stylolites can be used as stress gauges [52,53] enables reconstruction of the burial experienced by sedimentary rocks (e.g., [7,54–60]). These methods rely on the ubiquitous features of carbonate rocks, and we propose in this contribution to combine paleopiezometry and isotope-based thermometry as a new methodology widely applicable to carbonates in order to assess the timing and depth of burial/contraction during deformation history. We present an original application by reconstructing the burial-deformation history of the carbonate sedimentary sequence of the Cingoli Anticline, an arcuate fold in the Umbria-Marche Apennines Ridge (UMAR, Northern Apennines, central Italy; Figure 1). The Cingoli Anticline is an excellent case study for testing such new methodologies: (i) it is a rather simple symmetrical fold, mainly formed by carbonate rocks exhibiting pervasive fractures and stylolites [61]; (ii) there is an abundant literature that discusses the timing of folding, based on sedimentologic and tectonic studies [62–64]; (iii) the geothermal gradient is known [65]; and (iv) it belongs to a young mountain belt where the past mesostructures received relatively poor attention [60,61,66–70] compared to their active counterparts.

In this study, we aim at identifying, characterizing and dating the mesostructures related to the main stages of deformation, i.e., faults, fractures, and stylolites linked to layer-parallel shortening (LPS), folding, and late-stage fold tightening (LSFT). For this purpose, we first carry out a classical field-based mesostructural study. We further reconstruct the burial history of strata by building burial curves using present strata thickness corrected for physical and chemical compaction and by applying the roughness inversion technique to sedimentary stylolites in order to quantify the maximal vertical stress, and hence the maximum burial depth experienced by the sedimentary sequence prior to contraction and during exhumation.

The timing of the deformation is further constrained by the temperatures derived from isotopic data (^{18}O,^{13}C and clumped isotopes), showing a thermal equilibrium with the host rocks considering previously published geothermal gradients, and compared to previous studies carried out in nearby folds [60,71]. This original and transferable approach reveals the burial evolution, the timing of deformation, and the fluid system in the Cingoli Anticline during the Apennine contraction in eastern UMAR.

2. Geological Setting

2.1. The Umbria Marche Apennine Ridge (UMAR)

The Apennines fold-and-thrust belt extends from the Po Plain to the Calabrian Arc over a distance of 1500 km, and it is the result of the Eurasian and African Plates convergence [72,73]. The Apennine belt accommodated significant orogenic contraction, estimated up to 50% through cross section balancing [74], and recorded shortening rates ranging from 6 mm/year up to 15-50 mm/year [74,75]. From a structural point of view, the Apennine belt is characterized by a succession of asymmetrical anticlines with eastward vergence, separated by narrower and often asymmetrical synclines. The Apennines are commonly divided into two main arcs, the Northern and the Southern Apennines arcs, each associated with its own geological and structural characteristics [76–78]. Moreover, this curved belt, with eastward convexity, is increasingly younger from west to east (from Oligocene to Pleistocene; [72,79]). This is the result of a roughly eastward migration of the deformation front (and its associated successive foredeep basin), which is related to the eastward retreating subduction of the Adriatic Plate under the European Plate, superimposed by post-orogenic extension at the rear of the propagating orogenic belt [29,72,80,81].

The UMAR represents the central-southern part of the Northern Apennines arc developed during late Miocene-Pliocene when the area was involved in the Apennines build-up. It is about 450 km long and rises to over 2000 m above sea level [75]. The sedimentary succession observed in the UMAR can be divided into three main units as follows: (i) Upper Triassic evaporites, the thickness of which is estimated at 1000 m before deformations and considered as a décollement level. They unconformably overlie crystalline rocks of the basement, which is barely or not even observable at the outcrop [72]; (ii) the Umbria-Marche carbonate-dominated succession (~2500 m thick), divided in several formations deposited from the earliest Jurassic to Oligocene [72,82–84]; (iii) Miocene hemipelagites and turbidites, deposited above these carbonate rocks, which record the progressive eastward involvement of the Meso-Cenozoic succession into the fold-and-thrust belt [62,85]. Indeed, during the foredeep stage, more than 3000 m of turbidites were deposited ahead of the advancing fold-and-thrust belt.

Both thick- and thin-skinned structural styles of deformation have been proposed for the Apennines. The thin-skinned interpretation considers a disharmonic deformation of the crust with the sedimentary units detached along the Triassic evaporites [86–88]. The thin-skinned model is opposed to the thick-skinned which considers the involvement of the basement during compressional deformation (e.g., [89]) Furthermore, several studies suggest that many thrusts are rooted on inherited pre-orogenic structures, mostly pre-existing normal faults formed either during the evolution of the Mesozoic passive margin or during the foreland flexure [90–96].

The UMAR undergoes the following main stages of regional deformation: (i) the forebulge stage consisting of the foreland flexuring [8], dated from late Oligocene-early Miocene in the western part of the ridge (i.e., eastern Tuscany-Monte Nero) and from middle Miocene in the eastern part of the ridge (i.e., Gubbio, San Vicino, and Cingoli areas) [79]; (ii) the LPS event, a pre/early-folding compressional stage NE-SW-oriented related to the Apenninic contraction [66,97–99], occurred by early Miocene to the west, and by middle Pliocene to the east [79]; (iii) the folding stage, started by early Miocene in the western part of the UMAR, and by middle Miocene in the eastern part. This stage is characterized by a maximum stress trending parallel to regional shortening, i.e., NE-SW-oriented [29], and local extension perpendicular to fold axis and associated with strata curvature at fold hinge [60]; (iv) the LSFT, associated with a NE-SW contractional trend. This stage corresponds to the moment when shortening is no longer accommodated, by e.g., limb rotation [60]; (v) a post-orogenic extension, starting by Miocene times in the western UMAR [99] and by early Pliocene times in the eastern part of the ridge [62,63], and continuing today. This extensional stage is associated with NNW-SSE trending related normal faults causing the downfaulting of the fold succession [86,99,100].

2.2. *The Cingoli Anticline*

2.2.1. Structural Pattern

The Cingoli Anticline is located in the eastern part of the UMAR along the footwall of the Sibillini Mountains (Figure 1A). This anticline is characterized by an arcuate geometry; its main NW-SE axis evolves toward a N-S orientation in its southern part; it is characterized by gently dipping limbs and its hinge is relatively flat. The fold has a spatial extension of about 15 km from north to south, and about 5 km from west to east and it culminates ~400 m above the local land (770 m above sea level). An WNW-ESE oriented left-lateral fault and a NNE-SSW oriented right-lateral fault bound the Cingoli Anticline to the north and south, respectively (Figure 1A). In addition, [101] and [63] suggest that inherited pre-contractional structures striking ~N-S (i.e., the Jurassic rifting and the late Miocene foreland flexure) strongly controlled the subsequent contractional tectonic evolution of the area (Figure 1A,C).

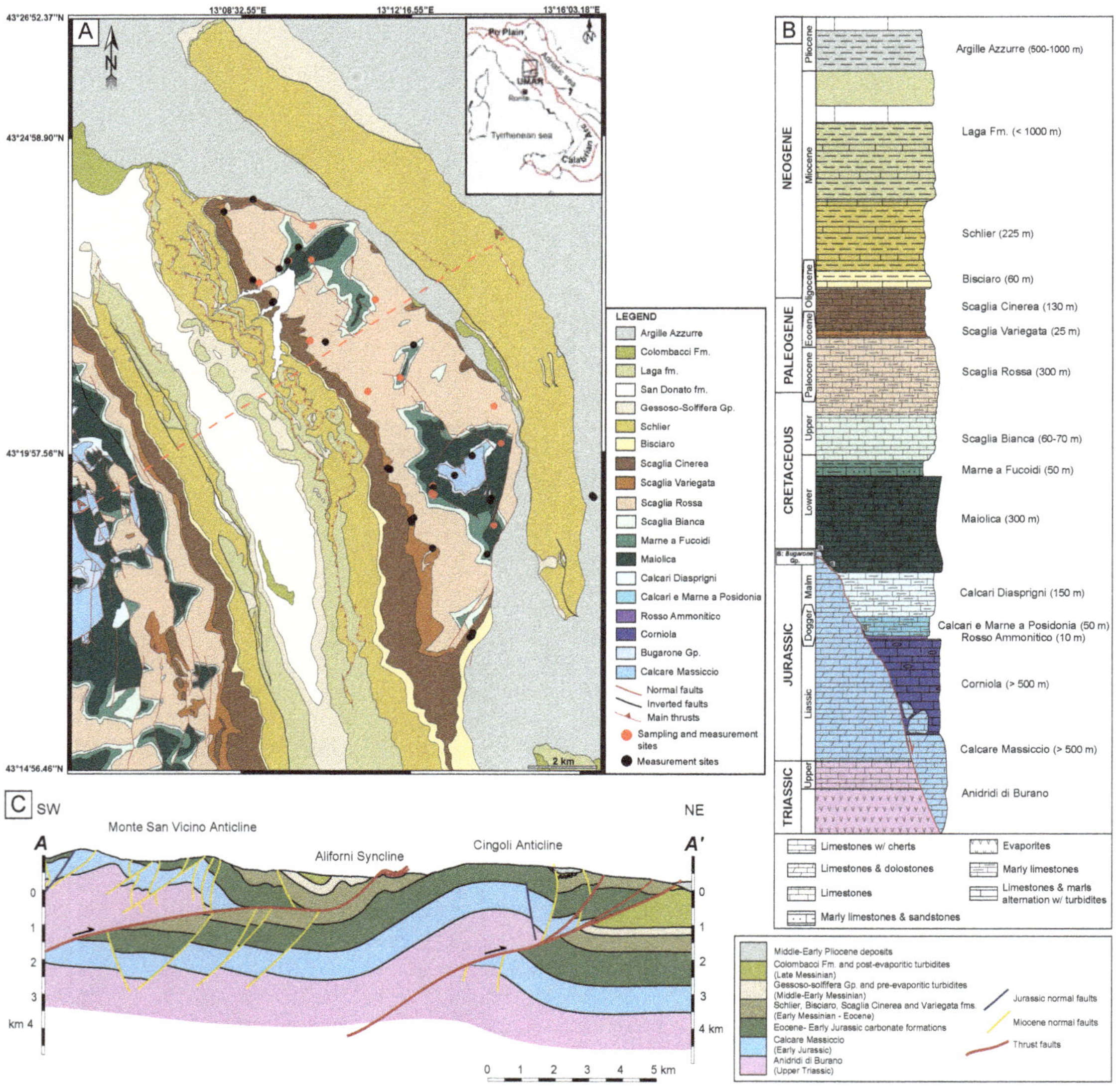

Figure 1. (**A**) Location and simplified geological map of the Cingoli Anticline. Red points represent the sampling sites for sedimentary stylolites analysis, and black points represent the measurement sites. (**B**) Stratigraphic column, not to scale, of the Umbria-Marche area, with thicknesses valid for the Cingoli Anticline area (modified from [61]). (**C**) SW-NE geological cross section through the northern part of the Cingoli Anticline [63].

2.2.2. Sedimentary Succession

Where not specified, the lithostratigraphic units described below are traditional and validated units [102], and correspond with formation-rank units.

The deformed units comprise Mesozoic and Cenozoic marine deposits, consisting of evaporites and platform carbonates at the base overlain by pelagic carbonates [63,64,101,103] (Figure 1B). The succession comprises:

(1) Upper Triassic anhydrites and dolostones, grouped in the Anidridi di Burano, unconformably deposited above the continental deposits of the Verrucano and the Hercynian basement; at the top of the Anidridi di Burano, euxinic interstratified marls are present.

(2) The Calcare Massiccio, formed by massive peridital limestones dated from Hettangian to Sinemurian (ca. 201–191 Ma).

(3) Four main Jurassic formations: (i) the Corniola, limestones with cherts beds (early Sinemurian–early Toarcian, ca. 199–183 Ma); (ii) the Rosso Ammonitico, nodular marly limestones dated Toarcian (ca. 183–174 Ma); (iii) the marls and cherty limestones of the Calcari e Marne a Posidonia (late Toarcian-early Bajocian, ca. 174–170 Ma); (iv) the Calcari Diasprigni, dominated by radiolarian-rich cherty limestones and cherts and, on top, by micritic limestones and marls bearing abundant Saccocoma sp. fragments (late Bajocian-early Tithonian, ca. 170–152 Ma). Because of the horst and graben structures related to the Jurassic extensional tectonics, these deposits accumulated in the hanging wall basins forming thick (hundreds of meters) "basinal" successions, while thin (up to few tens of meters), fossil-rich and condensed successions (Bugarone Group) accumulated on top of footwall blocks of Jurassic faults during the same time span (i.e., from early Pliensbachian to early Tithonian; ca. 191–152 Ma) [82,83,104–106].

(4) The Maiolica, micritic limestones associated with chert beds (Tithonian–earliest Aptian, ca. 152–124 Ma).

(5) Shales and marls of the Marne a Fucoidi (Aptian-Albian, ca. 124–100 Ma).

(6) The "Scaglia" group is composed of micritic limestones with cherts intercalations (late Aptian-Aquitanian, ca. 113–21 Ma), and divided into four formations: (i) the Scaglia Bianca (Cenomanian-earliest Turonian, ca. 100–94 Ma); (ii) the Scaglia Rossa (earliest Turonian-Lutetian, ca. 94–41 Ma), subdivided into three members; (iii) the Scaglia Variegata (Lutetian-Priabonian, ca. 48–34 Ma) and (iv) the Scaglia Cinerea (Rupelian-earliest Aquitanian, ca. 34–22 Ma).

(7) Bisciaro (Aquitanian-Burdigalian, ca. 22–16 Ma) and Schlier (Langhian-Tortonian, ca. 16–7 Ma) formations, hemipelagic limestones, marly limestones and marls.

(8) Siliciclastic foredeep deposits, grouped in two major sequences: (i) Messinian arenitic and pelitic turbidites (ca. 7–5 Ma), composed by the Laga formation and the Gessoso-Solfifera Group with the San Donato and Colombacci formations; (ii) arenites, pelites and fossiliferous marine clays and marls of the Argille Azzure, early to middle Pliocene in age (ca. 5–3 Ma). In this area, the Messinian and Pliocene deposits are continuous and widespread, and provide an almost complete record of the deformation history of the ridge's outer domains. In particular, growth strata observed in the San Donato and Colombacci formations within the Aliforni Syncline postdate flexural turbidites of the Laga formation, consequently broadly constraining fold growth to late Messinian-Zanclean (ca. 6–4 Ma) [62,63].

3. Materials and Methods

3.1. Fracture-Stylolite Network Characterization and Striated Fault Planes Analysis

The structural data collection and sampling sites are distributed along the whole anticline (Figure 1). The characterization of the fracture-stylolite network is based on field observations and measurements and analyses of joints, veins, striated faults, and tectonic stylolites. The dataset comprises more than 2300 orientations of mesostructures from the Cingoli Anticline. Furthermore, abutment and crosscutting relationships were carefully observed and analysed to establish the relative chronology of fracture sets. Fracture orientation data were projected on Schmidt stereodiagrams (lower hemisphere), in the current attitude of the strata (raw) and after unfolding (unfolded). In addition, the major sets of joints and/or veins (i.e., the most documented and representative at fold scale) were grouped and averaged by a Fisher statistical analysis, based on the following assumptions: (i) similar orientation considering natural variability (i.e., within 20°); (ii) deformation mode (e.g., opening, shearing, contraction), defined through thin sections observed with optical microscope; and (iii) chronological relationships.

Considering that stylolite peaks grow parallel to the main shortening direction [107], with respect to the distribution of dissolution gradients (i.e., non-soluble particles) [108], the orientation of the horizontal maximum principal stress (σ_1) was inferred from the maximum density of the peak orientation of tectonic stylolites. The early, syn-, and late folding sets of mesostructures were discriminated with the OpenStereo software [109]. Data were corrected for bedding attitude, by rotation about a horizontal axis to remove the dip angle of the strata, in order to investigate the relationships between fracturing and folding. Thus, fractures were grouped according to their geographical and structural position (respectively north/south and forelimb/backlimb), and according to the main stages of deformation.

Approximately 40 mesoscale striated faults were also measured to complement this mesostructural analysis, in 3 sites of measurements located in the northern and southern ends of the anticline. We used Angelier's inversion technique [110] which, under specific assumptions [111], allowed us to calculate paleostress orientations (i.e., local trend and plunge of principal stress axes) and stress ratios for each site of measurement.

3.2. Rock Mechanical Properties

To calculate the mechanical properties of the studied rocks, we used the Schmidt rebound hammer technique, which is a non-destructive method used for the estimation of the uniaxial compressive strength and Young modulus of concrete and natural rocks (e.g., [112]). It implies the use of a spring-loaded piston (the Schmidt hammer), pressed orthogonally against a surface of rock. The energy created by the resistance of the surface to the impact enables the piston to rebound. The distance traveled by the piston after the rebound is called the rebound value R, which is considered to be a proxy of the surface hardness [112], itself used to quantify uniaxial compressive strength and Young modulus of the rock [113]. A Silver Schmidt OS8200 (manufactured by PROCEQ) was used on 12 sites in various localities and sedimentary units of the fold. Each site consists of 50 to 90 rebounds performed perpendicularly to the surface, in most case lying flat (i.e., the hammer being vertical), from which rebound values were averaged as a single rebound value valid for the site. In order to ensure that this value is free from any outlier due to local heterogeneity, we calculated a moving average incremental mean until the mean rebound value stabilizes (supplementary material, Figure S1, Table S1).

3.3. Sedimentary Stylolite Roughness Inversion

Sedimentary stylolites are pressure-solution surfaces usually developed parallel to bedding in sub-horizontal strata during burial, i.e., when σ_1 was vertical. They are frequently observed in sedimentary rocks, and especially in carbonates. The 1D roughness of a track along these dissolution surfaces, i.e., the difference in height between two points along the track (sensu [53]), results from a competition between two forces [53]: (i) roughening forces, i.e., pinning on non-soluble particles in the rocks, and (ii) smoothing forces, associated with the surface energy at scale <1 mm and the elastic energy at scale >1 mm. Two main scaling regimes are discriminated by the stylolite growth model [56,107,108,114] depending on the predominant energy and the associated Hurst exponent (i.e., roughness exponent): the surface energy-controlled scale, characterized by a steep-slope and a roughness exponent of 1.1 ± 0.1, and the elastic energy-controlled scale, associated with a gentle slope and a Hurst exponent between 0.5 and 0.6. At the transition of these scale regimes, the change in roughness exponent is associated with a crossover length, estimated in mm by the signal processing approach. The authors in [53] directly link this crossover length L_c to the magnitude of prevalent mean stress σ_m and differential stress σ_d in the strata at the time the stylolite stopped to be an active dissolution surface, with Equation (1):

$$L_c = \frac{\gamma E}{\beta \sigma_m \sigma_d} \tag{1}$$

where L_c is the crossover length converted to m, E the Young modulus of the rock (in Pa), γ is the solid-fluid interfacial energy (in $J \cdot m^{-2}$), and β a dimensionless constant depending on the Poisson ratio (ν) and calculated with the relation:

$$\beta = \frac{\nu(1-2\nu)}{\pi} \tag{2}$$

The mean stress and the differential stress, are defined in Pa according to the Equations (3) and (4):

$$\sigma_m = \frac{\sigma_1 + \sigma_2 + \sigma_3}{3} \tag{3}$$

$$\sigma_d = \sigma_1 - \sigma_3 \tag{4}$$

Several studies successfully applied this approach, establishing the spectral analysis of the roughness of sedimentary stylolites as a robust paleopiezometry tool [54,57,59,60,108,115–118], especially because the final roughness is acquired quickly at the end of the stylolite's growth, thus depends only on stress and no longer on strain rate [108].

More than 100 sedimentary stylolites, with peaks perpendicular to the dissolution plane, were sampled in several localities of the Cingoli Anticline (Figure 1A) in the Maiolica, Scaglia Rossa, and Scaglia Variegata. The samples were cut and polished to better analyze the stylolite track. Each cut was made perpendicular to the plane of the stylolite and scanned with a resolution of 12,800 pixels per inches; the resulting file is an image on which the 1D track was hand drawn at magnifications 200% and 400% for greater precision. Then, each track was analyzed as a periodic signal. Usual analyses involve the Fourier Power Spectrum (FPS) and Average Wavelet Coefficient (AWC) methods [115]. We chose the method of Average Wavelet spectrum with Daubechies D4 wavelets [115,119], which is proven to be more stable and less sensitive to resolution effects. We used a non-linear regression with fixed Hurst coefficients of 0.5 and 1.1, corresponding to elastic and surface regimes, respectively (please refer to supplementary material, Figure S2, for the plots). The uncertainty for this regression approach to estimate L_c has been previously estimated to 23% [57]. To calculate the vertical stress magnitude, the horizontal stress was then considered as isotropic (i.e., uniaxial strain hypothesis, $\sigma_v > \sigma_h = \sigma_H$) in the case of sedimentary stylolites. Thus, the Equation (1) established by [53] is simplified as:

$$\sigma_v{}^2 = \frac{\gamma E}{\alpha L_c} \tag{5}$$

α being a constant defined as follows:

$$\alpha = \frac{(1-2\nu)(1+\nu)^2}{30\pi(1-\nu)^2} \tag{6}$$

To verify this assumption, sections perpendicular to the stylolite plane were cut for several samples and treated following the same inversion method, the isotropy of the horizontal stress implying a constant value of the crossover length regardless of the track direction [53,115].

While the solid-fluid interfacial energy γ is well-known and stable, fixed to 0.24 $J \cdot m^{-2}$ for dolomite and of 0.32 $J \cdot m^{-2}$ for calcite [120], and while the Poisson ratio ν is rather stable in carbonates and can be approximated to 0.25 ± 0.05, a major source of uncertainty lies in the values of the Young modulus E [57,121]. With known E, ν, and γ, the uncertainty on the calculated stress is 12% [57]. As in previous works [57,60,117,118] the calculated vertical stress $\sigma_v = \sigma_1$ was translated directly into the burial depth (z) of rocks using the relation

$$\sigma_v = \varrho g z \tag{7}$$

with ρ the average dry density of overlying rocks and g the gravity acceleration. Indeed, the stylolite roughness can be considered as unaffected by local fluid overpressure because the dissolution is located along a fluidic film [56,108,116], an assumption that remains valid until the system is fluidized [122]. For this reasons, ρ was considered as the average dry rock density for clastic and carbonate sediments (evaluated mean value at 2400 $kg \cdot m^{-3}$), without any additional assumption on the past thermal gradient or fluid pressure [7], and g the gravity acceleration, fixed at 9.8 $m \cdot s^{-2}$.

3.4. O-C Stable Isotopes

The analysis of stable oxygen and carbon isotopes, associated with the study of the diagenetic state of the rocks, allows for the identification and characterization of fluid generations at the origin of mineralization and vein filling within sedimentary rocks [20,123–129]. We focus hereinafter on the calcite cements filling the tectonic veins related either to LPS or to strata curvature at fold hinges. The mineralogy of some host-rock was checked with X-ray diffraction (supplementary material, Figure S3) using a Bruker D2 Phaser diffractometer from the ISTeP laboratory (Sorbonne Université) with X-Ray wavelength of 1.54056 Å. The resulting X-ray patterns showing mostly pure calcite with minor amounts of quartz. The diagenetic state of calcite veins was checked under cathodoluminescence microscopy and were performed on a cold cathode Cathodyne platform (CITL CCL 8200 Mk4) at stable vacuum of 60 mThor, a voltage of 12 kV, and a current of 200 μA, corresponding to the ideal voltage-current conditions to activate the luminescence of the carbonates.

We selected vein cements where the vein texture [130] and the diagenetic state support a single phase of filling, occurring at the same time or soon after fracture development (e.g., elongated blocky, Figure 2).

The isotope ratios of oxygen ($^{18}O/^{16}O$) and carbon ($^{13}C/^{12}C$) of samples collected in several localities of the Cingoli Anticline were obtained by Isotope-Ratio Mass Spectrometry (IRMS). The spectrometric equipment couples an automatic sample preparation line (KIEL IV) and an analysis section (DELTA V advantage) from Thermo Fisher Scientific at the ISTeP laboratory (Paris). We first selected veins of which (i) the cement witnessed a single growth step, unaltered by later diagenetic events, checked with cathodoluminescence microscopy; and (ii) the cement was likely related to vein opening, considering antitaxial or elongated blocky textures [130]. Thirty (30) to 50 μg of powder were sampled from the vein and the surrounding host-rocks as well, using a computer assisted micromill drill. The powder was reacted with anhydrous orthophosphoric acid at 70 °C to extract CO_2 gas, itself ionized in the spectrometer. The measured isotopic ratio reported in permil relative to the Vienna Pee Dee Belemnite (‰VPDB) are with an accuracy of 0.05‰ and 0.1‰ for carbon and oxygen, respectively.

3.5. Carbonate Clumped-Isotope Paleothermometry (Δ_{47})

All analyses were performed at the Laboratoire des Sciences du Climat et de l'Environnement (LSCE, Gif sur Yvette). Eight samples of homogenized carbonate powder were converted to CO_2 by anhydrous phosphoric acid reaction at 90 °C in a common, stirred acid bath for 15 minutes. Initial phosphoric acid concentration was 103% (1.91 g/cm^3) and each batch of acid was used for 7 days. After the cryogenic removal of the water, the evolved CO_2 was helium-flushed at 25 mL/mn through a purification column packed with Porapak Q (50/80 mesh, 1 m length, 2.1 mm ID) and held at −20 °C, and then quantitatively recollected by cryogenic trapping and transferred into an Isoprime 100 dual-inlet mass spectrometer equipped with six Faraday collectors (m/z 44–49). Each analysis took about 2.5 hours, during which the analyte gas and working reference gas were allowed to flow from matching, 10 mL reservoirs into the Nier-type ion source through deactivated fused silica capillaries (65 cm length, 110 μm ID). Every 20 minutes, gas pressures were adjusted to achieve m/z = 44 current of 80 nA, with differences between analyte gas and working gas generally below 0.1 nA. Pressure-dependent background current corrections were measured 12 times for each analysis. All background measurements from a given analytical session are then used to determine a mass-specific relationship linking background intensity (Zm), total m/z = 44 intensity (I44), and time (t):

$$Zm = a + bI44 + ct + dt^2 \quad (8)$$

Background-corrected ion current ratios (δ^{45} to δ^{49}) were converted to $\delta^{13}C$, $\delta^{18}O$, and "raw" Δ_{47} values as described by [131], using the IUPAC oxygen-17 correction parameters. The isotopic composition ($\delta^{13}C$, $\delta^{18}O$) of our working reference gas was computed based on the nominal isotopic composition of carbonate standard ETH–3 [132] and an oxygen-18 acid fractionation factor of 1.00813 [133]. Raw Δ_{47} values were then converted to the "absolute" Δ_{47} reference frame defined by the "ETH" carbonate standards [132] using regression methods detailed by [134]. Full analytical errors are derived from the external reproducibility of unknowns and standards (N_f = 78) and conservatively account for the uncertainties in raw Δ_{47} measurements as well as those associated with the conversion to the "absolute" Δ_{47} reference frame. The precipitation temperature was calculated using the calibration proposed by [135] and updated by [132].

3.6. Burial Model

The burial model associated with the Cingoli Anticline area was constructed from the open access well data collection of the ViDEPI project selected around the Cingoli Anticline (i.e., Misa1, Rosora1, Burano1, Treia1 wells). Thicknesses were sequentially uncompacted to obtain burial depths of the strata of interest through time. Two kinds of corrections were therefore applied, considering the effects of both physical and chemical compaction. The computer interface used to produce these burial curves is the Backstrip software, which performs 1D backstripping of sedimentary strata [136] and involves several parameters for modeling.

First, the layer thicknesses were corrected for chemical compaction, in order to be referenced in the software. Thickness information was provided by stratigraphic studies or well data (i.e., current formation thicknesses). In this case, thicknesses were corrected for chemical compaction, considering spacing and amplitudes of bedding-parallel stylolites (BPS) for each formation studied. The average number of sedimentary stylolites per meter was estimated from outcrop data. The height of the highest tooth (i.e., the height from tooth to base line) associated with each analyzed stylolite was also computed from the samples. Chemical compaction was then deduced from these two parameters, calculated as their product, and expressed as a percentage of bed thickness.

The second step consisted of defining parameters needed to evaluate physical compaction undergone by the different layers, related to the weight of the sedimentary column and possibly of the water column (according to the type of basin considered). These parameters, such as the dry density ρ and the porosity coefficient c, were defined on the basis of the work by [137,138], as follows: (i) the dry density ρ was chosen at 2700 kg/m^3 for carbonate rocks and 1800 kg/m^3 for other lithologies; (ii) the porosity coefficient c was calculated with the following equations and the porosity-depth curves established by [137,138]. Φ is the porosity at depth y, while Φ_0 the surface porosity, both given by the curves.

$$\Phi = \Phi_0 e^{-cy} \text{ with } c = \frac{1}{y} \ln\left(\frac{\Phi}{\Phi_0}\right) \quad (9)$$

Thus, c is equal to 0.58 for carbonate rocks and to 0.3 for other lithologies. Corrections related to the weight of sediments and water being significantly different [137,138], the type of basin was also defined (0 for a marine basin, 1 for a continental basin), in order not to introduce bias into the resulting burial curves.

4. Results

4.1. Fracture-Stylolite Network Characterization and Striated Fault Planes Analysis

The fracture deformation mode was defined through thin sections observed with optical microscope, either by considering the texture (i.e., elongate blocky or crack-seal), or the object shift in the matrix (Figure 2).

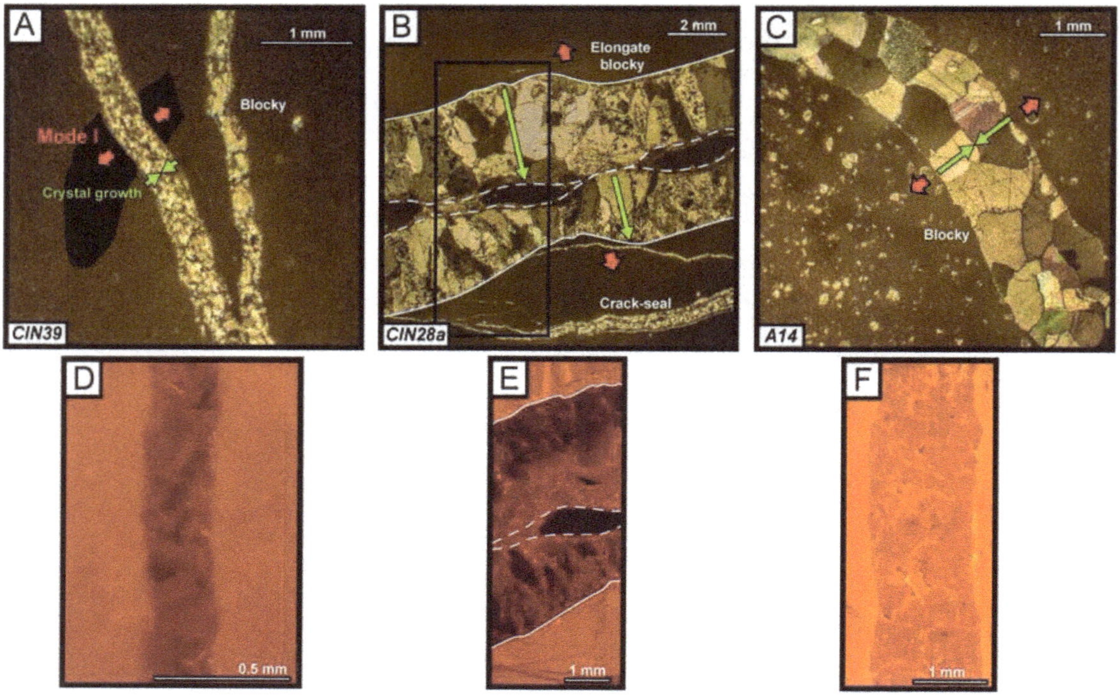

Figure 2. Observation of the different types of fractures in optical microscopy. (**A**) N-S fractures, (**B**) N045 fractures and (**C**) N135-160 fractures. Textures and shifts in the matrix were characterized, in order to verify the deformation mode, based on the classification of [130]. Red arrows represent the direction of the opening (mode I), and green arrows the direction of calcite crystal growth. (**D–F**) observation of these different sets in cathodoluminescence, indicating a single crystallization phase, synchronous with the mode I opening.

Three major sets of fractures were discriminated on the basis of their average orientation (Figures 3 and 4A,B) whereas their chronological sequence was established through abutment and crosscutting relationships at different scales (from outcrop to thin section):

- set I gathers bedding-perpendicular joints oriented N180 to N020 (after unfolding). This set is observed over the entire anticline, predates sets II and III, because it is intersected and abutted by a set of stylolites with peaks oriented N045, which are themselves intersected and abutted by the joints/veins of set II (Figure 4C).
- set II gathers joints and veins with N045 $\pm$ 10° orientation, present throughout the study area. They are perpendicular to bedding strike. This set postdates set I and predates set III.
- set III comprises bedding-perpendicular N130 to N160-oriented joints parallel to bedding strike (i.e., N135-140 in the North and N160 in the South). They are mainly parallel to the axis of the anticline and crosscut or abut all other joint/vein sets (Figure 4C).
- another set of E-W fractures, poorly represented at the scale of the anticline (i.e., only in the northern part, in three sites of measurements), includes N070 to N110-oriented joints (after unfolding) and perpendicular to the bedding, developed after set II.

Because of the low number of measurements (i.e., 25 of 3000 fractures analyzed) and because they systematically developed near faults (Figure 3), this family of fractures is considered as minor and of local meaning only, and therefore not affiliated to a major set. Consequently, it will not be interpreted thereafter.

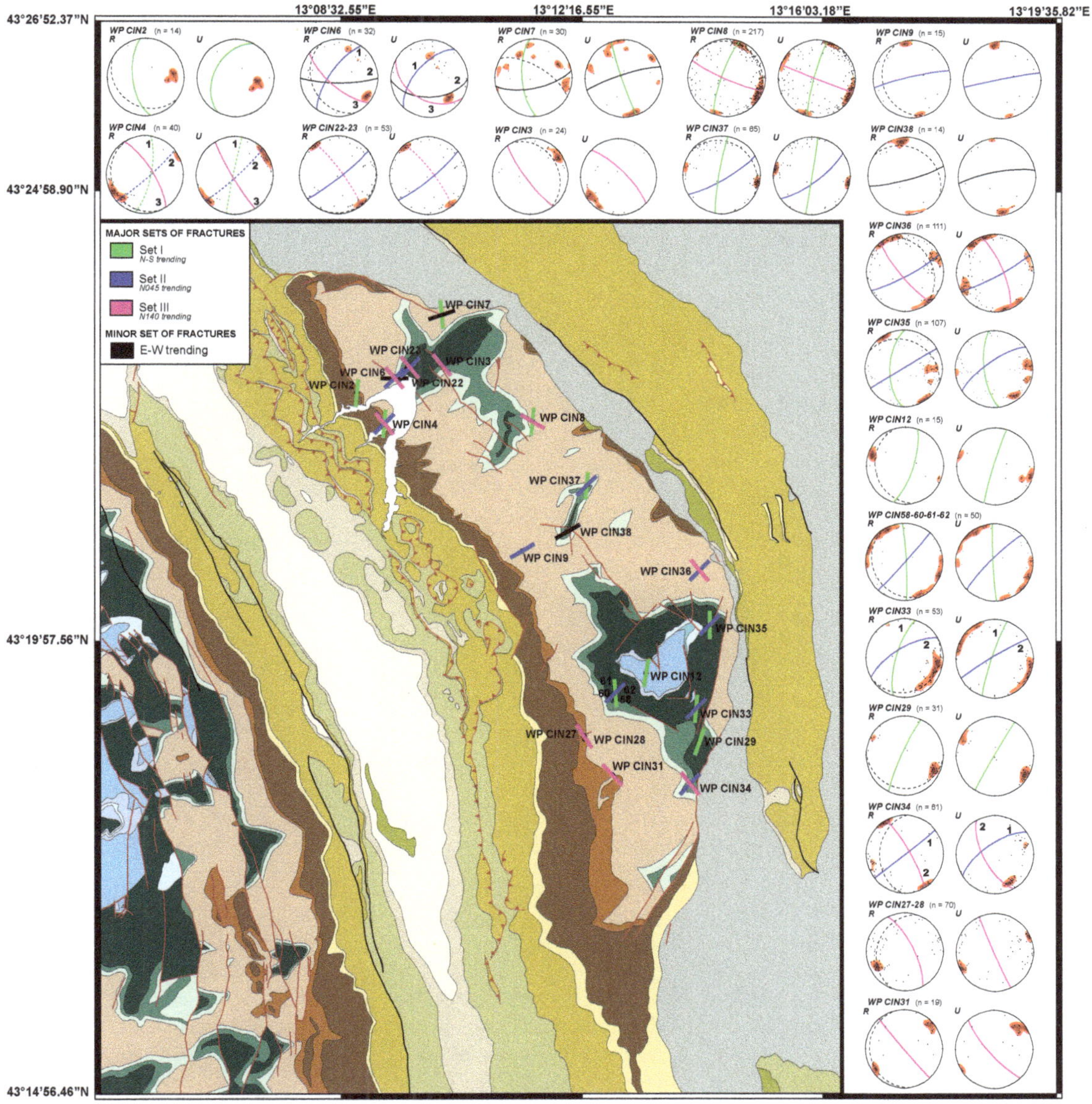

Figure 3. Location of fracture planes measured on the geological map (GPS locations are provided in the supplementary material, Table S2). Each measurement point is associated with two stereodiagrams (lower hemisphere), representing main fracture orientations in current (R) and unfolded attitude (U); on each stereodiagram, the bedding is reported as dashed lines, and fracture planes by solid-colored lines, each color relating to one of the three major fracture sets defined (green: set I, blue: set II, pink: set III).

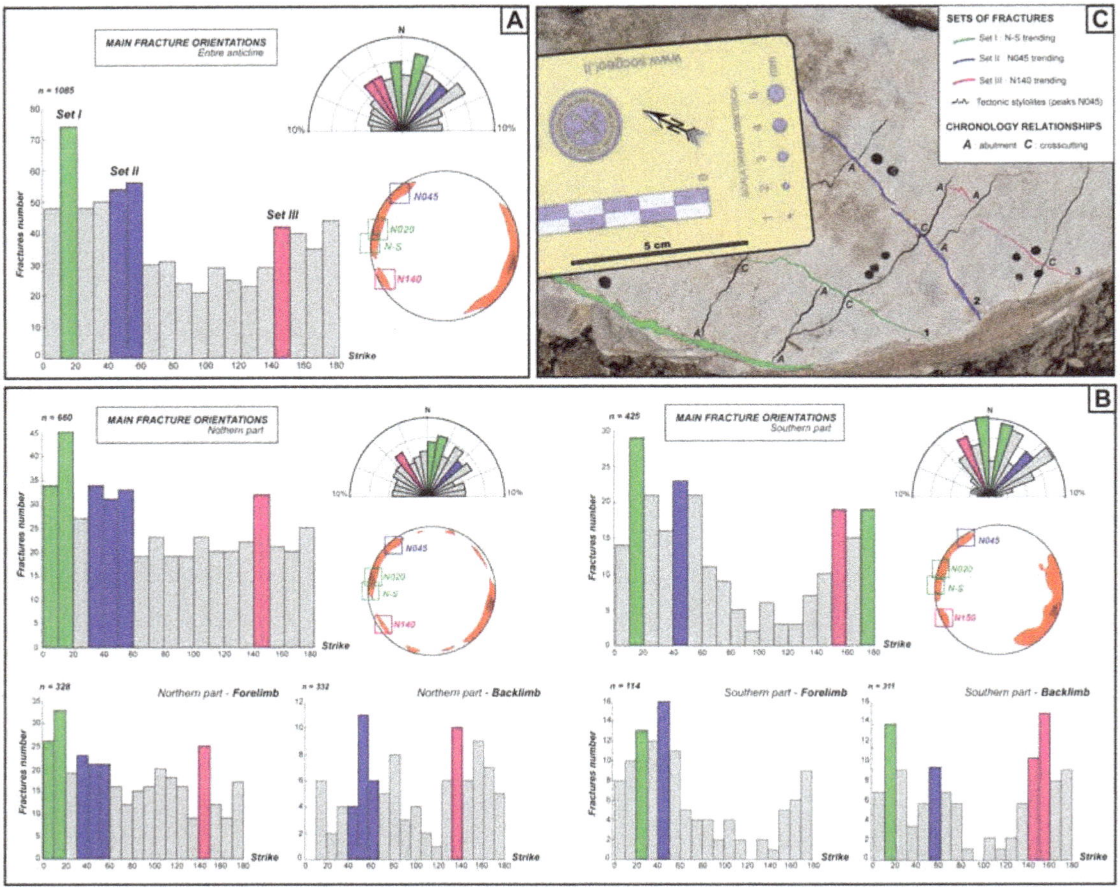

Figure 4. Main fractures orientations (after unfolding) plotted on histograms, stereograms and rose diagrams for (**A**) the whole anticline and (**B**) northern and southern parts of the anticline, discriminating forelimb and backlimb measurements. Three major sets of fractures discriminated according these orientations are represented with their specific color (green: set I, blue: set II, pink: set III). (**C**) Chronological relationships between fractures and stylolites (i.e., abutment and crosscutting), observed at mesoscopic scale.

Tectonic stylolite peaks are mostly oriented N045 (Figure 5). The continuous change in dip of the stylolite plane from vertical to oblique suggests that part of the tectonic stylolites developed before folding and other after folding. Few prefolding stylolites peaks are oriented N140 (WP CIN 37 and WP CIN 27–28) in the Maiolica and Scaglia in the southern backlimb, and N090 in Calcare Massiccio in the southern part of the anticline (WP CIN 12).

The inversion of striated faults for stress was carried out in few sites in the anticline (Figure 5):

- in the northern backlimb, conjugate NW-SE trending reverse faults reveal a compressional stress regime with a σ_1 axis roughly oriented N045;
- in the northern forelimb, N170–180-oriented normal faults indicate either an extensional regime with σ_3 oriented N045, or, more likely correspond to tilted oblique-slip reverse faults consistent with a pre-tilting N020 compression;
- in the southern backlimb, the few fault-slip data preclude any reliable stress tensor calculation. The dataset is however consistent with a post-tilting σ_1 oriented N045 and σ_3 oriented N135.

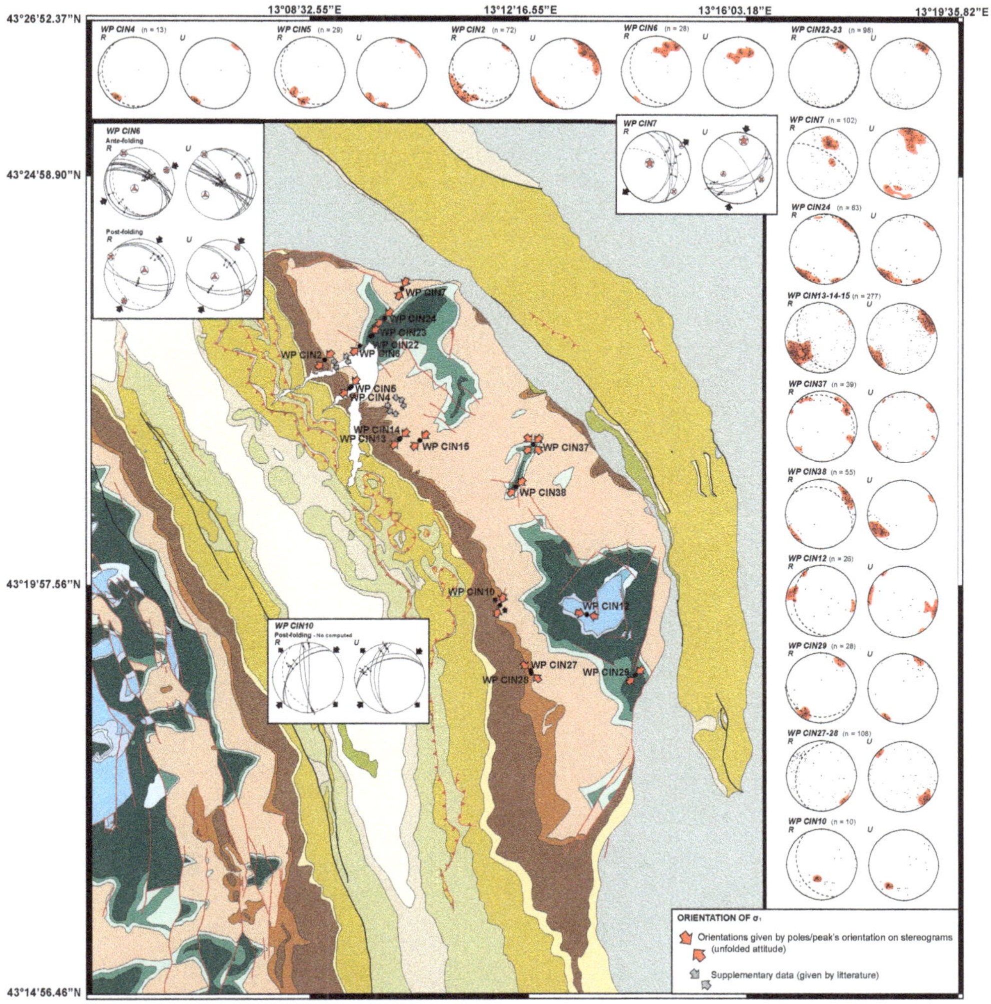

Figure 5. Location and plot of measured tectonic stylolites on the geological map. Each measurement point is associated with two stereodiagrams (lower hemisphere), representing main orientations of tectonic stylolites peaks measured (current and unfolded attitude). On each stereodiagram, the bedding is reported as dashed lines, and peaks orientation (i.e., σ_1 orientation) is given by high pole density zones. Stereodiagrams with fault-slip data and principal stress axes are also reported.

4.2. Young Modulus Estimate

Rock elastic properties were measured on flat homogeneous surfaces, for Maiolica, Scaglia Rossa and Scaglia Variegata, corresponding to 12 sites of measurement (n = 1063). For each site, the mean for rebound value R was represented as a function of the number of rebound incorporated in the mean calculation (supplementary material, Figure S1, Table S1); the stabilized R value (represented as a plateau on the graph) is then believed to be corrected from heterogenous effect and outliers, and so represent the rebound value R for the rock studied. Strikingly, the average of these representative R values is similar in the Maiolica, Scaglia Rossa and Scaglia Variegata formations, with values of 45 ± 8.4, 48 ± 5.8 and 46 ± 8.5, respectively. R values were further interpreted as Young moduli following the empiric relationship determined in [113] for sedimentary rocks and return a

E value similar for the 3 formations at about 20 GPa, very similar to the one reconstructed from stylolite inversion by [121] of 23 GPa.

4.3. Sedimentary Stylolite Roughness Inversion

The stylolite roughness inversion method was applied on 112 BPS sampled in the northern, central and southern parts of the Cingoli Anticline (Figure 1A), within the Cretaceous to Eocene carbonate formations. The inversion was successful (i.e., returning a value of crossover length L_c) on 77 BPS covering the anticline and distributed as follows: Maiolica (early Cretaceous, n = 56), Scaglia Rossa (late Cretaceous-early Eocene, n = 18) and Scaglia Variegata (middle to late Eocene, n = 3). For several stylolites, this paleopiezometric inversion was applied on two orthogonal tracks, in order to ensure that the stress on the horizontal plane was isotropic. L_c values are summarized in Table 1, and reported as an interval for each formation studied, considering an uncertainty of 23%:

- Maiolica: [0.27 ± 0.06; 1.76 ± 0.40] mm
- Scaglia Rossa: [0.36 ± 0.08; 1.17 ± 0.27] mm
- Scaglia Variegata: [0.75 ± 0.17; 1.75 ± 0.40] mm

Table 1. Results of stylolite roughness inversion, applied on bedding-parallel stylolites.

Sample	GPS	Formation	L_c (mm)	σ_v (MPa) *	Depth (m)
CIN13	58	Maiolica	0.44 ± 0.10	34	1440
			0.60 ± 0.14	29	1250
			0.77 ± 0.18	26	1100
			0.84 ± 0.19	24	1040
			0.30 ± 0.07	41	1740
			0.27 ± 0.06	43	1840
			0.31 ± 0.07	40	1700
CIN14	59	Maiolica	0.34 ± 0.08	38	1630
			0.63 ± 0.14	28	1200
CIN3	60	Maiolica	0.39 ± 0.09	36	1540
			0.54 ± 0.12	30	1300
			0.49 ± 0.11	32	1360
CIN6	60	Maiolica	0.54 ± 0.12	30	1300
			0.36 ± 0.08	37	1550
CIN8	60	Maiolica	0.53 ± 0.12	30	1300
			0.76 ± 0.17	26	1100
CIN9	60	Maiolica	0.39 ± 0.09	36	1540
CIN10	60	Maiolica	0.36 ± 0.08	37	1600
			0.31 ± 0.07	40	1700
			0.29 ± 0.07	42	1800
CIN15	61	Maiolica	0.49 ± 0.11	32	1360
			0.84± 0.19	25	1040
			0.38 ± 0.09	37	1550
CIN17	61	Maiolica	0.85 ± 0.20	25	1040
CIN18	61	Maiolica	0.45 ± 0.10	33	1400
			0.91 ± 0.21	23	1000
			1.09 ± 0.25	22	900
CIN33	64	Maiolica	0.73 ± 0.17	26	1100
			1.08 ± 0.25	22	900
			0.46 ± 0.11	33	1400
			0.34 ± 0.08	38	1630
CIN38	64	Maiolica	0.53 ± 0.12	30	1300
			0.55 ± 0.13	30	1300
CIN40	64	Maiolica	0.76 ± 0.17	26	1100
			0.46 ± 0.11	33	1400
			0.32 ± 0.07	40	1700
			0.56 ± 0.13	30	1300
			0.35 ± 0.08	38	1600

Table 1. *Cont.*

Sample	GPS	Formation	L_c (mm)	σ_v (MPa) *	Depth (m)
C3	WP_CIN3	Maiolica	1.76 ± 0.40	17	720
			1.38 ± 0.32	19	810
C51	WP_CIN23	Maiolica	1.18 ± 0.27	21	880
			0.67 ± 0.15	27	1150
C56	WP_CIN23	Maiolica	0.35 ± 0.08	38	1600
C67′	WP_CIN29	Maiolica	0.75 ± 0.17	26	1100
			0.58 ± 0.13	29	1250
C68	WP_CIN29	Maiolica	0.33 ± 0.08	39	1650
C69	WP_CIN29	Maiolica	0.74 ± 0.17	26	1100
			0.46 ± 0.11	33	1400
C70	WP_CIN29	Maiolica	0.35 ± 0.08	38	1600
			0.34 ± 0.08	38	1600
C71	WP_CIN29	Maiolica	0.38 ± 0.09	36	1540
C72	WP_CIN29	Maiolica	0.50 ± 0.12	32	1350
C86	WP_CIN38	Maiolica	1.33 ± 0.31	20	850
			0.39 ± 0.09	36	1540
			0.43 ± 0.10	34	1450
			1.11 ± 0.26	22	900
C79	WP_CIN36	Scaglia Rossa	0.38 ± 0.09	37	1550
C15	WP_CIN7	Scaglia Rossa	0.49 ± 0.11	32	1300
			0.51 ± 0.12	32	1300
C87	WP_CIN8	Scaglia Rossa	1.17 ± 0.27	21	880
			0.75 ± 0.17	26	1100
C21	WP_CIN9	Scaglia Rossa	0.59 ± 0.14	29	1250
			0.68 ± 0.16	27	1150
			0.45 ± 0.10	34	1420
C26	WP_CIN13	Scaglia Rossa	0.87 ± 0.20	24	1020
			0.85 ± 0.20	24	1020
			0.98 ± 0.23	23	960
C28	WP_CIN13	Scaglia Rossa	0.45 ± 0.10	33	1400
C29	WP_CIN13	Scaglia Rossa	0.63 ± 0.14	28	1200
			0.64 ± 0.15	28	1200
			0.84 ± 0.19	25	1040
C30	WP_CIN14	Scaglia Rossa	0.36 ± 0.08	38	1600
			0.60 ± 0.14	29	1230
			0.40 ± 0.09	35	1500
C2	WP_CIN2	Scaglia Variegata	0.75 ± 0.17	26	1100
			1.26 ± 0.29	20	850
			1.75 ± 0.40	17	720

* Crossover length given within 23% uncertainty. Vertical stress σ_v given within 12% uncertainty calculated according to Equation (5), considering a Young modulus E = 23 GPa ([139], this study), a Poisson ratio $\nu = 0.25$, and interfacial energy $\gamma = 0.32\ J \cdot m^{-2}$. Depth calculated using dry density of rock d = 2400 $g \cdot m^{-3}$, acceleration of gravity g = 9.81 $m \cdot s^{-2}$.

4.4. Burial Model

The burial curves resulting from the backstripping process are presented in Figure 6.

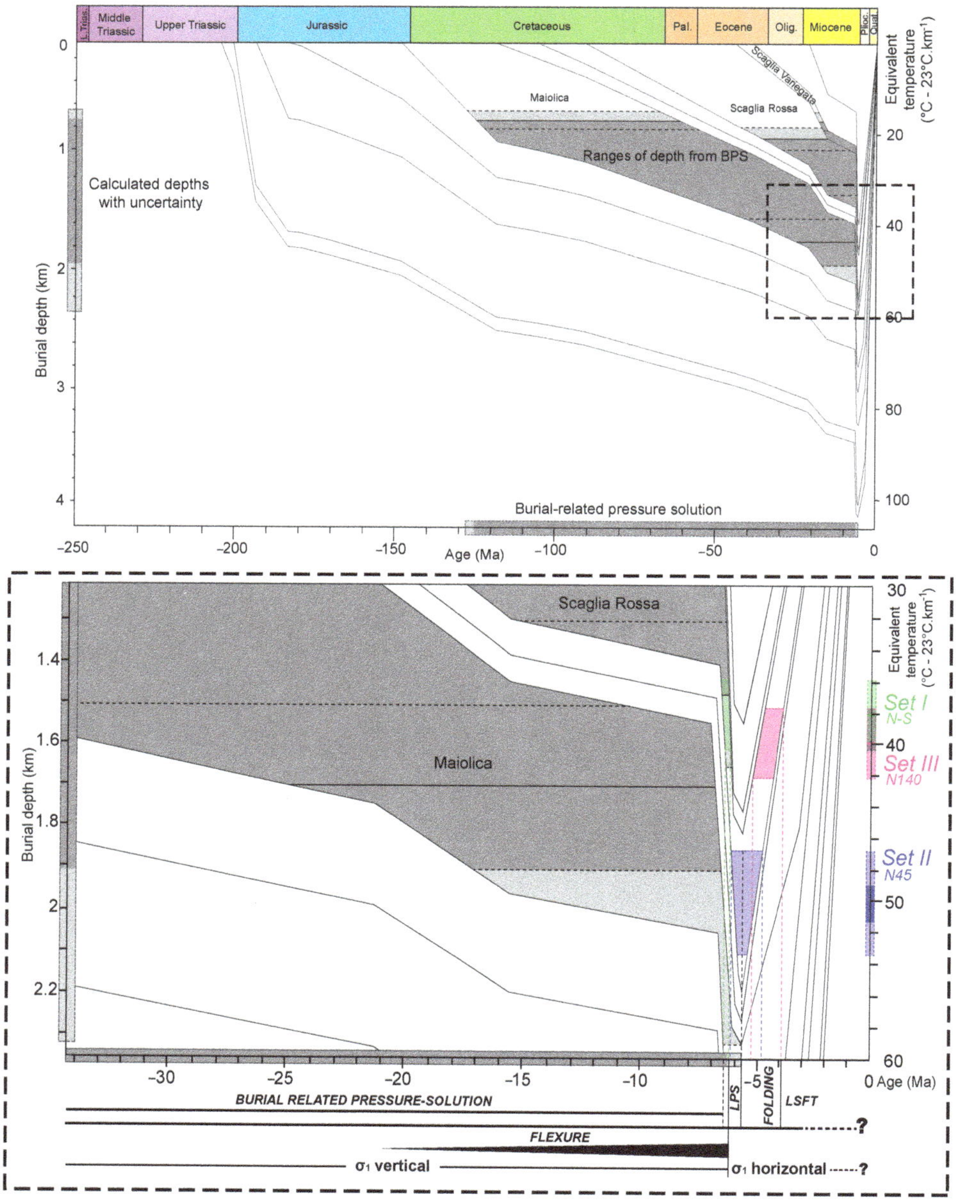

Figure 6. Burial model constructed considering thickness from stratigraphic and well data corrected for chemical and physical compaction. The range of depths reconstructed from BPS roughness inversion (with uncertainty shaded in light grey) are reported for each formation as grey levels. The corresponding timing and depth of active dissolution are reported on the x axis and left y axis, respectively. The results of clumped isotope analysis (i.e., temperatures of precipitation of vein cements at thermal equilibrium with the host rock) are reported on the right y axis. The timing of the deformation is reported on the right-hand side in the insert. Onsets of LPS, folding stage and LSFT are deduced from the results of the roughness inversion process applied on sedimentary stylolites, as well as from clumped isotope data.

They were reconstructed for the Triassic to Pliocene formations in the Cingoli area, considering: (i) the chemical compaction calculated at 8% for Maiolica, and 3% for Scaglia Rossa and Scaglia Variegata considering spacing and amplitude of BPS (following [108]); (ii) physical compaction by using the open-source software BackStrip [136]. The temperatures linked to these depths were calculated by considering a geothermal gradient of 23 °C·km-1 reconstructed in the outermost western part of the UMAR from organic matter thermal maturity [65] and clay minerals [140] (Figure 6). These curves illustrate a first phase of increasing burial, corresponding to the deepening of the Umbria-Marche basin and a second phase of exhumation since early Pliocene. The maximum burial depths computed for the formations of interest, and equivalent temperatures, can be deduced from the left and right y-axis of Figure 6, respectively. These curves are consistent with models established for the inner part of the belt in the area of the Monte Tancia thrust [71].

4.5. Oxygen and Carbon Stable Isotopes

Twenty-eight (28) vein calcite cements and surrounding calcite host-rocks from the Scaglia Rossa were analyzed for $\delta_{18}O$ and $\delta_{13}C$ (Table 2, Figure 7).

Table 2. Results of Stable Isotopic Analyses of Oxygen and Carbon Isotopes.

		Vein		*Host-Rock*	
Sample	**Set**	**$\delta^{13}C$ (‰VPDB)**	**$\delta^{18}O$ (‰VPDB) Calcite**	**$\delta^{13}C$ (‰VPDB)**	**$\delta^{18}O$ (‰VPDB) Calcite**
CIN23-V1	I	2.04	2.19	2.15	−2.95
CIN23-V2	I	1.90	−0.44	2.15	−2.95
CIN23-V3	I	1.91	−1.56	2.15	−2.95
CIN23-V4	I	1.98	−0.68	2.15	−2.95
CIN25-V1	I	2.06	1.79	2.10	−2.27
CIN25-V2	I	2.06	−0.97	2.10	−2.27
CIN25-V3	I	2.08	2.07	2.10	−2.27
CIN39-V1	I	0.05	−0.74	1.08	−1.59
CIN39-V2	I	0.10	−0.30	1.08	−1.59
CIN39-V3	I	0.31	−0.66	1.08	−1.59
CIN7-V1	II	2.48	1.37	2.47	−1.68
CIN7-V2	II	2.97	2.09	2.47	−1.68
CIN28a-V1	II	2.07	1.21	2.01	−1.82
CIN28a-V2	II	1.81	0.58	2.01	−1.82
CIN28a-V3	II	1.95	2.22	2.01	−1.82
CIN26-V1	II	2.00	2.59	2.02	−1.74
CIN26-V2	II	1.95	2.34	2.02	−1.74
CIN26-V3	II	2.14	1.47	2.02	−1.74
CIN26-V4	II	2.05	1.38	2.02	−1.74
CIN28b-V1	II	2.08	2.20	2.03	−1.64
CIN28b-V2	II	1.98	1.68	2.03	−1.64
A13-V1	III	3.23	0.79	3.08	−1.45
A14-V1	III	3.21	0.21	2.89	−1.53
A14-V2	III	3.19	0.61	2.89	−1.53
CIN37-V1	III	1.39	−0.59	1.09	−1.55
CIN37-V2	III	1.25	0.46	1.09	−1.55
CIN37-V3	III	1.10	−1.12	1.09	−1.55
CIN37-V3	III	1.10	−1.12	1.09	−1.55

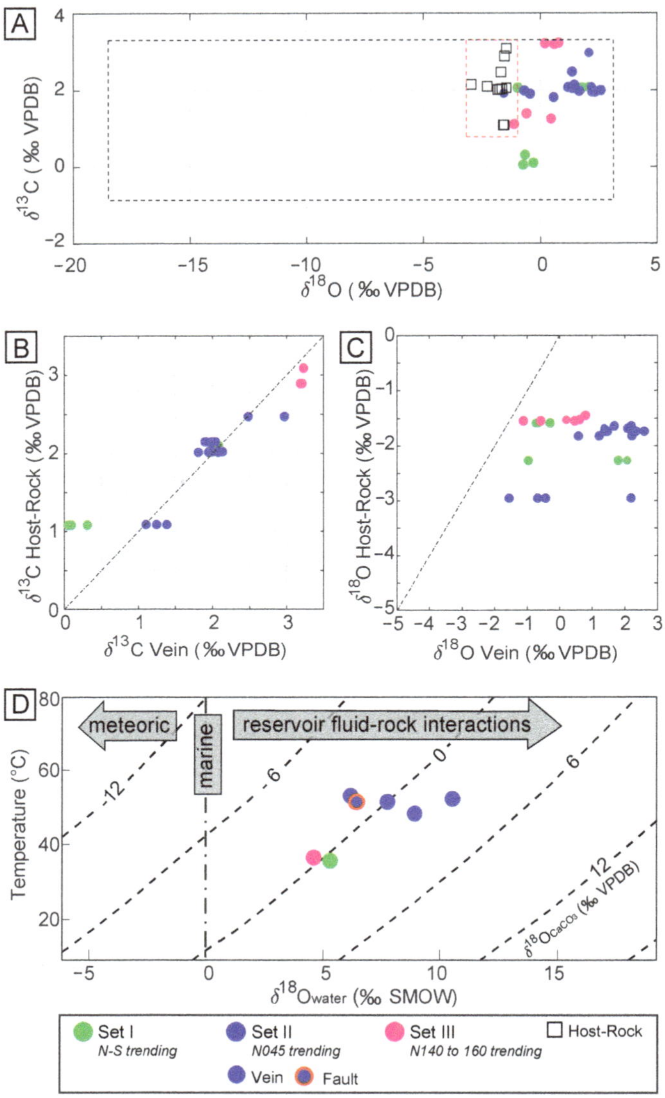

Figure 7. Isotopic data from tectonic veins, faults and host-rocks sampled in the Scaglia Rossa. (**A**) Oxygen versus Carbon stable isotopic ratio (‰VPDB) of host rocks (black squares), and vein cements according to the vein sets. Red dotted frame represents the range of isotopic values documented in the UMAR from the Hettangian to Aquitanian carbonates, black dotted frame represents the range of isotopic values documented in tectonic related fracture fillings at the scale of the range [60]. (**B**) $\delta^{13}C$ values of vein cements versus $\delta^{13}C$ values of the surrounding host rocks (‰VPDB), according to the vein sets. (**C**) $\delta^{18}O$ values of vein cements versus $\delta^{18}O$ values of the surrounding host rocks (‰VPDB), according to the vein sets. (**D**) $\Delta_{47}CO_2$ measured temperature of precipitation (°C) versus $\delta^{18}O$ values of calcite cements (‰VPDB, oblique dotted lines) and corresponding $\delta^{18}O$ values of the related fluids (‰SMOW) calculated from temperature-dependent fractionation equation $CaCO_3$-H_2O of [141]. For A-D, tectonic veins are reported as full circles of which color relates to the set they belong to (green: N-S, blue: N045, purple: N140). Note that the circle with red contour on D correspond to an LPS-related fault.

In the host-rock (n = 11), the $\delta^{18}O$ isotopic values range from −2.95 to −1.45‰ VPDB while the $\delta^{13}C$ isotopic values range from 1.08 to 3.08‰ VPDB. In the calcite veins (n = 28), the $\delta^{18}O$ isotopic values range from −1.56 to 2.59‰ VPDB while the $\delta^{13}C$ isotopic values range from 0.05 to 3.23‰ VPDB. The vein cements show variable isotopic values: for the set I (N-S, n = 10), $\delta^{18}O$ ratio ranges from −1.56 to 2.19‰ VPDB while $\delta^{13}C$ ratio ranges from 0.05 to 2.08‰ VPDB; for the set II (N045, n = 11), $\delta^{18}O$ ratio ranges from 0.58 to 2.59 ‰ VPDB while $\delta^{13}C$ ratio ranges from 1.81 to 2.97‰ VPDB; for the set III N140 (n = 7), $\delta^{18}O$ ratio ranges from −1.12 to 0.79‰ VPDB while $\delta^{13}C$ ratio ranges from 1.1 to 3.23‰ VPDB; $\delta^{18}O$ ratio ranges from −1.57 to 1.93‰ VPDB while $\delta^{13}C$ ratio ranges from 2.02 to 2.21‰ VPDB (Table 2, Figure 7A). In order to account for possible rock buffering effect, the isotopic values of the veins were plotted against isotopic values of the surrounding host-rock, for both carbon (Figure 7B) and oxygen (Figure 7C). Results show that most veins have a $\delta^{13}C$ value similar to their host rock, with a difference ranging from −0.50 to 0.25‰ VPDB in all sets except in the set I where this difference reaches −1.05‰ VPDB. Considering the difference in $\delta^{18}O$ values, the results are more scattered, ranging from −0.12 to 5.14‰ VPDB. Notably, the difference in the set II is higher than the one in the set III.

4.6. Carbonate Clumped-Isotope Paleothermometry (Δ47)

Eight of the 9 samples presented in the Supplementary Material were selected as being unambiguously related to a major fracture set. Consequently, 7 samples of vein cements and 1 sample of striated coating of fault plane were selected for Δ47 clumped isotope measurements (Table 3), with Δ47 values ranging from 0.593 ± 0.006‰ to 0.630 ± 0.006‰, (1SE) corresponding to precipitation temperature (T47) ranging from 38.30 ± 1.9 °C to 51.4 ± 2.2 °C (1SE). Veins belonging to different tectonic sets appear to yield distinct temperatures of precipitation, with T47 ranging from 48.7 ± 2.1 °C to 51.4 ± 2.2 °C for the set II (n = 5) and related fault cements, while vein cements from sets I and III have T47 ranging from 38.8 ± 2.0 °C to 45.1 ± 2.1 °C.

Table 3. Results of clumped isotope measurements, with Δ_{47} ratio and associated temperature of precipitation T_{47}.

Sample	N	Set	$\delta^{13}C$ (‰ VPDB)	$\delta^{18}O$ (‰VPDB, Calcite)	Δ_{47} (‰, 1σ)	T_{47} (°C ± 1σ)
NB_CIN23_V1	3	I	1.77	−0.40	0.6299 ± 0.0058	38.3 ± 1.9
NB_CIN25_V1prime	3	I	1.93	1.68	0.6006 ± 0.0058	48.7 ± 2.1
NB_CIN26_V1	3	II	1.92	2.13	0.5933 ± 0.0058	51.4 ± 2.2
NB_CIN28a_V1	3	II	1.73	−0.03	0.5968 ± 0.0057	50.0 ± 2.2
NB_CIN28b_V1	3	II	2.01	1.95	0.5944 ± 0.0058	51.0 ± 2.2
NB_CIN7_V1	3	II	2.44	0.86	0.6103 ± 0.0058	45.1 ± 2.1
NB_A14_V1	3	III	3.08	0.45	0.6283 ± 0.0058	38.8 ± 2.0
NB_CIN25_FAILLE	3	/	2.01	1.77	0.5985 ± 0.0058	49.4 ± 2.2

5. Interpretation of Results

5.1. Sequence of Mesostructures in Relation to Folding

The main stages of regional deformation, already described in the literature [29,60–63,68,121,142], were associated with sets of fracture-stylolite network identified in this domain of the UMAR.

Set I (N-S to N020-oriented fractures) is the oldest set encountered in the Cingoli Anticline. Because of its orientation and opening mode, we propose to interpret it as an along-strike joint set related to the flexure stage associated with forebulge development (sensu [8]).

Vertical, bedding, and fold-axis perpendicular set II joints/veins, associated with early folding stylolites with N045-oriented peaks likely reflects a stage of LPS with σ_1 striking perpendicular to the northern part of the UMAR structure axes. This is confirmed by reverse faulting associated with a N045 σ_1 after unfolding (Figure 5). Local complexities are interpreted as resulting from LPS related stress perturbation, resulting in a slight local stress rotation in the vicinity of local heterogeneities such as inherited faults (Figure 3, e.g., [23,143,144]). For instance, we interpret the N020 contraction in the northern part of the fold as a local rotation around the WNW-ESE fault. We also consider that stylolites with peaks-oriented E-W documented in the Calcare Massiccio relate to LPS perturbed by the reactivation of N-S striking inherited normal fault.

The joints/veins of set III postdate those of set II (Figure 4C) and are bedding-perpendicular and strike parallel to the local fold axis and bedding strike. We propose to relate this set to the folding stage, reflecting outer-arc extension associated to strata curvature at fold hinge. The ~20° variation of the orientation of this set between the north and south of the fold (N140 in the northern part and N160 in the southern part, Figure 3 and Figure 4) is consistent with the arcuate shape of the fold and then strengthen this interpretation.

Late folding, tectonic stylolites with horizontal peaks striking N045, along with post-tilting strike-slip faults (Figure 5) are interpreted to be related to horizontal NE-SW contraction affecting the strata after the fold was locked, corresponding to LSFT [8,15].

Our results therefore demonstrate that the N045 compression prevailed during the entire contractional history, i.e., from LPS to LSFT.

5.2. Evolution of the Burial Depth

The calculation of vertical stress involves the use of the following mechanical parameters: (i) crossover lengths L_c values, calculated by considering an uncertainty of 23% and reported in Table 1; (ii) mechanical and chemical parameters, defined in the literature and given above (i.e., Young modulus E, Poisson ratio ν and the solid-fluid interfacial energy γ). Then, the burial depths were calculated from each value of the vertical stress using Equation (2) and rounded to the closest 10 m (Table 1). The corresponding depth ranges for each formation are:

- Maiolica: from 720 ± 85 m to 1840 ± 220 m;
- Scaglia Rossa: from 880 ± 100 m to 1590 ± 190 m;
- Scaglia Variegata: from 720 ± 85 m to 1100 ± 130 m.

These ranges of burial depth correspond to the ranges of depth in which pressure solution along sedimentary stylolites was active, i.e., at the time vertical shortening (σ_1 vertical) was prevailing over horizontal shortening [60,145]. Figure 6 shows these ranges of depth reported on the burial model for comparison. The inversion and modeling data appear to be consistent as the maximum burial recorded by sedimentary stylolites never exceeds the maximum depth of the formation they belong to (Figure 6). In addition, the largest range of depths is associated with the Maiolica (i.e., the oldest formation); for the Scaglia Rossa and Variegata, intervals are overall narrower, and the more recent the formation, the narrower the depth range and the shallower the depth returned. In the case of the Scaglia Variegata (i.e., the youngest), the maximum burial recorded by the stylolites does not exceed 1200 m. Thus, BPS would not develop between 1200 and 1750 m, i.e., at the maximum burial values associated with the Scaglia Variegata (Figure 6). However, these data indicate that the burial was continuous from the Cretaceous to the late Miocene until the maximum burial depths of the sedimentary layers studied were reached. The reconstruction of the complete burial/exhumation history in relation to the deformation stages requires the combination of these data with isotope analyses.

5.3. Fluid System

The fluid system in the Cingoli Anticline can be partially characterized using stable O, C and clumped isotope dataset. The positive difference of isotopic values between vein cements and related host-rock, being low to null for $\delta^{13}C$ values (Figure 7B) yet significant for $\delta^{18}O$ values (Figure 7C), argues against rock buffered fluid precipitation as well as diagenesis related to burial. Instead, it strongly suggests that cements precipitated from local fluids originated from the studied sedimentary sequence (hence with identical $\delta^{13}C$ signature), with various but limited degrees of fluid-rock interaction likely related to migration, leading to an increase of the $\delta^{18}O$ ratios. The Δ_{47} results complement and support this interpretation as the combination of the temperature of precipitation T_{47} with the $\delta^{18}O$ value of the cement yields the $\delta^{18}O$ values of the precipitating fluid (Figure 7D) using the temperature dependent equation of fractionation of [141]. The reconstructed $\delta^{18}O$ values of the precipitating fluids range from 4.80 to 10.50‰ SMOW, irrespective of the tectonic vein sets where fluids precipitated. Positive $\delta^{18}O$ values points towards a more or less evolved brine origin for the fluid, which is consistent with a scenario involving a migration of Turonian-Lutetian marine fluids inside the host Scaglia Rossa and precipitating at thermal equilibrium within the host. It is noteworthy that these characteristics of the fluid system in the Cingoli Anticline are consistent with the fluid systems reconstructed in most of the other folds and thrusts of the UMAR, except for the Subasio Anticline [60] and Monte Tancia thrust [71]. Interestingly, our dataset does not document a rock buffering of external fluids as interpreted in the Monte Tancia thrust (southern part of the UMAR) by [71], and it does not reflect the late meteoric derived fluid infiltration documented there and related to the currently active extensional tectonics. When considering the tectonic vein sets, temperature of precipitation T_{47} differs significantly between set I, set II, and set III, supporting the interpretation of a thermal equilibrium between local fluids and the host rocks throughout the burial history. Moreover, the difference in $\delta^{18}O_{fluids}$ values between set II, related to LPS, and set III, related to local curvature of the strata at fold hinge, suggest that the degree of lateral fluid-rock interaction was higher during LPS than during folding, as is the case elsewhere [139].

6. Discussion

The results of mesostructural and isotopic analyses, together with inversion of the roughness of the sedimentary stylolites for maximum burial depth, have been combined in order to unravel the history of deformation in the Cingoli area. Moreover, because the fluid system appears to be at thermal equilibrium during deformation and that each fracture set has a specific T_{47} signature, it is possible to infer the timing of fracture development by comparing the range of precipitation temperatures T_{47} measured in the vein cements from each set with burial curves and maximum depths of active pressure-solution reconstructed from inversion of stylolite roughness. The stages of deformation (Figure 8), together with their absolute timing and sequence were therefore characterized, and compared with existing data for this study area [60,71], as well as in other localities of the belt, i.e., the anticlines of Monte Nero [121], Monte Catria [66] and Monte Conero [68]; the ages determined using the isotopic data are given with an uncertainty of 0.2 Ma, due to uncertainties on temperature ($\pm$ 2 °C) (Figure 6):

(i) pre-contractional stage, marked by burial, vertical compaction and dissolution along BPS recognized at the scale of the fold-and-thrust belt [60,61,66,68,121], under a vertical σ_1. It lasted until the early Messinian (ca. 6.4 Ma);
(ii) this pre-contractional stage is partly coeval with an E-W extension related to the flexure of the Adriatic foreland [63,101], the onset of which is set to early Burdigalian (ca. 21 Ma, as defined by the inflection point of the burial curves), and which ended by middle Messinian (ca. 6.3 Ma, Figure 8). This E-W extension would be at the origin of the development of a network of N-S fractures (set I). Pre-folding N-S striking joints have already been described in this anticline [61], and in other anticlines, Monte Nero [121] and Monte Catria [66], without being related to regional extension. In the Conero anticline, however, [68] related a set of N-S, high angle to bedding, joints and veins associated with normal faults to a flexural event. Further considerations provide the existence of polyphase syndepositional normal faulting: the Barremian [146–148] and late Cretaceous phases of stretching (e.g., [95,149,150], well known in the Umbria-Marche-Sabina area, could have developed set I joints in the Maiolica and Scaglia Rossa, as well as the development of stylolites in the Mesozoic rocks.
(iii) LPS stage, a pre/early-folding compressional stage with σ_1 NE-SW-oriented related to the Apenninic contraction [66,97–99]. The onset of this stage corresponds to the

switch from formerly vertical to horizontal σ_1, associated with a N045 compression marked by the fractures and stylolites of set II. This stage of deformation is consistent regionally as it has been documented in the Monte Nero [121], Conero [68], and Monte Catria [66] anticlines, and identified in numerous folds of the UMAR [52,54]. Based on T_{47} precipitation temperature and burial history, the LPS stage started since the middle Messinian (ca. 6.3 Ma). It has been reported in other case studies that the fold growth and associated underlying ramp activation is likely to be responsible for the uplift we reconstructed ca. 5.8 Ma [60]. Thus, we consider the LPS stage to have lasted from 6.3 Ma to 5.8 Ma;

(iv) fold growth stage, characterized by a compression parallel to regional shortening, i.e., NE-SW-oriented [29] and local extension perpendicular to fold axis and related with strata curvature at fold hinge [60]. Based on the dating of the LPS, fold growth started at 5.8 Ma yet the T_{47} points towards a related N135 striking joints/veins development during the latest Neogene-early Pliocene (ca. 5.2 to 3.9 Ma, Figure 8). That difference in timing suggests a 0.6 My long fault activity and strata tilting before curvature became high enough to developed outer-arc extension fractures;

(v) LSFT, post-dating the fold growth stage and still associated with a NE-SW contractional trend. At this stage shortening is no longer accommodated by e.g., limb rotation [60] and is associated with tectonic stylolites with N045-oriented peaks. The E-W fractures locally measured cannot be associated with this deformation stage, because no consistent chronological relationships with the syn-folding fracture sets were identified. Isotopic analyses (Figure 8) suggest the onset of the LSFT by the Pliocene (ca. 3.9 Ma). Despite the record of recent seismic activity in this northern part of the Apennines, linked to a post-orogenic NE-SW extension [151], the end of this deformation stage can precisely be determined neither from previous studies carried out in Cingoli [61] nor from data collected during this study.

This deformation scenario is in line to the one proposed by [61], that discriminates seven sets of stylolites, of which complexity can be related to more local effect of the fold evolution.

The fold growth duration in Cingoli as constrained by the above results (ca. 1.9 My, from 5.8 to 3.9 Ma) is consistent with that established by [62] using the age of foredeep deposits. The age of the end of the foreland flexure, evaluated at ca. 6.3 Ma in our study, is also consistent within uncertainties with the early Messinian age (ca. 7.2–6.5 Ma) derived from the sedimentary dating of the flexure-related normal faults [63,64]. The abrupt increase in the slope of the burial curves (Figure 6) likely reflects the initiation of this flexure at about 21 Ma, which is older than the late Burdigalian (ca. 16 Ma) age previously proposed on the basis of the distribution and variations in thickness of the Schlier (Aquitanian-Serravallian) marking the deepening of the basin [63], and older than Serravallian (ca. 13.8 Ma) proposed at the west of the Cingoli Anticline, in the western part of the belt [79]. The horizontal isotropy of the compaction-related sedimentary stylolites, along which dissolution was active until 7 Ma, however, suggests a very limited imprint of the flexure on the magnitude of the horizontal stresses in the Cretaceous-Paleogene rocks until the early Messinian, when the flexure became important enough to cause fractures and large-scale normal faults [63].

The folding duration, estimated in our work at ~2 My (from 5.8 to 3.9 Ma) is consistent with the results of previous studies carried out in the more internal areas of the belt, using absolute dating methods to reconstruct the duration and timing of the deformations [60,71]: the difference in timing is in line with [62], and the duration of the folding is consistent with [60,71], which respectively dated the folding from 8 to 5 Ma [60], and from 9 to 7 Ma [71] (i.e., a duration of 2 to 3 My).

This case study illustrates the potential of the combination of mesostructural, paleopiezometric, and isotopic analyses, which reveals a regionally consistent sequence and timing of deformation stages, despite multiple sources of uncertainties. Namely, the refining of the timing of the flexure expression on the sedimentary reservoir is an example of how the study of mesostructures can provide insights on the large-scale structures.

Another example of such upscaling lies in the interpretation of the arcuate shape of the Cingoli Anticline, that the distribution and timing of LPS related veins (set II) bounds to be a primary feature of the fold development, likely linked to the reactivation of an inherited N-S normal fault during folding (Figure 8).

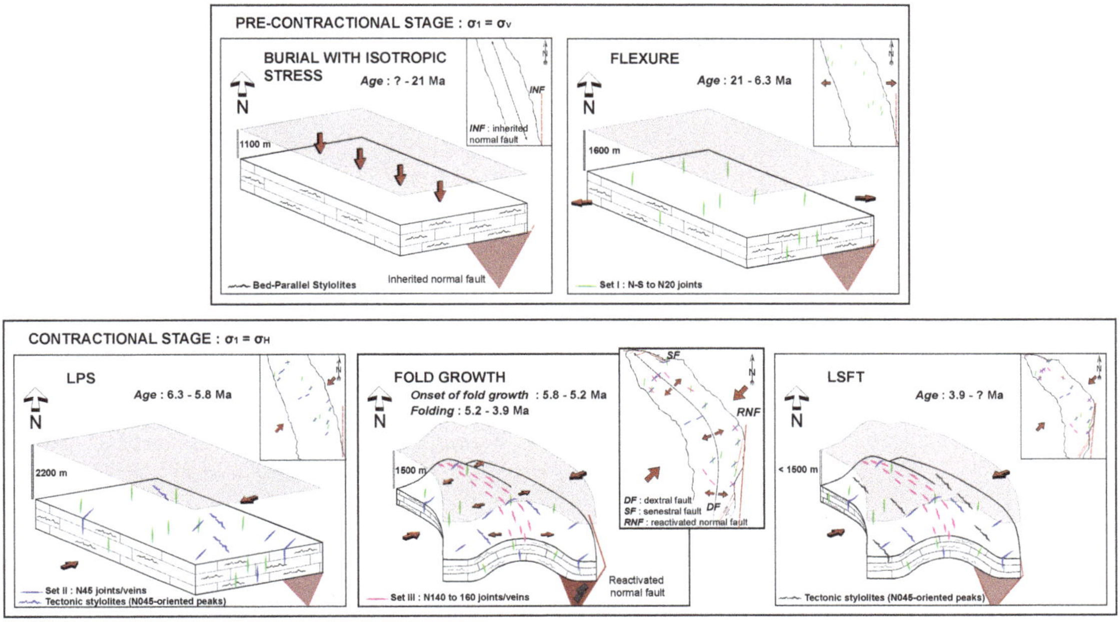

Figure 8. Interpretative model of the history of deformation in the Cingoli Anticline. The structural evolution of the area is represented in 3D and in map view. For each stage of deformation, the main principal stress σ_1 is represented as red arrows, as well as the associated mesostructures (green: set I, blue: set II, pink: set III) and burial depths recorded by the Scaglia Rossa (deduced from burial curves). Faults are represented by red lines in map view, dotted lines when inactive, and solid lines when active. The inherited normal fault is also represented on the 3D block, by a red plane in transparency when it is inactive.

7. Conclusions

This work, focused on the Cingoli Anticline in eastern UMAR, shows how the burial-deformation history of folded rocks can be unraveled using an original combination of ubiquitous features of carbonate rocks: fracture analysis, BPS paleopiezometry and vein cement geochemistry. The main conclusions are:

- different stages of deformation were recognized: (i) E-W extension related to foreland flexure (σ_1 vertical); (ii) N045 oriented LPS; (iii) fold growth; (iv) LSFT, under a horizontal N045 contraction. Mesostructural analyses also support that the arcuate geometry of the Cingoli Anticline is a primary feature, probably linked to the oblique reactivation of a N-S inherited normal fault.
- the burial history of strata was reconstructed with high resolution using roughness inversion applied to sedimentary stylolites. Our results highlight that this paleopiezometric technique yields consistent maximum depth estimates down to 2500 m, in agreement with previous studies in the western part of the UMAR.
- the timing of deformation, and particularly the duration of the Apenninic contractional stages, was reconstructed from combined paleopiezometric, isotopic and mesostructural data. Following foreland flexure (ca. 21.2 to 6.3 Ma), LPS was dated from middle Messinian to early Pliocene (ca. 6.3 to 5.8 Ma) and fold growth occurred between early

and middle Pliocene (ca. 5.8 to 3.9 Ma). The precise duration of LSFT remains out of reach. The duration of the fold growth phase is in line with previous estimates based on other proxies such as K-Ar and U-Pb absolute dating [71].

- the O and C stable isotope signatures and clumped isotopes of Δ_{47} of vein cements imply that the paleofluid system that prevailed during LPS and folding in this structure involve marine local fluids with limited interaction with the host rock, in agreement with earlier findings in the eastern UMAR.

Beyond regional implications, this study demonstrates the high potential of our new approach combining paleopiezometric, isotopic, and mesostructural data to reconstruct the sequence and to constrain the timing not only of local mesoscale deformation, but also of regional tectonic events in an orogenic system. Our results further confirm that the paleopiezometric inversion of the roughness of sedimentary stylolites for the vertical stresses is a reliable and powerful tool to unravel the amount and timing of burial without any assumption about the past geothermal gradient.

Supplementary Materials: The following are available online at https://www.mdpi.com/2076-3263/11/3/135/s1, Figure S1: graphical representation of the incremental mobile average value of the elastic rebound as a function of the number of measurements incorporated in the calculation, for each measurement site, Table S1: stabilized average rebound value R for the Maiolica, Scaglia Rossa and Scaglia Variegata. Standard deviation and number of measures are detailed for each site of measurement, Figure S2: Average Wavelet analysis of the stylolite roughness for all studied samples, Table S2: GPS coordinates of measurement and sampling sites, Figure S3: interpretated X-ray diffractometry spectrum of the Scaglia Rossa, Figure S4: Δ_{47} analysis report, detailing analysis and results.

Author Contributions: Conceptualization, A.L., N.E.B., O.L., and J.-P.C.; methodology, A.L., N.E.B. and S.K.; formal analysis, A.L., N.E.B., M.D., S.K. and L.E.; fieldwork, O.L., N.E.B., J.-P.C. and L.P.; writing—original draft preparation, A.L.; writing—review and editing, A.L., N.E.B., O.L., L.E., L.P., M.D. and J.-P.C.; project administration, N.E.B.; funding acquisition, N.E.B., O.L., J.-P.C. All authors have read and agreed to the published version of the manuscript.

Funding: This research was funded by the Sorbonne Université, grant number C14313, the Région Nouvelle- Aquitaine, and the Agence Nationale de la Recherche (Projet Investissement d'Avenir (programme E2S)).

Acknowledgments: A.L and N.E.B are funded through the ISITE programme E2S, supported by ANR PIA and Région Nouvelle-Aquitaine. Authors also thank two anonymous reviewers for insightful comments that improved the manuscript and Giancarlo Molli, Angelo Cipriani and Domenico Liotta, for their editorial work.

Conflicts of Interest: The authors declare no conflict of interest. The funders had no role in the design of the study; in the collection, analyses, or interpretation of data; in the writing of the manuscript, or in the decision to publish the results.

References

1. Barton, C.A.; Zoback, M.D. Stress perturbations associated with active faults penetrated by boreholes: Possible evidence for near-complete stress drop and a new technique for stress magnitude measurement. *J. Geophys. Res.* **1994**, *99*, 9373–9390. [CrossRef]
2. Zoback, M.L.; Zoback, M.D. Chapter 24: Tectonic stress field of the continental United States. *Mem. Geol. Soc. Am.* **1989**, *172*, 523–539. [CrossRef]
3. Sibson, R.H. Crustal stress, faulting and fluid flow. *Geol. Soc. Spec. Publ.* **1994**, *78*, 69–84. [CrossRef]
4. Sanderson, D.J.; Zhang, X. Stress-controlled localization of deformation and fluid flow in fractured rocks. *Geol. Soc. Spec. Publ.* **2004**, *231*, 299–314. [CrossRef]
5. Sanderson, D.J.; Zhang, X. Critical stress localization of flow associated with deformation of well-fractured rock masses, with implications for mineral deposits. *Geol. Soc. Spec. Publ.* **1999**, *155*, 69–81. [CrossRef]
6. Mourgues, R.; Gressier, J.B.; Bodet, L.; Bureau, D.; Gay, A. "Basin scale" versus "localized" pore pressure/stress coupling—Implications for trap integrity evaluation. *Mar. Pet. Geol.* **2011**, *28*, 1111–1121. [CrossRef]
7. Beaudoin, N.; Lacombe, O. Recent and future trends in paleopiezometry in the diagenetic domain: Insights into the tectonic paleostress and burial depth history of fold-and-thrust belts and sedimentary basins. *J. Struct. Geol.* **2018**, *114*, 357–365. [CrossRef]

8. Tavani, S.; Storti, F.; Lacombe, O.; Corradetti, A.; Muñoz, J.A.; Mazzoli, S. A review of deformation pattern templates in foreland basin systems and fold-and-thrust belts: Implications for the state of stress in the frontal regions of thrust wedges. *Earth-Sci. Rev.* **2015**, *141*, 82–104. [CrossRef]
9. Bergbauer, S.; Pollard, D.D. A new conceptual fold-fracture model including prefolding joints, based on the Emigrant Gap anticline, Wyoming. *Bull. Geol. Soc. Am.* **2004**, *116*, 294–307. [CrossRef]
10. Bellahsen, N.; Fiore, P.; Pollard, D.D. The role of fractures in the structural interpretation of Sheep Mountain Anticline, Wyoming. *J. Struct. Geol.* **2006**, *28*, 850–867. [CrossRef]
11. La Bruna, V.; Lamarche, J.; Agosta, F.; Rustichelli, A.; Giuffrida,, A.; Salardon, R.; Marie, L. Structural Diagenesis, Early Embrittlement and Fracture Setting in Shallow-Water Platform Carbonates (Monte Alpi Southern Apennines, Italy). In *Proceedings of the Fifth International Conference on Fault and Top Seals*; European Association of Geoscientists & Engineers: Houten, The Netherlands, 2019; Volume 2019, pp. 1–5.
12. Rocher, M.; Lacombe, O.; Angelier, J.; Chen, H.W. Mechanical twin sets in calcite as markers of recent collisional events in a fold-and-thrust belt: Evidence from the reefal limestones of southwestern Taiwan. *Tectonics* **1996**, *15*, 984–996. [CrossRef]
13. Lacombe, O.; Angelier, J.; Rocher, M.; Bergues, J.; Deffontaines, B.; Chu, H.T.; Hu, J.C.; Lee, J.C. Stress patterns associated with folding at the front of a collision belt: Example from the Pliocene reef limestones of Yutengping (Taiwan). *Bull. Soc. Geol. Fr.* **1996**, *167*, 361–374.
14. Rocher, M.; Lacombe, O.; Angelier, J.; Deffontaines, B.; Verdier, F. Cenozoic folding and faulting in the south Aquitaine Basin (France): Insights from combined structural and paleostress analyses. *J. Struct. Geol.* **2000**, *22*, 627–645. [CrossRef]
15. Amrouch, K.; Lacombe, O.; Bellahsen, N.; Daniel, J.M.; Callot, J.P. Stress and strain patterns, kinematics and deformation mechanisms in a basement-cored anticline: Sheep Mountain Anticline, Wyoming. *Tectonics* **2010**, *29*. [CrossRef]
16. Amrouch, K.; Beaudoin, N.; Lacombe, O.; Bellahsen, N.; Daniel, J.M. Paleostress magnitudes in folded sedimentary rocks. *Geophys. Res. Lett.* **2011**, *38*. [CrossRef]
17. Callot, J.P.; Breesch, L.; Guilhaumou, N.; Roure, F.; Swennen, R.; Vilasi, N. Paleo-fluids characterisation and fluid flow modelling along a regional transect in Northern United Arab Emirates (UAE). *Arab. J. Geosci.* **2010**, *3*, 413–437. [CrossRef]
18. Sassi, W.; Guiton, M.L.E.; Leroy, Y.M.; Daniel, J.M.; Callot, J.P. Constraints on bed scale fracture chronology with a FEM mechanical model of folding: The case of Split Mountain (Utah, USA). *Tectonophysics* **2012**, *576–577*, 197–215. [CrossRef]
19. Smart, K.J.; Ferrill, D.A.; Morris, A.P.; McGinnis, R.N. Geomechanical modeling of stress and strain evolution during contractional fault-related folding. *Tectonophysics* **2012**, *576–577*, 171–196. [CrossRef]
20. Beaudoin, N.; Bellahsen, N.; Lacombe, O.; Emmanuel, L. Fracture-controlled paleohydrogeology in a basement-cored, fault-related fold: Sheep Mountain Anticline, Wyoming, United States. *Geochem. Geophys. Geosyst.* **2011**, *12*. [CrossRef]
21. Saintot, A.; Stephenson, R.; Brem, A.; Stovba, S.; Privalov, V. Paleostress field reconstruction and revised tectonic history of the Donbas fold and thrust belt (Ukraine and Russia). *Tectonics* **2003**, *22*. [CrossRef]
22. Laubach, S.E.; Olson, J.E.; Gale, J.F.W. Are open fractures necessarily aligned with maximum horizontal stress? *Earth Planet. Sci. Lett.* **2004**, *222*, 191–195. [CrossRef]
23. Bellahsen, N.; Fiore, P.E.; Pollard, D.D. From spatial variation of fracture patterns to fold kinematics: A geomechanical approach. *Geophys. Res. Lett.* **2006**, *33*. [CrossRef]
24. Cooper, S.P.; Goodwin, L.B.; Lorenz, J.C. Fracture and fault patterns associated with basement-cored anticlines: The example of Teapot Dome, Wyoming. *Am. Assoc. Pet. Geol. Bull.* **2006**, *90*, 1903–1920. [CrossRef]
25. Ahmadhadi, F.; Lacombe, O.; Daniel, J.-M. Early Reactivation of Basement Faults in Central Zagros (SW Iran): Evidence from Pre-folding Fracture Populations in Asmari Formation and Lower Tertiary Paleogeography. In *Thrusts Belts and Foreland Basins. Frontiers in Earth Sciences*; Lacombe, O., Roure, F., Vergés, J., Eds.; Springer: Berlin/Heidelberg, Germany, 2007; pp. 205–228.
26. Ahmadhadi, F.; Daniel, J.M.; Azzizadeh, M.; Lacombe, O. Evidence for pre-folding vein development in the Oligo-Miocene Asmari Formation in the Central Zagros Fold Belt, Iran. *Tectonics* **2008**, *27*. [CrossRef]
27. Lacombe, O.; Bellahsen, N.; Mouthereau, F. Fracture patterns in the Zagros Simply Folded Belt (Fars, Iran): Constraints on early collisional tectonic history and role of basement faults. *Geol. Mag.* **2011**, *148*, 940–963. [CrossRef]
28. Casini, G.; Gillespie, P.A.; Vergés, J.; Romaire, I.; Fernández, N.; Casciello, E.; Saura, E.; Mehl, C.; Homke, S.; Embry, J.C.; et al. Sub-seismic fractures in foreland fold and thrust belts: Insight from the Lurestan Province, Zagros Mountains, Iran. *Pet. Geosci.* **2011**, *17*, 263–282. [CrossRef]
29. Tavani, S.; Storti, F.; Baus, J.; Muoz, J.A. Late thrusting extensional collapse at the mountain front of the northern Apennines (Italy). *Tectonics* **2012**, *31*. [CrossRef]
30. Beaudoin, N.; Leprêtre, R.; Bellahsen, N.; Lacombe, O.; Amrouch, K.; Callot, J.P.; Emmanuel, L.; Daniel, J.M. Structural and microstructural evolution of the Rattlesnake Mountain Anticline (Wyoming, USA): New insights into the Sevier and Laramide orogenic stress build-up in the Bighorn Basin. *Tectonophysics* **2012**, *576–577*, 20–45. [CrossRef]
31. Stearns, D.W.; Friedman, M. Reservoirs in Fractured Rock. In Stratigraphic Oil and Gas Fields—Classification, Exploration Methods, and Case Histories. In *Stratigraphic Oil and Gas Fields—Classification, Exploration Methods, and Case Histories*; AAPG Special Volumes; Goud, H.R., Ed.; The American Association of petroleum Geologist: Tulsa, OK, USA, 1972; Volume 10, pp. 82–106.
32. Engelder, T. *Joints and Shear Fractures in Rock*; Department of Geoscience, Pennsylvania State University: State College, PA, USA, 1987; pp. 27–69. ISBN 0-12-066265-5.

33. Laubach, S.E. Subsurface fractures and their relationship to stress history in East Texas basin sandstone. *Tectonophysics* **1988**, *156*, 37–49. [CrossRef]
34. Laubach, S.E. Paleostress directions from the preferred orientation of closed microfractures (fluid-inclusion planes) in sandstone, East Texas basin, U.S.A. *J. Struct. Geol.* **1989**, *11*, 603–611. [CrossRef]
35. Fischer, M.P.; Woodward, N.B.; Mitchell, M.M. The kinematics of break-thrust folds. *J. Struct. Geol.* **1992**, *14*, 451–460. [CrossRef]
36. Cooke, M.L. Fracture localization along faults with spatially varying friction. *J. Geophys. Res. Solid Earth* **1997**, *102*, 22425–22434. [CrossRef]
37. Aldega, L.; Carminati, E.; Scharf, A.; Mattern, F.; Al-Wardi, M. Estimating original thickness and extent of the Semail Ophiolite in the eastern Oman Mountains by paleothermal indicators. *Mar. Pet. Geol.* **2017**, *84*, 18–33. [CrossRef]
38. Aldega, L.; Bigi, S.; Carminati, E.; Trippetta, F.; Corrado, S.; Kavoosi, M.A. The Zagros fold-and-thrust belt in the Fars province (Iran): II. Thermal evolution. *Mar. Pet. Geol.* **2018**, *93*, 376–390. [CrossRef]
39. Balestra, M.; Corrado, S.; Aldega, L.; Morticelli, M.G.; Sulli, A.; Rudkiewicz, J.-L.; Sassi, W. Thermal and structural modeling of the Scillato wedge-top basin source-to-sink system: Insights into the Sicilian fold-and-thrust belt evolution (Italy). *Bulletin* **2019**, *131*, 1763–1782. [CrossRef]
40. Anders, M.H.; Laubach, S.E.; Scholz, C.H. Microfractures: A review. *J. Struct. Geol.* **2014**, *69*, 377–394. [CrossRef]
41. Becker, S.P.; Eichhubl, P.; Laubach, S.E.; Reed, R.M.; Lander, R.H.; Bodnar, R.J. A 48 m.y. history of fracture opening, temperature, and fluid pressure: Cretaceous Travis Peak Formation, East Texas basin. *Bull. Geol. Soc. Am.* **2010**, *122*, 1081–1093. [CrossRef]
42. Laubach, S.E.; Fall, A.; Copley, L.K.; Marrett, R.; Wilkins, S.J. Fracture porosity creation and persistence in a basement-involved Laramide fold, Upper Cretaceous Frontier Formation, Green River Basin, USA. *Geol. Mag.* **2016**, *153*, 887–910. [CrossRef]
43. Fall, A.; Eichhubl, P.; Cumella, S.P.; Bodnar, R.J.; Laubach, S.E.; Becker, S.P. Testing the basin-centered gas accumulation model using fluid inclusion observations:Southern Piceance Basin, Colorado. *Am. Assoc. Pet. Geol. Bull.* **2012**, *96*, 2297–2318. [CrossRef]
44. Lespinasse, M. Are fluid inclusion planes useful in structural geology? *J. Struct. Geol.* **1999**, *21*, 1237–1243. [CrossRef]
45. Corrado, S.; Invernizzi, C.; Aldega, L.; D'errico, M.; Di Leo, P.; Mazzoli, S.; Zattin, M. Testing the validity of organic and inorganic thermal indicators in different tectonic settings from continental subduction to collision: The case history of the Calabria–Lucania border (southern Apennines, Italy). *J. Geol. Soc. London.* **2010**, *167*, 985–999. [CrossRef]
46. Schito, A.; Andreucci, B.; Aldega, L.; Corrado, S.; Di Paolo, L.; Zattin, M.; Szaniawski, R.; Jankowski, L.; Mazzoli, S. Burial and exhumation of the western border of the Ukrainian Shield (Podolia): A multi-disciplinary approach. *Basin Res.* **2018**, *30*, 532–549. [CrossRef]
47. Beaudoin, N.; Lacombe, O.; Roberts, N.M.W.; Koehn, D. U-Pb dating of calcite veins reveals complex stress evolution and thrust sequence in the Bighorn Basin, Wyoming, USA. *Geology* **2018**, *46*, 1015–1018. [CrossRef]
48. Hoareau, G.; Claverie, F.; Pécheyran, C.; Paroissin, C.; Grignard, P.-A.; Motte, G.; Chailan, O.; Girard, J.-P. Direct U-Pb dating of carbonates from micron-scale femtosecond laser ablation inductively coupled plasma mass spectrometry images using robust regression. *Geochronology* **2021**, *3*, 67–87. [CrossRef]
49. Aldega, L.; Viola, G.; Casas-Sainz, A.; Marcén, M.; Román-Berdiel, T.; Lelij, R. Unraveling Multiple Thermotectonic Events Accommodated by Crustal-Scale Faults in Northern Iberia, Spain: Insights From K-Ar Dating of Clay Gouges. *Tectonics* **2019**, *38*, 3629–3651. [CrossRef]
50. Hoareau, G.; Crognier, N.; Lacroix, B.; Aubourg, C.; Roberts, N.M.W.; Niemi, N.; Branellec, M.; Beaudoin, N.; Ruiz, I.S. Combination of Δ47 and U-Pb dating in tectonic calcite veins unravel the last pulses related to the Pyrenean Shortening (Spain). *Earth Planet. Sci. Lett.* **2021**, *553*, 116636. [CrossRef]
51. Mangenot, X.; Bonifacie, M.; Gasparrini, M.; Götz, A.; Chaduteau, C.; Ader, M.; Rouchon, V. Coupling Δ47 and fluid inclusion thermometry on carbonate cements to precisely reconstruct the temperature, salinity and δ18O of paleo-groundwater in sedimentary basins. *Chem. Geol.* **2017**, *472*, 44–57. [CrossRef]
52. Renard, F.; Schmittbuhl, J.; Gratier, J.P.; Meakin, P.; Merino, E. Three-dimensional roughness of stylolites in limestones. *J. Geophys. Res.* **2004**, *109*. [CrossRef]
53. Schmittbuhl, J.; Renard, F.; Gratier, J.P.; Toussaint, R. Roughness of stylolites: Implications of 3D high resolution topography measurements. *Phys. Rev. Lett.* **2004**, *93*, 238501. [CrossRef]
54. Ebner, M.; Piazolo, S.; Renard, F.; Koehn, D. Stylolite interfaces and surrounding matrix material: Nature and role of heterogeneities in roughness and microstructural development. *J. Struct. Geol.* **2010**, *32*, 1070–1084. [CrossRef]
55. Ebner, M.; Toussaint, R.; Schmittbuhl, J.; Koehn, D.; Bons, P. Anisotropic scaling of tectonic stylolites: A fossilized signature of the stress field? *J. Geophys. Res.* **2010**, *115*, B06403. [CrossRef]
56. Rolland, A.; Toussaint, R.; Baud, P.; Schmittbuhl, J.; Conil, N.; Koehn, D.; Renard, F.; Gratier, J. Modeling the growth of stylolites in sedimentary rocks. *J. Geophys. Res. Solid Earth* **2012**, *117*, B6. [CrossRef]
57. Rolland, A.; Toussaint, R.; Baud, P.; Conil, N.; Landrein, P. Morphological analysis of stylolites for paleostress estimation in limestones. *Int. J. Rock Mech. Min. Sci.* **2014**, *67*, 212–225. [CrossRef]
58. Koehn, D.; Rood, M.P.; Beaudoin, N.; Chung, P.; Bons, P.D.; Gomez-Rivas, E. A new stylolite classification scheme to estimate compaction and local permeability variations. *Sediment. Geol.* **2016**, *346*, 60–71. [CrossRef]
59. Bertotti, G.; de Graaf, S.; Bisdom, K.; Oskam, B.; Vonhof, H.B.; Bezerra, F.H.R.; Reijmer, J.J.G.; Cazarin, L.C. Fracturing and fluid-flow during post-rift subsidence in carbonates of the Jandaíra Formation, Potiguar Basin, NE Brazil. *Basin Res.* **2017**, *29*, 836–853. [CrossRef]

60. Beaudoin, N.; Labeur, A.; Lacombe, O.; Koehn, D.; Billi, A.; Hoareau, G.; Boyce, A.; John, C.M.; Marchegiano, M.; Roberts, N.M.; et al. Regional-scale paleofluid system across the Tuscan Nappe–Umbria Marche Arcuate Ridge (northern Apennines) as revealed by mesostructural and isotopic analyses of stylolite-vein networks. *Solid Earth* **2020**, *11*, 1617–1641. [CrossRef]
61. Petracchini, L.; Antonellini, M.; Billi, A.; Scrocca, D. Fault development through fractured pelagic carbonates of the Cingoli anticline, Italy: Possible analog for subsurface fluid-conductive fractures. *J. Struct. Geol.* **2012**, *45*, 21–37. [CrossRef]
62. Calamita, F.; Cello, G.; Deiana, G.; Paltrinieri, W. Structural styles, chronology rates of deformation, and time-space relationships in the Umbria-Marche thrust system (central Apennines, Italy). *Tectonics* **1994**, *13*, 873–881. [CrossRef]
63. Mazzoli, S.; Deiana, G.; Galdenzi, S.; Cello, G. Miocene fault-controlled sedimentation and thrust propagation in the previously faulted external zones of the Umbria-Marche Apennines, Italy. *EGU Stephan Mueller Spec. Publ. Ser.* **2002**, *1*, 195–209. [CrossRef]
64. Calamita, F.; Cello, G.; Invernizzi, C.; Paltrinieri, W. Stile strutturale e crono- logia della deformazione lungo la traversa M. S. Vicino. *Stud. Geol. Camerti* **1990**, 69–86. [CrossRef]
65. Caricchi, C.; Aldega, L.; Corrado, S. Reconstruction of maximum burial along the Northern Apennines thrust wedge (Italy) by indicators of thermal exposure and modeling. *Bull. Geol. Soc. Am.* **2015**, *127*, 428–442. [CrossRef]
66. Tavani, S.; Storti, F.; Salvini, F.; Toscano, C. Stratigraphic versus structural control on the deformation pattern associated with the evolution of the Mt. Catria anticline, Italy. *J. Struct. Geol.* **2008**, *30*, 664–681. [CrossRef]
67. Petracchini, L.; Antonellini, M.; Billi, A.; Scrocca, D. Syn-thrusting polygonal normal faults exposed in the hinge of the Cingoli anticline, northern Apennines, Italy. *Front. Earth Sci.* **2015**, *3*, 67. [CrossRef]
68. Díaz General, E.N.; Mollema, P.N.; Antonellini, M. Fracture patterns and fault development in the pelagic limestones of the Monte Conero anticline (Italy). *Ital. J. Geosci.* **2015**, *134*, 495–512. [CrossRef]
69. Calamita, F.; Coppola, L.; Deiana, G.; Ivernizzi, C.; Mastrovincenzo, S. Le associazioni strutturali di Genge e M. Rotondo; un motivo ricorrente nella thrust belt umbro-marchigiana settentrionale. *Boll. Della Soc. Geol. Ital.* **1987**, *106*, 141–151.
70. Invernizzi, C. Low temperature deformation in naturally deformed marly limestones from the Umbria-Marche Apennines. *Ann. Tectonicae* **1994**, *8*, 119–133.
71. Curzi, M.; Aldega, L.; Bernasconi, S.M.; Berra, F.; Billi, A.; Boschi, C.; Franchini, S.; Van der Lelij, R.; Viola, G.; Carminati, E. Architecture and evolution of an extensionally-inverted thrust (Mt. Tancia Thrust, Central Apennines): Geological, structural, geochemical, and K–Ar geochronological constraints. *J. Struct. Geol.* **2020**, *136*, 104059. [CrossRef]
72. Lavecchia, G.; Minelli, G.; Pialli, G. The Umbria-Marche arcuate fold belt (Italy). *Tectonophysics* **1988**, *146*, 125–137. [CrossRef]
73. Elter, F.M.; Elter, P.; Eva, C.; Eva, E.; Kraus, R.K.; Padovano, M.; Solarino, S. An alternative model for the recent evolution of the Northern-Central Apennines (Italy). *J. Geodyn.* **2012**, *54*, 55–63. [CrossRef]
74. Tozer, R.S.J.; Butler, R.W.H.; Corrado, S. Comparing thin- and thick-skinned thrust tectonic models of the Central Apennines, Italy. *Stephan Mueller Spec. Publ. Ser.* **2002**, *1*, 181–194. [CrossRef]
75. Scisciani, V.; Agostini, S.; Calamita, F.; Pace, P.; Cilli, A.; Giori, I.; Paltrinieri, W. Positive inversion tectonics in foreland fold-and-thrust belts: A reappraisal of the Umbria-Marche Northern Apennines (Central Italy) by integrating geological and geophysical data. *Tectonophysics* **2014**, *637*, 218–237. [CrossRef]
76. Mazzoli, S.; Aldega, L.; Corrado, S.; Invernizzi, C.; Zattin, M. Pliocene-quaternary thrusting, syn-orogenic extension and tectonic exhumation in the Southern Apennines (Italy): Insights from the Monte Alpi area. *Spec. Pap. Soc. Am.* **2006**, *414*, 55.
77. Boccaletti, M.; Calamita, F.; Viandante, M.G. The lithospheric Apennine Neo-Chain developing since the Lower Pliocene as a result of the Africa-Europe convergence. *Boll. della Soc. Geol. Ital.* **2005**, *124*, 87–105.
78. Satolli, S.; Pace, P.; Viandante, M.; Calamita, F. Lateral variations in tectonic style across cross-strike discontinuities: An example from the Central Apennines belt (Italy). *Int. J. Earth Sci.* **2014**, *103*, 2301–2313. [CrossRef]
79. Brozzetti, F.; Cirillo, D.; Luchetti, L. Timing of Contractional Tectonics in the Miocene Foreland Basin System of the Umbria Pre-Apennines (Italy): An Updated Overview. *Geosciences* **2021**, *11*, 97. [CrossRef]
80. Cello, G.; Mazzoli, S.; Tondi, E.; Turco, E. Active tectonics in the central Apennines and possible implications for seismic hazard analysis in peninsular Italy. *Tectonophysics* **1997**, *272*, 43–68. [CrossRef]
81. Ghisetti, F.; Vezzani, L. Normal faulting, extension and uplift in the outer thrust belt of the central Apennines (Italy): Role of the Caramanico fault. *Basin Res.* **2002**, *14*, 225–236. [CrossRef]
82. Santantonio, M. Pelagic carbonate platforms in the geologic record: Their classification and sedimentary and paleotectonic evolution. *Am. Assoc. Pet. Geol. Bull.* **1994**, *78*, 122–141. [CrossRef]
83. Santantonio, M. Facies associations and evolution of pelagic carbonate platform/basin systems: Examples from the Italian Jurassic. *Sedimentology* **1993**, *40*, 1039–1067. [CrossRef]
84. Carminati, E.; Lustrino, M.; Cuffaro, M.; Doglioni, C. Tectonics, magmatism and geodynamics of Italy: What we know and what we imagine. *J. Virtual Explor.* **2010**, *36*, 10–3809. [CrossRef]
85. Bigi, S.; Milli, S.; Corrado, S.; Casero, P.; Aldega, L.; Botti, F.; Moscatelli, M.; Stanzione, O.; Falcini, F.; Marini, M.; et al. Stratigraphy, structural setting and burial history of the Messinian Laga basin in the context of Apennine foreland basin system. *J. Mediterr. Earth Sci.* **2009**, *1*, 61–84. [CrossRef]
86. Bally, A.W.; Burbi, W.; Cooper, J.C.; Ghelardoni, L. Balanced sections and seismic reflection profiles across the Central Apennines. *Mem. Soc. Geol. Ital.* **1986**, *107*, 109–130.
87. Ghisetti, F.; Vezzani, L. Geometric and kinematic complexities in the Marche-Abruzzi external zones (Central Apennines, Italy). *Geol. Rundschau* **1988**, *77*, 63–78. [CrossRef]

88. Hill, K.; Hayward, A. Structural constraints on the Tertiary plate tectonic evolution of Italy. *Mar. Pet. Geol.* **1988**, *5*, 2–16. [CrossRef]
89. Lacombe, O.; Bellahsen, N. Thick-skinned tectonics and basement-involved fold–thrust belts: Insights from selected Cenozoic orogens. *Geol. Mag.* **2016**, *153*, 763–810. [CrossRef]
90. Hippolyte, J.-C.; Angelier, J.; Barrier, E. Compressional and extensional tectonics in an arc system: Example of the Southern Apennines. *J. Struct. Geol.* **1995**, *17*, 1725–1740. [CrossRef]
91. Mazzoli, S.; Cello, G.; Deiana, G.; Galdenzi, S.; Gambini, R.; Mancinelli, A.; Mattioni, L.; Shiner, P.; Tondi, E. Modes of foreland deformation ahead of the Apennine thrust front. *J. Czech Geol. Soc.* **2000**, *45*, 246.
92. Scisciani, V.; Calamita, F.; Tavarnelli, E.; Rusciadelli, G.; Ori, G.-G.; Paltrinieri, W. Foreland-dipping normal faults in the iner edges of syn-orogenic basins: A case from the Central Apennines, Italy. *Tectonophysics* **2001**, *330*, 211–224. [CrossRef]
93. Calamita, F.; Paltrinieri, W.; Pelorosso, M.; Scisciani, V.; Tavarnelli, E. Inherited Mesozoic architecture of the Adria continental palaeomargin in the neogene central apennines orogenic system, Italy. *Boll. Della Soc. Geol. Ital.* **2003**, *122*, 307–318.
94. Rusciadelli, G.; Viandante, M.G.; Calamita, F.; Cook, A.C. Burial-exhumation history of the central Apennines (Italy), from the foreland to the chain building: Thermochronological and geological data. *Terra Nov.* **2005**, *17*, 560–572. [CrossRef]
95. Tavarnelli, E. Ancient synsedimentary structural control on thrust ramp development: An example from the Northern Apennines, Italy. *Terra Nov.* **2007**, *8*, 65–74. [CrossRef]
96. Pace, P.; Calamita, F. Push-up inversion structures v. fault-bend reactivation anticlines along oblique thrust ramps: Examples from the Apennines fold-and-thrust belt (Italy). *J. Geol. Soc. Lond.* **2014**, *171*, 227–238. [CrossRef]
97. Marshak, S.; Geiser, P.A.; Alvarez, W.; Engelder, T. Mesoscopic fault array of the northern Umbrian Appennine fold belt, Italy: Geometry of conjugate shear by pressure-solution slip. *Geol. Soc. Am. Bull.* **1982**, *93*, 1013–1022. [CrossRef]
98. Storti, F.; Rossetti, F.; Salvini, F. Structural architecture and displacement accomodation mechanisms at the termination of the Priestley Fault, northern Victoria Land, Antarctica. *Tectonophysics* **2001**, *341*, 141–161. [CrossRef]
99. Barchi, M.; Alvarez, W.; Shimabukuro, D.H. The Umbria-Marche Apennines as a double orogen: Observations and hypotheses. *Ital. J. Geosci.* **2012**, *131*, 258–271. [CrossRef]
100. Calamita, F.; Deiana, G. The arcuate shape of the Umbria-Marche-Sabina Apennines (Central Italy). *Tectonophysics* **1988**, *146*, 139–147. [CrossRef]
101. Deiana, G.; Cello, G.; Chiocchini, M.; Galdenzi, S.; Mazzoli, S.; Pistolesi, E.; Potetti, M.; Romano, A.; Turco, E.; Principi, M. Tectonic evolution of the external zones of the Umbria-Marche Apennines in the Monte San Vicino-Cingoli area. *Boll. Della Soc. Geol. Ital.* **2002**, *1*, 229–238.
102. Cita, M.B.; Abbate, E.; Ballini, M.; Conti, A.; Falorni, P.; Germani, D.; Petti, F.M. Carta Geologica d'Italia–1: 50.000, Catalogo delle Formazioni, Unità tradizionali. *Quad. Serv. Geol. D'It. Ser. III* **2007**, *7*, 382.
103. Menichetti, M. La sezione geologica Cingoli—M. Maggio-Tevere nell'Appennino umbro-marchigiano: Analisi cinematica e strutturale. *Stud. Geol. Camerti* **1991**, *special vo*, 315–328.
104. Centamore, E.; Chiocchini, M.; Deiana, G.; Micarelli, A.; Pieruccini, U. Contributo alla conoscenda del Giurassico dell'Appennino Umbro-Marchigiano. *Stud. Geol. Camerti* **1971**, *1*, 7–89. [CrossRef]
105. Farinacci, A. *Jurassic Sediments in the Umbro-Marchean Apennines: An Alternative Model*; Istituto di Geologia e Paleontologia: Roma, Italy, 1981.
106. Santantonio, M.; Carminati, E. Jurassic rifting evolution of the Apennines and Southern Alps (Italy): Parallels and differences. *Bulletin* **2011**, *123*, 468–484. [CrossRef]
107. Koehn, D.; Renard, F.; Toussaint, R.; Passchier, C.W. Growth of stylolite teeth patterns depending on normal stress and finite compaction. *Earth Planet. Sci. Lett.* **2007**, *257*, 582–595. [CrossRef]
108. Toussaint, R.; Aharonov, E.; Koehn, D.; Gratier, J.P.; Ebner, M.; Baud, P.; Rolland, A.; Renard, F. Stylolites: A review. *J. Struct. Geol.* **2018**, *114*, 163–195. [CrossRef]
109. Grohmann, C.H.; Campanha, G.A. OpenStereo: Open source, cross-platform software for structural geology analysis. *AGU Fall Meet. Abstr.* **2010**, *2010*, IN31C-06.
110. Angelier, J. Tectonic analysis of fault slip data sets. *J. Geophys. Res.* **1984**, *89*, 5835–5848. [CrossRef]
111. Lacombe, O. Do fault slip data inversions actually yield" paleostresses" that can be compared with contemporary stresses? A critical discussion. *Comptes Rendus-Geosci.* **2012**, *344*, 159–173. [CrossRef]
112. Aydin, A.; Basu, A. The Schmidt hammer in rock material characterization. *Eng. Geol.* **2005**, *81*, 1–14. [CrossRef]
113. Katz, O.; Reches, Z.; Roegiers, J.-C. Evaluation of mechanical rock properties using a Schmidt Hammer. *Int. J. Rock Mech. Min. Sci.* **2000**, *37*, 723–728. [CrossRef]
114. Ebner, M.; Koehn, D.; Toussaint, R.; Renard, F. The influence of rock heterogeneity on the scaling properties of simulated and natural stylolites. *J. Struct. Geol.* **2009**, *31*, 72–82. [CrossRef]
115. Ebner, M.; Koehn, D.; Toussaint, R.; Renard, F.; Schmittbuhl, J. Stress sensitivity of stylolite morphology. *Earth Planet. Sci. Lett.* **2009**, *277*, 394–398. [CrossRef]
116. Koehn, D.; Ebner, M.; Renard, F.; Toussaint, R.; Passchier, C.W. Modelling of stylolite geometries and stress scaling. *Earth Planet. Sci. Lett.* **2012**, *341–344*, 104–113. [CrossRef]

117. Beaudoin, N.; Gasparrini, M.; David, M.E.; Lacombe, O.; Koehn, D. Bedding-parallel stylolites as a tool to unravel maximum burial depth in sedimentary basins: Application to Middle Jurassic carbonate reservoirs in the Paris basin, France. *Bull. Geol. Soc. Am.* **2019**, *131*, 1239–1254. [CrossRef]
118. Beaudoin, N.; Lacombe, O.; David, M.E.; Koehn, D. Does stress transmission in forelands depend on structural style? Distinctive stress magnitudes during Sevier thin-skinned and Laramide thick-skinned layer-parallel shortening in the Bighorn Basin (USA) revealed by stylolite and calcite twinning paleopiez. *Terra Nov.* **2020**, *32*, 225–233. [CrossRef]
119. Simonsen, I.; Hansen, A.; Nes, O.M. Determination of the Hurst exponent by use of wavelet transforms. *Phys. Rev.* **1998**, *58*, 2779–2787. [CrossRef]
120. Wright, K.; Cygan, R.T.; Slater, B. Structure of the (1014) surfaces of calcite, dolomite and magnesite under wet and dry conditions. *Phys. Chem. Chem. Phys.* **2001**, *3*, 839–844. [CrossRef]
121. Beaudoin, N.; Koehn, D.; Lacombe, O.; Lecouty, A.; Billi, A.; Aharonov, E.; Parlangeau, C. Fingerprinting stress: Stylolite and calcite twinning paleopiezometry revealing the complexity of progressive stress patterns during folding—The case of the Monte Nero anticline in the Apennines, Italy. *Tectonics* **2016**, *34*, 1687–1712. [CrossRef]
122. Vass, A.; Koehn, D.; Toussaint, R.; Ghani, I.; Piazolo, S. The importance of fracture-healing on the deformation of fluid-filled layered systems. *J. Struct. Geol.* **2014**, *67*, 94–106. [CrossRef]
123. Dietrich, D.; McKenzie, J.A.; Song, H. Origin of calcite in syntectonic veins as determined from carbon- isotope ratios. *Geology* **1983**, *11*, 547–551. [CrossRef]
124. Shemesh, A.; Ron, H.; Erel, Y.; Kolodny, Y.; Nur, A. Isotopic composition of vein calcite and its fluid inclusions: Implication to paleohydrological systems, tectonic events and vein formation processes. *Chem. Geol.* **1992**, *94*, 307–314. [CrossRef]
125. Templeton, A.S. Fluids and the Heart Mountain Fault revisited. *Geology* **1995**, *23*, 929–932. [CrossRef]
126. Douglas, T.A.; Chamberlain, C.P.; Poage, M.A.; Abruzzese, M.; Shultz, S.; Henneberry, J.; Layer, P. Fluid flow and the heart mountain fault: A stable isotopic, fluid inclusion, and geochronologic study. *Geofluids* **2003**, *3*, 13–32. [CrossRef]
127. Vilasi, N.; Malandain, J.; Barrier, L.; Callot, J.P.; Amrouch, K.; Guilhaumou, N.; Lacombe, O.; Muska, K.; Roure, F.; Swennen, R. From outcrop and petrographic studies to basin-scale fluid flow modelling: The use of the Albanian natural laboratory for carbonate reservoir characterisation. *Tectonophysics* **2009**, *474*, 367–392. [CrossRef]
128. Barbier, M.; Hamon, Y.; Callot, J.P.; Floquet, M.; Daniel, J.M. Sedimentary and diagenetic controls on the multiscale fracturing pattern of a carbonate reservoir: The Madison Formation (Sheep Mountain, Wyoming, USA). *Mar. Pet. Geol.* **2012**, *29*, 50–67. [CrossRef]
129. Gasparrini, M.; Lacombe, O.; Rohais, S.; Belkacemi, M.; Euzen, T. Natural mineralized fractures from the Montney-Doig unconventional reservoirs (Western Canada Sedimentary Basin): Timing and controlling factors. *Mar. Pet. Geol.* **2021**, *124*, 104826. [CrossRef]
130. Bons, P.D.; Elburg, M.A.; Gomez-Rivas, E. A review of the formation of tectonic veins and their microstructures. *J. Struct. Geol.* **2012**, *43*, 33–62. [CrossRef]
131. Daëron, M.; Blamart, D.; Peral, M.; Affek, H.P. Absolute isotopic abundance ratios and the accuracy of Δ47 measurements. *Chem. Geol.* **2016**, *442*, 83–96. [CrossRef]
132. Bernasconi, S.M.; Müller, I.A.; Bergmann, K.D.; Breitenbach, S.F.M.; Fernandez, A.; Hodell, D.A.; Jaggi, M.; Meckler, A.N.; Millan, I.; Ziegler, M. Reducing uncertainties in carbonate clumped isotope analysis through consistent carbonate-based standardization. *Geochem. Geophys. Geosyst.* **2018**, *19*, 2895–2914. [CrossRef] [PubMed]
133. Kim, S.T.; Mucci, A.; Taylor, B.E. Phosphoric acid fractionation factors for calcite and aragonite between 25 and 75 °C: Revisited. *Chem. Geol.* **2007**, *246*, 135–146. [CrossRef]
134. Daëron, M. Full propagation of analytical uncertainties in D47 measurements. *Geochem. Geophys. Geosyst.* **2020**, 1–26. [CrossRef]
135. Kele, S.; Breitenbach, S.F.M.; Capezzuoli, E.; Meckler, A.N.; Ziegler, M.; Millan, I.M.; Kluge, T.; Deák, J.; Hanselmann, K.; John, C.M.; et al. Temperature dependence of oxygen- and clumped isotope fractionation in carbonates: A study of travertines and tufas in the 6-95°C temperature range. *Geochim. Cosmochim. Acta* **2015**, *168*, 172–192. [CrossRef]
136. Cardozo, N. Backtrip: Performs 1D "Airy type" Backstripping of Sedimentary Strata. 2011. Available online: https://downloadsafer.com/software/osxbackstrip/ (accessed on 30 January 2021).
137. Allen, P.A.; Allen, J.R. *Basin Analysis: Principles and Aplications, Blackwell Scientific Publication*, 1st ed.; Blackwell Scientific Publications: Oxford, UK; London, UK, 1990; ISBN 0632024224.
138. Watts, A.B. *Isostasy and Flexure of the Lithosphere*; Paperback; Cambridge University Press: Cambridge, UK, 2001; ISBN 9780521006002.
139. Beaudoin, N.; Bellahsen, N.; Lacombe, O.; Emmanuel, L.; Pironon, J. Crustal-scale fluid flow during the tectonic evolution of the Bighorn Basin (Wyoming, USA). *Basin Res.* **2014**, *26*, 403–435. [CrossRef]
140. Aldega, L.; Botti, F.; Corrado, S. Clay mineral assemblages and vitrinite reflectance in the Laga Basin (Central Apennines, Italy): What do they record? *Clays Clay Miner.* **2007**, *55*, 504–518. [CrossRef]
141. Kim, S.T.; O'Neil, J.R. Equilibrium and nonequilibrium oxygen isotope effects in synthetic carbonates. *Geochim. Cosmochim. Acta* **1997**, *61*, 3461–3475. [CrossRef]
142. Di Naccio, D.; Boncio, P.; Cirilli, S.; Casaglia, F.; Morettini, E.; Lavecchia, G.; Brozzetti, F. Role of mechanical stratigraphy on fracture development in carbonate reservoirs: Insights from outcropping shallow water carbonates in the Umbria-Marche Apennines, Italy. *J. Volcanol. Geotherm. Res.* **2005**, *148*, 98–115. [CrossRef]
143. Petit, J.; Barquins, M. Can natural faults propagate under mode II conditions? *Tectonics* **1988**, *7*, 1243–1256. [CrossRef]

144. Homberg, C.; Hu, J.C.; Angelier, J.; Bergerat, F.; Lacombe, O. Characterization of stress perturbations near major fault zones: Insights from 2-D distinct-element numerical modelling and field studies (Jura mountains). *J. Struct. Geol.* **1997**, *19*, 703–718. [CrossRef]
145. Beaudoin, N.; Lacombe, O.; Koehn, D.; David, M.E.; Farrell, N.; Healy, D. Vertical stress history and paleoburial in foreland basins unravelled by stylolite roughness paleopiezometry: Insights from bedding-parallel stylolites in the Bighorn Basin, Wyoming, USA. *J. Struct. Geol.* **2020**, *136*, 104061. [CrossRef]
146. Fabbi, S.; Citton, P.; Romano, M.; Cipriani, A. Detrital events within pelagic deposits of the Umbria-Marche Basin (Northern Apennines, Italy): Further evidence of Early Cretaceous tectonics. *J. Mediterr. Earth Sci.* **2016**, *8*, 39–52.
147. Cipriani, A.; Bottini, C. Early Cretaceous tectonic rejuvenation of an Early Jurassic margin in the Central Apennines: The "Mt. Cosce Breccia" . *Sediment. Geol.* **2019**, *387*, 57–74. [CrossRef]
148. Cipriani, A.; Bottini, C. Unconformities, neptunian dykes and mass-transport deposits as an evidence for Early Cretaceous syn-sedimentary tectonics: New insights from the Central Apennines. *Ital. J. Geosci.* **2019**, *138*, 333–354. [CrossRef]
149. Decandia, F.A. Geologia dei Monti di Spoleto (Provincia di Perugia). *Boll. Della Soc. Geol. Ital.* **1982**, *101*, 291–315.
150. Marchegiani, L.; Bertotti, G.; Cello, G.; Deiana, G.; Mazzoli, S.; Tondi, E. Pre-orogenic tectonics in the Umbria-Marche sector of the Afro-Adriatic continental margin. *Tectonophysics* **1999**, *315*, 123–143. [CrossRef]
151. Mazzoli, S.; Santini, S.; Macchiavelli, C.; Ascione, A. Active tectonics of the outer northern Apennines: Adriatic vs. Po Plain seismicity and stress fields. *J. Geodyn.* **2015**, *84*, 62–76. [CrossRef]

geosciences

MDPI

Article

Stratigraphic and Tectonic Setting of the Liguride Units Cropping Out along the Southeastern Side of the Agri Valley (Southern Apennines, Italy)

Giacomo Prosser [1], Giuseppe Palladino [2,3,*], Dario Avagliano [4], Francesco Coraggio [4], Eleonora Maria Bolla [5], Marcello Riva [5] and Daniele Enrico Catellani [5]

1 Dipartimento di Scienze, Università degli Studi della Basilicata, 85100 Potenza, Italy; giacomo.prosser@unibas.it
2 Department of Geology and Geophysics, School of Geosciences, University of Aberdeen, Aberdeen AB24 3UF, UK
3 Italconsult S.p.A., Viggiano, 85059 Potenza, Italy
4 ENI S.p.A., Viggiano, 85059 Potenza, Italy; dario.avagliano@eni.com (D.A.); francesco.coraggio@eni.com (F.C.)
5 ENI S.p.A., San Donato Milanese, 20097 Milano, Italy; eleonora.maria.bolla@eni.com (E.M.B.); marcello.riva@eni.com (M.R.); daniele.enrico.catellani@eni.com (D.E.C.)
* Correspondence: giuseppe.palladino@abdn.ac.uk

check for updates

Citation: Prosser, G.; Palladino, G.; Avagliano, D.; Coraggio, F.; Bolla, E.M.; Riva, M.; Catellani, D.E. Stratigraphic and Tectonic Setting of the Liguride Units Cropping Out along the Southeastern Side of the Agri Valley (Southern Apennines, Italy). *Geosciences* **2021**, *11*, 125. https://doi.org/10.3390/geosciences11030125

Academic Editors: Domenico Liotta, Giancarlo Molli and Jesus Martinez-Frias

Received: 30 December 2020
Accepted: 4 March 2021
Published: 9 March 2021

Publisher's Note: MDPI stays neutral with regard to jurisdictional claims in published maps and institutional affiliations.

Copyright: © 2021 by the authors. Licensee MDPI, Basel, Switzerland. This article is an open access article distributed under the terms and conditions of the Creative Commons Attribution (CC BY) license (https://creativecommons.org/licenses/by/4.0/).

Abstract: This paper shows the main results of a multidisciplinary study performed along the southeastern sector of the Agri Valley in Basilicata (Southern Italy), where Cenozoic units, crucial for constraining the progressive evolution of the Southern Apennine thrust and fold belt and, more in general, the geodynamic evolution of the Mediterranean area are widely exposed. In particular, we aimed at understanding the stratigraphic and tectonic setting of deep-sea, thrust-top Cenozoic units exposed immediately to north of Montemurro, between Costa Molina and Monte dell'Agresto. In the previous works different units, showing similar sedimentological characteristics but uncertain age attribution, have been reported in the study area. In our study, we focussed on the Albidona Formation, pertaining to the Liguride realm, which shows most significant uncertainties regarding the age and the stratigraphic setting. The study was based on a detailed field survey which led to a new geological map of the area. This was supported by new stratigraphic, biostratigraphic and structural analyses. Biostratigraphic analysis provided an age not older than the upper Ypresian and not younger than the early Priabonian. Recognition of marker stratigraphic horizons strongly helped in the understanding of the stratigraphy of the area. The study allowed a complete revision of the stratigraphy of the outcropping Cenozoic units, the recognition of until now unknown tectonic structures and the correlation between surface and subsurface geology.

Keywords: Albidona Formation; biostratigraphy; liguride units; Agri Valley; southern Apennines

1. Introduction

Understanding surface geology of a given area provides fundamental clues on the lithological, stratigraphic and structural setting of the subsurface, useful for exploitation of natural resources. In particular, recognition of key stratigraphic horizons is essential for the correct identification of major tectonic structures and the interpretation and correlation of geophysical and well data to define reliable 3D geological models. In addition, the recognition of regional faults and permeable stratigraphic units can be helpful in modelling fluid flow in subsurface. The Agri Valley represents a good example where a better definition of the surface geology provides significant improvements in the interpretation of subsurface data.

Here, the carbonate reservoir rocks are covered by a thick pile of allochthonous thrust sheets emplaced during the building of the southern Apennine thrust and fold belt (Figure 1).

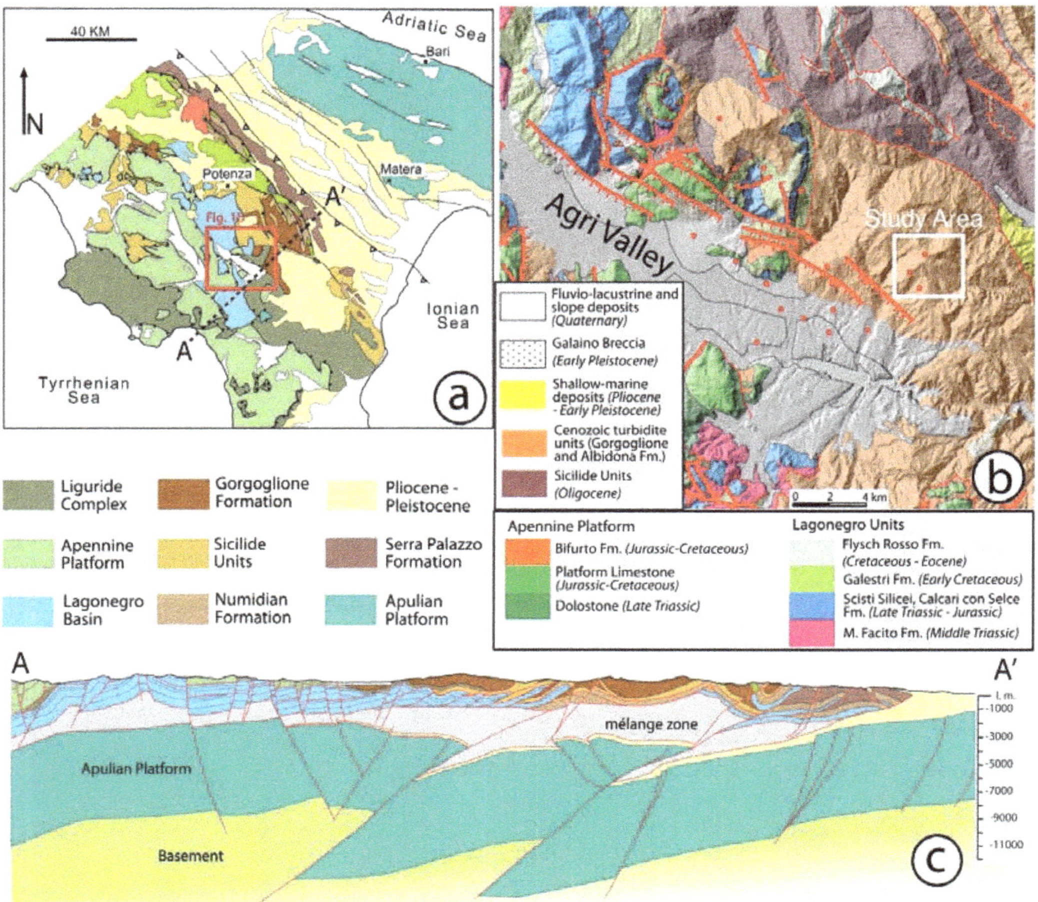

Figure 1. (**a**) Regional geological map of the Southern Apennines; (**b**) schematic sketch map of the Agri Valley graben showing the location of the study area; (**c**) schematic geological cross-section outlining the structural highs of the Internal Apulian Platform and the geometry of the overlying allochthonous units [1].

For this reason, the knowledge of the surface geology in this area is regarded as a priority. In the Agri Valley one of the major geological challenges consists in the distinction between different Cenozoic turbidite units, represented by the Albidona and the Gorgoglione formations sensu [2] and named "Albidona Formation" and "Gorgoglione Flysch" in the Italian Geological Map at 1:50,000 scale (CARG Project—e.g., Servizio Geologico d'Italia, 2005, 2009, 2014). The latter units are commonly interpreted as part of allochthonous thrust sheets. According to [3] the aforementioned formations commonly show similar lithological and sedimentological characteristics, which frequently hinder the correct identification of the vertical/lateral relationships. These issues, along with the scarce paleontological age determinations, led to stratigraphic misinterpretations and failure in recognizing important tectonic structures in the field that are fundamental when building a 3D geological model of the Agri Valley. Difficulty in discriminating between the

Albidona and the Gorgoglione formations is evident when comparing different geological maps of the Agri Valley area [3–6].

Although the Albidona and the Gorgoglione formations show similar lithological characteristics (in both the cases these formations consist of alternating turbiditic sandstone/conglomerates and clays) their correct attribution has significant geological, hydrogeological and environmental consequences. In addition, the correct mapping and the detailed definition of the boundaries (stratigraphic or tectonic) between the two aforementioned formations and a more precise definition of their age is also helpful for providing information on the evolution of the southern Apennine thrust and fold belt.

In this paper we show the main results of detailed geological mapping, accompanied by new stratigraphic and structural reconstructions, carried out in the Costa Molina—Tempa del Vento—La Rossa—Monte dell'Agresto area, located in the southeastern sector of the Agri Valley (Figure 2a,b). The study has been supported by new biostratigraphic data, performed on marker stratigraphic horizons, that have been used to precisely discriminate the Albidona and Gorgoglione formations. In addition, the new stratigraphic data provided evidence for important tectonic structures that have been recognized for the first time in the area.

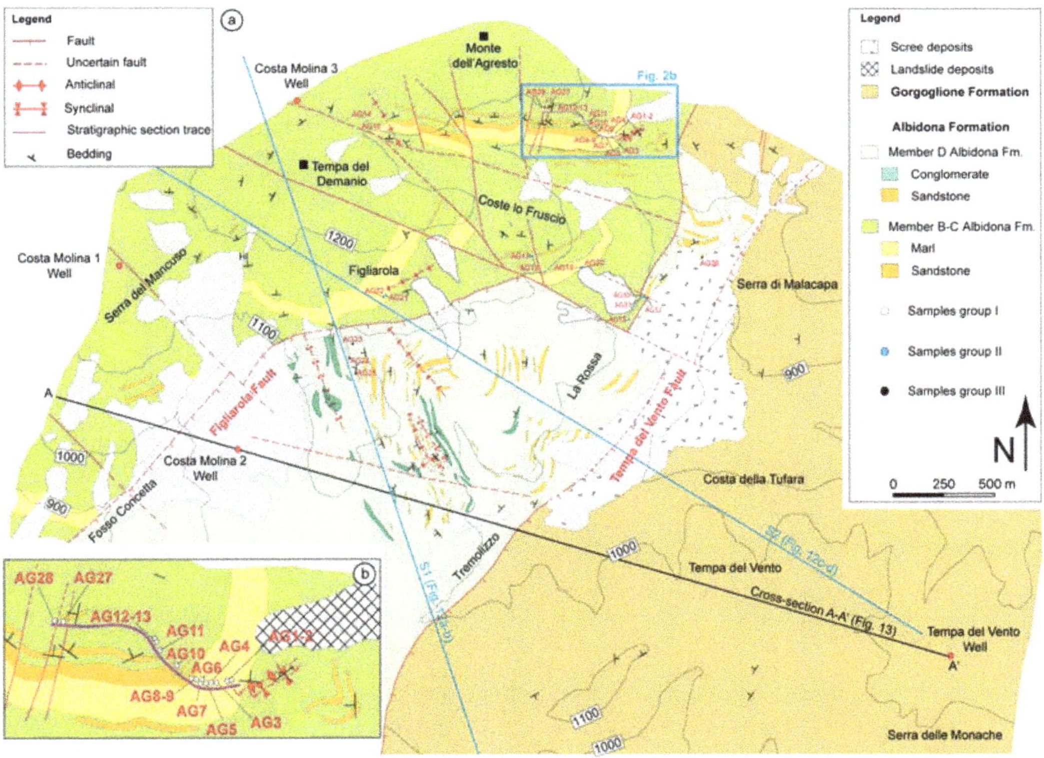

Figure 2. (**a**) Geological map of the Costa Molina—Monte dell'Agresto-Tempa del Vento area, showing (**b**) the location of the studied stratigraphic section, collected samples, seismic lines and geological cross section.

2. Geological Setting

The Agri Valley is a Quaternary tectonic depression, filled up with continental deposits, located in the axial sector of the southern Apennine thrust belt [7,8]. Its origin is mainly connected to the activity of Pliocene to Quaternary, NW-SE oriented, strike-slip and normal faults, which started to develop when thrusting was still active in the frontal sector of the belt [9,10]. In particular, in the southern Apennines extensional tectonics strictly reflects the eastward roll-back and crustal delamination of the of the Apulian slab, associated with the Tyrrhenian backarc basin opening [11–14]. Based on the available field and subsurface data, the Agri Valley is considered a Quaternary graben [15] controlled by two major bounding faults named, respectively, Eastern Agri Fault System (EAFS) and Monti della Maddalena fault system (MMFS) [16]. Active tectonics in the area is testified by recent continental deposits displaced by normal faults [7] and by the intense seismicity currently characterizing the Agri Valley area [17,18].

The Agri Valley graben developed on a pre-Pliocene substratum formed by a series of stacked allochthonous units forming the axial sector of the southern Apennine thrust belt. In the study sector, the architecture of the mountain belt can be observed on outcrops in the shoulders of the Agri Valley graben, or reconstructed by means of the abundant subsurface data (seismic lines and wells) acquired for hydrocarbon exploration [19]. The geometrically highest allochthonous unit is represented by the Liguride complex, consisting of minor fragments of oceanic crust, pertaining to Ligurian part of the Alpine Tethys realm [20], and the related sedimentary cover of Mesozoic to Cenozoic age [4]. This complex derives from a late Cretaceous–early Miocene accretionary wedge, formed on top of the NW-dipping subduction of the Ligurian Tethys [21,22]. All the underlying tectonic units pertain to the paleomargin of the African plate and are represented by the Apennine Platform carbonates, tectonically superimposed onto the deep-water deposits of the Lagonegro Basin, and the Apulian Platform carbonates that represent the lowest structural element of the southern Apennine tectonic pile. Emplacement of the allochthonous units took place during the middle to late Miocene and was accommodated by major low-angle thrusts, bounding at the base each tectonic unit [23,24]. However, severe internal deformation can be also recognized, as testified by folding, thrusting and local development of cleavage fabrics within the clay-rich intervals. This deformation style particularly well outlined in the Lagonegro Units, locally consisting of thick antiformal stacks, as documented by hydrocarbon well data [2]. The simple piling relationship between individual allochthonous units is frequently made more complex by out of sequence thrusting processes, produced during the latest deformation events of late Pliocene-early Pleistocene age [25,26].

The northeastward propagation of thrusting during formation of the Southern Apennines is recorded by a series of thrust-sheet top basins that rejuvenate progressively toward NE, according to the emplacement age of the allochthonous units. Typical thrust-sheet top deposits in the Agri Valley area, as indicated by previous literature [3], are represented by the Albidona and Gorgoglione formations. In particular, in the study area (Figure 2a) the Albidona Formation is stratigraphically covered by the Gorgoglione Formation by means a marked angular unconformity [2]. The latter formation has been deposited in two major NW-trending sub-basins, whose location and evolution were controlled by thrusting and folding within the underlying allochthonous units [27]. The study area corresponds to the western sub-basin, where the Gorgoglione Formation appears less involved by later thrusting and consequent tectonic load with respect to the eastern sub-basin [28,29]. According to the literature [30], the basal unconformity of the Gorgoglione Formation on the Albidona Formation has been intercepted by hydrocarbon wells drilled in the study area (i.e., the Costa Molina2 well). However, the same contact is less recognizable in the field because of striking lithological similarities between the Gorgoglione and Albidona formations. Although few doubts exist about the stratigraphic relationships between the two considered units, different interpretations still persist about the age and the significance of the Albidona Formation. The chronostratigraphic attribution of the Albidona Formation, the problems in discriminating it from the Gorgoglione Formation and the related tectonic and geodynamic implications will be discussed in the following section.

3. The Age and Significance of the Albidona Formation

The Albidona Formation mainly consists of siliciclastic turbidite, outlined by sandstone and conglomerate beds, alternating with intervals of shales, marls and silty marls, with a thickness of about 2000 m in the type locality of Albidona in Northern Calabria [31–33]. A typical feature is represented by the occurrence of about 50 m thick intervals of nearly homogeneous whitish marly limestones and marls, particularly frequent in the lower-intermediate part of the succession. The presence of both siciliclastic and calciclastic intervals suggests that sediments were supplied both from an active margin undergoing contractional deformation and from the marginal part of a carbonate platform [34]. According to Colella and Zuffa [35], the siliciclastic material as well as olistoliths of ophiolitic rocks [36] were fed by the Liguride accretionary wedge, located to the west, whereas the carbonate debris was supplied by the western marginal portion of the Apennine Platform, located to the east. Accordingly, the thick marly intervals were interpreted as carbonate turbidite megabeds [35]. The same intervals have been compared by Baruffini et al. ii [33] to the "homogenites" recognized in the deep-sea Holocene successions of the Mediterranean Sea [37], interpreted as tsunami-derived megaturbidites [38].

The age of the Albidona Formation is controversial. Selli [31], who described for the first time the formation, provided a Langhian age. On the other hand, several authors [39–44] obtained older Eocene ages by analysing planktonic foraminifera both in the type locality and in other sections of the southern Apennines. However, similar studies on the Albidona type section allowed to obtain Oligocene-early Burdigalian ages [32]. Younger ages (early–middle Burdigalian) were provided by Bonardi et al. ii [45] from the analysis of nannofossils in five samples collected in the type locality and in the Agri Valley. Based on a detailed analysis of calcareous nannofossils and palynomorphs from the Albidona type section, Baruffini et al. ii [33] reported again Eocene ages, similar to those already documented in the older studies and discussed also the Miocene ages indicated by the previous authors. In particular, they provided an early Ypresian to early Priabonian age and recognized a wide *hiatus* encompassing the early Lutetian—early Bartonian time span. Finally, sheets 505 Moliterno and 506 Sant'Arcangelo of the 1:50,000 official Italian geological map by ISPRA [5,6] indicated an early Miocene Age, based on data collected outside of the mapped areas. In the Trebisacce 1:50,000 sheet [46], which comprises the type-area of the Albidona Formation, an early Miocene age has been recognized, but only in the uppermost interval of the succession.

Difficulties in providing a consistent age of the Albidona Formation, combined with the paucity of key outcrops, have important consequences on the understanding of its paleogeographic significance as well as on the interpretation of its tectonic or stratigraphic relationships with the underlying units. According to a first interpretation [45,47], the basal contact of the Albidona Formation on the Apennine Platform and the Lagonegro Basin units, exposed in the southern sector of the high Agri Valley (Figure 1), has been indicated as an unconformity. An alternative interpretation [2] considers the same contact as a thrust surface and the Albidona Formation as deposited in a thrust-sheet top basin formed during the emplacement of the Liguride complex. Vezzani et al. ii [48] proposed a similar interpretation and considered the Albidona Formation as deposited in a thrust-sheet top basin that seals contractional structures affecting the Liguride accretionary wedge.

The first interpretation implies that at the time of deposition of the Albidona Formation the African paleomargin was already involved in the Apennine deformation. The second interpretation assumes that the Albidona thrust-sheet top basin formed on the Liguride accretionary wedge when the African paleomargin was still unaffected by contractional deformation. As the involvement of the African paleomargin in the Apennine chain took place during the early Miocene, according to the first deformation recorded in the Apennine

Platform and the Lagonegro Basin units [49,50], a more precise constrain on the age of the Albidona Formation would greatly help in solving this controversy and in providing new hints on the tectonic evolution of the southern Apennines. At this aim we analysed the stratigraphic organization of the Albidona Formation at Monte dell'Agresto, which is located along the southeastern side of the high Agri Valley.

4. Data and Methods

Detailed geological mapping at a 1:5000 scale of a 12.5 km^2 wide area, including the Costa Molina—Tempa del Vento—La Rossa—M. dell'Agresto localities, was performed in order to better understand the stratigraphy of the Albidona Formation and to analyse the structural setting of the area (Figure 2a). Mapping was focused on the identification of key stratigraphic horizons that have been correlated throughout the study area. After geological mapping, the most continuous and representative stratigraphic section (Figures 2b and 3) of the Albidona Formation, located at Coste dell'Agresto locality, was measured and sampled for biostratigraphic analysis on planktonic foraminifera and nannofossils. Other samples have been collected in significant outcrops of the area in order to better identify the age of the key horizons and to provide biostratigraphic constraints on some stratigraphic intervals of the Albidona Formation that were not represented in the measured stratigraphic section.

Analysis of calcareous nannofossils has been performed by means of standard smear slides prepared in the ENI laboratories of Bolgiano (San Donato Milanese). Semi-quantitative analyses of the assemblages were performed on a Zeiss Axioplan optical microscope. Okada and Bukry [51], Martini [52], Perch-Nielsen [53], Bown [54], Gradstein et al. ii [55], were used as standard references for the stratigraphic interpretation of nannofossil distribution.

Structural analysis has been focused on the description of fold geometry in the Albidona Formation and the documentation of the main fault zones. Large-scale folds affecting the Albidona Formation have been mapped by distinguishing the upright and overturned limbs and by identifying the trace of the fold axial surfaces. Overturned beds have been easily detected by means of classical way-up criteria used for turbidites. Brittle deformation has been analysed at Coste dell'Agresto locality, where the orientation of minor fault planes and fractures has been measured within one major fault zone of the study area. The number of fractures per unit length in the damage zones, defined as fracture intensity or P10 [56,57], has been estimated by means of linear scanlines.

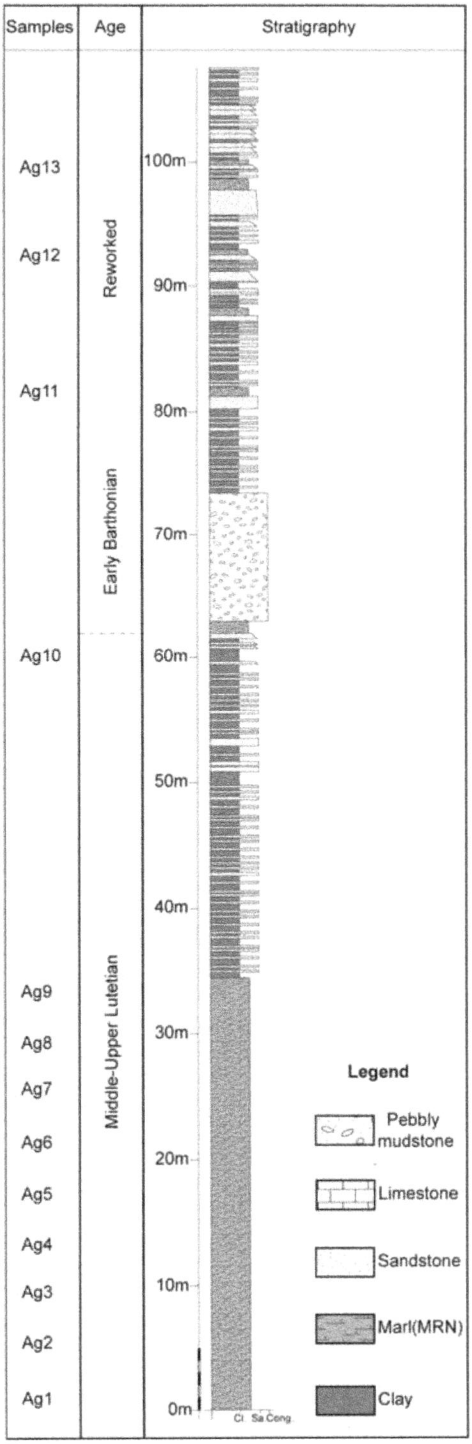

Figure 3. Stratigraphic section of the member B–C (Albidona Formation), measured in the Coste dell'Agresto locality.

5. Stratigraphy of the Monte Dell'agresto Area

We will describe the stratigraphic successions exposed at Monte dell'Agresto area (Figure 2a,b and Figure 3) adopting, for the Albidona Formation, the same stratigraphic divisions proposed by Baruffini et al. ii [33], defined in the Albidona type locality. The reason for this choice is the striking stratigraphical and chronostratigraphical similarities of the studied successions with the section exposed in the type locality. Baruffini et al. ii [33] identified four members in the Albidona formation: (i) member A, consisting of a turbiditic system made by coarse-grained sandstones and microconglomerates. Rare marly intervals locally alternate with the main lithologies; (ii) member B, made up of thick marly intervals (up to tens of metres), alternating with turbiditic sandstone beds and clay intervals. Thick conglomerate levels also occur; (iii) member C, consisting of marly clays and coarse-grained sandstones, with rare calciclastic levels; (iv) member D, mainly composed of coarse-grained turbiditic sandstones and conglomerates, with thin and rare calciclastic intervals.

In the Monte dell'Agresto area we recognized three of the four members described by Baruffini et al. ii [33]. However, the distinction between members B and C is not so straightforward as in the type-area; therefore, we decided to merge the two members together to form the member B–C. On the other hand, member D is easily distinguishable from the underlying members. Member A is not exposed in the study area.

5.1. Albidona Formation, Member B–C

The stratigraphic organization of the member B–C is visible in a natural section exposed along the southeastern slope of Monte dell'Agresto (Figure 3). Member B–C mainly consists of siliciclastic turbidites locally alternating with marly and calcareous intervals. The most common lithofacies, forming the background of the member B–C, is represented by alternating thin-bedded turbiditic sandstones and clays (Figures 3 and 4a).

Sandstones mainly consist of medium to fine—grained turbidites forming cm to dm-thick beds. They show marked erosional basal surfaces and sedimentary structures as well as the Ta-c intervals of the Bouma sequence [58]. The clays commonly form greyish to greenish m-thick, thinly laminated packages. Structureless sandstones in m-thick layers stand out at different stratigraphic heights within background sediments (Figure 4b). Often, the observed thickness of the sandstone beds is the result of amalgamation processes between different strata. Structureless sandstones commonly show lens-shaped geometry, marked erosional basal surfaces and clay chips, ripped up from the underlying clays. Further intercalations within the background deposits of the member B–C consist in three distinctive sedimentary bodies (Figures 3 and 4) that, for their uniqueness, represent very useful stratigraphic markers. These bodies form a triplet that can be traced for many kilometres throughout the study area.

The oldest is represented by a whitish marly interval, affected by cleavage, joints and calcite veins (Figure 4c), showing an anomalous maximum thickness of about 40 m. This interval represents not only an easily recognizable key bed but it also useful for biostratigraphic determinations, given its abundant faunal content (see the next section). Marly intervals occurring within the Albidona Formation have been interpreted as megaturbidites [35] or as "homogenites", possibly triggered by tsunami events [38].

Figure 4. Photographs illustrating peculiar stratigraphic intervals within member B–C and D of the Albidona Formation. (**a**) Thin-bedded turbidites and clays representing the background sedimentation of member B–C of the Albidona Formation; (**b**) Structureless sandstone intervals; (**c**) Marls affected by cleavage and calcite veins; (**d**) Pebbly mudstone exposed at Monte dell'Agresto. Note the occurrence of scattered basalt clasts within the clayey matrix. In the box, a close-up view of pillow lava fragments; (**e**) Pebbly sandstone. Scattered clasts mainly occur in the laminated lower portion.

A second key level, located about 30 m above the described marls, is represented by an about 10 m-thick pebbly mudstone (Figures 3 and 4d), consisting of angular to sub-rounded pebbles and boulders dispersed within a greenish to greyish silty clay. Clasts consist of magmatic, metamorphic and sedimentary rocks. Magmatic rocks are mainly represented by white and pink granites, porphyries, microgabbros and black to green basalts.

Locally, fragments of pillow lavas (Figure 4d), with minor inter-pillow sediments, have been recognized. Metamorphic clasts are commonly represented by phyllites, schists and paragneisses, possibly overprinted by contact metamorphism. Sedimentary clasts are represented by rare Paleocene-Eocene carbonate platform limestones, greenish and reddish cherts and sandstones. Microfacies of carbonate clasts, indicating a reefal environment, is not consistent with the typical Eocene limestones of the Apennine Platform, represented by the shelf-lagoonal facies of the Trentinara Formation [59]. In general, clasts composition is consistent with the provenance from the upper part of a continental crust, where phyllites and schists are intruded by granitoids, covered by porphyres and, finally, by platform carbonates. Moreover, the widespread occurrence of pillow lava fragments and microgabbro implies the presence of a second sedimentary source, consisting of an obducted ophiolitic complex. This is further supported by the concomitant occurrence of inter-pillow limestones and greenish to reddish cherts that typically occur in the upper part of an ophiolite section. Due to bad outcrop conditions pebbly mudstone did not show evident sedimentological characteristics and clear relationships with the surrounding strata. A possible interpretation, due to its structureless appearance, with clasts floating in a dominant clayey matrix, is that they result from the emplacement of a debris flow. This stratigraphic level might be correlated with similar intervals containing blocks of granitoids and mafic rocks exposed in various localities along the southwestern side of the Agri Valley [36,45,60].

The third marker bed is a pebbly sandstone (Figure 4e), which is located some metres above the pebbly mudstone. It consists of a 2–3 m-thick coarse-grained sandstone body showing well-rounded, a few centimetres in diameter, crystalline and metamorphic clasts distributed along the lower portion of the bed. Plane-parallel and cross-lamination typically occur in the basal portion of the bed, where the clasts are mostly present. Toward the top, clast frequency and lamination progressively decrease, and the sandstone become structureless or show a faint normal gradation. It shows a lens-shaped geometry and progressively pinches out eastward. In the opposite direction it can be followed for some kilometres, forming an easily recognizable marker bed. We interpreted the pebbly sandstone as the result of a turbidite flow or a debris flow evolving toward a turbidite flow.

5.2. Albidona Formation, Member D

Member D differs from member B–C mainly for two reasons: (i) fine to medium grained sandstones are replaced by coarse-grained sandstones, micro-conglomerates and conglomerates (Figure 5a,b); (ii) Marly intervals are thinner and commonly show higher clay content.

Figure 5. (**a**) Typical outcrops of the member D of the Albidona Formation represented by thick conglomerate/micro-conglomerate beds; (**b**) Close-up view of the conglomerates.

The conglomerate/micro-conglomerate beds show thicknesses ranging from few centimetres to some metres and can be followed laterally for hundreds of metres. They commonly show a lens-shaped geometry, dome-shaped tops and marked erosional basal surfaces. Conglomerate/micro-conglomerate beds display a matrix-supported texture with clast size ranging from few centimetres up to the decimetre. Clast roundness can vary from scarcely rounded to well rounded. Clast composition includes sedimentary, magmatic and metamorphic rocks. Among metamorphic rocks, well-rounded clasts of quartz veins and subrounded phyllite fragments are the most frequent, whereas magmatic rocks mainly consist of granite clasts. Unlike the member B–C, apparently they do not include ophiolite-derived debris.

Marls of member D show general features similar to those described for the underlying member B–C; however, they are characterized by a greater amount of clay and, consequently by paucity in faunal content. Marls are commonly organized in laterally continuous beds showing thicknesses ranging from some tens of centimetres to some meters and are generally affected by an intense cleavage, commonly parallel to the sedimentary layering.

5.3. Gorgoglione Formation

The Gorgoglione Formation (late Burdigalian—Tortonian according to [61]) mainly consists of arenaceous turbidites, variably alternated with matrix-supported conglomerate/microconglomerate layers and clay-rich intervals. In the study area arenaceous turbidites consist of 1–2 m thick coarse-grained graded sandstone, containing sparse clasts at the base and laminated finer-grained sandstone on top. Conglomerates generally consist of rounded clasts derived from sedimentary, metamorphic and igneous rocks, these latter showing a granite to granodiorite composition, within abundant matrix. Sandstones and conglomerates are generally characterized by a yellow color, may contain plant remains and are frequently organized in 10 to 50 m thick intervals. Minor cm- to dm-thick clayey intervals alternate with the previous lithologies. Frequently, thick sandstone and conglomerate intervals result from the amalgamation of individual m-thick turbidite levels. These coarse-grained intervals are laterally discontinuous and occur within a typical turbiditic sequence, consisting of dm-thick layers of graded sandstone intercalated within dominant clay or silty clay.

Interestingly, the Gorgoglione Formation never contains marly intervals, and this represented one of the main criteria that we used for distinguishing the Gorgoglione Formation from both B–C and D members of the Albidona Formation. Another distinctive feature is the exclusive presence of pebbles consisting of granitoids and minor metamorphic rocks in the coarse-grained facies, with the absence of ophiolitic material.

6. Biostratigraphy

In the Monte dell'Agresto area 31 samples were collected for the biostratigraphic analysis of the Albidona Formation succession. Samples have been mainly collected in marly intervals located at different stratigraphic heights, within both the B–C and the D members. They have been organized in three main groups representing, respectively, the intermediate and lower-intermediate intervals of the member B–C (Groups I and II; Tables 1 and 2), and the member D (Group III; Table 3). The samples are characterized by the significant occurrence of Cretaceous, Paleocene reworked forms. In addition, the constant occurrence of Eocene forms, not older than the Lutetian, has been ascertained by integrating calcareous nannoplankton and planktonic foraminifera. Although forms with a long-range distribution (i.e., *Chiloguembelina cubensis*, LO top Rupelian, *Sphenolithus praedistentus*, LO intra-Chattian) sometimes occur, we never recognized taxa with a short range distribution or a FO/FAD showing ages younger than lower Priabonian (i.e., *Cribrocentrum erbae*). Therefore, based on comparison between marker planktonic foraminifera and calcareous nannoplankton associations (Figure 6) we attributed of a Lutetian age for the member B–C and a Barthonian/early Priabonian age for the member D.

Table 1. Group I.

Samples	Markers (Calcareous Nannoplankton)	Markers (Plantktonic Foraminifera)
AG 28	*Sphenolithus radians*, *Girgisia gammation*, *Toweius pertusus*, *Discoaster barbadiensis* Sample biostratigraphic range: NP11–14	*Acarinina* cf. *topilensis*, *Chiloguembelina* sp., *Turborotalia* sp. Sample biostratigraphic range: E10–E11
AG 27	*S. radians*, *G. gammation*, *T. pertusus*, *D. barbadiensis* Sample biostratigraphic range: NP11–14	*Acarinina* cf. *boudreauxi*, *Morozovelloides* sp.? *Chi-loguembelina* sp., *Subbotina* spp., *Globigerinatheka* sp., *Morozovella* cf. *caucasica*, Globotruncanidae (rew.) Sample biostratigraphic range: E10–E11
AG 15	*S. radians*, *G. gammation*, *Toweius callosus*, *T. pertusus* Sample biostratigraphic range: NP11–14	Barren
AG 14	*S. radians*, *G. gammation*, *T. callosus*, *T. pertusus* Sample biostratigraphic range: NP11–14	Barren
AG 13	*S. radians*, *G. gammation*, *Sphenolithus orphanknollensis*, *Nannotetrina* sp., *D. barbadiensis* Sample biostratigraphic range: NP12–15	Barren
AG 12	*S. radians*, *G. gammation*, *S. orphanknollensis*, *Nannotetrina* sp., *D. barbadiensis* Sample biostratigraphic range: NP12–15	Very rare indeterminated planktonic foraminifera
AG 11	*S. radians*, *G. gammation*, *S. orphanknollensis*, *Nannotetrina* sp., *D. barbadiensis* Sample Biostratigraphic range: NP12–15	Barren
AG 10	*S. radians*, *G. gammation*, *S. orphanknollensis*, *Nannotetrina* sp., *D. barbadiensis* Sample biostratigraphic range: NP12–15	Rare indeterminated planktonic foraminifera
AG 9	*Sphenolithus predistentus*, *Helicosphaera lophota*, *G. gammation* Sample Biostratigraphic range: NP16	Barren
AG 8	*generic Eocene association*	*Acarinina* spp. and small indeterminated planktonic foraminifera
AG 7	*Sphenolithus spiniger*, *Sphenolithus furcatolithoides* Sample biostratigraphic range: NP15/16	*Morozovella aequa* (rew.), *Acarinina soldadoensis* (rew.) Sample biostratigraphic range: not younger than E7
AG 6	*S. radians*, *G. gammation*, *S. furcatolithoides*, *D. barbadiensis* Sample biostratigraphic range: NP15	*Acarinina bullbrooki*, *Acarinina* spp. Sample biostratigraphic range: E7–E11
AG 5	*S. radians*, *G. gammation*, *T. pertusus*, *T. callosus* Sample biostratigraphic range: NP11–14	*Acarenina* cf. *soldadoensis* (rew), *Morozovella aragonensis* Sample biostratigraphic range: E7–E11
AG 4	*S. radians*, *G. gammation*, *T. pertusus*, *T. callosus* Sample biostratigraphic range: NP11–14	*Acarinina* sp., *Morozovella* sp. Sample biostratigraphic range: not younger than E9
AG 3	*S. radians*, *G. gammation*, *T. pertusus*, *T. callosus* Sample biostratigraphic range: NP11–14	*Acarinina bullbrooki*, *Subbotina* spp., *Globanomalina* cf. *compressa* (rew.) Sample biostratigraphic range: E7–E11
AG 2	*S. radians*, *G. gammation*, *T. pertusus*, *T. callosus* Sample biostratigraphic range: NP11–14	*Acarinina* sp., *Morozovella* sp. Sample biostratigraphic range: not younger than E9
AG 1	*S. radians*, *G. gammation*, *T. pertusus*, *T. callosus* Sample biostratigraphic range: NP11–14	*Acarinina bullbrooki*, *Subbotina* spp., *Morozovella* spp. Sample biostratigraphic range: E7–E11

Table 2. Group II.

Samples	Markers (Calcareous Nannoplankton)	Markers (Plantktonic Foraminifera)
AG 18	*S. spiniger* Sample biostratigraphic range: NP16	*Acarinina* cf. *boudreauxi* Sample biostratigraphic range: E7–E9
AG 17	*S. spiniger* Sample biostratigraphic range: NP16	*Acarinina* spp., *A.* cf. *pentacamerata* (rew.), *A. soldadoensis* (rew.), *Chiloguembelina* cf. *cubensis* Sample biostratigraphic range: E10 or younger
AG 22	*Discoaster salisburgensis, S. radians, G. gammation* Sample biostratigraphic range: NP11–12	*Acarinina* spp., *A.* cf. *bullbrooki, A. soldadoensis* (rew), *Globanomalina* sp. (rew.) Sample biostratigraphic range: E7–E11
AG 21	*D. salisburgensis, S. radians, G. gammation* Sample Biostratigraphic range: NP11–12	*Acarinina* spp., *A.* cf. *boudreauxi, Acarinina soldadoensis* (rew), *Igorina* sp. (rew.), *Morozovella aequa* (rew.) Sample biostratigraphic range: E7–E11
AG 20	*S. radians, G. gammation, T. callosus, T. pertusus* Sample biostratigraphic range: NP11–14	*Acarinina* spp., *A.* cf. *boudreauxi, Acarinina soldadoensis* (rew), *Chiloguembelina crinita, Morozovelloides* sp.? *Subbotina* spp. Sample biostratigraphic range: E7–E13
AG 19	*S. radians, G. gammation, T. callosus, T. pertusus* Sample biostratigraphic range: NP11–14	*Acarinina* spp., *A.* cf. *boudreauxi, A. pentacamerata* (rew), *Chiloguembelina* spp., *Igorina pusilla* (rew.) Sample biostratigraphic range: E7–E9
AG 30	*S. radians, G. gammation, Campylosphaera dela, Discoaster saipanensis* *D. barbadiensis* Sample biostratigraphic range: NP14	Very rare small planktonic foraminifera fragments
AG 31	*S. radians, G. gammation, C. dela, D. saipanensis, D. barbadiensis* Sample biostratigraphic range: NP14	No sample
AG 32	*S. radians, G. gammation, T. callosus, T. pertusus* Sample biostratigraphic range: NP11–14	No sample
AG 33	*S. radians, G. gammation, T. callosus, T. pertusus* Sample biostratigraphic range: NP11–14	*Acarinina bullbrooki, A.* cf. *aspensis, Globigerinatheka* sp., Chiloguembelina sp., *Morozovella* spp., *M. aequa* (rew.), *Globanomalina compressa* (rew.), *Igorina albeari* (rew.) Sample biostratigraphic range: E7–E11

Table 3. Group III.

Samples	Markers (Calcareous Nannoplankton)	Markers (Plantktonic Foraminifera)
AG 26	*Cribrocentrum reticulatum, Dictyococcites bisectus, Reticulofenestra umbilicus, Sphenolithus obtusus, Cribrocentrum erbae* Sample biostratigraphic range: NP17	Barren
AG 25	*Cr. reticulatum, D. bisectus, R. umbilicus, S. obtusus, Cr. erbae, Blackites* sp. Sample biostratigraphic range: NP17	Barren
AG 24	*Cr. reticulatum, D. bisectus, R. umbilicus, S. obtusus, Cr. erbae, Blackites* sp. Sample biostratigraphic range: NP17	Barren
AG 23	*Cr. reticulatum, D. bisectus, R. umbilicus, S. obtusus, Cr. erbae, Blackites* sp. Sample biostratigraphic range: NP17	Barren

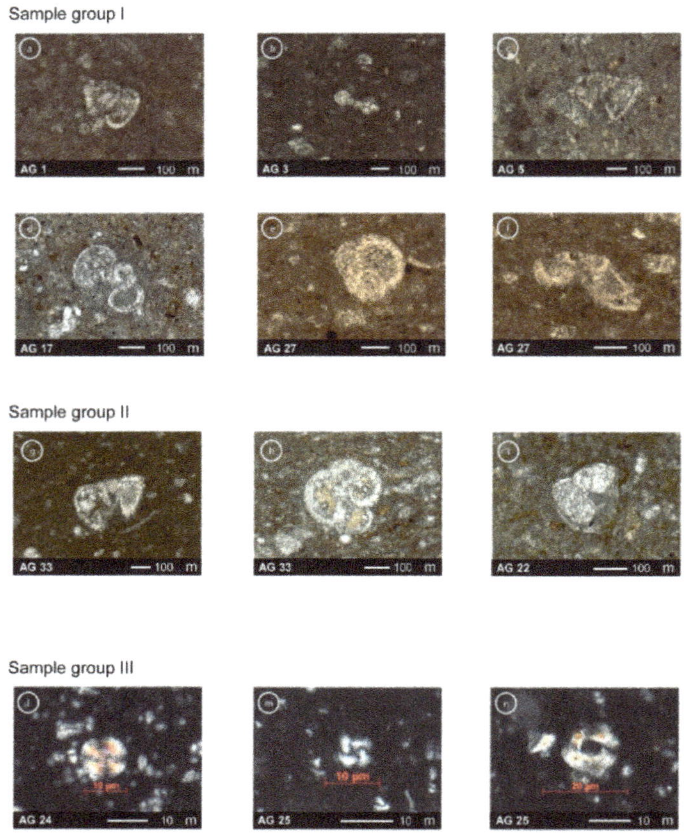

Figure 6. Sample group I: (**a**) *Acarinina bullbrooki;* (**b**) *Globanomalina* cf. *compressa* reworked; (**c**) *Morozovella aragonensis;* (**d**) *Acarinina soldadoensis* reworked; (**e**) *Globigerinatheka* sp.; (**f**) Globotruncanids (*Marginotruncana* sp.?) reworked; Sample group II: (**g**) *Acarinina bullbrooki;* (**h**) *Globigerinatheka* sp.; (**i**) *Turborotalia* sp. (juvenile); Sample group III: (**l**) *Dictyococcites bisectus;* (**m**) *Cribrocentrum reticulatum;* (**n**) *Reticulofenestra umbilicus.*

6.1. Group I

This group consists of a composite succession representative of the member B–C of the Albidona Formation, exposed northern sector of the study area (Figure 2). It includes samples from AG1 to AG13, collected at Coste dell'Agresto along the measured stratigraphic section (Figure 3), and samples AG14-15 and AG27-28 from minor marly intervals in the same area. In particular, the previously described marly key level (MRN interval in Figure 3) has been sampled at regular distances of 4–5 m (samples AG1-9; Figure 3). In the sampled marly intervals (Table 1), calcareous nannoplankton associations are abundant and well preserved. On the contrary planktonic foraminifera associations, analysed in thin sections, are scarcely represented and some of the samples resulted barren. In general, the existing forms are small, well preserved and iso-oriented. In some samples radiolarians are abundant. Information and marker forms for each sample are reported in Table 1. The most representative specimens recognized in this group are illustrated in Figure 6a–f.

Group I encompasses the middle-upper Lutetian (Biozones NP15-16) (see the chronostratigraphic scheme in Figure 7). The first form representative of the nannoplankton Biozone NP15 is *Sphenolithus furcatolithoides*, observed in the sample AG6. Furthermore, *Sphenolithus predistentus*, indicating the transition toward the Biozone NP16, has been recognized in sample AG9. A Lutetian age is also confirmed by the occurrence of planktonic foraminifera as well as *Acarinina* cf. *topilensis*, *Globigerinatheka* sp. recognized in the samples AG27 and AG28.

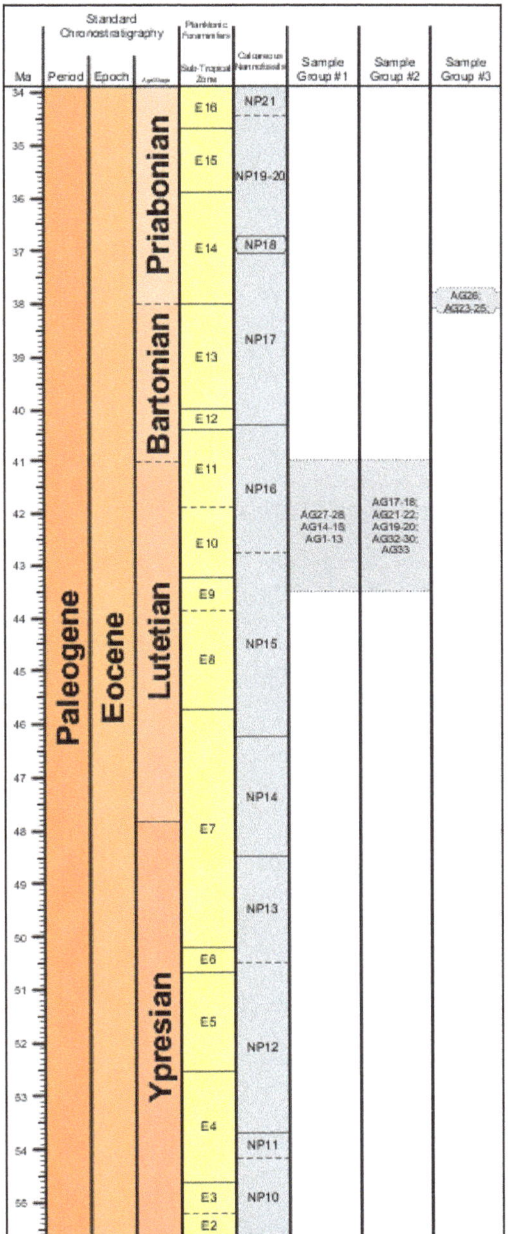

Figure 7. Final interpretative chronostratigraphic scheme reconstructed for the study succession. Reworked Ypresian and older bioevents are often present in samples.

In the other samples the age attribution is less clear since reworked forms are very common. In general, reworked planktonic foraminifera of Cretaceous, Paleocene and late Ypresian (i.e., *Acarinina soldadoensis*, *Morozovella aequa*) ages are present. Large range forms or markers ranging from nannoplankton biozones NP11 and NP24 also occur. In particular, the recognition of *Sphenolithus predistentus*, which ranges from biozone NP16 to biozones NP24 at least provided an age not older than biozone NP16. Although this latter form can easily reach the Oligocene we exclude this younger age because of the lack of a clear Oligocene faunal assemblage.

In more detail, samples AG1-AG5, collected at the base of the MRN interval, show fossil assemblages that do not allow to exclude a late Ypresian age. In fact, the occurrence of *Acarinina bullbrooki* in the sample AG1 provides an age not older than E7 biozone. Accordingly, calcareous nannoplankton associations indicate an age not younger than NP14 biozone. The occurrence of calcareous nannoplankton associations pertaining to the biozone NP15 in the sample AG6 might suggest that the underlying stratigraphic portions can be located in the upper part of the NP14 biozone and therefore in the Lutetian. The occurrence of *Sphenolithus predistensus* (total range NP16-NP24) allows to attribute the sample AG9 at the base of NP16 Biozone. In this sample, the fossil assemblage is also characterized by a group of taxa of lower Eocene affinity such as *Girgisia gammation* (total range NP11-NP14), always observed in the underlying samples. These latter forms could be interpreted as clearly reworked only from the sample AG6, due to the first occurrence of *Sphenolithus furcatulithoides* (total range NP15-16).

The aforementioned biostratigraphic reconstruction might suggest that the 34 m thick MRN interval could represent a condensed section ranging between the upper Ypresian? and the upper Lutetian (NP14-16). However, the MRN interval is included in a thick turbidite succession characterized by high sedimentation rates. Therefore, a prolonged deactivation of the clastic input (6–7 My) during deposition of the MRN interval seems unlikely. Moreover, it has to be noted that samples collected from marly levels overlying the MRN interval show again fossil associations with upper Ypresian—lower Lutetian affinities, similar to those recognized in the samples AG1-AG5. The occurrence of these intervals has been attributed to reworking processes, only occasionally preserving the primary biostratigraphic signal (samples AG 27—AG28). The recurring appearance of layers containing calcareous nannoplankton associations with an upper Ypresian—lower Lutetian (NP11-14) affinity, conflicting with their stratigraphic position in the Agresto succession, rises further doubts about the reliability of the age provided by samples AG1-AG5. For these reasons, the stratigraphic interval included between samples AG1 and AG5 has been assigned to the Lutetian (biozone NP15).

6.2. Group II

Group II consists of a composite succession representative of the member B–C of the Albidona Formation exposed central sector of the study area. It includes samples AG17-18-19-20-21-22-30-32-33 (Figure 2). Samples AG19-20-21-22 have been collected along the MRN marker interval (Figures 2 and 3). The most representative forms are listed in Table 2. Figure 6g–i show the most representative forms of this group.

Consistently with the samples of Group I, a Lutetian age is ascertained for Group II and abundant reworked forms of Cretaceous, Paleocene and Ypresian age have also been recognized. In particular, the reworked nature of early Eocene forms is similar to that discussed for the Group I.

Planktonic foraminifera represented by *Globigerinatheka* spp. recognized in the sample AG33 provided a Lutetian age not older than Biozone E8. Moreover, sample AG17, located above the MNR interval, provided and age not older than Biozone E10 (topmost NP15-NP16), because of the occurrence of *Chiloguembelina* cf. *cubensis*.

Summing up, we indicate an age comprised between the upper part of NP15 and the lower part of NP16 calcareous nannoplankton biozones, corresponding to E9-E10 planktonic foraminifera biozones for Group I and Group II samples (Figure 7). Therefore, the studied stratigraphic interval corresponds to a minor part of the entire Albidona member B–C, which is characterized by high sedimentation rates and a thickness of 1500 m in the type locality [33].

6.3. Group III

This group includes samples AG 23-24-25 (Figure 2), collected in a marly interval of the member D of the Albidona Formation. Additionally, sample AG26 has been attributed to this group, and referred to the same member, because of the strong affinities showed by the floral content with the other samples of Group III. Calcareous nannoplankton associations are abundant and very well preserved (Figure 6l–n). On the contrary, planktonic foraminifera content is almost absent. The occurrence of markers as *Blackites* sp., *Cribrocentrum erbae*, *Cribrocentrum reticulatum*, *Dictyococcites bisectus*, *Reticulofenestra umbilicus*, *Sphenolithus obtusus*, allowed to refer Group III to the topmost Barthonian/lower Priabonian (Biozone NP17) (Figure 7).

7. Structural Setting of the Monte dell'Agresto Area

The detailed stratigraphic reconstruction of the Albidona Formation and the identification of key stratigraphic intervals allowed a better definition of the overall geological features of the study area, which appear substantially upgraded when compared with the previous geological maps [3,5,6]. The most striking characteristics is that most of the previously mapped Gorgoglione Formation turned out to be the Albidona Formation (Figure 2). This information was crucial for identifying major tectonic contacts that separate the two formations in most of the study area. Another important improvement consists in the distinction of members B–C and D of the Albidona Formation, which allowed a more detailed mapping of this thick and sometimes monotonous succession. These advancements allowed the identification of two roughly SW-NE oriented major faults, crossing the entire study area, named, respectively, as the Figliarola and Tempa del Vento faults. These two faults crosscut older contractional structures, represented by folds, which formed during the building of the southern Apennine thrust belt. In the next sections we will describe the main contractional and extensional structures recognized in the study area.

7.1. Contractional Structures

Contractional tectonics in the study area is mainly outlined by the presence of folds at the meso- and macro-scale. Minor thrusts and reverse faults, generally characterized by a limited displacement (up to some metres), also occur. We observed a remarkable difference in the folding style between the members B–C and D. In particular, in the member B–C folds are more commonly open and upright, which is very likely controlled by the occurrence of more than 10 m thick layers, represented by the marly key interval and by thick sandstone beds. On the contrary, folds in the member D show a tighter geometry, related to the predominance of thinner sandstone or microconglomerate layers.

7.1.1. Geometry and Orientation of Folds in the Member B–C

Folds in the member B–C consist of roughly NW-SE oriented anticlines and synclines. Fold orientation has been deduced by plotting the poles of the bedding planes (stereoplot A and B; Figure 8) that, although dispersed, are consistent with a best fit axis dipping of about 45° to 300° N.

Coherent with this orientation is an open NW-trending anticline recognized along the southern side of Coste dell'Agresto locality (Figure 2), which is outlined by a thick sandstone interval and by the marly key bed (Figure 9a). In the same locality, overturned minor folds also affect the member B–C. These structures commonly show axes dipping toward W of 20° (Figure 2) and, due to their orientation and asymmetry, are not consistent with the dominant large-scale NW-trending folds detected within the member B–C. Consequently, these structures probably pertain to different deformation stage.

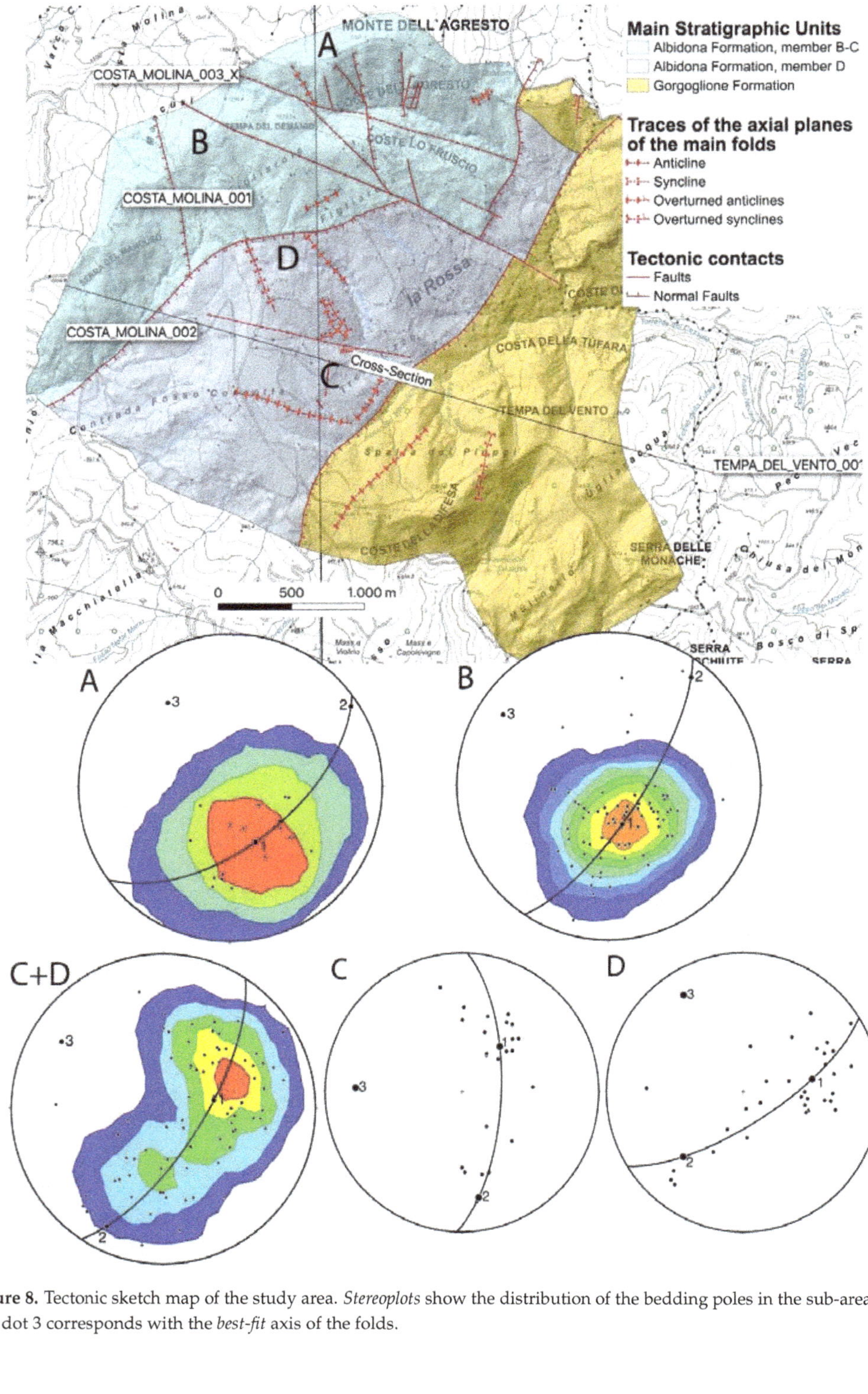

Figure 8. Tectonic sketch map of the study area. *Stereoplots* show the distribution of the bedding poles in the sub-areas **A**–**D**. The dot 3 corresponds with the *best-fit* axis of the folds.

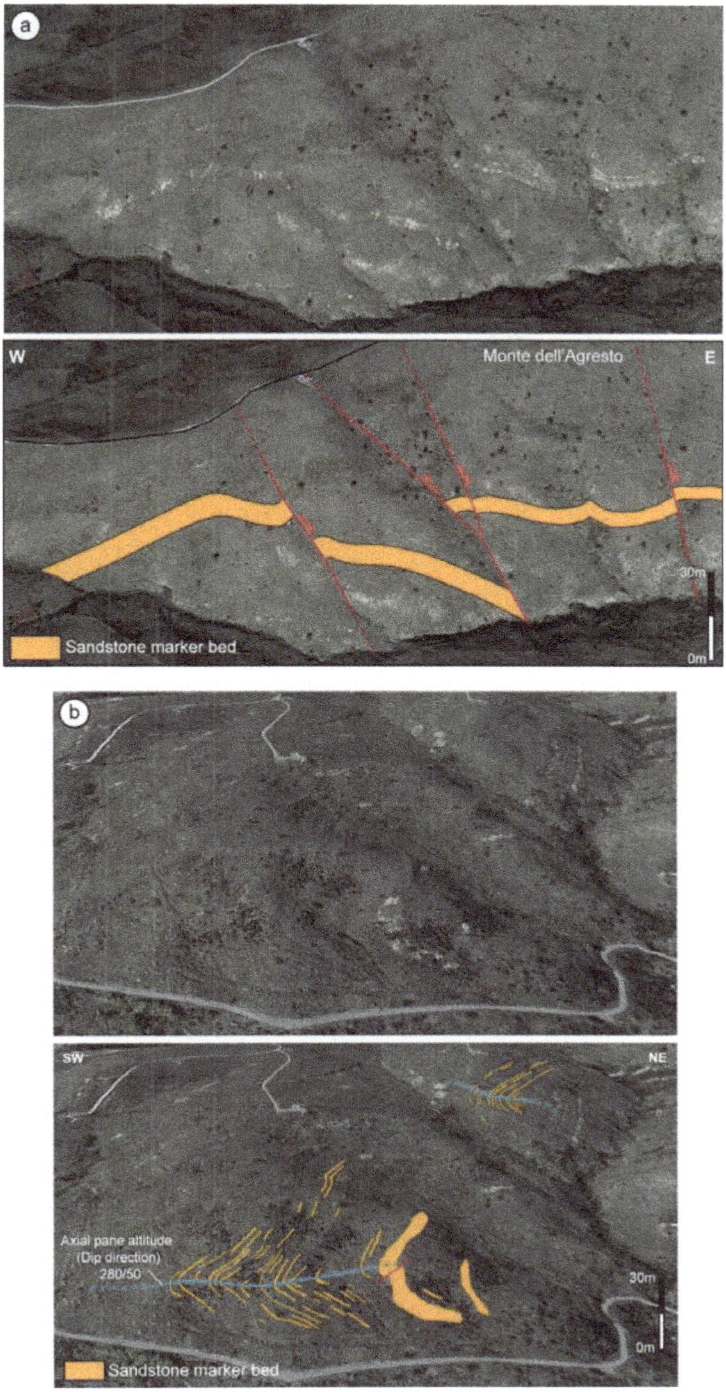

Figure 9. Examples of tectonic structures exposed in the M. dell'Agresto—C.da La Rossa area. (**a**) NW-trending anticline outlined by a thick sandstone bed at M. dell'Agresto; (**b**) Geometry of the E-trending Tremolizzo overturned anticline. Pictures from Google Earth.

7.1.2. Geometry and Orientation of Folds in the Member D

Similar to the previous case, folds in the member D consist of anticlines and synclines showing NW-SE and E-W hinge directions (stereoplot C and D in Figure 8). They generally consist of tight folds, often overturned, affecting interbedded clay, sandstone and microconglomerate. Hinge zones, where exposed, commonly show an angular to sub-angular geometry. A major E-W trending kilometre-scale overturned anticline crops out at Tremolizzo. The fold shows an axis plunging 20° towards N278. In the same locality, the presence of evident and angular fold hinges (Figure 9b) allows to appreciate an axial surface dipping of about 20–30° to the west (N280), consistent with the westward plunge of the fold axis. The same anticline can be mapped eastwards, between Tremolizzo and the Costa Molina, where the north-dipping overturned limb is well exposed (Figure 2). Interestingly, in the Costa Molina area the fold axial surface dips approximately to the east, indicating that the Tremolizzo anticline has been refolded by a gentle NW-trending syncline, generating a type 2 interference structure [62]. The overall geometry of the Tremolizzo anticline, characterized by an overturned southern limb, clearly indicates a southward-directed vergence, not consistent with the general northeastward thrust transport direction displayed by the Southern Apennine belt. Similar E-W trending asymmetric folds are present in other outcrops in the study area, where in some cases display hinge collapse phenomena and layer-parallel shearing.

Summarizing, interference structures in member D document that the Albidona Formation has been affected by at least two folding phases. The first phase (D1) consists of asymmetric and frequently overturned folds with southward transport direction. The second folding phase (D2) is generally characterized by NW-trending open folds (Figure 8; stereoplot D), with local overturned beds.

7.2. Extensional Structures

Three main fault sets, oriented, respectively, NNE-SSW, NW-SE and WNW-ESE have been recognized throughout the study area (Figure 2). Due to the poor outcrop conditions, in most cases fault have been inferred by the offset of marker levels or on the basis of stratigraphic inconsistency. In rarer cases, outcropping, meso-scale fault planes have been observed and measured. Considering the NE-SW set, two major kilometre-scale faults, named, respectively, Tempa del Vento and Figliarola faults, have been detected. Their occurrence has been mainly deduced on the basis of key beds offset and on the geometrical relationships between the mapped stratigraphic units. However, the presence of both faults is locally documented by exposures of the fault zones and is confirmed by the analysis of seismic lines acquired for hydrocarbon exploration in the area.

The Tempa del Vento Fault (Figure 2) is a NNE-SSW oriented and ESE-dipping normal fault, which separates the member D of the Albidona Formation at the footwall from the Gorgoglione Formation at the hangingwall. The main evidence for the presence of the fault comes from the geometrical relationships between the two formations and, particularly, from the attitude of the bedding planes in the Gorgoglione Formation, which dips constantly of 30° to towards Albidona Formation at the footwall. Tilting of the Gorgoglione Formation is consistent with the geometry of the downthrown block of a normal fault and supports the extensional kinematics of the Tempa del Vento Fault. The presence of the Gorgoglione Formation at the footwall of the fault in the northeastern sector of the study area suggests a gradual decrease of the vertical displacement towards the northeast.

The Figliarola Fault consists in two segments, oriented NE-SW and NNE-SSW, respectively, with a general dip towards the SE. This normal fault shows the deposits pertaining to the member B–C of the Albidona Formation at the footwall and the member D of the Albidona Formation plus the overlying Gorgoglione Formation at the hangingwall. The fault consists of a poorly exposed master fault plane and series of closely spaced, mostly antithetic minor faults. It shows a wide damage zone clearly exposed along the eastern side of Monte dell'Agresto. The damage zone affects a 15 m thick sandstone and microconglomerate layer pertaining to the Gorgoglione Formation, located at the hangingwall of the structure. Fracture measurement carried out in this outcrop are showed in Figure 10a,b Three main fracture sets have been recognized, two of them, showing a NNE-SSW strike and dipping, respectively, toward WNW and ESE, form a conjugate fracture system which is consistent with the orientation and the extensional kinematics of the Figliarola Fault. A third set strikes NNW-SSE and dips steeply towards the WSW. A scanline measured orthogonally to the conjugate fracture set, along a 16 m long exposure, allowed to evaluate the fracture intensity (P10) [56,57]. As showed in Figure 10c the fracture intensity progressively decreases in a direction opposite to the master fault, which is located on the on the left-hand side of the histogram, showing a good agreement with the distribution of fractures in damage zones [63,64]. Isolated peaks occurring, respectively, at 8, 12 and 15 m, coincides with the presence subsidiary faults within the damage zone.

Figure 10. Structural data from the damage zone of the Figliarola Fault. (**a**) close up outcrop photograph; (**b**) rose and contour plot diagrams of fracture data measured along the scanline shown in (**a**); (**c**) distribution of the fracture intensity along the scanline shown in (**a**). Master fault is located at the left end side of the histogram.

At the hangingwall of the Figliarola Fault a series of antithetic, meso-scale faults also occur in an outcrop located 400 m East from the master fault (Figure 2). These structures cut through the Gorgoglione Formation deposits which, in this area, consist of thick beds of yellowish sandstone alternating with thin clay intervals (Figure 11a). Faults are oriented 10° N–20° E, with an average dip angle of 60°, and show displacements ranging from few decimetres to a minimum of 3 m and spacing varying form few decimetres to some meters. Fracture intensity within the fault-bounded blocks has been analysed by means of two scanlines (Figure 11b) that have been oriented parallel to the two main sandstone beds. Histograms (Figure 11c) show high values of fracture intensity approaching the fault A, which is characterized by the maximum displacement, and in the central part of the outcrop, where the closely spaced faults C and D bound a narrow sandstone block. Fractures are oriented NNE-SSW in average, which is consistent with the orientation of the minor antithetic faults and of the Figliarola master fault and appear arranged into two conjugate sets, showing dip angles of 45–55° (Figure 11d). Less commonly, a third set showing steeper dip angles can be detected (Figure 11d).

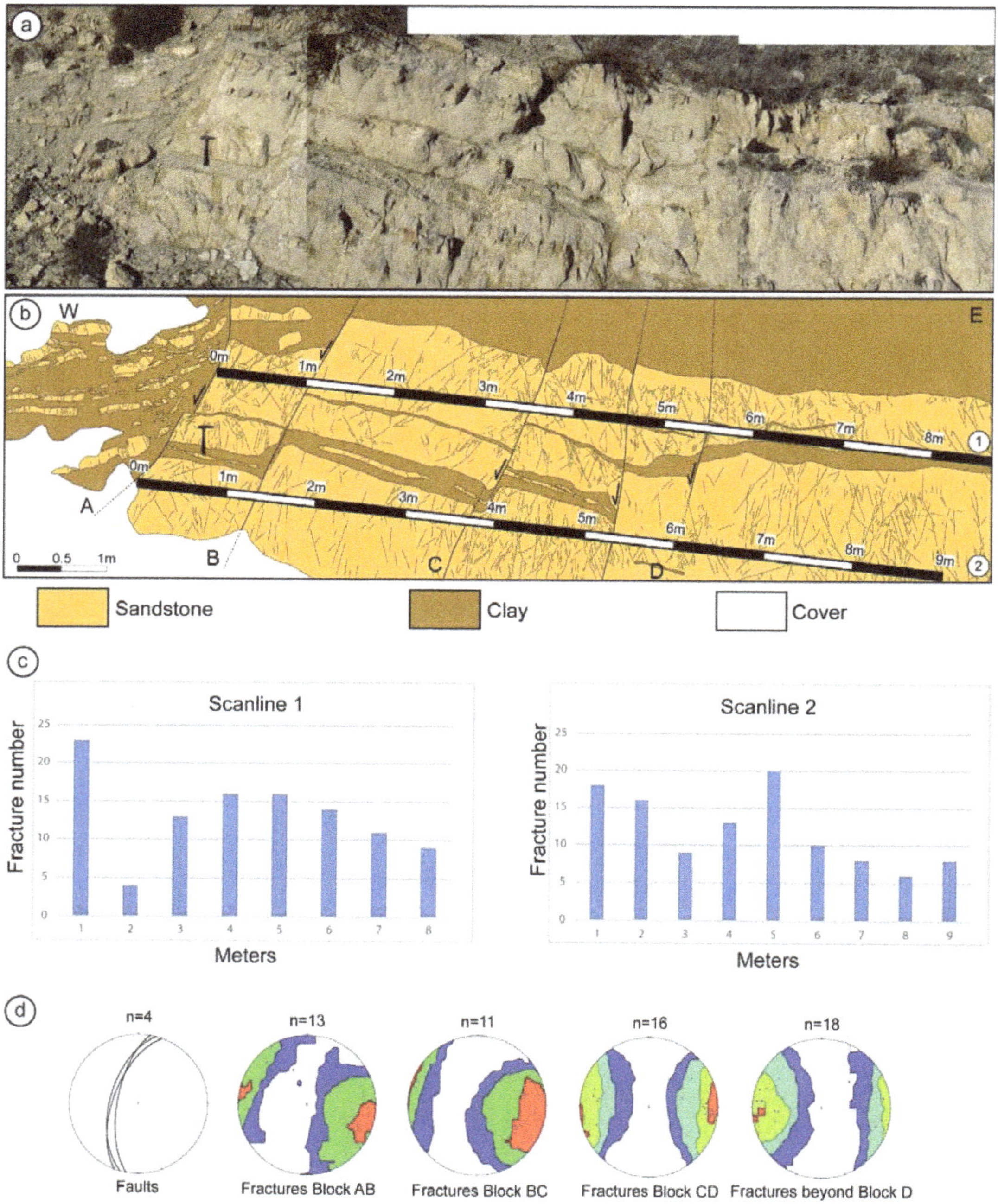

Figure 11. Minor antithetic faults, identified by letters A to D, in the Gorgoglione Formation exposed at the hangingwall of the Figliarola Fault. (**a**) outcrop photograph; (**b**) interpretation showing the position of two scanlines measured along two main sandstone beds; (**c**) diagrams of the fracture intensity measured along the two scanlines; (**d**) stereoplots showing the orientation of fault planes and fractures in each fault-bounded block shown in (**b**).

Seismic Profiles

A series of seismic profiles, acquired during exploration of the Val d'Agri oil field by ENI, provide support on the subsurface geometry of the Tempa del Vento and Figliarola faults. In particular, the geometrical features of the two faults are imaged in seismic lines S1 and S2 (Figure 12; see traces of seismic profiles in Figure 2). The two profiles can be interpreted in light of the stratigraphic data provided by the Costa Molina 2 and Tempa del Vento well logs [30] and VIDEPI web site. In both the seismic profiles the faults tend to join downwards while cutting through the tectonic contact separating the Liguride from the Lagonegro Units (Figure 12).

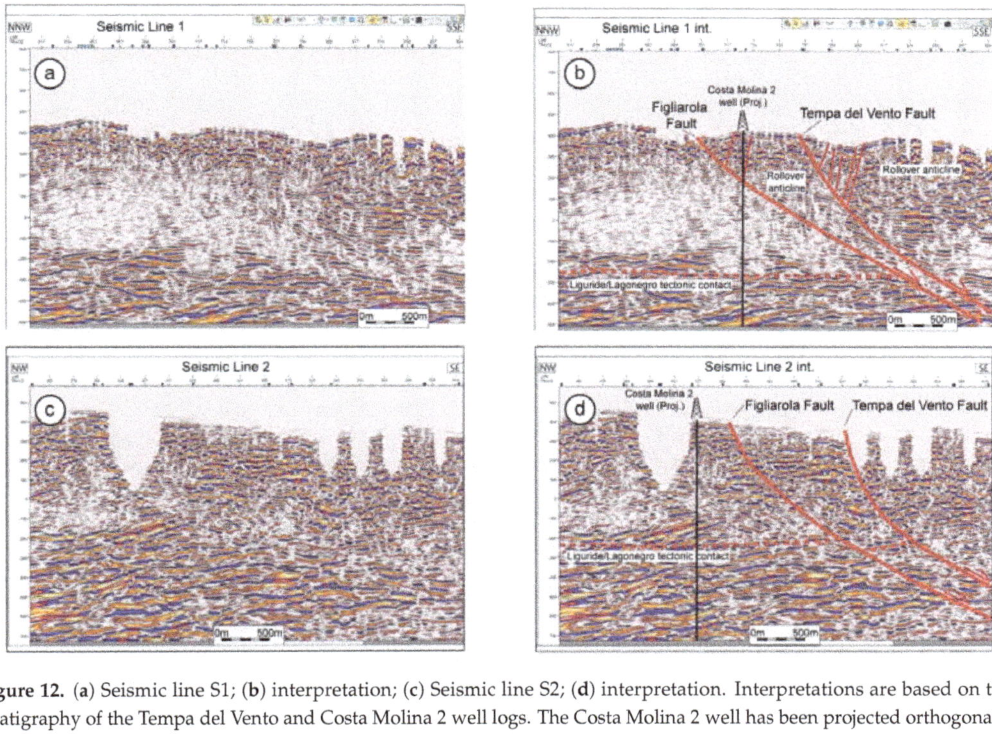

Figure 12. (**a**) Seismic line S1; (**b**) interpretation; (**c**) Seismic line S2; (**d**) interpretation. Interpretations are based on the stratigraphy of the Tempa del Vento and Costa Molina 2 well logs. The Costa Molina 2 well has been projected orthogonally to the seismic sections.

Moreover, line S1 clearly shows a rollover anticline at the hanging wall of the Tempa del Vento and Figliarola faults, which is consistent with a listric geometry of the fault trajectory. It also shows a series of antithetic normal faults affecting the rollover anticline, which likely resulted from crestal collapse [65] during slip along the listric fault. The curved geometry displayed by the two faults at depth is also consistent with a listric geometry, even though velocity effects connected to lithological differences between the Liguride and the Lagonegro Units cannot be excluded. It is important to note that the apparent dip angles of the fault planes decrease from line S2 to line S1, due to the different orientations of the seismic profiles with respect to the strike of the faults. The listric geometry and the tendency of the two structures to join at depth indicates that Figliarola and the Tempa del Vento faults are connected and rooted at a common detachment level.

8. Discussion

The detailed stratigraphic and structural analysis carried out along the southeastern sector of the Agri Valley, with the support of new biostratigraphic data, enabled us to redefine the geological setting of an important sector of the Agri Valley, where the distinction between different Cenozoic flyschoid units allowed the detection of important tectonic structures. In particular, mapping, correlation and age attribution of different marly intervals allowed us to extend the presence of the Albidona Formation in an area previously ascribed to the Gorgoglione Formation, with the NE-trending Tempa del Vento Fault tectonically separating the two formations. The new findings represent an important update for the available geological maps (sheets 505 and 506 [5,6]) and encourage further discussion about the age of the Albidona Formation. They also provide new evidence for understanding the relationships between tectonics and sedimentation in thrust-sheet top basins, which is crucial for reconstructing the tectonic evolution of the southern Apennines thrust and fold belt. The main implications will be discussed in the next sections.

8.1. Statigraphic Characteristics and Age of the Albidona Formation at the Monte dell'Agresto in Comparison with the Albidona Formation Type-Area

The stratigraphic succession exposed at Monte dell'Agresto share many characteristics with the Albidona Formation succession exposed in the type-area [33]. Lithological and stratigraphical similarities are represented by the common occurrence of a lower portion (member B–C) consisting of alternating sandstone, clay and marly intervals and an upper portion (member D) characterized by alternating coarse-grained sandstones and conglomerates. Another striking similarity is the middle-late Eocene age exhibited by both the study succession and the Albidona type-area succession, according to Baruffini et al. [33]. Nevertheless, regarding the age attribution, some significant discrepancies emerged as well. In fact, while the Barthonian/lower Priabonian age of member D is consistent with the type-area succession, the age of member B–C is different. Very interestingly, the time span encompassed by member B–C at Monte dell'Agresto includes the boundary between the biozones NP15 and NP16 that, in the Albidona type-area is not preserved because of an unconformity producing an important intraformational hiatus at the base of member D. This means that the studied succession shows a less pronounced unconformity between members B–C and D, probably indicating that the Albidona Formation of Monte dell'Agresto was less affected by synsedimentary tectonics and basin instability when compared to the type-area. In conclusion, the member B–C of the Albidona succession exposed at Monte dell'Agresto can be considered in part complementary with that exposed in the type-area.

The Eocene ages obtained in the present study might be useful for discussing the effective age of the Albidona Formation exposed in the Agri Valley area. In fact, it strongly contrasts with the early Miocene age indicated in the official geological maps [5,6]. In our opinion, this latter age attribution is not robustly supported by data. In fact, based on the descriptions accompanying the maps, this age has been inferred by stratigraphic intervals located outside of the mapped areas. In addition, the authors never considered the Agresto succession for age determinations, since it was originally attributed to the Gorgoglione Formation.

The Eocene age is also supported by subsurface data provided by hydrocarbon wells. In particular, the Costa Molina 2 and Costa Molina 3 wells (Figure 2) clearly report the occurrence of Eocene deposits attributable to the Albidona Formation. Moreover, the Tempa del Vento well (Figure 2) shows the occurrence of Eocene deposits right below the Gorgoglione basal unconformity. The Costa Molina 1 well log reports the occurrence of Miocene deposits in the first hundreds of meters. However, more recent stratigraphic reinterpretations carried out by ENI, have questioned this age.

Said that, we cannot exclude that the uppermost portion of the Albidona Formation can be attributed to the lower Miocene, as recently documented in the type-area [46]. However, our data indicate that a large portion of this formation was deposited during the Eocene. A similar age interpretation is provided in the stratigraphic scheme by Vezzani et al. [48], where the Albidona Formation encompasses the Eocene—early Miocene age interval.

8.2. Geodynamic Significance of the Albidona Formation

Clast composition of coarse-grained sediments pertaining to the Albidona Formation provides useful indications about the tectonically active source areas and consequently on the geodynamic scenario at the time of deposition. Obviously, particularly suited at this aim are the coarse-grained sediments represented by conglomerates and the level of pebbly mudstone. The study of these lithologies revealed that most of the clasts derived from a crystalline basement comparable to the uppermost portion of the continental crustal section exposed in the Sila Massif of Calabria [66], where low to medium grade metamorphic rocks are intruded by granodiorites and by later porphyritic dykes [67,68]. On the other hand, the ophiolitic material occurring in the pebbly mudstone, mainly represented by pillow basalts, microgabbro and reddish to greenish cherts, is consistent with the provenance of these clasts from the Liguride accretionary wedge, whose relics are currently exposed in the Pollino area of Basilicata and in Northern Calabria [4,22,69,70]. Therefore, the available data clearly suggest that the Calabrian Arc and the Liguride oceanic palaeodomain were strongly involved by active tectonics in the source area of the Albidona Formation, corresponding to the Liguride accredionary wedge. The search for evidence about a possible involvement in the deformation of the African paleomargin has been fruitless since no pebbles clearly derived from either the Apennine Platform or the Lagonegro Basin have been recognized. Possible sediments sourced from the Apennine Platform might be represented by the marly intervals. However, according to Colella and Zuffa [35], they could just represent very distal sediments supplied by the western margin of the platform located in the foreland setting of the Liguride accretionary wedge. Moreover, very fine grained megaturbidites, such as those generated by tsunami waves, might travel for long distances and transport large amount of carbonate mud, as indicated by Cita et al. ii [38] for the homogenites of the Mediterranean basin. In conclusion, our evidence about clast composition are in good agreement with the geodynamic model proposed by Patacca and Scandone [2] and Vezzani et al. ii [48] which suggests that the Albidona Formation was deposited in a thrust sheet top basin located above the Calabride/Liguride accretionary wedge, during subduction of the oceanic crust of the Ligurian Tethys. Deep subduction followed by fast exhumation of the Tethyan crust during Eocene is documented by radiometric ages and thermobarometry obtained for high-P bluschists from Northern Calabria [71,72].

8.3. Tectonic Setting of the Study Area

Study on fold axes orientations in the investigated area revealed the presence of at least two folding phases showing E-W and NW-SE hinge directions (stereoplot C and D in Figure 8). Occurrence of E-W hinge orientations represent an anomaly for the considered sector of the southern Apennines, where N- to NW-trending structures with an average NE-facing tectonic transport direction have been recognized [73]. This anomaly further supports the results obtained by the study of clast composition namely that the Albidona Formation was deposited in a thrust top basin on the Liguride accretionary wedge that underwent deformation during subduction of the Ligurian Tethys. In fact, the southwards vergence of the D1 folds recognized in the Albidona Formation can be compared with similar structures, showing a tectonic transport toward SW, recognized in the internal Units of the southern Apennines by Vitale and Ciarcia [74]. The same Authors refer the age of this early deformation stage to the early-middle Miocene. Folding stage D2 is related to later evolutionary stages of the southern Apennines thrust belt.

8.4. Significance and Interpretation of the NE-Trending Faults

NE-trending faults occurring in the southern Apennines are largely associated with the Pliocene to Pleistocene evolution of the thrust belt. These tectonic structures generally consist of either right or left—lateral strike slip faults offsetting pre-existing contractional structures, associated with the build-up of the chain, or forming lateral ramps accommodating the thrust emplacement [8,75]. Although predominant, NE-trending strike slip faults are not exclusive in the southern Apennines and an increasing number of NE-trending normal faults have been recognized. These latter faults have been associated to earlier extensional tectonic stages [76] or to the activation of low—angle normal faults (LANF) [77–79].

The NE-trending Figliarola and the Tempa del Vento faults represent the main extensional structures recognized in the study area. Standing on the geometrical characteristics described in the previous sections, such as the listric geometry observed in the seismic profiles, these structures can be interpreted as SE-dipping intermediate to low angle normal faults. A geological cross-section (Figure 13) allowed assessing the cumulative displacement of the Tempa del Vento and Figliarola faults.

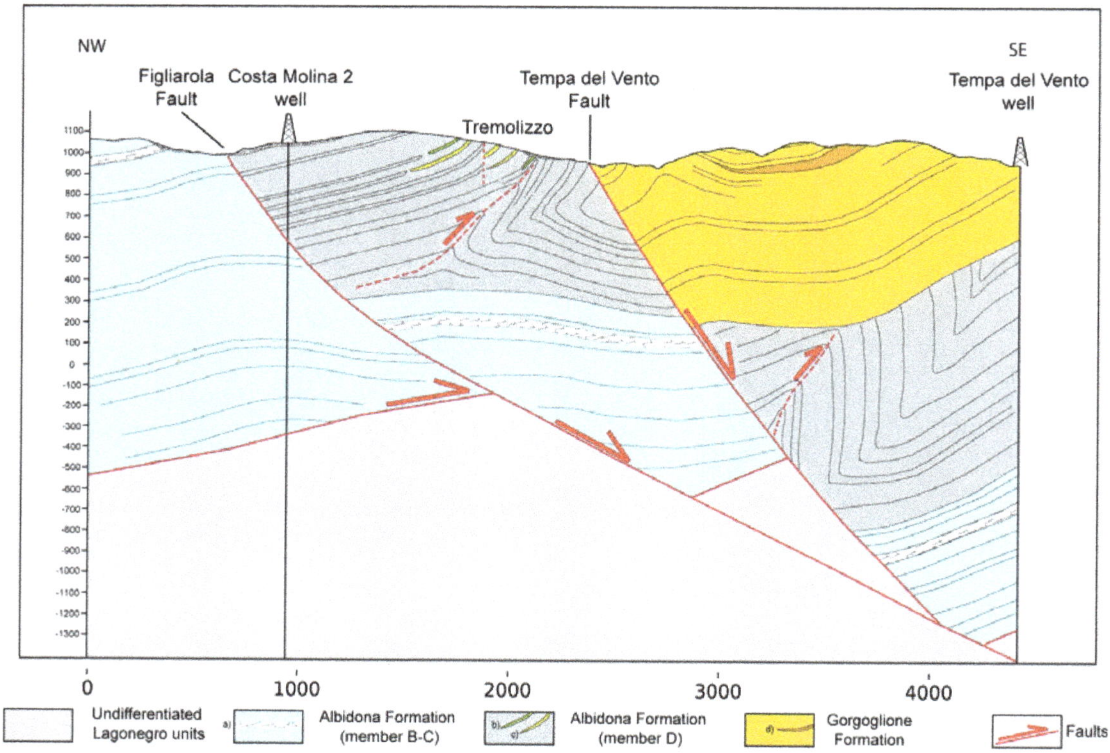

Figure 13. Interpretative cross-section of the M. dell'Agresto—Contrada La Rossa area, showing the overall geometry and displacement of the Figliarola—Tempa del Vento Fault system. Legend: (a) marly marker horizon; (b) microconglomerate; (c) sandstone; (d) conglomerate.

The interpretation presented in the geological cross-section relies on the aforementioned seismic lines and on the data obtained from the Costa Molina 2 and Tempa del Vento wells (Figures 2 and 13). Considering that both normal faults are probably rooted at a common detachment level, the Figliarola Fault can be interpreted as the major structure and the Tempa del Vento Fault as a secondary splay. The cumulative displacement of the two faults can be estimated by considering the marly marker horizon recognized by geological mapping at the footwall of the Figliarola Fault, within the Albidona member B–C, and a similar horizon recognized in the stratigraphic log of the Tempa del Vento well, located at a depth of 1550 m at the hangingwall of the Tempa del Vento Fault. By considering this marker bed, the cumulative displacement of the Figliarola and Tempa del Vento Faults can be estimated at about 1800 m.

8.5. Further Implications

Recognition of the Albidona Formation and previously unknown tectonic structures in the study area has important consequences on reservoir modelling in the Val d'Agri oil field. In particular, based on the present study, it will be possible to update the seismic velocity model, considering the petrophysical characteristics of the Albidona Formation instead of the Gorgoglione Formation. In addition, faults recognized for the first time in the study area will be considered in order to improve the 3D model of the Agri Valley.

The widespread presence of the Albidona Formation in the study area has also strong environmental implications regarding, in particular, the definition of the chemical composition of soils. Knowledge of the composition of soils developed on peculiar rock types is fundamental for differentiating the background concertation of elements connected to either geogenic or anthropogenic processes. In this context, the recognition of the pebbly mudstone interval of the Albidona Formation, containing clasts derived from ophiolitic, igneous and metamorphic source areas, can be helpful in better understanding anomalous concentrations of some elements in soils.

9. Conclusions

The present work stresses on the importance of performing an accurate field mapping for the understanding of surface and subsurface geology. At this aim we presented the case study of the Agri Valley in southern Italy where field survey allowed to revise the distribution of two major Cenozoic turbiditic units of the Southern Apennines (the Albidona and Gorgoglione Formations) and to identify three distinct stratigraphic marker intervals in the Albidona Formation that resulted very useful for stratigraphic correlations both in outcrop and subsurface. From the older to the younger the three intervals are (i) a 40 m thick marly interval, (ii) a pebbly mudstone and (iii) a sandstone interval.

Biostratigraphic analysis performed on the marly level and other similar intervals distributed at different stratigraphic highs in the study succession allowed the attribution of the Albidona Formation to the Eocene. Different facies characteristics and age determinations allowed the differentiation of the Albidona Formation in two members, with the older one, identified as member B–C, Lutetian in Age, consisting of alternating sandstones and clays and the younger one, Barthonian/Priabonian in age, identified as member D, consisting in alternating sandstones and conglomerates.

The geometrical relationships between the two members and with the Miocene Gorgoglione Formation allowed recognising two major NE-trending normal faults, documented for the first time in the study area. These structures, named, respectively, as Figliarola and Tempa del Vento faults, have also been recognized in the subsurface by means available seismic lines provided by ENI.

Recognition of the marker marly horizon in the stratigraphy of Costa Molina and Tempa del Vento wells allowed us to calculate the cumulative displacement of the Figliarola and Tempa del Vento Faults that can be estimated to about 1800 m.

Author Contributions: Conceptualization, G.P. (Giacomo Prosser) and G.P. (Giuseppe Palladino); Data curation, G.P. (Giacomo Prosser), G.P. (Giuseppe Palladino), E.M.B., M.R. and D.E.C.; Formal analysis, G.P. (Giacomo Prosser), G.P. (Giuseppe Palladino), E.M.B., M.R. and D.E.C.; Investigation, G.P. (Giacomo Prosser), G.P. (Giuseppe Palladino), E.M.B., M.R. and D.E.C.; Methodology, G.P. (Giacomo Prosser), E.M.B., M.R. and D.E.C.; Project administration, D.A. and F.C.; Resources, D.A.; Software, G.P. (Giacomo Prosser) and G.P. (Giuseppe Palladino); Supervision, G.P. (Giacomo Prosser), D.A., F.C., M.R. and D.E.C.; Validation, G.P. (Giacomo Prosser), D.A. and F.C.; Visualization, G.P. (Giacomo Prosser), G.P. (Giuseppe Palladino), E.M.B., M.R. and D.E.C.; Writing—original draft, G.P. (Giacomo Prosser), G.P. (Giuseppe Palladino), E.M.B., M.R. and D.E.C.; Writing—review & editing, G.P. (Giacomo Prosser), D.A. and F.C. All authors have read and agreed to the published version of the manuscript.

Funding: This research received no external funding.

Institutional Review Board Statement: Not applicable.

Informed Consent Statement: Not applicable.

Acknowledgments: We very grateful to the editor G. Wang and the guest editors D. Liotta, G. Molli and A. Cipriani for the opportunity to share our research in the Special Issue "The Apennines: Tectonics, Sedimentation, and Magmatism from the Palaeozoic to the Present". We are also grateful to two anonymous referees for their critical reviews that helped improve this paper.

Conflicts of Interest: The authors declare no conflict of interest.

References

1. Prosser, G.; Schiattarella, M.; Tramutoli, M.; Doglioni, C.; Harabaglia, P.; Bigozzi, A. Una sezione rappresentativa dell'Appennino meridionale. In Proceedings of the Conferenza Sulla Ricerca Scientifica in Basilicata, Potenza, Italy, 29 February–1 March 1996.
2. Patacca, E.; Scandone, P. Geology of Southern Apennines. Bollettino della Società Geologica Italiana. *Ital. J. Geosc.* **2007**, *7*, 75–119.
3. Carbone, S.; Catalano, S.; Lazzari, S.; Lentini, F.; Monaco, C. Presentazione della carta geologica del bacino del Fiume. *Agric. Mem. Soc. Geol. Ital.* **1991**, *47*, 129–143.
4. Bonardi, G.; Amore, F.O.; Ciampo, G.; de Capoa, P.; Miconnet, P.; Perrone, V. Il Complesso Liguride Auctorum: Stato delle conoscenze e problemi aperti sulla sua evoluzione pre-appenninca ed i suoi rapporti con l'arco calabro. *Mem. Soc. Geol. Ital.* **1988**, *41*, 17–35.
5. Sevizio Geologico d'Italia, ISPRA, 2014. Carta Geologica d'Italia Alla Scala 1:50,000, Foglio 505 "Moliterno". Available online: https://www.isprambiente.gov.it/Media/carg/505_MOLITERNO/Foglio.html (accessed on 30 December 2020).
6. Sevizio Geologico d'Italia, ISPRA, 2005. Carta Geologica d'Italia Alla Scala 1:50,000, Foglio 506 "Sant'Arcangelo". Available online: https://www.isprambiente.gov.it/Media/carg/506_SANT_ARCANGELO/Foglio.html (accessed on 30 December 2020).
7. Giano, S.I.; Maschio, L.; Alessio, M.; Ferranti, L.; Improta, S.; Schiattarella, M. Radiocarbon dating of active faulting in the Agri high valley, southern Italy. *J. Geodyn.* **2000**, *29*, 371–386. [CrossRef]
8. Cello, G.; Gambini, R.; Mazzoli, S.; Read, A.; Tondi, E.; Zucconi, V. Fault zone characteristics and scaling properties of the Val d'Agri Fault System (Southern Apennines, Italy). *J. Geodyn.* **2000**, *29*, 293–307. [CrossRef]
9. Hippolyte, J.-C.; Angelier, J.; Roure, F. A major geodynamic change revealed by Quaternary stress patterns in the southern Apennines (Italy). *Tectonophysics* **1994**, *230*, 199–210. [CrossRef]
10. Ferranti, L.; Oldow, J.S. Latest Miocene to Quaternary horizontal and vertical displacement rates during simultaneous contraction and extension in the Southern Apennines orogen, Italy. *Terra Nova* **2005**, *17*, 209–214. [CrossRef]
11. Malinverno, A.; Ryan, W.B.F. Extension in the Tyrrhenian Sea and shortening in the Apennines as result of arc migration driven by sinking of the lithosphere. *Tectonics* **1986**, *5*, 227–245. [CrossRef]
12. Royden, L.; Patacca, E.; Scandone, P. Segmentation and configuration of subducted lithosphere in Italy: An important control on thrust-belt and foredeep-basin evolution. *Geology* **1987**, *15*, 714–717. [CrossRef]
13. Patacca, E.; Sartori, R.; Scandone, P. Tyrrhenian basin and Apenninic arcs: Kinematic relations since Late Tortonian times. *Mem. Soc. Geol. Ital.* **1990**, *45*, 425–451.
14. Doglioni, C. A proposal for the kinematic modelling of W-dipping subductions—Possible applications to the Tyrrhenian-Apennines system. *Terra Nova* **1991**, *3*, 423–434. [CrossRef]
15. Morandi, S.; Ceragoli, E. Integrated interpretation of seismic and resistivity images across the «Val d'Agri» graben (Italy). *Ann. Geophys.* **2002**, *45*, 259–271.
16. Maschio, L.; Ferranti, L.; Burrato, P. Active extension in Val d'Agri area, Southern Apennines, Italy: Implications for the geometry of the seismogenic belt. *Geophys. J. Int.* **2005**, *162*, 591–609. [CrossRef]
17. Burrato, P.; Valensise, G. Rise and Fall of a Hypothesized Seismic Gap: Source Complexity in the Mw 7.0 16 December 1857 Southern Italy Earthquake. *Bull. Seism. Soc. Am.* **2008**, *98*, 139–148. [CrossRef]
18. Improta, L.; Ferranti, L.; De Martini, P.M.; Piscitelli, S.; Bruno, P.P.; Burrato, P.; Civico, R.; Giocoli, A.; Iorio, M.; D'Addezio, G.; et al. Detecting young, slow-slipping active faults by geologic and multidisciplinary high-resolution geophysical investigations: A case study from the Apennine seismic belt, Italy. *J. Geophys. Res. Space Phys.* **2010**, *115*, 11307. [CrossRef]
19. Shiner, P.; Beccacini, A.; Mazzoli, S. Thin-skinned versus thick-skinned structural models for Apulian carbonate reservoirs: Constraints from the Val d'Agri Fields, S Apennines, Italy. *Mar. Pet. Geol.* **2004**, *21*, 805–827. [CrossRef]
20. Handy, M.R.; Schmid, S.M.; Bousquet, R.; Kissling, E.; Bernoulli, D. Reconciling plate-tectonic reconstructions of Alpine Tethys with the geological–geophysical record of spreading and subduction in the Alps. *Earth-Sci. Rev.* **2010**, *102*, 121–158. [CrossRef]
21. Knott, S.D. The Liguride Complex of Southern Italy—A Cretaceous to Paleogene accretionary wedge. *Tectonophysics* **1987**, *142*, 217–226. [CrossRef]
22. Vitale, S.; Ciarcia, S.; Fedele, L.; Tramparulo, F.D. The Ligurian oceanic successions in southern Italy: The key to decrypting the first orogenic stages of the southern Apennines-Calabria chain system. *Tectonophysics* **2019**, *750*, 243–261. [CrossRef]
23. Scandone, P. Studi di geologia lucana: Carta dei terreni della serie calcareo-silico-marnosa e note illustrative. *Boll. Soc. Natur. Napoli* **1972**, *273*, 225–299.
24. Bucci, F.; Novellino, R.; Tavarnelli, E.; Prosser, G.; Guzzetti, G.; Cardinali, M.; Gueguen, E.; Guglielmi, P.; Adurno, I. Frontal collapse during thrust propagation in mountain belts: A case study in the Lucanian Apennines, Southern Italy. *J. Geol. Soc.* **2014**, *171*, 571–581. [CrossRef]

25. Patacca, E.; Scandone, P. Late thrust propagation and sedimentary response in the thrust-beltforedeep system of the Southern Apennines (Pliocene-Pleistocene). In *Anatomy of an Orogen: The Apennines and Adjacent Mediterranean Basins*; Vai, G.B., Martini, I.P., Eds.; Kluwer Academic Publishers: London, UK, 2001.
26. Noguera, A.; Rea, G. Deep structure of the Campanian–Lucanian Arc (Southern Apennine, Italy). *Tectonophysics* **2000**, *324*, 239–265. [CrossRef]
27. Boiano, U. Anatomy of a siliciclastic turbidite basin: The Gorgoglione Flysch, Upper Miocene, southern Italy: Physical stratigraphy, sedimentology and sequence-stratigraphic framework. *Sediment. Geol.* **1997**, *107*, 231–262. [CrossRef]
28. Cavalcante, F.; Fiore, S.; Lettino, A.; Piccarreta, G.; Tateo, F. Illite-smectite mixed layers in Sicilide shales and piggy-back deposits of the Gorgoglione Formation (Southern Apennines): Geological inference. *Boll. Soc. Geol. Ital.* **2007**, *126*, 241–254.
29. Cavalcante, F.; Prosser, G.; Agosta, F.; Belviso, C.; Corrado, G. Post-depositional history of the Miocene Gorgoglione Formation (southern Apennines, Italy): Inferences from mineralogical and structural analyses. *Bull. Soc. Géol. Fr.* **2015**, *186*, 243–256. [CrossRef]
30. Buttinelli, M.; Improta, L.; Bagh, S.; Chiarabba, C. Inversion of inherited thrusts by wastewater injection induced seismicity at the Val d'Agri oilfield (Italy). *Sci. Rep.* **2016**, *6*, 37165. [CrossRef]
31. Selli, R. Il Paleogene nel quadro della geologia dell'Italia centro-meridionale. *Mem. Soc. Geol. Ital.* **1962**, *3*, 737–789.
32. Zuppetta, A.; Russo, M.; Turco, E.; Gallo, L. Età e significato della Formazione di Albidona in Appennino Meridionale. *Boll. Soc. Geol. Ital.* **1984**, *103*, 159–170.
33. Baruffini, L.; Lottaroli, F.; Torricelli, S.; Lazzari, D. Stratigraphic revision of the Eocene Albidona Formation in the type locality (Calabria, southern Italy). *Riv. Ital. Paleontol. Stratigr.* **2000**, *106*, 73–98.
34. Critelli, S.; Muto, F.; Tripodi, V.; Perri, F. Link between thrust tectonics and sedimentation processes of stratigraphic sequences from the southern Apennines foreland basin system, Italy. *Rend. Online Soc. Geol. Ital.* **2013**, *25*, 21–42.
35. Colella, A.; Zuffa, G.G. Megastrati carbonatici e silicoclastici della Formazione di Albidona (Miocene, Appennino meridionale): Implicazioni paleogeografiche. *Mem. Soc. Geol. Ital.* **1988**, *41*, 791–807.
36. Dietrich, D.; Scandone, P. The position of basic and ultrabasic rocks in the tectonic units of the southern Apennines. *Atti dell'Accademia Pontiana* **1972**, *21*, 1–15.
37. Kastens, K.A.; Cita, M.B. Tsunami-induced sediment transport in the abyssal Mediterranean Sea. *GSA Bull.* **1981**, *92*, 845–857. [CrossRef]
38. Cita, M.B.; Camerlenghi, A.; Rimoldi, B. Deep-sea tsunami deposits in the eastern Mediterranean: New evidence and depositional models. *Sediment. Geol.* **1996**, *104*, 155–173. [CrossRef]
39. Pavan, G.; Pirini, C. Microfossili cretacici ed eocenici nella zona di Monte Falapato (Lucania). *Mem. Soc. Geol. Ital.* **1963**, *4*, 1105–1134.
40. Mostardini, F.; Pieri, M.; Pirini, C. Stratigrafia del foglio 212, Montalbano Jonico. *Boll. Serv. Geol. Ital.* **1966**, *87*, 541–547.
41. Ogniben, L. Schema introduttivo alla geologia del confine calabro-lucano. *Mem. Soc. Geol. Ital.* **1969**, *8*, 453–763.
42. Vezzani, L. Nota preliminare sulla stratigrafia della Formazione di Albidona. *Boll. Soc. Geol. Ital.* **1966**, *85*, 767–776.
43. Vezzani, L. Distribuzione, facies e stratigrafia della Formazione del Saraceno (Albiano—Daniano) nell'area compresa tra il Mare Jonio e il Torrente Frido. *Geol. Rom.* **1968**, *7*, 229–276.
44. Vezzani, L. Il Flysch di Albidona nell'area del confine tra Calabria e Lucania. *Geol. Rom.* **1970**, *9*, 101–126.
45. Bonardi, G.; Ciampo, G.; Perrone, V. La Formazione di Albidona nell'Appennino calabro-lucano: Ulteriori dati stratigrafici e relazioni con le unità esterne appenniniche. *Boll. Soc. Geol. Ital.* **1985**, *104*, 539–549.
46. Sevizio Geologico d'Italia, ISPRA, 2009. Carta Geologica d'Italia alla scala 1:50,000, Foglio 535 "Trebisacce". Available online: https://www.isprambiente.gov.it/Media/carg/535_TREBISACCE/Foglio.html (accessed on 30 December 2020).
47. Lentini, F.; Carbone, S.; Catalano, S.; Monaco, C. Confronti sedimentologico petrografici e posizione strutturale del Flysch di Albidona e di Gorgoglione nella media Val d'Agri (Appennino lucano). *Mem. Soc. Geol. Ital.* **1987**, *38*, 259–263.
48. Vezzani, L.; Festa, A.; Ghisetti, F.C. *Geology and Tectonic Evolution of the Central-Southern Apennines, Italy*; Geological Society of America: Boulder, CO, USA, 2010.
49. Patacca, E.; Scandone, P.; Bellatalla, M.; Perilli, N.; Santini, U. Numidian sand event in the Southern Apennines. *Mem. Soc. Geol. Padova* **1992**, *43*, 297–337.
50. Palladino, G.; Parente, M.; Prosser, G.; Di Staso, A. Tectonic control on the deposition of the Lower Miocene sediments of the Monti della Maddalena ridge (Southern Apennines): Synsedimentary extensional deformation in a foreland setting. *Boll. Soc. Geol. Ital.* **2008**, *127*, 317–335.
51. Okada, H.; Bukry, D. Supplementary modification and introduction of code numbers to the low-latitude coccolith biostratigraphic zonation (Bukry, 1973; 1975). *Mar. Micropaleontol.* **1980**, *5*, 321–325. [CrossRef]
52. Martini, E. Standard Tertiary and Quaternary calcareous nannoplankton zonation. In *Proceedings of the Second Planktonic Conference, Roma, 1970*; Tecnoscienza: Roma, Italy, 1971; pp. 739–765.
53. Perch-Nielsen, K. Cenozoic calcareous nannofossils. In *Plankton Stratigraphy*; Bolli, H.M., Saunders, J.B., PerchNielsen, K., Eds.; Cambridge University Press: Cambridge, UK, 1985; pp. 427–554.
54. Bown, P.R. *Calcareous Nannofossil Biostratigraphy*; Kluwer Academic Publishers: Norwell, MA, USA, 1998; pp. 1–315.
55. Gradstein, F.M.; Ogg, J.G.; Schmitz, M.D.; Ogg, G.M. *The Geological Time Scale 2012*; Elsevier: Amsterdam, The Netherlands, 2012.
56. Priest, S.D. *Discontinuity Analysis for Rock Engineering*; Springer Science and Business Media LLC.: Secaucus, NJ, USA, 1993.

57. Ortega, O.J.; Marrett, R.A.; Laubach, S.E. A scale-independent approach to fracture intensity and average spacing measurement. *AAPG Bull.* **2006**, *90*, 193–208. [CrossRef]
58. Bouma, A.H. *Sedimentology of Some Flysch Deposits*; Elsevier: Amsterdam, The Netherlands, 1962.
59. Sartoni, S.; Crescenti, U. *Ricerche Biostratigrafiche nel Mesozoico dell'Appennino Meridionale*; Museo Geologico Giovanni Capellini: Bologna, Italy, 1962; Volume 29, pp. 153–302.
60. Scandone, P.; Sgrosso, I. Flysch con Inocerami nella Valle del Cavolo presso Tramutola (Lucania). *Boll. Soc. Natur. Napoli* **1964**, *73*, 166–175.
61. Giannandrea, P.; LoIacono, F.; Maiorano, P.; Lirer, F.; Puglisi, D. Geological map of the eastern sector of the Gorgoglione Basin (southern Italy). *Ital. J. Geosci.* **2016**, *135*, 120–141. [CrossRef]
62. Ramsay, J.G. *Folding and Fracturing of Rocks*; McGraw-Hill: New York, NY, USA, 1967.
63. Kim, Y.-S.; Peacock, D.C.; Sanderson, D.J. Fault damage zones. *J. Struct. Geol.* **2004**, *26*, 503–517. [CrossRef]
64. Choi, J.-H.; Edwards, P.; Ko, K.; Kim, Y.-S. Definition and classification of fault damage zones: A review and a new methodological approach. *Earth-Sci. Rev.* **2016**, *152*, 70–87. [CrossRef]
65. McClay, K. Extensional fault systems in sedimentary basins: A review of analogue model studies. *Mar. Pet. Geol.* **1990**, *7*, 206–233. [CrossRef]
66. Caggianelli, A.; Prosser, G. An exposed cross-section of Late Hercynian upper and intermediate continental crust exposed in the Sila Nappe (Calabria, Southern Italy). *Period. Mineral.* **2001**, *70*, 275–301.
67. Langone, A.; Godard, G.; Prosser, G.; Caggianelli, A.; Rottura, A.; Tiepolo, M. P–T–t path of the Hercynian low-pressure rocks from the Mandatoriccio complex (Sila Massif, Calabria, Italy): New insights for crustal evolution. *J. Metamorph. Geol.* **2010**, *28*, 137–162. [CrossRef]
68. Festa, V.; Langone, A.; Caggianelli, A.; Rottura, A. Dike magmatism in the Sila Grande (Calabria, southern Italy): Evidence of Pennsylvanian–Early Permian exhumation. *Geosophy* **2010**, *6*, 549–566. [CrossRef]
69. Cavalcante, F.; Belviso, C.; Finizio, F.; Lettino, A.; Fiore, S. *Carta Geologica delle Unità Liguridi Dell'area del Pollino (Basilicata): Nuovi dati Geologici, Mineralogici e Petrografici*; Digilabs: Bari, Italy, 2009.
70. Liberi, F.; Morten, L.; Piluso, E. Geodynamic significance of ophiolites within the Calabrian Arc. *Isl. Arc* **2006**, *15*, 26–43. [CrossRef]
71. Shimabukuro, D.H.; Wakabayashi, J.; Alvarez, W.; Chang, S.-C. Cold and old: The rock record of subduction initiation beneath a continental margin, Calabria, southern Italy. *Lithosphere* **2012**, *4*, 524–532. [CrossRef]
72. Tursi, F.; Bianco, C.; Brogi, A.; Caggianelli, A.; Prosser, G.; Ruggieri, G.; Braschi, E. Cold subduction zone in northern Calabria (Italy) revealed by lawsonite–clinopyroxene blueschists. *J. Metamorph. Geol.* **2020**, *38*, 451–469. [CrossRef]
73. Mazzoli, S.; Barkham, S.; Cello, G.; Gambini, R.; Mattioni, L.; Shiner, P.; Tondi, E. Reconstruction of continental margin architecture deformed by the contraction of the Lagonegro Basin, southern Apennines, Italy. *J. Geol. Soc.* **2001**, *158*, 309–319. [CrossRef]
74. Vitale, S.; Ciarcia, S. Tectono-stratigraphic and kinematic evolution of the southern Apennines/Calabria–Peloritani Terrane system (Italy). *Tectonophysics* **2013**, *583*, 164–182. [CrossRef]
75. Bonini, M.; Sani, F. Pliocene-Quaternary transpressional evolution of the Anzi-Calvello and Northern S. Arcangelo basins (Basilicata, Southern Apennines, Italy) as a consequence of deep-seated fault reactivation. *Mar. Pet. Geol.* **2000**, *17*, 909–927. [CrossRef]
76. Bucci, F.; Novellino, R.; Guglielmi, P.; Prosser, G.; Tavarnelli, E. Geological map of the northeastern sector of the high Agri Valley, Southern Apennines (Basilicata, Italy). *J. Maps* **2012**, *8*, 282–292. [CrossRef]
77. Ferranti, L.; Oldow, J.S.; Sacchi, M. Pre-Quaternary orogen-parallel extension in the Southern Apennine belt, Italy. *Tectonophysics* **1996**, *260*, 325–347. [CrossRef]
78. Novellino, R.; Prosser, G.; Spiess, R.; Viti, C.; Agosta, F.; Tavarnelli, E.; Bucci, F. Dynamic weakening along incipient low-angle normal faults in pelagic limestones (Southern Apennines, Italy). *J. Geol. Soc.* **2015**, *172*, 283–286. [CrossRef]
79. Giano, S.I.; Pescatore, E.; Agosta, F.; Prosser, G. Geomorphic evidence of Quaternary tectonics within an underlap fault zone of southern Apennines, Italy. *Geomorphology* **2018**, *303*, 172–190. [CrossRef]

geosciences

MDPI

Article

Structural and Stratigraphic Setting of Campagna and Giffoni Tectonic Windows: New Insights on the Orogenic Evolution of the Southern Apennines (Italy)

Stefano Vitale [1], Ernesto Paolo Prinzi [1], Maria Monda [1], Francesco D'Assisi Tramparulo [2] and Sabatino Ciarcia [3,*]

1 Dipartimento di Scienze della Terra, dell'Ambiente e delle Risorse (DiSTAR), Università degli Studi di Napoli Federico II, 80126 Napoli, Italy; stefano.vitale@unina.it (S.V.); prinziernesto@gmail.com (E.P.P.); mariam.monda@gmail.com (M.M.)
2 Istituto Nazionale di Geofisica e Vulcanologia, Sezione di Napoli Osservatorio Vesuviano, 80124 Napoli, Italy; francesco.tramparulo@ingv.it
3 Dipartimento di Scienze e Tecnologie, Università degli Studi del Sannio, 82100 Benevento, Italy
* Correspondence: sabatino.ciarcia@unisannio.it

Received: 30 August 2020; Accepted: 7 October 2020; Published: 10 October 2020

Abstract: We present a structural study on the tectonic windows of Giffoni and Campagna, located in the western sector of the southern Apennines (Italy). We analyzed thrusts, folds, and related minor deformation structures. Here, a major in-sequence E-verging thrust fault juxtaposes Meso-Cenozoic successions of the Apennine Platform (Picentini Mts unit) and the Lagonegro-Molise Basin (Frigento unit). However, out-of-sequence thrusts duplicated the tectonic pile with the interposition of the upper Miocene wedge-top basin deposits of the Castelvetere Group. We reconstructed the orogenic evolution of these two tectonic windows, including five deformation phases. The first (D1) was related to the in-sequence thrusting with minor thrusts and folds, widespread both in the footwall and the hanging wall. A subsequent extension (D2) has formed normal faults crosscutting the D1 thrusts and folds. All structures were subsequently affected by two shortening stages (D3 and D4), which also deformed the upper Miocene wedge top basin deposits of the Castelvetere Group. We interpreted the D3–D4 structures as related to an out-of-sequence thrust system defined by a main frontal E-verging thrust and lateral ramps characterized by N and S vergences. Low-angle normal faults were formed in the hanging wall of the major thrusts. Out-of-sequence thrusts are observed in the whole southern Apennines, recording a crustal shortening event that occurred in the late Messinian–early Pliocene. Finally, we suggest that the two tectonic windows are the result of the formation of an E–W trending regional antiform, associated with a late S-verging back-thrust, that has been eroded and crosscut by normal faults (D5) in the Early Pleistocene.

Keywords: southern Apennines; out-of-sequence thrust; wedge-top basin; tectonic window

1. Introduction

The fold-and-thrust belt of the southern Apennines (Figure 1) is defined by low- and high-angle thrusts [1–9]. Thin-skinned tectonics, characterized by dominant low-angle thrusts, has been active mainly in the first part of the orogenic evolution, from the early Miocene to the early Pliocene, recording most of the orogenic shortening (e.g., [2,10]). On the other hand, thick-skinned tectonics, defined by high-angle thrusts, played a significative role in the crustal shortening during the Pliocene–Middle Pleistocene; they were nucleated at deeper structural levels showing smaller displacements (e.g., [1,4,7,8]). Thick-skinned tectonics has probably acted through the positive inversion of pre-existing normal faults located in the Permo-Triassic rocks of the downgoing Adria plate [3], likely triggered by the buttressing of the thick Apulian Mesozoic carbonates against the allochthonous wedge in the early Pliocene [6,11]. The fault inversion generated deep-seated envelopment thrusts [12,13] that crosscut upward the already formed orogenic wedge and presently appear as out-of-sequence structures (Figure 1b; [11,14,15]). In the axial sector of the southern Apennine chain, these ramp-dominated thrusts have formed several anticlines within the buried Apulian carbonates that originated hydrocarbon traps presently drilled and exploited (e.g., [16]). These envelopment thrusts are widespread in the southern Apennines (e.g., [11,14,15,17]), frequently deforming the upper Miocene–Middle Pleistocene wedge-top basin deposits both in the western and eastern sectors of the chain. However, few studies have provided information about the orientation, structural style, kinematics, and age of these out-of-sequence structures (e.g., [11,14,15]). In this work, we analyzed the tectonic windows of Giffoni and Campagna in the southern Apennines (Figure 1a), where in- and out-of-sequence thrusts and related structures are remarkably exposed. These two regional structures have been previously studied by different authors [18–22] and were mapped in the new official geological cartography [23,24]. We present a detailed study of the mesoscale deformation structures, with the aim to reconstruct the orogenic evolution of this key sector of the chain, where a crustal section of the allochthonous wedge is naturally exposed.

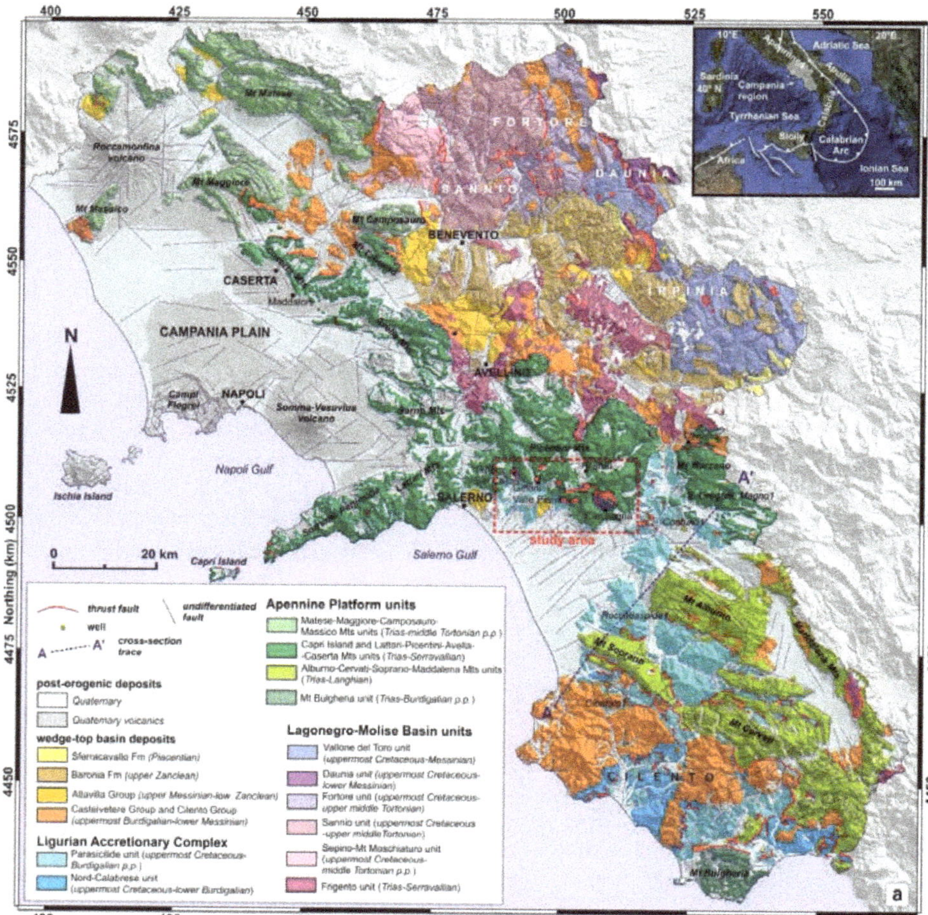

Figure 1. *Cont.*

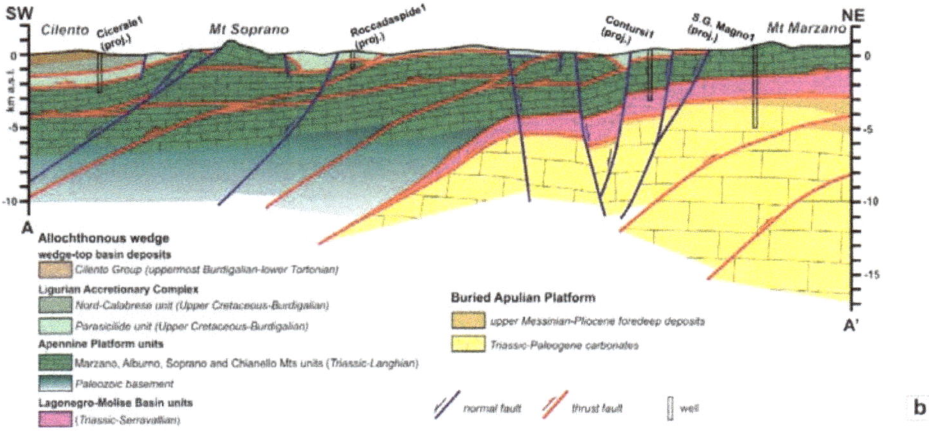

Figure 1. (**a**) Schematic geological map of the southern Apennines. (**b**) Geological cross-section. (**a**,**b**) modified after [8].

2. Geological Setting

The superposition of different thrust-sheets characterizes the study area (Figure 1a,b). The main tectonic units refer to three paleogeographic domains: (i) Ligurian Accretionary Complex (Parasicilide unit; [25–31]); (ii) Apennine Platform (Picentini Mts and Mt Croce units; [7,8,19–24,32–39]); (iii) Lagonegro-Molise Basin (Frigento unit; [7,8,10,40–44]).

The Parasicilide unit is widely exposed in the Sele River valley and the Salerno area (Figure 1a). It is made up of deep-basin deposits (Figure 2), including at the base of the uppermost Cretaceous–middle Eocene pelitic Argille Scagliose Fm, upward passing to upper Eocene–Aquitanian marly-calcareous M. S. Arcangelo Fm and varicolored clays of the Argille Varicolori Fm. Burdigalian foredeep sandstones of the Albanella Fm seal the Parasicilide succession.

The Picentini Mts unit mainly consists of a thick shallow-water succession (Figure 2) with at base Carnian–Norian dolostones and limestones, upward passing to Jurassic–Upper Cretaceous dolomitic limestones and limestones. Pelagic dolostones and limestones with cherts locally replace the shallow-water Norian–Lower Jurassic succession. At places, the Maastrichtian–Paleocene margin recrystallized calcareous breccias (Calcari Cristallini Fm) cover the Mesozoic succession.

The Monte Croce unit, exposed only in the Campagna tectonic window, is composed, from bottom to top (Figure 2), of Upper Cretaceous–lower Aquitanian margin recrystallized breccias, calcarenites, and scaglia-like reddish argillites (Fontana Frasci Fm) upward passing to Aquitanian–Serravallian deep basin, scaglia-like, pelitic, marly, and calciclastic turbidites and, finally, Serravallian foredeep sandstones (Fontana Porcellara Fm and Vallimala Flysch). The lower part of the succession is partially replaced by diagenetic dolomite.

The Frigento unit (Figure 2) consists of Ladinian–Carnian, slope to basin, calcilutites, and sandstones embedding boundstone bodies (Monte Facito Fm) at the base, covered by Carnian–Norian cherty calcarenites and calcilutites (Calcari con Selce Fm). The succession passes upward to Rhaetian–Jurassic radiolarites and reddish and greenish silicized argillites (Scisti Silicei Fm), and Lower Cretaceous dark siliceous shales with intercalations of calcilutites (Flysch Galestrino Fm). Finally, Upper Cretaceous–lower Miocene calcarenites, calcilutites, and varicolored clays (Flysch Rosso Fm), Langhian Numidian sandstones, and Serravallian foredeep sandstones of Serra Palazzo Fm cover the whole succession.

All tectonic units are sealed by the wedge-top basin deposit of the Castelvetere Group (CVTG), which in the Campagna tectonic windows is represented by the Fontana Frigine Fm. Other uppermost Messinian–Pliocene wedge-top basin deposits, with minor extension, unconformably cover the older successions. Finally, Lower–Middle Pleistocene alluvial and lacustrine clastic deposits, including the Eboli conglomerates, extensively cover the analyzed area.

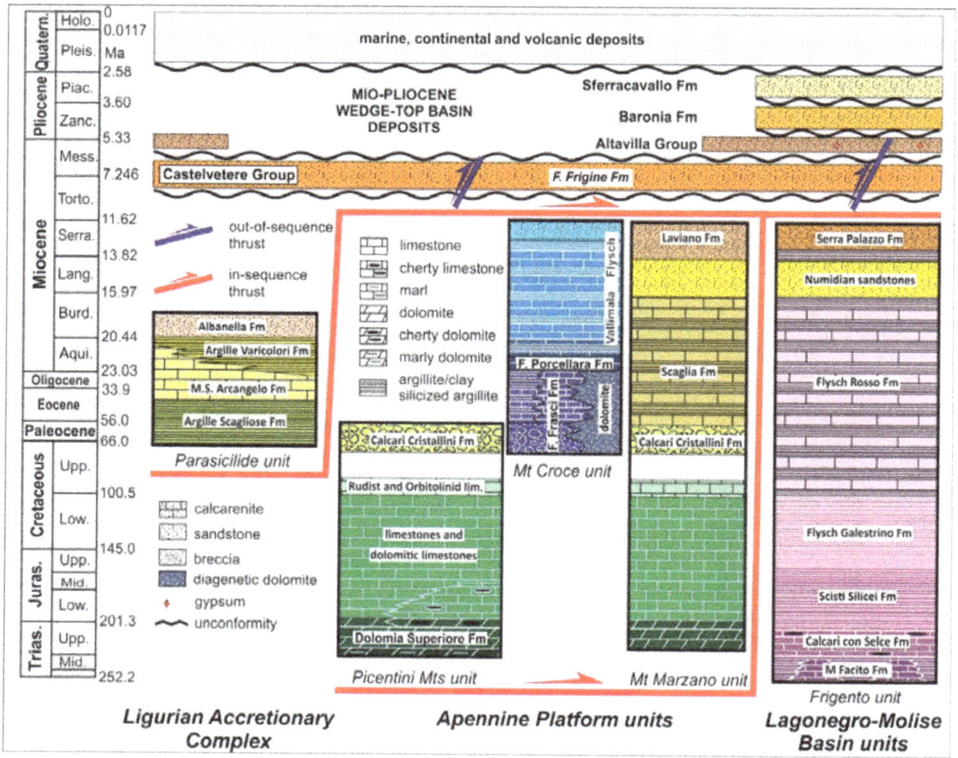

Figure 2. Schematic stratigraphic logs of the successions exposed in the Giffoni and Campagna tectonic windows (modified after [11]).

The Campagna tectonic window is a 12 km long E–W trending structure (Figure 3). Here, a complex thrust-sheet pile is tectonically exposed below the Apennine Platform succession (Picentini Mts unit). From top to bottom, the footwall consists of the Mt Croce unit (Apennine Platform margin), covered by the upper Miocene wedge-top basin of the Castelvetere Group (CVTG), sandwiched between the Frigento unit (Lagonegro-Molise Basin) thrust-sheets.

The Giffoni tectonic window (Figure 4) includes a larger structure (Giffoni Sei Casali) and two minor windows (Giffoni Valle Piana). Here, the Frigento unit is exposed, tectonically covered by the Picentini Mts unit. Both the Giffoni and Campagna areas host low-angle normal faults (LANFs; [22,23]) and high-angle normal faults. The latter structures frequently hide the exposition of the thrust faults. In particular, Lower Pleistocene SW-dipping normal faults, with hundreds of meters of displacement, lowered the SW sector (Salerno Plain), bringing near the highest thrust-sheet (Parasicilide unit) and the lowest succession (Triassic rocks) of the Picentini Mts unit. The volumetrically massive deposits of Eboli Conglomerates are associated with the activity of these normal faults [45].

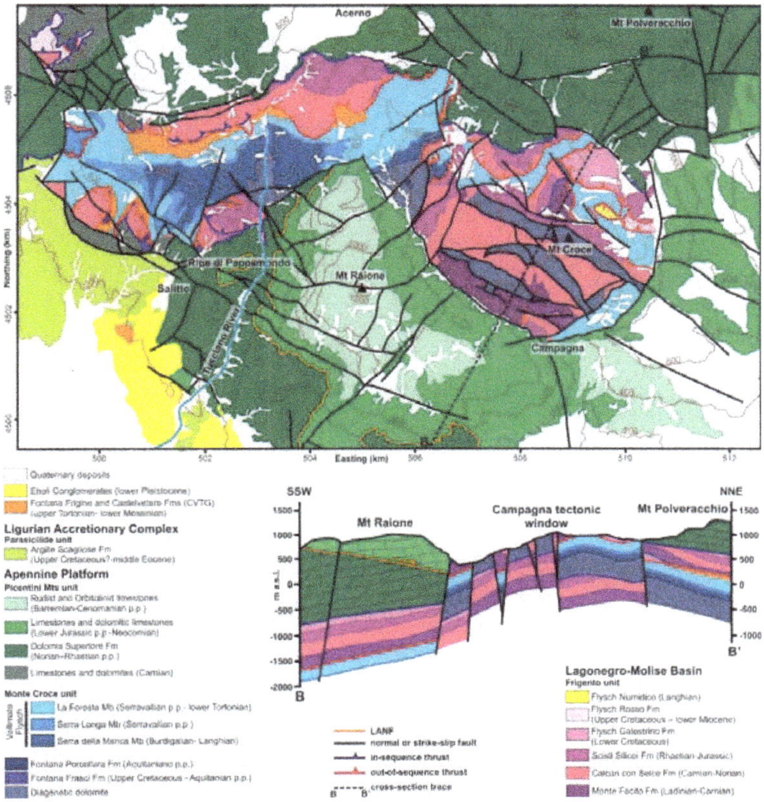

Figure 3. Geological map of the Campagna tectonic window (modified after [23,24]). WGS84-F33N Projection.

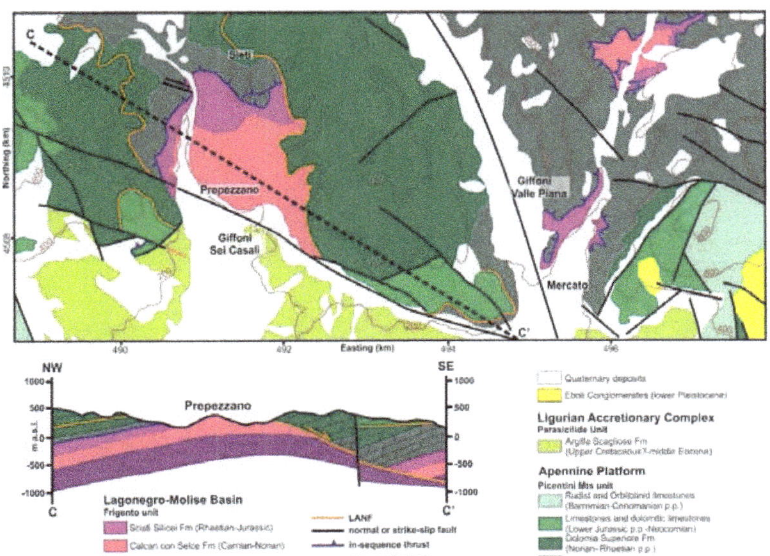

Figure 4. Geological map of the Giffoni tectonic window (modified after [23]). WGS84-F33N Projection.

3. Structures

In the following paragraphs, we illustrate the mesoscale structures observed in the tectonic windows of Campagna and Giffoni. To discriminate between out- and in-sequence thrusts, we used, as a criterion to identify the out-of-sequence structures, involvement in the thrusting of the wedge-top basin deposit of CVTG, whose sedimentation occurred within basins located on top of the already structured thrust-sheet prism. To reconstruct the chronology of the deformation structures, we analyzed the superposition relations between them. In not all areas, structures with different ages are sufficiently exposed. For this reason, we first illustrate the Salitto area because of the exposure completeness of the deformation structures that allowed us to reconstruct the temporal succession of the different deformation stages.

3.1. Campagna Tectonic Window

In the Salitto area (Figure 3), well-exposed structures occur in the footwall of the out-of-sequence thrust that places the Frigento unit above the Monte Croce unit with the interposition of the CVTG deposits. The oldest deformation structures are well-recorded in the Mt Croce unit. They are overturned tight folds (F1) associated with thrust faults (TF1; Figure 5a–c). These structures are cut by normal faults (NF2) frequently appearing as oblique faults (Figure 5a,c). Usually, drag folds are associated with these extensional faults (Figure 5c,d). F1, TF1, and NF2 structures are, in turn, deformed by late thrust faults and associated folds (Figure 5a,c–e, and Figure 6a,e,g). TF3 thrust faults are more developed in the Salitto area, frequently parallel to the NF2 faults (Figure 5d), at places forming conjugate sets (Figure 5e). In many instances, the calcareous beds of the Mt Croce unit host tectonic stylolites orthogonal to the bedding. Commonly, they form two orthogonal sets, suggesting E–W and N–S shortening directions, with the last one the youngest (Figure 5f).

TF3 thrust faults are locally tilted by folding associated with the late TF4 thrusts and presently appearing as strike-slip faults with a ramp-flat geometry and associated vertical folds (Figure 6a). At mesoscale, the two late thrusting events (TF3 and TF4) and associated fold sets (F3 and F4) form interference patterns such as that illustrated in Figure 6b where a TF3 and related F3 fold, verging to NE, are folded by an F4 fold, indicating an almost perpendicular NW–SE shortening. The two folds form a dome structure (type 1 of Ramsay's classification; [46]).

In the marly layers of the Mt Croce succession, the more developed F3 folding forms an axial plane cleavage (Figure 6c,d) dipping to the west. These two sets of late thrusts also deform the sandstones and conglomerates of the CVTG (Figure 6e–g). Minor thrusts (Figure 6e) and folds (Figure 6f) occur within the CVTG conglomerates with the development, in the marly layers, of an axial plane cleavage (Figure 6f). Finally, it is common to observe on the slickenside plane of thrust faults, hosted in the CVTG sandstones, two orthogonal fiber sets, E- and N-verging, with the latter generally superposed to the former one (Figure 6g).

Poles to bedding form a rough NE–SW-directed girdle (Figure 7a), with the main cluster suggesting a mean moderately SW-dipping bedding for the Salitto area. The projection of the fold axes (A1) of the folding set (F1) indicates dominant subhorizontal NW–SE-directed structures with some steep axes (Figure 7b). Poles to axial planes (AP1) suggest a dominant NE vergence (Figure 7c), as well as the few measured TF1 thrust faults (Figure 7d). Because the subsequent folds tilted older structures, such as the TF1, we restored them, reporting to horizontal the bedding. Hence, we rotated each tilted TF1 thrust faults (Figure 7e). We used as the rotation-axis, the direction line of the measured bedding crosscut by the thrust fault, and the bedding dip as the rotation angle. After the rotation, the data inversion indicates a NE–SW shortening (Figure 7f).

Figure 5. Campagna tectonic window (Salitto). Mt Croce unit: (**a**) early fold and thrusts (TF1) crosscut by normal faults (NF2) subsequently cut by a late thrust (TF3); (**b**) early thrust fault (TF1) verging to the east with an associated overturned syncline in the footwall, the fold host veins in the outer arc, and cleavage in the inner arc; (**c**) early normal faults (NF2) subsequently deformed by a late thrust fault (TF3); (**d**) normal (NF2) and thrust (TF3) faults; (**e**) conjugate thrust faults; (**f**) two orthogonal sets of stylolites.

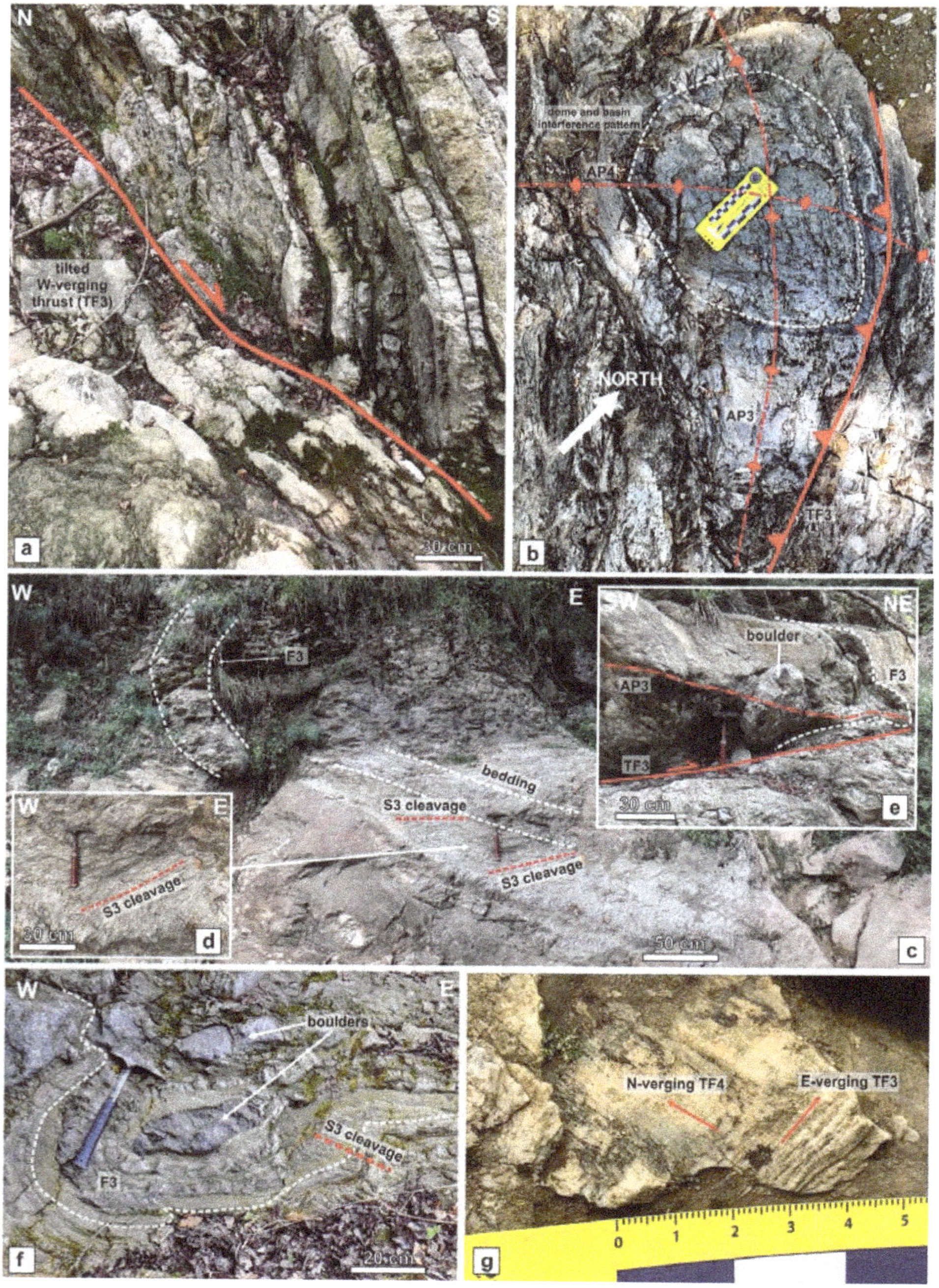

Figure 6. Campagna tectonic window (Salitto). Mt Croce unit: (**a**) W-verging thrust (TF3) subsequently tilted by an N–S shortening (D4 stage). (**b**) Dome-and-basin interference pattern between the two orthogonal folding events related to the D3 and D4 stages; (**c**) macro-scale overturned fold in the marls and sandstones; (**d**) fold axial plane cleavage associated with the macro-scale fold. CVTG: (**e**) minor thrust fault and hanging wall anticline in the CVTG. (**f**) Minor fold in the CVTG conglomerate; (**g**) two orthogonal sets of slickenside fibers along the same thrust plane.

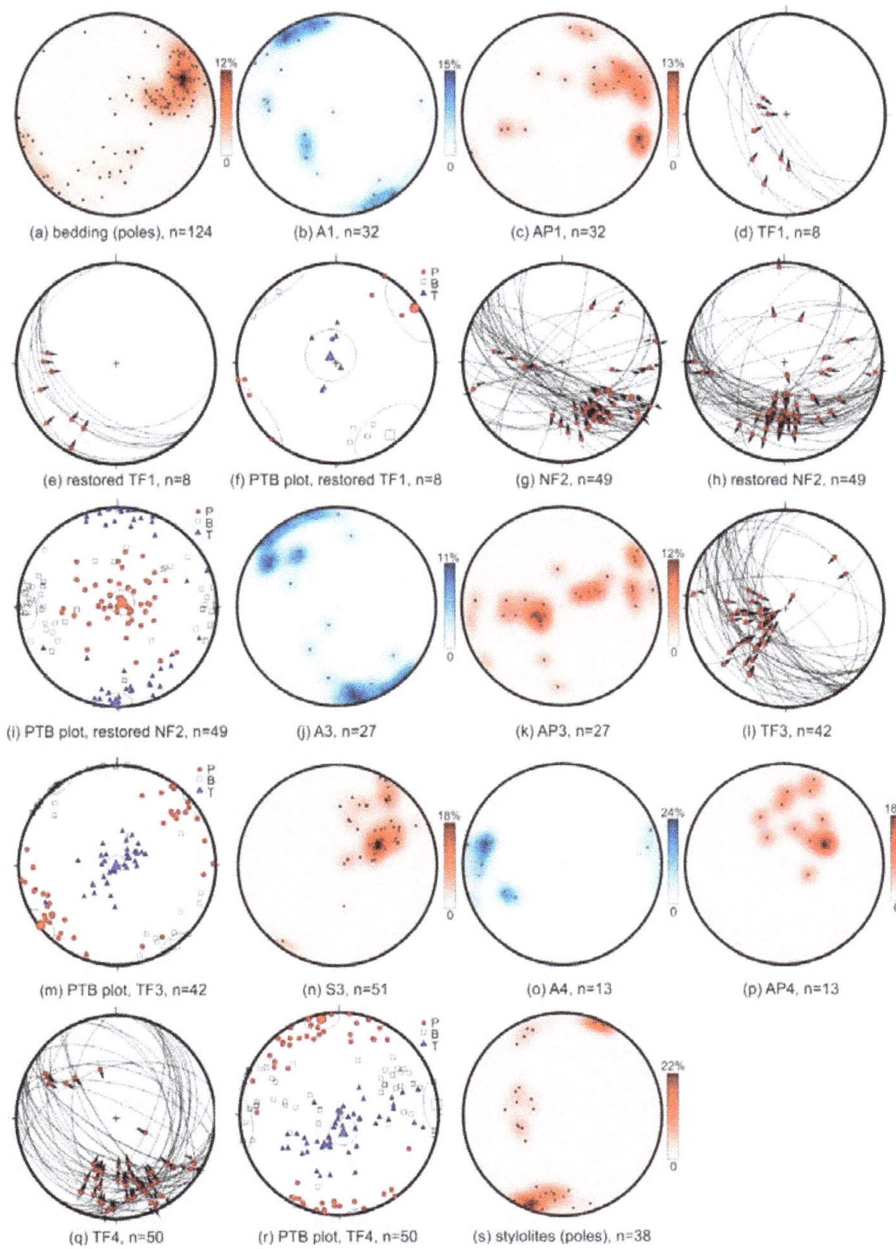

Figure 7. Salitto (Campagna tectonic window). (**a–s**) Stereographic projections (lower hemisphere, equiareal net) and PBT plots of the analyzed structures described in the text. A—fold axis; AP—axial plane; NF—normal fault; TF—thrust fault. S—tectonic foliation. Color bar indicates the density percentile of the contour plot. (**f**) θ = 30°; mean vectors; P: 055/04 (R = 85%); T: 143/11 (R = 83%); B: 324/82 (R = 87%). (**i**) θ = 44°; mean vectors; P: 076/86 (R = 74%); T: 268/09 (R = 65%); B: 181/03 (R = 75%). (**m**) θ = 54°; mean vectors; P: 232/03 (R = 75%); T: 144/02 (R = 80%); B: 316/88 (R = 87%). (**r**) θ = 40°; mean vectors; P: 349/06 (R = 70%); T: 085/01 (R = 50%); B: 166/76 (R = 68%).

We inverted fault kinematic data by the PBT method [47] that provides for every fault, defined by plane attitude, slip orientation, and kinematics, the axis of maximum shortening (P), maximum stretching (T), and the intermediate axis (B), orthogonal to the P–T plane. We used the software TectonicsFP (version 1.79; [48]), which allows one to calculate the best-fit angle (θ), minimizing the sum of all misfit angles between the measured slip direction and the calculated maximum shear stress. Furthermore, the software also provides the confidence cones of every axis expressed as a percentile (%R). In the case of lacking conjugate faults, we used a fixed value of angle θ equal to 30° for all fault-slip data.

NF2 faults show dominant strike-slip kinematics (Figure 7g); however, when restored (Figure 7h), they indicate dominant normal kinematics. The inversion of the NF2 data suggests an N–S extension (Figure 7i). A3 fold axes are mainly subhorizontal with a dominant NNW–SSE direction (Figure 7j), and poles to AP3 axial planes (Figure 7k) form a girdle suggesting an ENE–WSW shortening. TF3 thrust faults generally are moderately dipping to SW (Figure 7l), marking a NE–SW shortening (Figure 7m). In addition, the poles to S3 cleavage (Figure 7n) provide a mean moderately SW-dipping plane. A4 fold axes are generally gently dipping or subhorizontal with an E–W direction (Figure 7o), and AP4 poles indicate a dominant N-vergence (Figure 7p), as well as the TF4 thrust faults (Figure 7q), which inversion furnishes an N–S shortening (Figure 7r). Finally, poles to stylolites (Figure 7s) indicate two main shortening directions: NNE-SSW (main) and E–W (secondary).

In the Acerno-Tusciano River area (Figure 3), D3 and D4 structures are better-developed with respect to the D1 and D2 structures. Several mesoscale D4 back-thrusts affect the Mt Croce unit frequently with associated anticline and syncline in the hanging wall and footwall, respectively (Figure 8a). A decametric-sized shear zone is located in the footwall of a main FT4 thrust (Figure 8b). Here, shales and marly levels are characterized by widespread S-C structures, with shear planes (C) forming low-angles to the bedding (Figure 8b) and indicating a mean SSW vergence. F3 and F4 folds are from open to tight (Figure 8c–e). The F3 folds form an interference pattern of type 3 of Ramsay's classification [46] with the F1 folds (Figure 8d) and of type 2 with the F4 folds (Figure 8e). In the Campagna area, the hanging wall carbonates host several LANFs. The largest is localized in Mt Raione (Figure 3). Here, the western slope of the mountain is defined by the superposition of Jurassic onto Triassic rocks through a LANF. This structure is highlighted by the increase in the Jurassic succession thickness toward the NE. A segment of the Mt Raione LANF is exposed in the Ripe di Pappamondo (Figure 3); here, a major low-angle normal fault separates the Jurassic limestones in the hanging wall and Triassic dolomites in the footwall with a well-developed cataclasite (Figure 8f). Here, the gently dipping fault plane shows small ramps where slickenside striations occur, indicating normal kinematics (Figure 8g).

A3 fold axes are generally subhorizontal and N–S directed (Figure 9a). Poles to AP3 axial planes form an E–W girdle (Figure 9b), indicating both vergences to E and W, as well as the TF3 thrust faults (Figure 9c). The inversion of fault data suggests an ESE–WNW shortening (Figure 9d). A4 fold axes (Figure 9e) are about subhorizontal with a main WSW–ENE direction. AP4 poles form a girdle with a main cluster indicating dominant SSE dipping planes (Figure 9f). Finally, TF4 thrust faults show both vergences to N and S (Figure 9g), and the PBT plot suggests an N–S shortening (Figure 9h). S and C planes of the S-C structures (Figure 9i) generally dip to N/NE, with the C-planes showing lower dip angles. We calculated the attitude of the reverse slip vectors as that of the line forming an angle of 90° from the intersection between S and C planes. The projection of the C-shear planes and corresponding slip vectors (Figure 9i) indicates a prevalence of sense of shear top to the south. LANFs measured in the Ripe di Pappamondo area are characterized by planes and slip vectors dipping both to the north and south (Figure 9j). The inversion of LANF kinematic data indicates an N–S extension (Figure 9k).

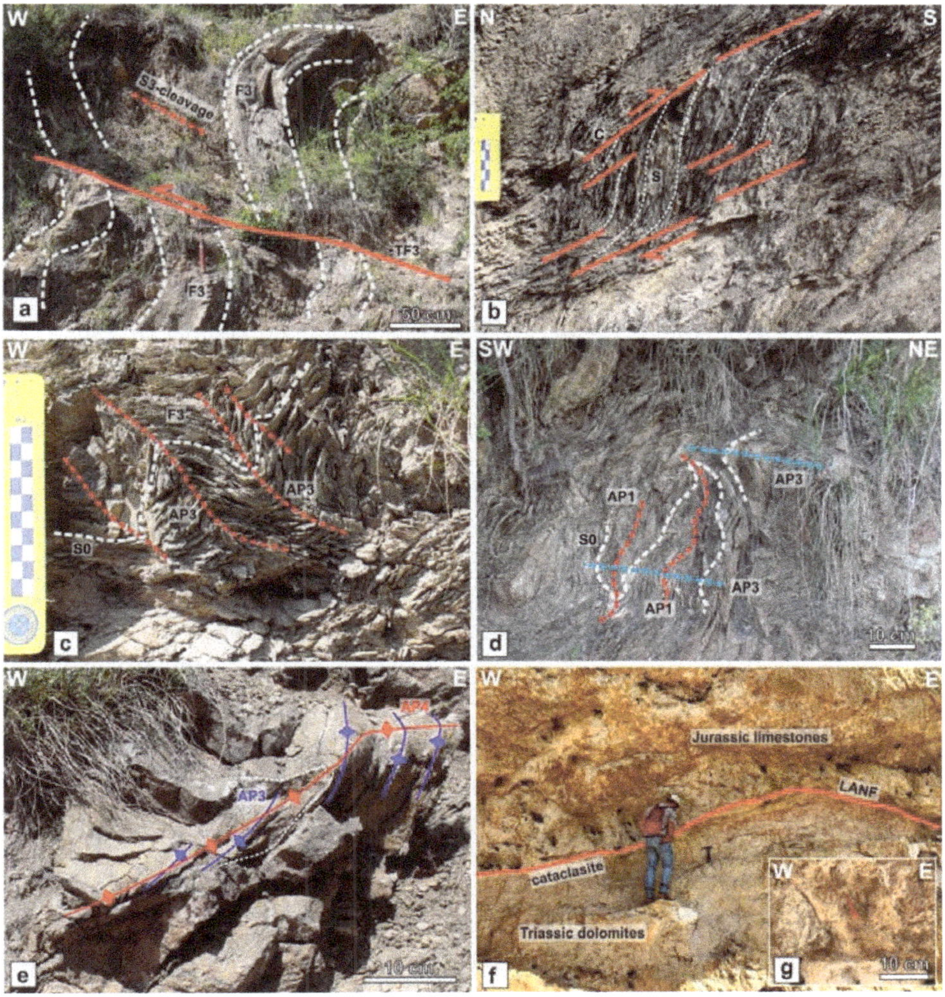

Figure 8. Campagna tectonic window (Acerno-Tusciano River). Mt Croce unit: (**a**) thrust fault (TF4) verging to SW with associated anticline and syncline in the hanging wall and footwall, respectively; (**b**) shear zone located in the footwall of a TF4 with S-C structures (S is the tectonic cleavage S1 associated to the D1 deformation); (**c**) F3 folds verging to W. Interference pattern between F3 and F4 folds: (**d**) type 3; (**e**) type 2. (**f**) Low-angle normal fault (LANF) located between Triassic dolomites (footwall) and Jurassic limestones (hanging wall) of the Apennine Platform (Ripe di Pappamondo). (**g**) Particulars of the LANF of the previous image, showing the gently dipping fault plane with slickenside striations.

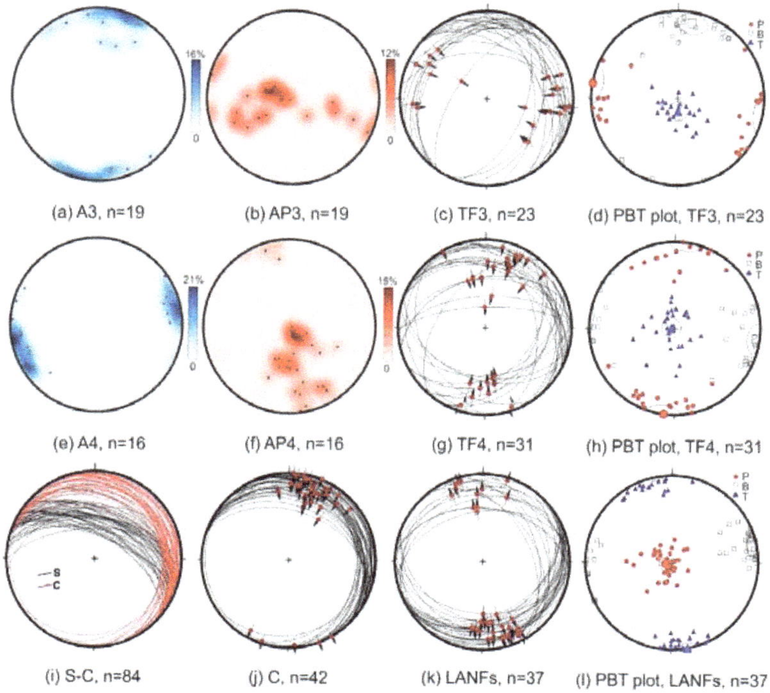

Figure 9. Acerno-Tusciano River (Campagna tectonic window). (**a–l**) Stereographic projections (lower hemisphere, equiareal net) and PBT plots of the analyzed structures described in the text. A—fold axis; AP—axial plane; TF—thrust fault. Color bar indicates the density percentile of the contour plot. (**d**) θ = 26°; mean vectors; P: 280/01 (R = 81%); T: 011/14 (R = 84%); B: 178/78 (R = 90%). (**h**) θ = 26°; mean vectors; P: 187/00 (R = 79%); T: 097/06 (R = 82%); B: 283/86 (R = 85%). (**l**) θ = 70°; mean vectors; P: 258/84 (R = 92%); T: 079/07 (R = 91%); B: 169/01 (R = 92%).

3.2. *Giffoni Tectonic Window*

The structural survey of the Giffoni Valle Piana and Giffoni Sei Casali areas was carried out in the localities of Sieti, Prepezzano, and Mercato villages (Figure 4). In-sequence thrust faults (TF1) are excellently exposed in these areas. Generally, Carnian dolomites of the Picentini Mts unit tectonically superpose on the Lower Jurassic silicized argillites of the Frigento unit (Figure 10a,b). In the Prepezzano locality, dolomites in the hanging wall are characterized by an intense fracturing, and frequently, a cataclasite made exclusively of dolomitic clasts occurs associated with the TF1 thrust (Figure 10b). Several Riedle shears crosscut the F1 thrust plane (Figure 10c), indicating a SE vergence. Rocks located in the footwall (Scisti Silicei Fm) host several deformation structures. The oldest ones are minor thrust faults (Figure 10d,e), probably associated with the in-sequence thrusts. These early structures are dislocated by normal faults subsequently tilted and now commonly appearing with reverse kinematics (Figure 10d–f). Similar early thrust faults are also hosted in the Calcari con Selce Fm (Figure 10g). All described structures are deformed by subsequent thrusting and folding stages. Similarly to the previous analyzed areas, we observed two sets of late thrusts (TF3-4). TF3 thrusts are defined by vergences both to E and W, whereas TF4 thrusts show dominant N-vergences. Riedel shears occur associated with both structures.

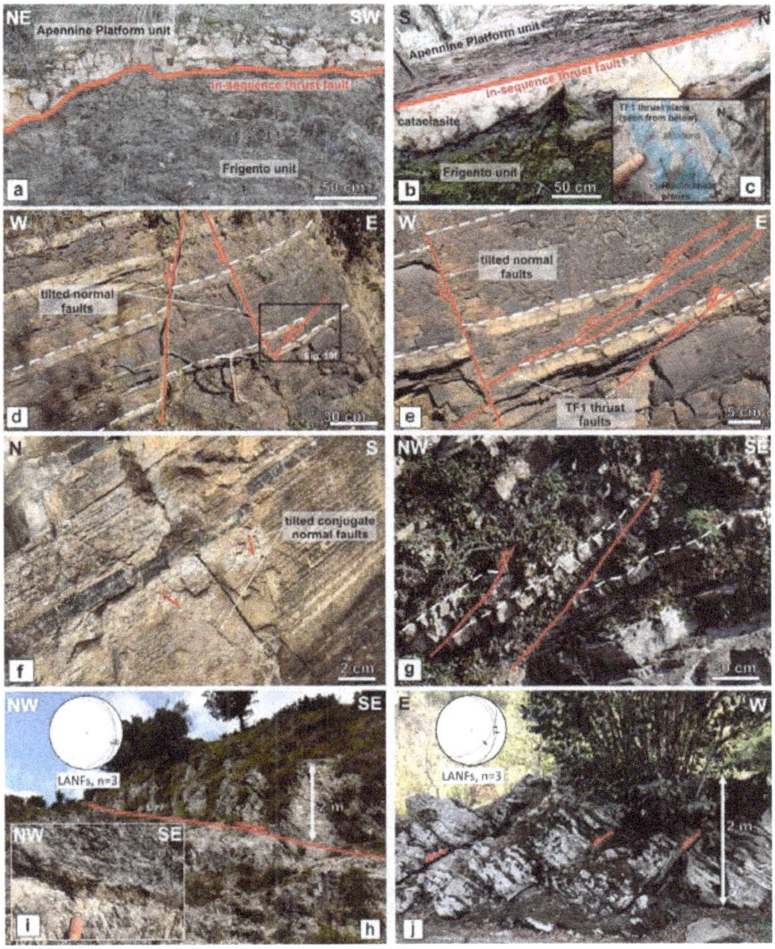

Figure 10. Giffoni tectonic window. (**a**,**b**) in-sequence thrust (TF1) between the Apennine Platform unit (Triassic dolomites) in the hanging wall and the Frigento unit (Scisti Silicei Fm) in the footwall (Prepezzano). (**c**) Riedle shears crosscutting the TF1 thrust fault plane. Frigento unit (Scisti Silicei Fm): (**d**,**e**) tilted conjugate normal faults (NF2) crosscutting thrust faults (TF1) (Prepezzano). (**f**) Tilted conjugate normal faults (Sieti). (**g**) Minor thrust faults (TF1) in Calcari con Selce Fm (Giffoni Valle Piana). (**h**) LANF in Rhaetian–Jurassic dolostones (Giffoni Valle Piana). (**i**) Fault gauge associated with the LANF of the previous picture. (**j**) minor LANFs in Carnian well-bedded limestones (Prepezzano).

As reported in Figure 4, the tectonic window is characterized by the occurrence of LANFs, such as the Campagna area. A well-exposed LANF occurs in the Rhaetian–Jurassic dolostones in the hanging wall of the Giffoni Valle Piana tectonic window (Figure 10h). A gauge is present along the fault plane (Figure 10i). This LANF dips to ESE with an angle of 30° (stereographic projection of Figure 10h). LANFs with minor displacements are widespread in the area, such as illustrated in Figure 10j, always showing a down to E/SE sense of shear.

Poles to bedding (Figure 11a) form a main cluster indicating a mean moderately W-dipping plane (mean value 283/29). A1 fold axes are subhorizontal with a NE–SW direction (Figure 11b). Poles to AP1 axial planes form an NW–SE girdle distribution (Figure 11c). The Riedel shears associated with the TF1 thrust faults (Figure 11d), when restored, indicate a mean ESE vergence (Figure 11e). Tilted NF2 normal faults (Figure 11f) when restored (Figure 11g) suggest an E–W extension (Figure 11h). TF3 thrust faults form conjugate sets, both verging to W and E (Figure 11i) with the PBT plot, marking an E–W shortening (Figure 11j). Riedel shears associated with TF3 thrusts indicate both W and E vergences (Figure 11k). Finally, TF4 thrust faults are mainly dipping to NNW, indicating an SSE vergence (Figure 11l). The PBT plot (Figure 11m) shows an NNW–SSE shortening. The few Riedel shears furnish an NNW vergence (Figure 11n).

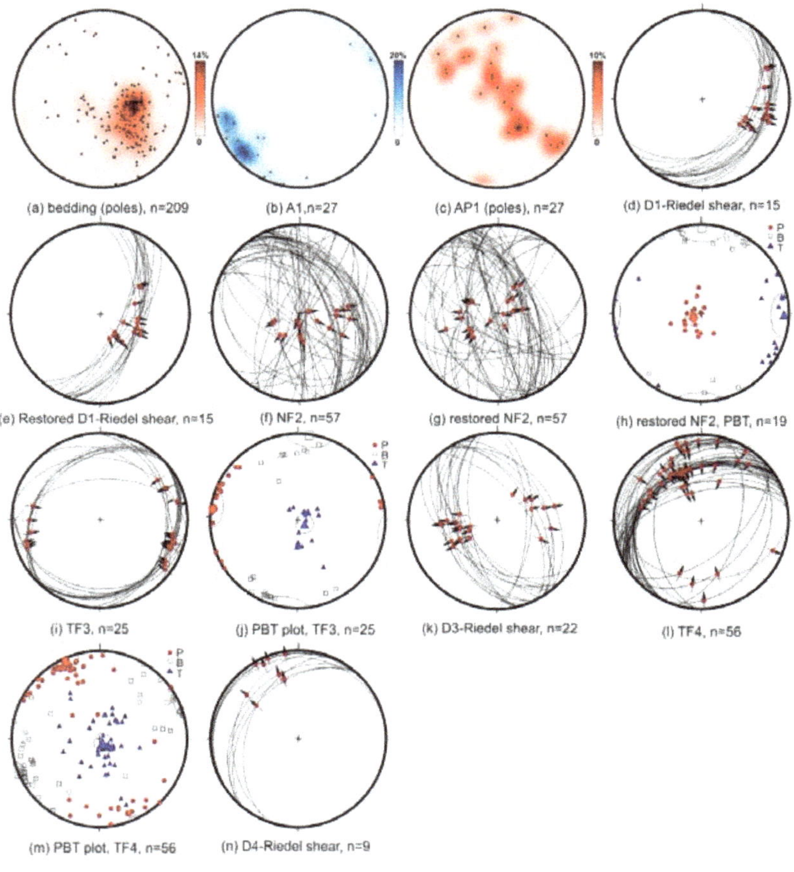

Figure 11. Giffoni tectonic window. (**a–n**) Stereographic projections (lower hemisphere, equiareal net) and PBT plots of the analyzed structures described in the text. A—fold axis; AP—axial plane; NF—normal fault; TF—thrust fault. (**h**) θ = 22°; mean vectors; P: 141/80 (R = 88%); T: 004/10 (R = 78%); B: 273/03 (R = 74%).(**j**) θ = 20°; mean vectors; P: 277/03 (R = 82%); T: 007/01 (R = 74%); B: 119/83 (R = 90%). (**m**) θ = 38°; mean vectors; P: 155/03 (R = 81%); T: 247/01 (R = 78%); B: 005/87 (R = 84%).

4. Discussion

The tectonic windows of Giffoni and Campagna provide the opportunity to study naturally exposed sections of the allochthonous wedge both in terms of stratigraphy and tectonics. The successions of the Picentini Mts unit (Apennine Platform) and the Frigento unit (Lagonegro-Molise Basin) are well-studied and are widespread in the whole southern Apennines [7,8,10,21,22,32–35,38–43]. On the contrary, the Mt Croce unit crops out only in the Campagna tectonic window, and it has been poorly studied [19,20,35–37]. This succession shows analogies with that exposed in the Laviano area, on the northern side of the Mt Marzano (Figures 1a and 2). The Laviano succession was recently studied by [49]. It is an excellent example of shallow-water Triassic–Upper Cretaceous carbonates covered by a margin to deep basin succession from Maastrichtian to Serravallian in age. The authors relate this deepening of the depositional environment with the Late Cretaceous–Paleogene abortive rifting that occurred in Adria, in analogy with that observed from Libya to southern Sicily [49]. A similar margin to slope succession characterizes the Mt Croce unit. It starts with Upper Cretaceous–Paleocene calcareous breccias, corresponding to the Calcari Cristallini Fm, a widespread recrystallized calcareous clastic deposit, marking the start of the platform sinking in several places of the southern Apennines [7,8,49]. The succession passes upward to a thick, scaglia-type slope calciclastics deposit and basin calcilutites, marls, and argillites. The succession ends with Serravallian foredeep sandstones corresponding to the similar sediments in the Laviano area. Finally, unconformable deposits of the upper Tortonian–lower Messinian CVTG cover both successions. This close analogy suggests a similar paleogeographic position in the uppermost Cretaceous, where they formed the eastern margin of the Apennine Platform. Furthermore, both margin successions were involved in the same out-of-sequence thrusting event [11].

The structural analysis presented in the previous paragraphs gives us some inputs to propose a possible reconstruction of the tectonic events of this key sector of the southern Apennines. The best place to observe the superposition between the different orogenic stages is the Salitto outcrop (Campagna tectonic window). Here, we have recognized the superposition of five deformation stages in the footwall of the out-of-sequence thrust, which juxtaposes the Frigento unit onto the Mt Croce unit, with the interposition of the CVTG rocks. The first (D1) includes regional thrusts and minor mesoscale thrusts with related folds, all defined by a vergence to the E/SE. This in-sequence thrusting event allowed the tectonic covering of the Apennine Platform rocks (Picentini Mts unit) onto the Lagonegro-Molise Basin deposits (Frigento unit). The second stage (D2) was recorded by normal faults, deformed by the subsequent shortening deformations. As suggested by [11,31], we associate this extensional event with the formation of several structural depressions where the CVTG sediments were deposited. A synorogenic extensional environment was also envisaged for the older Langhian–lower Tortonian wedge-top basin deposits of the Cilento Group [27,28,31]. A recent study [50] in the Cilento area (Figure 1a) about temperature-dependent clay minerals and vitrinite reflectance in the Cilento Group and Mt Sacro Fm (CVTG) indicates a basin evolution marked by two phases of severe subsidence, interpreted as the result of syn-orogenic extension at shallow crustal levels. Hence, this extensional event occurred in the southern Apennines with the deposition of CVTG sediments within structurally controlled depocenters. Furthermore, in several outcrops of the southern Apennines, including those localized in the Campagna tectonic window, the base of CVTG deposits is marked by carbonate conglomerates with an arenaceous matrix [14,15,40,51–53]. This feature could mark the erosion of carbonates placed in the footwall of the D2 normal faults.

The latter two stages (D3 and D4), defined by thrusts, folds, S-C structures, and tectonic stylolites, also affected the wedge-top basin deposits of CVTG, highlighting the out-of-sequence nature of these deformation events. The first two events (D1 and D2) are also well-recorded in the Giffoni area. Here, the Frigento unit (footwall) hosts early thrusts (D1) crosscut by normal faults (D2) and subsequently tilted. Furthermore, the younger deformations have been recorded only as minor thrusts. In the Acerno-Tusciano River area (Campagna tectonic window), the last two shortening stages (D3 and D4) were mainly recorded.

According to the orientation analysis, the D1 stage was characterized by dominant NE (Salitto) and E/SE (Giffoni) vergences. Hence, a mean tectonic transport to the east is inferred for the in-sequence thrusting (D1) in the study area. The D2 event was defined by N–S (Salitto) and E–W (Giffoni) extensions. These two orthogonal directions suggest a radial extensional strain field that characterized the D2-extension in the study area. The D3 stage was characterized by NE–SW (Salitto) and E–W (Acerno-Tusciano River and Giffoni) shortening directions with a prevalence of the NE vergence for the Salitto area and both vergences (NE and SW) for the Acerno-Tusciano River and Giffoni sectors. Finally, the D4 event with N–S (Salitto and Acerno-Tusciano River) and NNW–SSE (Giffoni) shortening directions had dominant vergences toward N (Salitto), S (Giffoni), or both N and S (Giffoni).

In order to carry out the mean shortening directions for the out-of-sequence thrusting events (D3-4), we joined all measurements of the two tectonic windows (Figure 12a,c) and performed the PBT plots (Figure 12b,d). The data inversion indicates the dispersion of the P-axes in both cases with mean shortening directions about WSW–ENE- and NNW–SSE-oriented for the D3 (Figure 12b) and D4 (Figure 12d) stages, respectively. It is worth noting as in both D3 and D4 events, thrusts and back-thrusts are equally present.

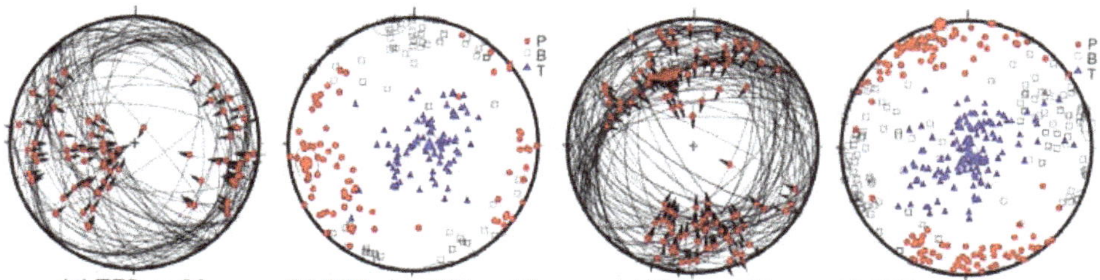

(a) TF3, n=90 (b) PBT plot, TF3, n=90 (c) TF4, n=137 (d) PBT plot, TF4, n=137

Figure 12. (**a**–**d**) Stereographic projections (lower hemisphere, equiareal net) and PBT plots of the total out-of-sequence thrust faults of D3 and D4 deformation stages measured in both tectonic windows. (**b**) θ = 32°; mean vectors; P: 260/15 (R = 63%); T: 349/04 (R = 66%); B: 086/80 (R = 81%). (**d**) θ = 30°; mean vectors; P: 348/00 (R = 74%); T: 080/02 (R = 64%); B: 1466/87 (R = 77%).

As concerns the two out-of-sequence thrusting stages (D3 and D4), as shown in previous papers [11,14,15], they form the most common shortening structures in the southern Apennines. The age of these events is late Messinian–early Pliocene as suggested by different features, including (i) these thrusts deform the upper Tortonian–lower Messinian deposits of CVTG; (ii) the analogy with similar structures described in the Central Apennines dated late Messinian–early Pliocene [54–57]; (iii) recent dating of the out-of-sequence thrust of the Mt Massico [17] to 5.1 ± 3.7 Ma (early Pliocene) through the U-Pb geochronology on synkinematic calcite fibers; and finally, (iv) the fission tracks analysis on Apatites, collected within sandstones of the Mt Croce unit [6], furnishes an age of 5.2 ± 0.9 Ma (early Pliocene) for the oldest tectonic exhumation event of this thrust-sheet.

In the Campagna tectonic window, the D3 and D4 thrusts indicate E–W and N–S shortening directions, such as previously described by [22] and observed in other areas of the southern Apennines [11]. According to [11,14,15], these out-of-sequence thrusts were formed by the positive inversion of normal faults located in the buried Apulian Platform. Some of them propagated upward, crosscutting the overlying allochthonous wedge, and, although they form as in-sequence thrusts at deeper structural levels with respect to the orogenic belt, they appear as out-of-sequence structures at surface. Such tectonic features are called envelopment thrusts [12–14].

All these geological features allowed us to reconstruct three geological sections, at a regional scale, crossing the two tectonic windows and the Mt Marzano-Laviano sector (Figure 13). To constraint the cross-sections, we used information from Acerno01, San Gregorio Magno01, Contursi01, and Nusco02 boreholes [58,59] and the seismic profile of CROP04 interpreted by [60].

The B″-B‴ cross-section (Figure 13b) starts from the Campagna sector up to the thrust front located on the north side of the Picentini Mts. Hence, in our reconstruction, the out-of-sequence thrust, which allowed the superposition of the Mt Croce unit onto the Frigento unit, joins to that exposed at the front of the Picentini Mts, such as suggested in [11]. This interpretation also suggests that the Campagna tectonic window is related to the formation of an E–W trending antiform associated with a blind late back-thrust rooted in the Apulian carbonates. Furthermore, we suggest that the margin succession of Mt Croce laterally passes northward to the pelagic succession of the Frigento unit.

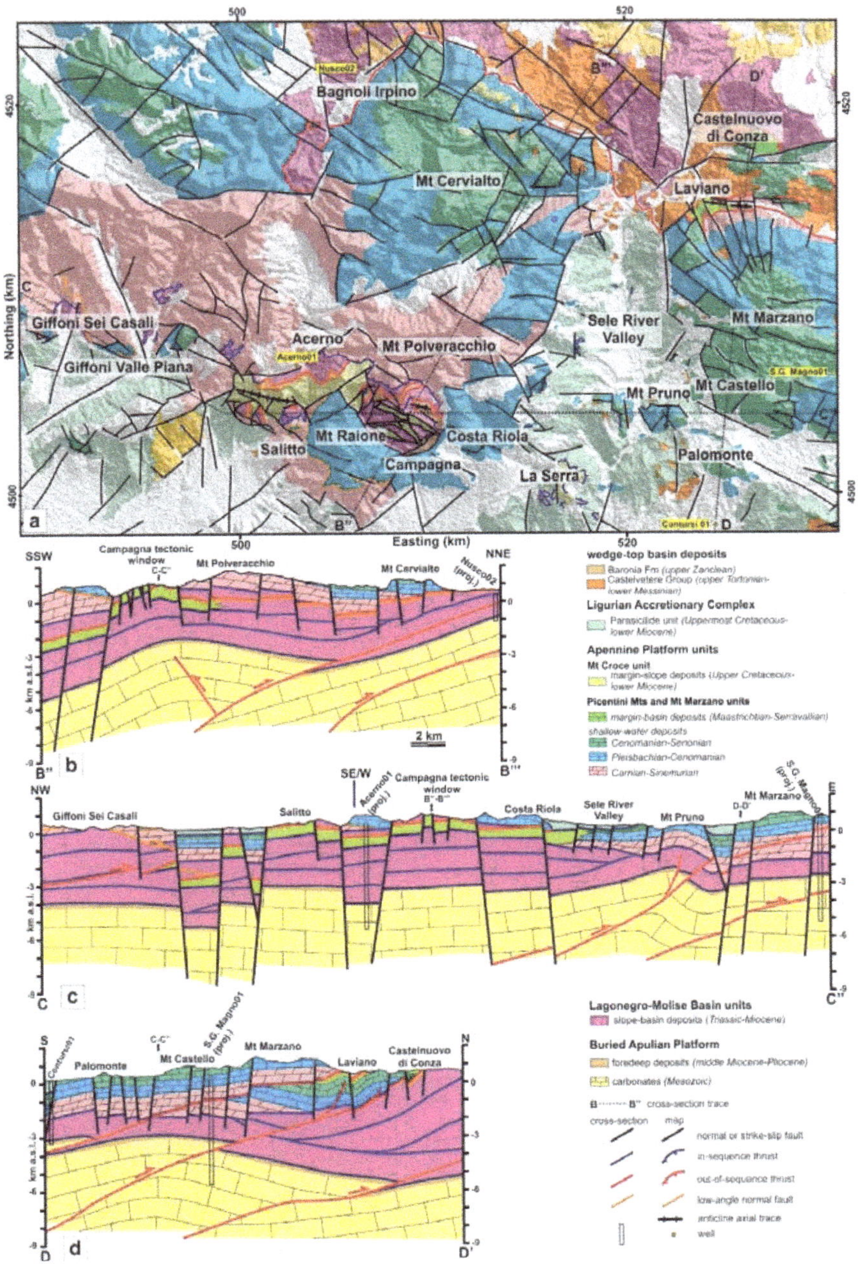

Figure 13. (**a**) Geological map of the Giffoni and Campagna tectonic windows, Picentini Mts, Sele River Valley, and Mt Marzano (modified after [8]). (**b–d**) Geological cross-sections.

The C-C″ cross-section (Figure 13c) runs through the two tectonic windows up to the Sele River Valley. In our interpretation, the out-of-sequence thrust exposed in the Campagna tectonic window is dipping to the west, forming a frontal anticline in the hanging wall (Costa Riola fold, Figure 13a) bounding the western side of the Sele River Valley. Furthermore, another thrust fault in the buried

Apulian carbonates propagates upward, breaching, and forming the antiform within the Apennine Platform carbonates (Mt Pruno fold, Figure 13a). As concerns the stratigraphy, we suggest a westward lateral variation of the Apennine Platform carbonates (Mt Marzano unit) to the margin succession of the Mt Croce unit. Such a hypothesis is supported by the exposition in the Sele River Valley of Calcari Cristallini Fm below the Parasicilide unit (La Serra, Figure 13a)

Finally, section D-D′ (Figure 13d) crosscuts the Mt Marzano up to the Laviano area. Here, differently from the Mt Croce unit, the Maastrichtian–Serravallian margin to the basin Laviano succession is covered by the shallow-water Mesozoic carbonates of the Apennine Platform (Mt Marzano unit). The cross-section shows as the Mt Marzano unit, rocks cover the Laviano succession with the interposition of the CVTG through an out-of-sequence thrust propagated by the buried Apulian carbonates. In turn, the Laviano succession overthrusts the Frigento unit again with the CVTG in the footwall. The latter thrust has two blind splays forming two large folds in the Laviano area, and the frontal ramp forms the anticline of Castelnuovo di Conza.

In the hanging wall of these thrusts, LANFs are present. The structural survey carried out in this study indicates structures with slip-vectors down to SE for the Giffoni tectonic window and both to north and south for the Campagna area. Different works studied these extensional structures in the Giffoni area. The authors in [22,61] related these faults with the synorogenic NW–SE trending extension parallel to the chain axis, whereas in [62], with the opening of the back-arc Marsili oceanic basin during the late Pliocene–Pleistocene. However, in the Campagna area, these structures indicate an N–S extension, not related to the previous directions indicated by [22,61]. Hence, we suggest relating the formation of the LANFs to the out-of-sequence thrusting events, which over-thickened the orogenic wedge, causing extension in the shallow level. This genetic relation is also observed elsewhere in the southern Apennines, such as in Mt Alpi, where deep-seated thrusts allowed the buried Apulian Platform to uplift, triggering LANFs in the overlying allochthonous wedge [63].

From a macro-scale point of view, the area around the two tectonic windows, from the Lattari to Picentini Mts, exposes the oldest (Carnian–Norian in age) rocks of the Apennine Platform. According to [22], we suggest that this feature is related to the formation of a very large E–W trending antiform (about 40 km in length) and its subsequent erosion in the crest. As illustrated in the cross-section B″-B‴ we related this antiform to an out-of-sequence blind thrust.

With this in mind, we reconstructed the succession of the tectono-stratigraphic events for the Campagna and Giffoni tectonic windows between Serravallian and Early Pleistocene shown in the scheme of Figure 14. After the Serravallian–middle Tortonian in-sequence E-verging thrusting (Figure 14a) that juxtaposed the Apennine Platform (Picentini Mts and Mt Croce units) onto the Lagonegro-Molise Basin (Frigento unit), the whole tectonic pile was affected by a radial extensional deformation field characterized by E–W and N–S main directions (Figure 14b). This extension caused a general dismembering of the tectonic prism. This feature triggered erosion within the Apennine Platform carbonates, which fed, together with the siliciclastic supply coming from the overriding plate and the tectonic wedge itself [64,65], the depocenters with the sedimentation of the upper Tortonian–lower Messinian CVTG (Figure 14b). Subsequently, in the late Messinian– early Pliocene, the out-of-sequence E-verging thrusting duplicated the tectonic pile with the interposition of CVTG deposits (Figure 14c). The thrust fault system was characterized by several lateral ramps, verging both to N and S. Synchronous with the out-of-sequence event LANFs developed (Figure 14c,d). In the Giffoni tectonic window, the WNW–ESE extension prevailed, whereas, in the Campagna sector, the N–S extension was dominant. A late out-of-sequence event produced blind-thrust faults that deformed the previously formed structures producing a regional E–W trending antiform (Figure 14d). Finally, with the starting of the Early Pleistocene, this antiform was eroded (Figure 14f) with the formation of a large amount of calcareous detritus, presently forming the Eboli conglomerates [66] widespread to SW of the tectonic windows (Figure 3). This event was associated with an extensional stage [45,66] that lowered the Salerno area with the formation of a significant tectonic depression bounded by normal faults with hundreds of meters of displacement. This severe erosion of the regional antiform allowed the lowermost part of the hanging wall (Triassic rocks) to be exposed, now surrounding the two tectonic windows.

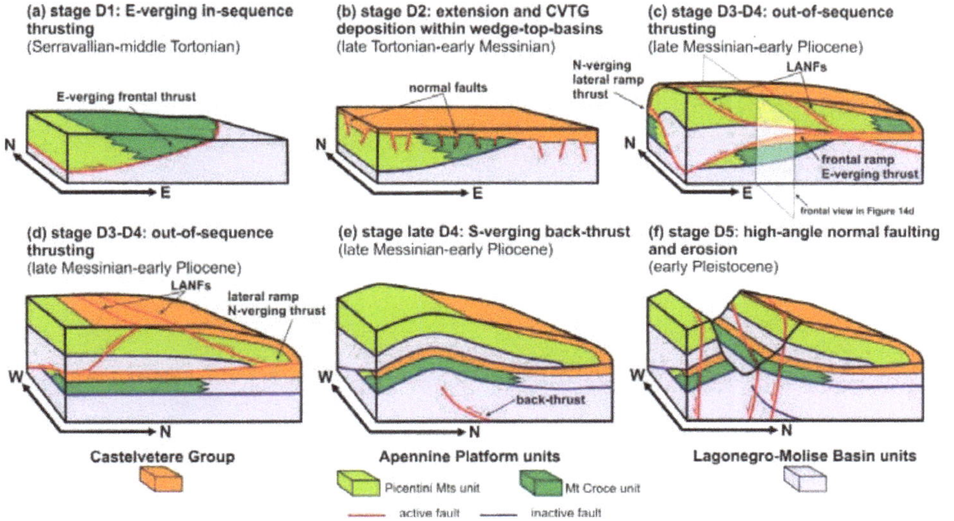

Figure 14. (**a**–**f**) Image showing the tectono-stratigraphic evolution from Serravallian to Early Pleistocene of the Campagna and Giffoni tectonic windows.

5. Conclusions

The detailed structural survey of the tectonic windows of Campagna and Giffoni allowed us to provide, for the first time, the reconstruction of the main orogenic stages that characterized this sector of the southern Apennines chain from the Serravallian to the Early Pleistocene. Furthermore, we clearly identified in- and out-of-sequence thrusting events and an extensional phase, interspersed between the crustal shortening pulses. Finally, we provided a comparison between the succession of Mt Croce, involved in the out-of-sequence thrusting, with that exposed in front of the chain (Laviano area).

The Campagna and Giffoni tectonic windows provide naturally exposed sections showing the superposition of several thrust-sheets. We reconstructed five main events that define the tectonic evolution of this area. The first orogenic pulse (D1) was recorded by an in-sequence thrusting that superposed the Apennine Platform (Picentini Mts and Mt Croce units) onto the Lagonegro-Molise Basin (Frigento unit), which occurred in the Serravallian. D1 structures are well-exposed in the footwall of the main regional compressive faults as mesoscale thrust and folds. A second stage (D2) was characterized by extension recorded as normal faults, tilted by the subsequent deformations. The deposition of the upper Tortonian–lower Messinian wedge-top basin CVTG sediments was related to this extensional event. The stages D3 and D4 were characterized by out-of-sequence thrusting involving the whole tectonic pile, included CVTG deposits that well-recorded these shortening deformations as mesoscale thrusts, folds, S-C, and pressure–solution structures. This thrusting doubled the allochthonous wedge, allowing the superposition of the Frigento unit onto the Mt Croce unit. The latter succession shows several tectono-stratigraphic analogies with that exposed in the Laviano area (located to NE of the two tectonic windows), both representing the eastern margin of the Apennine Platform. We suggest that the D3–D4 structures were associated with frontal E-verging thrusts and N/S lateral ramp thrusts, respectively, which occurred in the early Pliocene. As suggested by [11,14,15], these out-of-sequence thrusts have to be considered as envelopment thrusts, formed by the positive inversion of normal faults located at depth in the buried Apulian Platform and propagated upward within the overlying allochthonous wedge. Related to the out-of-sequence activity, LANFs developed in the hanging wall of both tectonic windows. The main thrusts, exposed in the two tectonic windows, are folded by a late event of out-of-sequence thrusting, forming a regional E–W trending antiform that was subsequently eroded, allowing the exposition of the oldest rocks of the Apennine Platform succession and the thrust-sheet pile in its footwall. The regional folding and erosion, with added severe normal faulting (stage D5) that occurred in the Early Pleistocene, produced a large amount of carbonate detritus, now forming the Eboli conglomerates widely exposed around the two tectonic windows.

Author Contributions: Conceptualization, S.V., S.C.; Supervision, S.V., S.C.; methodology, S.V., E.P.P., F.D.T.; software, E.P.P., F.D.T., M.M.; validation, S.V., S.C.; formal analysis, E.P.P., F.D.T.; investigation, S.V., E.P.P., F.D.T., S.C., M.M.; resources, S.V., S.C.; data curation, S.V., E.P.P., F.D.T., M.M.; writing—original draft preparation, S.V., S.C.; writing—review and editing, S.V., S.C.; visualization, S.V., E.P.P., M.M. All authors have read and agreed to the published version of the manuscript.

Funding: This research received no funding.

Acknowledgments: We thank the editor L. Lin and the guest editors D. Liotta, G. Molli, and A. Cipriani for the opportunity to share our research in the Special Issue "The Apennines: Tectonics, Sedimentation, and Magmatism from the Palaeozoic to the Present". We are grateful to the three anonymous reviewers for their useful suggestions and corrections that improved the final version of the manuscript. Finally, we thank L. Ferranti for useful discussions about the geology of the Giffoni tectonic window.

Conflicts of Interest: The authors declare no conflict of interest.

References

1. Noguera, A.; Rea, G. Deep structure of the Campanian–Lucanian Arc (Southern Apennine, Italy). *Tectonophysics* **2000**, *324*, 239–265. [CrossRef]
2. Mazzotti, A.; Stucchi, E.; Fradelizio, G.L.; Zanzi, L.; Scandone, P. Seismic exploration in complex terrains: A processing experience in the Southern Apennines. *Geophysics* **2000**, *65*, 1402–1417. [CrossRef]
3. Shiner, P.; Beccacini, A.; Mazzoli, S. Thin-skinned versus thick-skinned structural models for Apulian carbonate reservoirs: Constraints from the Val d'Agri Fields, S Apennines, Italy. *Mar. Pet. Geol.* **2004**, *21*, 805–827. [CrossRef]
4. Butler, R.W.H.; Mazzoli, S.; Corrado, S.; Dedonatis, M.; Dibucci, D.; Gambini, R.; Naso, G.; Nicolai, C.; Scrocca, D.; Shiner, P.; et al. Applying thickskinned tectonic models to the Apennine thrust belt of Italy: Limitations and implications. In *Thrust Tectonics and Hydrocarbon Systems*; McClay, K.R., Ed.; The American Association of Petroleum Geologists: Tulsa, OK, USA, 2004; pp. 647–667.
5. Scrocca, D.; Carminati, E.; Doglioni, C. Deep structure of the southern Apennines, Italy: Thin-skinned or thick-skinned? *Tectonics* **2005**, *24*. [CrossRef]
6. Mazzoli, S.; D'Errico, M.; Aldega, L.; Corrado, S.; Invernizzi, C.; Shiner, P.; Zattin, M. Tectonic burial and 'young' (<10 Ma) exhumation in the southern Apennines fold and thrust belt (Italy). *Geology* **2008**, *36*, 243–246. [CrossRef]
7. Vitale, S.; Ciarcia, S. Tectono-stratigraphic and kinematic evolution of the southern Apennines/Calabria–Peloritani Terrane system (Italy). *Tectonophysics* **2013**, *583*, 164–182. [CrossRef]
8. Vitale, S.; Ciarcia, S. Tectono-stratigraphic setting of the Campania region (southern Italy). *J. Maps* **2018**, *14*, 9–21. [CrossRef]
9. Sabbatino, M.; Vitale, S.; Tavani, S.; Consorti, L.; Corradetti, A.; Cipriani, A.; Arienzo, I.; Parente, M. Constraining the onset of flexural subsidence and peripheral bulge extension in the Miocene foreland of the southern Apennines (Italy) by Sr-isotope stratigraphy. *Sediment. Geol.* **2020**, *401*, 105634. [CrossRef]
10. Mostardini, F.; Merlini, S. Appennino centro-meridionale: Sezioni Geologiche e proposta di modello strutturale. *Mem. Soc. Geol. Ital.* **1986**, *35*, 177–202. (In Italian)
11. Vitale, S.; Prinzi, E.P.; Tramparulo, F.D.A.; De Paola, C.; Di Maio, R.; Piegari, E.; Sabbatino, M.; Natale, J.; Notaro, P.; Ciarcia, S. Late Miocene-Early Pliocene Out-of-Sequence Thrusting in the Southern Apennines (Italy). *Geosciences* **2020**, *10*, 301. [CrossRef]
12. Bally, A.W.; Gordy, P.L.; Stewart, G.A. Structure, seismic data and orogenic evolution of the southern Canadian Rocky Mountains. *Bull. Can. Pet. Geol.* **1966**, *14*, 337–381.
13. Ghisetti, F.; Barchi, M.; Bally, A.W.; Moretti, I.; Vezzani, L. Conflicting balanced structural sections across the Central Apennines (Italy): Problems and implications. In *Generation, Accumulation and Production of Europe's Hydrocarbons*; European Association of Petroleum Geoscientists, Special Publication; Spencer, A.M., Ed.; Springer: Berlin/Heidelberg, Germany, 1993; Volume 3, pp. 219–231.

14. Vitale, S.; Tramparulo, F.D.A.; Ciarcia, S.; Amore, F.O.; Prinzi, E.P.; Laiena, F. The northward tectonic transport in the southern Apennines: Examples from the Capri Island and western Sorrento Peninsula (Italy). *Int. J. Earth Sci.* **2017**, *106*, 97–113. [CrossRef]
15. Vitale, S.; Prinzi, E.P.; Ciarcia, S.; Sabbatino, M.; Tramparulo, F.D.A.; Verazzo, G. Polyphase out-of-sequence thrusting and occurrence of marble detritus within the wedge-top basin deposits in the Mt. Massico (southern Apennines): Insights into the late Miocene tectonic evolution of the central Mediterranean. *Int. J. Earth Sci.* **2019**, *108*, 501–519. [CrossRef]
16. Bertello, F.; Fantoni, R.; Franciosi, R.; Gatti, V.; Ghielmi, M.; Pugliese, A. From Thrust and-Fold Belt to Foreland: Hydrocarbon Occurrences in Italy. In Proceedings of the 7th Petroleum Geology Conference, London, UK, 30 March–2 April 2009; Vining, B.A., Pickering, S.C., Eds.; Geological Society of London: London, UK, 2010; pp. 113–126.
17. Smeraglia, L.; Aldega, L.; Billi, A.; Carminati, E.; Di Fiore, F.; Gerdes, A.; Albert, R.; Rossetti, F.; Vignaroli, G. Development of an Intrawedge Tectonic Mélange by Out-of-Sequence Thrusting, Buttressing, and Intraformational Rheological Contrast, Mt. Massico Ridge, Apennines, Italy. *Tectonics* **2019**, *38*, 1223–1249. [CrossRef]
18. Ietto, A. Su alcune particolari strutture connesse alla tettonica di sovrascorrimento dei Monti Picentini (Appennino Meridionale). *Boll. Soc. Nat. Napoli* **1965**, *74*, 65–85. (In Italian)
19. Scandone, P.; Sgrosso, I.; Vallario, A. Finestra tettonica nella serie calcareo-silico-marnosa lucana presso Campagna (Monti Picentini, Salerno). *Boll. Soc. Nat. Napoli* **1967**, *76*, 3–10. (In Italian)
20. Turco, E. La finestra tettonica di Campagna (M. Picentini, Salerno). *Boll. Soc. Nat. Napoli* **1976**, *85*, 639–665. (In Italian)
21. Pappone, G.; Ferranti, L. Thrust tectonics in the Picentini Mountains, Southern Apennines, Italy. *Tectonophysics* **1995**, *252*, 331–348. [CrossRef]
22. Ferranti, L.; Oldow, J.S. History and tectonic implications of low-angle detachment faults and orogen parallel extension, Picentini Mountains, Southern Apennines fold and thrust belt, Italy. *Tectonics* **1999**, *18*, 498–526. [CrossRef]
23. ISPRA. Carta Geologica d'Italia Alla Scala 1:50.000, Foglio Foglio 467 "Salerno". 2020. Available online: http://www.isprambiente.gov.it/Media/carg/467_SALERNO/Foglio.html (accessed on 2 August 2020).
24. ISPRA. Carta Geologica d'Italia Alla Scala 1:50.000, Foglio Foglio 468 "Eboli". 2020. Available online: http://www.isprambiente.gov.it/Media/carg/468_EBOLI/Foglio.html (accessed on 2 August 2020).
25. Ciarcia, S.; Vitale, S.; Di Staso, A.; Iannace, A.; Mazzoli, S.; Torre, M. Stratigraphy and tectonics of an Internal Unit of the southern Apennines: Implications for the geodynamic evolution of the peri-Tyrrhenian mountain belt. *Terra Nova* **2009**, *21*, 88–96. [CrossRef]
26. Ciarcia, S.; Mazzoli, S.; Vitale, S.; Zattin, M. On the tectonic evolution of the Ligurian accretionary complex in southern Italy. *Geol. Soc. Am. Bull.* **2012**, *124*, 463–483. [CrossRef]
27. Vitale, S.; Ciarcia, S.; Mazzoli, S.; Iannace, A.; Torre, M. Structural analysis of the 'Internal' Units of Cilento, Italy: New constraints on the Miocene tectonic evolution of the southern Apennine accretionary wedge. *Comptes Rendus Geosci.* **2010**, *342*, 475–482. [CrossRef]
28. Vitale, S.; Ciarcia, S.; Mazzoli, S.; Zaghloul, M. Tectonic evolution of the 'Liguride' accretionary wedge in the Cilento area, southern Italy: A record of early Apennine geodynamics. *J. Geodyn.* **2011**, *51*, 25–36. [CrossRef]
29. Vitale, S.; Fedele, L.; Tramparulo, F.D.A.; Ciarcia, S.; Mazzoli, S.; Novellino, A. Structural and petrological analyses of the Frido Unit (southern Italy): New insights into the early tectonic evolution of the southern Apennines–Calabrian Arc system. *Lithos* **2013**, 219–235. [CrossRef]
30. Vitale, S.; Ciarcia, S.; Tramparulo, F.D.A. Deformation and stratigraphic evolution of the Ligurian Accretionary Complex in the southern Apennines (Italy). *J. Geodyn.* **2013**, *66*, 120–133. [CrossRef]
31. Vitale, S.; Ciarcia, S.; Fedele, L.; Tramparulo, F.D.A. The Ligurian oceanic successions in southern Italy: The key to decrypting the first orogenic stages of the southern Apennines-Calabria chain system. *Tectonophysics* **2019**, *750*, 243–261. [CrossRef]
32. Scandone, P.; Sgrosso, I. Il Mesozoico nel gruppo montuoso dell'Accellica (M. Picentini-Salerno). *Mem. Soc. Geol. Ital.* **1964**, *4*, 1–8. (In Italian)
33. Scandone, P.; Sgrosso, I. Sulla paleogeografia della Penisola Sorrentina dal Cretacico superiore al Miocene. *Boll. Soc. Nat. Napoli* **1965**, *74*, 159–177. (In Italian)

34. Servizio Geologico d'Italia, Geological Map of Italy, Scale 1:100,000, Sheets 186 S. Angelo de' Lombardi and 198 Eboli. 1970. Available online: http://193.206.192.231/carta_geologica_italia/sud.htm (accessed on 2 August 2020).
35. D'Argenio, B.; Pescatore, T.; Scandone, P. Schema geologico dell'Appennino meridionale (Campania e Lucania). In *Atti del Conv. Moderne Vedute Sulla Geologia dell'Appennino*; Atti Accademia Nazionale dei Lincei: Rome, Italy, 1973; pp. 49–72. (In Italian)
36. Scandone, P.; Sgrosso, I. La successione miocenica dell'alta Vallimala nella finestra tettonica di Campagna (Monti Picentini). *Boll. Soc. Geol. Ital.* **1974**, *93*, 1043–1047. (In Italian)
37. Pagliaro, S. *Geologia della Finestra Tettonica di Campagna (Monti Picentini–Salerno)*; Arte Tipografica Editrice: Napoli, Italy, 2003; pp. 1–43.
38. Patacca, E. Stratigraphic constraints on the CROP-04 seismic line interpretation: San Fele 1, Monte Foi 1 and San Gregorio Magno 1 wells (Southern Apennines, Italy). *Boll. Soc. Geol. Ital.* **2007**, *7*, 185–239.
39. Bonardi, G.; Ciarcia, S.; Di Nocera, S.; Matano, F.; Sgrosso, I.; Torre, M. Carta delle principali unità cinematiche dell'Appennino meridionale. Nota illustrativa. *Ital. J. Geosci.* **2009**, *128*, 47–60.
40. Pescatore, T.; Sgrosso, I.; Torre, M. Lineamenti di tettonica e sedimentazione nel Miocene dell'Appennino campano-lucano. *Mem. Soc. Nat. Napoli* **1970**, *78*, 337–408. (In Italian)
41. Cocco, E.; Cravero, E.; Ortolani, F.; Pescatore, T.; Russo, M.; Torre, M.; Coppola, L. Le unità irpine nell'area a nord di Monte Marzano, Appennino Meridionale. *Mem. Soc. Geol. Ital.* **1974**, *13*, 607–654. (In Italian)
42. Dazzaro, L.; Di Nocera, S.; Pescatore, T.; Rapisardi, L.; Romeo, M.; Russo, B.; Senatore, M.R.; Torre, M. Geologia del margine della catena appenninica tra il F. Fortore ed il T. Calaggio (Monti della Daunia—Appennino Meridionale). *Mem. Soc. Geol. Ital.* **1988**, *41*, 411–422. (In Italian)
43. Basso, C.; Ciampo, G.; Ciarcia, S.; Di Nocera, S.; Matano, F.; Torre, M. Geologia del settore irpino-dauno dell'Appennino Meridionale: Unità meso-cenozoiche e vincoli stratigrafici nell'evoluzione tettonica mio-pliocenica. *Studi Geol. Camerti* **2002**, *1*, 7–27. (In Italian)
44. Mattioni, L.; Shiner, P.; Tondi, E.; Vitale, S.; Cello, G. The Argille Varicolori Unit of Lucania (Italy): A record of tectonic offscraping and gravity sliding in the Mesozoic-Tertiary Lagonegro Basin, southern Apennines. In *Geology and Active Tectonics of the Western Mediterranean Region and North Africa*; Special Publication; Moratti, G., Chalouan, A., Eds.; Geological Society of London: London, UK, 2006; Volume 262, pp. 277–288.
45. Hippolyte, J.-C.; Angelier, J.; Roure, F. A major geodynamic change revealed by Quaternary stress patterns in the southern Apennines (Italy). *Tectonophysics* **1994**, *230*, 199–210. [CrossRef]
46. Ramsay, J.G. *Folding and Fracturing of Rocks*; McGraw Hill: New York, NY, USA, 1967.
47. Angelier, J.; Mechler, P. Sur une méthode graphique de recherche des contraintes principales également utilisable en tectonique et en séismologie: La méthode des diédres droits. *Bull. Soc. Géol. Fr.* **1977**, *19*, 1309–1318. (In French) [CrossRef]
48. Reiter, F.; Acs, P. TectonicsFP. Software for Structural Geology. Innsbruck University, Austria, 1996–2020. Available online: https://github.com/freiter/TectonicsFP/releases (accessed on 2 August 2020).
49. Vitale, S.; Amore, O.F.; Ciarcia, S.; Fedele, L.; Grifa, C.; Prinzi, E.P.; Tavani, S.; Tramparulo, F.D.A. Structural, stratigraphic, and petrological clues for a Cretaceous-Paleogene abortive rift in the southern Adria domain (southern Apennines, Italy). *Geol. J.* **2017**, *53*, 660–681. [CrossRef]
50. Corrado, S.; Aldega, L.; Perri, F.; Critelli, S.; Muto, F.; Schito, A.; Tripodi, V. Detecting syn-orogenic extension and sediment provenance of the Cilento wedge top basin (southern Apennines, Italy): Mineralogy and geochemistry of fine-grained sediments and petrography of dispersed organic matter. *Tectonophysics* **2019**, *750*, 404–418. [CrossRef]
51. Ogniben, L. Flysch miocenico autoctono e parautoctono e argille scagliose alloctone nella zona di Caiazzo (Caserta). *Boll. Soc. Geol. Ital.* **1956**, *75*, 169–179. (In Italian)
52. Critelli, S.; Le Pera, E. La Formazione di Castelvetere nell'evoluzione petrostratigrafica dell'avanfossa del Tortoniano-Messiniano dell'Appennino Meridionale. *Boll. Soc. Geol. Ital.* **1995**, *114*, 615–634. (In Italian)
53. Santo, A. Le ruditi dei Monti Alburni nel quadro dell'evoluzione altomiocenica dell'Appennino campano. *Boll. Soc. Geol. Ital.* **1996**, *115*, 519–528. (In Italian)
54. Cerisola, R.; Montone, P. Analisi strutturale di un settore della catena dei Monti Ausoni–Aurunci (Lazio, Italia centrale). *Boll. Soc. Geol. Ital.* **1992**, *111*, 449–457. (In Italian)

55. Naso, G.; Tallini, M.; Tozzi, M. Caratteristiche geologico-strutturali dell'area di Miranda (Isernia): Un contributo alla comprensione dei rapporti tra le falde molisane e avanfossa del Messiniano-Pliocene inferiore. *Mem. Soc. Geol. Ital.* **1995**, *114*, 423–441. (In Italian)
56. Scrocca, D.; Tozzi, M.; Parotto, M. Assetto strutturale del settore compreso tra il Matese, Le Mainarde e l'Unita' di Frosolone. Implicazioni per l'evoluzione neogenica del sistema di sovrascorrimenti nell'Appennino centro-settentrionale. *Studi Geol. Camerti* **1995**, *2*, 407–418. (In Italian)
57. De Corso, S.; Scrocca, D.; Tozzi, M. Geologia dell'anticlinale del Matese e implicazioni per la tettonica dell'Appennino molisano. *Boll. Soc. Geol. Ital.* **1998**, *117*, 419–441. (In Italian)
58. Patacca, E.; Scandone, P. Geology of southern Apennines. *Boll. Soc. Geol. Ital.* **2007**, *7*, 75–119.
59. ViDEPI. 2020. Available online: https://www.videpi.com/videpi/pozzi/pozzi.asp (accessed on 2 August 2020).
60. Scrocca, D.; Sciamanna, S.; Di Luzio, E.; Tozzi, M.; Nicolai, C.; Gambini, R. Structural setting along the CROP-04 deep seismic profile (Southern Apennines—Italy). *Boll. Soc. Geol. Ital.* **2007**, *7*, 283–296.
61. Ferranti, L.; Oldow, J.S.; Sacchi, M. Pre-Quaternary orogen-parallel extension in the Southern Apennine belt, Italy. *Tectonophysics* **1996**, *260*, 325–347. [CrossRef]
62. Casciello, E.; Cesarano, M.; Pappone, G. Extensional detachment faulting on the Tyrrhenian margin of the southern Apennines contractional belt (Italy). *J. Geol. Soc.* **2006**, *163*, 617–629. [CrossRef]
63. Mazzoli, S.; Aldega, L.; Corrado, S.; Invernizzi, C.; Zattin, M. Pliocene-quaternary thrusting, syn-orogenic extension and tectonic exhumation in the Southern Apennines (Italy): Insights from the Monte Alpi area. In *Styles of Continental Contraction*; Special Paper; Mazzoli, S., Butler, R.W.H., Eds.; Geological Society of America: Boulder, CO, USA, 2006; Volume 414, pp. 55–77.
64. Carrara, C.; Serva, L. I ciottoli contenuti nei flysch cretacico-paleogenici e miocenici e nei depositi post-tortoniani dell'Appennino meridionale. Loro significato paleotettonico. *Boll. Soc. Geol. Ital.* **1982**, *101*, 441–496. (In Italian)
65. Critelli, S.; Le Pera, E. Tectonic Evolution of the Southern Apennines Thrust-Belt (Italy) as Reflected in Modal Compositions of Cenozoic Sandstone. *J. Geol.* **1995**, *103*, 95–105. [CrossRef]
66. Cinque, A.; Guida, F.; Russo, F.; Santangelo, N. Dati cronologici e stratigrafici su alcuni depositi continentali della piana Piana del F. Sele: I Conglomerati di Eboli. *Geogr. Fis. Din. Quat.* **1988**, *1*, 39–44. (In Italian)

© 2020 by the authors. Licensee MDPI, Basel, Switzerland. This article is an open access article distributed under the terms and conditions of the Creative Commons Attribution (CC BY) license (http://creativecommons.org/licenses/by/4.0/).

geosciences

MDPI

Article

Late Miocene-Early Pliocene Out-of-Sequence Thrusting in the Southern Apennines (Italy)

Stefano Vitale [1,*], Ernesto Paolo Prinzi [1], Francesco D'Assisi Tramparulo [2], Claudio De Paola [1,3], Rosa Di Maio [1], Ester Piegari [1], Monia Sabbatino [1], Jacopo Natale [1], Pasqualino Notaro [4] and Sabatino Ciarcia [5]

1 Dipartimento di Scienze della Terra, dell'Ambiente e delle Risorse (DiSTAR), Università degli Studi di Napoli Federico II, 80126 Napoli, Italy; prinziernesto@gmail.com (E.P.P.); Claudio.DePaola@socotec.com (C.D.P.); rosa.dimaio@unina.it (R.D.M.); ester.piegari@unina.it (E.P.); monia.sabbatino@unina.it (M.S.); jacopo.natale@unina.it (J.N.)

2 Istituto Nazionale di Geofisica e Vulcanologia, sezione di Napoli Osservatorio Vesuviano, 80124 Napoli, Italy; francesco.tramparulo@ingv.it

3 SOCOTEC Italia s.r.l., Lainate, 20045 Milano, Italy

4 EalCUBO (Environment, Earth, Engineering) Soc. Coop., Arcavacata, 87063 Cosenza, Italy; pasqualino.notaro@gmail.com

5 Dipartimento di Scienze e Tecnologie, Università degli Studi del Sannio, 82100 Benevento, Italy; sabatino.ciarcia@unisannio.it

* Correspondence: stefano.vitale@unina.it

Received: 4 July 2020; Accepted: 3 August 2020; Published: 6 August 2020

Abstract: We present a structural study on late Miocene-early Pliocene out-of-sequence thrusts affecting the southern Apennine orogenic belt. The analyzed structures are exposed in the Campania region (southern Italy). Here, thrusts bound the N-NE side of the carbonate ridges that form the regional mountain backbone. In several outcrops, the Mesozoic carbonates are superposed onto the unconformable wedge-top basin deposits of the upper Miocene Castelvetere Group, providing constraints to the age of the activity of this thrusting event. Moreover, a 4-km-long N-S oriented electrical resistivity tomography profile, carried out along the Caserta mountains, sheds light on the structure of this thrust system in an area where it is not exposed. Further information was carried out from a tunnel excavation that allowed us to study some secondary fault splays. The kinematic analysis of out-of-sequence major and minor structures hosted both in the hanging wall (Apennine Platform carbonates) and footwall (Castelvetere Group deposits and Lagonegro-Molise Basin units) indicates the occurrence of two superposed shortening directions, about E-W and N-S, respectively. We associated these compressive structures to an out-of-sequence thrusting event defined by frontal thrusts verging to the east and lateral ramp thrusts verging to the north and south. We related the out-of-sequence thrusting episode to the positive inversion of inherited normal faults located in the Paleozoic basement. These envelopments thrust upward to crosscut the allochthonous wedge, including, in the western zone of the chain, the upper Miocene wedge-top basin deposits.

Keywords: southern Apennines; out-of-sequence thrust; wedge-top basin; electrical resistivity tomography

---

1. Introduction

The study of the kinematics and structural style of thrust sheets, as well as dating their emplacement, is crucial for understanding the tectonic architecture and evolution of orogenic chains. This task can be a challenge when the primary structures are not well exposed, such as the case of the southern Apennines, which has experienced a severe post-orogenic high-angle normal and strike-slip faulting starting from the Pleistocene [1–5]. This event produced the deposition at the toe of the mountains of a massive amount of talus during the Early-Middle Pleistocene glacial–interglacial cycles [6–9]. Furthermore, the carbonate ridges, bounding the Campanian Plain (Figure 1a), were locally covered by a considerable amount of pyroclastic deposits (up to tens of meters thick) during the Middle Pleistocene and Holocene activity of the Roccamonfina, Campi Flegrei and Somma-Vesuvius volcanoes (e.g., [10] and references therein), making more and more infrequent the exposure of older structures (i.e., thrust faults).

Recently, many studies have provided temporal constraints to the fault activity in several fold-and-thrust belts worldwide through different geochronological methods. They include isotopic K-Ar and Ar-Ar dating for cataclasites containing silicate minerals (e.g., [11–13]) and U-Th/He, U-Pb dating for the synkinematic carbonate tectonites (e.g., [14,15]). Different authors have applied these methodological approaches also to the study of thrusts in the Apennines (e.g., [16]), and in some cases being able to date multiple reactivations of the thrust faults especially in the case of positive inversion that makes difficult the interpretation of the previous fault kinematics and deformation history (e.g., [17]).

This study focuses on the Campania region in southern Italy (Figure 1a). We have chosen this area because it exposes a large segment of the Apennine orogen [18]. Furthermore, a complete cartography has been recently provided [19] that makes easier the correlations between the different regional structures. Only in recent years, the systematic study of the thrust fault kinematics allowed the full-definition of an out-of-sequence thrusting stage affecting several sectors of the southern Apennines (e.g., [19,20]). This tectonic event profoundly changed the original orogenic architecture, made of several thrust sheets, forming a further tectonic imbrication of the Meso-Cenozoic successions (Apennine Platform and Lagonegro-Molise Basin units). This crustal shortening produced a complex regional thrust system with the hanging wall ramp formed by the carbonate ridges of the Marzano-Picentini Mts to the south, up to the Avella-Caserta and Taburno Mts to the north (Figure 1a). The main flat thrust is further exposed in the Campagna tectonic window (Figure 1a) located in the western sector of the chain (e.g., [21]). This study aims to reconstruct the out-of-sequence thrust system geometry and its kinematics through a detailed structural survey and analysis of the principal and secondary structures, and the construction of a 4-km-long geological cross-section. For this purpose, to define the geometry of the thrust system, we used an electrical resistivity tomography survey across the Maddaloni-Durazzano ridge (Figure 1a), and information came out during a tunnel excavation in that area (Mt Aglio access window).

The southern Apennines are a fold-and-thrust belt formed by the superposition of several thrust sheets, including oceanic and continental deep basin to shallow-water successions [18,22–31]. This orogenic chain results from the subduction of the Ligurian oceanic lithosphere beneath the Europe/AlKaPeCa plate, starting from the Paleocene-Eocene and the following Miocene-Middle Pleistocene thrusting and folding affecting the Adria plate (e.g., [18,32] and references therein). The sinking of the oceanic lithosphere triggered the fast eastward migration of the slab hinge (roll-back mechanism, [23]). In the lower Miocene, an accretionary wedge made up of oceanic deep basin sediments developed (Ligurian Accretionary Complex; [24,33,34]). During the Miocene, the Ligurian Accretionary Complex overrode the Adria domain forming a new accretionary prism consisting of thrust sheets of the Apennine Platform (basin to shallow-water) and Lagonegro-Molise Basin (slope to pelagic) successions. From the Messinian to Middle Pleistocene, the allochthonous pile superposed onto the Apulian Platform domain. During the orogenic crustal shortening stage, from lower Miocene to Middle Pleistocene, the orogenic prism was the locus of clastic sedimentation in several wedge-top basins. The unconformable deposits include the uppermost Burdigalian-lower Tortonian Cilento Group, the upper Tortonian-lower Messinian Castelvetere Group (CVTG), the upper Messinian-lowermost Zanclean Altavilla Group (AG), the upper Zanclean Baronia Fm and the Piacentian Sferracavallo Fm [19,35]. At depth, the tectonic pile superposes onto a buried para-autochthon imbricate belt (Figure 1b) consisting of the Apulian Platform succession hosting ramp-dominated thrust faults [36], which upward crosscut the allochthonous wedge producing out-of-sequence structures, frequently deforming the upper Miocene-lower Pliocene wedge-top basin deposits [20]. In this work, we studied the segment of the southern Apennines exposed in the Campania region (Figure 1a). The highest mountains include some NW-SE directed ridges made of Meso-Cenozoic carbonates. The tectonic pile is further crosscut by Quaternary high-angle faults with normal to strike-slip kinematics, at places producing large structural depressions filled by marine, transitional and volcanic deposits, such as the Campania Plain (Figure 1a). This study, for the first time, provides a systematic analysis of thrust faults in a vast area of the southern Apennines, which sheds light on the intricate orogenic evolution that characterizes this segment of the circum-Mediterranean Alpine chains.

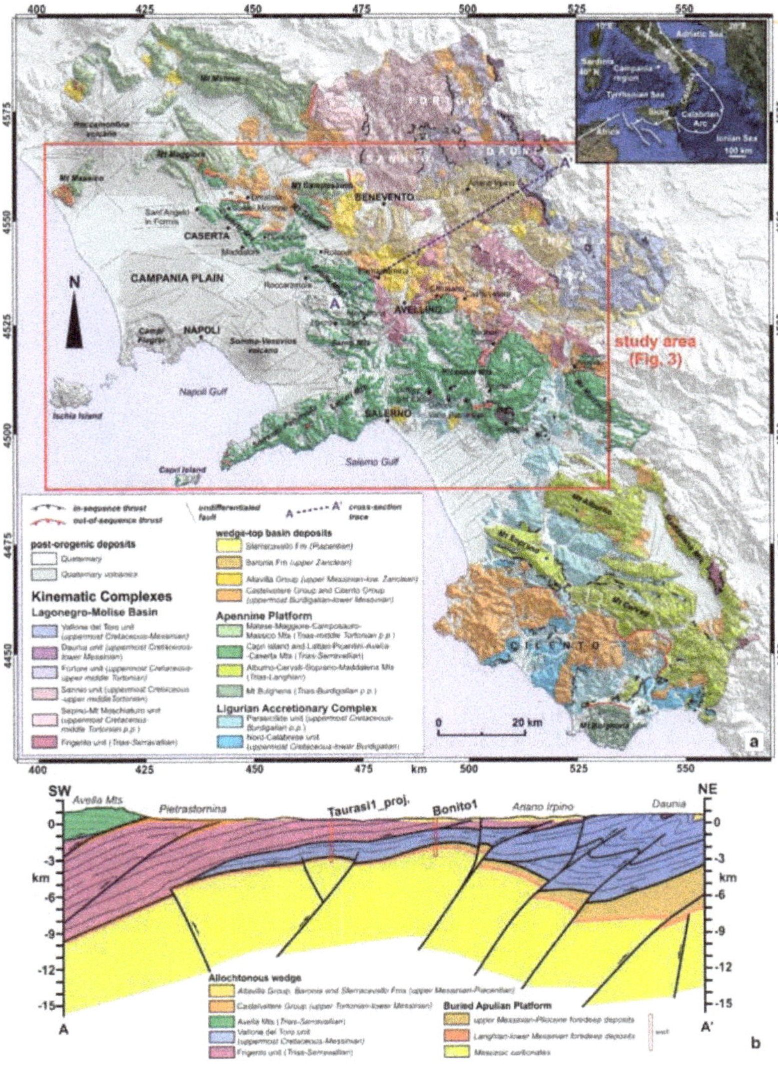

Figure 1. (**a**) Tectonic scheme of the Campania region (southern Italy), inset map of southern Italy from Google Earth. (**b**) Cross-section from Avella Mts to the Daunia sector. Modified after [19].

2. Geological Setting

Regional Stratigraphy

The sedimentary successions, exposed in the analyzed area (Figures 1a and 2), form different thrust sheets covered by Miocene-Pliocene unconformable wedge-top basin deposits. The tectonic nappes derive from three paleogeographic domains ([18,19,25]; Figure 2): (i) the Ligurian Accretionary Complex; (ii) the Apennine Platform and (iii) the Lagonegro-Molise Basin. The Ligurian Accretionary Complex units (Figure 1a) are represented by the deep-basin Parasicilide succession [25,37], cropping out only in the Sele River Valley, and some limited areas of the Salerno sector. The Parasicilide unit includes uppermost Cretaceous-Eocene p.p. shales (Argille Scagliose Fm), covered by Eocene p.p.-Aquitanian limestones (M.S. Arcangelo Fm), marls and varicolored clays (Argille Varicolori Fm) and finally by Burdigalian foredeep sandstones (Albanella Fm).

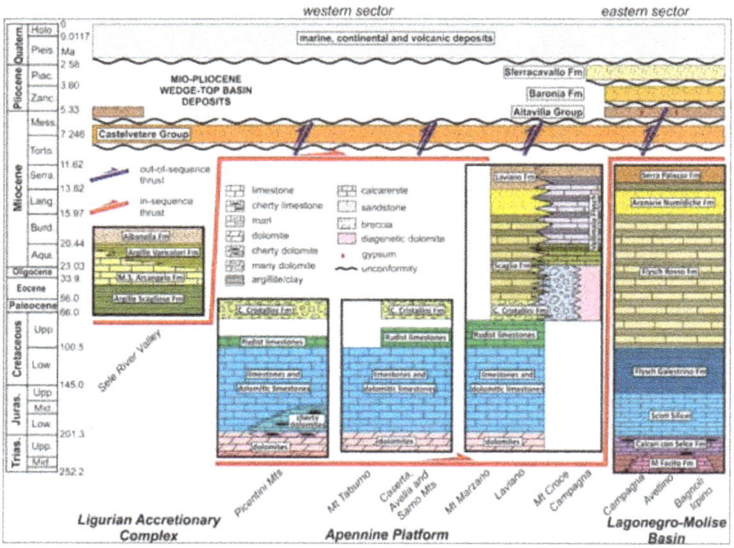

Figure 2. Schematic stratigraphic logs of the analyzed successions.

The Apennine Platform units consist of a dominant carbonate succession ([19,27]; Figure 2) cropping out in the Picentini, Taburno and Caserta Mts. It includes shallow-water rocks made up from the base to the top by Carnian–Hettangian p.p. dolomites and Jurassic–Lower Cretaceous limestones and dolomitic limestones; some lateral heteropies of slope-to-basin cherty carbonates at Picentini Mts occur. The succession continues upward with Upper Cretaceous shallow-water Rudist limestones. At some localities, the Cretaceous shallow-water carbonates pass upward to the uppermost Cretaceous-Paleocene margin recrystallized calcareous breccias (Calcari Cristallini Fm). In the Laviano area, the latter carbonates evolve upward to Eocene–lower Miocene slope-to-basin Scaglia Fm deposits, including cherty limestones, marls and shales, covered by Serravallian foredeep calcarenites and sandstones of Laviano Fm [38]. The Campagna tectonic window exposes a carbonate platform margin succession, known as the Mt Croce unit, consisting of dominant clastic rocks (Vallimala Flysch). Finally, the Lagonegro-Molise Basin units [31,39,40] are represented by the Frigento succession ([18,25]; Figure 2), cropping out eastward of the Apennine Platform ridge and in the Campagna tectonic window [41–43]. The succession includes at the base Anisian–Carnian shallow-water to basin rocks (Monte Facito Fm), evolving upward to Carnian–Lower Cretaceous deep basin deposits (Calcari con Selce, Scisti Silicei and Flysch Galestrino Fms), Upper Cretaceous-Burdigalian slope to basin deposits

(Flysch Rosso Fm), Langhian sandstones (Flysch Numidico Fm), Serravallian marls and foredeep turbidites (Serra Palazzo Fm).

The upper Tortonian–lower Messinian CVTG is constituted at the base by calcirudites including clasts of Cretaceous-Tertiary limestones, with intercalated layers and lenses of sandstones and residual clays and carbonate clasts upward passing to turbiditic sandstones with intercalations of clays, marls and olistostromes of varicolored clays (Ligurian Accretionary Complex unit) and carbonate olistoliths. These deposits unconformably cover the Ligurian Accretionary Complex, the Apennine Platform and the Lagonegro-Molise Basin units. In the study area, the younger wedge-top basin deposits include the upper Messinian-lowermost Pliocene Altavilla Group, the upper Zanclean Baronia Fm and the Piacenzian Sferracavallo Fm, all characterized by dominant silico- and calciclastic rocks [19].

3. Methods and Materials

3.1. P-B-T Method

In order to analyze fault slip data, collected in the field, we applied the P-B-T technique [44]. This method furnishes the orientations of the three principal paleostress axes $\sigma 1$, $\sigma 2$ and $\sigma 3$ (with $\sigma 1 > \sigma 2 > \sigma 3$) and the stress ratio $R = (\sigma 2 - \sigma 3)/(\sigma 1 - \sigma 3)$ [45]. The P-B-T technique provides for every single fault, characterized by well-defined plane attitude, slip orientation and kinematics, the direction of maximum shortening (P), the direction of maximum stretching (T) and the intermediate axis (B), orthogonal to the P-T plane. We used the software TectonicsFP 1.7.8 [46], which allows one to calculate the best-fit angle (θ), minimizing the sum of all misfit angles between the measured slip direction and the calculated maximum shear stress. In the case of lacking conjugate faults, we used a fixed value of angle θ equal to 30° for all fault-slip data.

3.2. Nannoplankton Content Analysis

We performed a nannoplankton content analysis on marls and calcareous clay levels of the CVTG deposits sampled in the footwall of the analyzed thrusts to give temporal constraints on the deformational events. We collected 38 samples in the localities of Acerno, Campagna, Castelvetere, Chiusano, Mt Taburno, Maddaloni, Pietra Maula (Lauro), Mt Faggeto (Monteforte Irpino) and Sant'Angelo in Formis (Figure 1a). To obtain optimal results for each type of material, the coccolith specimens were prepared using the centrifugation and simple smear slide techniques, following the procedures described in [47]. We observed the nannoplankton content through the polarized light microscope Zeiss Axioscop equipped with an ×100 oil immersion objective lens. We performed a qualitative evaluation of the assemblages on all the samples, but only 19 of them proved to be fossiliferous, while the other ones resulted barren or very poorly fossiliferous. Dating this material was not a trivial task both because some nannoplankton contents resulted strongly altered by diagenesis, which usually partially or entirely hides the primary morphological characteristics, and because of the intense reworking of specimens. Indeed, many older specimens have been found with very few biostratigraphic markers. However, specimens have been identified at the species level, and biostratigraphic markers have been used for dating.

3.3. Electrical Resistivity Tomography (ERT)

A 2D ERT profile is a geophysical prospecting technique that provides an electrical resistivity cross-section of the shallow earth structure until depths of a few tens of hundreds of meters. Among the many applications of the ERT profiling in mining, environmental, engineering and archaeological research [48], it is to emphasize its success in modeling the structural setting of relatively complex geological areas thanks to the capability to define, in terms of electrical resistivity contrasts, the contact between different geological formations and the flexure structure produced by faulting (e.g., [49–55]). The acquired apparent resistivity data were collected by the IRIS SYSCAL Pro SWITCH 96 resistivitymeter in multi-electrode configuration, employing cables with 96 electrodes interspaced at 10 m, according to the so-called roll-along technique (e.g., [56,57]), which is commonly used when the measured profile length is greater than the cable length, as for our case study. Specifically, seven roll-along sequences, with overlapping of 48 electrodes between two consecutive rolls, were performed, which allowed us to collect, also thanks to the high 2D coverage offered by the chosen pole–dipole electrode configuration, 27.256 data points along the investigated section and an exploration depth of about 200 m below the ground level (b.g.l.). All the acquired apparent resistivity values were deployed in a single 2D model, which takes into account the real profile topography, and then processed by the *ERTlab*™ software (version 64, Geostudi Astier S.r.l., Livorno, Italy) based on a finite-element inversion algorithm.

4. Structural Analysis and Data Presentation

We performed a geological survey in different areas of the southern Apennines, where main out-of-sequence thrusts and related minor structures are exposed. Results are presented starting from the south (Marzano-Picentini Mts) to the north (Caserta Mts).

4.1. Marzano-Picentini Mts

We analyzed the Laviano area located along the northern side of the Mt Marzano (Figure 3). Here, the Apennine Platform succession hosts a major thrust that also deforms the wedge-top basin sediments of the CVTG. To the north of the analyzed area, the Lagonegro-Molise Basin unit is exposed, covered by the CVTG. Several secondary structures are associated with the main thrust, such as minor thrust faults within the Serravallian calcarenites (Laviano Fm; Figure 4a), several S-C structures (Figure 4b) and pre-buckle thrusts marked by the cherty layers (Figure 4c) within the Paleogene-lower Miocene Scaglia Fm (Figure 2). In the footwall of the main thrust, the CVTG deposits cover the carbonates, filling graben where the former normal faults are cut by the younger thrust faults (Figure 4d,e). The plot of thrust faults (Figure 5a) suggests a prevalence of planes moderately dipping to SSW, and the related PBT diagram indicates an NNE-SSW shortening.

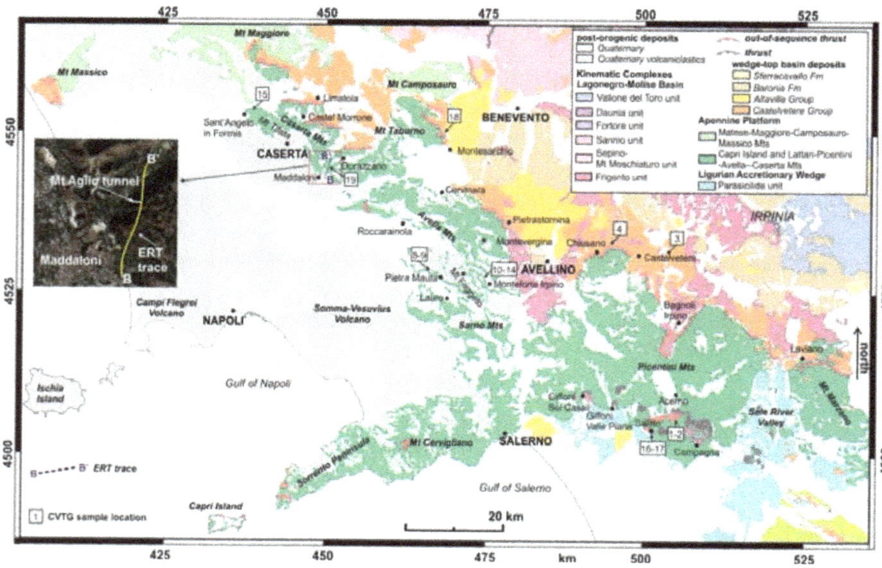

Figure 3. Tectonic map of the study area showing the analyzed sites and the trace of the electrical resistivity tomography (ERT) profile and tunnel excavation.

Figure 4. Laviano: (**a**) thrust fault in calcarenites of the Laviano Fm.; (**b**) S-C structures associated to a north-verging thrust within the gray Scaglia Fm; (**c**) pre-buckle thrust in red chert within the Scaglia Rossa Fm; (**d**,**e**): normal fault plane crosscut by north-verging thrust faults. Bagnoli Irpino: (**f**) back-thrust within the Cretaceous succession; (**g**,**h**) striations and steps.

The mountain area surrounded by the Laviano, Avellino, Salerno and Campagna towns (Figure 3), includes the carbonate ridge of the Picentini Mts, facing northward to the Irpinia area, dominated by the Frigento unit (Lagonegro-Molise Basin) covered by the wedge-top basin deposits of CVTG, Baronia and Sferracavallo Fms. The Picentini Mts confine to the east with the Sele River Valley, where the Ligurian Accretionary Complex is widely exposed [37]. Here, the out-of-sequence thrust between the carbonates and Frigento unit, with the interposition of the CVTG, is exposed only in a few localities. In the areas where the main thrust is hidden, several secondary structures allowed us to reconstruct the geometry and kinematics, such as the Bagnoli Irpino area (Figure 3). Here, a few deformation structures are exposed; amongst others, a good example is the back-thrust within the Mesozoic limestones

(Figure 4f), accompanied by several secondary slickenside planes with striations and steps (Figure 4g,h). These structures indicate a vergence to the south (Figure 5b). In the Chiusano-Castelvetere area (Figure 3), the Upper Cretaceous limestones tectonically cover the CVTG deposits through a flat-lying ramp (Figure 6a). The frontal thrust form an anticline ramp evidenced by the high-angle bedding of carbonates and overlying CVTG deposits. An instructive example is in the quarry located close to the Chiusano village; here, the almost vertical strata show also parasitic folds (Figure 6b). In this case, the well-cemented Pleistocene slope talus preserved these structures, presently exposed by the digging activity. In a dismissed quarry, a back-thrust verging to the south occurs within the carbonates with a drag fold in the footwall (Figure 6c). Only in one locality, close to the Castelvetere village (Figure 3), the superposition of carbonates onto clays and sandstones of Castelvetere is exposed (Figure 6d,e). The siliciclastic rocks form lenses within a cataclasite, subsequently cut by normal faults (Figure 6d). The measured fault planes (Figure 5c) dip to the north and the east; however, the PBT diagram indicates an E-W shortening.

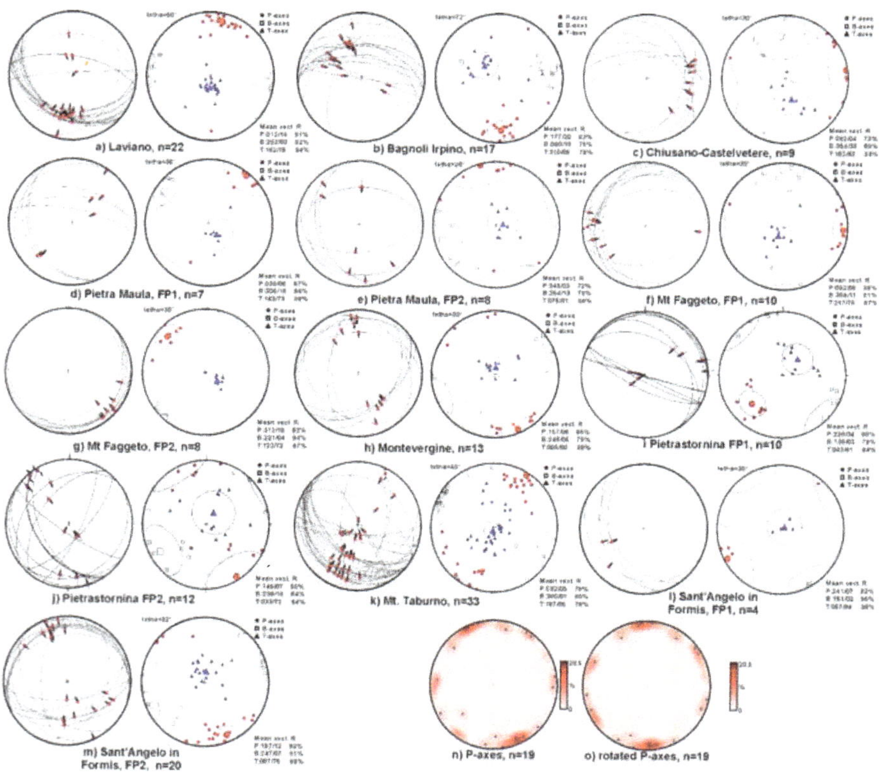

Figure 5. Stereographic projections of (**a–m**) faults and corresponding PBT plots; (**n**) T-axes; (**o**) T-axes rotated 20° counterclockwise.

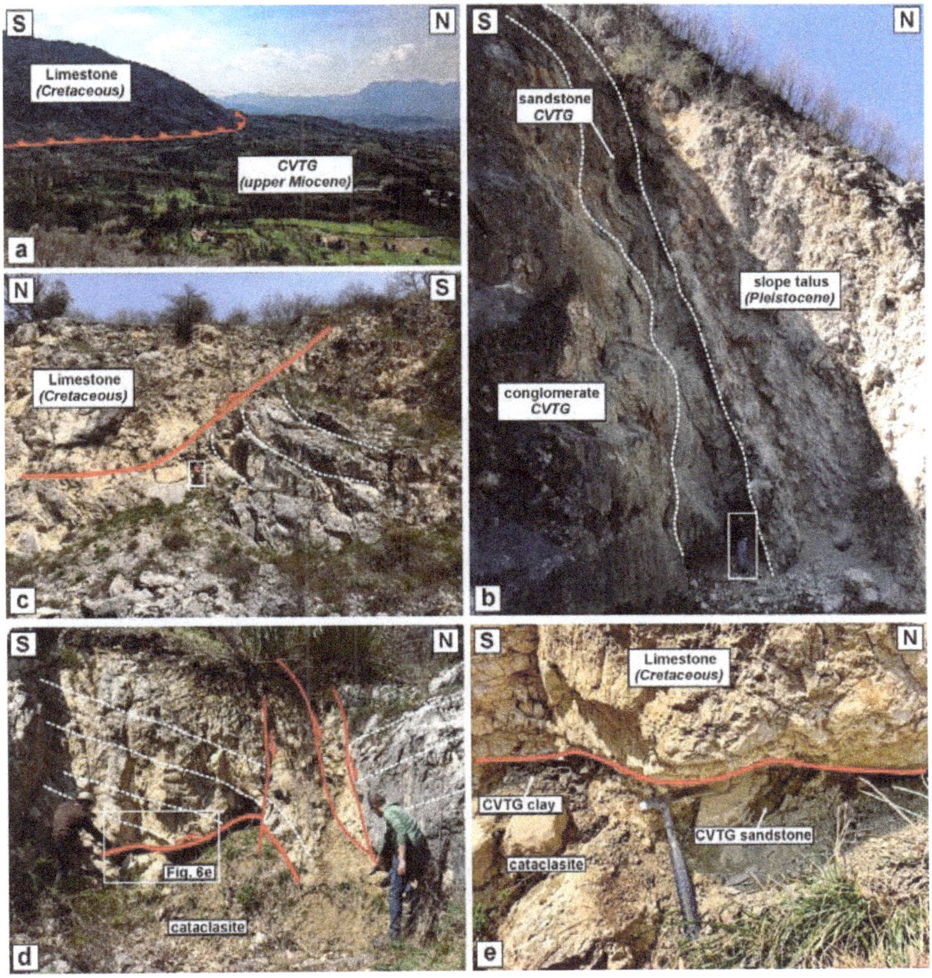

Figure 6. Chiusano area: (**a**) panoramic view of the frontal thrust; (**b**) highly dipping forelimb of the hanging wall anticline associated to the frontal thrust formed by Upper Cretaceous carbonates covered by beds of Castelvetere Group (CVTG) conglomerates and sandstones; the succession is covered by the Middle Pleistocene slope talus (Belvedere quarry); (**c**) back-thrust within the Upper Cretaceous limestones with an associated footwall syncline. Castelvetere area: (**d**) the main thrust crosscut by late normal faults; (**e**) particulars of the cataclasite in the footwall made of lenses of clays and sandstones of the CVTG.

4.2. *Avella and Taburno Mts*

In this area, the Apennine Platform and Lagonegro-Molise Basin units crop out, unconformably covered by CVTG, Altavilla Group and Baronia Fm. Along the E-W ridge of Roccarainola (Figure 3), [58] reported the occurrence of a major thrust. The authors indicate an NNW tectonic vergence of the thrust that juxtaposes the Jurassic carbonates on the Cretaceous rocks with an associated hanging wall overturned anticline. In this work, we analyzed further outcrops, including the Lauro, Monteforte and Montevergine areas (Figure 3). Along the road connecting Lauro with the town of Avellino, deposits of the lower part of the CVTG are exposed, covering the Upper Cretaceous carbonates.

At Pietra Maula (north of Lauro) and Mt Faggeto (northwest of Monteforte Irpino), a 20 m thick CVTG succession crops out (Figure 3) consisting of calcareous conglomerates with lenses of sandstones and red and grey clays (Figure 7a–d). Clasts are mainly made of the Eocene Trentinara Fm (e.g., [19]), not exposed in the analyzed area. The succession hosts two sets of thrusts and folds. In the Pietra Maula outcrop, the main thrust superposes the Upper Cretaceous carbonates onto the upper Miocene conglomerates, sandstones and clays (Figure 7a). Minor thrust faults deform the clay lenses (Figure 7a), in particular, it is common to find along the same fault plane two sets of slickenside striations and steps (Figure 7b) indicating an early ENE-WSW shortening, with associated folds especially in the clay and sand lenses, and a later NNW-SSE shortening. Two sets of thrust faults are also observed in Mt Faggeto. However, the NW-verging thrust is the best-developed (Figure 7e). At Pietra Maula, fault planes of the first out-of-sequence deformation (Figure 5d) form a conjugate system with a statistical NE-SW shortening direction, whereas the structures related to the second deformation stage (Figure 5e) suggest a NNW-SSE shortening. At Mt Faggeto, the thrust faults are W-dipping with an E-W shortening direction for the first deformation stage (Figure 5f), whereas the fault planes, related to the second stage, are gently dipping to SE, indicating an NW-SE shortening (Figure 5g). On the top of the Montevergine mountain, the Apennine Platform Jurassic carbonates overthrust the CVTG (Figure 7f), the latter made of calcareous conglomerates with a sandy matrix. Only in the upper part of these clastic deposits, clasts are made of crystalline rocks. Thrusts and related folds occur in the Apennine Platform Jurassic succession (Figure 7g) as well as the presence of pressure solution planes orthogonal to the bedding (Figure 7h). Thrust faults form a conjugate system, of which the PBT diagram provides an NNW-SSE shortening direction (Figure 5h). The carbonate ridge from Montevergine to Cervinara (Figure 3) is defined by Apennine Platform Cretaceous limestones covered by the CVTG deposits forming an anticline ramp with high-angle strata along the northern side. The best exposure is in the Pietrastornina area (Figure 3), where CVTG beds are steep and locally overturned (Figure 8a,c). In a few localities, the thrust is exposed, indicating a double vergence, with shales and sandstones in the footwall (Figure 8b). Generally, the CVTG conglomerate shows indented clasts (Figure 8d). In the footwall of the main thrust, the Frigento succession and the overlying CVTG deposits are deformed by folds and thrusts. Remarkable examples are the ramp anticlines within the slope carbonates of the Flysch Rosso Fm (Frigento unit) cropping out in different places such as that exposed in the quarry north of Pietrastornina (Figure 8e). Here, the vertical beds of the Flysch Rosso Fm and CVTG deposits are cut by sub-horizontal thrusts with small displacements and associated drag folds in the footwall (Figure 8e). Measured thrust faults mark two shortening directions: NE-SW and NW-SE (Figure 5i,j). Another place to observe the major structure is at the toe of Triassic-Jurassic carbonates of the Mt Taburno located north of the Avella Mts (Figure 3). Along the NE-side of this mountain, the main thrust crops out. It is defined by a flat-lying fault (Figure 8g) with highly deformed clays and sandstones in the footwall (Figure 8h). Striations and steps indicate a N vergence. Several secondary thrust faults are present (Figure 8g). Such as the previous areas, here the main thrust is cut by younger normal faults. Associated with this thrust, some folds in Flysch Rosso Fm in the footwall occur, the largest is a thigh overturned fold verging to the north (Figure 8f). Poles to fold limbs indicate a theoretical fold axis of 302/24 (Figure 8f). Thrust planes moderately dip to SE, and the PBT diagram suggests an NE-SW shortening direction (Figure 5k).

4.3. Caserta Mts

The Caserta Mts (Figure 3) are formed by three E-W trending carbonate ridges of Maddaloni, Castel Morrone and Limatola Mts, and the NW-SE trending Mt Tifata (Sant'Angelo in Formis). A few are the sites in the Limatola ridge where secondary thrusts are exposed. The best outcrop is located along the road connecting the Castel Morrone and Limatola towns (Figure 3). Here, a major back-thrust occurs (Figure 9a), which brings the Apennine Platform Jurassic limestones to overthrust the CVTG deposits. Minor thrust faults occur, suggesting an N-S shortening. The area of Sant'Angelo in Formis was the subject of a few studies in the last century (e.g., [59,60]). Reference [60] describes several thrust faults between Mesozoic Apennine Platform carbonates (hanging wall) and CVTG in the footwall. We have not found the thrusts described by these authors. However, we observed several minor structures, such as mesoscale thrusts and related folds. Two sets of secondary thrusts are present, the first furnishes an ENE-WSW shortening direction (Figure 5l), whereas the second set indicates an NNW-SSE shortening (Figure 5m).

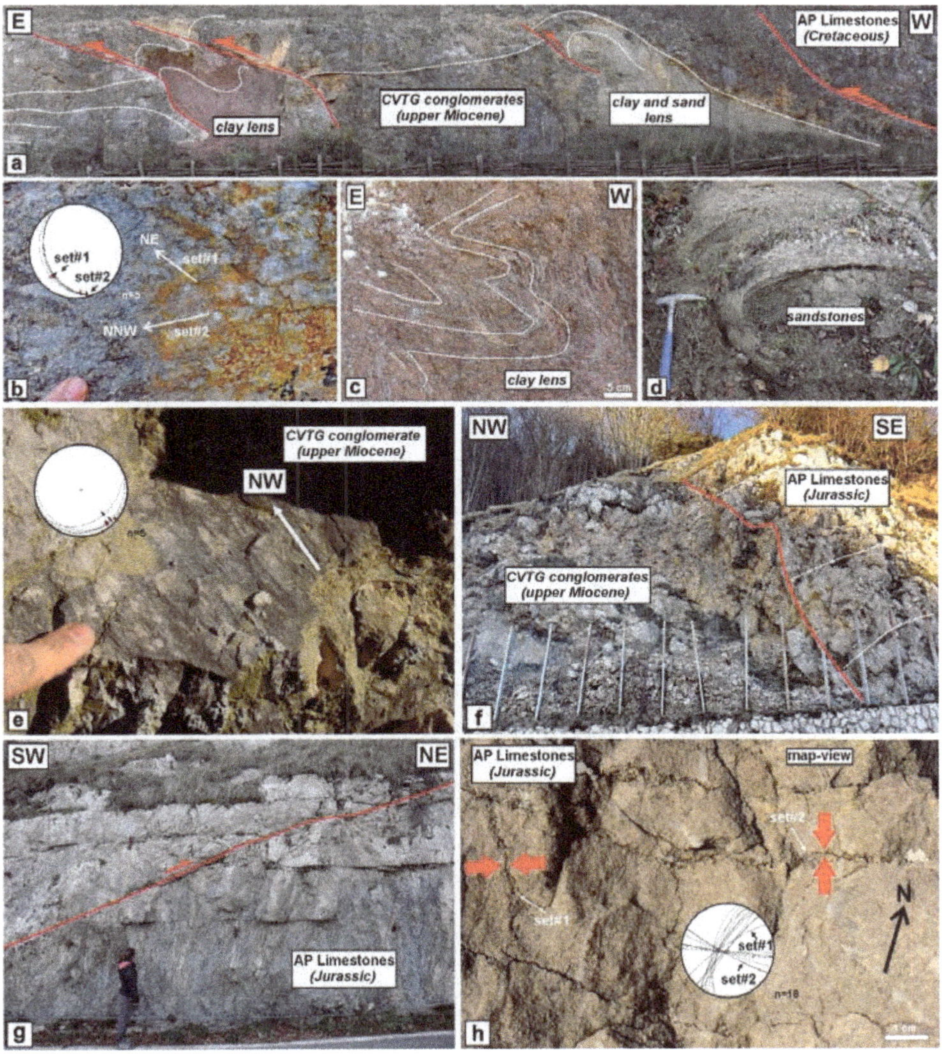

Figure 7. Pietra Maula: (**a**) imbricate fan of thrust sheets superposing Apennine Platform (AP) limestones onto CVTG conglomerates, sandstones and clays; (**b**) thrust fault plane showing two sets of slickenside striations and steps; (**c**) folds in CVTG clays; (**d**) fold in the CVTG sandstones. (**e**) Thrust plane showing striations and steps indicating an N vergence (Mt Faggeto). Montevergine: (**g** high-angle thrust of AP Jurassic limestones onto CVTG; (**h**) two orthogonal stylolite sets in AP Jurassic oolitic limestones.

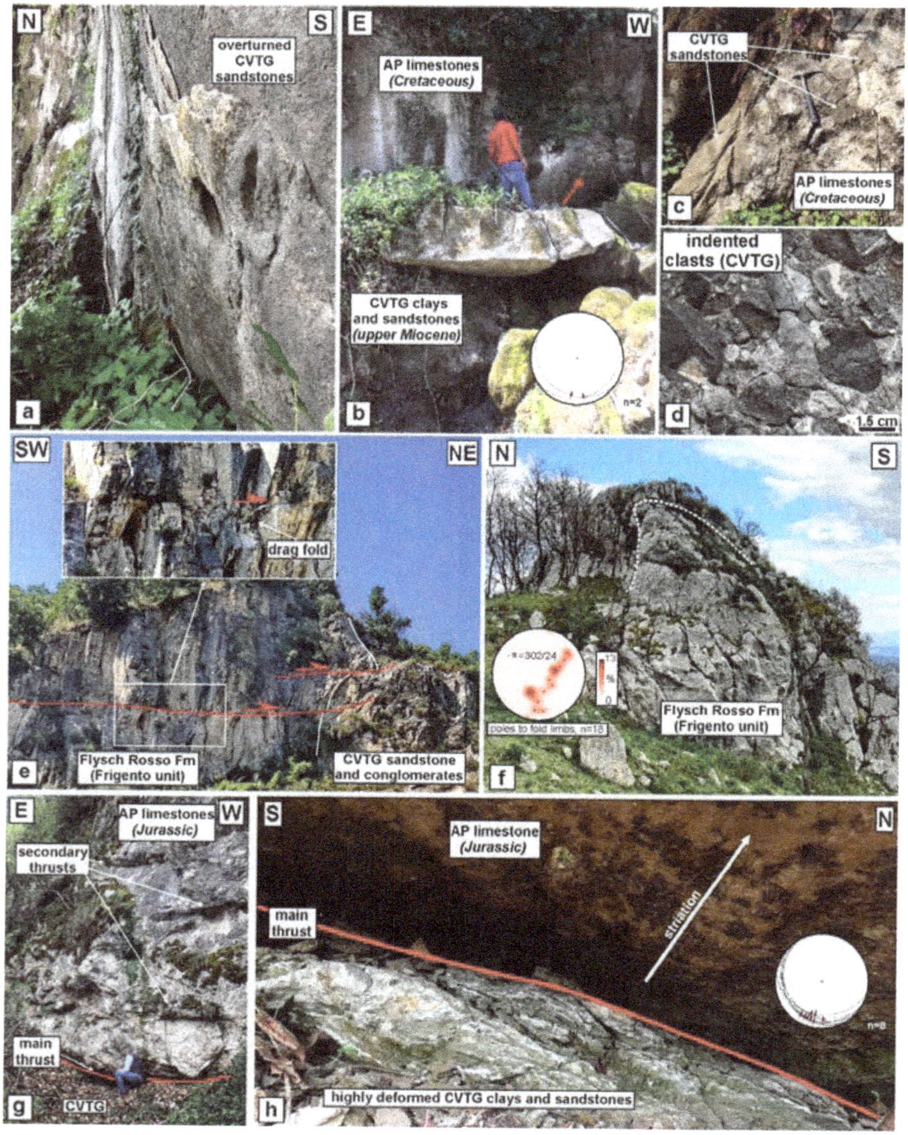

Figure 8. Pietrastornina: (**a**) overturned CVTG beds; (**b**) main thrust fault of Apennine Plattform (AP) Cretaceous limestones onto shales and sandstones of CVTG; (**c**) CVTG sandstone covering AP Cretaceous limestones; (**d**) indented calcareous clasts in CVTG; (**e**) vertical beds of Flysch Rosso Fm crosscut by a sub-horizontal thrust verging to NE with associated drag folds in the footwall. Mt Taburno: (**f**) overturned fold in Flysch Rosso Fm verging to the north; (**g**,**h**) the main thrust between AP Jurassic carbonates onto highly deformed clays of CVTG.

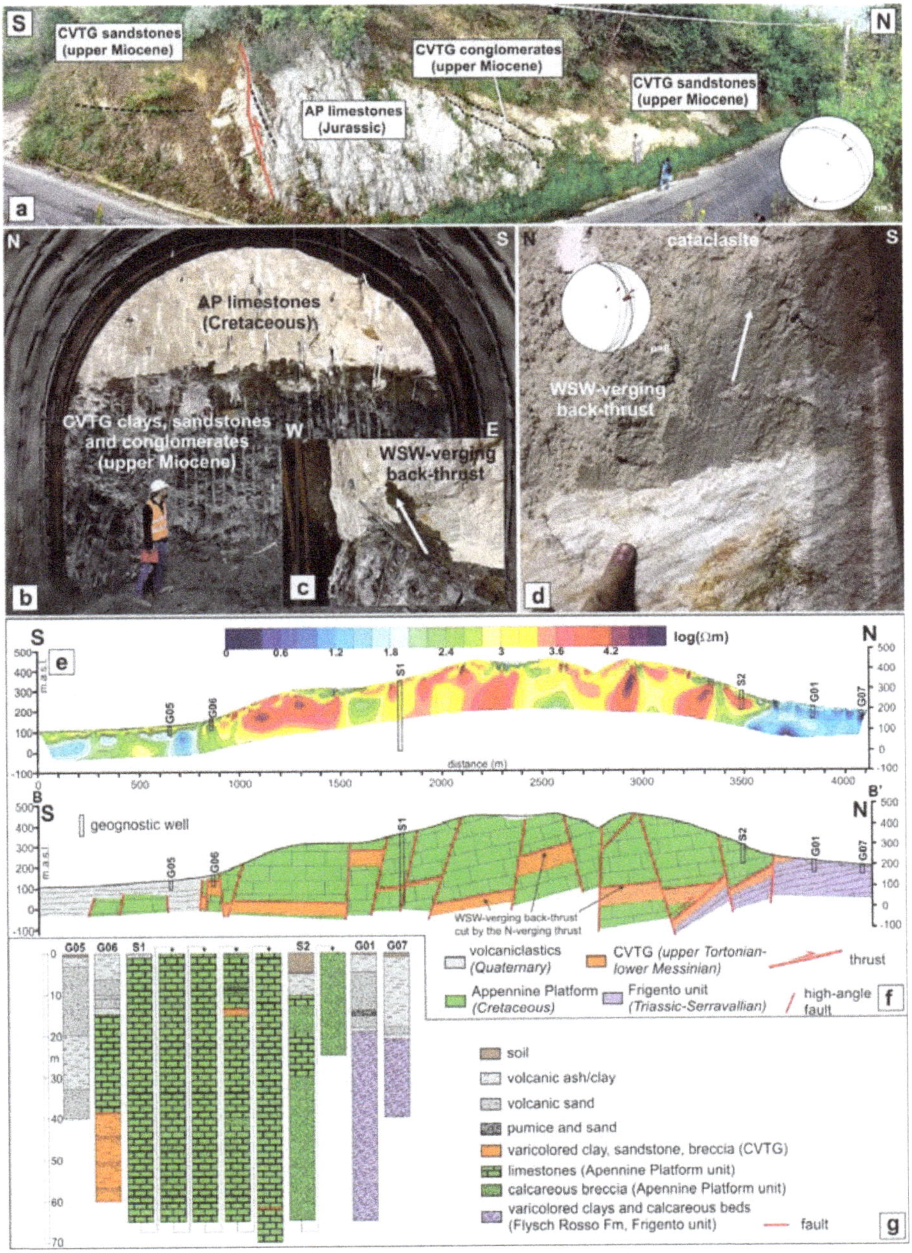

Figure 9. (**a**) Back-thrust of Apennine Platform (AP) Jurassic limestones onto CVTG sandstones (Limatola). Maddaloni Tunnel: (**b**) excavation front showing the tectonic superposition of the AP Cretaceous carbonates onto the CVTG; (**c**) particulars of the back-thrust plane; (**d**) slickenside striations and stereographic projections of thrust faults. (**e**) ERT profile; (**f**) cross-section; (**g**) well logs.

4.3.1. Maddaloni Carbonate Ridge

The northern slope of the Maddaloni ridge is E-W oriented (Figure 3). Here, the carbonate ridge is bounded by normal faults that hide the main thrust. To investigate the tectonic architecture in depth, we used an ERT profile 4 km long, and N-S directed, crosscutting the western side of the ridge, from the Maddaloni town to the north (Figure 3). Furthermore, we followed the excavation of an E-W directed tunnel within the carbonate succession orthogonal to the ERT profile (Figure 3). The structural survey performed on the excavation fronts furnished useful information about the geometry and kinematics of thrust faults hosted in the Apennine Platform Meso-Cenozoic succession. The best structure observed is defined by the superposition of Cretaceous carbonates onto CVTG varicolored clays, sandstones and conglomerates (Figure 9b). The footwall is highly deformed with well-marked foliation and some overturned folds. Along the thrust plane dipping to west, we observed striations and steps (Figure 9c,d) indicating a vergence to WSW. Finally, we reconstructed a geological cross-section (B-B′; Figure 9f), integrating data from some geognostic wells (Figure 9g), the ERT profile (Figure 9e) and information gathered in the tunnel. It is interesting to note that in the well G06 (Figure 9g), the Cretaceous limestones tectonically cover the CVTG clays. Further, in the well S1, a cataclasite with a varicolored clayey matrix is present at -275 m below the field surface.

ERT-Cross Section

The ERT profile used to reconstruct the architecture of the Maddaloni ridge fault zone is indicated in Figure 3 as B-B′. It extends approximately S-N in a straight line for 4100 m. The inverted resistivity section (Figure 9e) was obtained for a root mean square misfit error lower than 10% to the field data. The resistivity model shows a generally resistive pattern abruptly broken by two prominent conductive zones at the ends of the profile. The architecture of the resistive zone appears rather complex: sectors characterized by resistivity values of a few thousand of Ωm (shades of yellow color in Figure 9e), which are locally interrupted by very high-resistivity nuclei (>10^4 Ωm, shades of red color in Figure 9e), overtop areas with resistivities of the order of some hundreds of Ωm (deep green color zones in Figure 9e). Both the amplitudes and inhomogeneous distribution of the observed high-resistivity values (yellow and red sectors in Figure 9e) are consistent with strongly fractured calcareous rocks, likely interested by karstic phenomena. Different water contents along fault/fracture systems and/or degraded portions of the carbonate structure can, therefore, justify this wide range in the observed resistivity variations.

Conversely, the deep relatively conductive zones can well describe, both for the measured resistivity values and for their shape and in-depth position, clay lenses, sandstones and conglomerates, such as those highlighted by the tunnel excavation (Figure 9c). Finally, the abrupt lowering of the resistivity values observed at the two ends of the ERT profile can instead be attributed to the contact of the calcareous formation with lithologically different formations. In particular, we note that the resistivity values observed in the southern edge of the profile are compatible with pyroclastic deposits, while the very low resistivity values in the northern part (< 10^2 Ωm, shades of blue color in Figure 9e) well correlate with clayey materials characterized by different degrees of compaction.

In order to validate the interpretative hypotheses coming from the ERT prospection, we have compared the modeled cross-section with available geological evidence from geognostic surveys located along the ERT profile (Figure 9e). In particular, the S1 and S2 stratigraphic logs (Figure 9g) show a very thin layer of topsoil followed by altered/degraded limestone from weakly to strongly fractured, locally rather vacuolar. The other well, drilling a calcareous succession, is G06 (Figure 9g). The stratigraphy shows an upper interval of tectonized limestones overlying highly deformed clays similar to that observed in the tunnel excavation that we ascribed to the CVTG. Completely different geological sequences, instead, come from the three geognostic wells placed at both ends of the ERT cross-section. The G05 stratigraphic column (Figure 9g) is constituted down to the maximum exploration depth by horizons of silts (or sandy-silts), which alternate with levels of sands (or silty-sands) from fine to coarse grain size with the inclusion of weakly cemented pyroclasts.

Conversely, the stratigraphy provided by the G01 and G07 wells (Figure 9g) shows an alternation of pyroclastic layers that differ in size (clay, silts, sands and gravel) and saturation degree only for the shallowest portion of the investigated subsurface. Starting at a depth of about 20 m b.g.l., varicolored shales of the Frigento unit are found, which are characterized by the lowest values of the electrical resistivity observed along the whole profile. The ERT model, combined with geological field data, has been used to reconstruct the cross-section in Figure 9f, where high-angle faults are recognized by sharp decreases in resistivity due to the increase in permeability values associated with a higher density of fractures. The sharpest resistivity contrast in the northernmost part of the profile well marks the main thrust fault.

5. Nannoplankton Content Analysis

The nannoplankton analysis carried out on the CVTG samples (Table 1) indicates mainly an upper Miocene age for the fossiliferous samples, although they are affected by intense reworking. The samples from CVTG5 to CVTG14 collected at Pietra Maula (Lauro) and Mt Faggeto (Monteforte Irpino) are characterized by the presence of biostratigraphic markers as *Discoaster brouweri* (Figure 10a), *D. surculus*, *D. variabilis* (Figure 10b), *Helicosphaera wallichii* (Figure 10c) and *Sphaenolithus abies*. In the samples CVTG1-2 and CVTG16-17 collected in the areas of Acerno and Salitto, we found the markers *Discoaster variabilis*, *Helicospaera wallichii* and *Sphaenolithus abies*. The samples CVTG3-4 collected in the Castelvetere and Chiusano localities contain the markers *Discoaster brouweri*, *D. variabilis* and *Sphaenolithus abies* (Figure 10d) in association with other specimens as *Reticulofenestra pseudoumbilicus* (Figure 10e; Table 1). CVTG18 sampled in the Mt Taburno area revealed the presence of the markers *Discoaster loeblichii* (Figure 10f) and *D. variabilis*. We found the markers *Helicospaera wallichii*, and *Sphaenolithus abies* in the sample CVTG19 collected in the Maddaloni tunnel. Finally, the sample CVT15 collected at Sant'Angelo in Formis did not reveal the presence of any biostratigraphic markers. However, we found an association of specimens that is comparable with most of the other samples. Based on the first occurrence datum of these markers found in the CVTG, we can consider these deposits as no older than the upper Tortonian. Due to the occurrence of the markers *D. loeblichii* and *D. surculus*, we can constrain these deposits as ranging from the lower part of the nannofossil zone NN11 of [61], which corresponds to the base of CN9A of [62] or CNM16 of [63], to the upper part of the NN11 of [61], corresponding to CN9B of [62] or CNM18 of [63]. We cannot define the last occurrence datum for all of the formation due to the intense reworking of the specimens and also because not all the samples contain markers as *D. loeblichii* with a very narrow time range of distribution (e.g., for Mt Taburno area). However, if we consider that all these deposits are coeval and belong to the CVTG, their age can be constrained to the upper Tortonian-lower Messinian. This result also agrees with the age of CVTG recently dated in Mt Massico and Sorrento Peninsula ([20,64]) and with the literature on the wedge-top basin deposits of Castelvetere Fm [18,25,65,66].

Table 1. Summary of biostratigraphic data (calcareous nannofossils). X: occurrence of specimen; R: reworked specimen; M: marker specimen.

CVTG SAMPLE	1	2	3	4	5	6	7	8	9	10	11	12	13	14	15	16	17	18	19
Braarudosphaera bigelowii				X					X	X			X						
Carinaster sp.													X						
Calcidiscus leptoporus			X	X								X	X	X	X			X	X
Calcidiscus macintyrei			X	X														X	
Coccolithus formosus					R	R			R							R			R
Coccolithus pelagicus	X	X	X	X			X	X	X	X	X	X	X	X	X	X	X	X	X
Coccolithus miopelagicus													R						
Coccolithus tenuiforatus				X															
Cruciplacolithus sp.												X		X					
Cyclicargolithus abisectus									R										
Cyclicargolithus floridanus	R	R		R			R	R	R		R	R	R			R			
Discoaster barbadiensis									R	R									
Discoaster brouweri				M									M						
Discoaster deflandrei									R										
Discoaster druggii										R									
Discoaster loeblichii																		M	
Discoaster multiradiatus				R	R														
Discoaster saipanensis						R			R								R		
Discoaster salisburgensis										R									
Discoaster surculus													M						
Diascoaster variabilis				M									M				M	M	
Helicosphaera carteri	X	X							X		X	X	X		X				
Helicosphaera recta		R																	
Helicosphaera walbersdorfensis									R										
Helicosphaera wallichii		M							M										M
Microrhabdulus decoratus									R										
Nannotetrina sp.	R											R	R			R			
Pontosphaera discopora													X						X
Pontosphaera multipora													X						X
Reticulofenestra bisecta	R	R		R	R		R	R	R		R	R			R				
Reticulofenestra dictyoda									R										
Reticulofenestra haqii													X						
Reticulofenestra minuta				X								X	X			X	X	X	
Reticulofenestra pseudoumbilicus	X	X		X					X			X	X	X		X	X		
Sphenolithus abies		M	M	M				M									M		M
Sphenolithus ciperoensis										R									
Sphenolithus conicus				R				R								R			
Sphenolithus delphix									R										
Sphenolithus dissimilis		R																	
Sphenolithus heteromorphus									R							R	R		
Sphenolithus moriformis		X		X			X	X	X			X		X	X	X	X		X
Sphenolithus neoabies													X						
Sphenolithus radians				R				R	R			R							
Sphenolithus procerus				R				R											
Transversopontis sp.										R									
Watznaueria barnese									R	R									
Zygrhablithus bijugatus	R		R						R				R			R			

UTM (33T, WGS84) coordinates (meter) of the listed samples: CVTG1 and CVTG2 (Acerno): 503794E, 4505896N; CVTG3 (Castelvetere): 503508E, 4530468N; CVTG4 (Chiusano): 494276E, 4532154N; CVTG5 to CVTG9 (Pietra Maula): 469301E, 4527138 N; CVTG10 to CVTG14 (Monteforte Irpino): 33T 472491E, 4527607N; CVTG15 (Sant'Angelo in Formis): 437645E, 4552581N; CVTG16 and CVTG17 (Salitto): 501632E, 4504105N; CVTG18 (Mt Taburno): 468531E, 4549402N; CVTG19 (Maddaloni, Tunnel: 450290E, 4545491N.

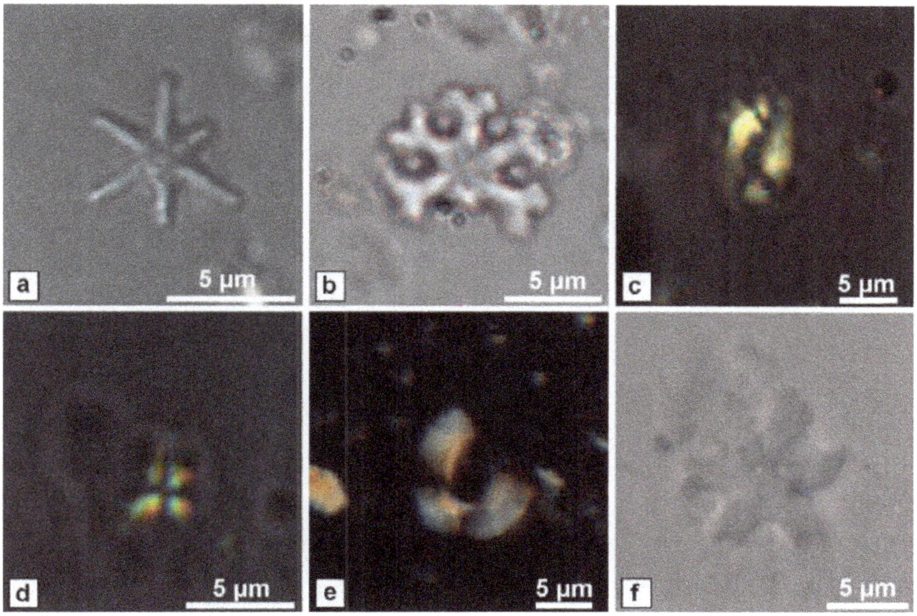

Figure 10. Microphotographs of the calcareous nannofossil species: (**a**) *Discoaster brouweri*, parallel nicols (sample CVTG 13); (**b**) *Discoaster variabilis*, parallel nicols (sample CVTG 13); (**c**) *Helicosphaera wallichii*, cross nicols (sample CVTG 9); (**d**) *Sphenolithus abies*, cross nicols (sample CVTG 3); (**e**) *Rericulofenestra pseudoumbilicus*, cross nicols (sample CVTG 4); (**f**) *Discoaster loeblichii*, parallel nicols (sample CVTG 18).

6. Discussion

The structural survey, carried out in several sites of the Apennine chain in the Campania region, allowed us to collect a considerable dataset of attitude and kinematic measurements of structures related to an out-of-sequence thrusting event (Figure 11). Primary and secondary structures, such as thrust faults, folds and pressure solution planes, suggest a deformation event that occurred after the deposition of the CVTG. In most of the analyzed outcrops, we observed structures associated with an NW-SE/NNE-SSW shortening (Figure 11). However, in the localities Sant'Angelo in Formis, Campagna, Pietrastornina, Pietra Maula, Mt Faggeto and Maddaloni ridge, we recognized that this deformation superposes onto structures related to an E-W/ESE-WNW directed shortening with a dominant tectonic vergence to the east (Figure 11). These two thrusting stages can be considered as out-of-sequence deformation pulses because they deform the upper Miocene CVTG deposits, whose sedimentation occurred within basins located on top of the already structured thrust sheet prism. Similar structures were also observed elsewhere in the Campania region, such as at Capri Island and the western Sorrento Peninsula [64] and Mt Massico [20]. In particular, in the first two localities, a single N-verging thrusting stage was recorded, whereas at Mt Massico, two shortening directions occur (ca. E-W and N-S), with the N-verging as the later deformation stage (Figure 11).

Further evidence of such a thrusting event is found in a tunnel excavation and by the ERT survey along the Maddaloni carbonate ridge that allowed us to reconstruct the thrust geometry in an area where it is not exposed. We interpreted the Maddaloni structure as the superposition of the out-of-sequence imbrication of different thrust sheets made of Mesozoic carbonates and CVTG deposits verging to WSW (back-thrusts), crosscut by an N-verging thrust (Figure 9f). To obtain mean values of shortening directions of the two out-of-sequence deformation stages at regional scale, we gathered all data from the PBT plots (Figure 5a–m) and those calculated for Capri Island, Sorrento Peninsula and Mt Massico (data from [20,64]). The resulting plot (Figure 5n) indicates the occurrence of a main cluster of the σ1, with a direction of NNW-SSE and a secondary direction scattered between NE-SW and SE-NW. Hence, taking into account the counterclockwise rotation of ca. 20° (e.g., [67]) recorded in the Plio-Pleistocene for this sector of the southern Apennines, the shortening directions related to the two out-of-sequence episodes are ca. E-W and N-S (Figure 5o), with the latter as the more developed widespread shortening direction in the western sector of the chain.

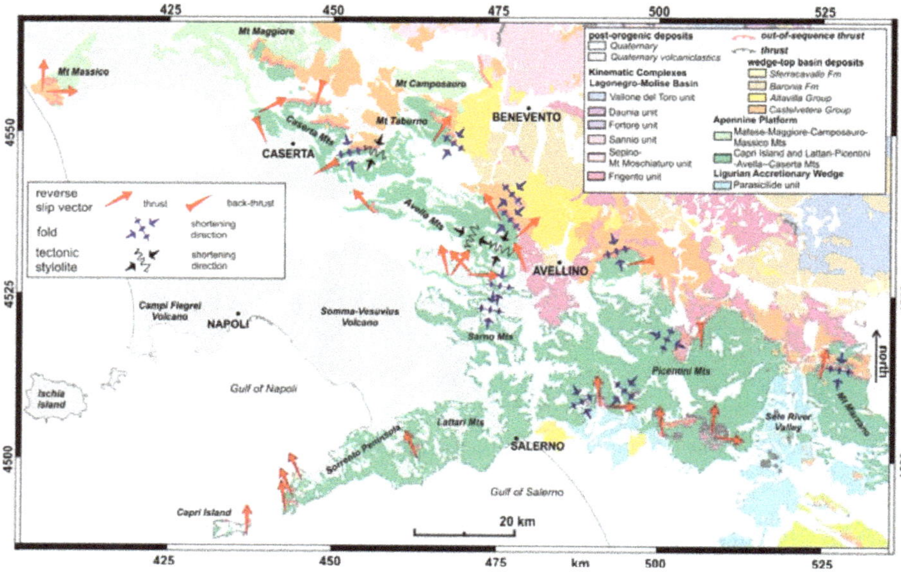

Figure 11. Tectonic map of the study area showing tectonic vergences and shortening directions for the studied out-of-sequence deformation structures.

Several papers described similar structures also in the central Apennines (e.g., [68–72]). However, a long-lived debate exists about the tectonic meaning of these structures because they frequently juxtapose younger-over-older rocks. In fact, they were interpreted as out-of-sequence thrusts (e.g., [68–74]) or low-angle normal faults (e.g., [28,75]). Contrarily to the previous orogenic area, in the southern Apennines, the occurrence of the upper Miocene CVTG deposits in the footwall of Mesozoic rocks clearly demonstrates their out-of-sequence origin. Furthermore, the exposure of these structures in several areas of the chain suggests a regional deformation event that affected a large part of the orogenic belt and not only a limited area. As concerns the age of this event, stratigraphic constraints indicate that the out-of-sequence thrusting was active following the early Messinian, the youngest age of the CVTG. We exclude that the analyzed out-of-sequence thrusting event has had an older activity because of the evidence of an extensional setting before the sedimentation of the CVTG such as what has been documented in other areas of the region (e.g., Mt Massico, [20]; Cilento, [76]) and in this study (e.g., Laviano area). It is worth noting a close analogy with the central Apennines where Tortonian-lower Messinian clastic deposits (Brecce della Renga Fm), corresponding to the CVTG rocks, were sedimented in an extensional environment before their involvement in the out-of-sequence thrusting [74].

Recently, [14] dated the E-verging out-of-sequence thrust of Mt Massico (Figure 3) to the early Pliocene (5.1 Ma) by using U-Pb geochronology on synkinematic calcite fibers. Hence, we can infer that this out-of-sequence deformation stage was successive to the deposition of CVTG (ca. 10-7 Ma) and probably active in the late Messinian-Zanclean, such as suggested for similar structures elsewhere located in the central Apennines and Matese area (e.g., [68–72,77]). East-verging out-of-sequence thrusts have also been reported in the eastern sector of the Campania region. In particular, these structures deform the upper Messinian-lowermost Pliocene Altavilla Group deposits (e.g., [18]) generally sealed by the lower Pliocene Baronia Fm [78]. However, in the easternmost sector of the chain (e.g., Daunia area, Figure 1), also the wedge-top basin deposits of Baronia Fm and middle Pliocene Sferracavallo Fm are deformed by such east-verging structures [79]. The eastward out-of-sequence thrusting observed in several localities is consistent with this E-W shortening that ruled the superposition of the Apennine Platform onto the Lagonegro-Molise Basin domain [20,64]. The complete closure of the Lagonegro-Molise Basin occurred with the overthrusting of the easternmost sector (Vallone del Toro succession, [25]) onto the Apulian Platform in the late Messinian-early Pliocene [19]. Subsequently, the E-W shortening, which acted, until that time, mainly through the thin-skinned tectonics, affected the buried Apulian carbonates through thick-skinned tectonics. Here, the deformation was recorded by ramp-dominate thrusts, now forming a buried imbricate belt [80,81]. This thick-skinned thrusting was probably triggered by the buttressing of the thick Apulian Mesozoic carbonates against the allochthonous wedge (e.g., [29]) with the positive inversion of pre-existing normal faults located within the Permo-Triassic basement (e.g., [36]). Such an interpretation has also been considered for the central and northern Apennines, where deep-rooted normal faults underwent positive inversion during the Neogene compression and thrust–fold development (e.g., [82–84]). With this in mind, the eastward out-of-sequence thrusts have to be considered as in-sequence structures enucleated at depth and crosscutting the already formed orogenic chain. These types of structures are called envelopment thrusts [64,85,86]. As concerns the N-S shortening direction, we suggest that it could be considered as related to lateral ramps of the main frontal E-verging thrusts. Figure 12 shows a tectono-stratigraphic evolution scheme of the analyzed area in the Serravallian-early Pliocene interval. From the Serravallian-early Tortonian, the Apennine Platform overthrusted onto the Lagonegro-Molise Basin successions, forming an imbricate fan of thrust sheets in the footwall (Figure 12a). Subsequently, in the late Tortonian-early Messinian, the CVTG sediments deposited within wedge-top basins on top of the tectonic pile, sealing the in-sequence tectonic contacts (Figure 12b). The sedimentation was preceded by an extensional event recorded by normal faults. In the late Messinian-early Pliocene, out-of-sequence thrusts have developed as the positive inversion of deep-seated normal faults (Figure 12c,d) with a main eastward tectonic transport and the formation of lateral ramps with N and S vergences.

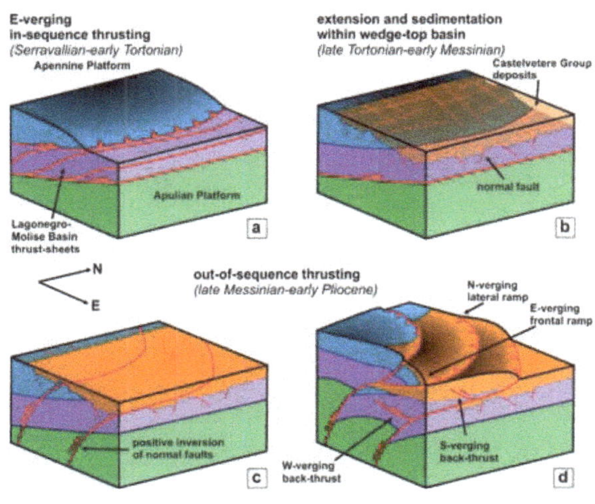

Figure 12. (**a**–**d**) Cartoon showing the tectono-stratigraphic evolution scheme of the analyzed area between the Serravallian and early Pliocene.

7. Conclusions

Our study reports the occurrence of a regional tectonic event during the late Messinian-early Pliocene defined by out-of-sequence thrusting. The main thrust, related to this event, is exposed along the N-NE flank of the carbonate ridges forming the mountain backbone of the Campania region. This out-of-sequence thrust juxtaposed the Meso-Cenozoic Apennine Platform carbonates onto its margin (e.g., Laviano and Mt Croce successions) or onto the Frigento unit (Lagonegro-Molise Basin unit), with the interposition, in both cases, of the upper Miocene wedge-top basin deposits of the CVTG. The main thrust is marked by a ramp anticline exposed in few localities such as the areas of Chiusano-Castelvetere and Pietrastornina. We studied the main thrust only in few places, whereas several secondary structures are common both in the hanging wall and footwall. The structural analysis indicates two main shortening directions: ca. E-W and N-S. Thanks to the use of the electrical resistivity tomography and information obtained following a tunnel excavation, the evidence of such a main thrusting event is given in an area where the thrust is not exposed. The age of this tectonic event is younger than the age of the CVTG sediments, whose dating of upper Tortonian-lower Messinian is confirmed by the nannoplankton analysis performed in this study. Similar structures described in other studies in the whole southern and central Apennines have been dated to the late Messinian-early Pliocene. These out-of-sequence thrust faults are the surficial expression of deep-seated ramp thrusts formed as a positive inversion of normal faults located in the Permo-Triassic basement. These envelopment thrusts developed from west to east of the chain, involving younger and younger wedge-top basin deposits, from the late Miocene to the late Pliocene, respectively.

Author Contributions: Conceptualization, S.V., R.D.M., S.C.; methodology, S.V., E.P.P., F.D.A., C.D.P., R.D.M., E.P., M.S.; software, E.P.P., F.D.T., C.D.P., R.D.M., E.P.; validation, S.V., S.C.; formal analysis, E.P.P., F.D.T., C.D.P., R.D.M., E.P., M.S.; investigation, S.V., E.P.P., F.D.T., C.D.P., P.N., S.C.; resources, S.V, S.C., R.D.M.; data curation, E.P.P., F.D.T., C.D.P., R.D.M., E.P., M.S.; writing—original draft preparation, S.V., R.D.M., M.S., J.N., S.C.; writing—review and editing, S.V., R.D.M., M.S., S.C.; visualization, S.V., E.P.P., M.S. All authors have read and agreed to the published version of the manuscript.

Funding: This research received no funding.

Acknowledgments: We thank the editor G. Wang and the guest editors D. Liotta, G. Molli and A. Cipriani for the opportunity to share our research in the Special Issue "The Apennines: Tectonics, Sedimentation, and Magmatism from the Palaeozoic to the Present". We are grateful to the two reviewers L. Aldega and P. Pace, for the useful suggestions and corrections that improved the final version of the manuscript. We thank the Consorzio CANCELLO FRASSO TELESINO (CFT) for the permission to publish the ERT and well log data, access to the tunnel excavation and use photos of the front-excavation. Finally, we thank G. Vespasiano and F. Ferraro for the help in the field.

Conflicts of Interest: The author declares no conflict of interest.

References

1. Hippolyte, J.C.; Angelier, J.; Roure, F. A major geodynamic change revealed by Quaternary stress patterns in the southern Apennines (Italy). *Tectonophysics* **1994**, *230*, 199–210. [CrossRef]
2. Schiattarella, M. Quaternary tectonics of the Pollino Ridge, Calabria-Lucania boundary, Southern Italy. In *Continental Transpressional and Transtensional Tectonics*; Holdsworth, R.E., Strachan, R.A., Dewey, J.F., Eds.; Geological Society of London: London, UK, 1998; Volume 135, pp. 341–354.
3. Ascione, A.; Cinque, A.; Improta, L.; Villani, F. Late quaternary faulting within the southern Apennines seismic belt: New data from Mt. Marzano area (southern Italy). *Quat. Int.* **2003**, *101*, 27–41. [CrossRef]
4. Ferranti, L.; Santoro, E.; Mazzella, M.E.; Monaco, C.; Morelli, D. Active transpression in the northern Calabria Apennines, southern Italy. *Tectonophysics* **2009**, *476*, 226–251. [CrossRef]
5. Brozzetti, F. The Campania-Lucania Extensional Fault System, southern Italy: A suggestion for a uniform model of active extension in the Italian Apennines. *Tectonics* **2011**, *30*, TC5009. [CrossRef]
6. Capaldi, G.; Cinque, A.; Romano, P. Ricostruzione di sequenze morfoevolutive nei Picentini meridionali (Campania, Appennino meridionale). *Geogr. Fis. Din. Quat.* **1988**, *1*, 207–222.
7. Cinque, A.; Guida, F.; Russo, F.; Santangelo, N. Dati cronologici e stratigrafici su alcuni depositi continentali della Piana del F. Sele: I Conglomerati di Eboli. *Geogr. Fis. Din. Quat.* **1988**, *1*, 39–44.

8. Coltorti, M.; Dramis, F. The chronology of Upper Pleistocene stratified slope-waste-deposits in Central Italy. *Permafr. Periglac. Processes* **1995**, *6*, 235–242. [CrossRef]
9. Santangelo, N.; Di Donato, V.; Lebreton, V.; Romano, P.; Russo Ermolli, E. Palaeolandscapes of Southern Apennines during the late Early and the Middle Pleistocene. *Quat. Int.* **2012**, *267*, 20–29. [CrossRef]
10. Cioni, R.; Isaia, R.; Sulpizio, R. The city of Napoli and its active volcanoes. *Geol. Field Trips* **2019**. [CrossRef]
11. Haines, S.; Lynch, E.; Mulch, A.; Valley, J.W.; Van der Pluijm, B. Meteoric fluid infiltration in crustalscale normal fault systems as indicated by δ18O and δ2H geochemistry and 40Ar/39Ar dating of neoformed clays in brittle fault rocks. *Lithosphere* **2016**, *8*, 587–600. [CrossRef]
12. Viola, G.; Torgersen, E.; Mazzarini, F.; Musumeci, G.; Van der Lelij, R.; Schönenberger, J.; Garofalo, P.S. New Constraints on the Evolution of the Inner Northern Apennines by K-Ar Dating of Late Miocene-Early Pliocene Compression on the Island of Elba, Italy. *Tectonics* **2018**, *37*, 3229–3243. [CrossRef]
13. Aldega, L.; Viola, G.; Casas-Sainz, A.; Marcén, M.; Román-Berdiel, T.; Lelij, R. Unraveling Multiple Thermotectonic Events Accommodated by Crustal-Scale Faults in Northern Iberia, Spain: Insights From K-Ar Dating of Clay Gouges. *Tectonics* **2019**, *38*, 1–23. [CrossRef]
14. Smeraglia, L.; Aldega, L.; Billi, A.; Carminati, E.; Di Fiore, F.; Gerdes, A.; Rossetti, F.; Vignaroli, G. Development of an Intrawedge Tectonic Mélange by Out-of-Sequence Thrusting, Buttressing, and Intraformational Rheological Contrast, Mt. Massico Ridge, Apennines, Italy. *Tectonics* **2019**, *38*, 1223–1249. [CrossRef]
15. Carminati, E.; Aldega, L.; Smeraglia, L.; Scharf, A.; Mattern, F.; Albert, R.; Gerdes, A. Tectonic evolution of the Northern Oman Mountains, part of the Strait of Hormuz Syntaxis: New structural and paleothermal analyses and U-Pb dating of synkinematic calcite. *Tectonics* **2020**, *39*, e2019TC005936. [CrossRef]
16. Carboni, F.; Viola, G.; Aldega, L.; Van der Lelij, R.; Brozzetti, F.; Barchi, M.R. K-Ar fault gouge dating of Neogene thrusting: The case of the siliciclastic deposits of the Trasimeno Tectonic Wedge (Northern Apennines, Italy). *Ital. J. Geosci.* **2020**, *139*, 300–308. [CrossRef]
17. Curzi, M.; Aldega, L.; Bernasconi, S.M.; Berra, F.; Billi, A.; Boschi, C.; Franchini, S.; Van der Lelij, R.; Viola, G.; Carminati, E. Architecture and evolution of an extensionally-inverted thrust (Mt. Tancia Thrust, Central Apennines): Geological, structural, geochemical, and K–Ar geochronological constraints. *J. Struct. Geol.* **2020**, *136*, 104059. [CrossRef]
18. Vitale, S.; Ciarcia, S. Tectono-stratigraphic and kinematic evolution of the southern Apennines/Calabria-Peloritani Terrane system (Italy). *Tectonophysics* **2013**, *583*, 164–182. [CrossRef]
19. Vitale, S.; Ciarcia, S. Tectono-stratigraphic setting of the Campania region (southern Italy). *J. Maps* **2018**, *14*, 9–21. [CrossRef]
20. Vitale, S.; Prinzi, E.P.; Ciarcia, S.; Sabbatino, M.; Tramparulo, F.D.A.; Verazzo, G. Polyphase out-of-sequence thrusting and occurrence of marble detritus within the wedge-top basin deposits in the Mt. Massico (southern Apennines): Insights into the late Miocene tectonic evolution of the central Mediterranean. *Int. J. Earth Sci.* **2019**, *108*, 501–519. [CrossRef]
21. Pappone, G.; Ferranti, L. Thrust tectonics in the Picentini Mountains, Southern Apennines, Italy. *Tectonophysics* **1995**, *252*, 331–348. [CrossRef]
22. Carminati, E.; Lustrino, M.; Doglioni, C. Geodynamic evolution of the central and western Mediterranean: Tectonics vs. igneous petrology constraints. *Tectonophysics* **2012**, *579*, 173–192. [CrossRef]
23. Malinverno, A.; Ryan, W.B.F. Extension in the Tyrrhenian Sea and shortening in the Apennines as result of arc migration driven by sinking of the lithosphere. *Tectonics* **1986**, *5*, 227–245. [CrossRef]
24. Vitale, S.; Ciarcia, S.; Fedele, L.; Tramparulo, F.D.A. The Ligurian oceanic successions in southern Italy: The key to decrypting the first orogenic stages of the southern Apennines-Calabria chain system. *Tectonophysics* **2019**, *750*, 243–261. [CrossRef]
25. Bonardi, G.; Ciarcia, S.; Di Nocera, S.; Matano, F.; Sgrosso, I.; Torre, M. Carta delle principali unità cinematiche dell'Appennino meridionale. Nota illustrativa. *Ital. J. Geosci.* **2009**, *128*, 47–60.
26. Vitale, S.; Dati, F.; Mazzoli, S.; Ciarcia, S.; Guerriero, V.; Iannace, A. Modes and timing of fracture network development in poly-deformed carbonate reservoir analogues, Mt. Chianello, southern Italy. *J. Struct. Geol.* **2012**, *37*, 223–235. [CrossRef]
27. D'Argenio, B.; Pescatore, T.; Scandone, P. Schema geologico dell'Appennino meridionale (Campania e Lucania). Atti del Conv. Moderne vedute sulla geologia dell'Appennino. *Atti Accad. Naz. Lincei* **1973**, *182*, 49–72.

28. Mostardini, F.; Merlini, S. Appennino centro-meridionale: Sezioni Geologiche e proposta di modello strutturale. *Mem. Soc. Geol. Ital.* **1986**, *35*, 177–202.
29. Mazzoli, S.; D'Errico, M.; Aldega, L.; Corrado, S.; Invernizzi, C.; Shiner, P.; Zattin, M. Tectonic burial and 'young' (<10 Ma) exhumation in the southern Apennines fold and thrust belt (Italy). *Geology* **2008**, *36*, 243–246.
30. Sabbatino, M.; Vitale, S.; Tavani, S.; Consorti, L.; Corradetti, A.; Cipriani, A.; Arienzo, I.; Parente, M. Constraining the onset of flexural subsidence and peripheral bulge extension in the Miocene foreland of the southern Apennines (Italy) by Sr-isotope stratigraphy. *Sediment. Geol.* **2020**, *401*, 105634. [CrossRef]
31. Mattioni, L.; Shiner, P.; Tondi, E.; Vitale, S.; Cello, G. The Argille Varicolori Unit of Lucania (Italy): A record of tectonic offscraping and gravity sliding in the Mesozoic-Tertiary Lagonegro Basin, southern Apennines. In *Geology and Active Tectonics of the Western Mediterranean Region and North Africa*; Moratti, G., Chalouan, A., Eds.; Geological Society of London: London, UK, 2006; Volume 262, pp. 277–288.
32. Handy, M.R.; Schmid, S.M.; Bousquet, R.; Kissling, E.; Bernoulli, D. Reconciling platetectonic reconstructions of Alpine Tethys with the geological–geophysical record of spreading and subduction in the Alps. *Earth-Science Rev.* **2010**, *102*, 121–158. [CrossRef]
33. Ciarcia, S.; Mazzoli, S.; Vitale, S.; Zattin, M. On the tectonic evolution of the Ligurian accretionary complex in southern Italy. *GSA Bull* **2012**, *124*, 463–483. [CrossRef]
34. Vitale, S.; Ciarcia, S.; Tramparulo, F.D.A. Deformation and stratigraphic evolution of the Ligurian Accretionary Complex in the southern Apennines (Italy). *J. Geodyn.* **2013**, *66*, 120–133. [CrossRef]
35. Ciarcia, S.; Vitale, S. Sedimentology, stratigraphy and tectonics of evolving wedge-top depozone: Ariano Basin, southern Apennines, Italy. *Sediment. Geol.* **2013**, *290*, 27–46. [CrossRef]
36. Shiner, P.; Beccacini, A.; Mazzoli, S. Thin-skinned versus thick-skinned structural models for Apulian Carbonate Reservoirs: Constraints from the Val D'Agri Fields. *Mar. Pet. Geol.* **2004**, *21*, 805–827. [CrossRef]
37. Ciarcia, S.; Vitale, S.; Di Staso, A.; Iannace, A.; Mazzoli, S.; Torre, M. Stratigraphy and tectonics of an internal unit of the Southern Apennines: Implications for the geodynamic evolution of the peri-Tyrrhenian mountain belt. *Terra Nova* **2009**, *21*, 88–96. [CrossRef]
38. Vitale, S.; Amore, O.F.; Ciarcia, S.; Fedele, L.; Grifa, C.; Prinzi, E.P.; Tavani, S.; Tramparulo, F.D.A. Structural, stratigraphic and petrological clues for a Cretaceous-Paleogene abortive rift in the southern Adria domain (southern Apennines, Italy). *Geol. J.* **2018**, *53*, 660–681. [CrossRef]
39. Scandone, P. Studi di geologia lucana; la serie calcareo silico marnosa e i suoi rapporti con l'Appennino calcareo. *Boll. Soc. Nat. Napoli* **1967**, *76*, 301–469.
40. Scandone, P. Studi di Geologia lucana: Carta dei terreni della serie calcareo-silico-marnosa e note illustrative. *Boll. Soc. Nat. Napoli* **1972**, *81*, 225–300.
41. Scandone, P.; Sgrosso, I.; Vallario, A. Finestra tettonica nella serie calcareo-silico-marnosa lucana presso Campagna (Monti Picentini, Salerno). *Boll. Soc. Nat. Napoli* **1967**, *76*, 3–10.
42. Turco, E. La finestra tettonica di Campagna (M. Picentini, Salerno). *Boll. Soc. Nat. Napoli* **1976**, *85*, 639–665.
43. ISPRA. Carta Geologica d'Italia Alla Scala 1:50.000, Foglio 468 "Eboli". Available online: http://www.isprambiente.gov.it/Media/carg/468_EBOLI/Foglio.html (accessed on 4 July 2020).
44. Angelier, J.; Mechler, P. Sur une méthode graphique de recherche des contraintes principales également utilisable en tectonique et en séismologie: La méthode des diédres droits. *Bull. Soc. Geol. Fr.* **1977**, *19*, 1309–1318. [CrossRef]
45. Angelier, J. Determination of the mean principal directions of stresses for a given fault population. *Tectonophysics* **1979**, *56*, T17–T26. [CrossRef]
46. Reiter, F.; Acs, P.; Tectonics, F.P. Software for Structural Geology. Innsbruck University: Innsbruck, Austria, 1996–2017. Available online: http://www.tectonicsfp.com (accessed on 4 July 2020).
47. Bown, P.R.; Young, J.R. Techniques. In *Calcareous Nannofossil Biostratigraphy*; Bown, P.R., Ed.; Springer: Dordrecht, The Netherlands, 1998; pp. 16–28.
48. Reynolds, J.M. *An Introduction to Applied and Environmental Geophysics*, 2nd ed.; Wiley: Chichester, UK, 2011; p. 688. ISBN 978-0-471-48535-3.
49. Di Maio, R.; Patella, D.; Petrillo, Z.; Siniscalchi, A.; Cecere, G.; De Martino, P. Application of electric and electromagnetic methods to the definition of the Campi Flegrei caldera (Italy). *Ann. Geophys.* **2000**, *43*, 375–390.

50. Park, S.K.; Wernicke, B. Electrical conductivity images of quaternary faults and tertiary detachments in the California basin and range. *Tectonics* **2003**, *22*, 1030. [CrossRef]
51. Koukadaki, M.A.; Karatzas, G.P.; Papadopoulou, M.P.; Vafidis, A. Identification of the saline zone in a coastal aquifer using electrical tomography data and simulation. *Water Resour. Manag.* **2007**, *21*, 1881–1898. [CrossRef]
52. Vanneste, K.; Verbeeck, K.; Petermans, T. Pseudo-3D imaging of a low-slip-rate, active normal fault using shallow geophysical methods: The Geleen fault in the Belgian Maas River valley. *Geophysics* **2008**, *73*, B1–B9. [CrossRef]
53. Giocoli, A.; Galli, P.; Giaccio, B.; Lapenna, V.; Messina, P.; Peronace, E.; Romano, G.; Piscitelli, S. Electrical resistivity tomography across the Paganica-San Demetrio fault system (L'Aquila 2009 earthquake). *Boll. Geofis. Teor. Appl.* **2011**, *52*, 457–469.
54. Giocoli, A.; Magrì, C.; Vannoli, P.; Piscitelli, S.; Rizzo, E.; Siniscalchi, A.; Basso, C.; Di Nocera, S. Electrical Resistivity Tomography investigations in the Ufita Valley (Southern Italy). *Ann. Geophys.* **2008**, *51*, 213–223.
55. Seminsky, K.Z.; Zaripov, R.M.; Olenchenko, V.V. Interpretation of shallow electrical resistivity images of faults: Tectonophysical approach. *Russ. Geol. Geophys.* **2016**, *57*, 1349–1358. [CrossRef]
56. Lecocq, T.; Camelbeeck, T. Electrical resistivity tomography data across the Hockai Fault Zone (Ardenne, Belgium). *Data in Brief* **2017**, *11*, 1–4. [CrossRef]
57. Loke, M.H. *Tutorial: 2-D and 3-D Electrical Imaging Surveys*; Geotomo Software: Penang, Malaysia, 2001.
58. Bravi, S.; Civile, D.; Martino, C.C.; Putignano, M.L. La struttura da interferenza nei carbonati mesozoici della dorsale di Monte Fellino (Appennino Campano). *Boll. Soc. Geol. Ital.* **2006**, *125*, 105–116.
59. ISPRA. Servizio Geologico d'Italia. Geological Map of Italy, Scale 1:100,000, Sheet 171 Gaeta. Available online: http://193.206.192.231/carta_geologica_italia/tavoletta.php?foglio=171 (accessed on 4 July 2020).
60. Pescatore, T.; Sgrosso, I. I rapporti tra la Piattaforma Campano-Lucana e la Piattaforma Abruzzese-Campana nel casertano. *Boll. Soc. Geol. Ital.* **1973**, *92*, 925–938.
61. Martini, E. Standard Tertiary and Quaternary calcareous nannoplankton zonation. In *Proceedings of the Second Planktonic Conference*; Farinacci, A., Ed.; Tecnoscienza: Roma, Italy, 1971; Volume 2, pp. 739–765.
62. Okada, H.; Bukry, D. Supplementary modification and introduction of code numbers to the low-latitude coccolith biostratigraphic zonation (Bukry 1973, 1975). *Mar. Micropaleontol.* **1980**, *5*, 321–325. [CrossRef]
63. Backman, J.; Raffi, I.; Rio, D.; Fornaciari, E.; Pälike, H. Biozonation and biochronology of Miocene through Pleistocene calcareous nannofossils from low and middle latitudes. *Newsletters Strat.* **2012**. [CrossRef]
64. Vitale, S.; Tramparulo, F.D.A.; Ciarcia, S.; Amore, F.O.; Prinzi, E.P.; Laiena, F. The northward tectonic transport in the southern Apennines: Examples from the Capri Island and western Sorrento Peninsula (Italy). *Int. J. Earth Sci.* **2017**, *106*, 97–113. [CrossRef]
65. Pescatore, T.; Sgrosso, I.; Torre, M. Lineamenti di tettonica e sedimentazione nel Miocene dell'Appennino campano-lucano. *Mem. Soc. Nat. Napoli* **1970**, *78*, 337–408.
66. Critelli, S.; Le Pera, E. Tectonic evolution of the southern Apennines Thrust-Belt (Italy) as reflected in modal compositions of Cenozoic sandstone. *J. Geol.* **1995**, *103*, 95–105. [CrossRef]
67. Gattacceca, J.; Speranza, F. Paleomagnetic constraints for the tectonic evolution of the southern Apennines belt (Italy). *Boll. Soc. Geol. Ital.* **2007**, *7*, 39–46.
68. Cerisola, R.; Montone, P. Analisi strutturale di un settore della catena dei Monti Ausoni–Aurunci (Lazio, Italia centrale). *Boll. Soc. Geol. Ital.* **1992**, *111*, 449–457.
69. Naso, G.; Tallini, M.; Tozzi, M. Caratteristiche geologico-strutturali dell'area di Miranda (Isernia): Un contributo alla comprensione dei rapporti tra le falde molisane e avanfossa del Messiniano-Pliocene inferiore. *Mem. Soc. Geol. Ital.* **1995**, *114*, 423–441.
70. Scrocca, D.; Tozzi, M.; Parotto, M. Assetto strutturale del settore compreso tra il Matese, le Mainarde e l'Unità di Frosolone. Implicazioni per l'evoluzione neogenica del sistema di sovrascorrimento nell'Appennino centro-meridionale. *Studi Geol. Camerti* **1995**, *2*, 407–418.
71. De Corso, S.; Scrocca, D.; Tozzi, M. Geologia dell'anticlinale del Matese e implicazioni per la tettonica dell'Appennino molisano. *Boll. Soc. Geol. Ital.* **1998**, *117*, 419–441.
72. Sani, F.; Del Ventisette, C.; Montanari, D.; Coli, M.; Nafissi, P.; Piazzini, A. Tectonic evolution of the internal sector of the Central Apennines, Italy. *Mar. Pet. Geol.* **2004**, *21*, 1235–1254. [CrossRef]
73. Ghisetti, F.; Vezzani, L. Thrust belt development in the Central Apennines: Northward polarity of thrusting and out-of-sequence deformations in the Gran Sasso chain (Italy). *Tectonics* **1991**, *10*, 904–919. [CrossRef]

74. Carminati, E.; Fabbi, S.; Santantonio, M. Slab bending, syn-subduction normal faulting, and out-of-sequence thrusting in the Central Apennines. *Tectonics* **2014**, *33*, 530–551. [CrossRef]
75. Pace, P.; Domenica, A.D.; Calamita, F. Summit low-angle faults in the Central Apennines of Italy: Younger on-older thrusts or rotated normal faults? Constraints for defining the tectonic style of thrust belts. *Tectonics* **2014**, *33*, 756–785. [CrossRef]
76. Corrado, S.; Aldega, L.; Perri, F.; Critelli, S.; Muto, F.; Schito, A.; Tripodi, V. Detecting syn-orogenic extension and sediment provenance of the Cilento wedge top basin (southern Apennines, Italy): Mineralogy and geochemistry of fine-grained sediments and petrography of dispersed organic matter. *Tectonophysics* **2019**, *750*, 404–418. [CrossRef]
77. Scrocca, D.; Tozzi, M. Tettogenesi mio-pliocenica dell'Appennino molisano. *Boll. Soc. Geol. Ital.* **1999**, *118*, 255–286.
78. ISPRA. Carta Geologica d'Italia Alla Scala 1:50.000, Foglio 433 "Ariano Irpino". Available online: http://www.isprambiente.gov.it/Media/carg/433_ARIANO_IRPINO/Foglio.html (accessed on 4 July 2020).
79. Dazzaro, L.; Di Nocera, S.; Pescatore, T.; Rapisardi, L.; Romeo, M.; Russo, B.; Senatore, M.R.; Torre, M. Geologia del margine della catena appenninica tra il F. Fortore ed il T. Calaggio (Monti della Daunia-Appennino Meridionale). *Mem. Soc. Geol. Ital.* **1988**, *41*, 411–422.
80. Mazzoli, S.; Barkham, S.; Cello, G.; Gambini, R.; Mattioni, L.; Shiner, P.; Tondi, E. Reconstruction of continental margin architecture deformed by the contraction of the Lagonegro Basin, southern Apennines, Italy. *J. Geol. Soc. Lond.* **2001**, *158*, 309–319. [CrossRef]
81. Ascione, A.; Ciarcia, S.; Di Donato, V.; Mazzoli, S.; Vitale, S. The Pliocene-Quaternary wedge-top basins of southern Italy: An expression of propagating lateral slab tear beneath the Apennines. *Basin Res.* **2012**, *24*, 456–474. [CrossRef]
82. Pace, P.; Calamita, F. Push-up inversion structures v. fault-bend reactivation anticlines along oblique thrust ramps: Examples from the Apennines fold-and-thrust belt (Italy). *J. Geol. Soc. Lond* **2014**, *171*, 227–238. [CrossRef]
83. Scisciani, V.; Patruno, S.; Tavarnelli, E.; Calamita, F.; Pace, P.; Iacopini, D. Multi-phase reactivations and inversions of Paleozoic–Mesozoic extensional basins during the Wilson cycle: Case studies from the North Sea (UK) and the Northern Apennines (Italy). In *Fifty Years of the Wilson Cycle Concept in Plate Tectonics*; Wilson, R.W., Houseman, G.A., Mccaffrey, K.J.W., Doré, A.G., Buiter, S.J.H., Eds.; Geological Society of London: London, UK, 2019; Volume 470, pp. 205–243.
84. Scisciani, V.; Agostini, S.; Calamita, F.; Pace, P.; Cilli, A.; Giori, I.; Paltrinieri, W. Positive inversion tectonics in foreland fold-and-thrust belts: A reappraisal of the Umbria–Marche Northern Apennines (Central Italy) by integrating geological and geophysical data. *Tectonophysics* **2014**, *637*, 218–237. [CrossRef]
85. Bally, A.W.; Gordy, P.L.; Stewart, G.A. Structure, seismic data and orogenic evolution of the southern Canadian Rocky Mountains. *Bull. Can. Pet. Geol.* **1966**, *14*, 337–381.
86. Ghisetti, F.; Barchi, M.; Bally, A.W.; Moretti, I.; Vezzani, L. Conflicting balanced structural sections across the Central Apennines (Italy): Problems and implications. In *Generation, Accumulation and Production of Europe's Hydrocarbons*; Spencer, A.M., Ed.; Springer: Berlin/Heidelberg, Germany, 1993; Volume 3, pp. 219–231.

© 2020 by the authors. Licensee MDPI, Basel, Switzerland. This article is an open access article distributed under the terms and conditions of the Creative Commons Attribution (CC BY) license (http://creativecommons.org/licenses/by/4.0/).

geosciences

MDPI

Article

The Capo Castello Shear Zone (Eastern Elba Island): Deformation at the Contact between Oceanic and Continent Tectonic Units

Caterina Bianco

Department of Earth and Geo-Environmental Sciences, University of Bari, Piazza Umberto I, 1, 70121 Bari BA, Italy; biancocate0@gmail.com

Received: 19 August 2020; Accepted: 7 September 2020; Published: 10 September 2020

Abstract: Low-grade mylonitic shear zones are commonly characterized by strain partitioning, with alternating low strain protomylonite and high strain mylonite and ultramylonite, where the shearing is most significant. In this paper the capo Castello shear zone is analyzed. It has developed along the contact between continental quartzo-feldspathic, in the footwall, and oceanic ophiolitic units, in the hangingwall. The shear zone shows, mostly within the serpentinites, a heterogeneous strain localization, characterized by an alternation of mylonites and ultramylonites, without a continuous strain gradient moving from the protolith (i.e., the undeformed host rock) to the main tectonic contact between the two units. The significance of this mylonitic shear zone is examined in terms of the dominant deformation mechanisms, and its regional tectonic frame. The combination of the ultramafic protolith metamorphic processes and infiltration of derived fluids caused strain softening by syntectonic metamorphic reactions and dissolution–precipitation processes, leading to the final formation of low strength mineral phases. It is concluded that the strain localization, is mainly controlled by the rock-fluid interactions within the ophiolitic level of the Capo Castello shear zone. Regarding the regional setting, this shear zone can be considered as an analogue of the initial stage of the post-collisional extensional fault, of which mature stage is visible along the Zuccale fault zone, a regional structure affecting eastern Elba Island.

Keywords: strain partitioning; shear zone; quartzo-feldspathic rocks; serpentinite; Elba Island

1. Introduction

Mylonites are within shear zones where high strain conditions developed during ductile deformation. Some of the causes of their nucleation can be attributed to: (i) pre-existing mechanical discontinuities [1,2]; (ii) high shear heating [3]; (iii) strain softening caused by fluid-rock interaction and grain size reduction [4,5]. This latter condition favors crystallization mechanisms [6,7], mostly linked to the dissolution–precipitation and diffusive mass-transfer [8–10]. There are several cases of strain partitioning in various tectonic settings and at different scales [10–16]. In the context of collisional and post-collisional evolution of an orogeny, the partitioning of the strain is well evident at the tectonic contact between continental and oceanic units, characterized by quartzo-feldspathic and ultramafic rocks, respectively. This contact commonly develops deformation at temperature conditions up to about 400 °C during collision [17,18]. Furthermore, the presence of syn-tectonic fluids can trigger metamorphic reactions and pressure–solution processes [19], determining the growth of new weak minerals, such as micas, in quartzo-feldspathic rocks and talc, serpentine polymorphs and chlorite, in ultramafic rocks [20–22]. Such processes occur in preferential levels where strain tends to be localized ([23,24], even at lower temperature values: 300 °C–500 °C, [25]). In particular, the role of serpentinite alteration tends to deeply weaken ultramafic rocks strength [25–27] up to make soft an original highly competent level, leading to the final formation of a phyllosilicate network (for example, formed by talc, muscovite and kaolinite) with a very low frictional strength [28–31].

In this large context, a further example is from the Capo Castello shear zone, here presented. This shear zone is exposed in the eastern part of Elba Island (Italy) delimiting the contact (Figures 1 and 2) between the oceanic Ligurian units (Oceanic unit 1, in [32,33]) and the continental units (Continental units 2–3, in [32,33]). It involves ultramaphics, from which insights on the weakening process and role of talc formation comes out.

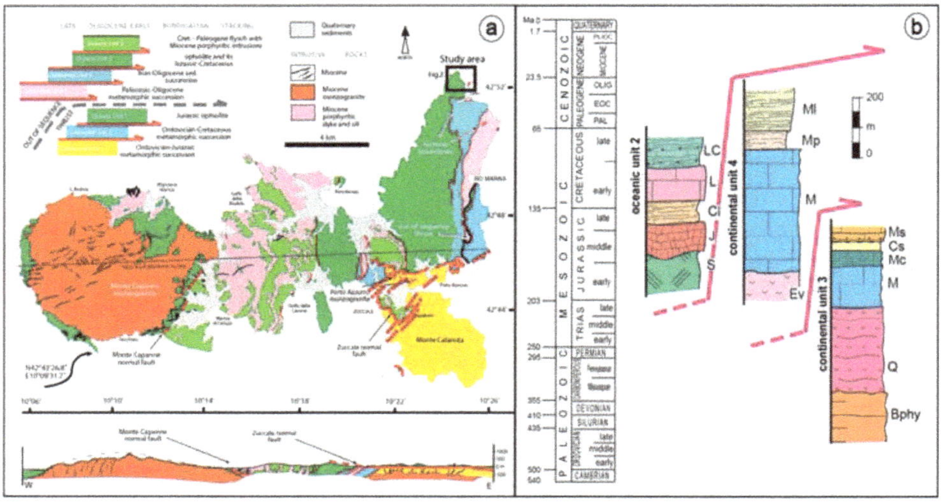

Figure 1. (**a**) Geological sketch map of Elba Island showing the seven tectonic units belonging to the collisional tectonic pile. The black solid square indicates the study area, enlarged Figure 2; (**b**) Tectono-stratigraphic columns in the study area: Oceanic Unit 2: S = ophiolite; J = radiolarite (Mt. AlpeChert Fm.); Cl = calcilutite and cherty limestone (Nisportino Fm.); L = cherty limestone (Calpionella Limestone Fm.); LC = limestone and shale (Palombini Shale Fm.). Continental Unit 4: Ev = evaporite (Calcare Cavernoso Fm.); M = massive and cherty limestone and dolostone (Pania di Corfino Fm., Mt. Cetona Fm., Calcare Massiccio Fm., Grotta Giusti Limestone, Rosso Ammonitico Fm., Limano cherty Limestone Fm.); Mp = marls (Posidonia Marlstone Fm.); Ml = Varicolored Shales (Cavo Fm.). Continental Unit3: Bphy = black phyllite (Rio Marina Fm.); Q = quartzite and phyllite (Verruca Fm., Mt. Serra quartzite Fm.); M = marble (Valle Giove Limestone Fm.; Capo Pero Limestone Fm.; Capo Castello Calcschist Fm.); Mc = cherty marble; Cs = calcschist and phyllite (Varicoloured Sericitic Schist Fm.); Ms = metasandstone and phyllite (Pseudomacigno Fm.) Modified, after [32,33].

Although the Capo Castello shear zone shows a low-grade metamorphism, its architecture does not show the clear continuous strain gradient typical of mylonitic low-grade shear zone [34,35], whereas a from edge-to-center alternation of mylonites and ultramylonites is displayed [36].

Then, in the frame of Elba tectonic evolution, some authors account for a contractional and/or pulsing tectonic regime, active from the Cretaceous to Pliocene–Pleistocene times [37–42]. In this context, the study of shear zones represents a key-factor in order to contribute to this issue. The reader is addressed to papers such as [43–47] where this topic is widely treated by means of a discussion on the reasons why a continuous extensional framework can better explain the regional geological features of the inner Northern Apennines, where Elba is structurally located.

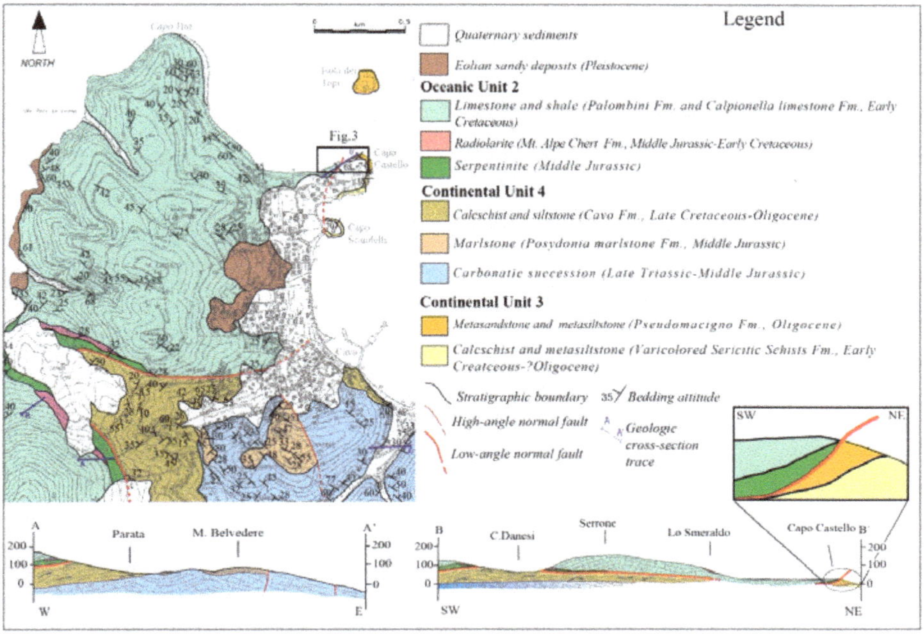

Figure 2. Geologic map and cross sections of the study area. The black square indicates the location Capo Castello shear zone. A schematic and not-to-scale cross section of the study shear zone (grey shade) is also showed, corresponding to the zoomed area within the BB′ cross section.

2. Geological Framework

The northern Apennines is a collisional belt that originated from the convergence-and-collision (Cretaceous–Early Miocene) between the Adria microplate and the European plate, the latter represented by the Sardinia–Corsica block [48,49]. This geodynamic process determined the eastward stacking of several tectonic units derived from the oceanic (Ligurian Domain), to transitional (Sub-Ligurian Domain) and continental palaeogeographic domains (Tuscan Domain and Umbro–Marchean Domain) of the northern Apennines [50].

Since the early-middle Miocene to present, the eastward migrating extension [51–53] affected the inner part of the northern Apennines and was characterized by [47] Miocene low-angle normal faults, successively cross cut by Pliocene–present high-angle normal faults. Then, since the late Miocene, extension was accompanied by widespread exhumation and magmatism with an eastward younging direction [54].

Elba Island is characterized by seven main tectonic units belonging both to continental and oceanic domains (Figure 1a). Its tectonic evolution is framed in the one of the northern Apennines, where (i) a compressional stage (late Oligocene/early Miocene) determined the overthrusting of the oceanic and continental units, including out-sequence thrusts [55]; (ii) a Miocene–Pliocene extensional stage characterized by low angle normal faults, e.g., the Zuccale fault [56–58], at least, developed during and after the emplacement of the Monte Capanne (about 7.0 Ma, [59]) and Porto Azzurro (5.9 Ma, [60]) granitic complexes, respectively, located to the west and the east of the Island.

The interaction with faulting and magmatism favored exhumation and uplift of the plutons and their hosting mid-crustal rocks [58]. The Miocene–Pliocene extensional structures are cross-cut by high-angle normal faults, interpreted as transfer faults, responsible for the hydrothermal fluid flow that is linked to the cooling of the pluton [46].

In the study area, the superimposition of oceanic units onto the continental ones is clearly evident (Figures 1b and 2). In detail, the units of continental pertinence are characterized by carbonate and quartzo-feldspathic metasedimentary rocks (Cretaceous–Oligocene, [61]; Continental Unit 3 in Figure 1b) and by calcareous and marly pelagic sediments (Jurassic to Cretaceous–Oligocene [61], Continental Unit 4 in Figure 1b), locally characterized by intense deformation. The oceanic unit is composed of remnants of the Jurassic ophiolite succession and its sedimentary cover, made up of Jurassic radiolarite and Cretaceous–Eocene calcareous sediments, with the latter largely cropping out.

3. The Capo Castello Shear Zone

The hangingwall of the Capo Castello shear zone is characterized by grey limestone, calcilutite and shale (Palombini shale Fm., in [61], belonging to the oceanic sedimentary cover) and by low-grade serpentinite (oceanic basement) [62], interested by a pervasive retrograde serpentinization, but unaffected by the Pliocene thermo-metamorphism induced by the Porto Azzurro pluton [63]. The maximum metamorphic temperature is in fact less than 350 °C, comparable with the temperature for the serpentinite [62] above Continental unit II (Figure 1). Conversely, the footwall is defined by phyllite, calcschist (Varicolored Sericitic Schist Fm., [61]) and metasandstone with intercalation of dark phyllite (Pseudomacigno Fm., [59]), belonging to the continental domain (Tuscan Domain). These rocks are part of the epimetamorphic Tuscan-type succession [63] and experienced metamorphic peak temperatures of less than 300 °C, thus indicating a low regional metamorphic grade [63], ascribed to greenschist/Qtz-Alb-Ms-Chlsubfacies by [64].

The structure of the Capo Castello shear zone, on the whole gently dipping to the SW (Figure 2), was analyzed in a 97 m long section, SSW–NNE oriented where the mylonitic rocks have an apparent thickness of 47.7 m (Figure 3 and Table 1) and are delimited by two different protoliths. The protolith (i.e., the rock outside the shear zone), at its eastern side (Figure 3), is made up of phyllite and calcschist (Varicolored Sericitic Schist Fm., in [61]) belonging to the continental unit III, and showing an intense internal deformation, with isoclinal re-folded folds [61]. Locally, these rocks show intercalated and metamorphosed dark grey limestones. Differently, the western protolith crops out for 49.3 m (Figure 3) and it is represented by SW dipping grey limestone and calcilutite, belonging to the oceanic unit (Palombini shale Fm., [61]). The western edge of the shear zone is characterized by protomylonites, about 5 m thick, affecting the ophiolite sedimentary cover (Palombini shale Fm., [61]), while the eastern one is defined by about 10 m thick continental metasandstone (Pseudomacigno Fm., [61]) protomylonites (Figure 3). Both eastern and western protomylonites grade to mylonite and ultramylonite serpentinite domains, where the bulk of the deformation is localized (Figure 3).

The differentiation in protomylonite-mylonite and ultramylonite has been carried out taking into account the distribution and size of porphyroclasts and their percentage areal distribution. Their semi-quantitative estimation was obtained by the free software SXM v2.5 (http://www.ImageSXM.org.uk), taking into account four scanned thin sections representative of low and high strain domains (Figure 4).

In the following section, the full range of structures are described, starting from the protomylonitic to the mylonitic and ultramylonitic rocks. In addition, a section on later brittle structures affecting the whole shear zone, is presented, in order to better constrain the different deformation phases of the study shear zone and to frame the shear zone in the Elba Island tectonic evolution.

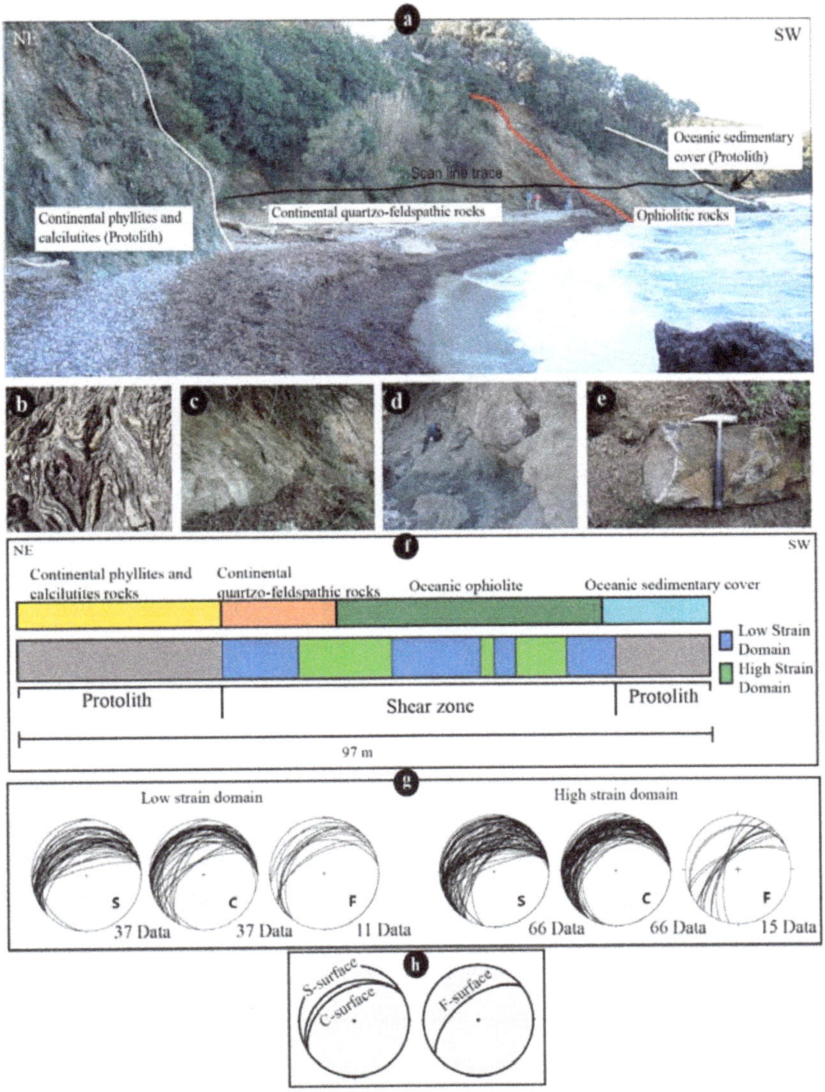

Figure 3. Capo Castello analyzed shear zone involving the quartzo-feldspathic rocks belonging to the Continental Unit 3 and the ophiolite-bearing rocks, with their sedimentary cover, belonging the Oceanic Unit 2. (**a**) Panoramic view of the study area with the boundary between different rock formations involved in the faulting and the scan line trace in black; (**b**) Protolith (i.e., the rock undeformed by the study shear zone) of calcschist and phyllite characterized by isoclinal folds; (**c**) metasandstones involved in the shear zone; (**d**) serpentinite level in the shear zone; (**e**) protolith with limestone and shale, structurally located over the serpentinite; (**f**) scan-line showing the lithology, and the distribution of the strain domains; (**g**) Structural data (lower hemisphere, Schmidt diagram) of the low and high strain domains. Stereoplots indicate cyclographic traces of C and S surfaces and of the brittle west-dipping structures F see the text for description; (**h**) Average stereoplots showing top-to-the ESE shear sense (to the right) and west dipping brittle structures (to the left).Structural and kinematic data are plotted in stereographic diagrams, lower hemisphere, equiareal projection. Stereograms were obtained by OSX Stereonet, from: http://www.geo.cornell.edu/geology/faculty/RWA/RWA.html.

Table 1. Descriptive strain intensity Capo Castello shear zone domains within quartzo-feldspathic rocks and ophiolite, excluding the small level of ophiolite sedimentary cover.

Lithology	Strain Intensity	Features
Quartzofeldspathic rock (metasandstone)	Low strain domain	Protomylonitic fabric Weak crystallized matrix (< 50%) Weak foliation Low frequency distribution of S/C structures Absence of shear bands (C′/C″) Minor grainsize reduction Minor sintectonic mineral growth with weak preferred orientation Area covered by porphyroclasts: 0.8%
	High strain domain	Mylonitic fabric Dark grey recrystallized matrix (between 50% and 90%) Intensely foliated High frequency of S/C structures Presence of C/C″ shear bands Significantly grainsize reduction Moderate sintectonic mineral growth with moderate preferred orientation Porphyroclast with σ-shape, rare θ-type Area covered by porphyroclasts: 10%
Ophiolite (serpentinite)	Low Strain domain	Mylonitic fabric Dark crystallized matrix (between 50% and 90%) Intensely foliated Moderate frequency distribution of S/C structures Absence of shear bands (C′/C″) Moderate grainsize reduction Moderate sintectonic mineral growth with strong preferred orientation Porphyroclast with σ and ∂-shape Area covered by porphyroclasts: 13%
	High strain domain	Ultramylonitic fabric Almost completely recrystallized matrix (<90%) Intensely foliated High frequency of S/C structures Presence of C/C″ shear bands Intense grainsize reduction Pervasive sin-tectonic mineral growth with strong preferred orientation Porphyroclast ∂ and θ-shape area covered by porphyroclasts: 2%

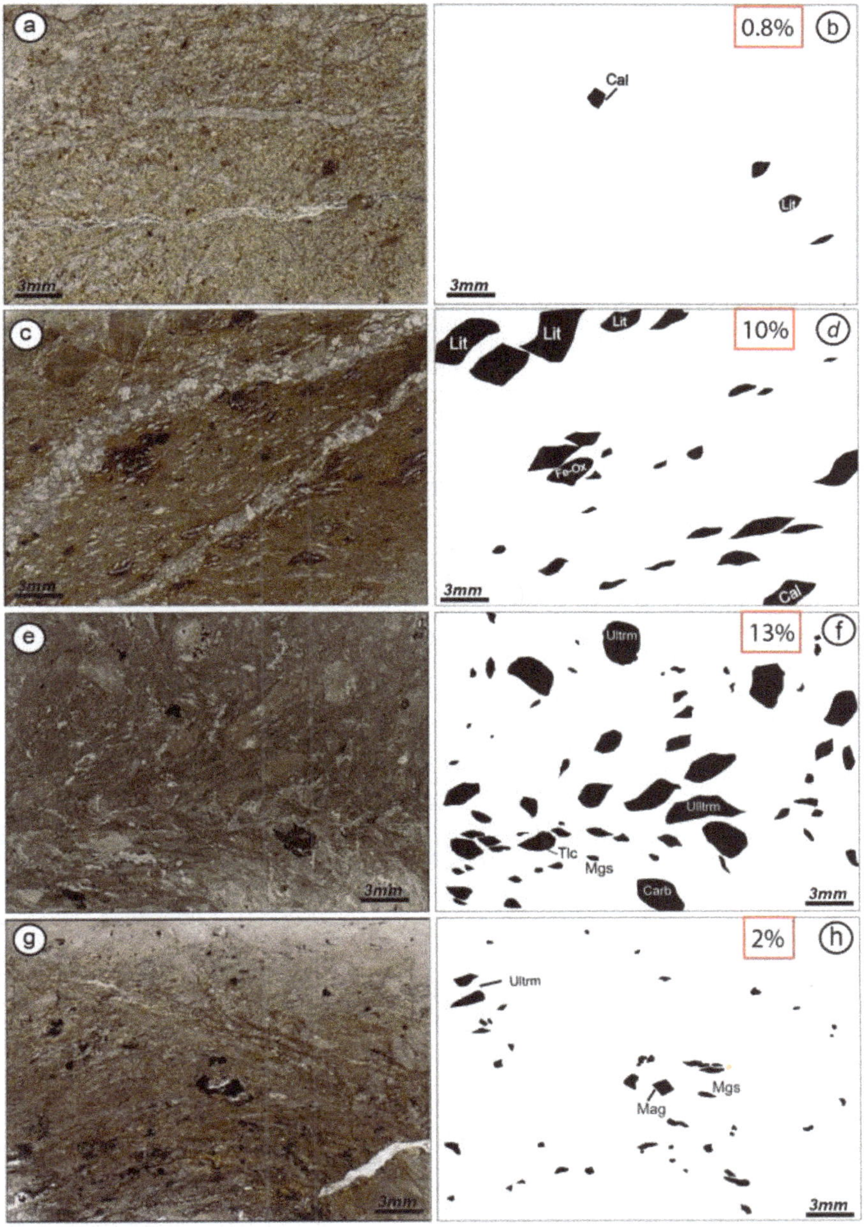

Figure 4. Area percentage occupied by porphyroclasts, as computed using image analysis. (**a**,**b**) quartzo-feldspathic protomylonite domain; (**c**,**d**) quartzo-feldspathic mylonitic domain; (**e**,**f**) myloniticserpentinite; (**g**,**h**) ultramylonite in serpentinite level. Srp: serpentine; Lit: lithic clast; Carb: carbonate clast; Mag: magnetite; Mgs: magnesite; Tlc: talc; Ultrm: ultramafic clast.

3.1. Protomylonites

On the SW side of the shear zone, the protomylonite is composed by metasandstone with sericitic-chloritic matrix and less amount of calcitic cement. This level is weakly foliated with rare S/C structures (Figures 3g and 5a). The matrix results only partly crystallized (< 50%) and the original mineral assemblage is still recognizable, made up of Qtz + Kfs + Ms ± Cal + Bt ± Chl + Fe − Ox. The grain size did not experience a drastic reduction and syntectonic minerals with a preferred orientation are rare. The porphyroclasts are scarce and often represented by lithics with common microlithic textures. The image analysis displays porphyroclasts covering an area of about 1% (Figure 4a,b). Micas, together with elongated and stretched magnesite and quartz ribbons, are mainly located along the S-foliation (Figure 5b,c). The contact between the metasandstoneprotomylonite and the metasandstonemylonites is gradual and marked by an eastward increase in the intensity of foliation and number of porphyroclasts in the matrix.

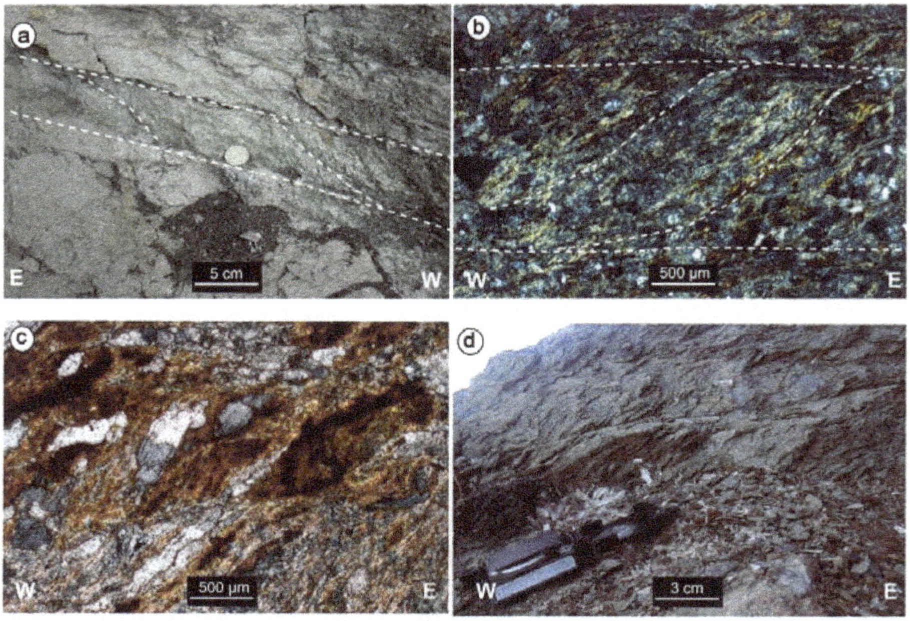

Figure 5. Protomylonitic domains in the quartzo-feldspathic rocks (**a**–**c**) and in the (**d**) calcilutite of oceanic pertinence. (**a**) S/C structures with a top-to-the-east shear sense and related line drawing; (**b**) microscale S/C structures by micas and Fe-oxides and related line drawing; (**c**) crossed polars micrograph showing S surface underlined by magnesite ribbons, micas and Fe-oxides; (**d**); top-to-the-east S/C structures in the calcilutite.

On the NE side of the shear zone, the protomylonite is characterized by the sedimentary cover of the serpentinite composed of calcilutite with interbeddedshales. This domain is less deformed with respect to the quartzo-feldspathic protomylonite and the matrix is weakly crystallized. The clayey layers, comprised between isolated calcilutite blocks, record a relatively intermediate deformation, indicated by the occurrence of a centimeter to decimeter scale top-to-the-east S/C structures, with angle of about 45° between S and C foliations, thus indicating a scarce rotational strain with respect to the one observed in the serpentine mylonites and ultramylonites (Figure 5d). The end of the protomylonitic fabric matches the lithologic passage to the basement cover, represented by the underlying serpentine.

3.2. Mylonites

Metasandstonemylonites differ from the protomylonite by the higher proportion of dark grey crystallized matrix and a larger amount of porhyroclasts and foliation intensity. The latter is indicated by an increase in the alignment of micas and by a stretching and flattening of quartz and magnesite crystals. Foliations are here represented by S/C structures, generally making angles of less than 40° (Figure 3g). These, both in oriented thin-sections and at the outcrop scale, show a general top-to-the-ESE sense of shear (Figure 3g,h and Figure 6a).

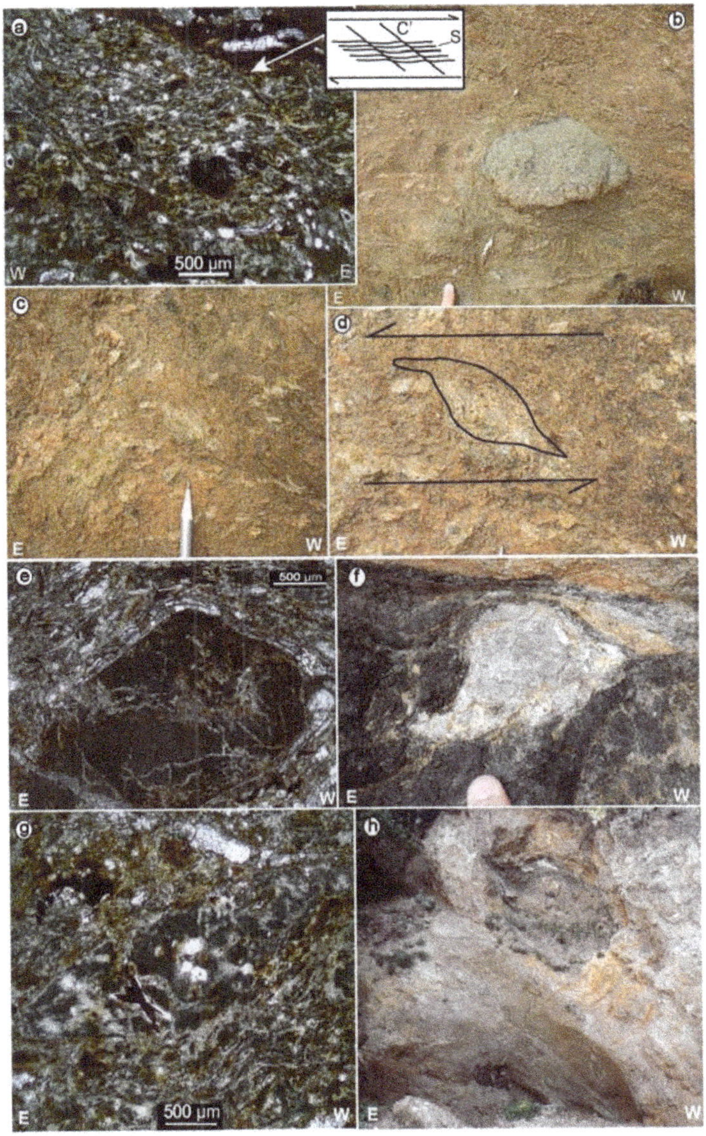

Figure 6. Mylonitic domain within quartzofeldspathic rocks. (**a**) crossed polars micrograph of a highly foliated matrix with small porhyroclasts and shear bands; (**b**) isolated carbonatic porphyroclast; (**c**,**d**) sigmoidal carbonate porhyroclasts and its detail with the line drawing, pointing to a top-to-the east shear sense; (**e**) crossed polars micrographs displaying an altered lithic porphyroclast, affected by small carbonate veins and wrapped by magnesite ribbons; magnesite is also dispersed in the highly foliated matrix; (**f**) carbonate porphyroclast in a darker crystallized matrix; (**g**) crossed polars micrographs showing lithic porhyroclast dispersed in an intensely foliated and crystallized matrix. (**h**) large δ-type ultramafic porphyroclast pointing to a top-to-the east shear sense.

In the matrix, dark and very fine-grained extensional shear bands (C′/C″) are also recognizable (Figure 6a). Commonly, the porphyroclasts contained in the matrix, show small size (up to 6 × 4 cm on average) and are characterized by qtz-lithics and/or carbonate and Fe-Ox clasts (Figure 6b–g). The only exception is represented by two large (ca. 1.5 × 3 m and 90 × 50 cm) ultramafic porphyroclasts (Figure 6h), including smaller (ca. 30 × 20 cm and 15 × 5 cm, respectively) carbonate clasts.

Many porhyroclasts are parallel to the foliation or, more often, they display, σ-shape features (Figure 6c,d), locally showing strain shadows prevalently filled by white micas. Such σ-shape features point to a general top-to-the-ESE sense of shear, coherently with the S/C structures, as previously described. Occasionally, porphyroclasts are also rounded or naked-θ-type wrapped by magnesite ribbons and partly crystallized. On the whole, the average area covered by porhyroclasts is about 10% (Figure 4c,d) of the total area.

Within the myloniticserpentinite domain, the intensity of the foliation increases. The S/C fabric is well developed and porphyroclasts of ultramafic and carbonate composition (Figure 7), ranging from 5 × 8 cm to 70 × 30 cm in size (Figure 7), are abundant, covering about 13% (Figure 4e,f) of the total area. Commonly, these porphyroclasts are aligned along the main foliation (Figure 7a–c). When they have experienced rotation, they show asymmetric tails and pressure shadows, forming σ-shape structures (Figure 7b–e). With the increasing of rotation, ∂-type clasts are also generated (Figure 7f). Again, the asymmetry of porphyroclasts displays a top-to-the-east sense of shear.

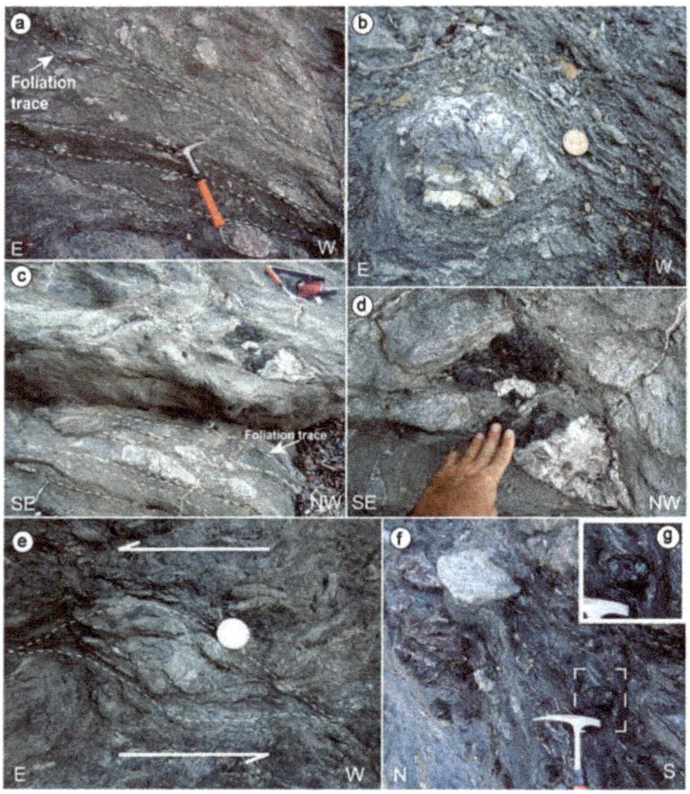

Figure 7. Mylonitic domain in serpentinite rocks. (**a**–**c**) well expressed schistosity, in crystallized matrix with large porphyroclasts (predominantly of carbonate composition), aligned along the main foliation; (**d**,**e**) top-to-the-east carbonate and ultramafic porhyroclasts; (**f**) top-to-the-east δ-type ultramafic (enlarged in g) and carbonate porphyroclasts.

Where the mylonitic serpentine passes into the ultramylonite domain, the contact is gradational and the strain intensity rises up. This is marked by an increase in the intensity of foliation, definition of S/C fabric, and a gradual decrease in the presence of porphyroclasts and their size.

3.3. Ultramylonite

Ultramylonitic serpentinites are characterized by dark-green to black crystallized matrix (<90%) and intense grain-size reduction (Figure 8a–c). These rocks show a strong planar fabric and a drastic reduction of the amount and size of the porphyroclasts. These are mainly represented by ultramafic, carbonate and magnesite clasts, with size up to 5 × 2 cm. Clasts are dispersed within a highly foliated matrix (Figure 8b–g). By the image analysis, porphyroblasts occupy an area of about 2% (Figure 4g,h), having a σ and δ-shape with pressure shadows filled by serpentine and talc. Locally, naked θ-shape porphyroclasts, typical of ultramylonites, can occur, testifying an increase in rotation, a shearing thinning and/or a complete crystallization of the porphyroclast tails. The mylonitic foliation is well preserved and it is almost parallel to the SW dipping C-plane (Figure 8a–c). This latter, coupled with the attitude of the S-planes, generates small angles (<25°), indicating a main top-to-the-SE sense of shear (Figure 3g–h), in agreement with the asymmetric shape of the porhyroclasts and the geometry of their asymmetric pressure shadow tails (Figure 8d,e,g).

Figure 8. Ultramyloniticserpentinite domain. (**a**,**b**) closely spaced top-to-the-east S/C and shear bands structures, respectively; (**c**) highly foliated matrix with tiny porphyroclasts aligned along the main schistosity; (**d**) Crossed polars micrograph showing an asymmetric magnetite porphyroclast with well evident strain shadow, mainly composed of talc and serpentine and pointing to a top-to-the-east shear sense; (**e**) crossed polars micrograph showing an asymmetric (top-to-the-east shear sense) magnesiteporphyroclast with a well evident strain shadow, mainly composed of talc; (**f**) rounded ultramafic porphyroclasts dispersed in a highly foliated matrix; (**g**) crossed polars micrograph showing an euhedral magnetite porphyroclast with a well-developed strain shadow, mainly composed of talc and serpentine.

3.4. Petrographic Features of the Mylonitic and Ultramylonitic Serpentine

Since the serpentinite level played an important role in controlling the deformation along the shear zone, to deepen the knowledge of their mineralogical composition and microstructures, key-samples have been analyzed by SEM/EDS technique and Raman spectroscopy. The SEM/EDS operating conditions were of 15 Kv and 15 nA (Table 2), while the Raman measurements were performed using a Labram Micro-Raman spectrometer by Horiba, equipped with a He-Ne laser source at 632.8 nm (nominal output power 18 mW). The spectral region from 150 to 1100 cm^{-1} was investigated since these include the lattice vibration modes of the serpentine species [65]. The detected bands show low temperature serpentine minerals, such as lizardite [66], locally mixed with antigorite and in minor amount, chrysotile (Figure 9), often present with chlorite in late veins. The peaks around 800 testify the presence of olivine mesh texture. Spectra were likened to those obtained in [65,67–69], with comparable results.

Table 2. Representative mineral analyses in serpentinite gouge rock. Srp = serpentine; Tlc = talc; Mgs = magnesite; Mag = magnetite; Mg-Chr = magnesium-chromite; Cal = calcite; Dol = dolomite; Chl = chlorite, four types of chlorite (Penninite, Pycnochlorite, Chlinochlore and Sheridanite) have been detected. Structural formulae of Mag and Mg-Chromite have been calculated on the basis of 4 oxygens and 3 cations. Apart from the last two oxides, for all the minerals the total amount of Fe has been considered only as $Fe_2{}^{+}CO_3$ has been calculated.

	Mgs	Cal	Dol	Tlc	Srp	Chl-Penninite	Chl-Oycnochlorite	Chl-Chlinochlore	Chl-Sheridanite	Mag	Mg-Chr
SiO_2	0.000	0.000	0.000	62.360	41.960	35.060	34.960	35.640	31.200	0.000	0.000
TiO_2	0.000	0.000	0.000	0.000	0.000	0.000	0.000	0.000	0.000	0.000	0.370
Al_2O_3	0.000	0.000	0.000	0.470	2.210	16.440	20.070	24.300	27.200	0.000	4.900
Cr_2O_3	0.000	0.000	0.000	0.000	0.410	0.000	0.000	0.000	0.000	0.000	32.400
FeO	8.520	0.870	0.770	1.910	3.340	11.200	19.790	1.870	3.500	92.410	46.380
MnO	0.260	0.230	0.370	0.000	0.000	0.000	0.000	0.000	0.000	0.000	0.000
MgO	38.970	2.710	16.070	29.320	38.440	30.250	19.860	33.700	31.080	0.410	14.450
CaO	0.980	45.490	27.510	0.000	0.000	0.080	0.360	0.360	0.360	0.180	0.000
Na_2O	0.000	0.000	0.000	0.000	0.000	0.830	0.580	0.780	0.780	0.000	0.000
K_2O	0.000	0.000	0.000	0.000	0.000	0.120	0.160	0.180	0.180	0.000	0.000
CO_2	48.700	39.335	39.838								
TOTAL	97.430	88.635	84.558	94.060	86.360	93.980	95.780	96.830	94.300	93.000	98.500
Si	0.000	0.000	0.000	8.030	1.979	6.349	6.414	5.965	5.426	0.000	0.000
Al	0.000	0.000	0.000	0.071	0.123	3.509	4.340	4.793	5.575	0.000	0.009
Ti	0.000	0.000	0.000	0.000	0.000	0.000	0.000	0.000	0.000	0.000	0.189
Cr	0.000	0.000	0.000	0.000	0.015	0.000	0.000	0.000	0.000	0.000	0.835
Fe^{3+}	0.000	0.000	0.000	0.000	0.000	0.000	0.000	0.000	0.000	2.000	0.958
Fe^{2+}	0.214	0.027	0.023	0.206	0.132	1.696	3.037	0.262	0.509	0.969	0.307
Mn	0.007	0.009	0.011	0.000	0.000	0.000	0.000	0.000	0.000	0.000	0.000
Mg	1.747	0.153	0.891	5.628	2.703	8.167	5.432	8.408	8.058	0.024	0.702
Ca	0.032	1.811	1.076	0.016	0.000	0.016	0.071	0.065	0.067	0.007	0.000
Na	0.000	0.000	0.000	0.291	0.000	0.291	0.206	0.253	0.263	0.000	0.000
K	0.000	0.000	0.000	0.028	0.000	0.028	0.037	0.038	0.040	0.000	0.000
Σcations	2.000	2.000	2.000	13.935	4.952	20.056	19.538	19.784	19.938	3.000	3.000
CO_3	2.000	2.000	2.000								
X_{Mg}	0.874	0.077	0.445	0.965	0.954	0.828	0.641	0.970	0.941	0.024	0.696

The mineralogical paragenesis, characterizing the matrix, is composed by talc-serpentine minerals-chlorite-magnesite-calcite-pyroxene-spinel-oxides and small amount of dolomite and quartz. On the whole, this mineral assemblage points to a sub-green-schistsfacies [4]. Porphyroclasts are mainly represented by lizardite-rich serpentinite fragments, talc, magnesite and calcite (Figure 10a,b). Often, talc porphyroclasts contain chrysotile vein textures (Figure 10c). There are no textures indicating a prograde alteration of lizardite to antigorite.

Crystal habits of relict orthopyroxene and olivine (Figure 10f–h), suggesting a harzburgite composition of the ultramafic protolith, are sometimes preserved, although no pure composition has been detected. However, the original orthopyroxene, is indicated by the presence of bastite (serpentine + fine-grained carbonate) and by the texture and striations disposed along the cleavage planes (Figure 10f,g). Bastite lamellae are generally colorless and with a lamellar extinction pattern.

Sometimes relic pyroxene is dissected by carbonate veins (Figure 10f). Serpentinized olivine although sometimes replaced by talc, is indicated by the presence of mesh and hourglass texture with variable internal configuration, that may enclose relicts of olivine crystals, evident under optical microscope (Figure 10h). The great majority of these pseudomorphic textures are prevalently composed by micro-crystalline lizardite lamellae and minor chrysotile and antigorite, suggesting that serpentine mainly crystallized at the boundary Opx-Ol, and mostly at the expense of olivine.

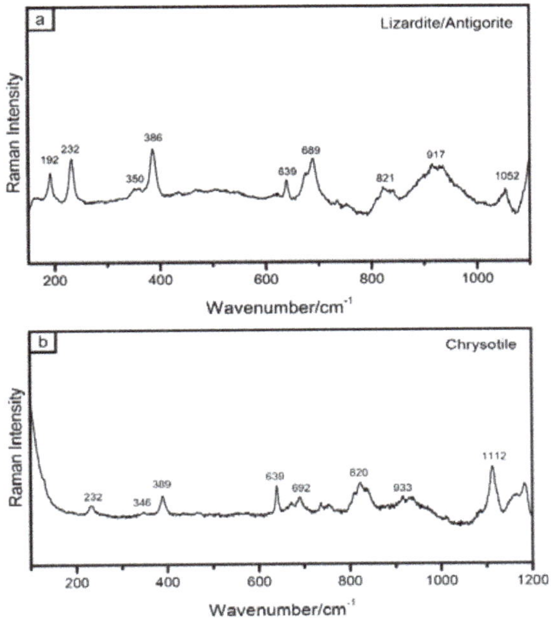

Figure 9. Raman spectra in the low-wave number region of (**a**) mixed lizardite and antigorite and (**b**) chrysotile.

Magnesite is abundant, occurring as porphyroclast wrapped by talc and serpentine (Figure 10b), and as fine aggregates in the matrix. Often magnesite is encountered as monomineralic and cryptocrystalline veinlets, with serrated or sharp contacts in the enclosing serpentinite and talc (Figure 10i). The presence of magnesite, both in the matrix and as porphyroclasts, suggests that the alteration of the protolith to foliated serpentinite predates the mylonitic event.

Magnetite, and subordinately magnesium-chromite, occurs as prismatic, sometimes brecciated and rarely elliptical. In some cases, the oxide strain shadows are characterized by talc and serpentine. During serpentinization, magnetite can precipitate during the alteration and oxidation of olivine and orthopyroxene [70,71]. In this case, magnetite forms fine clusters of tiny opaque crystals, surrounding the altered primary minerals, concentrating at the mesh rims or resulting dispersed within the serpentinite texture and deformation zone. (Figure 10h).

Within the highly deformed rock, the mylonitic and phyllonitic fabric is underlined by the presence of lizardite, chrysotile, talc, Mg-chlorite (penninite, clynochlore, Table 2) and carbonate clasts. A strongly oriented pervasive foliation, characterized by lizardite and talc (Figure 10b,d,e), is locally outlined by oxide alignment (Figure 10a), determined by the spinel Mg-chromite and magnetite (Table 2). Often, both talc and lizardite wrap porhyroclasts composed of magnesite and serpentinite fragments (Figure 10a,b,d) and occur in their strain shadows, indicating a syntectonic crystallization, coeval with the grain-size reduction, in the phyllonitic levels.

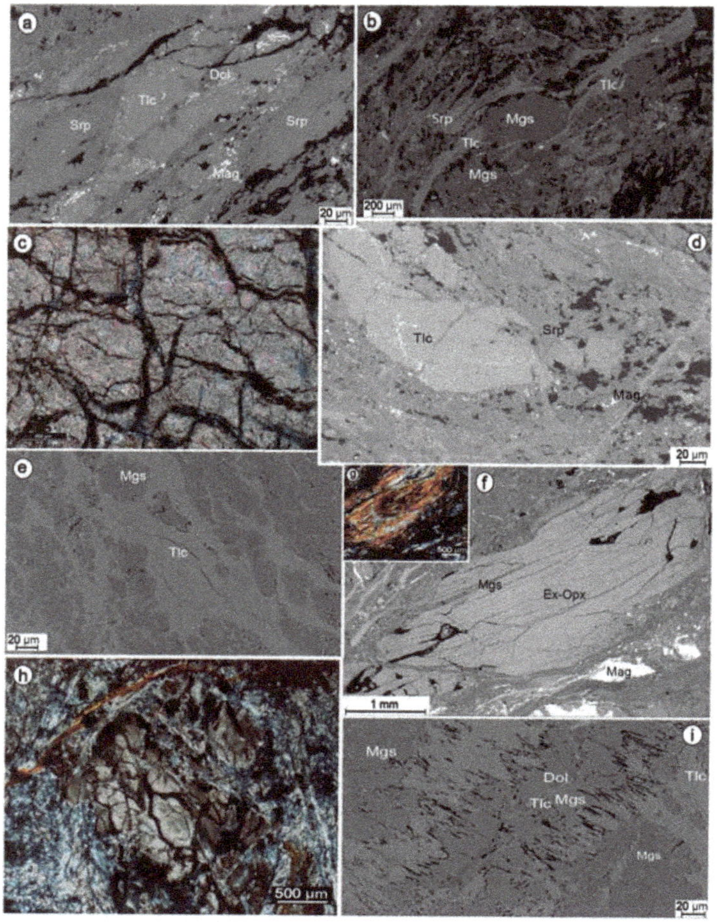

Figure 10. (**a**) sigmoids of serpentine and talc in the mylonitic serpentinite domain; tiny magnetite crystals are aligned along the main schistosity. Within the talc, small dolomite minerals are recognizable; (**b**) magnesite δ-type porphyroclast with strain shadows filled by talc; (**c**) chrysotileveinlets cross cutting talc crystal; (**d**) microstructures pointing to plastic deformation accommodated by a dissolution–precipitation mechanism in the presence of weak minerals as talc and serpentine; (**e**) Interconnected network of talc and magnesite; (**f**) bastite texture (lizardite + fine-grained carbonate) with lamellar extinction and striations disposed along the cleavage planes of the original orthopyroxene in SEM/BSE image; (**g**) the same pseudomorphic texture under optical microscope (crossed nicols); (**h**) deformed mesh boundaries picked out by magnetite and characterizing the substitution of olivine crystal with serpentine (lizardite) and talc (optical microscope, crossed nicols); (i) vein filled with magnesite and having serrated boundaries indicating a reaction processes producing talc. Along the central axis of the vein, dolomite and talc in small amounts indicate dissolution–precipitation reactions producing talc.

3.5. Brittle Structures

The rocks involved in the shear zone are also affected by later west dipping faults cross-cutting the previously formed S/C structures (Figure 11a,b and Figure 12a,b) and porphyroclasts (Figure 12c,d). Often these faults show shear veins on their slip surfaces with calcite mineralization (Figure 12b). Their average dip-direction is N220, with a plunge of about 50° (Figure 3g,h). These structures are also recognized at the microscale, determining the slicing of porhyroclasts (Figure 11c,d and Figure 12e–g). Locally, these grain-scale faults are also subsequent to carbonate extensional veins (Figure 11e,f). The later brittle event is also testified by occurrence of centimeter to millimeter thick E–W trending extensional fibrous antitaxial calcite and magnesite veins (Figures 11e–g and 12h) and by Fe-Oxides and hydroxydes veins cross-cutting porphyroclasts. These veins become more abundant toward the domain where the continental metasandstone crops out. In most cases, the veins follow the foliation defined by the C-planes. Within the metasandstone, fibrous magnesite veins also occur, cross-cutting the main foliation. Locally, also magnesite blocky veins are found, showing well developed magnesite crystal faces and cemented by a microcrystalline calcite (Figure 11h).

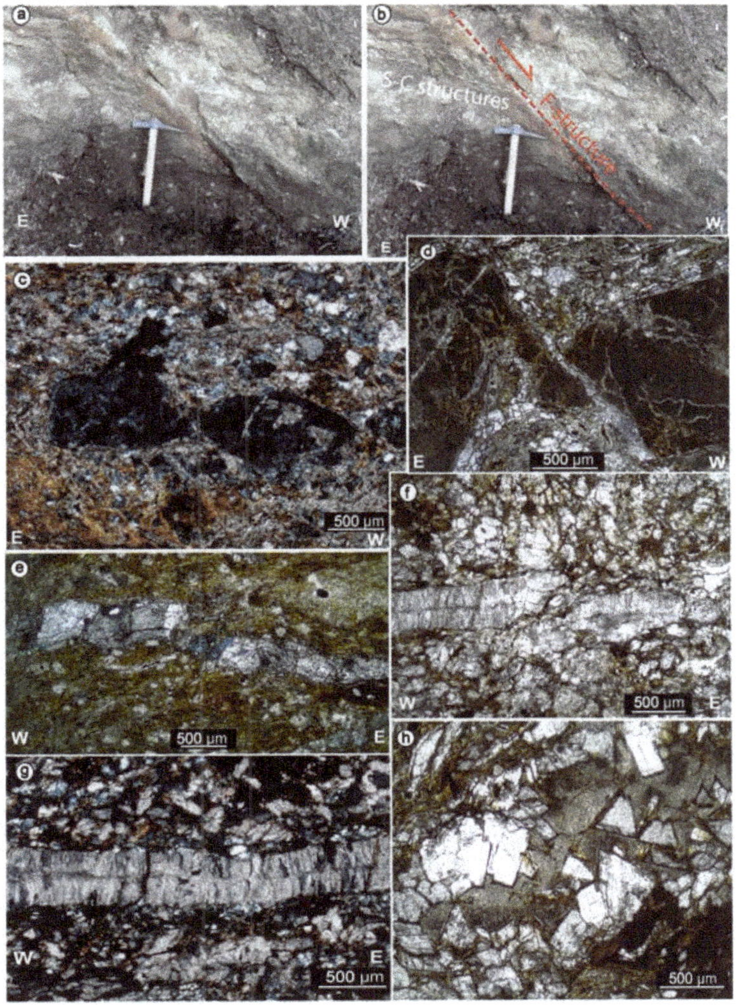

Figure 11. Brittle structures in the quartzo-feldspathic rocks. (**a**,**b**) West dipping faults dissecting the S/C structures and its corresponding line-drawing; (**c**,**d**) micrographs showing lithic porphyroclasts cross cut by small scale faults (crossed nicols); in (**d**) the porphyroclast is also wrapped by magnesite ribbons; (**e**) magnesite vein affected by fractures; (**f**,**g**) micrographs of fibrous, antitaxial extensional calcite; (**h**) blocky magnesite.

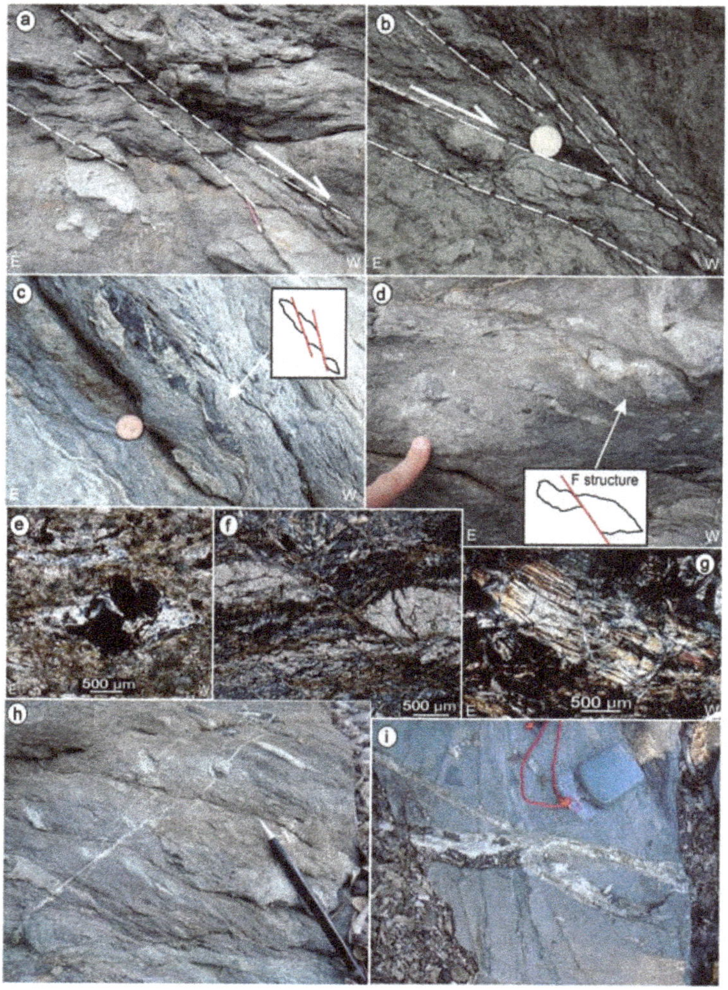

Figure 12. Later brittle structures in ophiolites. (**a**) west-dipping normal fault zone; (**b**) west-dipping faults with calcite mineralization on their slip-planes; (**c**,**d**) carbonate and ultramafic porphyroclasts dislocated by west-dipping fault (F-structure); (**e**) crossed polars micrograph showing a sliced up magnetite porphyroclast; (**f**) crossed polars micrographs showing a magnesite porhyroclast cross cut by a west-dipping fault; (**g**) crossed polars micrograph showing a fragmented pyroxene cross cut by magnesite veins; (**h**) calcite veins overprinting previous brittle structures cross cutting porhyroclasts; (**i**) calcilutite cross cut by calcite veins with internal small regular elements, testifying a possible contribute of hydraulic pressure during the fracturing process.

In addition, the bulk of the carbonate beds belonging to the oceanic sediments are affected by calcite veins with thickness up to about 5 cm (Figure 12i). Often, these veins are characterized by small regular clasts, accounting for a contribute of the hydraulic pressure during the fracturing process. Locally, these veins also cross-cut the west-dipping brittle structures.

4. Discussion

This section discusses the deformation of the two main lithotypes controlling the evolution of the Capo Castello shear zone: the serpentinite and the quartzo-feldspathic sedimentary rocks and the implications of the study shear zone in Elba geological framework.

It is well known the principal mineral phase of serpentinite, the serpentine, is characterized by different polymorphs, reflecting different elastic [72,73] and rheological properties [26,27]. To constrain the PT to which serpentinite underwent the characterization of the serpentine polymorphs is a crucial information. Commonly, lizardite and chrysotile are the dominant varieties occurring in low-grade metamorphic ophiolites [74]. In addition, experimental studies ([68] and natural sample analyses [71], show that under 300 °C and P < 4 kbar (sub-greenschist facies) the lizardite and secondarily the chrysotile are the stable polymorphs, while between 320 °C and 380 °C and at a higher pressure (4 < P < 24 kbar, blueschists facies) the lizardite is gradually replaced by antigorite that became the only serpentine variety occurring above 380 °C.

As regards the case under study, Raman spectroscopy analysis displays that the most abundant serpentine mineral, both in the matrix and in the pseudomorphic meshes, is the low-grade lizardite and in minor amount chrysotile (Figure 9). These two polymorphs commonly occur in the sub-green-schists facies [4]. However, some spectrums show a weakly mixing of lizardite and antigorite, suggesting a transition reaction from these two polymorphs. Therefore, comparing the Capo Castello serpentinite with other experimental and natural cases [26,68,74–76], the hypothesized PT conditions at which the study shear zone was active are T ≈ 320 °C and P < 5 kbar, thus related to the upper part of the greenschist facies (Figure 13).

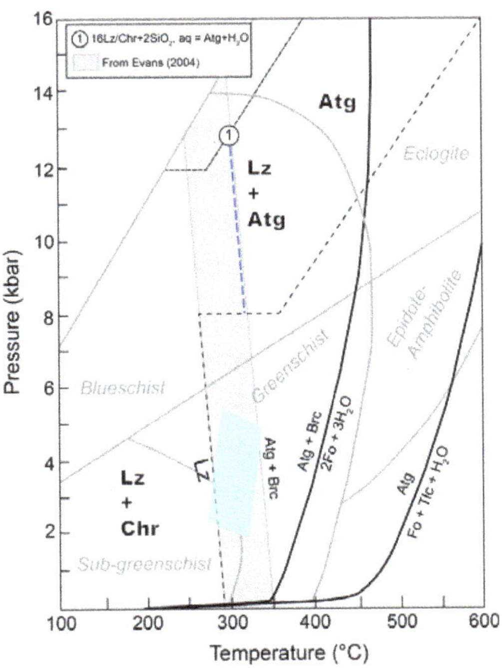

Figure 13. Phase diagram of antigorite and lizardite (from [69], modified). Symbols: Lz-lizardite, Atg-antigorite, Chr-chrysotile, Fo-forsterite, Tlc-talc, Brc-brucite. Reaction (1) is from [68] and occurs in the presence of SiO2-rich fluids by dissolution–precipitation processes. The pale blue area represents the coexisting lizardite and antigorite in the Capo Castello shear zone stability field.

Serpentine minerals are also at the base of Capo Castello shear zone deformation mechanism, because of their low frictional coefficient due to the weak interactions among OH groups and silicate layers [26,77] mostly within the lizardite and chrysotile polymorphs [78,79]. Therefore, given the easy gliding along their basal plane [80], the deformation within the Capo Castello harzburgite protolithis favored and its serpentinization can be considered an efficient mechanism to reduce rock strength, thus permitting easy slip processes. In this context, [29] demonstrate that during serpentinization, the friction coefficient is lowered from $\mu = 0.6$ to 0.3. According to its mineralogical assemblage, the Capo Castello serpentinite developed at low temperature (T < 320 °C), in an open system that favored the infiltration of fluids, [81] demonstrated that meteoric groundwater can be the main factor in serpentinization, forming lizardite and chrysotile minerals. In addition, as documented in several fault zones [5,30,82–84], the intense foliation can also derive from an earlier influx of hydrous fluids, favoring reactions in the initial stage of the cataclasite formation, thus leading to softening of the ultramafic rock and triggering fluid assisted dissolution–precipitation and pressure–solution processes (Figure 10a,d).

The high strain domain along the shear zone (Figure 10e) is also a consequence of the presence of talc and minor talc-carbonates, distributed along the serpentinite foliation [83]. Summing up, and taking into account the textural and petrographic observation, different metamorphic reactions, leading to the formation of low strength mineral phases, can be reasonably considered in the following order [85]:

$$3Fe_2SiO_4 + 2H_2O = 2Fe_3O_4 + 3SiO_2 + 2H \quad (1)$$

$$3Mg_2SiO_4 + SiO_2 + 2H_2O = 2Mg_3Si_2O_5(OH)_4 \quad (2)$$

These first reactions lead to the serpentinization of olivine, as suggested by the pseudomorphic olivine texture (Figure 10h). These reactions produce the initial weakening of the shear zone. Fayalite is oxidized in magnetite, releasing SiO_2 fluids, in turn consumed in the second reaction, to form serpentine from the forsterite component. The results assemblage is serpentine (chrysotile/lizardite), magnetite and some left unreacted olivine [83], as testified by the detected mineral phases within the serpentinite samples.

The following reaction is given by [83]:

$$Mg_3Si_2O_5(OH)_4 + SiO_2 = Mg_3Si_4O_{10}(OH)_2 + H_2O \quad (3)$$

Serpentine, in the presence of silica-rich fluids, can then react to provide talc, a further low strength mineralogy phase. Silica-rich fluids originate from the interaction with the neighboring quartz-rich sediments, suggesting an open system within the ultramafic rock.

The system opening, then, can bring to other, even coeval, metamorphic reactions, here referred to as (4) [86]. Magnesite, being present both along the foliation and within later veins (Figure 11e,h), can be considered as developing from the initial-to-mature stages of shearing, based on the following reaction:

$$4Mg_2SiO_4 + 2H_2O + 5CO_2 = Mg_3Si_4O_{10}(OH)_2 + 5MgCO_3 \quad (4)$$

This is enhanced by the interaction with CO_2 rich fluids, coming from atmosphere and the surrounding carbonate rocks [87,88], i.e., the ophiolitic sedimentary cover, enhancing the alteration processes of the ultramafic protolith [89,90].

Furthermore, magnesite, can result from the reaction of tremolite, that is normally present in ophiolite. In our study area, however, this amphibole was never detected, thus suggesting its total involvement in the following reaction as (5) [91], producing talc, too:

$$Ca_2Mg_5Si_8O_{22}(OH)_2 + 4CO_2 = Mg_3Si_4O_{10}(OH)_2 + 2CaMg(CO_3) + 4SiO_2 \quad (5)$$

It can be concluded that the strain localization and partitioning, is mainly controlled by the occurrence of infiltrated fluids when serpentine is already formed, thus influencing the mechanical behavior of the fault rocks, causing a strain softening process by syntectonic metamorphic reactions and formation of phyllosilicate networks. This process is therefore envisaged to modify low to high strain zones [86], since dissolution–precipitation processes are efficient even in low strain domains, as demonstrated in several upper crustal fault zones [92,93]. As already described, in the quartzo-feldspathic rocks, the low strain domain is represented by protomylonitic fabrics. It shows a partly crystallized matrix (<50%), a slight grain size reduction (Figure 5a,e) and rare newly forming minerals, with a weak preferred orientation. The foliation is well expressed and porphyroclasts are scarce. Differently, the higher strain domain, characterized by a mylonitic fabrics, displays an increased proportion of crystallized matrix, a higher amount of porhyroclasts and foliation intensity, diffuse top-to-the-east S/C structures, alignment of micas, and stretching and flattening of quartz and magnesite crystals (Figure 6). The occurrence of magnesite testifies the presence of syntectonic magnesium-bearing solutions within the quartzo-feldspathic rocks. Fluids, in fact, became Ca^- and Mg^- enriched, passing firstly through the ophiolitecarbonatic cover and then through the serpentinite levels, structurally located at the top of the quartzo-feldspathic rocks. At the 200–300 °C temperature interval the presence of fluids activates pressure–solution processes [88], becoming the dominant deformation mechanism. As a consequence, the production of relative lower strength mineral phases, as biotite and white mica and their segregation in particular layers, led to a softening reaction process and volume reduction [19].On the whole, along the Capo Castello shear zone, the serpentinite domain results the most deformed with respect to the metasandstone; this indicates that during the shear zone activity the crystallization and fluid flow were strongly localized within the serpentinite rocks.

As regards the frame of the Elba island regional geology, the structural and petrographic features of the Capo Castello shear zone can be compared with the Zuccale normal fault (Figure 1), an extensional regional structure affecting Elba Island and locally juxtaposing ophiolite onto quartzitic rocks [30,55–58,84]. Both shear zones, in fact, are characterized by S/C and C/C′ type shear structures, having a general consistent top-to-the-east sense of shear (Figure 14).

According to the Zuccale fault zoning, proposed by [56], their highly foliated levels (L2–L3) with serpentine-tremolite-talc-chlorite schists and chlorite phyllonite record the same deformational mechanisms of the Capo Castello serpentinite level. The main difference between the two structures is that, the Zuccale fault is characterized by a concentrated level of low strength mineral phases, whereas the latter is more widely distributed in the Capo Castello shear zone, probably a consequence of the original composition of the involved ultramaphics and/or to a minor degree of flattening. Another difference between the Zuccale and Capo Castello shear zone can be recognized in the origin of silica-rich fluids. These are in fact related to a magmatic origin at Zuccale, due to the proximity with the Porto Azzurro Pluton. Differently, the Capo Castello area is far from any magma source [46] and therefore the origin of silica rich fluids should be related to fluid circulation from nearest quartzo-feldspathic rocks. Finally, from a tectonic point of view, both structures determined a tectonic omission in the original stacked pile, although at different stages of evolution (Figure 15). By this, the Capo Castello shear zone, located in the hanging wall of the Zuccale fault, can be considered as a part of the deformation associated to an older, exhumed, regional detachment, representative of the initial stage of deformation, now visible at Zuccale in its mature stage. Regarding kinematics, at Capo Castello shear zone, it was observed that the micro and meso-structures generally point to a top-to-the-east sense of shear (Figure 14), while the main shear zone attitude is dipping to the west (Figure 2), as all the later brittle structures cross cutting the S/C pairs and the asymmetric porphyroclasts. This feature suggests a passive rotation of the fault hangingwall as a consequence of gravitational processes occurred during the lateral crustal segmentation that generated megaboudin structures, as already documented in other part of the Tuscany ([41,49,50]). In this context, the kinematic indicators of the study shear zone were also rotated and the present top-to-ESE kinematic indication (Figure 3) is reasonably influenced by this latter process.

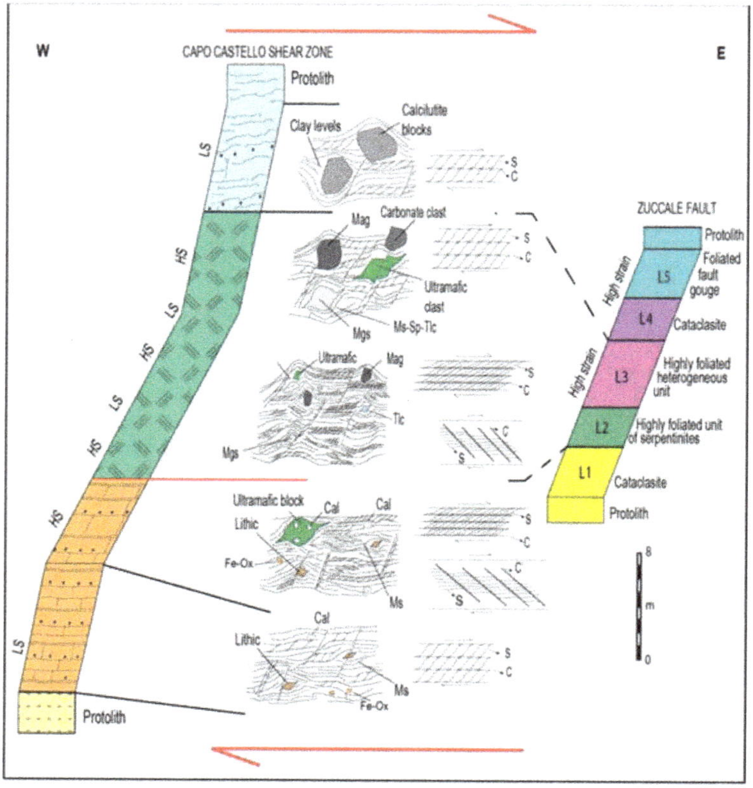

Figure 14. Sketch comparing the Capo Castello shear zone and the Zuccale normal fault stratigraphy. Along the Capo Castello shear zone log, the different strain domains are represented. The Zuccale normal fault log is reconstructed taking into account [56]. **L1**: cataclasite, locally with protomylonitic basement clasts, in a carbonate-quartz-chlorite matrix; **L2**: highly foliated unit of serpentinite plus tremolite-talc-chlorite schist; **L3**: highly heterogeneous unit of chlorite phyllonites with lenses of carbonate mylonite; **L4**: carbonate vein-rich unit set in a cataclasite with carbonate and ultramafic lenses; **L5**: foliated fault gouge and fault breccia. Deformation structures and deformation mechanisms of **L2–L3** units are similar to the ones of the serpentinite levels in the Capo Castello shear zone.

Finally, even if no dating is available in this area, the activity of the Capo Castello shear zone can be encompassed between 19.8 ± 1.4 Ma ([33], age of HP event) and late Miocene–Early Pliocene (age of Zuccale activity, [59]). In this view, the west dipping F-structures (i.e., the latest brittle deformation), accommodated the passive rotation of the hangingwall.

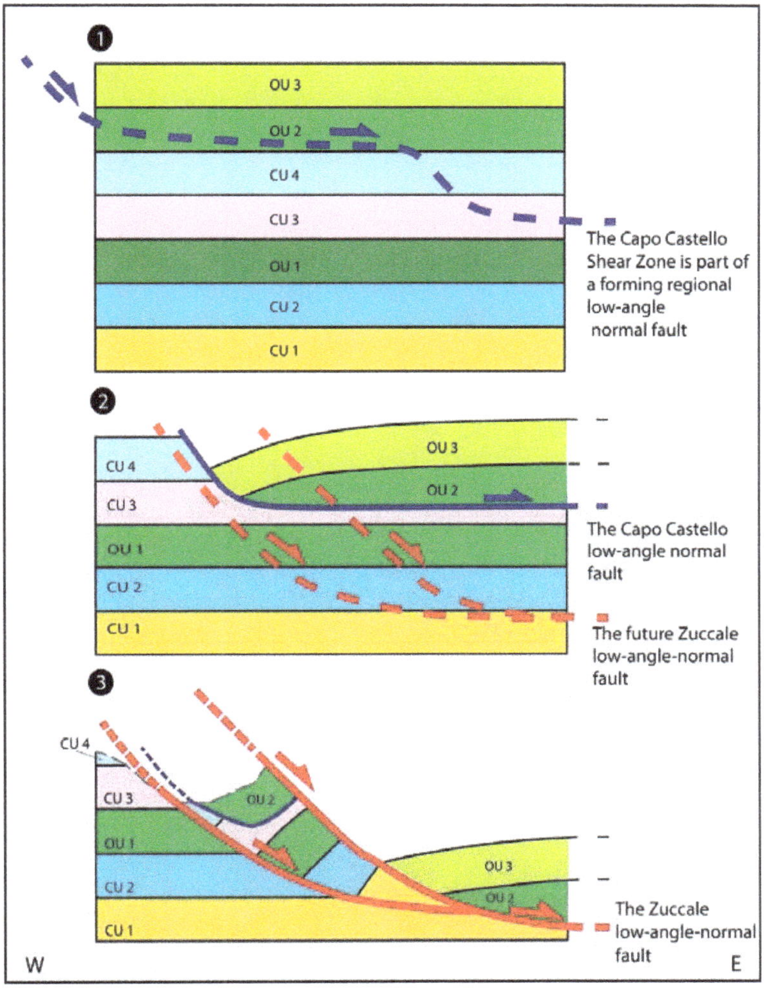

Figure 15. Tectonic scheme, not to scale, showing the tectonic omission, caused by the low angle normal fault to which the Capo Castello shear zone belongs, and its relationship with the low-angle Zuccale normal fault, in the stacked pile. From the top to the bottom: (**1**) the original Elba Island stacked pile linked to the Late Oligocene–early Miocene collisional phase; (**2**) the tectonic omission linked to the activity of the low angle normal fault to which the Capo Castello shear zone is associated; the Oceanic Unit 2 is in direct contact with the continental unit 3; (**3**) the tectonic omission generated by the Zuccale normal fault along which the Oceanic unit 3, the highest unit in the tectonic pile, overlay the deepest continental unit 1. Symbols indicate the Oceanic (OU) and the Continental Units (CU) stacked during the orogenesis. See also Figure 1.

5. Conclusions

The study of the geometry, kinematics, textural and mineral chemistry of the Capo Castello shear zone allowed to draw the following conclusions:

Along the shear zone the heterogeneous strain effects are prevalently influenced by lithology. Within the quartzofeldspathic rocks the low strain domains are expressed by a protomylonitic fabric, generally characterized by a poorly crystallized matrix and a weak grain size reduction and foliation. Within the higher strain domain, the metasandostones show a mylonitic fabric with a dark grey crystallized matrix and a higher amount of porhyroclasts and foliation intensity, represented by a top-to-the-east S/C structure, locally accompanied by C′/C″ shear bands. Apart from the ophiolitic sedimentary cover showing a weak protomylonitic fabric, the serpentinite level records a more intense deformation. In fact, the low strain domains are characterized by a mylonitic fabric, while the higher strain domains by an ultramylonitic one, both of them displaying an intense foliation marked by S/C structures. These two domains show an alternation along the shear zone without a clear strain gradient from the edge to the center of the structure. The passage from the mylonites zone to the ultramylonites is underlined by increasing the proportion of crystallized matrix, of the grain-size reduction and of the number and size of the porphyroclasts. Within the ophiolitic rocks, the partitioning of the strain is expressed as the development of the mylonitic fabric in the low strain domains, and of the ultramylonitic fabric in the high strain ones.

The serpentinite level, due to the peculiar mineralogy of the serpentine minerals, acts as a lubricant along the shear zone, accommodating most of the deformation. The presence of the infiltrated fluid influences the mechanical behavior of the serpentinite, causing strain softening processes by sintectonic metamorphic reactions and dissolution–precipitation processes leading to the final formation of a low strength phyllosilicate networks dominated by talc.

As a maximum, deformation ceased at $T \cong 320$ °C and $P < 5$ kbar (upper greenschistsfacies) as defined by the stability field of lizardite + chrysotile, the dominant serpentine polymorph along the Capo Castello shear zone, and by the antigorite mixing testifying the previous transition to lizardite.

In the frame of the extensional tectonics continuously characterizing the inner northern Apennines since the Miocene, the study shear zone is considered an older, exhumed analogue of the Zuccale low-angle normal fault. According to this interpretation, this structure is ascribed to an extensional detachment acted in the time interval between the high-pressure event (19.8 ± 1.4 Ma) and the activity of the Zuccale normal fault (Late Miocene–Early Pliocene). The main novelty of this work is to have recognized a shear zone, being part of an older, exhumed structure comparable to the low-angle Zuccale regional fault. This implies that the extensional tectonics continuously developed through time, joined with regional uplift and exhumation.

Funding: The research leading to these results has received funding from the European Community's Seventh Framework Program under grant agreement No. 608553 (Project IMAGE).

Acknowledgments: Two anonymous reviewers, by their constructive criticism and suggestions, helped me to improve the original manuscript.

Conflicts of Interest: The author declares no conflict of interest.

References

1. Pennacchioni, G.; Mancktelow, N.S. Nucleation and initial growth of a shear zone network within compositionally and structurally heterogeneous granitoids under amphibolite facies conditions. *J. Struct. Geol.* **2007**, *29*, 1757–1780. [CrossRef]
2. Van der Zee, W.; Wibberley, C.A.; Urai, J.L. The influence of layering and pre-existing joints on the development of internal structure in normal fault zones: The Lodève basin, France. *Geol. Soc. Lond. Spec. Publ.* **2008**, *299*, 57–74. [CrossRef]
3. Regenauer-Lieb, K.; Weinberg, R.; Rosenbaum, G. The effect of energy feedbacks on continental strength. *Nature* **2006**, *442*, 67–70. [CrossRef] [PubMed]
4. Barnes, I.; O'Neil, J.R. The relationship between fluids in some fresh alpine-type ultramafics and possible modern serpentinization, western United States. *Geol. Soc. Am. Bull.* **1969**, *80*, 1947–1960. [CrossRef]
5. Evans, J.P.; Chester, F.M. Fluid-rock interaction in faults of the San Andreas system: Inferences from San Gabriel fault rock geochemistry and microstructures. *J. Geophys. Res. Solid Earth* **1995**, *100*, 13007–13020. [CrossRef]
6. White, S.H.; Burrows, S.E.; Carreras, J.; Shaw, N.D.; Humphreys, F.J. On mylonites in ductile shear zones. *J. Struct. Geol.* **1980**, *2*, 175–187. [CrossRef]

7. Platt, J.P.; Behr, W.M. Grain size evolution in ductile shear zones: Implications for strain localization and the strength of the lithosphere. *J. Struct. Geol.* **2011**, *33*, 537–550. [CrossRef]
8. Menegon, L.; Stünitz, H.; Nasipuri, P. Partitioning of metamorphism and deformation at the nappe scale and implications for nappe-stacking mechanisms: The example of the Kalaknappe complex (north-Norwegian Caledonides). In Proceedings of the EGU General Assembly Conference Abstracts, Vienna, Austria, 22–27 April 2012.
9. Ruiz-Agudo, E.; Putnis, C.V.; Putnis, A. Coupled dissolution and precipitation at mineral–fluid interfaces. *Chem. Geol.* **2014**, *383*, 132–146. [CrossRef]
10. Gratier, J.P.; Thouvenot, F.; Jenatton, L.; Tourette, A.; Doan, M.L.; Renard, F. Geological control of the partitioning between seismic and aseismic sliding behaviours in active faults: Evidence from the Western Alps, France. *Tectonophys.* **2013**, *600*, 226–242. [CrossRef]
11. Silver, C.R.; Murphy, M.A.; Taylor, M.H.; Gosse, J.; Baltz, T. Neotectonics of the western Nepal fault system: Implications for Himalayan strain partitioning. *Tectonics* **2015**, *34*, 2494–2513. [CrossRef]
12. Liu, Q.; van der Hilst, R.D.; Li, Y.; Yao, H.; Chen, J.; Guo, B.; Qi, S.; Wang, J.; Huang, H.; Li, S. Eastward expansion of the Tibetan Plateau by crustal flow and strain partitioning across faults. *Nat. Geosci.* **2014**, *7*, 361–365. [CrossRef]
13. Leever, K.A.; Gabrielsen, R.H.; Sokoutis, D.; Willingshofer, E. The effect of convergence angle on the kinematic evolution of strain partitioning in transpressional brittle wedges: Insight from analog modeling and high-resolution digital image analysis. *Tectonics* **2011**, *30*, TC2013. [CrossRef]
14. Lee, P.E.; Jessup, M.J.; Shaw, C.A.; Hicks III, G.L.; Allen, J.L. Strain partitioning in the mid-crust of a transpressional shear zone system: Insights from the Homestake and Slide Lake shear zones, central Colorado. *J. Struct. Geol.* **2012**, *39*, 237–252. [CrossRef]
15. Zhang, B.; Zhang, J.; Zhong, D. Structure, kinematics and ages of transpression during strain-partitioning in the Chongshan shear zone, western Yunnan, China. *J. Struct. Geol.* **2010**, *32*, 445–463. [CrossRef]
16. Holdsworth, R.E.; Tavarnelli, E.; Clegg, P.; Pinheiro, R.V.L.; Jones, R.R.; McCaffrey, K.J.W. Domain deformation patterns and strain partitioning during transpression: An example from the Southern Uplands terrane, Scotland. *J. Geol. Soc.* **2002**, *159*, 401–415. [CrossRef]
17. Beltrando, M.; Rubatto, D.; Manatschal, G. From passive margins to orogens: The link between ocean-continent transition zones and (ultra) high-pressure metamorphism. *Geology* **2010**, *38*, 559–562. [CrossRef]
18. Molli, G.; Tribuzio, R.; Marquer, D. Deformation and metamorphism at the eastern border of the Tenda Massif (NE Corsica): A record of subduction and exhumation of continental crust. *J. Struct. Geol.* **2006**, *28*, 1748–1766. [CrossRef]
19. Wintsch, R.P.; Yi, K. Dissolution and replacement creep: A significant deformation mechanism in mid-crustal rocks. *J. Struct. Geol.* **2002**, *24*, 1179–1193. [CrossRef]
20. Guillot, S.; Schwartz, S.; Reynard, B.; Agard, P.; Prigent, C. Tectonic significance of serpentinites. *Tectonophysics* **2015**, *646*, 1–19. [CrossRef]
21. Chernak, L.J.; Hirth, G. Deformation of antigorite serpentinite at high temperature and pressure. *Earth Planet. Sci. Lett.* **2010**, *296*, 23–33. [CrossRef]
22. Schroeder, T.; John, B.E. Strain localization on an oceanic detachment fault system, Atlantis Massif, 30N, Mid-Atlantic Ridge. *Geochem. Geophys. Geosyst.* **2004**, *5*, Q11007. [CrossRef]
23. Passchier, C.W.; Simpson, C. Porphyroclast systems as kinematic indicators. *J. Struct. Geol.* **1986**, *8*, 831–843. [CrossRef]
24. Mancktelow, N.S. How ductile are ductile shear zones? *Geology* **2006**, *34*, 345–348. [CrossRef]
25. Raleigh, C.B.; Paterson, M.S. Experimental deformation of serpentinite and its tectonic implications. *J. Geophys. Res.* **1965**, *70*, 3965–3985. [CrossRef]
26. Amiguet, E.; Van De Moortèle, B.; Cordier, P.; Hilairet, N.; Reynard, B. Deformation mechanisms and rheology of serpentines in experiments and in nature. *J. Geophys. Res. Solid Earth* **2014**, *119*, 4640–4655. [CrossRef]
27. Hirauchi, K.I.; Katayama, I. Rheological contrast between serpentine species and implications for slab–mantle wedge decoupling. *Tectonophysics* **2013**, *608*, 545–551. [CrossRef]
28. Morrow, C.A.; Moore, D.E.; Lockner, D.A. The effect of mineral bond strength and adsorbed water on fault gouge frictional strength. *Geophys. Res. Lett.* **2000**, *27*, 815–818. [CrossRef]

29. Escartin, J.; Hirth, G.; Evans, B. Nondilatant brittle deformation of serpentinites: Implications for Mohr-Coulomb theory and the strength of faults. *J. Geophys. Res. Solid Earth* **1997**, *102*, 2897–2913. [CrossRef]
30. Collettini, C.; Viti, C.; Smith, S.A.F.; Holdsworth, R.E. Development of interconnected talc networks and weakening of continental low-angle normal faults. *Geology* **2009**, *37*, 567–570. [CrossRef]
31. Viti, C.; Hirose, T. Dehydration reactions and micro/nanostructures in experimentally-deformed serpentinites. *Contrib. Mineral. Petrol.* **2009**, *157*, 327–338. [CrossRef]
32. Bianco, C.; Brogi, A.; Caggianelli, A.; Giorgetti, G.; Liotta, D.; Meccheri, M. HP-LT metamorphism in Elba Island: Implications for the geodynamic evolution of the inner Northern Apennines (Italy). *J. Geodyn.* **2015**, *91*, 13–25. [CrossRef]
33. Bianco, C.; Godard, G.; Halton, A.; Brogi, A.; Liotta, D.; Caggianelli, A. The lawsonite-glaucophaneblueschists of Elba Island (Italy). *Lithos* **2019**, *348–349*, 105198. [CrossRef]
34. Ramsay, J.G.; Graham, R.H. Strain variation in shear belts. *Can. J. Earth Sci.* **1970**, *7*, 786–813. [CrossRef]
35. Sibson, R.H. Transient discontinuities in ductile shear zones. *J. Struct. Geol.* **1980**, *2*, 165–171. [CrossRef]
36. Vitale, S.; Mazzoli, S. Heterogeneous shear zone evolution: The role of shear strain hardening/softening. *J. Struct. Geol.* **2008**, *30*, 1383–1395. [CrossRef]
37. Bonini, M.; Cerrina Feroni, A.; Martinelli, P.; Moratti, G.; Valleri, G.; Certini, L. The intramessinian angular unconformity within the Radicondoli syncline (Siena, Tuscany, Italy): Structural and biostratigraphical preliminary data. *Mem. Soc. Geol. Ital.* **1994**, *48*, 501–507.
38. Bonini, M.; Moratti, G. Evoluzione tettonica del bacino neogenico di Radiocondoli-Volterra (Toscana meridionale). *Boll. Soc. Geol. Ital.* **1995**, *114*, 549–573.
39. Moratti, G.; Bonini, M. Structural development of the NeogeneRadicondoli–Volterra and adjoining hinterland basins in Western Tuscany (Northern Apennines, Italy). *Geol. J.* **1998**, *33*, 223–241. [CrossRef]
40. Sani, F.; Bonini, M.; Cerina Feroni, A.; Mazzarini, F.; Moratti, G.; Musumeci, G.; Corti, G. Iatta, F.; Ellero, A. Messinian-Early Pliocene crustal shortening along the Tyrrhenian margin of Tuscany, Italy. *Boll. Soc. Geol. Ital.* **2009**, *128*, 593–604.
41. Musumeci, G.; Vaselli, L. Neogene deformation and granite emplacement in the metamorphic units of northern Apennines (Italy): Insights from mylonitic marbles in the Porto Azzurro pluton contact aureole (Elba Island). *Geosphere* **2012**, *8*, 470–490. [CrossRef]
42. Bonini, M.; Moratti, G.; Sani, F.; Balestrieri, M.L. Compression-to-extension record in the Late Pliocene-Pleistocene Upper Valdarno Basin (Northern Apennines, Italy): Structural and thermochronological constraints. *Ital. J. Geosci.* **2013**, *132*, 54–80. [CrossRef]
43. Brogi, A. Contractional structures as relicts of the Northern Apennines collisional stage recorded in the Tuscan nappe of the Mt. Amiata geothermal area (Italy). *Boll. Soc. Geol. Ital.* **2005**, *4*, 53–64.
44. Brogi, A.; Lazzarotto, A.; Liotta, D.; Ranalli, G.; CROP18 Working Group. Crustal structures in the geothermal areas of southern Tuscany (Italy): Insights from the CROP 18 deep seismic reflection lines. *J. Volcanol. Geotherm. Res.* **2005**, *148*, 60–80. [CrossRef]
45. Brogi, A.; Fidolini, F.; Liotta, D. Tectonic and sedimentary evolution of the Upper Valdarno Basin: New insights from the lacustrine S. Barbara Basin. *Ital. J. Geosci.* **2013**, *132*, 81–97.
46. Liotta, D.; Brogi, A.; Meccheri, M.; Dini, A.; Bianco, C.; Ruggieri, G. Coexistence of low-angle normal and high-angle strike-to oblique-slip faults during Late Miocene mineralization in eastern Elba Island (Italy). *Tectonophysics* **2015**, *660*, 17–34. [CrossRef]
47. Brogi, A.; Liotta, D. Highly extended terrains, lateral segmentation of the substratum, and basin development: The middle-late Miocene Radicondoli Basin (inner northern Apennines, Italy). *Tectonics* **2008**, *27*, 1–20. [CrossRef]
48. Boccaletti, M.; Elter, P.; Guazzone, G.P. Plate tectonic models for the development of the western Alps and Northern Apennines. *Nature* **1971**, *234*, 108–111. [CrossRef]
49. Scandone, P. Origin of the Tyrrhenian Sea and Calabrian Arc. *Boll. Soc. Geol. Ital.* **1979**, *98*, 27–34.
50. Molli, G. Northern Apennines-Corsica orogenic system: An updated overview. Siegesmund, S., Fügenschuh, B., Froitzheim N. (Eds.), Tectonic aspects of the Alpine–Dinaride–Carpathian system. *Geol. Soc. London Spec. Publ.* **2008**, *298*, 413–442. [CrossRef]

51. Carmignani, L.; Decandia, F.A.; Disperati, L.; Fantozzi, P.L.; Lazzarotto, A.; Liotta, D.; Oggiano, G. Relationships between the Sardinia–Corsica–Provencal Domain and the Northern Apennines. *Terra Nova* **1995**, *7*, 128–137. [CrossRef]
52. Barchi, M.; Beltrando, M. The Neogene-Quaternary evolution of the Northern Apennines: Crustal structure, style of deformation and seismicity. *J. Virtual Explor.* **2010**, *36*. [CrossRef]
53. Rossetti, F.; Glodny, J.; Theye, T.; Maggi, M. Pressure–temperature–deformation–time of the ductile Alpine shearing in Corsica: From orogenic construction to collapse. *Lithos* **2015**, *218*, 99–116. [CrossRef]
54. Serri, G.; Innocenti, F.; Manetti, P. Geochemical and petrological evidence of the subduction of delaminated Adriatic continental lithosphere in the genesis of the Neogene Quaternary magmatism of central Italy. *Tectonophysics* **1993**, *223*, 117–147. [CrossRef]
55. Keller, J.V.A.; Pialli, G. Tectonics of the island of Elba: A reappraisal. *Boll. Soc. Geol. Ital.* **1990**, *109*, 413–425.
56. Collettini, C.; Holdsworth, R.E. Fault zone weakening and character of slip along low-angle normal faults: Insights from the Zuccale fault, Elba, Italy. *J. Geol. Soc.* **2004**, *161*, 1039–1051. [CrossRef]
57. Smith, S.A.F.; Holdsworth, R.E.; Collettini, C. Interactions between Low-Angle normal faults and plutonism in the upper crust: Insights from the Island of Elba, Italy. *Bull. Geol. Soc. Am.* **2011**, *123*, 329–346. [CrossRef]
58. Westerman, D.S.; Dini, A.; Innocenti, F.; Rocchi, S. Rise and fall of a nested Christmas-tree laccolith complex, Elba Island, Italy. *Geol. Soc. Lond. Spec. Publ.* **2004**, *234*, 195–213. [CrossRef]
59. Dini, A.; Innocenti, F.; Rocchi, S.; Tonarini, S.; Westerman, D.S. The magmatic evolution of the late Miocene laccolith–pluton–dyke granitic complex of Elba Island, Italy. *Geol. Mag.* **2002**, *139*, 257–279. [CrossRef]
60. Maineri, C.; Benvenuti, M.; Costagliola, P.; Dini, A.; Lattanzi, P.; Ruggieri, G.; Villa, I.M. Sericitic alteration at the La Crocetta deposit (Elba Island, Italy): Interplay between magmatism, tectonics and hydrothermal activity. *Miner. Depos.* **2003**, *38*, 67–86. [CrossRef]
61. Bortolotti, V.; Fazzuoli, M.; Pandeli, F.; Principi, G.; Babbini, A.; Corti, S. Geology of Central and Eastern Elba Island Italy. *Ofioliti* **2001**, *26*, 97–150.
62. Duranti, S.; Palmeri, R.; Pertusati, P.C.; Ricci, C.A. Geological evolution and metamorphic petrology of the basal sequences of eastern Elba (complex II). *ActaVulcanol.* **1992**, *2*, 213–229.
63. Pandeli, E.; Bortolotti, V.; Principi, G. La successione toscana epimetamorfica di Capo Castello (Cavo, Isola d'Elba nord-orientale). *Atti Ticinesi Scienze Terra* **1995**, *38*, 171–191.
64. Elter, F.M.; Pandeli, E. Structural evolution of anchi-/epimetamorphic units of Central and Eastern Elba (Ortano, Acquadolce, Monticiano-Roccastrada and Grassera Units). *Ofioliti* **2001**, *26*, 219–228.
65. Rinaudo, C.; Gastaldi, D.; Belluso, E. Characterization of chrysotile, antigorite and lizardite by FT-Raman spectroscopy. *Can. Mineral.* **2003**, *41*, 883–890. [CrossRef]
66. Viti, C.; Mellini, M. Contrasting chemical compositions in associated lizardite and chrysotile in veins from Elba, Italy. *Eur. J. Mineral.* **1997**, *9*, 585–596. [CrossRef]
67. Petriglieri, J.R.; Salvioli Mariani, E.; Mantovani, L.; Tribaudino, M.; Lottici, P.P.; Laporte-Magoni, C.; Bersani, D. Micro-Raman mapping of the polymorphs of serpentine. *J. Raman Spectrosc.* **2015**, *46*, 953–958. [CrossRef]
68. Groppo, C.; Rinaudo, C.; Cairo, S.; Gastaldi, D.; Compagnoni, R. Micro-Raman spectroscopy for a quick and reliable identification of serpentine minerals from ultramafics. *Eur. J. Miner.* **2006**, *18*, 319–329. [CrossRef]
69. Schwartz, S.; Guillot, S.; Reynard, B.; Lafay, R.; Debret, B.; Nicollet, C.; Lanari, P.; Auzende, A.L. Pressure–temperature estimates of the lizardite/antigorite transition in high pressure serpentinites. *Lithos* **2013**, *178*, 197–210. [CrossRef]
70. Andréani, M.; Boullier, A.M.; Gratier, J.P. Development of schistosity by dissolution–crystallization in a Californian serpentinite gouge. *J. Struct. Geol.* **2005**, *27*, 2256–2267.
71. Viti, C.; Mellini, M. Mesh textures and bastites in the Elba retrograde serpentinites. *Eur. J. Mineral.* **1998**, *10*, 1341–1359. [CrossRef]
72. Christensen, N.I. Serpentinites, peridotites, and seismology. *Int. Geol. Rev.* **2004**, *46*, 795–816. [CrossRef]
73. Boudier, F.; Baronnet, A.; Mainprice, D. Serpentine mineral replacements of natural olivine and their seismic implications: Oceanic lizardite versus subduction-related antigorite. *J. Petrol.* **2009**, *51*, 495–512. [CrossRef]
74. Evans, B.W. The serpentinite multisystem revisited: Chrysotile is metastable. *Int. Geol. Rev.* **2004**, *46*, 479–506. [CrossRef]
75. O'Hanley, D.S. *Serpentinites, Records of Tectonic and Petrological History*; Oxford University Press: New York, NY, USA, 1996; p. 277.

76. Nicolas, A.; Meshi, A.; Boudier, F.; Jousselin, D.; Muceku, B. Mylonites in ophiolite of Mirdita (Albania): Oceanic detachment shear zone. *Geosphere* **2017**, *13*, 136–154. [CrossRef]
77. Reinen, L.A.; Weeks, J.D.; Tullis, T.E. The frictional behavior of lizardite and antigorite serpentinites: Experiments, constitutive models, and implications for natural faults. *Pure Appl. Geophys.* **1994**, *143*, 317–358. [CrossRef]
78. Moore, D.E.; Lockner, D.A.; Ma, S.; Summers, R.; Byerlee, J.D. Strengths of serpentinite gouges at elevated temperatures. *J. Geophys. Res. Solid Earth* **1997**, *102*, 14787–14801. [CrossRef]
79. Behnsen, J.; Faulkner, D.R. The effect of mineralogy and effective normal stress on frictional strength of sheet silicates. *J. Struct. Geol.* **2012**, *42*, 49–61. [CrossRef]
80. Hirauchi, K.I.; Michibayashi, K.; Ueda, H.; Katayama, I. Spatial variations in antigorite fabric across a serpentinite subduction channel: Insights from the Ohmachi Seamount, Izu-Bonin frontal arc. *Earth Planet. Sci. Lett.* **2010**, *299*, 196–206. [CrossRef]
81. Wenner, D.B.; Taylor, H.P. Temperatures of serpentinization of ultramafic rocks based on O^{18}/O^{16} fractionation between coexisting serpentine and magnetite. *Contrib. Miner. Petrol.* **1971**, *32*, 165–185. [CrossRef]
82. Holdsworth, R.E.; Stewart, M.; Imber, J.; Strachan, R.A. The structure and rheological evolution of reactivated continental fault zones: A review and case study. *Geol. Soc. Lond. Spec. Publ.* **2001**, *184*, 115–137. [CrossRef]
83. Moore, D.E.; Rymer, M.J. Talc-bearing serpentinite and the creeping section of the San Andreas fault. *Nature* **2007**, *448*, 795–797. [CrossRef] [PubMed]
84. Collettini, C.; Niemeijer, A.; Viti, C.; Marone, C. Fault zone fabric and fault weakness. *Nature* **2009**, *462*, 907–910. [CrossRef] [PubMed]
85. Bucher, K.; Grapes, R. *Petrogenesis of Metamorphic Rocks: Metamorphism of Ultramafic Rocks*; Springer: Berlin/Heidelberg, Germany, 2011; pp. 191–338.
86. Kelemen, P.B.; Hirth, G. Reaction-driven cracking during retrograde metamorphism: Olivine hydration and carbonation. *Earth Planet. Sci. Lett.* **2012**, *345*, 81–89. [CrossRef]
87. Klein, F.; Garrido, C.J. Thermodynamic constraints on mineral carbonation of serpentinizedperidotite. *Lithos* **2011**, *126*, 147–160. [CrossRef]
88. Quesnel, B.; Gautier, P.; Boulvais, P.; Cathelineau, M.; Maurizot, P.; Cluzel, D.; Ulrich, M.; Guillot, S.; Lesimple, S.; Couteau, C. Syn-tectonic, meteoric water-derived carbonation of the New Caledonia peridotitenappe. *Geology* **2013**, *41*, 1063–1066. [CrossRef]
89. Abu-Jaber, N.S.; Kimberley, M.M. Origin of ultramafic-hosted vein magnesite deposits. *Ore Geol. Rev.* **1992**, *7*, 155–191. [CrossRef]
90. Boschi, C.; Dini, A.; Dallai, L.; Ruggieri, G.; Gianelli, G. Enhanced CO2-mineral sequestration by cyclic hydraulic fracturing and Si-rich fluid infiltration into serpentinites at Malentrata (Tuscany, Italy). *Chem. Geol.* **2009**, *265*, 209–226. [CrossRef]
91. El-Sharkawy, M.F. Talc mineralization of ultramafic affinity in the Eastern Desert of Egypt. *Miner. Depos.* **2000**, *35*, 346–363. [CrossRef]
92. Mancktelow, N.S.; Pennacchioni, G. The control of precursor brittle fracture and fluid—Rock interaction on the development of single and paired ductile shear zones. *J. Struct. Geol.* **2005**, *27*, 645–661. [CrossRef]
93. Stewart, M.; Holdsworth, R.E.; Strachan, R.A. Deformation processes and weakening mechanisms within the frictional—Viscous transition zone of major crustal-scale faults: Insights from the Great Glen Fault Zone, Scotland. *J. Struct. Geol.* **2000**, *22*, 543–560. [CrossRef]

© 2020 by the author. Licensee MDPI, Basel, Switzerland. This article is an open access article distributed under the terms and conditions of the Creative Commons Attribution (CC BY) license (http://creativecommons.org/licenses/by/4.0/).

geosciences

Article

The Gavorrano Monzogranite (Northern Apennines): An Updated Review of Host Rock Protoliths, Thermal Metamorphism and Tectonic Setting

Andrea Brogi [1,2,*], Alfredo Caggianelli [1], Domenico Liotta [1,2], Martina Zucchi [1], Amalia Spina [3], Enrico Capezzuoli [4], Alessandra Casini [5,6] and Elena Buracchi [5,6]

1 Department of Earth and Geoenvironmental Sciences, University of Bari, 70125 Bari, Italy; alfredo.caggianelli@uniba.it (A.C.); domenico.liotta@uniba.it (D.L.); martina.zucchi@uniba.it (M.Z.)
2 CNR-IGG Pisa, Institute of Geosciences and Earth Resources, 56127 Pisa, Italy
3 Department of Physics and Geology, University of Perugia, 06123 Perugia, Italy; amalia.spina@unipg.it
4 Department of Earth Sciences, University of Florence, 50121 Firenze, Italy; enrico.capezzuoli@unifi.it
5 Tuscan Mining UNESCO Global Geopark, Parco Tecnologico Archeologico delle Colline Metallifere Grossetane, 58100 Gavorrano, Italy; direttore@parcocollinemetallifere.it (A.C.); geology@parcocollinemetallifere.it (E.B.)
6 Department of Physics, Earth and Environmental Sciences, University of Siena, 53100 Siena, Italy
* Correspondence: andrea.brogi@uniba.it; Tel.: +39-080-544-2576

check for updates

Citation: Brogi, A.; Caggianelli, A.; Liotta, D.; Zucchi, M.; Spina, A.; Capezzuoli, E.; Casini, A.; Buracchi, E. The Gavorrano Monzogranite (Northern Apennines): An Updated Review of Host Rock Protoliths, Thermal Metamorphism and Tectonic Setting. *Geosciences* **2021**, *11*, 124. https://doi.org/10.3390/geosciences11030124

Academic Editors: Domenico Liotta, Giancarlo Molli, Angelo Cipriani, Rodolfo Carosi and Jesus Martinez-Frias

Received: 28 December 2020
Accepted: 1 March 2021
Published: 8 March 2021

Publisher's Note: MDPI stays neutral with regard to jurisdictional claims in published maps and institutional affiliations.

Copyright: © 2021 by the authors. Licensee MDPI, Basel, Switzerland. This article is an open access article distributed under the terms and conditions of the Creative Commons Attribution (CC BY) license (https://creativecommons.org/licenses/by/4.0/).

Abstract: We review and refine the geological setting of an area located nearby the Tyrrhenian seacoast, in the inner zone of the Northern Apennines (southern Tuscany), where a Neogene monzogranite body (estimated in about 3 km long, 1.5 km wide, and 0.7 km thick) emplaced during early Pliocene. This magmatic intrusion, known as the Gavorrano pluton, is partially exposed in a ridge bounded by regional faults delimiting broad structural depressions. A widespread circulation of geothermal fluids accompanied the cooling of the magmatic body and gave rise to an extensive Fe-ore deposit (mainly pyrite) exploited during the past century. The tectonic setting which favoured the emplacement and exhumation of the Gavorrano pluton is strongly debated with fallouts on the comprehension of the Neogene evolution of this sector of the inner Northern Apennines. Data from a new fieldwork dataset, integrated with information from the mining activity, have been integrated to refine the geological setting of the whole crustal sector where the Gavorrano monzogranite was emplaced and exhumed. Our review, implemented by new palynological, petrological and structural data pointed out that: (i) the age of the Palaeozoic phyllite (hosting rocks) is middle-late Permian, thus resulting younger than previously described (i.e., pre-Carboniferous); (ii) the conditions at which the metamorphic aureole developed are estimated at a temperature of c. 660 °C and at a depth lower than c. 6 km; (iii) the tectonic evolution which determined the emplacement and exhumation of the monzogranite is constrained in a transfer zone, in the frame of the extensional tectonics affecting the area continuously since Miocene.

Keywords: magmatism; extensional tectonics; contact metamorphism; Gavorrano pluton; palynomorphs

1. Introduction

The inner Northern Apennines (i.e., northern Tyrrhenian Sea and southern Tuscany), after having experienced HP/LT metamorphism during late Oligocene-early Miocene [1–3] was affected by extension since Burdigalian [4]. The clearest evidence of this process is the opening of the Tyrrhenian Basin [5] and the present 20–26 and 30–50 km crustal and lithospheric thickness, respectively [6–8]. Extension favoured partial melting in the lower crust and in the mantle, thus generating crustal and hybrid magmas (Tuscan Magmatic Province: [9,10] for a review). Igneous activity and extensional tectonics migrated eastwards [11–13]: hence, the first magmatic evidence occurred during middle Miocene in Corsica (Sisco lamproite, 14 Ma), and then, in the Tuscan archipelago (Late Miocene-Pliocene) and southern Tuscany (Pliocene-Pleistocene), up to the presently cooling magmatic bodies [14–18]. These are producing relevant geothermal anomalies [19], as it is the case for the Monte Amiata [14,20] and Larderello [15,18,21] areas [22,23]. In the inner Northern Apennines (Figure 1), intrusive bodies, emplaced at <10 km depth, are partially exposed in the Tuscan Archipelago and southern Tuscany, where crustal uplift and extensional tectonics induced their exhumation ([24] with references therein).

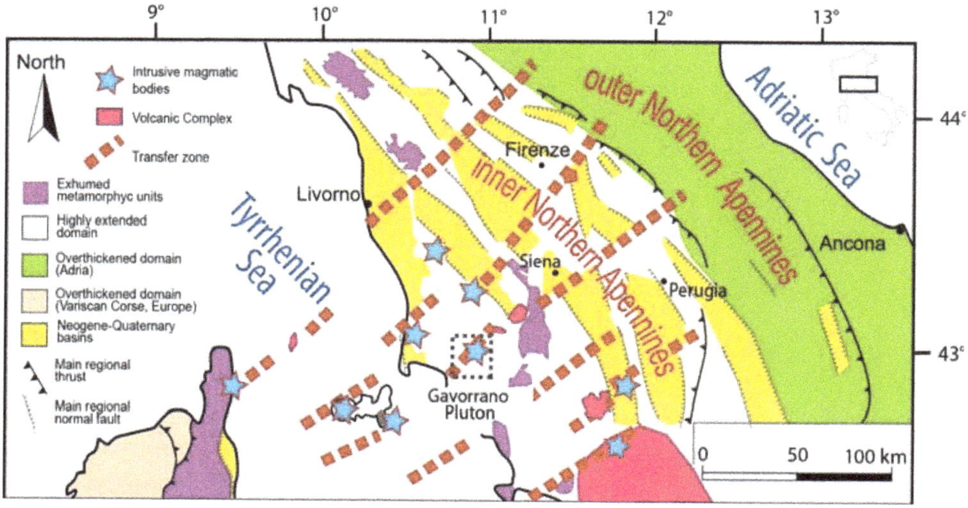

Figure 1. Structural sketch map of the Northern Apennines and Northern Tyrrhenian Sea showing inner and outer zones. The main Pliocene–Quaternary basins, transfer zones, Neogene–Quaternary Volcanic Complexes and Intrusive magmatic bodies and metamorphic units are highlighted. Location of the Gavorrano pluton (the study area) is also indicated.

The Gavorrano pluton is an example of this process [25]. Such a pluton is an about 3 km^3 laccolith [26], dated at 4.9 Ma [27] and partially exposed few kms to the east of the Tyrrhenian seacoast (Figure 1). It consists of cordierite-bearing monzogranite [28] with references therein with K-feldspar phenocrysts (up to 10 cm long), intruded by tourmaline-rich microgranite, porphyritic and aplitic dykes [25]. This magmatic intrusion and its contact aureole were mined from the last decades of the 19th century up to the 1981, to exploit sulphide (mainly pyrite) ore deposit, mostly occurring at the boundary between the igneous and host rocks, and in fault zones [29]. Although numerous studies were dedicated to this pluton, with the aim to reconstruct genesis and setting of the ore deposits (e.g., [26,29–31]), contrasting interpretations still remain with regards to: (i) the nature and age of the quartzitic-phyllite hosting rocks, contrastingly referred to Permian [32] or pre-Carboniferous [33]; (ii) the thermal conditions across the contact aureole and the related P-T peak conditions in the contact aureole, pointing to significantly different emplacement depths (cfr. [25,26,34]); (iii) the tectonic evolution of the Gavorrano area that was explained in extensional (e.g., [25,29]), transtensional [35] or compressional framework [34,36]. The compressional setting was also taken into account by [37,38] to explain the emplacement of the Gavorrano pluton, assumed to be contemporaneous to Pliocene regional thrusts and associated roof-anticlines. In this scenario, these authors considered the Gavorrano pluton as a key example for explaining the pluton emplacement in a compressional scenario, basically active since the Cretaceous in the inner Northern Apennines and northern Tyrrhenian sea.

In this paper, the state of the art on these themes, the contrasting interpretations, and hypotheses are discussed in the frame of new datasets. As a main conclusion: (i) we document the Permian age of the quartzitic-phyllite hosting rocks; (ii) we point to a peak temperature of c. 660 °C at a maximum pressure of 150 MPa for the metamorphic conditions in the contact aureole; and iii) we reconstruct the deformation within that sector of a Neogene regional transfer zone, which controlled the emplacement and exhumation of the Gavorrano pluton in the extensional framework characterizing inner Northern Apennines.

2. Geological Outline

The Gavorrano pluton intruded the lower part of the Tuscan Unit, following the main foliations and lithological boundaries (Figure 2) in the Palaeozoic-Triassic quartzite and phyllite, Triassic metacarbonate and late Triassic evaporite successions [25,29,39]. These rocks experienced, therefore, LP-metamorphism making particularly problematic the age attribution of the quartzitic-phyllite hosting rocks, contrastingly referred to Permian [32] or pre-Carboniferous [33], with different fallouts on the palaeogeography and context in which the overlying Triassic succession took place.

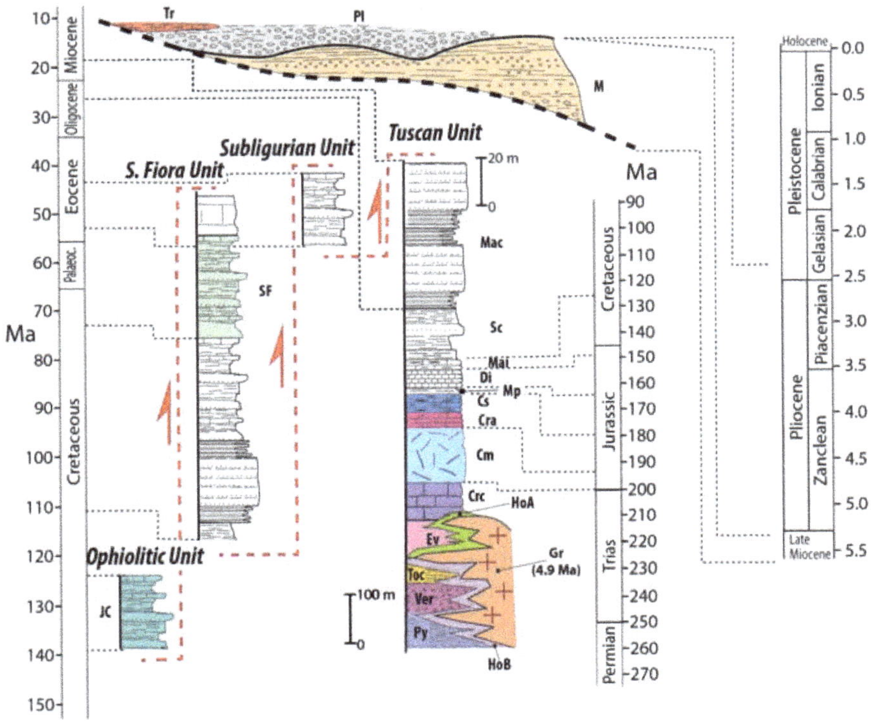

Figure 2. Stratigraphic logs of the tectonic units exposed in the study areas. The successions that are exposed are indicated by colours. Symbols: Tr—continental carbonate consisting of travertine and lacustrine limestone; Ple—fluvio-lacustrine sediments consisting of pebbly sand and sandy clay with interbedded pebble layers; M—Late Messinian polygenic reddish sandy-conglomerate and clay level (Montebamboli conglomerate Auctt.); JC—Argille a Palombini Fm: siliceous calcilutite, calcarenite, shale and marl; SF—Santa Fiora Fm: limestone, sandstone and shale passing to marl and silty-marl at the top; Mac—Macigno Fm: quartz-feldspar sandstone and shale; Sc—Scaglia Toscana Group: shale, limestone, marl, calcarenite and biocalcirudite; Mai—Maiolica Fm: cherty limestone, calcilutite; Di—Diaspri Fm: radiolarite and shale; Mp—Marne a Posidonia Fm: marl and marly limestone; Cs—Calcare Selcifero Fm: cherty limestone, marl and shale; Cra—Rosso Ammonitico Fm: reddish nodular limestone and shale; Cm—Calcare Massiccio Fm: massive limestone; Crc—Calcari a *Rhaetavicula contorta* Fm: bedded limestone and marl. Ev—Burano Fm: dolostone and gypsum/anhidryte layers; Toc—Tocchi Fm: metacarbonate and phyllite; Ver—Verrucano Group: quartz-metaconglomerate, metasandstone and phyllite. Py—Palaeozoic phyllite-quartzite Group: organic-matter bearing phyllite and metasandstone. HoA: carbonate hornfels; Hob: pelitic hornfels; Gr: Gavorrano magmatic complex.

The initial studies on the intrusive rocks were carried out by [40–44]. Marocchi [44] firstly described the Gavorrano granite as a magmatic complex formed by a porphyritic granite, a tourmaline-bearing microgranite and mica-bearing microgranite. Furthermore, Martelli [45] presented a geochemical and crystallographic study of both magmatic rocks and pyrite, hence describing, for the first time, the habitus and morphology of the pyrite and K-feldspar. However, the most complete paper dealing with the Gavorrano intrusion was published by [25], who defined the porphyritic granite as a quartz-monzonite, crossed by tourmaline-rich microgranite and aplitic dykes. Barberi et al. [46] implemented the study of the pluton, in the meantime dated at 4.9 Ma by K/Ar radiometric data [27].

The laccolithic shape of the magmatic intrusion was constrained by data from the underground mining activity [29] and finally defined by [47] as a body with a maximum length of 3 km, a width of 1.7 km and a thickness of 0.7 km.

The depth of the Gavorrano granite emplacement was estimated by [26] at a maximum value of 2–2.5 km, corresponding to a lithostatic pressure lower than 100 MPa. Differently, [47] indicate a maximum depth ranging between 4 and 5 km, corresponding to a lithostatic pressure lower than 200 MPa.

Magma cooling was accompanied by a significant hydrothermal process that led to pyrite ore deposits. Mining activity was carried out nearby the partially exposed monzogranite (Figure 3a,b).

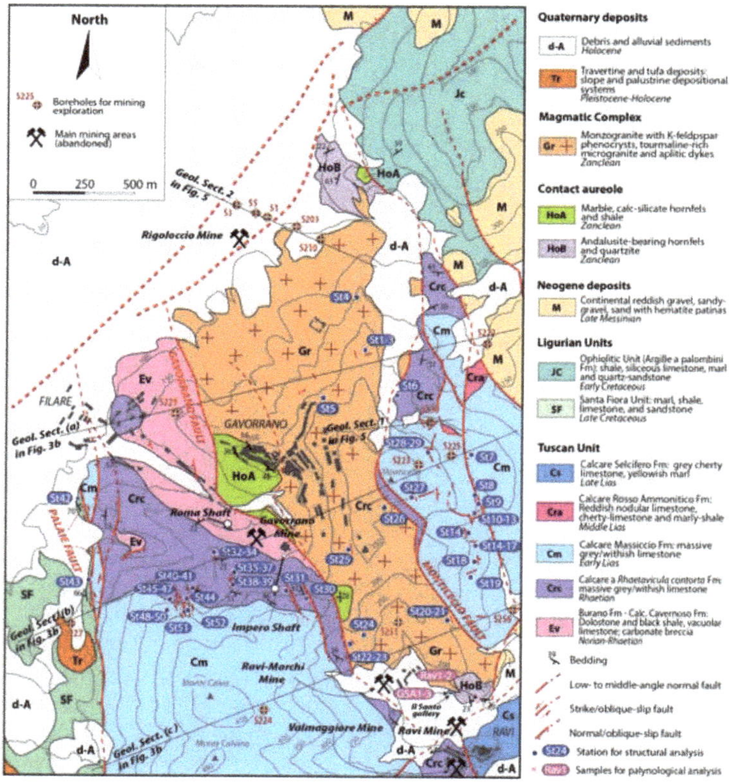

(a)

Figure 3. *Cont.*

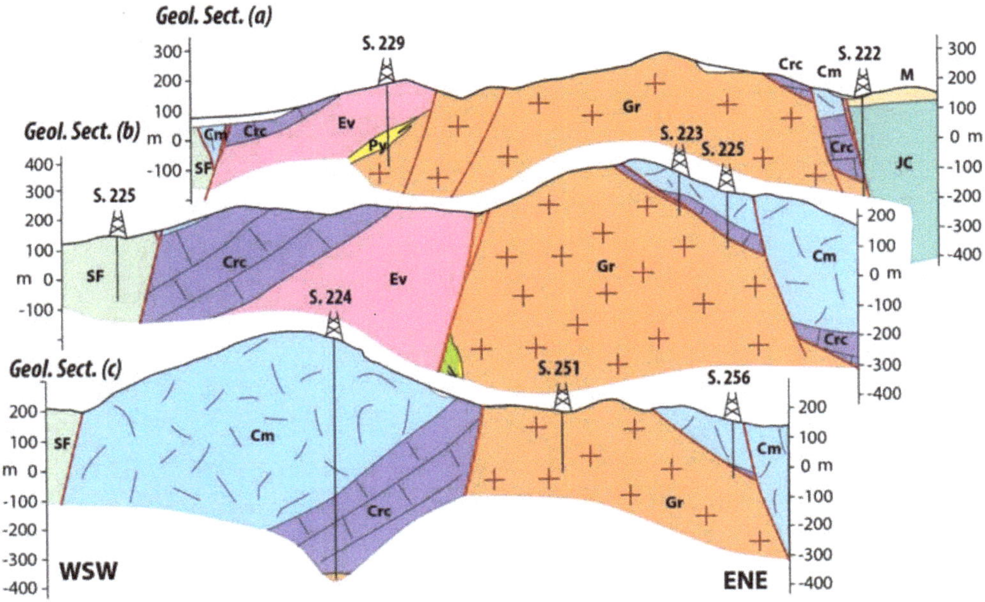

(b)

Figure 3. (**a**) Geological map of the Gavorrano mining district. Location of the abandoned mines and boreholes drilled during the mining exploration are also indicated. (**b**) Geological sections across the Gavorrano area. Their traces are indicated in (**a**).

Exploitation was encompassed between the end of nineteenth century and the last decades of the twentieth century, having produced about 25 million tons of pyrite [26]. Pyrite formed within a quartz-calcite gangue, with minor content of galena, sphalerite, chalcopyrite, fluorite, marcasite, barite, realgar and stibnite [26,29]. Three main pyrite bodies (named as Rigoloccio, Massa Boccheggiano and Valmaggiore) were distinguished, and exploited in an interconnected network of more than 30 km long tunnels belonging to five distinct mines [29]: Rigoloccio, Gavoranno, Ravi Marchi, Ravi Montecatini, Valmaggiore (Figure 3a). Several studies were addressed to the origin of the ore body and to its geometrical setting and attitude [26,29–31,43,48–65]. On the basis of these results, two main contrasting models were proposed for the genesis of the ore deposit: (i) an epigenetic origin related to the emplacement and cooling of the magmatic intrusion (e.g., [25,26,29,49,51,52,55,60,65–67]) and (ii) a sedimentary/metamorphic origin (e.g., [61,62,68,69]). Geochemical and isotopic studies from [70–76] and [65] described the distribution of minor elements within the pyrite and sphalerite, as well as the chemistry of the igneous and hosting metamorphic rocks, thus providing information on the origin of the mineralization; accordingly, the origin of the sulfur is partly referred to the magmatic source and partly to the remobilization of sedimentary rocks, by means of geothermal fluids. These latter were strictly controlled by permeable volumes in fault zones, active during the geothermal circulation [29] and by lithological discontinuities hydraulically connected with the fault zones. Ore bodies are in fact located: (i) at the boundary between the monzogranite and hosting rocks (Figure 4); (ii) along the contact separating the Tuscan metamorphic unit from the overlying late Triassic evaporite [26,29] (Figures 3 and 4); (iii) along normal faults juxtaposing sedimentary rocks with the magmatic intrusion (e.g., Monticello Fault, Figure 5).

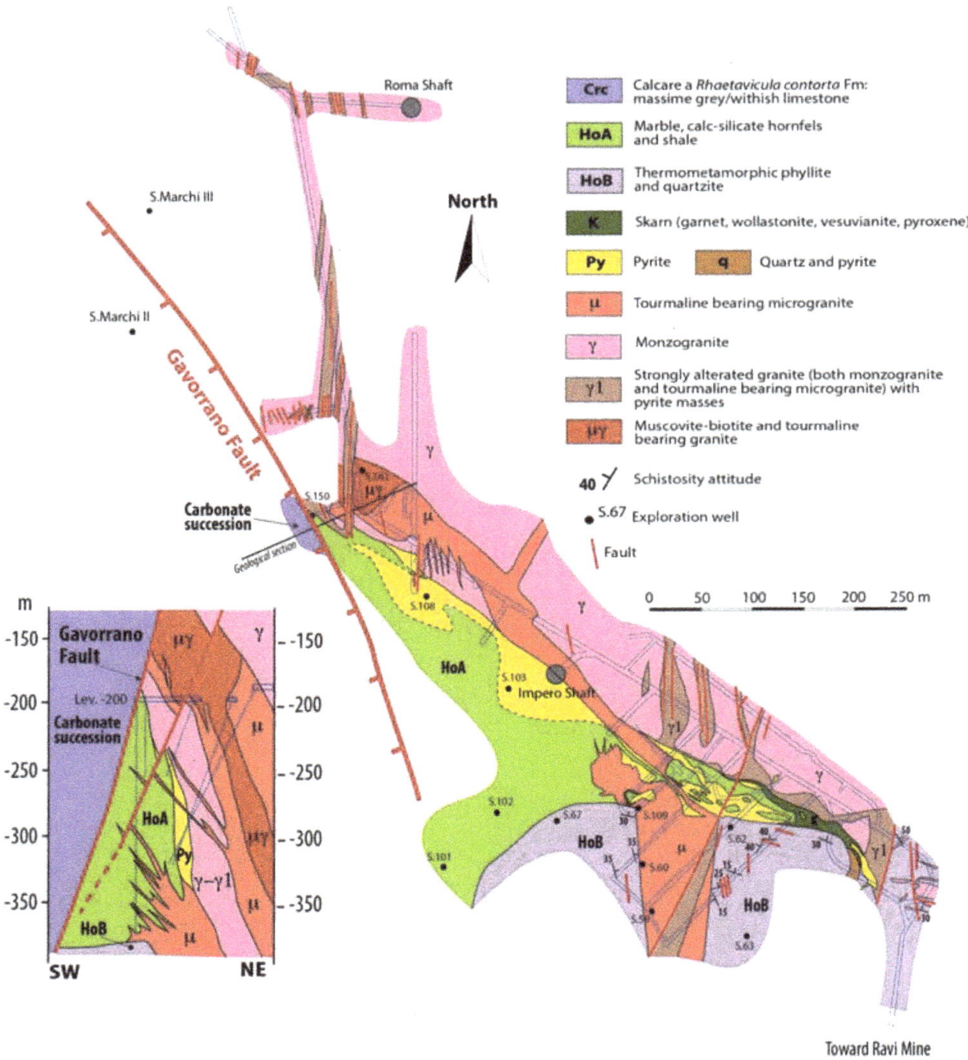

Figure 4. Geological map of the −200 m a.s.l. level of the Gavorrano mine (see Figure 3a for the location) and related geological section, redrawn from original documents stored in the archive of the Tuscan Mining Geopark (www.parcocollinemetallifere.it, Gavorrano; accessed date 4 March 2021).

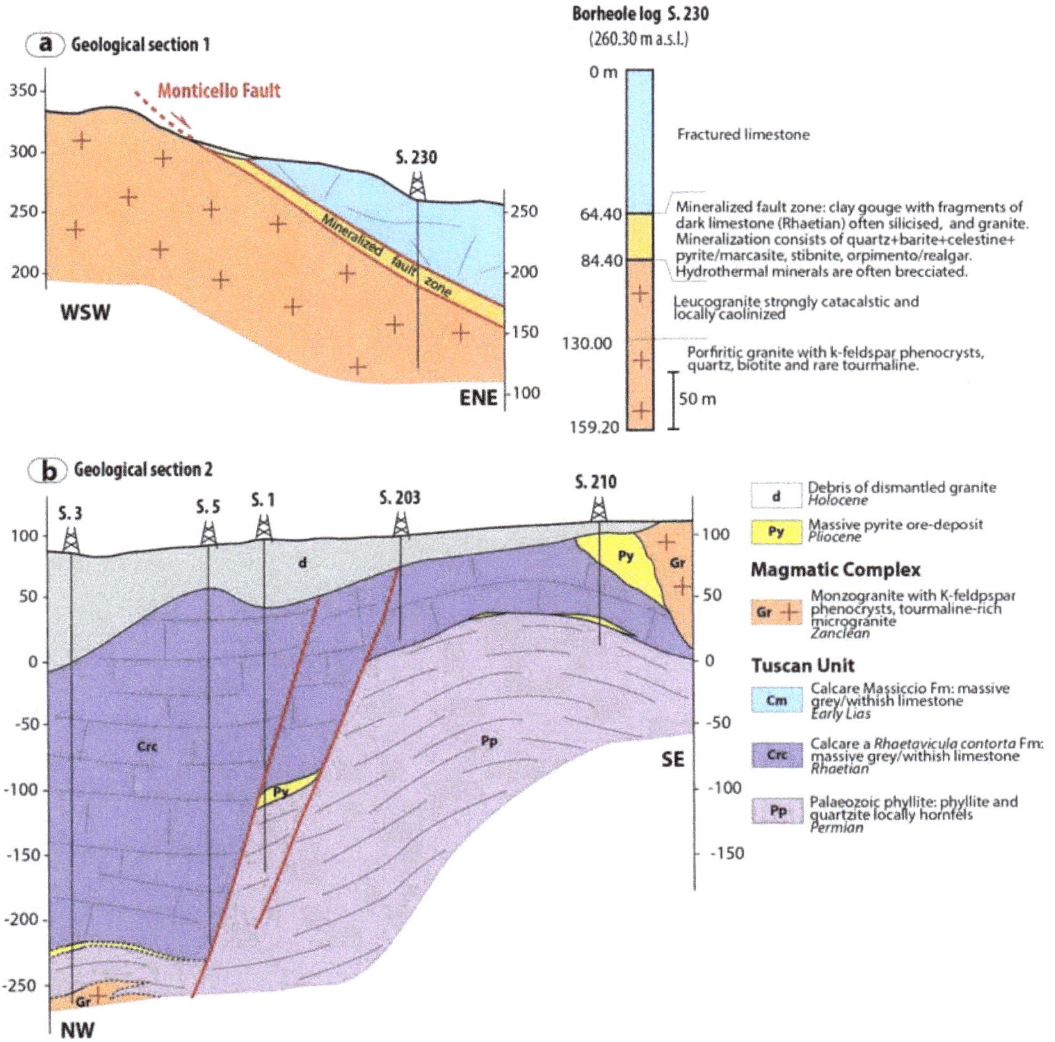

Figure 5. Geological cross-sections across the mining areas, redrawn from original mine documents stored in the archive of the Tuscan Mining Geopark (www.parcocollinemetallifere.it, Gavorrano; accessed date 4 March 2021). (**a**) Geological section across the Monticello Fault (the trace is indicated in Figure 3a), exploited during the mining activity: the main hydrothermal parageneses are indicated on the right of the borehole log. (**b**) Geological section across the Rigoloccio mine (the trace is indicated in Figure 3a) highlighting the two NE-striking fault segments dissecting the mineralisation. Note how the pyrite ore body has been found at the boundary between the magmatic and hosting rocks, and at the contact between the Palaeozoic-Triassic phyllite and quartzite and late Triassic carbonate successions.

Today, the intrusive rocks are partially exposed at surface or were tunneled at shallow depth in tunnels dug during the mining activity (Figure 3a,b). Their exhumation was controlled by normal faults, well-constrained in terms of geometry and displacements by means of surface and mining data [26,29,35].

Main faults were named as Gavorrano (NNW-SSE striking) and Monticello (N-S striking) faults, delimiting the western and eastern margins of the pluton, respectively (Figure 3a,b). The Gavorrano Fault is described as a west dipping high angle normal fault (60–70°) and with an arcuate geometry [29]. Its total offset exceeds 600 m. The Monticello

Fault is a middle-angle (35–50°) normal fault dipping to the east and is characterized by a total offset of about 1000 m [29]. Mining data highlight that the Gavorrano and Monticello Faults intersect each other in proximity of the Ravi village (Figure 3a). Both faults are mineralized although with different hydrothermal parageneses: the Gavorrano Fault hosts pyrite-ore bodies associated to minor content of galena, chalcopyrite and blend [29]. Differently, the Monticello Fault was mineralized by a hydrothermal mineral paragenesis made up of quartz, barite, celestine, pyrite/marcasite, stibnite, fluorite, orpiment/realgar (Figure 5).

The northern margin of the Gavorrano magmatic body is delimited by a SW-NE trending fault system, interpreted by some authors as the continuation of the Gavorrano Fault (e.g., [26,29]). Despite the significant role, this SW-NE trending fault is not mentioned by [34,36,38,47], although its occurrence is well documented by the mining data from the Rigoloccio mine (Figures 3 and 5) and described in several previously published geological maps and structural sketches [26,29,35].

Another N-S striking fault, named as the Palaie Fault (Figure 3a), was considered associated to the Gavorrano Fault, being almost parallel to this latter (cf. [29] with references therein). This structure delimits the western slope of the Monte Calvo [29] and was not interested by mining exploration. Nevertheless, this fault and the fault system delimiting to the east the monzogranite was investigated by [35] who presented a structural and kinematic dataset documenting a dominant strike- to oblique-slip kinematics. On the other hand, [47] account for a normal component of the Palaie, Gavorrano, and Monticello Faults, whereas [34] hypothesize a reverse/transpressive kinematics at least for the Palaie Fault. This view was later implemented by [36] who reported two adjunctive NW-SE trending faults, up to 2 km long (named as the Monte Calvo and Rigoloccio Faults: Figure 2 in [36]), and interpreted as cartographic scale reverse faults.

3. Age of Hosting Rocks

Protoliths of the LP-metamorphic rocks forming the contact aureole, consisting of metacarbonate and metapelite, are referred to the Tuscan Unit [57]. Dallegno et al. [26], Lotti [49], De Launay and Gites [52], Lotti [43] interpreted the dominantly metacarbonate succession exposed in the NW side of the magmatic intrusion and exploited at depth, as a part of the Late Triassic succession (i.e., Burano and Calcare a *Rhaetavicula contorta* formations; black limestone, In [77]). Part of this succession, tunneled in the Gavorrano mine, was considered by [25] as the transition from the late Triassic carbonate/evaporite to the Triassic metasiliciclastic succession of the Verrucano Group, later defined as the Tocchi Fm [78,79], never documented before in the Gavorrano area. Marinelli [25], Lotti [49], De Wukerslooth [57], Lotti [43] referred the andalusite-bearing metapelite exposed north and south of the monzogranite (Figure 2), to the Palaeozoic succession underlining the late Triassic carbonate one. Marinelli [25], Arisi Rota and Vighi [29] considered this succession as part of the Filladi di Boccheggiano Fm, attributed to the Permian or pre-Sudetian by [32,33], respectively. Dallegno et al. [26] agreed with the interpretations of the previously mentioned authors about the interpretation of the outcrops exposed at south of the monzogranite, in proximity of the Ravi village (Figure 3a); furthermore, Dallegno et al. [26] proposed an alternative hypothesis regarding the northern exposure (at north of the Gavorrano village, Figure 3a) where the exposed pelitic hornfel and metaquartzite (mainly consisting of metasandstone and quartz-metaconglomerate) were related to the Triassic Verrucano Group [33,80], on the basis of their textural and compositional features, as well as the occurrence of tourmalinolite and red porphyry clasts. In order to better constrain the age of this discussed metapelite succession, we have analysed key samples from: (i) the exposures along the main road in proximity of the Ravi village, and (ii) the mining tunnel, named as Il Santo gallery, not so far from the previous exposure (Figure 3a). Since LP-metamorphism reasonably obliterated the fossil contents, making any age determination impossible, we applied the study of palynological content, a useful methodology due to the fact that the wall of sporomorphs is characterized by a sporopollenin, a biopolymer of complex and not-completely known structure very resistant up to high temperatures (e.g., [81–85]) and provide a good chronological resolution (e.g., [86]). We collected key samples of spotted black metapelite and phyllitic-quartzite with high organic matter content. In particular, 2 samples have been collected in the exposures at north of the Ravi village (Rav 1 and Rav 2) and 3 samples (GSA 1–3) have been collected in the Il Santo gallery belonging to the Ravi mine (Figure 3a). Samples were treated with HCl (37%) and HF (50%) to destroy the carbonate and siliciclastic component. Boiling HCl (30%) was then used to remove the insoluble fluorosilicate. The organic residue was sieved with a 20 mm filter. The yield of the sample was treated repeatedly with Schultz solution, due to strongly high degree of thermal alteration preventing the identification of black-colour (graphitized?) palynomorphs. Light microscope observations were made on palynological slides using a Leica DM1000 microscope with differential interference contrast technique in transmitted light. Images were captured using the digital camera connected to the microscope and strongly corrected for brightness and contrast and colour using the open-source Gimp software. Palynological slides are stored at the Sedimentary Organic Matter Laboratory of the Department of Physics and Geology, University of Perugia, Italy. Samples GSA1-3 resulted almost barren in terms of palynomorph content. The yield of the samples mainly consists of large opaque phytoclasts such as inertinite (ligneous fragments completely oxidised) and some indeterminate black organic microfossils. On the contrary, in the samples Rav1 and Rav2, despite the low preservation grade prevents the recognition of almost all microfloristic elements, some sporomorphs were identified (Figure 6).

These consist of taeniate bisaccate pollen grains as *Distriatites insolitus, Hamiapollenites* spp. and non taeniate as *Alisporites sp.* cf. *splendens* and *Alisporites* spp. Spores as *Horriditriletes ramosus, Vallatisporites* sp. cf. *arcuatus, Thymospora opaqua, Densoisporites* sp. and *Kraeuselisporites* sp. also occur in assemblage with the incertae sedis organic microfossil *Reduviasporonites chalastus*. This microflora association shows close similarities with the one yielded from other pre-Triassic successions cropping out in Tuscany (Arenarie di Poggio al Carpino, Arenarie del Monte Argentario and Farma formations; [87–89], as well as in the Elba Island (Rio Marina and Mt. Calamita formations; [86]) and attributed to the Guadalupian-Lopingian time interval (middle-late Permian). In Southern Alps, along the praeparvus Zone of Changxingian age (late Lopingian), [90] documented an analogous palynoflora also characterized by the occurrence of *Alisporites* sp. cf. *splendens, Densoisporites* sp., *Kraeuselisporites* sp. and *Reduviasporonites chalastus*. This incertae sedis organic microfossil relatively long range from Capitanian (late Guadalupian) to early Triassic and widespread occur in different countries as well as Australia, Greenland, South China (Meishan, GSSP of Permian-Triassic boundary), Russian Platform, UK, Austria, USA, South Africa and other (e.g., [91–95]). In the Guadalupian of Northern Gondwana regions (e.g., Turkey; Oman, Saudi Arabia, Iran, Pakistan) R. chalastus was documented in assemblage with *Thymospora opaqua, Distriatites insolitus, Hamiapollenites* spp., *Kraeuselisporites* sp. *Densoisporites* sp. (e.g., [96–101]). On the basis of sporomorphs stratigraphic range, the microfloristic assemblage from the analysed samples is attributed to Guadalupian. The presence of forms as *Horriditriletes ramosus* and *Vallatisporites* sp. cf. *arcuatus* occurring in the early Permian of both palaeotemperate (e.g., Norway; [102]; and palaeotropical latitudes (e.g., Oman and Saudi Arabia; [96]), is here interpreted as probably reworked contents from older metasediments.

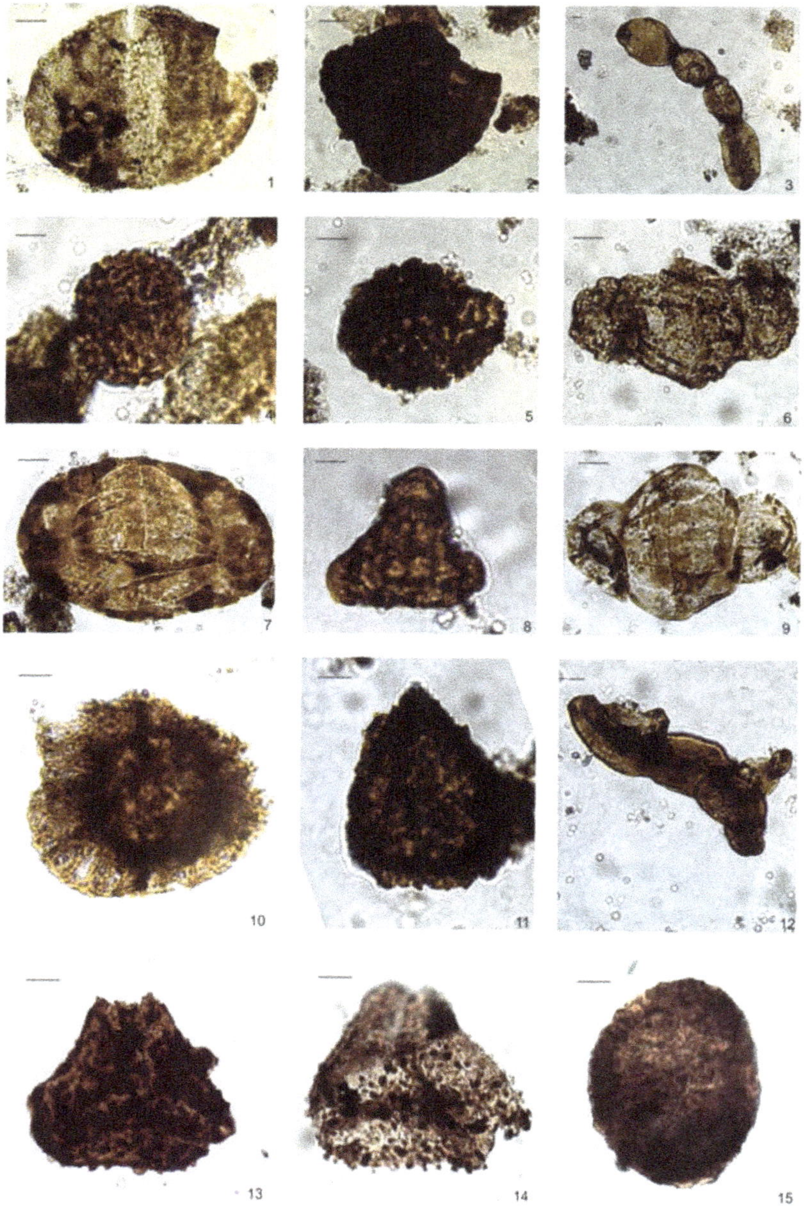

Figure 6. Sporomorphs from samples collected in the Gavorrano area (scale bar indicates 10 μm): (1) *Alisporites* sp. cf. *splendens* (Leschik) Foster 1979 (slide: RAV-1); (2) Indeterminate ornamented spore tetrad (slide: RAV-1); (3;12) *Reduviasporonites chalastus* (Foster) Elsik 1999 (slide RAV-1); (4–5) *Thymospora opaqua* Singh 1964 (slide RAV-2); (6,9) *Hamiapollenites* spp. (slide RAV-2); (7) *Distriatites insolitus* Bharadwaj and Salujha 1964 (slide RAV-1), (9) Indeterminate trilete spore (slide RAV-1); *Vallatisporites* sp. cf. *arcuatus* (Marques–Toigo) Archangelsky and Gamerro 1979 (slide RAV-2); (11) *Kraeuselisporites* sp. (slide RAV-1); (13,14) *Horriditriletes ramosus* (Balme and Hennelly) Bharadwaj and Salujah 1964 (slide RAV-1); (15) *Densoisporites* sp.

4. The Contact Aureole

The emplacement of the Gavorrano pluton produced LP-metamorphism on the host rocks resulting in a narrow contact aureole with a thickness of 200–300 m [25,26]. LP-metamorphism superimposed on the regional metamorphism which affected the pre-evaporitic metamorphic "basement" mainly represented by the Palaeozoic-Triassic phyllitic-quartzite units (i.e., dominantly pelitic successions), and the late Triassic carbonate rocks, producing hornfelses with different mineral assemblages, as firstly described by [25]. Concerning the pelitic rocks, [25,26] document a mineralogical assemblage made up of quartz + muscovite + K-feldspar + andalusite and chlorite + biotite + cordierite in Mg^- and Fe-bearing phyllite. Differently, [47] describe quartz + plagioclase + K-feldspar + andalusite and blasts replaced by fine-grained white mica they interpret as relicts of cordierite. [25] also describes corundum and green spinel, replaced by biotite and plagioclase, found within xenoliths collected in the Gavorrano mine. Differently, in the carbonate rocks, calc-silicate hornfels, partially replaced by skarn, shows a mineral assemblage mainly formed by garnet + epidote + spinel + wollastonite + diopside + forsterite + scapolite + quartz + calcite + vesuvianite [25,26]. At depth, the contact between granite and hornfels was described at −50 m, −200 m and −250 m [26,46] and wollastonite + calcite + quartz and diopside + forsterite + calcite mineral assemblages, with local levels enriched of garnet + vesuvianite + scapolite, have been found [46]. In the deepest levels of the Gavorrano mine (−200 m depth b.s.l.), [26] document dolomitic marble characterised by centimetric calcite and dolomite crystals, intimately associated to calc-silicate hornfels. Similarly, at the contact with the monzogranite, the same authors describe 1–2 m thick mineral assemblages consisting of: (i) diopside + garnet + dolomite + calcite approaching the hornfels, and (ii) epidote + tremolite + diopside + scapolite + calcite + garnet approaching the monzogranite. Diopside + tremolite veins, classified as replacement skarn [103], have also been documented in veins that cut the hornfels; similarly, narrow bands of phlogopite + tremolite (± actinolite) composition have also been described at the boundary between hornfels and skarn. No data are available for the mineralogical assemblage of the pelitic rocks, at the depth where observations were carried out. LP-metamorphism was followed by a subsequent hydrothermal event which produced, among the Fe-ore deposit [26,29,65], the alteration of the forsterite and diopside into serpentine, tremolite, talc, and chlorite, and the formation of veins filled by quartz + adularia + epidote + sulphides± calcite ± albite ± tremolite indicating temperature of about 250–300 °C [26]. Speculation of maximum temperature of about 175 °C was proposed for the last hydrothermal circulation by [104] analysing goethite and clay minerals at the Rigoloccio Mine (Figure 3a), derived from the hydrothermal alteration of the monzogranite and pyrite body.

We have implemented the existing dataset by analysing key samples of pelitic and carbonate hornfels from some key outcrops nearby the Ravi mine (Figure 3a) and from underground. These latter samples have been collected at: (i) the level −50 m b.s.l. of the Gavorrano mine; (ii) samples collected in the mining dump and possibly coming from the level −200 m of the Gavorrano mine. On the whole, our data agree with those reported by the previous Authors and provide additional information on the pelitic hornfels, particularly from the deep part of the Gavorrano mine.

The analysed pelitic and semipelitic rocks grade from spotted schist to hornfels (Figure 7). A compositional layering is generally recognisable being highlighted by an alternation of quartz- and mica-rich levels. In several cases, an intense deformation is observed in the form of serrated microfolds and winged d-porphyroclast (Figure 7a). In the spotted schist, the mineral assemblage is typically made up of quartz + biotite + muscovite + andalusite + tourmaline. Tiny elliptical cloudy spots are observed, probably derived from original cordierite (Figure 7b,c). In the hornfels from the deep level of the Gavorrano mine, muscovite-out conditions were reached as testified by the presence of K-feldspars and locally of corundum. Quartz crystals display variable grains size and are commonly characterised by polygonal shapes. In some cases, quartz shows lobate grain boundaries suggesting that a dynamic recrystallization took place. Biotite flakes increase in abundance from spotted schist to hornfels where they show orange-brown colour when oriented parallel to the lower polarizer. Andalusite porphyroblasts commonly show euhedral habit, with elongated and square diamond shapes (Figure 7d–f). The latter, usually contain the cross-shaped dark inclusion pattern typical of chiastolite (Figure 7d,e) as also described by [25]. Corundum is abundant and well recognisable at microscope scale in the form of spots made up of isolated crystals or aggregates within the biotite-rich levels devoid of quartz (Figure 7g). It shows a polygonal shape and a corona made up of K-feldspar, rare muscovite ± rutile (Figure 7h,j). It often displays a pale blue colour typical of sapphire variety. Tourmaline is zoned with brown to cyan colours being of dravite type and is mostly found within biotite-rich levels (Figure 7k). Among the accessory phases, zircon and opaque minerals are always present, whereas rutile is found in corundum-bearing hornfels.

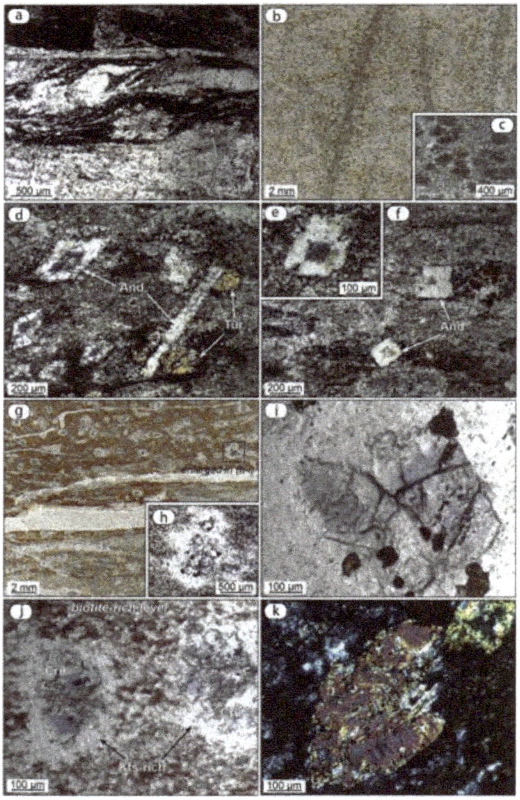

Figure 7. Micrographs of the pelitic and semipelitic rocks affected by contact metamorphism. (**a**) d-porphyroclast completely replaced by sericite (plane polarized light). (**b**,**c**) Scanned thin section (plane polarized light in (**b**) with a detail (crossed polars in (**c**) showing the elliptical cloudy spots. (**d**–**f**) Andalusite crystals (plane polarized light) that, in some cases, show dark inclusions arranged in geometrical pattern typical of chiastolite. Tourmaline crystals adjacent to elongated andalusite (**d**) can be observed. (**g**) Scanned thin section and detail (**h**) of corundum crystals characterised by a corona made up of K-feldspar and minor amount of muscovite flakes within biotite-rich level (plane polarized light). (**i**,**j**) Details (plane polarized light) of pale blue corundum of sapphire variety and rutile (**i**). (**k**) Detail of tourmaline crystal with interference colours up to second order blue (crossed polars). Mineral abbreviation from [105].

The analysed carbonate rocks collected in the Gavorrano mine (level—50m b.s.l.) consist of marbles with a variable grain size. In most cases, they contain olivine (Figure 8a,b) without diopside suggesting that they derive from carbonatic-silica-pure protolith. Locally, in the fine-grained type a polygonal fabric of calcite can be recognised, indicating a static recrystallization (Figure 8a). In some cases, olivine-rich levels show diffuse serpentinization, with few olivine relicts still present (Figure 8c,d), justified by [25,26] as the effect of a later hydrothermal fluid flowing through the thermal aureole.

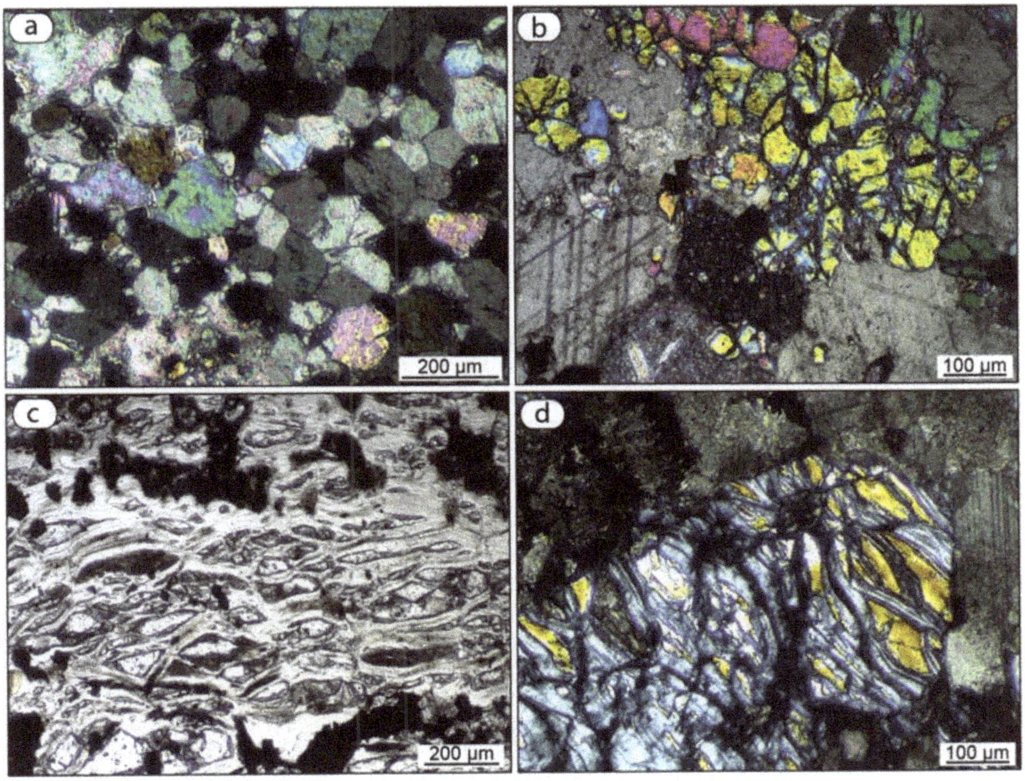

Figure 8. Micrographs of marbles. (**a**) Calcite grains in fine-grained olivine bearing marble with a typical polygonal fabric indicating static recrystallization (crossed polars). (**b**) Detail of olivine bearing marble showing tabular thick twins in calcite [106] (crossed polars). (**c**) Olivine-rich level in marble affected by intense serpentinization (plane polarized light). (**d**) Detail of olivine crystal almost totally replaced by serpentine, locally showing a mesh fabric (crossed polars).

Some considerations can be provided on the peak P-T conditions reached in the thermal aureole. In the pelitic hornfels, recording the maximum temperature in the contact aureole, muscovite-out conditions were reached through the reaction:

$$\text{Ms} + \text{Qtz} = \text{And} + \text{Kfs} + \text{H}_2\text{O} \quad (1)$$

Alternatively, in silica-poor domains, the genesis of corundum could be promoted by reaction:

$$\text{Ms} = \text{Crn} + \text{Kfs} + \text{H}_2\text{O} \quad (2)$$

After muscovite disappearance, corundum could be produced by reaction provided by Pattison and Harte [107]:

$$2\ \text{Bt} + 15\ \text{Al}_2\text{SiO}_5 = 9\ \text{Crn} + 3\ \text{Crd} + 2\ \text{Kfs} + 0.5\ \text{H}_2\text{O} \quad (3)$$

In the analysed samples, there is no evidence for the simultaneous blastesis of corundum and cordierite. Thus, reaction (2) is preferred for the genesis of corundum.

In order to constrain P-T conditions for the contact metamorphism, a look at a simple P-T grid is practical. The diagram in Figure 9, in addition to reaction curves (1) and (2), shows the wet solidus curve for granite and the andalusite—sillimanite equilibrium line. The absence in the hornfels of sillimanite and of microstructures indicative of partial melting indicate that the andalusite-sillimanite equilibrium line and the granite solidus curve were not crossed during the heating phase. On the other hand, the presence of corundum allows to constrain the metamorphic peak beyond reaction (2), within the grey area. A maximum limit for the pressure, provided by the intersection of reaction (2) with the andalusite-sillimanite equilibrium is of c. 170 MPa, corresponding to a temperature of c. 640 °C. At lower pressures, higher temperatures for the thermal peak are possible.

Quantitative estimates of the temperature were attempted by the Ti-in-biotite thermometer by Wu et al. [108]. This was calibrated for pelitic rocks containing a Ti-rich phase such as ilmenite or rutile at pressure higher of 100 MPa, thus being appropriate for the present case. On the basis of 7 biotite analyses of a corundum-bearing hornfels sample (Table A1), a mean value of c. 660 °C was obtained at a pressure of 170 MPa and of c. 650 °C at a pressure of 100 MPa. A check on the compatibility of these numerical results with the P-T extent of the grey area in the diagram of Figure 9, suggests a pressure value lower than c. 150 MPa, corresponding to a depth lower than c. 6 km, assuming an average density of 2650 kg/m^3 for the upper crust. However, considering the error of the thermometer, this latter P limit should be verified through more refined petrological methods and/or geological constraints.

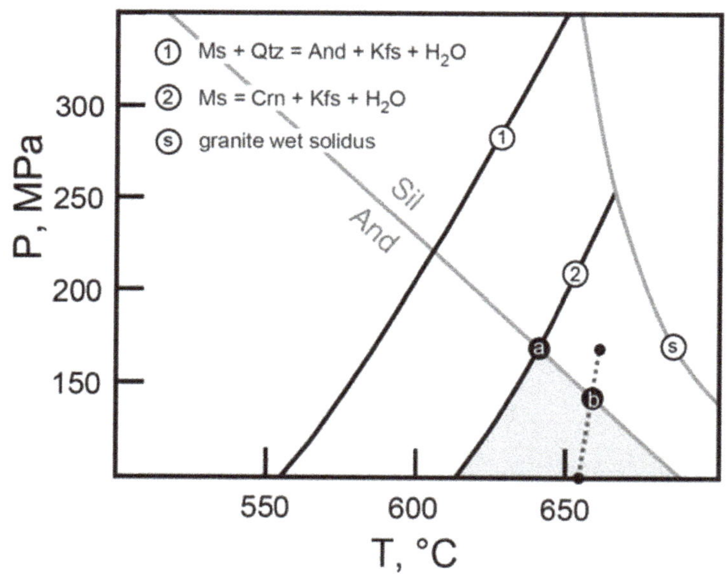

Figure 9. P-T grid from [109] here adopted to constrain conditions for the peak of contact metamorphism. Muscovite breakdown curves at $P_{H2O} = P_{total}$ are from [110], the granite wet solidus curve is from [111] and the andalusite-sillimanite equilibrium line from [112–114]. The grey area indicates peak P-T region compatible with the presence of andalusite + K-feldspar and, in silica-poor domains, of corundum + K-feldspar. Point (a) indicates maximum estimate for pressure on the basis of the corundum + K-feldspar presence, resulting in a value of 170 MPa. The dotted line connects points related to temperature estimates by Ti-in-biotite thermometer at 170 and 100 MPa, respectively. Point (b) indicates maximum estimate for pressure on the basis of the Ti-in-biotite thermometer, resulting in a value of c. 150 MPa.

5. Structural and Kinematic Data

The geological setting was already reconstructed by the large amount of mining data as reported in several papers (e.g., [25,26,29,65]). Nevertheless, still contrasting hypotheses are provided by different authors on the tectonic evolution that accompanied the pluton emplacement and its exhumation (cf. [26,35,36,115]).

In order to contribute to this issue, existing mining documents and a new dataset of structural and kinematic data have been integrated. Figure 3a shows the location of the stations where the structural analysis has been carried out. The results are shown in the stereographic diagrams, reported in the Annex 1.

Both cartographic and outcrop-scale evidence highlight superposed faulting events that can be categorized In (i) low- to middle-angle (<50° of dipping value) normal faults, affecting both granite and the carbonate succession; (ii) high angle (>50° of dipping value) strike-slip faults coexisting with the low- to middle-angle normal faults; (iii) high angle normal faults displacing the previous formed structures (Figure 3a,b).

As it regards the low- to middle-angle faults, the best example is the Monticello Fault (Figures 3a and 5), which decouples the monzogranite from the overlying sedimentary cover, by an almost ten-meter thick mineralized cataclastic zone, as it is well documented by the mining data (Figure 5a). Therefore, the consideration of the mining data changes the view of the Monticello Fault, previously interpreted as a high-angle normal fault, parallel to the Gavorrano Fault although dipping in the opposite direction and delimiting the monzogranite to the east [26,29,35,47].

By the new integration of data, the Monticello Fault assumes the role of an already existing fault decoupling the magmatic intrusion from the hosting rocks and contributing to the exhumation of the monzogranite. Such a structure was later affected by high-angle faults to which the Gavorrano Fault belongs (Figure 3a,b). It is worth to note that, on the basis on the mining data form the Ravi mine (located in the southern part of the Monticello fault), Marinelli [25] accounts for a shear zone, separating the magmatic intrusion from the hosting rocks, similarly to what is observed along the Monticello Fault.

Low-angle faults affecting the carbonate succession (Figures 3a and 10), also occur in the hanging wall of the Monticello Fault (Figure 3a). These are well-exposed in the abandoned quarries on the northern slope of the Monte Calvo area and are arranged in sub-parallel and anastomosed segments that define decameters-thick sheared and delaminated volumes with conjugated fault segments forming lozenge-shape geometries and meter-/decameter-scale extensional horses (Figures 11 and 12). Fault segments are characterized by kinematic indicators consisting of calcite fibres and steps, indicating normal, mostly top-to-the E-NE sense of shear (Figure 11, Figure A1). All these data contrast with the kinematic interpretation proposed by [36], although conducted in the same outcrops (cfr. Figure 7a,b in [36]). These authors, in fact, support a top-to-the west reverse kinematics of these faults, notwithstanding the fact that kinematic indicators clearly indicate a normal movement (Figure 11b,f). Furthermore, it is worth to underline that this particular kinematics is in agreement with the data collected in the whole Gavorrano area (Figure A1) and with the geometrical setting of the low-angle faults, as visible in the quarry exposures (Figure 12).

Figure 10. Google Earth photograph of the northern slope of the Monte Calvo, where late Triassic and early Jurassic carbonate succession is exposed in abandoned quarries. The location of the photographs shown in the indicated figures are also reported.

Figure 11. *Cont.*

Figure 11. Structural and kinematic features of low- to middle-angle normal faults affecting the carbonate succession (Calcari a *Rhaetavicula contorta* Fm, late Trias). See Figure 10 for the location of the pictures. (**a**) Panoramic view of a saw-cut quarry wall where sub-parallel and anastomosed low-angle normal fault segments occur. (**b**) Detail of the inset shown in (**a**), where the kinematic data have been collected and reported in the stereographic diagram (lower hemisphere, equiareal projection); note that this is the same detail on which [36] reported their analyses, in contrast with these ones. (**c**) Detail of the inset shown in (**a**) where the kinematic data have been collected and reported in the stereographic diagram (lower hemisphere, equiareal projection). (**d**) Detail of the inset shown in (**c**) where kinematic indicators consisting of mechanical striation and associated step strongly support for a normal (top-to-the NE) kinematics of the fault. (**e**) Panoramic view of a part of the saw-cut quarry wall where anastomosed low-angle normal fault segments are dissected by high-angle faults to which the reported stereographic diagram (lower hemisphere, equiareal projection) is referred. (**f**) Detail of the inset shown in (**e**), where the location of the kinematic indicators shown in (**g**,**h**) is shown; note that this is the same detail on which [36] reported their analyses accounting for a thrust surface. (**g**,**h**) Details of the inset shown in (**f**) where the kinematic indicators consisting of calcite fibers and fiber steps have been recognized. These account for polyphase movements on the same fault plane with a dominant normal component with a top-to-the NE-E sense of shear.

Figure 12. *Cont.*

Figure 12. Extensional structures exposed in the quarry saw-cut walls. See Figure 10 for the location of the pictures. (**a**) Decameter-size carbonate levels segmented by top-to-the NE extensional detachments. (**b**) Panoramic view of the abandoned quarry where the detail shown in figure (**c**) is indicated. (**c**) Meter-size extensional horses related to top-to-the NE extensional detachments.

Low-angle faults affecting the monzogranite (Figure 13a,b) have also been recognized. Here, these show striated slip-surfaces (Figure 13c) bounded by a centimeter-thick core zone with ultra-comminuted grains (Figure 13d) and centimeters-thick level of foliated monzogranite, showing s-c structures, with a top-to-the west sense of shear (Figure 13e,f). Although faults exposure in granite are limited, their setting accounts for a lozenge-shaped geometry (Figure 13b), thus explaining the occurrence of both top to E-ENE (dominant) and top to W-WSW sense of shear on their slip planes, respectively.

Figure 13. *Cont.*

Figure 13. Structural and kinematic features of faults affecting the monzogranite. (**a**) Panoramic view of analyzed outcrop with indicated the location of insets shown in (**b**,**g**). (**b**,**c**) Top-to the west low- to middle-angle normal faults cut by high-angle normal faults. (**d**) Detail of the low-angle normal fault zone highlighting a cm-thick comminuted cataclasite and S-C structures affecting the hangingwall. (**e**,**f**) Striated surfaces indicating a clear normal movement. (**g**) Detail of the high-angle normal fault cutting the low-to middle-angle normal faults. The inset shows the mechanical striation on the fault surface indicating a normal movement.

Concerning the high angle faults, N-S and SW-NE strike-slip faults occur in the whole area (Figure 3a). The best exposures (especially for the N-S striking faults) were recognized in the quarries, north of the Monte Calvo (Figure 10) and in the western part of the study area (i.e., Palaie Fault, Figure 3a).

In the abandoned quarries, these faults define decameters-thick vertical brittle shear zones (Figure 14) formed by sub-parallel and conjugate fault planes (Figure 14a–c), surrounded by well-developed damage zones. Left-lateral strike to oblique-slip kinematics is then suggested by indicators locally preserved on the slip-surfaces and consisting of calcite slicken-fibres and steps (Figure 14d,e). In some cases, syn-kinematic cm- to dm-thick banded calcite veins formed along the fault planes, or in extensional jogs (Figure 14f).

Figure 14. Strike-slip faults exposed in the quarry saw-cut walls. See Figure 10 for the location of the pictures. (**a**) Panoramic view of analyzed outcrop with indicated the location of insets shown in (**b**,**c**). (**b**,**c**) Detail of the fault zones showing a sub-vertical attitude and well distributed anastomosed fault segments forming the damage zones. (**d**,**e**) Kinematic indicators recognized on the fault planes, consisting of calcite fibers and fiber-steps clearly indicating a slightly left-lateral oblique-slip movement. (**f**) Mineralized fault zone forming extensional-jogs filled by banded-calcite veins.

This attests the role of such faults in controlling the hydrothermal fluid paths from the late magmatic events onwards, at least. This is in fact attested by the several S-N and SW-NE oriented microgranite dykes intruding both the monzogranite and the hosting rocks in fault zones, as documented in the outcrops (Figure 15) and by the underground mining data (Figure 4). Thus, a local strike-slip regime is supposed to have controlled the deformation in the Gavorrano area, and probably the pluton emplacement. Nevertheless, although the interplay between the low-angle normal faults and the S-N to SW-NE striking strike-slip faults has not been directly documented in the field, it is reasonable to assume that the transcurrent faults were contemporaneously active with the low-angle normal faults, since both fault systems are affected by syn-tectonic hydrothermal circulation. Their contemporaneity is also confirmed by the inversion of the kinematic data collected on fault-slip surfaces of both low-angle normal faults and strike-slip faults: it highlights a strong kinematic compatibility, as shown by the orientation of the main kinematic axes (Figure 16a–c). We can therefore assume that these faults were active under a common stress field: in this view, the low-angle normal faults developed as a consequence of the crustal thinning, having triggered magmatism and favoured the development of SW-NE striking km-thick sub-vertical brittle shear zone (i.e., transfer zone: [116–118]) of which the Gavorrano area is a part.

Figure 15. Examples of tourmaline-bearing microgranite dykes intruded within (**a**) monzogranite and the (**b**) phyllitic-quartzite (Verrucano Group) forming the monzogranite hosting rock.

In the context of the deformation induced by a transfer zone, the N-S striking left-lateral and SW-NE striking right-lateral strike-slip faults are thus framed in the same setting, as indicated by their kinematic compatibility (Figure 16c,d). Consequently, those are interpreted as minor faults linking the SW-NE striking main structures, in a common left-lateral strike-slip shear zone.

Concluding, we can depict a tectonic evolution where low-angle normal faults and strike-slip faults (N-S striking left-lateral strike-slip, SW-NE striking left- and right-lateral strike-slip faults) coexisted during the emplacement and exhumation of the monzogranite, as sketched in the conceptual model of Figure 17a,b.

The Palaie Fault (Figure 3a) has been described by several authors as a strike-slip fault [35] or a transpressive fault, by the kinematics reconstructed in a single outcrop [36]. Our data (Figure 18) highlight that what today is recognizable along the western slope of the Monte Calvo is the result of two superposed faulting events, at least: strike-slip fault segments are in fact preserved within lithons delimited by sub-parallel west-dipping normal faults (Figure 3a). In other words, the western slope of the Monte Calvo is delimited by a normal fault system partly reactivating and dissecting older N-S striking left-lateral strike-slip faults, thus determining lithons of which original attitude is reasonably modified. This can explain the singularity of the Palaie Fault, the single structure with visible kinematic indicators contrasting the general framework.

High angle normal faults, NNW-SSE and N-S striking are the youngest structures. These dissect the previous formed low-angle faults (Figure 19a,b) and are characterized by oblique-slip to normal movements (Figure 19c). Fault zones are with meters-thick damage zones (Figure 19a) where well-organized minor fractures affect both their hanging wall and footwall (Figure 19c). Kinematic indicators mainly consist of groove and mechanical striations developed on the fault surfaces.

Inversion of kinematics data collected on the normal faults (Figure 16e) show a kinematic compatibility with the low-angle normal faults (Figure 16a), thus supporting a stable E-NE trending extensional regime from the emplacement of the monzogranite until its exhumation (Figure 17c).

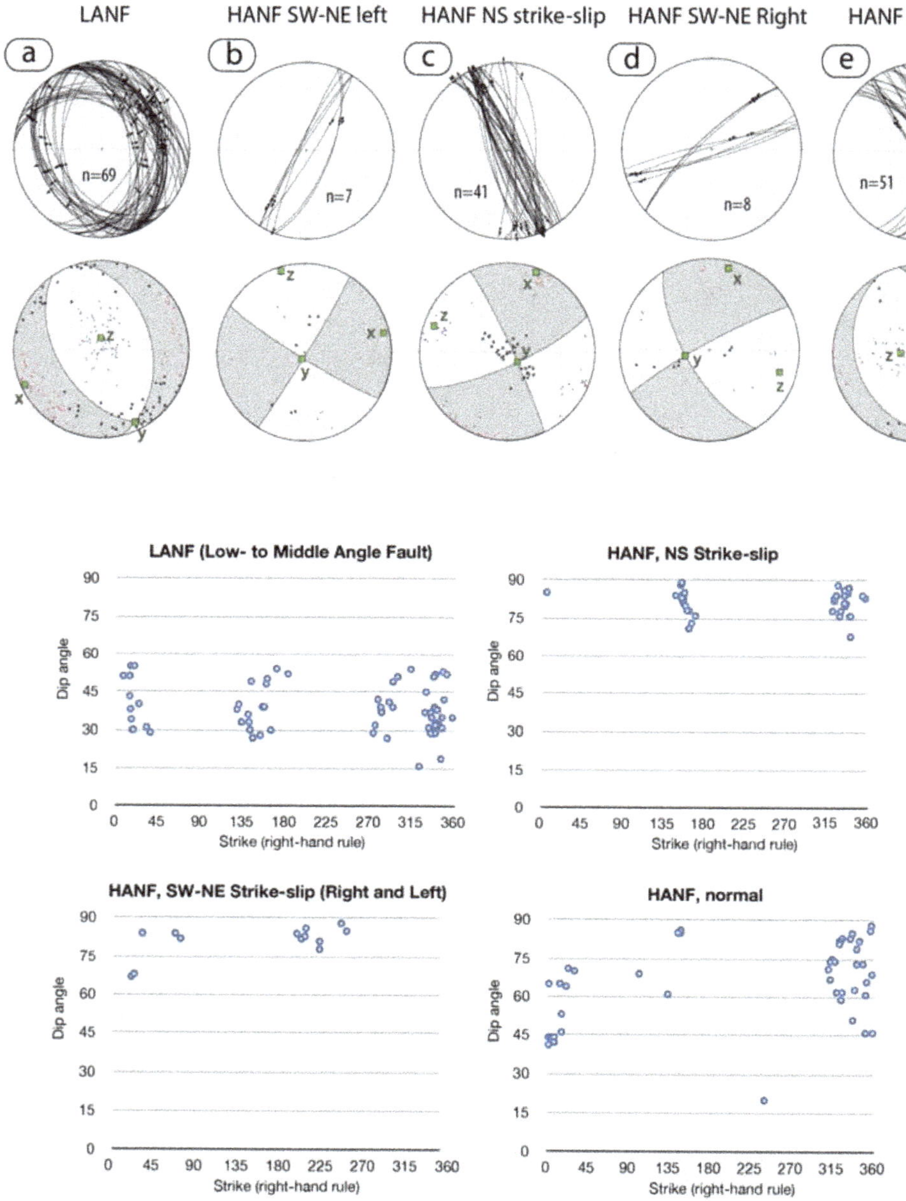

Figure 16. Upper part: fault-and-striae and fault plane solutions diagrams divided per systems from the inversion of kinematic data collected on fault-slip surfaces (Stereographic diagrams were performed using the software FAULTKIN, available at: http://www.geo.cornell.edu/geology/faculty/RWA/ accessed on 20 July 2019). Red and blue dots indicate P and T axes, respectively; x indicate neutral axis. Green squares show the orientation of the x, y and z strain axes. See the text for more indication. Lower part: strike vs. dip diagrams of the indicated fault categories.

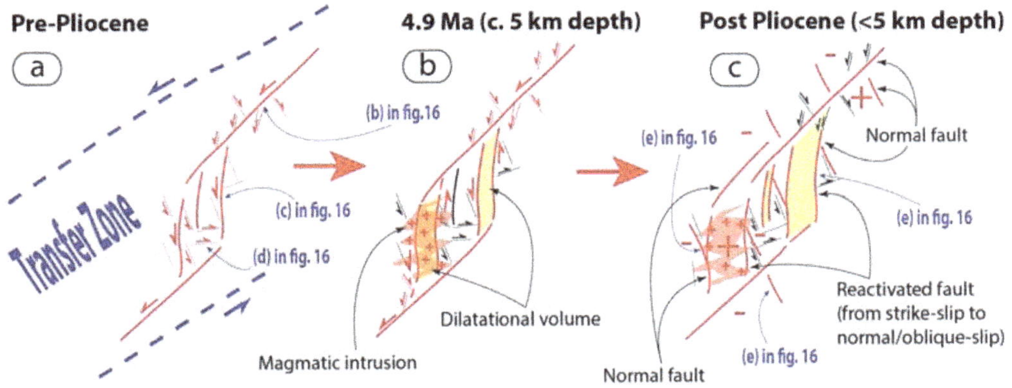

Figure 17. Conceptual model illustrating the relationships between faulting and magma intrusion/exhumation. SW-NE striking left-lateral regional transfer zone enucleated in a wide area including that one where the Gavorrano monzogranite is exposed, today. The transfer zone was active contemporaneously with top-to-the ENE e WSW low- to middle-angle normal faults, during the extensional evolution of the inner Northern Apennines. (**a**) The transfer zone gave rise to SW-NE striking left-lateral strike-slip faults linked by N-S striking faults in releasing step-over zones. Minor faults (NNW-left-lateral and WNW-striking right-lateral strike-slip faults) are associated with the N-S striking first-order faults. (**b**) The shearing evolution within the transfer zone formed vertical highly permeable volumes centred on the N-S striking faults. Magma was channeled within the permeable volume and intruded at the base of the late Triassic evaporite level, within the Permo-Triassic succession, in a depth interval comprised between 6.3 and 5.2 km. (**c**) Normal faults followed the magmatic emplacement and were active in the same regional stress field that was active at the time of pluton emplacement. These normal faults contributed to the exhumation of the monzogranite and the present configuration of the whole Gavorrano area.

Figure 18. Structural and kinematic features of Palaie Fault (see the Figure 3a for the location). The fault plane is exposed just in restricted exposures. (**a**) Panoramic view of the fault plane. (**b**) Enlarged area of the fault surface where mechanical striations can be observed. The fault plane was tilted during the activity of the normal fault systems bonding the western side of the Monte Calvo. Rotation therefore gave rise to an apparent east-dipping attitude of the Palaie Fault. Kinematic indicators, in the present coordinates, indicate left-lateral oblique-slip kinematics in agreement with the whole kinematic data collected in the Gavorrano area.

Figure 19. Structural and kinematic features of the normal faults affecting the carbonate succession. (**a**) Meters-thick fault zone affecting the late Triassic carbonate succession. (**b**) Detail of a WNW-dipping fault cutting an E-dipping middle-angle normal fault. (**c**) Well-developed damage zone consisting of sub-vertical fractures associated to a east-dipping normal fault with a slight oblique-slip movement.

6. Conclusive Remarks

On the basis of the new dataset integrated with the pre-existing data we can state the following points:

- The laccolithic monzongranite emplaced within the upper part of the Tuscan metamorphic succession, at the base of the Late Triassic carbonate succession. The exposed contact aureole at north of Ravi village is referred to the phyllitic-quartzite succession, similar to part of that one exposed at north of the Gavorrano village, underlying the metasandstone and quartz-metaconglomerate of the Triassic Verrucano Group. The succession exposed in the Gavorrano village and neighbourhood is referred to a transitional succession (i.e., Tocchi Fm) interposed between the Verrucano and late Triassic evaporite. The Palaeozoic succession hosting the monzogranite has a middle-late Permian age and can be referred to the coeval successions documented in the surroundings (e.g., Poggio al Carpino Fm, Filladi di Boccheggiano and Quarziti del Torrente Mersino formations). The decimetres-thick quartz-metasandstone levels interbedded within black phyllite, recognised in the Il Santo gallery, make this succession similar to the Poggio al Carpino Fm. On the contrary, the dominant black phyllite exposed at surface better corresponds to the Filladi di Boccheggiano and Quarziti del Torrente Mersino formations.
- The thermo-metamorphic paragenesis and Ti-in-biotite geothermometer point to a peak Temperature of c. 660 °C at a depth probably lower than 6 km. Dynamic recrystallisation of LP paragenesis suggests a syn-kinematic evolution of the contact aureole, in agreement with the active tectonic setting that assisted the magma emplacement, cooling and exhumation.
- We do not confirm the occurrence of regional and/or cartographic scale reverse faults, or thrust-related roof-anticline triggering the magma emplacement and hosting the magmatic intrusion, since those previously proposed interpretations contrast with

field data evidence. The pluton emplacement was coeval with coexisting strike-slip and extensional tectonics that continued also after the magma cooling and produced the exhumation of the magmatic system and of its contact aureole. The tectonic setting did not change through time: strike-slip and normal faults coexisted at least since the early Pliocene (age of the monzogranite emplacement). The Gavorrano pluton emplaced within a SW-NE trending sub-vertical strike-slip brittle shear zone (i.e., transfer zone) that accompanied the development of low-to middle-angle normal faults formed in a E-NE trending extensional setting. SW-NE striking strike-slip faults were mainly linked by NS striking strike-slip faults in releasing step-over zones, favouring the development of sub-vertical dilatational volumes with enough permeability to channel the magma from the deeper to upper crustal levels.

Author Contributions: Conceptualization, A.B.; methodology, A.B., A.C. (Alfredo Caggianelli), D.L., A.S., and M.Z.; software, A.B., A.C. (Alfredo Caggianelli), D.L., and M.Z.; validation, A.B. and D.L.; formal analysis, A.B., A.C. (Alfredo Caggianelli), D.L., A.S., and M.Z.; investigation, A.B., E.C., A.C. (Alfredo Caggianelli), D.L., and M.Z.; data curation, A.B., E.B., E.C., A.C. (Alfredo Caggianelli), A.C. (Alessandra Casini), D.L., A.S., and M.Z.; writing—A.B.; writing—review and editing, A.B., E.C., A.C. (Alfredo Caggianelli), A.C. (Alessandra Casini), D.L., A.S., and M.Z.; visualization, A.B.; project administration, A.B. and A.C. (Alessandra Casini); funding acquisition, A.B., D.L., and A.C. (Alessandra Casini). All authors have read and agreed to the published version of the manuscript.

Funding: This research was funded by: (1) Department of Physics and Geology (University of Perugia) 'REDAB—REDiscovering the Apennine Basement: a multidisciplinary correlation among the Northern Apennine Upper Paleozoic–Triassic successions (CAPEZBASE2017) (Amalia Spina and Enrico Capezzuoli); (2) European Community's Seventh Framework Programme under grant agreement No. 608553 (Project IMAGE) (Domenico Liotta, Andrea Brogi, Martina Zucchi); (3) FFABR grants (financial support for the base-research activity) from the MIUR (Ministry of Education and University) (Andrea Brogi). (4) Department of Earth and Geoenvironmental Sciences (University of Bari), University Project: Cooling of intra-crustal magmatic bodies and geothermal fluids circulation: comparison between late Hercinian and Plio-Quaternary granitoids from Calabria and southern Tuscany (Alfredo Caggianelli).

Informed Consent Statement: Informed consent was obtained from all subjects involved in the study.

Acknowledgments: We are indebted with Andrea Dini who supported us in discussing the evolution of the magmatic system and geological setting of the study area. We are grateful to the two anonymous reviewers for their helpful comments and valuable suggestions.

Conflicts of Interest: The authors declare no conflict of interest.

Appendix A

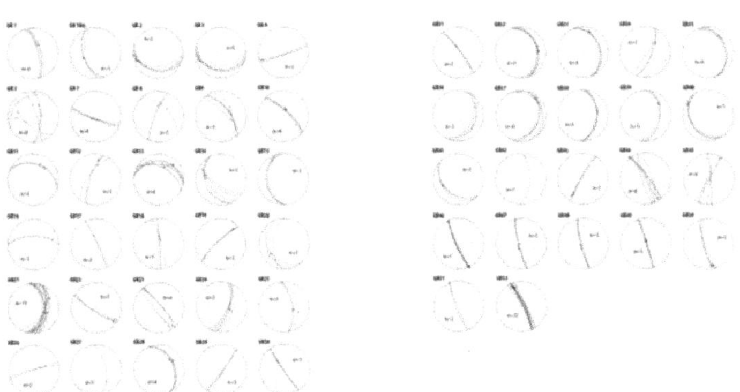

Figure A1. Structural stations: fault and striae from data collected on the indicated structural stations. Their location is indicated in the Figure 3a (lower hemisphere, equiareal diagram).

Table A1. Biotite analyses of corundum-bearing hornfels used for the application of the Ti-in-biotite geothermometer by [109]. Mineral formulae, calculated according to the method by [119].

	Bt 42	Bt 43	Bt 44	Bt 45	Bt 46	Bt 47	Bt 48
SiO_2	33.25	33.19	34.24	34.98	33.63	34.64	35.57
TiO_2	3.57	3.58	3.36	3.69	3.57	3.57	3.60
Al_2O_3	20.78	20.58	20.11	20.64	20.86	20.96	21.73
FeO	17.42	17.42	17.98	17.91	17.39	17.27	15.88
MnO	0.34	0.27	0.44	0.39		0.29	
MgO	8.36	8.36	8.51	8.79	8.18	8.62	8.87
Na_2O	0.18	0.23		0.18	0.32	0.30	0.18
K_2O	9.92	10.17	10.10	10.21	10.14	10.08	9.93
Tot	93.82	93.80	94.74	96.79	94.09	95.73	95.76
Si	2.57	2.57	2.62	2.62	2.59	2.61	2.64
Ti	0.21	0.21	0.19	0.21	0.21	0.20	0.20
Al^{IV}	1.43	1.43	1.38	1.38	1.41	1.39	1.36
Al^{VI}	0.46	0.45	0.44	0.44	0.48	0.47	0.55
Fe	1.12	1.13	1.15	1.12	1.12	1.09	0.99
Mn	0.02	0.02	0.03	0.02		0.02	
Mg	0.96	0.96	0.97	0.98	0.94	0.97	0.98
Na	0.03	0.04		0.03	0.05	0.04	0.03
K	0.98	1.00	0.99	0.97	0.99	0.97	0.94
$\sum$cat	7.78	7.81	7.77	7.77	7.79	7.76	7.69
$\sum$O	11.00	11.01	11.00	11.01	11.02	11.00	11.00
Fe^{3+}	0.03	0.06	0.04	0.05	0.11	0.10	0.16
Fe^{2+}	1.09	1.07	1.11	1.07	1.01	0.99	0.83
T, °C *	657	657	641	652	668	659	679

* Temperature estimate by Ti-in-biotite geothermometer [109] at P = 150 MPa.

References

1. Brunet, C.; Monié, P.; Jolivet, L.; Cadet, J.P. Migration of compression and extension in the Tyrrhenian Sea, insights from $^{40}Ar/^{39}Ar$ ages on micas along a transect from Corsica to Tuscany. *Tectonophysics* **2000**, *321*, 127–155. [CrossRef]
2. Rossetti, F.; Faccenna, C.; Jolivet, L.; Goffé, B.; Funiciello, R. Structural signature and exhumation P-T–t paths of the blueschist units exposed in the interior of the Northern Apennine chain, tectonic implication. *Boll. Soc. Geol. It.* **2002**, *1*, 829–842.
3. Bianco, C.; Godard, G.; Halton, A.; Brogi, A.; Liotta, D.; Caggianelli, A. The lawsonite-glaucophane blueschists of Elba Island (Italy). *Lithos* **2019**, 348–349. [CrossRef]
4. Carmignani, L.; Decandia, F.A.; Disperati, L.; Fantozzi, P.L.; Lazzarotto, A.; Liotta, D.; Oggiano, G. Relationships between the Tertiary structural evolution of the Sardinia-Corsica-Provencal Domain and the northern Apennines. *Terra Nova* **1995**, *7*, 128–137. [CrossRef]
5. Bartole, R. The North Tyrrhenian-Northern Apennines post- collisional system: Constraints for a geodynamic model. *Terra Nova* **1995**, *7*, 7–30. [CrossRef]
6. Calcagnile, G.; Panza, G.F. The main characteristics of the lithosphere–asthenosphere system in Italy and surrounding regions. *Pure Appl. Geophys.* **1980**, *119*, 865–879. [CrossRef]
7. Locardi, E.; Nicolich, R. Geodinamica del Tirreno e dell'Appennino centro-meridionale: La nuova carta della Moho. *Mem. Soc. Geol. It.* **1992**, *41*, 121–140.
8. Di Stefano, R.; Bianchi, I.; Ciaccio, M.G.; Carrara, G.; Kissling, E. Three-dimensional Moho topography in Italy: New constraints from receiver functions and controlled source seismology. *Geochem. Geophys. Geosyst.* **2011**, *12*, 1–15. [CrossRef]
9. Marinelli, G. Genèse des magmas du volcanisme Plio-Quaternaire des Apennines. *Geol. Rdsch.* **1967**, *57*, 127–141. [CrossRef]
10. Peccerillo, A. Plio-Quaternary magmatism in Italy. *Episodes* **2003**, *26*, 222–226. [CrossRef]
11. Elter, P.; Giglia, G.; Tongiorgi, M.; Trevisan, L. Tensional and compressional areas in the recent (Tortonian to Present) evolution of north Apennines. *Boll. Geofis. Teor. Appl.* **1975**, *17*, 3–18.
12. Serri, G.; Innocenti, F.; Manetti, P. Geochemical and petrological evidence of the subduction of delaminated Adriatic continental lithosphere in the genesis of the Neogene-Quaternary magmatism of central Italy. *Tectonophysics* **1993**, *223*, 117–214. [CrossRef]
13. Carmignani, L.; Decandia, F.A.; Disperati, L.; Fantozzi, P.L.; Kligfield, R.; Lazzarotto, A.; Liotta, D.; Meccheri, M. Inner northern Apennines. In *Anatomy of an Orogen: The Apennines and Adjacent Mediterranean Basins*; Vai, G.B., Martini, I.P., Eds.; Kluwer Academic: Dordrecht, The Netherlands, 2001; pp. 197–213.
14. Gianelli, G.; Puxeddu, M.; Batini, F.; Bertini, G.; Dini, I.; Pandeli, E.; Nicolich, R. Geological model of a young volcano-plutonic system: The geothermal region of Monte Amiata (Tuscany, Italy). *Geothermics* **1988**, *17*, 719–734. [CrossRef]

15. Dini, A.; Gianelli, G.; Puxeddu, M.; Ruggieri, G. Origin and evolution of Pliocene-Pleistocene granites from the Larderello geothermal field (Tuscan Magmatic Province, Italy). *Lithos* **2005**, *81*, 1–31. [CrossRef]
16. Brogi, A.; Lazzarotto, A.; Liotta, D.; Ranalli, G. Crustal structures in the geothermal areas of southern Tuscany (Italy): Insights from the CROP 18 deep seismic reflection. *J. Volcanol. Geotherm. Res.* **2005**, *148*, 60–80. [CrossRef]
17. Gola, G.; Bertini, G.; Bonini, M.; Botteghi, S.; Brogi, A.; De Franco, R.; Dini, A.; Donato, A.; Gianelli, G.; Liotta, D.; et al. Data integration and conceptual modelling of the Larderello geothermal area, Italy. *Energy Proc.* **2017**, *125*, 300–309. [CrossRef]
18. Rochira, F.; Caggianelli, A.; De Lorenzo, S. Regional thermo-rheological field related to granite emplacement in the upper crust: Implications for the Larderello area (Tuscany, Italy). *Geodin. Acta* **2018**, *30*, 225–240. [CrossRef]
19. Della Vedova, B.; Bellani, S.; Pellis, G.; Squarci, P. Deep temperatures and surface heat flow distribution. In *Anatomy of an Orogen: The Apennines and Adjacent Mediterranean Basins*; Vai, G.B., Martini, I.P., Eds.; Kluwer Academic: Dordrecht, The Netherlands, 2001; pp. 65–76.
20. Brogi, A. Kinematics and geometry of Miocene low-angle detachments and exhumation of the metamorphic units in the hinterland of the Northern Apennines (Italy). *J. Struct. Geol.* **2008**, *30*, 2–20. [CrossRef]
21. Liotta, D.; Brogi, A. Pliocene-Quaternary fault kinematics in the Larderello geothermal area (Italy): Insights for the interpretation of the present stress field. *Geothermics* **2020**, *83*, 101714. [CrossRef]
22. Gianelli, G.; Manzella, A.; Puxeddu, M. Crustal models of the geothermal areas of Southern Tuscany. *Tectonophysics* **1997**, *281*, 221–239. [CrossRef]
23. Batini, F.; Brogi, A.; Lazzarotto, A.; Liotta, D.; Pandeli, E. Geological features of the Larderello–Travale and Mt. Amiata geothermal areas (southern Tuscany, Italy). *Episodes* **2003**, *26*, 239–244. [CrossRef]
24. Dallmeyer, R.D.; Liotta, D. Extension, uplift of rocks cooling ages in thinned continental provinces: The Larderello geothermal area (inner Northern Apennines, Italy). *Geol. Mag.* **1998**, *135*, 193–202. [CrossRef]
25. Marinelli, G. L'intrusione terziaria di Gavorrano. *Atti. Soc. Tosc. Sci. Nat.* **1961**, *68*, 117–194.
26. Dallegno, A.; Gianelli, G.; Lattanzi, P.; Tanelli, G. Pyrite deposits of the Gavorrano area, Grosseto. *Atti. Soc. Tosc. Sci. Nat. Mem.* **1979**, *86*, 127–165.
27. Borsi, S.; Ferrara, G.; Tongiorgi, E. Determinazione con il metodo K/Ar della età delle rocce magmatiche della Toscana. *Boll. Soc. Geol. It.* **1967**, *86*, 403–410.
28. Serri, G.; Innocenti, F.; Manetti, P. Magmatism from Mesozoic to present: Petrogenesis, time-space distribution and geodynamic implications. In *Anatomy of an Orogen: The Apennines and Adjacent Mediterranean Basins*; Vai, G.B., Martini, I.P., Eds.; Kluwer Academic: Dordrecht, The Netherlands, 2001; pp. 77–103.
29. Arisi Rota, F.; Vighi, L. Le mineralizzazioni a pirite ed a solfuri misti. *Rend. Soc. Ital. Mineral. Petrol.* **1971**, *27*, 370–420.
30. Lotti, B. Sulla geologia del Gruppo di Gavorrano. *Boll. R. Com. Geol. Ital.* **1877**, *8*, 53–63.
31. Bencini, A.; Duchi, V.; Minissale, A.; Tanelli, G. Distribuzione di alcuni elementi metallici nelle rocce intrusive e carbonatiche associate alle mineralizzazioni a pirite di Gavorrano (Grosseto). *Rend. Soc. Ital. Mineral. Petrol.* **1980**, *36*, 599–609.
32. Cocozza, T.; Lazzarotto, A.; Vai, G.B. Flysch e molassa ercinici del Torrente Farma. *Boll. Soc. Geol. It.* **1974**, *93*, 115–128.
33. Bagnoli, G.; Tongiorgi, M. The Tuscan Paleozoic: A critical review. In *Report on the Tuscan Paleozoic Basement*; Tongiorgi, M., Ed.; Consiglio Nazionale delle Ricerche: Rome, Italy, 1978; pp. 9–26.
34. Musumeci, G.; Mazzarini, F.; Corti, G.; Barsella, M.; Montanari, D. Magma emplacement in a thrust ramp anticline: The Gavorrano Granite (northern Apennines, Italy). *Tectonics* **2005**, *24*, 1–17. [CrossRef]
35. Rossetti, F.; Faccenna, C.; Funiciello, R.; Pascucci, V.; Pietrini, M.; Sandrelli, F. Neogene strike-slip faulting and pluton emplacement in the colline metallifere region (Southern Tuscany, Italy): The Gavorrano-Capanne Vecchie area. *Boll. Soc. Geol. It.* **2001**, *120*, 15–30.
36. Musumeci, G.; Mazzarini, F.; Barsella, M. Pliocene crustal shortening on the Tyrrhenian side of the northern Apennines: Evidence from the Gavorrano antiform (southern Tuscany, Italy). *J. Geol. Soc. Lond.* **2008**, *165*, 105–114. [CrossRef]
37. Sani, F.; Bonini, M.; Cerrina Feroni, A.; Mazzarini, F.; Moratti, G.; Musumeci, G.; Corti, G.; Iatta, F.; Ellero, A. Messinian-Early Pliocene crustal shortening along the Tyrrhenian margin of Tuscany, Italy. *Ital. J. Geosci.* **2009**, *128*, 593–604. [CrossRef]
38. Montanari, D.; Corti, G.; Sani, F.; Del Ventisette, C.; Bonini, M.; Moratti, G. Experimental investigation on granite emplacement during shortening. *Tectonophysics* **2010**, *484*, 147–155. [CrossRef]
39. Lotti, B. *I Depositi Minerari Metalliferi*; Ind. Min.: Roma, Italy, 1928; p. 236.
40. Santi, G. *Viaggio Terzo per le Due Provincie Senesi che Forma il Seguito del Viaggio al Montamiata*; Ranieri Prosperi Stamp: Pisa, Italy, 1806; 182p.
41. Savi, P.; Meneghini, G. Considerazioni sulla geologia stratigrafica della Toscana. In *Memorie Sulla Struttura Geologica delle Alpi, degli Appennini e dei Carpazi*; Sir Murchison, R.E., Ed.; Stamperia Granducale: Firenze, Italy, 1851; pp. 279–522.
42. Lotti, B. Sull'età e origine dei graniti toscani. *Boll. R. Com. Geol. It.* **1886**, *1–15*.
43. Lotti, B. Sul giacimento di pirite di Gavorrano in Toscana. *Rass. Min.* **1901**, *1–8*, 116.
44. Marocchi, E. Studio sul granito di Gavorrano. *Atti. Soc. Tosc. Sco Nat. Mem.* **1897**, *15*, 171–187.
45. Martelli, A. Ricerche petrografiche e chimiche sulle formazioni granitiche di Gavorrano. *Atti. R. Acc. Lincei Rend.* **1909**, *18*, 1–7.
46. Barberi, F.; Innocenti, F.; Ricci, C.A. Il magmatismo. *Rend. Soc. Ital. Mineral. Petral.* **1971**, *27*, 169–210.

47. Mazzarini, F.; Corti, G.; Musumeci, G.; Innocenti, F. Tectonic control on laccolith emplacement in the northern Apennines fold-thrust belt: The Gavorrano intrusion (southern Tuscany, Italy). In *Physical Geology of High-Level Magmatic Systems*; Geological; Breitkreez, C., Petford, N., Eds.; Society, Special Publications: London, UK, 2004; Volume 234, pp. 151–161. [CrossRef]
48. Lotti, B. Osservazioni sulla memoria di L. de Launay "La métallogénie de l'Italie". *Boll. R. Com. Geol. d'Italia* **1907**, *38*, 3–23.
49. Lotti, B. Geologia della Toscana. *Mem. Descritt. Carta Geol. Ital.* **1910**, *13*, 484.
50. Lotti, B. Sui rapporti genetico-tettonici del giacimento di pirite dell'Isola del Giglio con quello di Gavorrano. *La Min. Ital.* **1930**, *14*, 206–209.
51. Toso, P. Sul modo di formazione dei principali giacimenti metalliferi aventi forma di irregolari ammassi o di strati, coltivati in Toscana ed in altre regioni d'Italia. *Boll. R. Com. Geol. d'Italia* **1912**, *3*, 113–233.
52. De Launay, L. Gites minéraux et métallifères. In *Traité de Métallogénie. Gites Minéraux et Métallifères. Gisements, Recherche, Production et Commerce des Minéraux Utiles et Minerais, Description des Principales Mines*; Librairie Polytechnique Ch. Béranger: Paris, France, 1913; Volume 1, pp. 350–353.
53. Sappa, G. Breve descrizione geologico-mineraria del giacimento di pirite di Gavorrano-Ravi. *Boll. Soc. Geol. Ital.* **1921**, *40*, 68–69.
54. Pompei, A. *Giacimenti di Pirite, di Fe e di Fosfati*; Ministero Economia Nazionale: Roma, Italy, 1927.
55. De Wijkerslooth, P. The mineralization of the Tuscan Mountains in connection with their tectonic evolution. *Proc. Konink. Akad. Wetensch.* **1930**, *33*, 557–564.
56. De Wukerslooth, P. Der Deckenbau Süd-Toskanas. *Proc. Konink. Akad. Wetensch.* **1930**, *33*, 1189–1200.
57. De Wukerslooth, P. *Bau und Entwicklung des Apennins besonders der Gebirge Toskanas*; Geologisch Instituut: Amsterdam, The Netherlands, 1934; 426p.
58. Turacchi, A. Pirite e miniere di pirite della Maremma. *Ind. Min.* **1954**, *5*, 15–26.
59. Trefzger, E.F. Ueber die Schwefelkies Lagerstatten der Toskanischen Maremma. *N. Jahrb. Min. Mon.* **1954**, *1*, 73–95.
60. Cavinato, A. *Giacimenti Minerari*; UTET: Torino, Italy, 1964; 686p.
61. Arnold, M. Etude préliminaire des sulfures des gisements de Niccioleta et de Gavorrano (Toscane). *C.R. Acad. Sci. Paris* **1973**, *276*, 445–447.
62. Natale, P. Relitti di bassa temperatura nelle piriti di alcuni giacimenti della Toscana. *Boll. Ass. Min. Subalp.* **1974**, *2*, 1–12.
63. Leonardelli, A. Toscana: Pirite e solfuri misti. In *Memoria Illustrativa della Carta Mineraria d'Italia*; Castaldo, G., Stampanoni, G., Eds.; Servizio Geologico d'Italia: Roma, Italy, 1975.
64. Bencini, A.; Martini, M.; Rebezzi, P.; Tanelli, G. Osservazioni preliminari sulla distribuzione di Fe, Mn, Ti, Li, Cu, Pb, Zn nelle rocce magmatiche del Campigliese (Campiglia Marittima, Toscana). *Rend. Soc. It. Min. Petr.* **1978**, *34*, 197.
65. Cortecci, G.; Lattanzi, P.; Leone, G.; Pochini, A.; Tanelli, G. Gli isotopi dello zolfo dei giacimenti a piriti di Niccioleta, Gavorrano, Boccheggiano e Ritorto, Toscana Meridionale. *Rend. Soc. It. Min. Petr.* **1980**, *36*, 261–277.
66. Jenks, W.E. Origin of some massive pyritic ore deposits of Western Europe. *Econ. Geol.* **1975**, *70*, 488–498. [CrossRef]
67. Rocchi, S.; Dini, A.; Mazzarini, F.; Poli, G. Campiglia Marittima and Gavorrano intrusive magmatism. *Per. Mineral.* **2003**, *72*, 127–132.
68. Bodechtel, J. Zur Genesis der Eisenerze der Toskana und der Insel Elba. *N. Mineral. Abh.* **1965**, *103*, 147–162.
69. Zuffardi, P. La metallogenesi italiana in relazione alla evoluzione del Mediterraneo. *Mem. Soc. Geol. It.* **1974**, *13*, 359–365.
70. Minguzzi, C. Dosatura spettrografica dell'oro in piriti italiane. *Atti. Soc. Tosc. Soc. Nat.* **1947**, *54*, 210–243.
71. Minguzzi, C.; Talluri, A. Indagini e considerazioni sulla presenza e sulla distribuzione di costituenti minori nelle piriti. *Atti. Soc. Tosc. Soc. Nat.* **1951**, *58*, 89–102.
72. Talluri, A. Studio genetico e geochimico del giacimento di pirite di Ravi Marchi. *Rend. Soc. Min. It.* **1953**, *9*, 264–266.
73. Caglioti, V.; Bettinali, C.; Giardini-Guidoni, A.; Mele, A. Isotopic abundance of some native sulphur and sulphide minerals. In Proceedings of the Summer Course on Nuclear Geology (CNEN 1960), Ravenna, Italy, 11 August 1960; pp. 202–213.
74. Ancarani-Rossiello, L.; Bettinali, C.; Faina, G. Sulla composizione isotopica dello zolfo nelle piriti della Maremma Toscana e dell'Isola d'Elba. *Per. Mineral.* **1962**, *31*, 369–373.
75. Garavelli, C.L. Contenuto in Fe e temperatura di formazione di blende italiane. *Atti. Soc. Tosc. Sci. Nat.* **1962**, *69*, 52–96.
76. Bralia, A.; Sabatini, G.; Troja, F. A revaluation of the Co/Ni ratio in pyrite as geochemical tool in ore genesis problems. Evidences from Southern Tuscany pyritic deposit. *Miner. Depos.* **1979**, *14*, 353–374. [CrossRef]
77. Zia, R. *Carta Geologica di Scarlino e Gavorrano (Grosseto)*; Centro Studi Geologia Appennino C.N.R.: Pisa, Italy, 1954.
78. Costantini, A.; Gandin, A.; Lazzarotto, A.; Mazzanti, R.; Sandrelli, F. Neotettonica dei Fogli 111-Livorno, 112-Volterra, 113-Castelfiorentino, 119-Massa Marittima, 120-Siena, 121-Montepulciano, 126-Isola d'Elba, 127-Piombino, 128-Grosseto, 129-S. Fiora. *Contr. Prelim. Realizz. Carta Neotett. Ital.* **1980**, *356*, 1075–1186.
79. Costantini, A.; Decandia, F.A.; Lazzarotto, A.; Sandrelli, F. L'unità di Monticiano-Roccastrada fra la Montagnola senese ed il Monte Leoni (Toscana meridionale). *Atti. Tic. Sc. Terra* **1988**, *31*, 382–420.
80. Rau, A.; Tongiorgi, M. Geologia dei Monti Pisani a Sud-Est della Valle del Guappero. *Mem. Soc. Geol. It.* **1974**, *13*, 227–408.
81. Pacltová, B. Palynology of metamorphic rocks (methodological study). *Rev. Palaeobot. Palynol.* **1986**, *48*, 347–356. [CrossRef]
82. Hanel, M.; Montenari, M.; Kalt, A. Determining sedimentation ages of high-grade metamorphic gneisses by their palynological record: A case study in the northern Schwarzwald (Variscan Belt, Germany). *Int. J. Earth Sci.* **1999**, *88*, 49–59. [CrossRef]

83. Schito, A.; Corrado, S.; Trolese, M.; Aldega, L.; Caricchi, C.; Cirilli, S.; Grigo, D.; Guedes, A.; Romano, C.; Spina, A.; et al. Assessment of thermal evolution of Paleozoic successions of the Holy Cross Mountains (Poland). *Mar. Pet. Geol.* **2017**, *80*, 112–132. [CrossRef]
84. Schito, A.; Spina, A.; Corrado, S.; Cirilli, S.; Romano, C. Comparing optical and Raman spectroscopic investigations of phytoclasts and sporomorphs for thermal maturity assessment: The case study of Hettangian continental facies in the Holy cross Mts. (central Poland). *Mar. Pet. Geol.* **2019**, *104*, 331–345. [CrossRef]
85. Spina, A.; Vecoli, M.; Riboulleau, A.; Clayton, G.; Cirilli, S.; Di Michele, A.; Marcogiuseppe, A.; Rettori, R.; Sassi, P.; Servais, T.; et al. Application of Palynomorph Darkness Index (PDI) to assess the thermal maturity of palynomorphs: A case study from North Africa. *Int. J. Coal Geol.* **2018**, *188*, 64–78. [CrossRef]
86. Spina, A.; Capezzuoli, E.; Brogi, A.; Cirilli, S.; Liotta, D. Middle-late Permian microfloristic evidences in the metamorphic successions of Northern Apennines: Insights for age constraining and palaeogeographic correlations. *J. Geol. Soc.* **2019**, *176*, 1262–1272. [CrossRef]
87. Cirilli, S.; Decandia, F.A.; Lazzarotto, A.; Pandeli, E.; Rettori, R.; Sandrelli, F.; Spina, A. Stratigraphy and depositional environment of the Mt. Argentario sandstones (southern Tuscany, Italy). *Boll. Soc. Geol. It.* **2002**, *1*, 489–498.
88. Lazzarotto, A.; Aldinucci, M.; Cirilli, S.; Costantini, A.; Decandia, F.A.; Pandeli, E.; Sandrelli, F.; Spina, A. Stratigraphic correlation of the Upper Palaeozoic-Triassic successions in southern Tuscany, Italy. *Boll. Soc. Geol. It.* **2003**, *2*, 25–35.
89. Aldinucci, M.; Brogi, A.; Spina, A. Middle-Late Permian sporomorphs from the Farma Formation (Monticiano-Roccastrada Ridge, southern Tuscany): New constraints for the tectono-sedimentary history of the Tuscan Domain. *Ital. J. Geosci.* **2008**, *127*, 581–597.
90. Spina, A.; Cirilli, S.; Utting, J.; Jansonius, J. Palynology of the Permian and Triassic of the Tesero and Bulla sections (Western Dolomites, Italy) and consideration about the enigmatic species Reduviasporonites chalastus. *Rev. Palaeobot. Palynol.* **2015**, *218*, 3–14. [CrossRef]
91. Visscher, H.; Brinkhuis, H.; Dilcher, D.L.; Elsik, W.C.; Looy, C.V.; Rampino, M.R.; Traverse, A. The terminal Paleozoic fungal event: Evidence of terrestrial ecosystem destabilization and collapse. *PNAS USA* **1996**, *93*, 2155–2158. [CrossRef]
92. Foster, C.B.; Stephenson, M.H.; Marshall, C.; Logan, G.A.; Greenwood, P.F. A revision of Reduviasporonites Wilson 1962: Description, illustration, comparison and biological affinities. *Palynology* **2002**, *26*, 35–58. [CrossRef]
93. Steiner, M.B.; Eshet, Y.; Rampino, M.R.; Schwindt, D.M. Fungal abundance spike and the Permian–Triassic boundary in the Karoo Supergroup (South Africa). *Palaeogeogr. Palaeoclimatol. Palaeoecol.* **2003**, *194*, 405–414. [CrossRef]
94. Galasso, F.; Fernandes, P.; Montesi, G.; Marques, J.; Spina, A.; Pereira, Z. Thermal history and basin evolution of the Moatize—Minjova Coal Basin (N'Condédzi sub-basin, Mozambique) constrained by organic maturation levels. *J. Afr. Earth Sci.* **2019**, *153*, 219–238. [CrossRef]
95. Galasso, F.; Pereira, Z.; Fernandes, P.; Spina, A.; Marques, J. First record of Permo-Triassic palynomorphs of the N'Condédzi sub-basin, Moatize-Minjova Coal Basin, Karoo Supergroup, Mozambique. *Rev. Micropaleontol.* **2019**, *64*. [CrossRef]
96. Stephenson, M.H.; Osterloff, P.L.; Filatoff, J. Palynological biozonation of the Permian of Oman and Saudi Arabia: Progress and challenges. *GeoArabia* **2003**, *8*, 467–496.
97. Stephenson, M.H. Stratigraphic Note: Update of the standard Arabian Permian palynological biozonation; definition and description of OSPZ5 and 6. *GeoArabia* **2006**, *11*, 173–178.
98. Stephenson, M.H. Spores and pollen from the middle and upper Gharif members (Permian) of Oman. *Palynology* **2008**, *32*, 157–182. [CrossRef]
99. Stephenson, M.H. Permian palynostratigraphy: A global overview. In *The Permian Timescale, Lucas, S.G., Shen, S.Z., Eds.;* Geology Society: London, UK, 2018; Volume 450, pp. 321–347. [CrossRef]
100. Spina, A.; Stephenson, M.H.; Cirilli, S.; Aria-Nasab, M.; Rettori, R. Palynostratigraphy of the Permian Faraghan Formation in the Zagros Basin, southern Iran. *Riv. Ital. Paleontol. Strat.* **2018**, *124*, 573–595.
101. Spina, A.; Cirilli, S.; Ghorbani, M.; Rettori, R.; Sorci, A.; Servais, T. Middle-late Cambrian acritarchs of the Zagros Basin, southwestern Iran. *Palynology* **2020**. [CrossRef]
102. Lindström, S. Palynology of Permian shale, clay and sandstone clasts from the Basen till in northern Vestfjella, Dronning Maud Land. *Antarct. Sci.* **2005**, *17*, 87–96. [CrossRef]
103. Burt, D.M. Mineralogy and petrology of skarn deposits. *Rend. Soc. Ital. Mineral. Petrol.* **1977**, *33*, 859–874.
104. Roy, R.; Osborn, E.F. The system Al_2O_3–SiO_2–H_2O. *Am. Mineral.* **1954**, *39*, 853–885.
105. Kretz, R. Symbols for rock-forming minerals. *Am. Mineral.* **1983**, *68*, 277–279.
106. Ferril, D.A.; Morris, A.P.; Evans, M.A.; Burkhard, M.; Groshong, J.R.H.; Onasch, C.M. Calcite twin morphology: A low-temperature deformation geothermometer. *J. Struct. Geol.* **2004**, *26*, 1521–1529. [CrossRef]
107. Pattison, D.R.M.; Harte, B. Petrography and mineral chemistry of pelites. In *Equilibrium and Kinetics in Contact Metamorphism: The Ballachulish Igneous Complex and Its Aureole*; Voll, G., Topel, J., Pattison, D.R.M., Seifert, F., Eds.; Springer: Berlin/Heidelberg, Germany, 1991; pp. 135–179. [CrossRef]
108. Wu, C.-M.; Chen, H.-X. Revised Ti-in-biotite geothermometer for ilmenite- or rutile-bearing crustal metapelites. *Sci. Bull.* **2015**, *60*, 116–121. [CrossRef]
109. Ferguson, C.C.; Al-Ameen, S.I. Muscovite breakdown and corundum growth at anomalously low *f*H2O: A study of contact metamorphism and convective fluid movement around the Omey granite, Connemara, Ireland. *Mineral. Mag.* **1985**, *49*, 505–514. [CrossRef]

110. Chatterjee, N.D.; Johannes, W. Thermal stability and standard thermodynamic properties of synthetic 2M muscovite, $KAl_2AlSi_3O_{10}(OH)_2$. *Contrib. Mineral. Petrol.* **1974**, *48*, 89–114. [CrossRef]
111. Boettcher, A.L.; Wyllie, P.J. Melting of granite with excess water to 30 kilobars pressure. *J. Geol.* **1968**, *76*, 235–244. [CrossRef]
112. Holdaway, M.J. Stability of andalusite and the aluminum silicate phase diagram. *Am. J. Sci.* **1971**, *271*, 97–131. [CrossRef]
113. Pattison, D.R. Stability of andalusite and sillimanite and the Al_2SiO_5 Triple point: Constraints from the Ballachulish aureole, Scotland. *J. Geol.* **1992**, *100*, 423–446. [CrossRef]
114. Cesare, B.; Marchesi, C.; Hermann, J.; Gómez-Pugnaire, M. Primary melt inclusions in andalusite from anatectic graphitic metapelites: Implications for the position of the Al_2SiO_5 triple point. *Geology* **2003**, *31*, 573. [CrossRef]
115. Acocella, V.; Rossetti, F. The role of extensional tectonics at different crustal levels on granite ascent and emplacement: An example from Tuscany (Italy). *Tectonophysics* **2002**, *354*, 71–83. [CrossRef]
116. Liotta, D. The Arbia-Val Marecchia line, Northern Apennines. *Eclogae Geol. Helv.* **1991**, *84*, 413–430.
117. Dini, A.; Westerman, D.S.; Innocenti, F.; Rocchi, S. Magma emplacement in a transfer zone: The Miocene mafic Orano dyke swarm of Elba Island, Tuscany, Italy. *Geol. Soc. Lond. Spec. Publ.* **2008**, *302*, 131–148. [CrossRef]
118. Liotta, D.; Brogi, A.; Meccheri, M.; Dini, A.; Bianco, C.; Ruggieri, G. Coexistence of low-angle normal and high-angle strike- to oblique-slip faults during Late Miocene mineralization in eastern Elba Island (Italy). *Tectonophysics* **2015**, *660*, 17–34. [CrossRef]
119. Dymek, R.F. Titanium, aluminum and interlayer cation substitutions in biotite from high-grade gneisses, West Greenland. *Am. Mineral.* **1983**, *68*, 880–899.

geosciences

Article

Lifecycle of an Intermontane Plio-Pleistocene Fluvial Valley of the Northern Apennines: From Marine-Driven Incision to Tectonic Segmentation and Infill

Massimiliano Ghinassi [1,*], Mauro Aldinucci [2], Valeria Bianchi [3], Andrea Brogi [4], Enrico Capezzuoli [5], Tsai-Luen Yu [6,7] and Chuan-Chou Shen [6,8]

1 Department of Geosciences, University of Padova, 35131 Padova, Italy
2 Eni México S. de R.L. de C.V., Mexico City 11000, Mexico; mauro.aldinucci@eni.no
3 School of Earth and Environmental Sciences, The University of Queensland, Brisbane 4072, Australia; valeria.bianchi2504@gmail.com
4 Department of Earth and Geoenviromental Sciences, University of Bari, 70125 Bari, Italy; andrea.brogi@uniba.it
5 Department of Earth Sciences, University of Florence, 50121 Florence, Italy; enrico.capezzuoli@unifi.it
6 High-Precision Mass Spectrometry and Environment Change Laboratory (HISPEC), Department of Geosciences, National Taiwan University, Taipei 10617, Taiwan; d00224009@g.ntu.edu.tw (T.-L.Y.); river@ntu.edu.tw (C.-C.S.)
7 Marine Industry and Engineering Research Center, National Academy of Marine Research, Kaohsiung City 806614, Taiwan
8 Research Center for Future Earth, National Taiwan University, Taipei 10617, Taiwan
* Correspondence: massimiliano.ghinassi@unipd.it

check for updates

Citation: Ghinassi, M.; Aldinucci, M.; Bianchi, V.; Brogi, A.; Capezzuoli, E.; Yu, T.-L.; Shen, C.-C. Lifecycle of an Intermontane Plio-Pleistocene Fluvial Valley of the Northern Apennines: From Marine-Driven Incision to Tectonic Segmentation and Infill. *Geosciences* **2021**, *11*, 141. https://doi.org/10.3390/geosciences11030141

Academic Editors: Angelo Cipriani and Jesus Martinez-Frias

Received: 30 January 2021
Accepted: 15 March 2021
Published: 18 March 2021

Publisher's Note: MDPI stays neutral with regard to jurisdictional claims in published maps and institutional affiliations.

Copyright: © 2021 by the authors. Licensee MDPI, Basel, Switzerland. This article is an open access article distributed under the terms and conditions of the Creative Commons Attribution (CC BY) license (https://creativecommons.org/licenses/by/4.0/).

Abstract: Downcutting and infill of incised valley systems is mostly controlled by relative sea-level changes, and studies on valley-fill successions accumulated independently from relative sea-level or lake-level oscillations are limited. This study focuses on the Plio-Pleistocene evolution of a fluvial drainage system developed in Southern Tuscany (Italy) following a regional marine forced regression at the end of Piacentian. Subsequent in-valley aggradation was not influenced by any relative sea-level rise, and valley morphological and depositional history mainly resulted from interaction between sediment supply and tectonic activity, which caused segmentation of the major valley trunk into localized subsiding depocenters separated by upwarping blocks. Fluvial sedimentation occurred until late Calabrian time, when the major river abandoned that valley, where minor fluvio-lacustrine depocenters allowed accumulation of siliciclastic and carbonate deposits. The present study demonstrates that the infill of the valley was not controlled by the forcing that caused its incision. Accumulation of the fluvial succession is discussed here in relation with localized, tectonic-controlled base levels, which commonly prevent from establishing of a clear downdip stratigraphic correlations. Chronological reconstruction of the study depositional dynamics provides solid constrains to frame them in the tectono-sedimentary evolution of the Northern Apennines.

Keywords: fluvial sedimentology; incised valley; palaeodrainage; Southern Tuscany

1. Introduction

Fluvial valley systems are key geomorphic elements that act as sediment transfers from source to sink areas. Occasionally, they allow accumulation of thick sedimentary successions [1–7], which might host relevant volumes of hydrocarbons or fossil water. In the downstream valley reaches, relative sea-level changes are dominant forcings on in-valley fluvial dynamics [1,8], whereas tectonic and climate are the main controls in upstream sectors [9,10]. Deposits accumulated in the downstream valley reaches commonly exhibit evidence of marine influence, in terms of adjustment of alluvial dynamics (e.g., tidal influence) as well as direct contribution to sedimentation (e.g., estuarine deposits). Oscillations of relative-sea level play a major role in triggering aggradation and degradation processes in downstream valley reaches [1]. Deposits accumulated in the upstream valley reaches have mainly fluvial origin [1,11] and processes of aggradational and degradation result from the complex interaction between tectonics, climate and drainage evolution [9,10,12]. Downstream valley-reach deposits are much more documented in comparison to the upstream counterpart, since the latter one has a low potential of preservation or can be hardly distinguishable from the surrounding extra-valley deposits [12].

In intermontane basins valleys play a key role in transferring sediment from the orogen to the surrounding basins [13–15]. The complex tectono-morphological configuration of extensional intermontane basins, like those of Northern and Central Apennines [16,17], allows coexistence of differently-subsiding areas separated by more stable substrate ridges. Beyond relative-sea level effects and tectono-climate forcings, this tectono-morphological configuration causes rivers cutting through post-orogenic landscapes to experience a number of interconnected processes [17–20] including: (i) variation in accommodations triggered by the occurrence of localized base levels; (ii) relocation due to piracy processes and (iii) variation in sediment supply and water discharge associated with evolution of the related drainage.

Major rivers of the inner Northern Apennines (e.g., Arno, Sieve, Ombrone) modelled the present-day landscape connecting extensional intermontane basins by cutting deep and narrow gorges on pre-Neogene rocks [18,21]. During Pliocene to Pleistocene time, tectono-geomorphic processes assisted these rivers to interact with a marine base level and to shift between adjacent basins [21], accumulating a wide spectrum of deposits along the related palaeovalleys. Although detection of these valley-fill succession is commonly hindered by paucity of outcrops, they are a precious archive to investigate the temporal and spatial evolution of the palaeo-drainage, and the associated processes.

This study focuses on a Pliocene–Pleistocene palaeovalley developed in Central Tuscany (Northern Apennines, Italy; Figure 1). Developed across a rocky ridge that separated three major intermontane basins, it experienced complex geomorphic and depositional dynamics, since its marine-driven incision to the final piracy-driven abandonment.

This study refines previously published data [22,23] and integrates new field-based information in a larger area of interest. The paper aims at depicting the depositional history of this palaeovalley through a multidisciplinary approach, which includes field mapping, sedimentary facies analyses, radiometric datings, palaeocurrent studies, and tectono-stratigraphic investigations. Beyond providing new insights to the regional geology, results from this study will be discussed in comparison with current facies models for valley-fill successions and also to highlight implication for the tectono-stratigraphic evolution of the Northern Apennines [24–26]. This work adopts the International Chronostratigraphic Chart v2020/03 nomenclature.

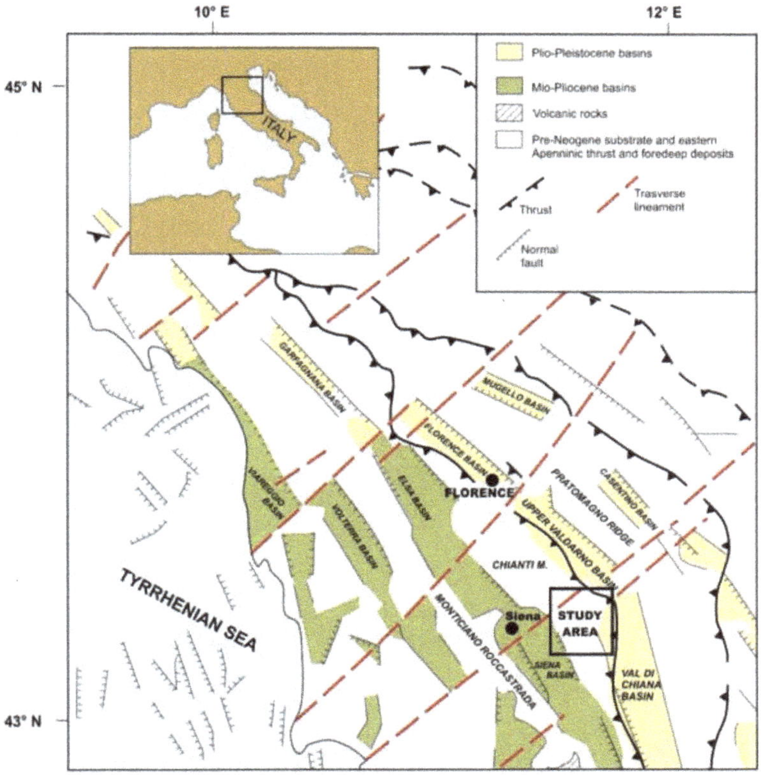

Figure 1. Tectonic map of the Northern Apennines, with indication of the main Neogene–Quaternary intermontane basins (Modified after [27]). Location of the study area is shown.

2. Geological Setting

After the collisional event that gave rise to the Northern Apennines (Late Oligocene-Early Miocene), a widespread extensional event produced the thinning of the previously overthickened continental crust [24,25] favoring the development of a Basins-and-Ranges upper-crustal architecture, especially in the inner zone of the belt (i.e., northern Tyrrhenian Sea and southern Tuscany) [16]. Extensional tectonics and related structural depressions (i.e., basins) have been migrating from west to east, thus the inland westernmost basins are the oldest ones [26]. In this view, the progressive eastward migration produced the development of intermontane basins, which evolution was characterized by an extensive tectonic activity conditioning the sedimentation [28,29].

The study area is drained by hydrographic systems pertaining to three major intermontane basins, namely the Upper Valdarno, Siena and Val di Chiana Basin (Figures 1–3). This area forms a narrow, elongated belt, that, for sake of simplicity, has been divided into three sectors based on the occurrence of three major faults (faults F1 to F3 in Figures 2 and 3, sectors are visible in Figure 3). Sector 1 and 2 are sited within the modern drainage system pertaining to the Upper Valdarno and Siena Basins, respectively, whereas sector 3 is sited in the drainage of the modern Val di Chiana Basin. Sector 1 is drained by the modern Ambra River (Figure 3), which after flowing southward for a few kilometers bends northward to join the Arno River in the Upper Valdarno Basin. Sector 2 is drained by the southward-flowing Ombrone River, which deeply cuts Pliocene marine deposits of the Siena Basin (Figure 3). The southern part of Sector 2 is characterized by the occurrence of a flat morphological surface (Pian di Bari and Bestina plain) that contrasts with the hilly landscape generated by the erosion of marine sediments, and caps the study deposits. Sector 3 is drained by the Foenna Creek (Figure 3), located at the bottom of a wide plain within the Sentino Basin (Figure 3). The Foenna Creek is sourced from the Rapolano area and drains eastward toward the Val di Chiana Basin (Figure 3).

The Upper Valdarno Basin is located about 35 km SE of Florence, between the Chianti Mountains and the Pratomagno Ridge, and it is filled with a few hundred meters of palustrine, lacustrine and alluvial deposits [30–33]. The basin fill succession consists of three major unconformity-bounded stratigraphic units [33] accumulated between Piacentian and Chibanian. The Calabrian to Chibanian depositional history of the Valdarno Basin is well-constrained by integrated magnetostratigraphic [31,33] and palaeontological data [34], which reveal the entrance of the Arno River into the basin [21,35]. The Arno River flowed from the Casentino Basin (Figure 1) southward to the Val di Chiana Basin since Pliocene time [21]. A piracy event caused the northward deflection into the Upper Valdarno Basin just before the Matuyama/Brunhes geomagnetic polarity transition [33,36]. In the basin fill succession, this geomorphic event is documented by: (i) a basin scale unconformity [36]; (ii) the increase of depth in channel of the axial drainage systems [33] and (iii) the occurrence of distinctive calcareous clasts from the Casentino area [35] in the axial fluvial deposits.

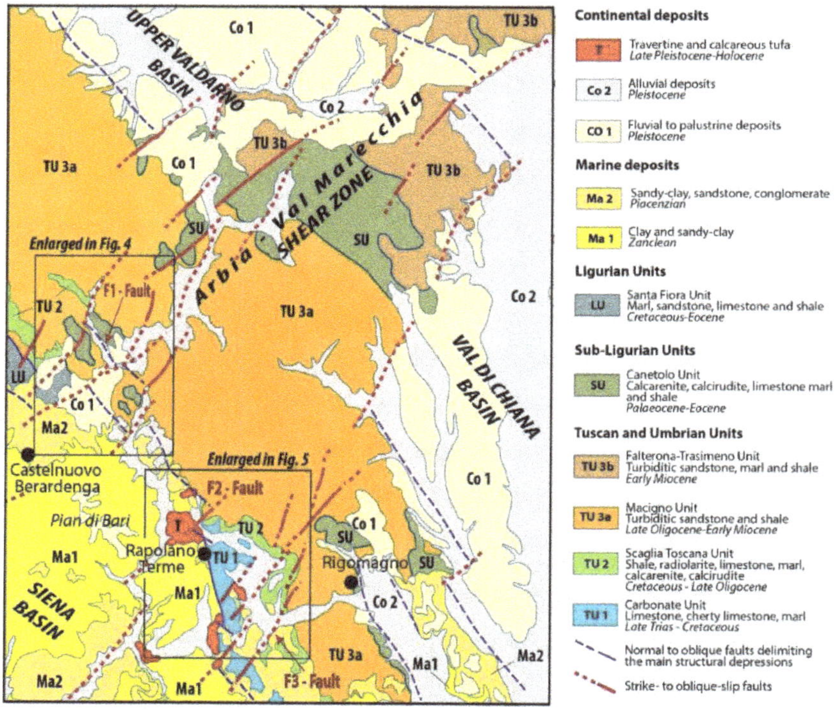

Figure 2. Geological map of the study area with indicated the enlarged areas in Figures 4 and 6. The indicated F1–F3 faults are mentioned in the text.

The Siena Basin is located about 15 km SE of the Upper Valdarno Basin, along the western side of the Chianti Ridge, and is bounded westward by the Monticiano-Roccastrada range (Figures 1 and 2). The basin-fill succession is ca. 100 m thick and consists of Miocene continental deposits covered by a Pliocene marine succession [37,38]. Marine sedimentation occurred since early Zanclean and it is well-documented along the Chianti margin, in the Castelnuovo Berardenga surroundings [39–41]. In this area, coastal marine deposits present three major relative sea-level changes [41] occurred during Piacentian time. In the Siena Basin, marine sedimentation ended up in the latest Piacentian [39–41], although recent studies pointed out the occurrence of latest Gelasian deposits [42]. This marine succession forms the youngest substrate of the investigated succession, which appears to be the only record of deposition in the area during Gelasian to Calabrian time ([22,23] Figure 4)

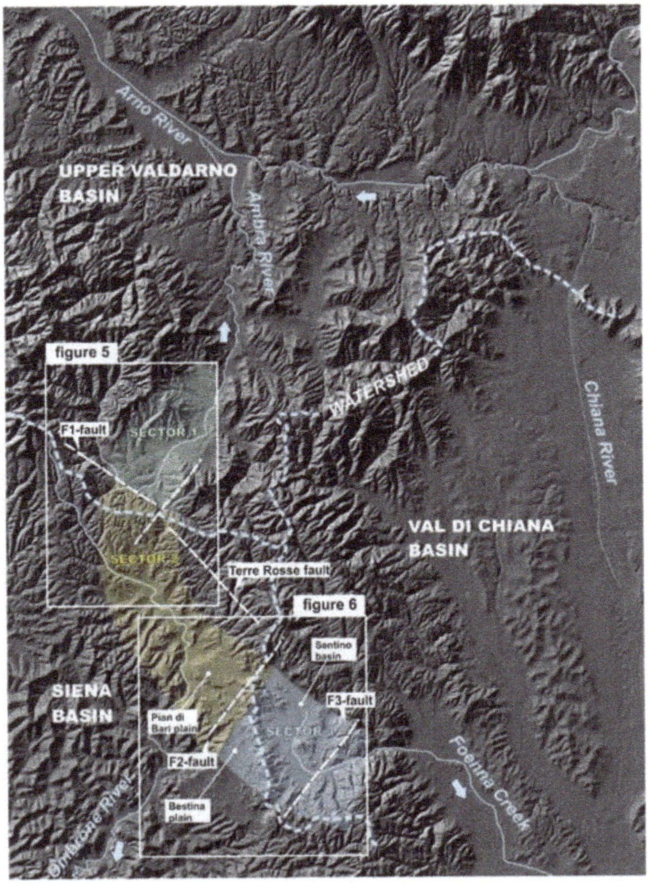

Figure 3. Geomorphological features of the study area. Watersheds separating the major drainage domains are highlighted, along with the key structural features associated with the present study (F1 to 3 faults).

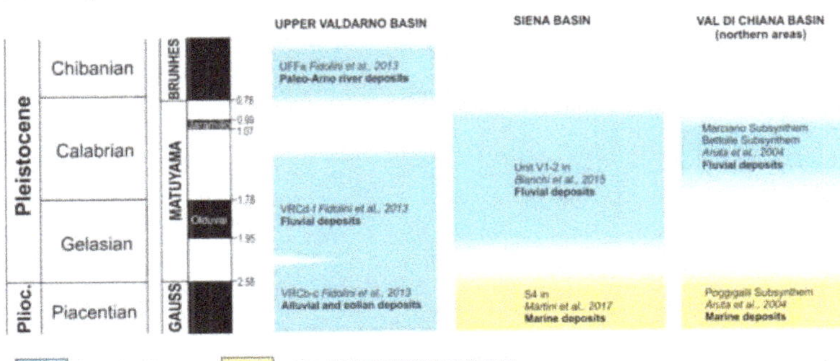

Figure 4. Age of deposits filling the Upper Valdarno, Siena and Val di Chiana Basins during Plio-Pleistocene time.

The Val di Chiana Basin represents the southern extension of the Upper Valdarno Basin, and it is separated from the latter one by a 10-km-wide rocky ridge (Figures 1–3). The Val di Chiana Basin is filled up with a 1500–2000 m thick sedimentary succession [43,44], consisting of marine and alluvial deposits accumulated between Pliocene and Pleistocene time. The stratigraphy of the Northern Val di Chiana Basin fill is still poorly known, although it was recently improved by [45]. In this sector of the basin, marine sedimentation occurred in the basin until late Piacentian [44,45], and it was replaced by alluvial sedimentation during Gelasian to Calabrian time [45]. The alluvial succession of the Val di Chiana Basin includes two major unconformity-bounded units (Bettolle and Marciano Synthems in [45]) consisting of sand and mud accumulated by a southward-directed palaeo-drainage. The Marciano Synthem is constrained to Calabrian by well-known mammal faunal associations [46].

The structural setting of the study area is defined by contractional structures related to the orogenesis of the Northern Apennines (i.e., fold-and-thrust system, [47] for a review) and by later structures related to the development of the Neogene-Quaternary basins ([37] for a review). Regarding the later ones, these were characterized by a prolonged activity from Late Serravallian to late Quaternary ([47] with references therein). The Pliocene-Pleistocene faults (Figure 2) controlled the configuration of the current structural depressions (i.e., Siena, Val di Chiana and Valdarno Basins) and their infill. In the study area (Figure 2), these faults can be categorized in three main systems depending on the age, the geometry and the kinematics: (i) Zanclean-Piacenzian N-S normal faults; (ii) Neogene-Quaternary NE-striking faults; and (iii) Neogene-Quaternary NW-striking faults. Zanclean N-S normal faults are mainly represented by the Rapolano Fault, a regional structure that bounds the eastern margin of the Siena Basin, partially covered by Piacenzian marine sediments [48,49]. It juxtaposes, in several places, the pre-Neogene succession of the Chianti- Cetona Mt. ridge (Tuscan Nappe, [50]) with the Zanclean-Piacenzian marine sediments filling the Siena Basin. Similar features characterize also minor faults associated to the Rapolano Fault (Figure 2). The maximum vertical offset of the Rapolano Fault was estimated in about 400 m on the basis of the interpretation of the seismic profiles [38]. Slickensides on minor faults mainly affecting the pre-Neogene Jurassic carbonate and siliceous units indicate a dominant dip–slip movement with minor left-lateral component. These faults are interrupted by NE- and NW-striking faults. These latter ones dissect the Piacenzian marine sediments and Late Pleistocene-Holocene travertine deposits [51–53], supporting the hypothesis of active tectonicsin the whole area. This is also confirmed by the prehistoric and historical seismicity mainly recorded by the travertine deposits [53,54]. Furthermore, NE- and NW-striking faults play a role in controlling the geothermal fluid circulation, as well as the location of thermal springs and travertine deposits [55–57]. These faults controlled also the late geomorphological evolution of the study area, mainly characterized by the Ambra Valley and Sentino Basin. In particular, the Ambra Valley area was mostly controlled by NE-striking fault system that is part of the so-called "Arbia-Valmarecchia Line" ([58] with references therein). This consists of a brittle shear zone characterized by anastomosed fault segments [52], which interfered with the continental sedimentation during the Quaternary [22]. The Sentino Basin developed in response of the activation of several NW- and NE-trending faults. Their interplay gave rise to the peculiar zig-zag shaped structural depression, controlling the continental sedimentation as illustrated in the next paragraph. Both NW- and NE-striking faults are characterized by a dominant left- and right-lateral oblique-slip kinematics.

3. Materials and Methods

The stratigraphy and most of the structural data from sector 1 and the northern part of sector 2 were investigated in previous studies [22], which outlined the occurrence of an incised valley in the northern margin of the Siena basin. The evidence was integrated with new sedimentological and structural data obtained expanding the study area toward the south and including field activities in sector 3 and in the southern part of sector 2.

The spatial extent of the studied deposits was mapped on a 1:10,000 scale topographic base (Figures 5 and 6). The sedimentary units were characterized and ascribed to specific depositional environments following basin principles of facies analyses. The main palaeodrainage routes were calculated from palaeocurrent measurements, which were obtained from dip of foresets from cross-bedded strata and a(t)b(i) clast imbrications.

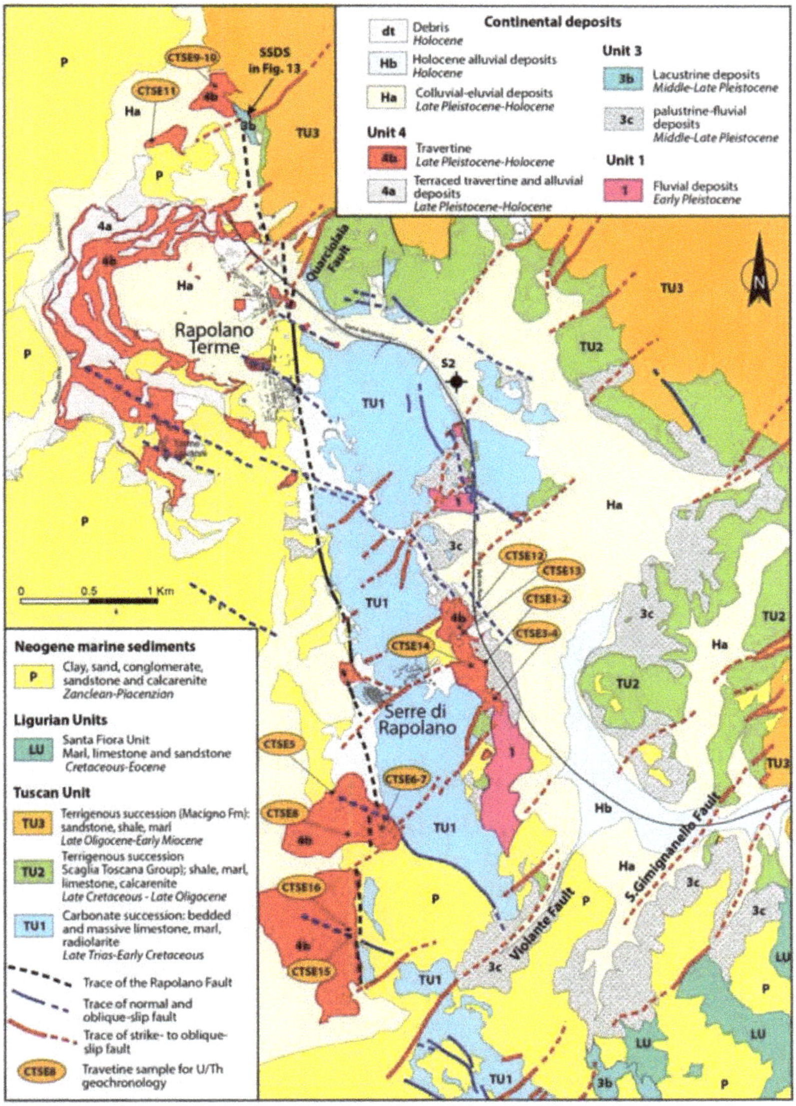

Figure 5. Geological map of the Sentino Basin (after [58], improved and expanded toward south). Querciolaia Fault corresponds to the F2 mentioned in the text and indicated in Figure 2; Violante and San Gimignanello Faults correspond to the F3 mentioned in the text and indicated in Figure 2.

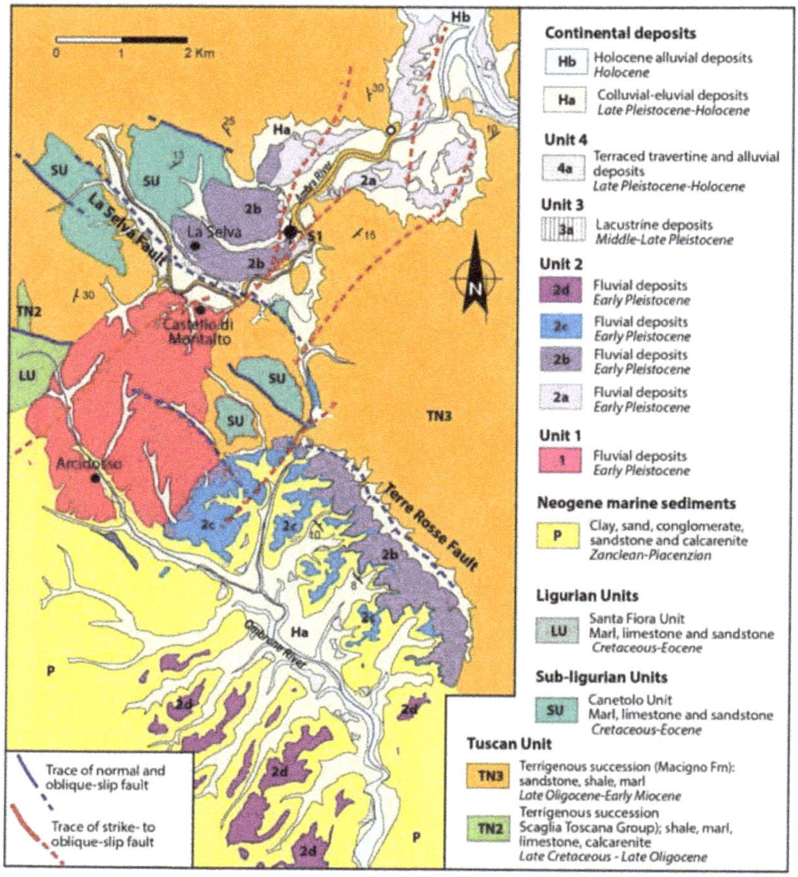

Figure 6. Geological map of the Ambra Valley (from [38], modified). La Selva Fault corresponds to the F1 mentioned in the text and indicated in Figures 2 and 3.

Structural analyses were focused on high-angle normal faults, which are thought to represent the youngest regional structures affecting both unconsolidated Pliocene–Pleistocene deposits and pre-Neogene bedrock. Structural and kinematic data represent discrete clusters (structural stations), along exposed meso-fault surfaces.

U–Th dating was carried out for travertines exposed in the southern part of the study area (Figure 5). The dating procedure was conducted at the High-Precision Mass Spectrometry and Environment Change Laboratory (HISPEC), Department of Geosciences, National Taiwan University [59,60] following the procedure already described in [53]. Determinations of all U-Th isotopic compositions and concentrations were made on a Thermo-Finnigan NEPTUNE multicollector inductively coupled plasma mass spectrometer (MC-ICPMS) [59]. Off-line data reduction methods followed [61]. Half-lives of U–Th nuclides used for age calculation was given by [62]. Isotopic and age errors given are two standard deviations of the mean and two standard deviations, respectively, unless otherwise noted.

Well logs and geophysical data reported by [22] were integrated with field evidence and a new borehole dataset in sector 3, in order to estimate thickness variability of the studied sedimentary entities, with particular focus on definition of the valley floor depth.

The studied deposits form a 1 to 3-km-wide and ca. 40-km-long entity, which crossed the three major sectors forming the study area. The succession was divided into four units, which are separated by erosional surfaces of abrupt changes in facies assemblages (e.g., change from fluvial to lacustrine deposits).

4. Results

4.1. Structural Features

Structural features are described for each sector indicated in Figure 3. Sector 1 (Figures 2 and 6) is part of the main brittle shear zone corresponding with the Arbia-Valmarecchia shear zone (Figure 2, cf. [63]). This is mainly characterized by dominant NE-striking fault segments that are linked by NW-striking ones (Figure 4). Fault 1 (La Selva Fault, Figure 4) is part of this fault system. N-S- striking faults pre-date the NW- and NE-striking ones, as indicated by their cross-cutting relationships. These faults affected only the bedrock mainly represented by Late Oligocene-Early Miocene sandstone (Macigno Fm), whereas NE- and NW-striking faults affected both the fluvial deposits and the bedrock (cf. [22,62]). Faults in bedrock are characterized by centimetre-thick core zones and damage zones wide up to 5 m; meso-faults display well-developed sets of fractures, often with a en-echelon geometry concentrated near the slip surface. These characteristics weakened large volumes of rock, creating greater erodibility along the NE and NW fault trend. This process, enhanced by the hydrogeological pattern, favored the development of interconnected valleys mainly parallel to the NE-striking faults. The offset of these surfaces does not exceed a few meters. N-S, NE- and NW-striking fault planes show a number of kinematic indicators consisting of mechanical striation and calcite fiber-steps. In the damaged zones, arrays of extensional jogs and T-fractures mainly occur. Kinematics of N-S striking faults indicate a dominant normal displacement. NE- and NW-striking faults (Figure 7) are characterized by dominant left-lateral oblique-slip movements. A synthesis of the collected data is reported in Figure 8.

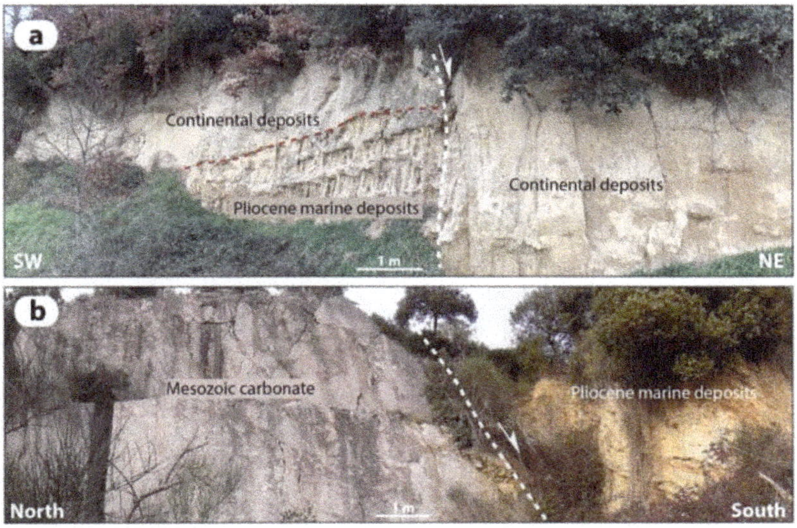

Figure 7. Panoramic view of faults affecting the successions filling the Sentino Basin and surroundings: (**a**) NW-striking normal fault dissecting Late Pleistocene continental sediments; (**b**) NE-striking left-lateral oblique-slip fault juxtaposing Piacenzian marine deposits to Mesozoic carbonate.

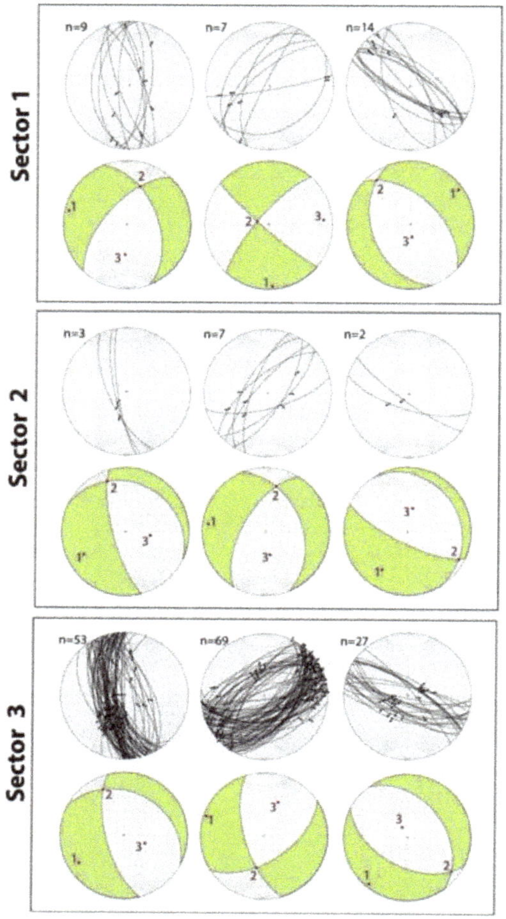

Figure 8. Stereographic diagrams (equiareal diagrams, lower hemisphere) illustrating cumulative kinematic data collected in different structural stations and divided per fault systems: N-S, NW- and NE-striking faults. For each fault system, faults and striae (upper diagrams) and fault plane solutions diagrams with kinematic axes from the inversion of kinematic data are shown. Location of the indicated sectors are reported in Figure 3. Kinematics data from area 1 have been integrated with those published in [38]; Kinematics data from area 3 have been integrated with those published in [38].

Sector 2 (Figures 2, 5 and 6) is mainly defined by a main cartographic-scale fault named as the Terre Rosse Fault (Figure 6). It dips south-westward and it delimits, for at least 3 km, the easternmost Pliocene sediments filling the Siena Basin from the bedrock composed of the Macigno Fm. The fault trace is buried by Quaternary deposits, denying direct observation and kinematic analyses on the main fault surface. Its maximum offset can be estimated at several tens of meters. The Terre Rosse Fault is interrupted to the north-west by the NE-striking fault. Kinematics analyses on minor structures in the footwall allow to recognise three fault systems, similarly to the Area 1: N-S, NW- and NE-striking (Figure 8). All faults show dominant normal to oblique-slip movements. (Figure 8).

Sector 3 (Figures 2 and 5) includes several cartographic scale faults (Figure 2). The most important ones bound the Sector 3, namely the Querciolaia Fault (F2 Fault in Figure 2) and Violante-Sangimignanello faults (F3 Fault in Figure 2) both NE-striking (Figures 2 and 6).

These are characterized by fault traces exceeding 1 km, by several tens of meters off-sets and they are associated with contemporaneous NW-striking faults (Figure 6). Both NW- and NE-striking faults were recognised affecting the bedrock, represented by Jurassic-Cretaceous carbonate succession, Cretaceous-Early Miocene shale/carbonate succession and the Pliocene-Quaternary deposits (cf. [50]). Faults in bedrock are characterized by centimetre-thick core zones and damage zones wide up to 20 m; meso-faults display sets of fractures and a number of kinematic indicators, mainly consisting of mechanical striation and calcite or quartz fiber-steps. All faults show dominant normal to oblique-slip movements (Figure 8). NE- and SW-striking faults dissected the N-S striking Rapolano Fault, which activity was from Zanclean to Piacenzian [38].

4.2. Stratigraphy

The stratigraphy of the studied area is summarized in Figure 9, by means of an idealized cross-section along the valley axis.

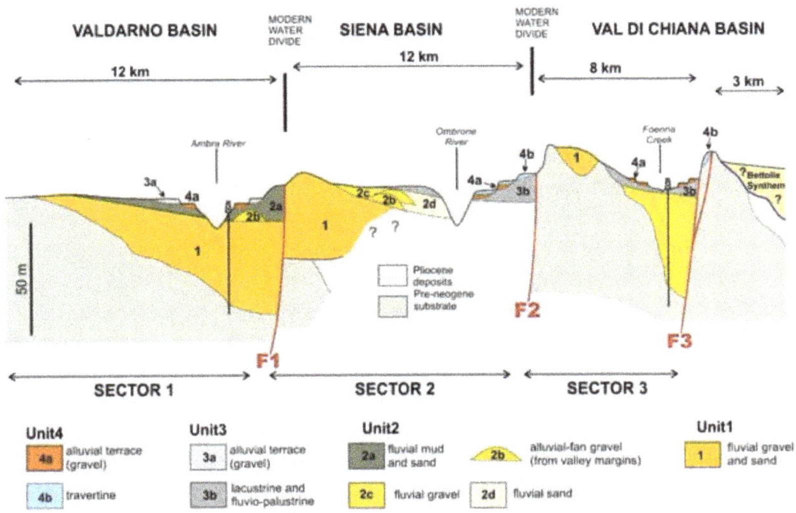

Figure 9. Schematic cross section along the axis of the study sedimentary body.

4.2.1. Unit 1

Sector 1

Sector 1 is characterized by a widespread cover of Holocene deposits (Figure 6). Deposits of Unit 1 do not occur in the most proximal reach of the sector 1, where pre-Neogene bedrock is exposed in the thalweg of the modern Ambra River, but borehole and geophysical data [22] show that 60-m-thick gravels occur just upstream of the fault F1 at the bottom of the modern valley (Figures 9 and 10). These gravelly deposits are confined in a 0.8–1.5 km wide palaeovalley (unit V1 in [22]; unit Q and FL in [64]), which wanders with an overall NNE–SSW trend and it cut the pre-Neogene bedrock. In the southernmost part of the sector 1, these deposits are locally exposed, and they consist of amalgamated and poorly-organized pebble to cobbles with scattered boulders.

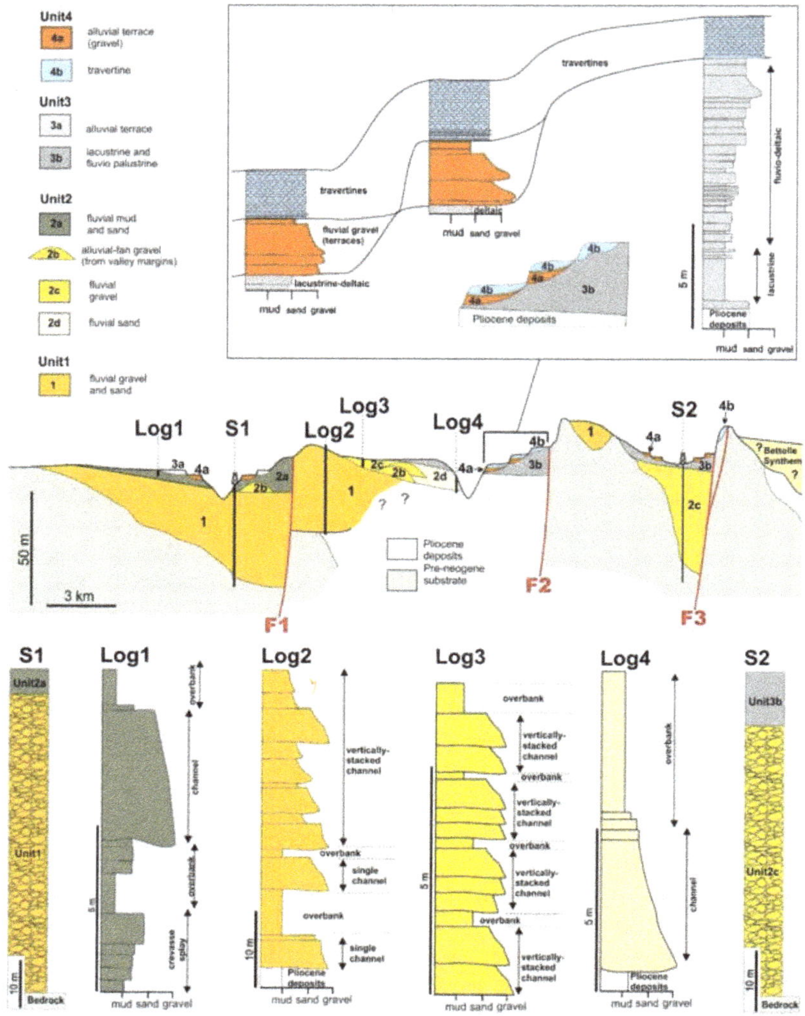

Figure 10. Simplified sedimentological logs and core data from the study deposits. Their location in the idealized longitudinal cross section shown in Figure 9 is shown.

Sector 2

Deposits of Unit 1 (unit V1 in [22]; unit Q and FL in [64]) are largely exposed in the northern part of the sector 2 (Figures 9, 10 and 11A). In the Castello di Montalto area, they cut the pre-Neogene bedrock, forming a sedimentary body wide ca 1.2 km and thick 60 to 70 m. In the Arcidosso area, they cut the Pliocene marine deposits and they show a similar thickness but increased width to ca 2.5 km. These deposits consist of gravels with subordinate sand and mud. Gravels consist of multilaterally arranged channelized bodies, which are up to 4.5-m thick and several tens of meters wide in sections perpendicular to the palaeoflow direction (Figure 11A). Gravels form large-scale inclined beds, which are mainly made of clast-supported pebbles and sandy pebbles. Pebbly beds are up to 50-cm thick and they show a well-developed plane-parallel to planar-cross stratification, which is commonly highlighted by alternation of matrix-free and matrix-rich strata. Sandstone pebbles and cobbles from the Macigno Fm are deeply weathered. Channelized sandy deposits are less common, and they mainly occur in the upper part of the unit in the sector 2. Sand beds are normally graded, erosively based and they exhibit plane parallel-stratification and ripple cross-lamination. Massive overbank mud is exposed in the proximal reach of the sector 2 (Figure 11A). These muds draped the whole valley and they are up to 5–6 m thick. They include organic-rich layers and pedogenic horizons with scattered root traces and carbonate concretions.

Figure 11. Unit 1 deposits. (**A**) Channelized gravel and overlying overbank mud exposed in the Arcidosso area. (**B**) Channelized sand overlying marine Pliocene deposits in the Sentino Basin area. (**C**) Poorly organized gravel in the Sentino Basin area.

Sector 3

Deposits of Unit 1 occurs in the sector 3 as an isolated, N-S trending lithosome overlying the pre-Neogene bedrock and Pliocene marine deposits in the east to Rapolano Terme. This sedimentary body is thick at least 20 m and it consists of clast-supported gravels grading southward into sand with subordinate gravels. Gravels occur above Mesozoic bedrock and they consist of moderately to well-rounded boulders of metric size? (Figure 11C). They are poorly organized exhibiting a gravelly to sandy matrix. Sandstone clasts from the Macigno Fm are deeply weathered, as those occurring in the sector 2. Sandy deposits occur above Pliocene marine sand and they are 4-m-thick channelized bodies (Figure 11B). These bodies are made of medium-grained sand with sets of large-scale inclined beds dipping at 5 to 20°. Inclined beds are characterized by a plane parallel and ripple cross-lamination. These beds are floored by channel lag gravels, which include deeply weathered sandstone pebbles of the Macigno Fm.

Palaeoflow Data

In the sector 1, deposits of Unit 1 are not exposed, but subsurface investigations show that they are enclosed in a N-S trending depression up to a few hundred meters wide (Figure 5). In the sector 2, deposits of Unit 2 form a N-S trending, elongated body that bends eastward, where alluvial deposits overlie Pliocene sediments. Palaeoflow of these deposits (Figure 12) is consistent with the orientation of the elongated body and it changes from southward to eastward, where the studied deposits cover the Pliocene sediments. In the sector 3, Unit 1 forms a N-S trending body and the palaeoflow is directed southward in both gravelly and sandy facies.

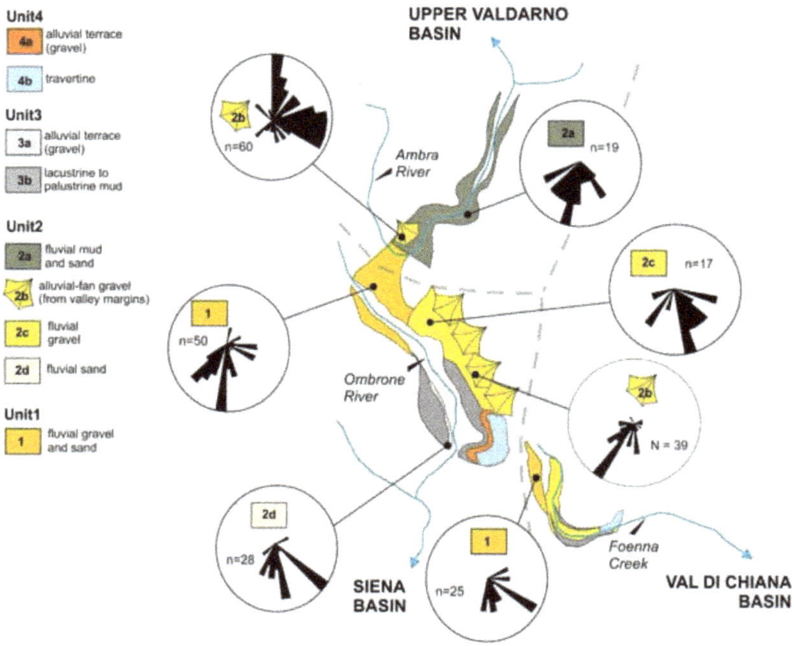

Figure 12. Paleocurrents distribution within the study deposits. Grey dashed line represents the watershed displayed in Figure 3.

Age

Organic-rich portions of overbank mud exposed in the Arcidosso area bear gastropod shells, including terrestrial *Pomatia elegans* and *Retinella* sp. [64]. The age distribution of this fauna spans from the earliest Gelasian to the Recent [65]. Palaeomagnetic studies revealed that a reverse magnetic polarity turned into a normal one at ca 25 m from the base of the unit [23].

4.2.2. Unit 2

Sector 1

Deposits of Unit 2 are confined within the same elongated palaeovalley that encloses the Unit 1 gravels occurring on top of them. Unit 2 deposits consist of mud with subordinate sand (V2fla in [22]), and interfingers with pebble to cobble-sized gravel in La Selva area (V2af in [22]). Mud is grey to dark grey and it is bedded or form massive tabular strata (Figure 13A). Greyish mud contains root traces and commonly root-bioturbated, whereas dark-grey mud is significantly enriched in organic matter and plant debris. Mud includes sandy deposits, which form tabular or channelized bodies. Tabular bodies thick up to 1.5 m consist of massive to plane parallel-stratified, medium-grained to fine-grained sand. Channelized bodies are up to 4 to 5 m thick and at least 50 to 60 m wide along-strike. Channelized sand consists of fining-upward bedsets of pebbly, coarse-grained sand grading into mud. These sandy bodies feature metric-scale inclined strata dipping at 5 to 10°. Inclined beds show plane parallel and trough cross-stratification and ripple cross-lamination. Gravels occurring in La Selva area consist of clast-supported, moderately to well-rounded, pebbles to boulders with subordinate sandy intercalations. Gravels form channelized units, up to 1.2–2 m thick. These units show an overall fining-upward trend and a range from massive at the base to crudely plane-parallel stratified at the top. In plain view, these gravels are organized in a fan shape, which has its apex on the eastern flank of the modern valley.

Figure 13. Deposits of Unit 2 and 3. (**A**) Massive overbank mud of Unit 2 exposed upstream of fault F1. (**B**) Fluvial gravels of Unit 2 overlying Pliocene marine deposits just downstream of fault F1. (**C**) Alluvial-fan gravel and sand (Unit 2) sourced from the bedrock forming the footwall block of Terre Rosse fault (Sector 2). (**D**) Sandy bar deposits (Unit 2) of the Pian di Bari area. (**E**) Scheme showing deposits of Units 2 and 3 in the Pian di Bari outcrop. (**F**) Fluvial gravelly sand of Unit 2 erosionally overlying Pliocene marine deposits. (**G**,**H**) Horizontally bedded palustrine deposits of Unit 3.

Sector 2

Deposits of Unit 2, are exposed in the northern part of sector 2 and consists of fluvial (V2flb in [22]) and alluvial fan (V2af in [22]) deposits. Fluvial deposits consist of gravels grading eastward into sand (Figure 13B) and are confined within a 2–2.5 m wide and N-S trending palaeovalley. Gravelly fluvial deposits are ca. 15 m thick and from 1 to 2 m thick, multi-storey channel bodies, commonly floored by a(t)b(i)-imbricated pebbles and scattered, angular boulder-sized clasts. Gravel ranges from plane-parallel to cross-stratified. Fluvial sand is characterized by multilateral channel bodies up to 5–6 m thick (Figure 13E). These bodies are made of medium-grained sand floored by fine pebbles (Figure 13F), which grade upward into pedogenized fine-grained sand and mud. Channelized sand is structured into sets of large-scale inclined beds dipping at 5 to 20° (Figure 13D). Inclined beds are characterized by a plane parallel, trough cross-stratification and ripple cross-lamination. The overlying mud displays evidence of pedogenesis. Alluvial-fan de-posits are sited along the eastern side of the palaeovalley and are sourced by the relief generated by the Terre Rosse Fault. Alluvial-fan deposits mainly consist of pedogenized sand that passes upward into channelized gravels (Figure 13C). Sandy deposits consist of vertically stacked, tabular beds which range from massive to slightly plane-parallel stratified. Channelized gravels are commonly floored by a(t)b(i)-imbricated cobbles with pebbles organized in plane-parallel to cross-stratification.

Sector 3

Deposits of Unit 2 do not occur in the eastward reach of the sector 3, where pre-Neogene bedrock crops out in the thalweg of the Foenna Creek. The western part of the sector 3 (i.e., Sentino Basin) is characterized by a widespread cover of Holocene deposits (Figure 5), but evidence for the occurrence of unit 2 arise from new borehole data (Figure 10). These data show that in this area the Unit 2 occurs below the Holocene alluvium of the Foenna Creek and consists of ca. 60 m thick gravel and sand.

Palaeoflow Data

Palaeocurrents from fluvial deposits in the sectors 1 and 2 point to an overall palaeoflow directed toward south (Figure 12). Alluvial-fan systems are associated with transverse supplies, which derive from west and east in the sector 1 and 2, respectively. No evidence of palaeoflow direction can be detected in the sector 3, since Unit 2 deposits have only been detected in the subsurface.

Age

No age constraints exist for deposits forming this unit.

4.2.3. Unit 3

Sector 1

The Unit 3 in sector 1 is represented by alluvial terraces of the Ambra River (Figure 6). These deposits are up to 2 m thick and are largely covered by Holocene colluvium. They mainly consist of channelizes sand covering a gravelly lag.

Sector 2

The Unit 3 is up to 15 m thick and it mainly occurs in the southern part of the sector. It is mainly represented by lacustrine and palustrine mud, with subordinate sandy intercalation, which are locally common in the upper part of the unit. In the Pian di Bari area, it is a few meters thick and cover overbank mud of Unit 2 (Figure 13E,G). In this area, it consists of palustrine, organic-rich mud (Figure 13H) with sandy intercalations containing abundant freshwater molluscs. In the southernmost part of the sector 2, the Unit 3 uncomfortably overlays Pliocene marine deposits and it consists of lacustrine mud grading upward into deltaic sandy beds. Lacustrine mud is tightly laminated including layers of well-sorted fine sand. These lacustrine layers are intensely deformed by dewatering structures and they are cut by syn-sedimentary normal faults (Figure 14). Overlying deltaic sandy beds are mainly tabular and characterized by widespread plane-parallel stratifications. Locally, they 0.5 to 1.5 m thick channelized bodies, which are paved by gravelly lags; they developed as distributary channels of small-scale deltaic systems locally sourced from the Rapolano Terme area. In the southernmost part of the sector 2, the flat morphological surface capping the lacustrine deposits defines a N-S oriented flange, that is ca. 6 km long and 0.5–1.5 km m wide. Such a flange is currently drained by the southward-flowing Bestina Creek, a 1–2 m wide rill that has been reclaimed by agricultural activities. No outcrops occur along the creek, but the modern soil flooring the flange includes abundant gravelly deposits.

Figure 14. Soft-sediment deformation structures in lacustrine sandy silt (Unit 5b); (**a**) detached highly-non cylindrical fold; (**b**) shear bands and related deformed laminations; (**c**,**d**) slump sheets; (**e**) slump sheets and deformed laminations; (**f**) clastic dykes.

Sector 3

In this sector, Unit 3 is up to 25–30 m thick but is poorly exposed. It mainly consists of mud with subordinate sands. Muddy deposits are massive, with organic-rich layers and local evidence of pedogenic processes (e.g., caliche). Sandy layers are up to 1.5 m thick and they show erosive bases floored by fine-grained gravels. The geometry of these sandy bodies cannot be defined because of paucity of outcrops.

Palaeoflow Data

In sector 1, the Unit 3 is represented by terraced deposits of the Ambra River, indicating a northward-directed palaeoflow, in agreement with the course of the modern Ambra River. In sectors 2 and 3, no palaeoflow data are available for Unit 3 deposits mainly due to their fine-grained nature and paucity of exposures.

Age

No age constraints have been detected for the deposits forming this unit.

4.2.4. Unit 4

Sector 1

Unit 4 in sector 1 is represented by alluvial terraces of the Ambra River (Figure 6). These consists of are 1–2 m thick sand and are largely covered by Holocene colluvium, which prevents from a detailed description. A precise boundary between terraces of Unit 3 and 4 cannot be defined.

Sector 2

The area is characterized by the presence of thick successions (up to 50 m) of continental carbonates (Figure 15A,C) and subordinate fluvial terraces (Figure 15B). These carbonates (mostly travertine, secondarily calcareous tufa and lacustrine deposits [55]) are widely diffused along all the basin margins (Figure 5). They were dated through U–Th isotopic composition (cf. [66,67]) and originated from several thermal springs in the area [68,69]. Facies association [51–56,70–75] evidences a large variability in their depositional environments, generally evolving from proximal systems (typically close to vents, like mounds and fissure ridges), to intermedi-ate (slopes and channels) up to distal (marsh, shallow lakes and transitional to alluvial plains) [76]. Local sections show the direct superposition of the carbonates to alluvial/fluvial clastic deposits (Figure 10). In some areas, the carbonate deposits are arranged in different quotes corresponding to at least three depositional terraces (Figure 10).

Figure 15. Deposits of Units 3 and 4. (**A**) Travertine of Unit 3 unconformably overlying lacustrine sand of Unit 3. (**B**) Fluvial terraces of the Ombrone River (Unit 4) covering Pliocene marine mud. (**C**) Travertines of Unit 4 in the Rapolano area.

Sector 3

Differently to Sector 2, the Sector 3 shows a patchy distribution of the continental carbonates (dominantly travertine) deposited in similar depositional environments, but with reduced extent (Figure 5) and thickness (up to 15 m). Also in this sector, travertines rest on fluvial/alluvial deposits (sands and conglomerates) and they evidence a distribution in at least three depositional terraces.

Palaeoflow Data

The alluvial deposits belonging to Unit 4 were accumulated by a drainage system configured similarly to that one of the present-day drainage. Likewise, the travertine deposition shows a dominant W-SW direction in the Sector 2 (from the basinal shoulder towards the Ombrone River hydrographical pattern) and an E-SE direction in the Sector 3 (towards the Sentino Basin and Foenna Creek)

Age

Sixteen travertine samples were collected in the area for radiometric analysis (Table 1 and Figure 16). All samples derived from macrocrystalline low porous material, avoiding the presence of secondary cements.

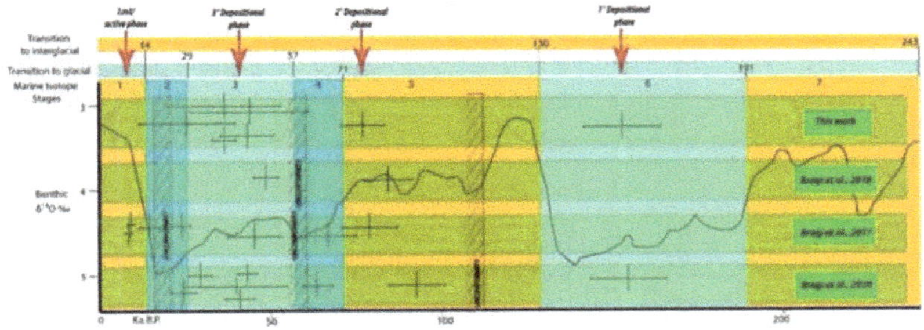

Figure 16. Comparison between available paleoclimate and travertine data of Rapolano Terme area. The new radiometric data here achieved (first line) are compared with previous ones [51,53,54] and recognized unconformities. Travertine data are reported with their standard deviation. Black line is the SPECMAP marine paleoclimatic δ18O record during late Pleistocene time [66]. It is evidenced the presence of at least three previous depositional phases. See text for details.

Due to different U and Th-concentrations, three samples resulted impossible to date, one not useful (infinite age error) and six samples with a very high error range. Combining the resulting data with stratigraphic and radiometric data, already published on the same travertine [51,53,54,74], it is possible to evidence a cluster of four main depositional phases separated by unconformities and succeeding from the latest Middle Pleistocene up to now.

Table 1. U–Th isotopic compositions and contents and ages by MC-ICPMS. For location of samples CTSE1–16, see Figure 6. a: $[^{238}U] = [^{235}U] \times 137.818$ (±0.65‰) [67]; $d^{234}U = ([^{234}U/^{238}U]$ activity − 1) × 1000. b: d^{234}Uinitial corrected was calculated based on ^{230}Th age (T), i.e., d^{234}Uinitial = d^{234}Umeasured X el^{234}*T, and T is corrected age. c: $[^{230}Th/^{238}U]$ activity = 1 − $e\text{-}l^{230}T$ + (d^{234}Umeasured/1000)[l230/(l230 − l234)](1 − e-(l230 − l234) T), where T is the age. Decay constants are 9.1705×10^{-6} year^{-1} for ^{230}Th, 2.8221×10^{-6} year^{-1} for ^{234}U [68,69], and 1.55125×10^{-10} year^{-1} for ^{238}U [70]. d: Age corrections, relative to chemistry date on 8 November 2018, were calculated using an estimated atomic $^{230}Th/^{232}Th$ ratio of 2 (±2) × 10^{-6}.

Sample ID	Weight g	^{238}U ppb [a]	^{232}Th ppt	δ^{234}U Measured [a]	$[^{230}Th/^{238}U]$ Activity [c]	$^{230}Th/^{232}Th$ Atomic (×10^{-6})	Age (yr Ago) Uncorrected	Age (year Ago) Corrected [c,d]	Age (year BP) Relative to 1950 AD	$\delta^{234}U_{initial}$ Corrected [b]
CTSE-1	0.04874	35.054 ± 0.075	27,753 ± 136	362.3 ± 6.2	0.519 ± 0.017	10.82 ± 0.35	51,144 ± 2069	43,474 ± 8222	43,405 ± 8222	410 ± 12
CTSE-2	0.05378	94.70 ± 0.18	708,677 ± 19,326	194.1 ± 3.4	0.806 ± 0.088	1.78 ± 0.20	117,345 ± 22,358	−33,993 —	−34,062 —	176 ± 94
CTSE-3	0.05536	45.624 ± 0.088	173,796 ± 2278	299.6 ± 4.9	0.653 ± 0.047	2.83 ± 0.20	73,685 ± 7309	27,761 ± 68,463	27,692 ± 68,463	324 ± 45
CTSE-4	0.05735	34.005 ± 0.067	43,933 ± 292	340.2 ± 5.1	0.416 ± 0.017	5.31 ± 0.23	39,862 ± 1997	26,651 ± 14,333	26,583 ± 14,333	367 ± 15
CTSE-5	0.05517	4.611 ± 0.019	51,737 ± 368	71 ± 23	0.123 ± 0.039	0.181 ± 0.057	13,297 ± 4496	—	—	—
CTSE-6	0.07041	3.205 ± 0.010	8376 ± 24	312 ± 22	0.773 ± 0.034	4.88 ± 0.22	92,641 ± 6725	64,393 ± 34,439	64,324 ± 34,439	374 ± 41
CTSE-7	0.04816	16.914 ± 0.049	53,671 ± 381	248.5 ± 7.8	0.812 ± 0.044	4.22 ± 0.23	109,143 ± 9762	71,545 ± 50,131	71,477 ± 50,131	304 ± 36
CTSE-8	0.06769	45.21 ± 0.11	32,168 ± 164	427.8 ± 5.6	0.792 ± 0.017	18.35 ± 0.39	83,872 ± 2582	77,570 ± 6962	77,501 ± 6962	532 ± 12
CTSE-9	0.05669	10.124 ± 0.029	14,796 ± 57	292.8 ± 9.9	0.503 ± 0.023	5.68 ± 0.26	52,667 ± 3085	37,095 ± 17,221	37,026 ± 17,221	325 ± 18
CTSE-10	0.05010	36.926 ± 0.064	12,819 ± 48	612.3 ± 6.3	0.5119 ± 0.0099	24.31 ± 0.48	40,602 ± 952	37,841 ± 2952	37,772 ± 2952	681 ± 9.0
CTSE-11	0.16055	92.95 ± 0.20	130,077 ± 1453	248.4 ± 3.9	0.538 ± 0.022	6.34 ± 0.27	60,234 ± 3249	44,822 ± 17,044	44,753 ± 17,044	282 ± 13
CTSE-12	0.05824	41.637 ± 0.081	251,379 ± 3270	220.3 ± 5.8	1.079 ± 0.057	2.95 ± 0.16	204,833 ± 30,052	118,921 ± 84,188	118,852 ± 84,188	308 ± 86
CTSE-13	0.0938	37.936 ± 0.064	169,548 ± 2313	169.9 ± 2.2	1.304 ± 0.062	4.81 ± 0.24	—	—	—	—
CTSE-14	0.0529	49.854 ± 0.089	174,281 ± 2521	167.1 ± 2.6	1.194 ± 0.063	5.63 ± 0.31	393,522 ± ∞	356,719 ± ∞	356,649 ± ∞	457.1 ± ∞
CTSE-15	0.0708	30.408 ± 0.058	31,495 ± 208	154.3 ± 2.8	0.931 ± 0.027	14.82 ± 0.44	167,748 ± 11,008	156,134 ± 15,886	156,064 ± 15,886	239.7 ± 11.6
CTSE-16	0.0541	378.76 ± 0.58	921,367 ± 32,039	206.6 ± 2.1	0.731 ± 0.043	4.95 ± 0.34	97,950 ± 9113	68,881 ± 35,800	68,812 ± 35,800	250.9 ± 22.3

5. Discussion

5.1. Palaeo-Drainage Evolution and Age

The confinement of the fluvial deposits along a narrow N-S trending depression, along with the dominance of palaeoflow parallel to the axis of this depression provide solid insights to confirm the occurrence of an ancient southward-draining valley in the study area [45]. This palaeovalley entered the western side of the Val di Chiana Basin in the Rigomagno area (Figure 2), where Gelasian to Calabrian valley-fill deposits were described [45].

The complex spatial distribution of the valley-fill deposits and the remarkable changes in their thickness, are characteristics explained here as a complex interplay between tectonics and sedimentation (Figure 17). The depositional history of this palaeovalley is reconstructed, with particular emphasis on the role played by the major faults (F1 to F3 in Figure 2) in controlling the aggradation and the avulsion of the related fluvial systems.

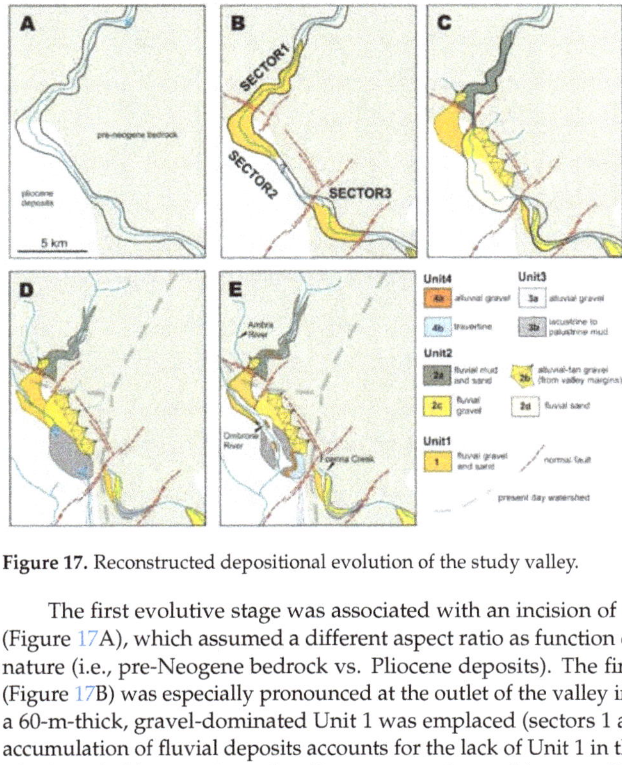

Figure 17. Reconstructed depositional evolution of the study valley.

The first evolutive stage was associated with an incision of the main valley system (Figure 17A), which assumed a different aspect ratio as function of the hosting substrate nature (i.e., pre-Neogene bedrock vs. Pliocene deposits). The first aggradational phase (Figure 17B) was especially pronounced at the outlet of the valley in the Siena Basin, where a 60-m-thick, gravel-dominated Unit 1 was emplaced (sectors 1 and 2). Such a localized accumulation of fluvial deposits accounts for the lack of Unit 1 in the distal part of sector 2, which probably experienced sediment starvation and bypass. The occurrence of Unit 1 in the proximal part of sector 3, suggests that the river outflowed from the Siena Basin dismantling its eastern rocky shoulder and producing an exceeding sediment load that locally congested the valley.

The onset of deposition of Unit 2 (Figure 17C) occurred as a consequence of the activations of F1-F3 faults, which strongly modified the valley bottom profile and caused the development of localized depocenters and valley shift in the proximal part of sector 2. The movement of the F1 fault caused the subsidence in sector 1, inducing an overall decrease in transport capability of the valley, with accumulation of overbank-dominated fluvial deposits within the valley and gravel deposition restricted to tributary inlets along the valley flanks. As the area sited downstream of F1 fault underwent uplift, the axis of fluvial incision within the valley progressively shifted towards the hydrographic left, a process ascribed to the increase of the F1 fault displacement towards the south-east. The progressive dismantling of this uplifted block caused aggradation of gravelly and sandy deposits in the proximal and distal parts of the sector 2, respectively. In sector 2, the shift of the major river occurred in parallel with the activation of the Terre Rosse Fault (Figures 2 and 4), that caused a development of alluvial fans along the left-hand side of the valley flank/north flank of the valley? Although exposures are limited at the present day, similar processes occurred in association with fault F2 and F3. Specifically, the coarse sediments produced by the dismantling of the F2 footwall were trapped in the subsiding area developed upstream of F3 fault, as attested by borehole data.

The deposition of Unit 3 was associated with a dramatic reorganization of the river network (Figure 17D). The major river abandoned the valley, and in sector 1 the Ambra River started to flow northward following the establishment of the modern watershed between the Upper Valdarno Basin and the Siena Basin. This caused incision and terracing of Unit 2 mud in the modern Ambra River valley. The Ombrone River was the only tributary of the major valley, that was flooded in sector2 as a consequence of damming due to F2 fault. The lake had probably an emissary in its southern termination, as suggested by the gravelly-floored abandoned valley now drained by the Bestina Creek. The fragmentation of the valley caused also a development of palustrine conditions in sector 3, that was dammed by F3 fault. Intense tectonic activity during this phase is also testified by the occurrence of syn-sedimentary faults and dewatering structures in the lacustrine deposits of the NE to Rapolano Terme.

The last depositional stage was associated with the establishment of the present-day drainage (Figure 17E). The Ambra River kept terracing its modern valley, mainly following entrenching of the Arno River in the Valdarno Basin [22]. The lake developed in sector 2 was emptied following a capture from a SE draining watercourse. This event caused a deflection of the Ombrone River toward SW along with the abandonment of the lake emissary that nowadays is drained by the Bestina Creek. The fluvial incision and the terracing of sector 2 was followed by accumulation of travertines, which were formed from precipitation of thermal water sourced by fault F2 and F3.

The proposed reconstruction frames the evolution of the studied palaeovalley between the regional marine regression and the accumulation of travertine and coeval fluvial terraced deposits (i.e., Unit 4), that occurred at 2.5 my and 40–150 kyrs, respectively. The diversion of the Ambra River toward north, at the transition between Unit 2 and 3, followed the entrance of the Arno River in the Upper Valdarno Basin just before the Matuyama/Brunhes transition [31,33,36] at 0.78 Ma. In this frame, the change from reverse to normal polarity detected by [22] in the middle part of the Unit 1 could be ascribed to the base of either Olduvai or Jaramillo sub-chrons. However, if marine sedimentation persisted in the Siena Basin until latest Gelasian, as suggested by [42], such a reverse polarity would correspond to the base of the Jaramillo sub-chron.

The thickness of the channel bars in the fluvial deposits of Unit 1 and 2 indicate that the palaeo-channels draining the valley were up to 4–5 m deep. Using a palaeo-hydrological approach, [22] estimated a bankfull discharge of ca. 266 $m^3 s^{-1}$ for a palaeo-channel of Unit 1. Although the river discharge strongly depends by local runoff, the value is comparable with that one of the modern Arno River at its entrance in the Upper Valdarno Basin, suggesting that the studied palaeovalley was drained by a major fluvial system. Nevertheless, further geological investigations and detailed provenance studies are required to reconstruct the origin of this fluvial system.

5.2. *Lifecycle of an Intermontane Fluvial Valley*

The incision of the studied valley system occurred after a tectonically-driven, regional uplift that caused a marine forced regression in the whole inner sector of Northern Apennines [16]. Therefore, the development of the valley matches with classical stratigraphic models, that relate the incision of valleys in coastal areas to episodes of base level fall and a dramatic subtraction of accommodation space [1]. These models also predict a wide spectrum of processes to fill incised valleys in coastal areas, although all of them required a relative sea-level rise, which allowed an increase of available accommodation space within the valley [3], heading its progressive backfilling according with the available sediment supply [1]. The infill of the valley cannot be ascribed to a relative sea-level rise since regional doming, in combination with the occurrence of inherited Neogene morpho-structural sills, prevented the sea from reentering southern Tuscany after the late Piacentian forced regression [16,77]. The aggradation within the valley can be ascribed to an unsteady sediment balance, and specifically to a progressive increase of sediment influx in comparison with the ebb [1]. This unsteadiness in sediment balance was differently manifested during accumulation of Unit 1–4. The tectonic uplift of the Chianti Ridge (i.e., one of the source area of the study palaeo-drainage) occurred at the Piacentian to Gelasian transition and it is represented by progressive unconformities in the basin-fill succession of the neighboring Upper Valdarno Basin [78]. Such an uplift enhanced the amount of sediment supplied into the valley and it triggered aggradation of Unit 1, especially at the entrance of the valley in the Siena Basin (sectors 1 and 2), where an abrupt decrease of gradient occurred. In sector 2, the valley-scale muddy overbank deposits might have been accumulated in response to an upstream gravel trapping, due to the localized decrease in the valley gradient associated with the activation of the F1 fault system. After dropping most of its load at the entrance of the Siena Basin, the river entrained new bedload crossing the F3 fault zone and it dropped it just downstream following a further morphotectonic-induced gradient decrease. Localized sediment trapping was furtherly enhanced during accumulation of Unit 2. At this stage the rate of F1–3 footwall uplift outpaced that one of the river embanking, forming localized barriers that strongly impacted on river transport capability and fluvial styles (cf. [75]). The desertion of the major river from the valley enhanced modifications of depositional dynamics during the accumulation of Unit 3 and 4, when the sector 1 hosted a reversed (i.e., northward) and shrank fluvial drainage (i.e., Ambra River), and sectors 2–3 acted as small fluvio-palustrine basins.

The reconstructed depositional history shows that incision and filling of the valley were, therefore, linked to a relative-sea level drop and morpho-tectonic forcings, respectively. This marks the difference between the studied succession and most of the documented valley fills, highlighting that forcings causing incision and filling of a valley do not have to be genetically related [79]. Additionally, the occurrence of fluvial, lacustrine and travertines deposits of Units 3 and 4 shows that the process of tectonic segmentation of the major valley might cause a large spatial differentiation of depositional dynamics which contributes to accumulate successions that are hardly identified in the rock record as the infill of ancient valleys. The wide spectrum of deposits filling the study valley is also a precious archive of climate and tectonic events affecting the area since latest Piacentian.

The three main depositional phases recognized in travertine and terraced deposits of Unit 4 assist in the understanding of the complex relationship between local tectonic uplift, fluvial morphodynamics and climate. The travertine deposition, that is often related to humid periods allowing the recharge of thermal reservoir, is an important evidence in this complexity. This situation was illustrated in other Italian travertine examples [80–83] evidencing how such recharge is directly related to the climatic conditions of the odd-numbered Marine Isotope Stages. Also in Rapolano Terme area, the sum of the available data evidence this relation and how, on the short period, the climate exerts a fundamental control on travertine deposition and on the development of alluvial/aggradational phases.

During the Pleistocene tectonic evolution of the inner Northern Apennines, relationships between fault activities and the evolution of the continental sedimentation strongly support a sedimentary response related to the tectonic pulses during the Quaternary. In particular, the activity of the Querciolaia and Violante-San Gimignanello faults (Figure 2), as well as the whole fault segments forming the "Arbia-Valmarecchia" shear zone, controlled the drainage and fluvio-lacustrine sedimentation during the Pleistocene and Holocene, thus attesting their primary role in controlling the continental sedimentation and the hydrographic pattern from Middle Pleistocene to Holocene. This reinforces the fact that, according to [52], the study area was affected by significant tectonic activity during Pleistocene-Holocene times, in response to a transtensive (mainly left-lateral) stress field [51,53,55]. Soft-sediment deformation structures (i.e., seismites) triggered by ground shacking present in the lacustrine sediments of the Unit 3 (Figure 13), indicate that the tectonic activity was accompanied by seismic events during the Middle Pleistocene. This completes the seismotectonic framework bringing back in time the beginning of the seismic activity in the study area, before referred to an oldest age of 84 ± 8–48.92 ± 5.08 ka [54]. This adds fundamental information for the better understanding of the seismotectonic framework and it sheds lights on the seismic potential of this sector of Tuscany, considered of modest interest with respect to seismic hazard and risk issues.

6. Conclusions

A Pleistocene palaeovalley and related deposits were recognized and mapped over a distance of ca. 35 km at the border between three major extensional intermontane basins of the Northern Apennines (Upper Valdarno, Siena and Val di Chiana Basin). This palaeovalley was cut following the regional late Piacentian marine forced regression, and it was filled since Gelasian following the intense tectonic activity that caused a progressive increase of sediment influx in comparison to the ebb. The major results produced from this work can be summarized as follow:

- Forcings causing incision and filling of a valley do not have to be genetically related. Although the study valley was cut as consequence of a relative sea-level fall, alluvial aggradation valley was not driven by a relative sea-level rise but resulted from an increase in sediment supply and a tectonic-driven sediment trapping within the valley.
- Tectonic segmentation of a valley can cause a wide spectrum of depositional environments within the major valley trunk. The occurrence of fluvial, lacustrine and travertines deposits in the studied valley-fill succession account for this process and highlight that such a large spatial variability of sedimentary facies can hinder detection of ancient valleys in the rock record.
- The wide spectrum of deposits filling the tectonically-segmented valley is an archive for tectono-climatic events. Specifically, it arises that the inner sector of Norther Apennines was affected by a significant tectonic activity during Pleistocene-Holocene times, in response to a transtensive stress field.

Author Contributions: Conceptualization, M.G., A.B., E.C., M.A., V.B.; formal analysis, T.-L.Y., C.-C.S.; investigation, M.G., A.B., E.C., M.A., V.B.; writing—original draft preparation, M.G., A.B., E.C.; writing—review and editing, M.G., A.B., E.C., M.A., V.B. All authors have read and agreed to the published version of the manuscript.

Funding: This research was funded by University of Padova (DOR2018 funds M.G.).

Data Availability Statement: Data are available following request to corresponding author.

Acknowledgments: M.G., A.B. and E.C. authors are indebted with R. Pancino for the remarkable support in the field. Marta Cosma and two anonymous reviewers are acknowledged for reading and commenting an early version of this work. U-Th dating in the HISPEC was supported by grants from the Science Vanguard Research Program of the Ministry of Science and Technology (MOST), Taiwan, ROC (109-2123-M-002-001 to C.-C.S.), the Higher Education Sprout Project of the Ministry of Education, Taiwan, ROC (109L901001 to C.-C.S.), the National Taiwan University (110L8907 to C.-C.S.)

Conflicts of Interest: The authors declare no conflict of interest.

References

1. Zaitlin, B.A.; Dalrymple, R.W.; Boyd, R. The Stratigraphic Organization of Incised-Valley Systems Associated with Relative Sea-Level Change. In *Incised-Valley Systems: Origin and Sedimentary Sequences*; Dalrymple, R., Boyd, R., Zaitlin, B.A., Eds.; SEPM Society for Sedimentary Geology: Tulsa, OK, USA, 1994; Volume SEPM SP 51, pp. 45–60. ISBN 9781565760905.

2. Corner, G.D. A Transgressive-Regressive Model of Fjord-Valley Fill: Stratigraphy, Facies and Depositional Controls. In *Incised Valleys in Time and Space*; Dalrymple, R., Leckie, D., Tillman, R., Eds.; SEPM Society for Sedimentary Geology: Tulsa, OK, USA, 2006; Volume SEPM SP 85, pp. 161–178, ISBN 1565761227.
3. Simms, A.R.; Anderson, J.B.; Taha, Z.P.; Rodriguez, A.B. Overfilled versus Underfilled Incised Valleys: Examples from the Quaternary Gulf of Mexico. In *Incised Valleys in Time and Space*; Dalrymple, R., Leckie, D., Tillman, R., Eds.; SEPM Society for Sedimentary Geology: Tulsa, OK, USA, 2006; Volume SEPM SP 85, pp. 117–139, ISBN 1565761227.
4. Deibert, J.E.; Camilleri, P.A. Sedimentologic and tectonic origin of an incised-valley-fill sequence along an extensional marginal-lacustrine system in the Basin and Range province, United States: Implications for predictive models of the location of incised valleys. *Am. Assoc. Pet. Geol. Bull.* **2006**, *90*, 209–235. [CrossRef]
5. Garrison, J.R., Jr.; van den Bergh, T.C.V. Effects of Sedimentation Rate, Rate of Relative Rise in Sea Level, and Duration of Sea-Level Cycle on the Filling of Incised Valleys: Examples of Filled and overfilled Incised Valleys From the Upper Ferron Sandstone, Last Chance Delta, East-Central Utah. In *Incised Valleys in Time and Space*; Dalrymple, R., Leckie, D., Tillman, R., Eds.; Special Publications of SEPM: Tulsa, OK, USA, 2006; Volume SEPM SP 85, pp. 239–279.
6. Plink-Björklund, P.; Steel, R.J. Incised Valleys On An Eocene Coastal Plain And Shelf, Spitsbergen—Part of A Linked Shelf–Slope System. In *Incised Valleys in Time and Space*; Dalrymple, R., Leckie, D., Tillman, R., Eds.; SEPM Society for Sedimentary Geology: Tulsa, OK, USA, 2006; Volume SEPM SP 85, pp. 281–307, ISBN 1565761227.
7. Gobo, K.; Ghinassi, M.; Nemec, W.; Sjursen, E. Development of an incised valley-fill at an evolving rift margin: Pleistocene eustasy and tectonics on the southern side of the Gulf of Corinth, Greece. *Sedimentology* **2014**, *61*, 1086–1119. [CrossRef]
8. Boyd, R.; Dalrymple, R.W.; Zaitlin, B.A. Estuarine and Incised-Valley Facies Models. In *Facies Models Revisited*; Walker, R.G., Posamentier, H., Eds.; SEPM Society for Sedimentary Geology: Tulsa, OK, USA, 2006; Volume SEPM SP 84, pp. 171–235. ISBN 9781565763302.
9. Blum, M.D.; Törnqvist, T.E. Fluvial responses to climate and sea-level change: A review and look forward. *Sedimentology* **2000**, *47*, 2–48. [CrossRef]
10. Holbrook, J. Origin, genetic interrelationships, and stratigraphy over the continuum of fluvial channel-form bounding surfaces: An illustration from middle Cretaceous strata, Southeastern Colorado. *Sediment. Geol.* **2001**, *144*, 179–222. [CrossRef]
11. Vincent, S.J. The Sis palaeovalley: A record of proximal fluvial sedimentation and drainage basin development in response to Pyrenean mountain building. *Sedimentology* **2001**, *48*, 1235–1276. [CrossRef]
12. Gibling, M.R.; Fielding, C.R.; Sinha, R. Alluvial Valleys and Alluvial Sequences:Towards a Geomorphic Assessment. In *From River to Rock Record: The Preservation of Fluvial Sediments and Their Subsequent Interpretation*; Davidson, S.K., Leleu, S., North, C.P., Eds.; SEPM Society for Sedimentary Geology: Tulsa, OK, USA, 2011; Volume SEPM SP 97, pp. 423–447, ISBN 9781565763050.
13. Hain, M.P.; Strecker, M.R.; Bookhagen, B.; Alonso, R.N.; Pingel, H.; Schmitt, A.K. Neogene to Quaternary broken foreland formation and sedimentation dynamics in the Andes of NW Argentina (25° S). *Tectonics* **2011**, *30*, 1–27. [CrossRef]
14. Mey, J.; Scherler, D.; Zeilinger, G.; Strecker, M.R. Estimating the fill thickness and bedrock topography in intermontane valleys using artificial neural networks. *J. Geophys. Res. Earth Surf.* **2015**, *120*, 1301–1320. [CrossRef]
15. Wang, P.; Scherler, D.; Liu-Zeng, J.; Mey, J.; Avouac, J.-P.; Zhang, Y.; Shi, D. Tectonic control of Yarlung Tsangpo Gorge revealed by a buried canyon in Southern Tibet. *Science* **2014**, *346*, 978–981. [CrossRef]
16. Martini, I.P.; Sagri, M. Tectono-sedimentary characteristics of Late Miocene-Quaternary extensional basins of the Northern Apennines, Italy. *Earth-Sci. Rev.* **1993**, *34*, 197–233. [CrossRef]
17. Cosentino, D.; Asti, R.; Nocentini, M.; Gliozzi, E.; Kotsakis, T.; Mattei, M.; Esu, D.; Spadi, M.; Tallini, M.; Cifelli, F.; et al. New insights into the onset and evolution of the central Apennine extensional intermontane basins based on the tectonically active L'Aquila Basin (central Italy). *Bull. Geol. Soc. Am.* **2017**, *129*, 1314–1336. [CrossRef]
18. Bartolini, C.; D'Agostino, N.; Dramis, F. Topography, exhumation, and drainage network evolution of the Apennines. *Episodes* **2003**, *26*, 212–216. [CrossRef]
19. D'Agostino, N.; Jackson, J.A.; Dramis, F.; Funiciello, R. Interactions between mantle upwelling, drainage evolution and active normal faulting: An example from the Central Appennines (Italy). *Geophys. J. Int.* **2001**, *147*, 475–497. [CrossRef]
20. Piacentini, T.; Miccadei, E. The role of drainage systems and intermontane basins in the Quaternary landscape of the Central Apennines chain (Italy). *Rend. Lincei* **2014**, *25*, 139–150. [CrossRef]
21. Bartolini, C.; Pranzini, G. Plio-Quaternary evolution of the Arno basin drainage. *Z. Geomorphol. Suppl. Stuttg.* **1981**, *40*, 77–91.
22. Bianchi, V.; Ghinassi, M.; Aldinucci, M.; Boaga, J.; Brogi, A.; Deiana, R. Tectonically driven deposition and landscape evolution within upland incised valleys: Ambra Valley fill, Pliocene-Pleistocene, Tuscany, Italy. *Sedimentology* **2015**, *62*, 897–927. [CrossRef]
23. Bianchi, V.; Salles, T.; Ghinassi, M.; Billi, P.; Dallanave, E.; Duclaux, G. Numerical modeling of tectonically driven river dynamics and deposition in an upland incised valley. *Geomorphology* **2015**, *241*, 353–370. [CrossRef]
24. Liotta, D.; Cernobori, L.; Nicolicl, R. Restricted rifting and its coexistence with compressional structures: Results from the CROP 3 traverse (Northern Apennines, Italy). *Terra Nova* **1998**, *10*, 16–20. [CrossRef]
25. Carmignani, L.; Decandia, F.A.; Disperati, L.; Fantozzi, P.L.; Kligfield, R.; Lazzarotto, A.; Liotta, D.; Meccheri, M. Inner Northern Apennines. In *Anatomy of an Orogen: The Apennines and Adjacent Mediterranean Basins*; Vai, G.B., Martini, I.P., Eds.; Springer: Dordrecht, The Netherlands, 2001; pp. 197–213, ISBN 978-94-015-9829-3.

26. Carmignani, L.; Decandia, F.A.; Dispera, L.; Fantozzi, P.L.; Lazzarotto, A.; Oggiano, G.; Angioj, V.C.M. Relationships between the Tertiary structural evolution of the Sardinia-Corsica-Provençal Domain and the Northern Apennines. *Terra Nova* **1995**, *7*, 128–137. [CrossRef]
27. Nalin, R.; Ghinassi, M.; Foresi, L.M.; Dallanave, E. Carbonate deposition in restricted basins: A Pliocene case study from the central Mediterranean (Northwestern Apennines), Italy. *J. Sediment. Res.* **2016**, *86*, 236–267. [CrossRef]
28. Brogi, A.; Liotta, D. Highly extended terrains, lateral segmentation of the substratum, and basin development: The middle-late Miocene Radicondoli Basin (inner northern Apennines, Italy). *Tectonics* **2008**, *27*, 1–20. [CrossRef]
29. Barchi, M.R. The Neogene-Quaternary evolution of the Northern Apennines: Crustal structure, style of deformation and seismicity. *J. Virtual Explor.* **2010**, *36*. [CrossRef]
30. Sagri, M.; Magi, M. Il Bacino del Valdarno Superiore. *Soc. Geol. Ital. L'Appennino Settentr. Guid. Alle Escursioni Post-Congr.* **1992**, *201*, 226.
31. Albianelli, A.; Bertini, A.; Magi, M.; Napoleone, G.; Sagri, M. Il bacino Plio-Pleistocenico del Valdarno Superiore: Eventi deposizionali, paleomagnetici e paleoclimatici. *Quaternario* **1995**, *8*, 11–18.
32. Ghinassi, M.; Magi, M. Variazioni climatiche, tettonica e sedimentazione al passaggio Pliocene medio-Pliocene superiore nel bacino del Valdarno Superiore (Appennino Settentrionale). *Ital. J. Geosci.* **2004**, *123*, 301–310.
33. Fidolini, F.; Ghinassi, M.; Magi, M.; Papini, M.; Sagri, M. The Plio-Pleistocene fluvio-lacustrine Upper Valdarno Basin (Central Italy): Stratigraphy and Basin fill evolution. *Ital. J. Geosci.* **2013**, *132*, 13–32. [CrossRef]
34. Rook, L.; Croitor, R.; Delfino, M.; Ferretti, M.P.; Gallai, G.; Pavia, M. The Upper Valdarno Plio-Pleistocene vertebrate record: An historical overview, with notes on palaeobiology and stratigraphic significance of some important taxa. *Ital. J. Geosci.* **2013**, *132*, 104–125. [CrossRef]
35. Billi, P.; Magi, M.; Sagri, M. Coarse-Grained Low-Sinuosity River Deposits: Example from Plio-Pleistocene Valdarno Basin, Italy. In *Recent Developments in Fluvial Sedimentolology*; SEPM Society for Sedimentary Geology: Tulsa, OK, USA, 1987; pp. 197–203.
36. Fidolini, F.; Ghinassi, M.; Aldinucci, M.; Billi, P.; Boaga, J.; Deiana, R.; Brivio, L. Fault-sourced alluvial fans and their interaction with axial fluvial drainage: An example from the Plio-Pleistocene Upper Valdarno Basin (Tuscany, Italy). *Sediment. Geol.* **2013**, *289*, 19–39. [CrossRef]
37. Bonini, M.; Sani, F. Extension and compression in the Northern Apennines (Italy) hinterland: Evidence from the late Miocene-Pliocene Siena-Radicofani Basin and relations with basement structures. *Tectonics* **2002**, *21*, 1–35. [CrossRef]
38. Brogi, A. Bowl-shaped basin related to low-angle detachment during continental extension: The case of the controversial Neogene Siena Basin (central Italy, Northern Apennines). *Tectonophysics* **2011**, *499*, 54–76. [CrossRef]
39. Martini, I.; Aldinucci, M.; Foresi, L.M.; Mazzei, R.; Sandrelli, F. Geological map of the Pliocene succession of the Northern Siena Basin (Tuscany, Italy). *J. Maps* **2011**, *7*, 193–205. [CrossRef]
40. Martini, I.; Arragoni, S.; Aldinucci, M.; Foresi, L.M.; Bambini, A.M.; Sandrelli, F. Detection of detached forced-regressive nearshore wedges: A case study from the central-southern Siena Basin (Northern Apennines, Italy). *Int. J. Earth Sci.* **2013**, *102*, 1467–1489. [CrossRef]
41. Martini, I.; Aldinucci, M. Sedimentation and basin-fill history of the pliocene succession exposed in the Northern Siena-Radicofani Basin (Tuscany, Italy): A sequence-stratigraphic approach. *Riv. Ital. Paleontol. Stratigr.* **2017**, *123*, 407–432. [CrossRef]
42. Baldanza, A.; Bizzarri, R.; Famiani, F.; Garassino, A.; Hyžný, M.; Pasini, G. The bathyal decapod crustacean community from the Poggio i Sodi quarries (Siena Basin, Tuscany, Italy). *Bol. Soc. Geol. Mex.* **2013**, *65*, 335–353. [CrossRef]
43. Barchi, M.; Minelli, G.; Magnani, B.; Mazzotti, A. Line CROP 03: Northern Apennines. *Mem. Descr. Cart. Geol. d'It.* **2003**, *LXII*, 127–136.
44. Bizzarri, R.; Baldanza, A. Integrated stratigraphy of the marine Early Pleistocene in Umbria. *Geosciences* **2020**, *10*, 371. [CrossRef]
45. Aruta, G.; Borgia, A.; Bruni, P.; Cecchi, G.; Cipriani, N.; Tredici, Y. Pliocene and Pleistocene unconformity bounded stratigraphic units (UBSU) in Val di Chiana. In *The "Regione Toscana" Project of Geological Mapping: Case Histories and Data Acquisition*; Morini, D., Bruni, P., Eds.; Tipografia Martinelli: Bagno a Ripoli, Italy, 2004; pp. 133–136.
46. Rook, L.; Martínez-Navarro, B. Villafranchian: The long story of a Plio-Pleistocene European large mammal biochronologic unit. *Quat. Int.* **2010**, *219*, 134–144. [CrossRef]
47. Brogi, A. Late evolution of the inner Northern Apennines from the structure of the Monti del Chianti-Monte Cetona ridge (Tuscany, Italy). *J. Struct. Geol.* **2020**, *141*, 104205. [CrossRef]
48. Liotta, D. Analisi del settore centro-meridionale del bacino pliocenico di Radicofani (Toscana meridionale). *Boll. Soc. Geol. Ital.* **1996**, *115*, 115–143.
49. Pascucci, V.; Martini, I.P.; Sagri, M.; Sandrelli, F. Effects of Transverse Structural Lineaments on the Neogene–Quaternary Basins of Tuscany (Inner Northern Apennines, Italy). In *Sedimentary Processes, Environments and Basins: A Tribute to Peter Friend*; Nichols, G., Williams, E., Paola, C., Eds.; John Wiley & Sons: Hoboken, NJ, USA, 2007; pp. 155–182, ISBN 9781405179225.
50. Bambini, A.M.; Brogi, A.; Cornamusini, G.; Costantini, A.; Foresi, L.M.; Lazzarotto, A. Geologia dell'area di Rapolano Terme in provincia di Siena (Appennino Settentrionale). *Ital. J. Geosci.* **2010**, *129*, 457–495. [CrossRef]
51. Brogi, A.; Capezzuoli, E.; Aqué, R.; Branca, M.; Voltaggio, M. Studying travertines for neotectonics investigations: Middle-Late Pleistocene syn-tectonic travertine deposition at Serre di Rapolano (Northern Apennines, Italy). *Int. J. Earth Sci.* **2010**, *99*, 1383–1398. [CrossRef]

52. Brogi, A.; Capezzuoli, E.; Martini, I.; Picozzi, M.; Sandrelli, F. Late quaternary tectonics in the inner Northern apennines (Siena Basin, southern Tuscany, Italy) and their seismotectonic implication. *J. Geodyn.* **2014**, *76*, 25–45. [CrossRef]
53. Brogi, A.; Capezzuoli, E.; Kele, S.; Baykara, M.O.; Shen, C.C. Key travertine tectofacies for neotectonics and palaeoseismicity reconstruction: Effects of hydrothermal overpressured fluid injection. *J. Geol. Soc.* **2017**, *174*, 679–699. [CrossRef]
54. Brogi, A.; Capezzuoli, E.; Moretti, M.; Olvera-García, E.; Matera, P.F.; Garduno-Monroy, V.H.; Mancini, A. Earthquake-triggered soft-sediment deformation structures (seismites) in travertine deposits. *Tectonophysics* **2018**, *745*, 349–365. [CrossRef]
55. Brogi, A. Faults linkage, damage rocks and hydrothermal fluid circulation: Tectonic interpretation of the Rapolano Terme travertines (southern Tuscany, Italy) in the context of Northern Apennines Neogene-Quaternary extension. *Eclogae Geol. Helv.* **2004**, *97*, 307–320. [CrossRef]
56. Brogi, A.; Capezzuoli, E. Travertine deposition and faulting: The fault-related travertine fissure-ridge at Terme S. Giovanni, Rapolano Terme (Italy). *Int. J. Earth Sci.* **2009**, *98*, 931–947. [CrossRef]
57. Brogi, A.; Capezzuoli, E. Earthquake impact on fissure-ridge type travertine deposition. *Geol. Mag.* **2014**, *151*, 1135–1143. [CrossRef]
58. Liotta, D. The Arbia-Val Marecchia line, Northern Apennines. *Eclogae Geol. Helv.* **1991**, *84*, 413–430.
59. Shen, C.C.; Cheng, H.; Edwards, R.L.; Moran, S.B.; Edmonds, H.N.; Hoff, J.A.; Thomas, R.B. Measurement of attogram quantities of 231Pa in dissolved and particulate fractions of seawater by isotope dilution thermal ionization mass spectroscopy. *Anal. Chem.* **2003**, *75*, 1075–1079. [CrossRef]
60. Shen, C.C.; Wu, C.C.; Cheng, H.; Lawrence Edwards, R.; Hsieh, Y.T.; Gallet, S.; Chang, C.C.; Li, T.Y.; Lam, D.D.; Kano, A.; et al. High-precision and high-resolution carbonate ^{230}Th dating by MC-ICP-MS with SEM protocols. *Geochim. Cosmochim. Acta* **2012**, *99*, 71–86. [CrossRef]
61. Shen, C.C.; Lawrence Edwards, R.; Cheng, H.; Dorale, J.A.; Thomas, R.B.; Bradley Moran, S.; Weinstein, S.E.; Edmonds, H.N. Uranium and thorium isotopic and concentration measurements by magnetic sector inductively coupled plasma mass spectrometry. *Chem. Geol.* **2002**, *185*, 165–178. [CrossRef]
62. Cheng, H.; Edwards, R.L.; Shen, C.C.; Polyak, V.J.; Asmerom, Y.; Woodhead, J.; Hellstrom, J.; Wang, Y.; Kong, X.; Spötl, C.; et al. Improvements in ^{230}Th dating, ^{230}Th and ^{234}U half-life values, and U-Th isotopic measurements by multi-collector inductively coupled plasma mass spectrometry. *Earth Planet. Sci. Lett.* **2013**, *371–372*, 82–91. [CrossRef]
63. Brogi, A.; Fidolini, F.; Liotta, D. Tectonic and sedimentary evolution of the Upper Valdarno Basin: New insights from the lacustrine S. Barbara Basin. *Ital. J. Geosci.* **2013**, *132*, 81–97. [CrossRef]
64. Aldinucci, M.; Ghinassi, M.; Sandrelli, F. Climatic and tectonic signature in the fluvial infill of a late Pliocene valley (Siena Basin, Northern Apennines, Italy). *J. Sediment. Res.* **2007**, *77*, 398–414. [CrossRef]
65. Esu, D.; Girotti, O. Late Pliocene and Pleistocene assemblages of continental molluscs in Italy. A survey. *Quaternario* **1991**, *4*, 137–150.
66. Martinson, D.G.; Pisias, N.G.; Hays, J.D.; Imbrie, J.; Moore, T.C.; Shackleton, N.J. Age dating and the orbital theory of the ice ages: Development of a high-resolution 0 to 300,000-year chronostratigraphy. *Quat. Res.* **1987**, *27*, 1–29. [CrossRef]
67. Hiess, J.; Condon, D.J.; McLean, N.; Noble, S.R. ^{238}U/^{235}U systematics in terrestrial uranium-bearing minerals. *Science* **2012**, *335*, 1610–1615. [CrossRef] [PubMed]
68. Minissale, A.; Vaselli, O.; Tassi, F.; Magro, G.; Grechi, G.P. Fluid mixing in carbonate aquifer near Rapolano (central Italy): Chemical and isotopic constraints. *Appl. Geochem.* **2002**, *17*, 1329–1342. [CrossRef]
69. Minissale, A. Origin, transport and discharge of CO_2 in central Italy. *Earth-Sci. Rev.* **2004**, *66*, 89–141. [CrossRef]
70. Jaffey, A.H.; Flynn, K.F.; Glendenin, L.E.; Bentley, W.C.; Essling, A.M. Precision measurement of half-lives and specific activities of ^{235}U and ^{238}U. *Phys. Rev. C* **1971**, *4*, 1889–1906. [CrossRef]
71. Capezzuoli, E.; Gandin, A.; Pedley, H.M. Travertines and calcareous tufa in southern Tuscany (central Italy). In *27th IAS Meeting of Sedimentology, Alghero, Italy. Fieldtrip Guidebook*; Pascucci, V., Andreucci, S., Eds.; EDES: Alghero, Italy, 2009; pp. 129–158, ISBN 9788865280652.
72. Guo, L.; Riding, R. Hot-spring travertine facies and sequences, Late Pleistocene, Rapolano Terme, Italy. *Sedimentology* **1998**, *45*, 163–180. [CrossRef]
73. Guo, L.; Riding, R. Rapid facies changes in Holocene fissure ridge hot spring travertines, Rapolano Terme, Italy. *Sedimentology* **1999**, *46*, 1145–1158. [CrossRef]
74. Carrara, C.; Ciuffarella, L.; Paganin, G. Inquadramento geomorfologico e climatico ambientale dei travertini di Rapolano Terme (SI). *Ital. J. Quat. Sci.* **1998**, *11*, 319–329.
75. Mancini, A.; Capezzuoli, E.; Erthal, M.; Swennen, R. Hierarchical approach to define travertine depositional systems: 3D conceptual morphological model and possible applications. *Mar. Pet. Geol.* **2019**, *103*, 549–563. [CrossRef]
76. Capezzuoli, E.; Gandin, A.; Pedley, M. Decoding tufa and travertine (fresh water carbonates) in the sedimentary record: The state of the art. *Sedimentology* **2014**, *61*, 1–21. [CrossRef]
77. Bossio, A.; Costantini, A.; Lazzarotto, A.; Liotta, D.; Mazzanti, R.; Salvatorini, G.; Sandrelli, F. Rassegna delle conoscenze sulla stratigrafia del Neoautoctono toscano. *Mem. Soc. Geol. Ital.* **1993**, *49*, 17–98.
78. Ghinassi, M.; Abbazzi, L.; Esu, D.; Gaudant, J.; Girotti, O. Facies analysis, stratigraphy and palaeontology (Molluscs and Vertebrates) in the Upper Pliocene sandy flood-basin deposits of the Upper Valdarno Basin (Northern Apennines). *Riv. Ital. Paleontol. Stratigr.* **2005**, *111*, 467–487. [CrossRef]

79. Holbrook, J.; Schumm, S.A. Geomorphic and sedimentary response of rivers to tectonic deformation: A brief review and critique of a tool for recognizing subtle epeirogenic deformation in modern and ancient settings. *Tectonophysics* **1999**, *305*, 287–306. [CrossRef]
80. Soligo, M.; Tuccimei, P.; Barberi, R.; Delitala, M.C.; Miccadei, E.; Taddeucci, A. U/Th dating of freshwater travertine from Middle Velino Valley (Central Italy): Paleoclimatic and geological implications. *Palaeogeogr. Palaeoclimatol. Palaeoecol.* **2002**, *184*, 147–161. [CrossRef]
81. Faccenna, C.; Soligo, M.; Billi, A.; De Filippis, L.; Funiciello, R.; Rossetti, C.; Tuccimei, P. Late Pleistocene depositional cycles of the Lapis Tiburtinus travertine (Tivoli, Central Italy): Possible influence of climate and fault activity. *Glob. Planet. Chang.* **2008**, *63*, 299–308. [CrossRef]
82. Janssens, N.; Capezzuoli, E.; Claes, H.; Muchez, P.; Yu, T.L.; Shen, C.C.; Ellam, R.M.; Swennen, R. Fossil travertine system and its palaeofluid provenance, migration and evolution through time: Example from the geothermal area of Acquasanta Terme (Central Italy). *Sediment. Geol.* **2020**, *398*, 105580. [CrossRef]
83. Mancini, A.; Della Porta, G.; Swennen, R.; Capezzuoli, E. 3D reconstruction of the Lapis Tiburtinus (Tivoli, Central Italy): Glacio-eustasy variations affecting travertine deposition. *Basin Res.* **2021**. accepted.

geosciences

MDPI

Article

New Chronological Constraints from Hypogean Deposits for Late Pliocene to Recent Morphotectonic History of the Alpi Apuane (NW Tuscany, Italy)

Ilaria Isola [1,2,*], Francesco Mazzarini [1], Giancarlo Molli [3], Leonardo Piccini [4], Elena Zanella [5], Giovanni Zanchetta [3], Russell Drysdale [6], John Hellstrom [7], Jon Woodhead [7], Adriano Roncioni [8], Flavio Milazzo [9], Diego Pieruccioni [10] and Eleonora Regattieri [2]

1 Istituto Nazionale di Geofisica e Vulcanologia Sezione di Pisa, 56125 Pisa, Italy; francesco.mazzarini@ingv.it
2 Istituto Geoscienze e Georisorse CNR, 56127 Pisa, Italy; eleonora.regattieri@igg.cnr.it
3 Dipartimento di Scienze della Terra, University of Pisa, 56126 Pisa, Italy; giancarlo.molli@unipi.it (G.M.); giovanni.zanchetta@unipi.it (G.Z.)
4 Department of Earth Science, University of Florence, 50121 Firenze, Italy; leonardo.piccini@unifi.it
5 Dipartimento di Scienze della Terra, University of Torino, 10125 Torino, Italy; elena.zanella@unito.it
6 School of Geography, The University of Melbourne, 3053 Melbourne, Australia; rnd@unimelb.edu.au
7 School of Earth Sciences, The University of Melbourne, 3053 Melbourne, Australia; j.hellstrom@unimelb.edu.au (J.H.); jdwood@unimelb.edu.au (J.W.)
8 Gruppo Speleologico Lucchese, 55100 Lucca, Italy; a.roncioni59@gmail.com
9 Dipartimento di Scienze Chimiche della Vita e della Sostenibilità Ambientale, University of Parma, 43124 Parma, Italy; flavio.milazzo@studenti.unipr.it
10 Dipartimento per il Servizio Geologico d'Italia, ISPRA, 00144 Roma, Italy; diego.pieruccioni@isprambiente.it
* Correspondence: ilaria.isola@ingv.it

check for updates

Citation: Isola, I.; Mazzarini, F.; Molli, G.; Piccini, L.; Zanella, E.; Zanchetta, G.; Drysdale, R.; Hellstrom, J.; Woodhead, J.; Roncioni, A.; et al. New Chronological Constraints from Hypogean Deposits for Late Pliocene to Recent Morphotectonic History of the Alpi Apuane (NW Tuscany, Italy). *Geosciences* **2021**, *11*, 65. https://doi.org/10.3390/geosciences11020065

Academic Editor: Jesus Martinez-Frias
Received: 18 December 2020
Accepted: 28 January 2021
Published: 2 February 2021

Publisher's Note: MDPI stays neutral with regard to jurisdictional claims in published maps and institutional affiliations.

Copyright: © 2021 by the authors. Licensee MDPI, Basel, Switzerland. This article is an open access article distributed under the terms and conditions of the Creative Commons Attribution (CC BY) license (https://creativecommons.org/licenses/by/4.0/).

Abstract: A sedimentary sequence of fluvial deposits preserved in the Corchia Cave (Alpi Apuane) provides new chronological constraints for the evolution of the cave system and the timing and rate of uplift of this sector of the Alpi Apuane since the late Pliocene. Supported by magnetostratigraphic analysis performed on fine-grained fluvial deposits, and by radiometric dating of speleothems, we suggest that the deposition of fluvial sediments occurred between ~1.6–1.2 Ma. This implies that the host volume of rock was already located close to the local base level, adding key information about the recent tectonic evolution of the Alpi Apuane. A few before ~1 Ma, an erosive phase occurred due to the base-level lowering, followed by continuous speleothem deposition since at least 0.97 Ma. From that time, Monte Corchia uplifted at a maximum rate of ~0.5 mm/year, which is consistent with isostatic uplift mainly driven by erosional unloading. The petrographical study of the fluvial deposits highlights the presence of material derived from the erosion of rocks that today are absent in the cave's catchment area, suggesting a different surface morphology during the Early Pleistocene. This study highlights the potential of cave sediments as archives for reconstructing the uplift history of mountain ranges.

Keywords: geochronology; karst; magnetostratigraphy; Corchia Cave; Alpi Apuane

1. Introduction

The Alpi Apuane (northwestern Italy) is a key area for understanding the morphological and tectonic evolution of the inner northern Apennines. A number of studies investigated the complex tectonic evolution of the region (e.g., [1–3] and reference therein), including the exhumation history and its relationship with the orogenesis of this sector of the Apennine chain, mainly through thermochronology methods (e.g., [4–8]). Thermochronology is the quantitative study of the thermal history of rocks using temperature-sensitive radiometric dating [9]. It is widely used to reconstruct rock exhumation rates and the tectonic evolution of mountain belts.

Other chronological constraints on paleotectonic evolution of mountain ranges can be provided by alluvial deposits preserved in river terraces located at different altitudes, but their formation is often strongly affected by climate variations, and so they are not directly related to tectonically induced base-level changes and their chronology is, however, difficult to be defined in detail. Conversely, cave systems are less climate dependent than river systems because their evolution is usually slower and less controlled by surface processes [10]. Therefore, caves can provide additional quantitative constraints on the vertical displacement of a subterranean hydrologic system relative to its base level [11–14]. The erosional forms of cave passages, as well as the occurrence and nature of infill deposits, provide valuable information about the processes of their formation [15,16]. Moreover, since caves are one of the most stable continental environments, they can preserve depositional records for long periods.

Horizontal/sub-horizontal levels of passages not controlled by litho-structural horizons are considered the product of a stationary paleo-water table, while vertical sections are commonly formed as a consequence of a base-level fall [10,17,18]. The ages of these developmental stages can be inferred by dating the hypogean deposits using, for example, uranium-series geochronology or paleomagnetism, from which rock valley incision rates and regional paleogeographic evolution can be inferred ([19–21] and reference therein, [13–22]).

The Alpi Apuane are a prominent massif that reaches elevations of almost 2000 m a.s.l., only a few kilometres inland from the coastline, creating a barrier to the humid westerly air masses of North Atlantic provenance and generating precipitation exceeding 2500 mm/year [23]. These characteristics, along with the presence of extensive carbonate rock outcrops, support intense karst processes, allowing the formation of hundreds of caves [24]. The Corchia Cave karst system is the longest and most complex cave in the Alpi Apuane [25], and it developed in several sub-horizontal levels connected by pits or deep canyon-like passages, revealing at least three phases of base-level stillstand [25,26]. We describe here fluvial deposits preserved within the Galleria delle Stalattiti (GdS), which belongs to one of the major horizontal passage systems of the Corchia Cave complex. U-Pb ages and paleomagnetic data acquired on samples of these deposits provide further constraints for the recent morphotectonic history of this sector of the Alpi Apuane.

2. Geological Setting

2.1. The Alpi Apuane and the Monte Corchia

The Apennine chain is a fold-and-thrust belt formed by thrusting from west to east of part of the Ligurian–Piedmont Ocean (Ligurian and Sub-Ligurian domain) over the Adria continental margin plate (Tuscan domain) during the Tertiary [27,28]. The deepest part of the Northern Apennine chain is exposed in the tectonic window of the Alpi Apuane ([1,29,30] and reference therein) (Figure 1), where two main tectono-metamorphic units, the Massa and the Apuane, outcrop. The former, exposed in the western part, experienced a higher degree of metamorphism and includes a Variscan basement and an upper Permian–Upper Triassic cover [3]. The latter, metamorphosed to greenschist facies, consists of a Paleozoic basement and metavolcanics of the early Permian age [31,32], unconformably covered by a Triassic-to-Oligocene sequence. The metasedimentary succession begins with Triassic continental to shallow-water deposits (Verrucano) followed by an Upper Triassic–Lower Jurassic carbonate sequence consisting of metadolostone (Grezzoni), metabreccia (Brecce di Seravezza), dolomitic marble, and marble. The sequence continues upward with Lower Jurassic—Lower Cretaceous cherty metalimestones, cherts, and calcschists. During the Lower Cretaceous–lower Oligocene, a progressive drowning of the carbonate platform allowed the sedimentation of phyllites and calcschists with marble interbeds. The sequence ends in the Oligocene–early Miocene with the deposition of turbiditic metasandstone (Pseudomacigno) ([33] and reference therein).

The Alpi Apuane are surrounded and structurally overlain by the Tuscan Unit, a Late Triassic–early Miocene carbonate and graywacke sequence, which experienced anchizonal metamorphism [3,34], and tectonically overlain by the non-metamorphic Ligurian Units composed of Jurassic ophiolites and Jurassic–Paleogene sedimentary rocks. The structure of the Alpi Apuane is commonly interpreted as being due to two main tectono-metamorphic regional events: the D1 event, leading to a large-scale northeast-facing isoclinal fold stacking at the metamorphic peak conditions [1,35], and the following D2, which was responsible for the progressive uplift and exhumation of the dome-like structure and the development of low-angle detachments [1–3]. The latest stages of D2 were associated with high-angle brittle faults mainly localized at the boundary between the Alpi Apuane and the surrounding Lunigiana/Versilia and Garfagnana tectonic depressions. Brittle deformation within the metamorphic core is expressed by localized and low-displacement, high-angle strike-slip to normal faults [3,36,37].

Monte Corchia is located in the southeast part of the Alpi Apuane Metamorphic Complex. According to Carmignani and Giglia [38], the Mt. Corchia structure is the result of a polyphase deformation during which an originally non-cylindrical eastward-facing D1 overtuned syncline was refolded up to the present geometry, giving a locally downward-younging direction.

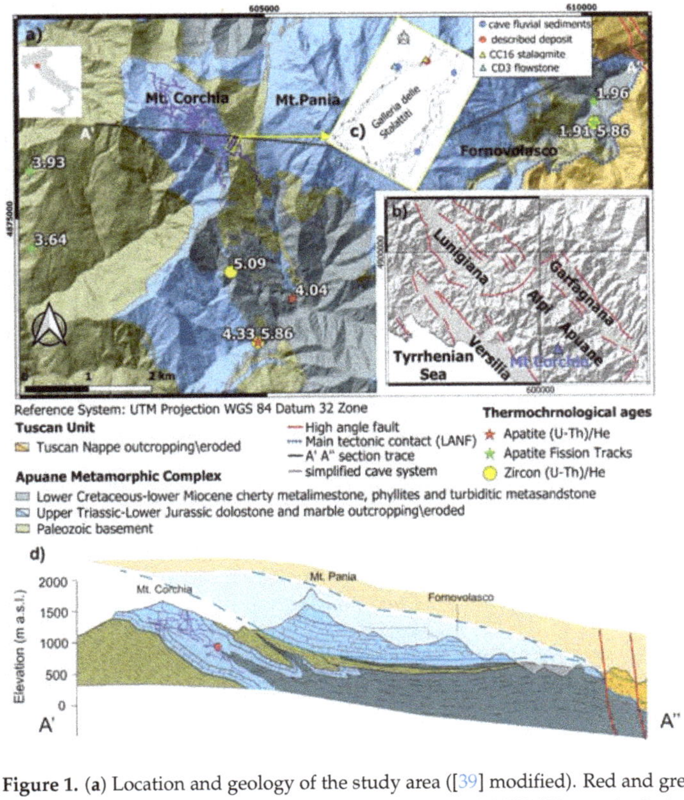

Figure 1. (**a**) Location and geology of the study area ([39] modified). Red and green stars and yellow circles represent the location of samples used for Apatite (U-Th)/He, apatite fission tracks, and Zircon (U-Th)/He thermochronology ages, respectively [8]. (**b**) Simplified tectonic sketch of the Garfagnana and Lunigiana valleys. (**c**) Galleria delle Stalattiti plan map ([40] modified). Circles and triangles show the location of fluvial and calcite deposits, respectively. (**d**) Geological cross section between A′ and A″. Dashed and continuous blue lines represent low-angle normal faults (LANFs) (former thrusts reactivated as LANFs: see [3,41] for further tectonic information). The red dot represents the Galleria delle Stalattiti (GdS) location.

The most complete low-temperature thermochronologic study, based on apatite and zircon fission tracks (AFT, ZFT) combined with apatite and zircon (U-Th)/He ages (AHE, ZHe), derived exhumation rates and processes related to the Alpi Apuane tectonic history [8]. The study suggested that the earlier and late Miocene high exhumation rates (up to >1.4 mm/year) can be interpreted as related to events of tectonic exhumation, whereas the low exhumation rates (<0.6 mm/year) of the last 4 Ma may instead be due to predominantly erosional exhumation.

2.2. The Corchia Cave System

The Corchia Cave system, more than 70 km long and about 1200 m deep, is mainly carved in the Upper Triassic to Lower Jurassic carbonate core of the Mt. Corchia syncline, confined by the phyllites and metavolcanics (Lower phyllites and Porphyroids, [42]) of the Paleozoic basement. Structural surfaces have guided the cave development, so the system is mainly elongated NW–SE and gently dipping to the SE, parallel to the syncline axial surface [25] (Figure 1). The horizontal passages follow, therefore, the main syn-metamorphic litho-structural surfaces of the metacarbonate core (Figure 1b), whereas most of the vertical passages mainly developed along low-displacement high-angle faults and/or fracture systems [26,43]. The Mt. Corchia groundwater network is drained mainly by a spring located at 175 m a.s.l. that flows to the Ligurian Sea via the Vezza River [25,44]. The cave system has 21 entrances at different elevations, the highest opening being close to the top of the mountain (1678 m a.s.l.) at 1637 m a.s.l. Most of the horizontal or sub-horizontal phreatic or epiphreatic cave passages are organized in different levels, the major ones of which are at around 1400, 1100–1200, and 900 m a.s.l., and are connected by steep to vertical passages [26]. This structure suggests different base-level stillstands, allowing the development of three major sub-horizontal levels, interrupted by relatively faster lowering of the base level. Local breakdown/collapse blocks and allochthonous fluvial sediments are the prevalent clastic deposits in the cave.

Our study focuses on the sediments occurring in the GdS, a sub-horizontal passage with floor elevation ranging between 860 and 870 m a.s.l. and vertically overlain by ~400 m of rock (mainly phyllites of the Paleozoic basement). It is part of the lower of the three abovementioned sub-horizontal major levels [26]. The gallery is partially filled by chemical (speleothems) and clastic deposits both autochthonous (breakdown/collapse blocks) and allogenic (due to stream transport). The latter are locally exposed along the sidewalls at different heights with different extents and thickness.

3. Materials and Methods

To overcome the intrinsic variability of the hypogean environment, we studied the GdS sedimentary section where stratigraphy is clear and continuous (~9 m thick and ~12 m laterally extended). We collected six samples to study the thin-section grain composition and to infer the possible source area of the allogenic recharge. The fluvial succession was also sampled for paleomagnetism at eight different levels. We sampled the finer-grain-size layers by collecting oriented samples for each stratum. At the laboratory, these samples were further reduced to standard cylindric specimens. All the measurements were performed at the Centro Interuniversitario di Magnetismo Naturale (CIMaN-ALP) Laboratory (Peveragno, Cuneo). For each specimen, we measured the natural remanent magnetization (NRM) using an AGICO JR6 spinner magnetometer. The specimens were then step-by-step Alternating Field (AF)-demagnetized by an ASC D-2000 device up to 100 mT. Isothermal remanent magnetization (IRM) acquisition and back field were performed on one specimen for each layer.

4. Results

4.1. Clastic Sediment Stratigraphy

The studied sequence (Figure 2) is mostly composed of clastic material transported and deposited by running water and lithified by calcite cementation. It directly lays on the carbonate bedrock and begins with a 1-cm-thick flowstone characterized by a very high detrital content. This is overlain by 0.9 m of fluvial sediment consisting of alternations of clast-supported conglomerate and medium- to fine-grained calcite-cemented arenite. In several points, the conglomerate layers show clasts imbrication and erosional troughs. The sequence continues with 5 m of centimetric to decimetric layers of arenite with calcite cement, passing via erosive contact to a 1.5-m-thick, centimetric to decimetric well-rounded heterometric conglomerate. An unconsolidated to partly consolidated 1.5 m layer of breakdown breccia covers the conglomerate and closes the clastic sequence. It consists of unsorted boulders and cobbles, with a prevalence of rock fall materials, ranging from centimeter- to meter-size clasts and overlain by a few decimeters of unconsolidated carbonate sand draping the irregular surface. A thin calcite flowstone, with several stalagmites grown on it, overlays the succession paraconformibly.

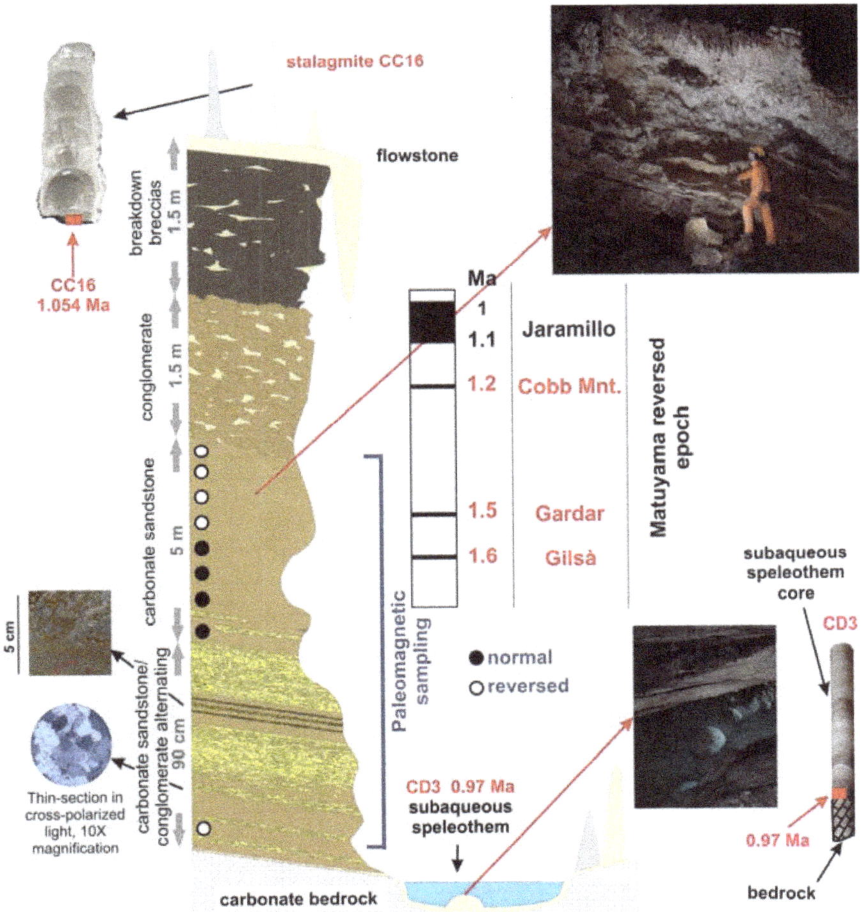

Figure 2. Stratigraphic column of the studied fluvial section and its relationship with speleothems. Vertical axes are not at scale, and the layer thickness is expressed by gray arrows.

At thin-section scale, the conglomerate has a fine carbonate matrix and sub-rounded clasts consisting of mineral grains (calcite, quartz, mica, feldspar, and epidote), sandstone, and metamorphic rocks. Sandstone clasts vary from medium to very fine in size and are well sorted and sub-rounded. They comprise minerals (calcite, quartz, mica, chlorite, feldspar, and oxides) and metamorphic and siliciclastic lithic components. Modal analysis carried out on sandstone samples and plotted on a ternary diagram (Figure 3a) following Dickinson [45] show a clustering of samples in the litharenite zone. To better understand the provenance of the detrital grains in the analyzed samples, we plotted them in a Metamorphic-Volcanic-Sedimentary ternary diagram (Figure 3b). Most of the lithics have metamorphic origin, but the presence of lithics of sedimentary origin is not negligible.

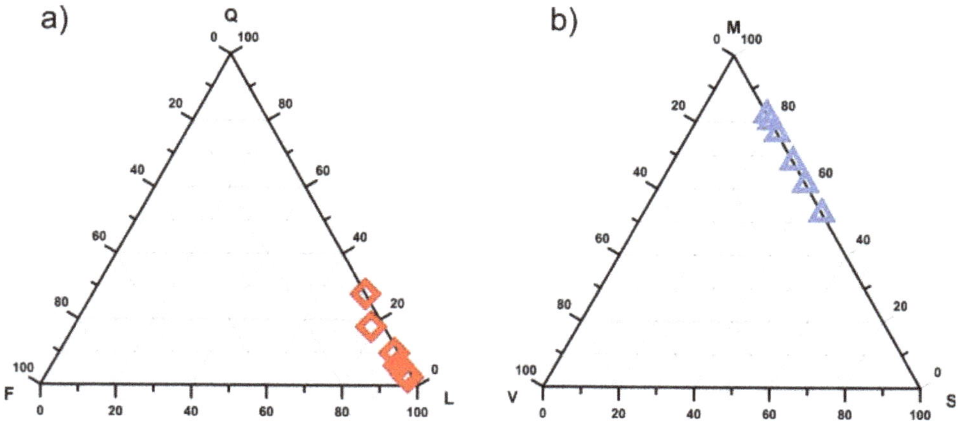

Figure 3. (**a**) Quartz-Feldspar-Lithic plot summarizing the sample compositions [45]. (**b**) Ternary diagram Metamorphic-Volcanic-Sedimentary of the lithic grains in the analyzed samples: notable is the presence of grains of sedimentary origin.

4.2. Paleomagnetic Analysis

Examples of the results of the demagnetization behavior of the specimens are shown in Figure 4a. The isothermal experiments confirm the occurrence of a low-coercivity phase in each layer, saturated by applying 0.3 to 0.5 T fields. The coercivity is mostly around 20 mT. This indicates that the main remanence carrier is magnetite. The intensity of the magnetization is rather low, ranging from 1×10^{-4} to 1×10^{-2} Am^{-1}, but mostly of ca 3×10^{-3} Am^{-1} (Figure 4b). Zijderveld diagrams [46] reveal the occurrence of two magnetization components, one of low stability, removable by applying a field of 10 to 15 mT and interpreted as a viscous component, and a high-stability component of both normal and reverse polarity, pointing to the origin of the diagram and interpreted as the Characteristic Remanent Magnetization (ChRM) (Figure 4c). The ChRM quality is expressed by the maximum angular deviation (MAD), which mostly ranges from 4.5 to 16. In the layers with low magnetization intensity (of the order of 10^{-4} Am^{-1}), MADs show higher values of around 20 to 25. Although these values are large, they have been considered acceptable because in the Zijderveld diagrams it is still possible to recover the magnetic polarity [47]. The mean magnetic polarity was computed using Fisher's Statistics [48]. The normal magnetic polarity is $D = 357.2°$, $I = -43.4$, alpha95 = 28.2°, Virtual Geomagnetic Pole (VGP) Lat N 71°, Long E 198°; its confidence limit is very high due to both the small number of normal polarity observations ($n = 10$) and the large dispersion of declinations. The mean reverse polarity is $D = 186.6$, $I = -41.7$, alpha95 = 11.6°, VPG Lat S 69°, Long E 173° (Figure 5).

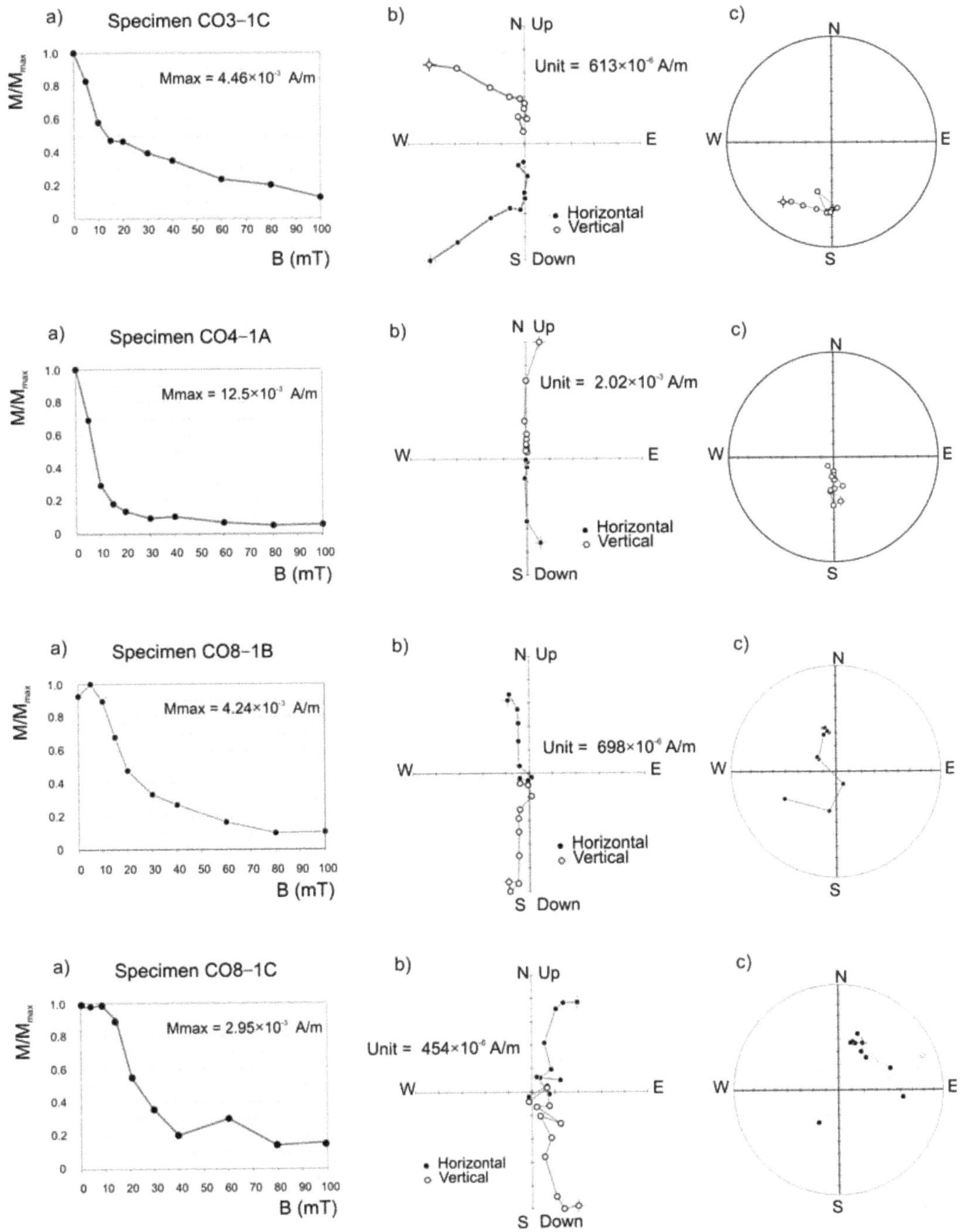

Figure 4. Examples of the magnetization behavior of reversed (first two rows) and normally polarized (last two rows) specimens during Alternating Field (AF) cleaning. For each specimen: (**a**) the decay of the normalized magnetization intensity versus the applied field; (**b**) the Zijderveld diagrams (black/white dots are declination/apparent inclination, respectively); and (**c**) equal-area projection of magnetic direction (black/white dots are positive/negative inclination, respectively).

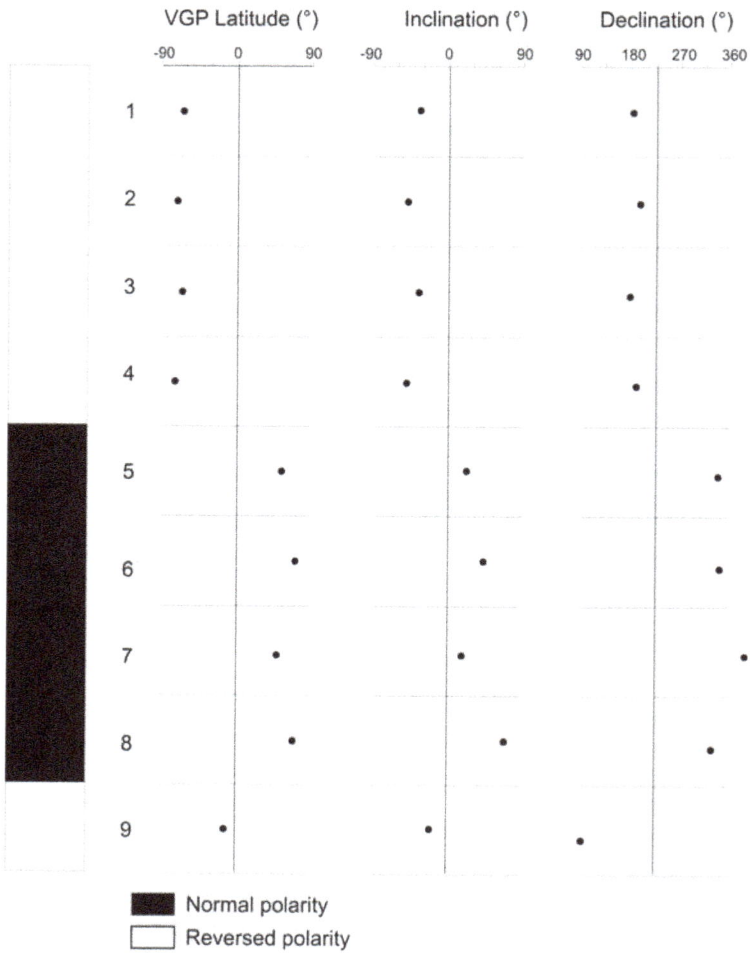

Figure 5. Paleomagnetic record. From left to right: polarity succession, latitude of the Virtual Geomagnetic Pole (VGP), magnetic Inclination and declination.

5. Discussion

5.1. Speleothem Ages

The Corchia Cave and, in particular, the GdS have been the object of wide-ranging paleoclimate research for the past 20 years [49–60]. Tens of speleothems have been dated, with more than 400 very precise radiometric ages [54,61]. The calcite of speleothems from the GdS has two important qualities: it is exceptionally free from detrital components, and it is rich in uranium content. In its deepest part, the GdS is overlain by about 400 m of rock, so the seepage water responsible for speleothem formation has lost most of its detrital load, leading to the precipitation of exceptionally pure calcite. The Brecce di Seravezza formation probably is the main source of the relatively high uranium content that allows very precise radiometric dating using both conventional U-series and the more novel U-Pb techniques [62,63]. Among the hundreds of ages obtained until now, the oldest comes from CC16, one of the stalagmites topping the described succession. The bottom of the stalagmite has been dated by Woodhead et al. [63] at 1.054 Ma ($2\sigma = 0.024$) using a methodology that allows producing U-Pb ages of high quality. Moreover, in the lower section of this gallery, a small and shallow lake, with walls covered by calcite flowstones, is present. On the floor, the $CaCO_3$ deposition has formed a subaqueous drapery with a cave-cloud shape, with lily-pad structures created where calcite crystals grow radially from projecting substrata. U-Pb dating shows that the basal age of the subaqueous calcite deposit reaches 0.967 + 0.018/−0.006 Ma [59].

5.2. The Galleria Delle Stalattiti (GdS) Evolution

Reconstructing the evolution of the GdS requires considering some uncertainties. A chronological interpretation of cave sedimentation is often complex. Waterways can be continuously obstructed and reopened, the sedimentation rate can be extremely variable, and deposits can be diachronous.

According to our stratigraphic and chronological reconstruction, the GdS was firstly formed as a phreatic tube, whose original morphology is preserved in the ceiling shape. Successively, the tube was affected by downcutting in epiphreatic to vadose conditions. Deposition is probably related to the transition from phreatic to vadose conditions and to a phase with high-solid-load transport. The grain sizes of the fluvial deposits and the occurrence of erosional surfaces testify to changes in water energy that could have also a climatic meaning. Otherwise, the fluvial deposits facies testify to a transition from low- to high-energy sediments (sand to pebble and gravel) progressively filling the passages. An increase of energy in the stream regime is also marked by the upper erosional surface of carbonate arenite, followed by coarse conglomerate deposition. A subsequent depositional phase, which was concurrent with widespread roof breakdown, was followed by an erosional phase cutting the entire sequence. A phase of calcite precipitation topped the detrital succession with a thin flowstone and stalagmites at least since 1.05 ± 0.2 Ma [63]. After the erosive phase ended, a small lake formed at the bottom of the gallery, where subaqueous speleothems have been growing directly on the rock substratum since ~0.97 Ma [59], indicating the definitive interruption of vadose flow.

The age of the carbonate arenite in the GdS can be constrained by coupling paleomagnetic data from fluvial deposits and radiometric ages from speleothems. At the base of the deposits, the magnetic zones (Figures 2 and 5) start with a reversed polarity record (older deposition phase: layer 9) followed by a normal polarity record (layers 8 to 5) and then by a reverse polarity record at the top (younger deposition phase: layers 4 to 1). Stalagmite CC16 caps the stratigraphic succession, providing an age of 1.05 Ma [63]. This means that the fluvial deposition reasonably started within the Matuyama Chron before the Jaramillo subchron (1071–991 ka), [64]. During the Matuyama, several short geomagnetic reversal events have been documented (see [64] and reference therein for details). Many of these are still controversial because defining the occurrence of a short reversal requires a high-precision reconstruction of a paleomagnetic field, which is dependent upon the temporal resolution of the archives and the geological constraints. Besides, not all the reported reversals have been recorded at all studied sites (see [65] for details). For these reasons, we have considered only the events reported by Simon et al. [64]. These have been detected by the relative paleo-intensity (geomagnetic dipole moment low) and Be-ratio records, and are commonly reported in the literature [65,66]. We identified three different events/subchrons that could match the normal polarity record from layers 8 to 5 in the cave sediment, namely the Cobb Mountains (1176/1204 ± 5 ka), the Gandar (1459 ± 9 ka), and the Gilsa (1587 ± 8 ka). This means that the fluvial deposition started at least before the normal Cobb Mountains subchron zone in the Matuyama reverse epoch, dated at about 1.2 Ma [64]. A local base-level fall may have occurred during the following reverse subchron prior to 1.05 Ma, causing the incision phase that cut the sediments underlying the CC16 stalagmite; this erosion phase had already expired at ~0.97 Ma when the lake had formed, and the calcite deposition took place. Since ~1 Ma, the scenario probably remained unchanged, as testified by the continuous and undisturbed deposition of the subaqueous speleothem [59]. The time lapse occurred between the end of the clastic deposition and the beginning of CC16 formation is unknown: the most conservative hypothesis is to consider this interval negligible.

The clastic deposition within the GdS probably occurred when this horizontal passage was close to the local base level. Furthermore, the erosion of coarse fluvial deposits requires a hydraulic gradient, therefore implying that the GdS was above local, and even more above the global base level, before speleothem formation, but its past real elevation is unknown. Presently, the base level of this karst system is at 175 m a.s.l., in the Vezza River valley, close to the contact between the Paleozoic basement and dolostone, that is, ~700 m below the present GdS elevation. The presence in the Corchia Cave system of relict phreatic passages at ~1600 m a.s.l. [67], that is, ~700 m above the GdS, implies that during GdS formation, there were at least 700-m-thick previously exhumed rocks. In summary, radiometric and paleomagnetic data indicate that at least by the Early Pleistocene (1.6–1.2 Ma), fluvial sediments were deposited in the GdS, indicating that this portion of Mt. Corchia was necessarily above the sea level at that time.

5.3. Exhumation/Uplift History from Thermochronological Dating for the Alpi Apuane

Since the first application of thermochronologic methods in the Alpi Apuane [68], various authors have tried to derive exhumation and rock-uplift rates from them [4–8]. Most of these authors agree on an exhumation rate ranging from 1.3 to 1.8 mm/year between about 6 and 4 Ma (late Miocene–early Pliocene), followed by a lower rate ranging from 0.6 to 0.9 mm/year, since early Pliocene to the present (Table 1).

Table 1. Summary of the Alpi Apuane exhumation rate estimated in previous studies.

Ages	Exhumation Rate mm/Year	Method	Reference
6–4 Ma	1.4		[8]
4–0 Ma	0.6	AFT, ZFT, AHe, ZHe	[8]
6–4 Ma	1.3–1.8	AFT	[7]
4–0 Ma	0.6–0.9	AFT	[7]
middle Pliocene	0.8	AFT, AHe, ZHe	[6]
6–2 Ma	0.8–1.7	AFT, ZFT	[5]

AFT: Apatite Fission Tracks; ZFT: Zircon Fission Tracks; AHE: Apatite (U-Th)/He; ZHe: Zircon (U-Th)/He. The ages represent the time periods relevant to the exhumation rate estimates.

Fellin et al. [8] suggested a high exhumation rate (>1.4 mm/year) during the late Miocene, followed by a decrease to ≤0.6 mm/year in the Pliocene. The only inconsistency with this interpretation, as noted by the authors, is related to some samples of a small region in the easternmost Alpi Apuane (i.e., Fornovolasco area, 7 km east of our studied site; see Figure 1) where AFT ages as young as 1.9 Ma were obtained [5]. Close to Mt. Corchia, AFT ages of around 3.8 Ma are reported [8]. The difference in age between the data at Fornovolasco and the data west of Mt. Corchia could either indicate a real differential exhumation or it could relate to overscatter of the data. As clearly shown in Figure 1a,d (see also [2,31,69,70]), no first-order high-angle faults with a high displacement throw (hundreds to thousands of meters) occurs just east of Mt. Corchia, justifying a significant differential uplift. No splitting of the metamorphic core in two sub-domains—Mt. Corchia in the west and Fornovolasco in the east—is also feasible, and therefore a differential exhumation history between them may be ruled out.

Notably, the highest of the three horizontal/sub-horizontal levels of the Corchia Cave system, about 500 m above the GdS, hosted the CC17-f2 stalagmite dated by Engel et al. [71] about 1.9 Ma. This age testifies a Corchia karst system well developed, at least in its higher part, with galleries already formed and speleothems growing. This scenario is inconsistent with AFT ages of about 1.9 Ma, indicating a depth of at least 3000 m just a few kilometers far. On this basis, we exclude the AFT ages from Fornovolasco and we consider only the AFT ages derived from the nearby Paleozoic rocks [5,8], west of the Corchia Cave, as representative of the whole Mt. Corchia area and of the eastern side of the Alpi Apuane core.

5.4. Rock Uplift Rates from Cave Evolution and Hypogean Deposits

Rock uplift rates can be inferred from exhumation rates, assuming that uplift triggers a similar denudation rate at regional scale. Otherwise, cave system evolution can furnish a reliable indication of the lowering rate of the local base level, which is usually strictly dependent on river incision. So, also in this different approach, we must presume that river incision acts as a quick-response process to tectonic uplift. These assumptions are quite acceptable in mountain regions, where surface processes are very active and the erosion rate is usually higher than uplift rate. Despite this intrinsic uncertainness of the two approaches, a rough estimate of uplift rates can provide useful information to reconstruct the tectonic dynamics of a mountain range.

We use the new chronological data for the GdS combined with the available thermochronology data to provide revised estimates of the uplift rates of this sector of the Alpi Apuane since the late Pliocene. The approximations previously discussed imply that the rates we are going to propose could underestimate the true uplift rate. For all these reasons, hereafter, we refer to the uplift rate computed between the closure of the apatite geothermal system and the deposition of CC16 as a minimum uplift rate. We take as major local constraints (i) the U-Pb speleothems ages; (ii) the magnetostratigraphic chronologies considering the three possible normal subchrons older than 1 Ma (Cobb Mountain, Gardar, Gilsa; Table 2); and (iii) the gallery morphology and the reconstruction of the cave infill phases (see Table 2).

Table 2. Ages used to estimate the uplift rates.

Sample	Present Elevation (m a.s.l.)	Age (Ma)	2σ	Dating Method	Lithology Formation Unit	Reference
CD3	870	0.97	0.010	U-Pb	Subaqueous flowstone	[59]
CC16	875	1.05	0.024	U-Pb	Stalagmite	[63]
CO5-8	865–874	1.176/1.204	0.005	Paleomagnetic chronology Cobb Mountain	Hypogean deposit	This work (using [64])
CO5-8	865–874	1.459	0.009	Paleomagnetic chronology Gardar	Hypogean deposit	This work (using [64])
CO5-8	865–874	1.587	0.005	Paleomagnetic chronology Gilsa	Hypogean deposit	This work (using [64])
CP1	675	3.93	0.36	AFT	Phyllite Paleozoic basement	[5]
CP3(4)	650	3.64	0.71	AFT	Phyllite Paleozoic basement	[5]

Considering the ages of the GdS deposits, the minimum local base level allowing gallery formation and fluvial sediment infilling, and the AFT closure depths (i.e., the depth of the 110 °C isotherm), a minimum rock uplift rate can be inferred from late Pliocene to Early Pleistocene.

To obtain a more reliable estimate, we must consider that at the time of GdS filling deposition, the rock where samples for AFT analysis are located were still buried below a thickness of rock successively eroded (Figure 6). In the simplest hypothesis that valley slopes have maintained a constant profile, this thickness should be of the same order of the valley incision and so of the base-level lowering, that is, ~700 m. Considering that the most elevated sections of the Alpi Apuane ridge consists of carbonate formations, we can suppose a slower denudation rate for summit areas and consequently a progressive increasing of slope gradient due to river incision. In this more realistic framework, the residual burial over AFT-dated rock samples at GdS deposition time would be around 500 m (see Figure 6). In other words, exhumation/uplift rates from the time of AFT closure to GDS sediment formation must be calculated on a thickness that is roughly 500 m lower than that calculated based only on geothermal gradients.

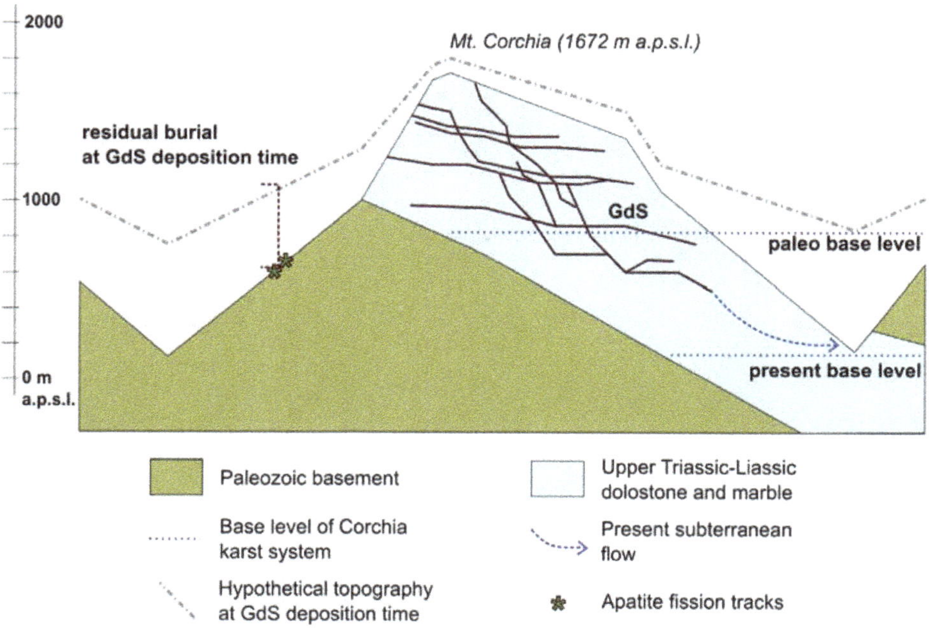

Figure 6. Schematic representation of the geomorphic evolution of Mt. Corchia in the last 1.2–1.6 Ma, considering the hypothesis that valleys were affected by an erosion rate higher than carbonate ridges.

We have estimated the exhumation/uplift rates considering two different geothermal gradients, 25 °C/km and 30 °C/km, and three hypothetical ages for basal GdS sediments, ~1.2, ~1.5, and ~1.6 Ma, obtaining rates ranging from ~1.6 to ~1.1 mm/year (Table 3).

Table 3. Rock uplift rates of Mt. Corchia constrained by AFTs (from [5]) and the magnetostratigraphic dating.

Minimum Rock Uplift Rate from Closure AFT to GdS Clastic Deposition				**Maximum Rock Uplift Rate from GdS Clastic Deposition to Present**	
Subchrone	Age (Ma)	Geothermal Gradient 25 °C/km (mm/year)	Geothermal Gradient 30 °C/km (mm/year)	Subchrone	(mm/year)
Cobb Mountain	1.2	1.3	1.1	COBB MNT	0.6
Gardar	1.5	1.5	1.2	GARDAR	0.5
Gilsa	1.6	1.6	1.3	GILSA	0.4

The phase of clastic deposition in the GdS was followed by a relative lowering of the local base level, which was responsible for an erosive phase cutting throughout the entire fluvial deposits until the carbonate bedrock. The migration of the cave stream toward lower passages allowed the formation of the Laghetto basso, a small pool mostly fed by dripping, where subaqueous speleothems have been continuously growing during about the last million years. This suggests that during this time, the uplift of the area was uniform, with no significant changes that could have disturbed or interrupted the seepage patterns and the lake feeding.

Moreover, assuming that a minimum hydraulic gradient was present during cave filling and considering the modern elevation of the sampling site at ca. 870 m a.s.l., the maximum uplift rate in the last 1.2–1.6 Ma in the Corchia area is estimated around 0.4/0.6 mm/year (Table 3). These values agree with the Middle–Late Pleistocene incision rates estimated by Piccini et al. [20] in the nearby Frigido River basin and with the vertical component of motion derived for the area by Bennett et al. [72] using the Common-mode

Signal Reduced time series of GPS networks and consistent with an uplift mainly driven by erosion.

5.5. Sediment Sources

Petrographic analysis and lithic components of the sampled fluvial deposits furnish information about the sediment sources. The presence of Paleozoic basement and marble clasts, as well as of other minerals typical of the Corchia area, indicates that at least a part of the source rocks would have been in a catchment area very close to the deposition site. Notably, non-metamorphic sandstones are also present as lithic components (Figure 3a,b), despite the absence of such rocks in the areas surrounding Mt. Corchia at the present time (Figure 1a,d). Alluvial infilling with pebbles of non-metamorphic rocks was already described by Piccini [26,67,73] in the higher levels of the Corchia system and dubitatively attributed to remnants of Ligurian Units in the catchment area during their deposition. Looking at the regional geology, non-metamorphic sandstones are present in the Tuscan Unit and the Ligurian Units or as secondary sediments in Neogene deposits. Presently, the closest outcrops of these units are several kilometers far from the Corchia Cave and at lower elevations than the upper entrances of the karst system. Consequently, the source of the studied deposits had to be completely removed (see Figure 1a,d) during the erosion-mediated exhumation of the metamorphic units activated by high-angle normal faulting that started in the region 4–5 Ma (e.g., [8,37]). We can so suppose that the allogenic components of sediment deposed in the GdS derive from the erosion of these materials deposed in the higher levels of the karst system during a previous deposition phase.

A schematic reconstruction of the Corchia Cave system and its catchment area development, from 2.5 Ma to the present, is proposed in Figure 7. AFT constrain the Apuane Metamorphic Complex, as covered by the Tuscan and Ligurian Units already juxtaposed by low-angle normal faults (LANFs) [3,41] still at a depth of 3.5–2.8 km (with a 25 °C or 30 °C geothermal gradient, respectively) at about 3.8 Ma. The Apuane metamorphic core, including the zone hosting the Corchia Cave system, rose from depth to surface with at least three significant stillstand phases, as testified by the three major horizontal/sub-horizontal levels of galleries at around 1400, 1200–1100, and 900 m above the present sea level [26], the higher of which was carved, and partially filled by alluvial sediments of non-metamorphic carbonate sandstone [26], before 1.9 Ma [71]. This testifies that before that time, surface waters were drained through the Corchia system coming from a catchment area partially formed by the Tuscan and Ligurian Units overlaying the metamorphic ones. This presently eroded shallow crustal section fed the cave fluvial deposits. The lowest of the main levels, where the GdS is found, was carved at least before ~1.6–1.2 Ma, when the deposition of the alluvial sediments occurred. Afterward, the base-level lowering led to stream erosion of the hypogean infill until the basal bedrock. The development of the small lake with submerged speleothems growing continuously since ~1.0 Ma testifies to the stability of this cave sector and possibly a continuous, uniform uplift in the area.

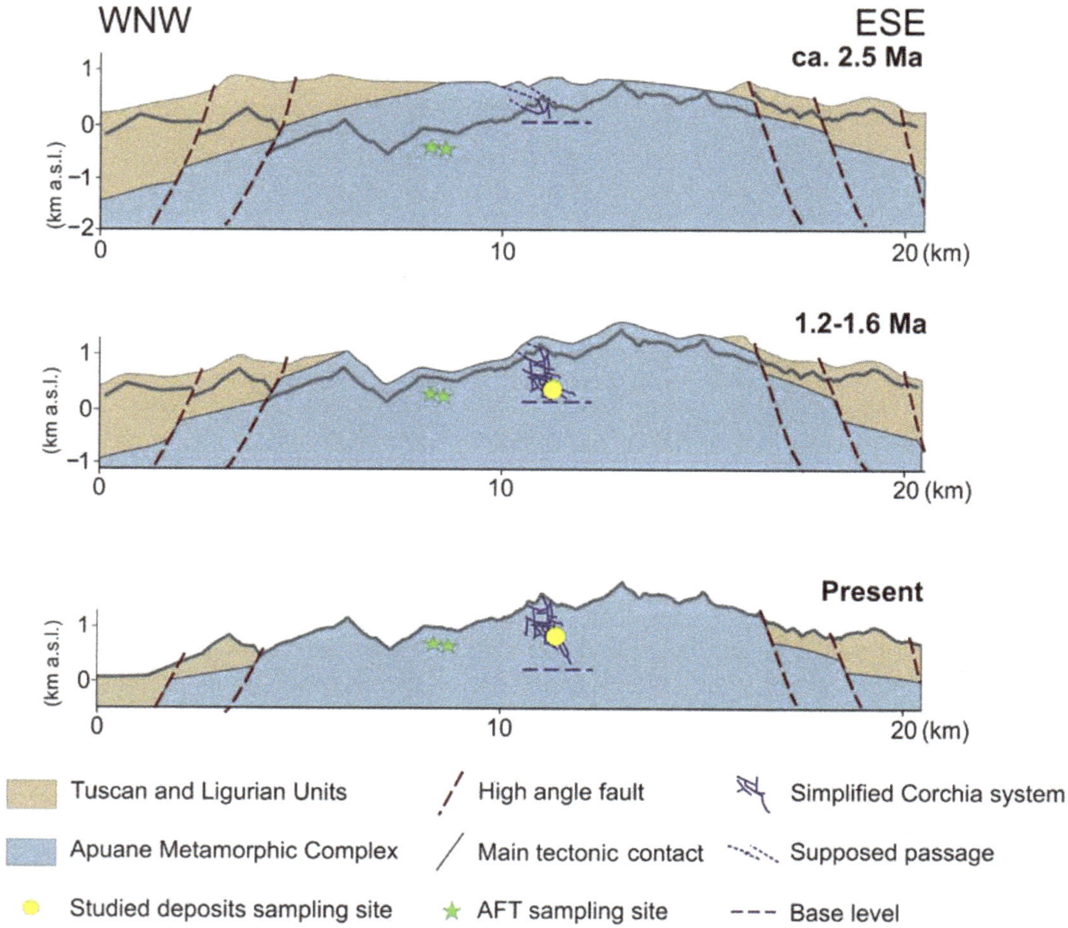

Figure 7. Schematic model showing the proposed chronology of the Corchia Cave system evolution and Alpi Apuane uplift over the last 2.5 Ma based on hypogean radiometric and paleomagnetic analyses and AFT thermochronology from the literature (Figure 1 and Table 2). 2.5 Ma: the highest of the sub-horizontal passages of the Corchia Cave system were already carved, while epigean streams sank and fed the sedimentary deposits with material eroded from metamorphic and non-metamorphic units. The AFT data locations are still buried at about 1.5 km depth. 1.2–1.6 Ma: the two higher karst levels of the Corchia Cave were completely developed, the lowest one was already carved near the base level, and the clastic infilling was occurring in the GdS. Present: the GdS is at about 870 m a.s.l., and a new phreatic level is forming around the modern base level.

6. Conclusions

Using robust dating from independent hypogean deposits in the Corchia Cave, we provide new information about the tectonic uplift of the Alpi Apuane since the lower Pliocene to the present. In particular, the evolution of the GdS is reconstructed by combining morphology observations, radiometric ages from speleothems, and magnetostratigraphic records from fluvial hypogean deposits (Figure 7). The ages suggest the GdS was carved as a phreatic tube and successively filled by fluvial deposits before the beginning of one of the normal subchrons comprised within the Matuyana reverse epoch, that is, about ~1.6/1.2 Ma ago. Between 1.6/1.2 and 1.0 Ma, a strong erosive fluvial phase testified for a local base-level lowering. This phase ended before 1.0 Ma, when the continuous and undisturbed deposition of subaqueous speleothems began. Based on these data, we estimate a minimum rock uplift rate of the Alpi Apuane since the late Pliocene to Early Pleistocene (1.6/1.2 Ma) of ~1.6/1.3 and 1.3/1.1 mm/year (for 25 and 30 °C/km geothermal gradient values, respectively). This phase was followed by a decrease to ~0.4/0.6 mm/year, in good agreement with the modern vertical velocity estimated by Bennett et al. [72] from continuous GPS station analysis (0.5 mm/year) and congruent with rock uplift driven mainly by erosion. The overall tectonic implications of these new estimated exhumation rates and rock uplift are outside the scope of this paper. Nevertheless, we would like to highlight the relevance of tectonics and of a high exhumation rate until the Early Pleistocene (not the Pliocene, as previously assumed) with a decreasing rate since at least 1 Ma. Further works, considering in a regional perspective thermochronologic ages and other constrains as hypogean deposits or stable oxygen isotope paleoaltimetry (e.g., [73]), will allow a better understanding of the vertical movements that affected the Apennine chain.

The petrographic study of the fluvial sediments highlights the presence of lithic components sourced from rocks no longer present in the modern Corchia catchment area. This implies an inversion of relief probably occurred during the Early Pleistocene, suggesting that the Corchia allogenic catchment area was a valley floor draining material from a cover since eroded. During an erosion-mediated exhumation history, probably activated by high-angle normal faulting, the carbonate valley floor became a ridge, as is the case in the present landscape. Further investigations in the higher cave levels could improve time constraints, allowing a more detailed reconstruction of tectonic uplift versus stillstand phases during the Pliocene–Early Pleistocene. Our results thus highlight the importance of several independent proxies for estimating exhumation or rock uplift rates in orogenic settings, because this process, even over time spans of less than 5 Myr, is not steady state. A comprehensive study of hypogean sediments in old karst terrains, coupled with dating, could therefore provide helpful tools to better understand and constrain the exhumation history of recent and active mountain belts.

Author Contributions: Conceptualization, I.I., F.M. (Francesco Mazzarini), G.M., G.Z., and R.D.; investigation, I.I., F.M. (Francesco Mazzarini), E.Z., J.H., J.W., A.R., F.M. (Flavio Milazzo), and D.P.; writing—original draft, I.I., F.M. (Francesco Mazzarini), G.M., L.P., E.Z., G.Z., and R.D.; and writing—review and editing, I.I., F.M. (Francesco Mazzarini), G.M., L.P., E.Z., G.Z., R.D., D.P., and E.R. All authors have read and agreed to the published version of the manuscript.

Funding: The speleothem research performed in this study was funded by the Australian Research Grants DP110102185 (to RND, JDW, and GZ) and DP0664621 (to JDW).

Institutional Review Board Statement: Not applicable.

Informed Consent Statement: Not applicable.

Data Availability Statement: Not applicable.

Acknowledgments: We are grateful to Federico Andreetto and Claudio Robustelli for their collaboration in paleomagnetic analysis, to the Parco Regionale delle Alpi Apuane for allowing the sampling in the Corchia Cave, to the Coop. Sviluppo e Futuro Levigliani for allowing us free admission to the touristic path, and to the Federazione Speleologica Toscana for speleological support. We are grateful to the reviewers for their comments and suggestions, which helped in improving and clarifying the text.

Conflicts of Interest: The authors declare no conflict of interest. The funders had no role in the design of the study; in the collection, analyses, or interpretation of data; in the writing of the manuscript; or in the decision to publish the results.

References

1. Carmignani, L.; Kligfield, R. Crustal extension in the northern Apennines: The transition from compression to extension in the Apuan Alps core complex. *Tectonics* **1990**, *9*, 1275–1303. [CrossRef]
2. Carmignani, L.; Decandia, F.A.; Fantozzi, P.L.; Lazzarotto, A.; Liotta, D.; Meccheri, M. Tertiary extensional tectonics in Tuscany (northern Apennines, Italy). *Tectonophysics* **1994**, *238*, 295–315. [CrossRef]

3. Molli, G.; Carlini, M.; Vescovi, P.; Artoni, A.; Balsamo, F.; Camurri, F. Neogene 3-D structural architecture of the north-west Apennines: The role of the low-angle normal faults and basement thrusts. *Tectonics* **2018**, *37*, 2165–2196. [CrossRef]
4. Coli, M. Time and mode of uplift of the Apuane Alps metamorphic complex. *Atti Tic. Sci. Terra* **1989**, *32*, 47–56.
5. Abbate, E.; Balestrieri, M.L.; Bigazzi, G.; Norelli, P.; Quercioli, C. Fission-track datings and recent rapid denudation in northern Apennines, Italy. *Mem. Soc. Geol. Ital.* **1994**, *48*, 579–585.
6. Brandon, M.; Abbate, E.; Balestrieri, M.L.; Bernet, M.; Bigazzi, G.; Picotti, V.; Reiners, P.; Zattin, M.; Zuffa, G.G. *Quantifying Exhumation History across the Northern Apennines*; Washington EOS Transactions American Geophysical Union: San Francisco, CA, USA; Volume 82.
7. Balestrieri, M.L.; Bernet, M.; Brandon, M.T.; Picotti, V.; Reiners, P.; Zattin, M. Pliocene and Pleistocene exhumation and uplift of two key areas of the Northern Apennines. *Quat. Int.* **2003**, *101*, 67–73. [CrossRef]
8. Fellin, M.L.; Reiner, P.W.; Brandon, M.T.; Wütrich, E.; Balestrieri, M.L.; Molli, G. Thermochronologic evidence for the exhumational history of the Apuan Alps metamorphic core complex, northern Apennines, Italy. *Tectonics* **2007**, *26*. [CrossRef]
9. Berger, G.W.; York, D. Geothermometry from dating experiments. *Geochim. Cosmochim. Acta* **1981**, *45*, 795–811. [CrossRef]
10. Ford, D.; Williams, P.D. *Karst Hydrogeology and Geomorphology*; Wiley: Chichester, UK, 2007; p. 562.
11. Granger, D.E.; Fabel, D.; Palmer, A.N. Pliocene—Pleistocene incision of the Green River, Kentucky, determined from radioactive decay of cosmogenic 26Al and 10Be in Mammoth Cave sediments. *Geol. Soc. Am. Bull.* **2001**, *113*, 825–836. [CrossRef]
12. Häuselmann, P.; Granger, D.E.; Jeannin, P.Y.; Lauritzen, S.E. Abrupt glacial valley incision at 0.8 Ma dated from cave deposits in Switzerland. *Geology* **2007**, *35*, 143–146. [CrossRef]
13. Strasser, M.; Strasser, A.; Pelz, K.; Seyfried, H. A mid Miocene to early Pleistocene multi-level cave as a gauge for tectonic uplift of the Swabian Alb (Southwest Germany). *Geomorphology* **2009**, *106*, 130–141. [CrossRef]
14. Calvet, M.; Gunnell, Y.; Braucher, R.; Hez, G.; Bourles, D.; Guillou, V.; Delmas, M.; ASTER Team. Cave levels as proxies for measuring post-orogenic uplift: Evidence from cosmogenic dating of alluvium-filled caves in the French Pyrenees. *Geomorphology* **2015**, *246*, 617–633. [CrossRef]
15. Springer, G.S. Clastic sediments in caves. In *Encyclopedia of Caves*; Culver, D., White, W., Eds.; Elsevier: Cambridge, MA, USA, 2005; pp. 102–108.
16. Martini, I. Cave clastic sediments and implications for speleogenesis: New insights from the Mugnano Cave (Montagnola Senese, Northern Apennines, Italy). *Geomorphology* **2011**, *134*, 452–460. [CrossRef]
17. Bögli, A. *Karst Hydrology and Physical Speleology*; Springer: Berlin, Germany, 1980.
18. Palmer, A.N. Cave levels and their interpretation. *Bul. Natl. Speleol. Soc. Am.* **1987**, *49*, 50–66.
19. Farrant, A.R.; Smart, P.L.; Whitaker, F.F.; Tarling, D.H. Long-term Quaternary uplift rates inferred from limestone caves in Sarawak, Malaysia. *Geology* **1995**, *23*, 357–360. [CrossRef]
20. Piccini, L.; Drysdale, R.; Heijnis, H. Karst morphology and cave sediments as indicators of the uplift history in the Alpi Apuane (Tuscany, Italy). *Quat. Int.* **2003**, *101*, 219–227. [CrossRef]
21. Richards, D.A.; Dorale, J.A. Uranium-series chronology and environmental applications of speleothems. *Rev. Mineral. Geochem.* **2003**, *52*, 407–460. [CrossRef]
22. Ballesteros, D.; Giralt, S.; García-Sansegundo, J.; Jiménez-Sánchez, M. Quaternary regional evolution based on karst cave geomorphology in Picos de Europa (Atlantic Margin of the Iberian Peninsula). *Geomorphology* **2019**, *336*, 133–151. [CrossRef]
23. Rapetti, F.; Vittorini, S. *Carta Climatica Della Toscana Centro-Settentrionale*; Pacini Editore: Pisa, Italy, 1994.
24. Catasto on Line Delle Grotte Della Regione Toscana 2009. Federazione Speleologica Toscana. Available online: http://speleotoscana.it/programmi_php/catasto/menu.php (accessed on 14 May 2020).
25. Piccini, L.; Zanchetta, G.; Drysdale, R.N.; Hellstrom, J.; Isola, I.; Fallick, A.E.; Leone, G.; Doveri, M.; Mussi, M.; Mantelli, F.; et al. The environmental features of the Monte Corchia cave system (Apuan Alps, Central Italy). *Int. J. Speleol.* **2008**, *37*, 153–172. [CrossRef]
26. Piccini, L. Speleogenesis in highly geodynamic contexts: The Quaternary evolution of Monte Corchia multi-level karst system (Apuan Alps, Italy). *Geomorphology* **2011**, *134*, 49–61. [CrossRef]
27. Boccaletti, M.; Elter, P.; Guazzone, G. Plate tectonic models for the development of the western Alps and northern Apennines. *Nature* **1971**, *234*, 108–111. [CrossRef]
28. Carmignani, L.; Decandia, A.; Disperati, L.; Fantozzi, L.; Lazzarotto, A.; Liotta, D.; Oggiano, G. Relationships between the Tertiary structural evolution of the Sardinia-Corsica-Provensal Domain and the Northern Apennines. *Terra Nova* **1995**, *7*, 128–137. [CrossRef]
29. Elter, P. Introduction à la géologie de l'Appennin septentrional. *Bull. Soc. Géol. Fr.* **1975**, *7*, 956–962. [CrossRef]
30. Molli, G.; Meccheri, M. Structural inheritance and style of reactivation at mid-crustal levels: A case study from the Apuan Alps (Tuscany, Italy). *Tectonophysics* **2012**, *579*, 74–87. [CrossRef]
31. Pieruccioni, D.; Galanti, Y.; Biagioni, C.; Molli, G. Geology and tectonic setting of the Fornovolasco area, Alpi Apuane (Tuscany, Italy). *J. Maps* **2018**, *14*, 357–367. [CrossRef]
32. Vezzoni, S.; Biagioni, C.; D'Orazio, M.; Pieruccioni, D.; Galanti, Y.; Petrelli, M.; Molli, G. Evidence of Permian magmatism in the Alpi Apuane metamorphic complex (Northern Apennines, Italy): New hints for the geological evolution of the basement of the Adria plate. *Lithos* **2018**, *318–319*, 104–123. [CrossRef]

33. Patacca, E.; Scandone, P.; Conti, P.; Mancini, S.; Massa, G. Ligurian-derived olistostrome in the pseudomacigno formation of the stazzema zone (Apuan Alps, Italy). Geological implications at regional scale. *Ital. J. Geosci.* **2013**, *132*, 463–476. [CrossRef]
34. Cerrina Feroni, A.; Plesi, G.; Fanelli, G.; Leoni, L.; Martinelli, P. Contributo alla conoscenza dei processi di grado molto basso (anchimetamorfismo) a carico della Falda Toscana nell'area del ricoprimento apuano. *Boll. Soc. Geol. Ital.* **1983**, *102*, 269–280.
35. Kligfield, R.; Hunziker, J.; Dallmeyer, R.D.; Schamel, S. Dating of deformation phases using K-Ar and 40Ar/39Ar techniques: Results from the Northern Apennines. *J. Struct. Geol.* **1986**, *8*, 781–798. [CrossRef]
36. Molli, G.; Cortecci, G.; Vaselli, L.; Ottria, G.; Cortopassi, A.; Dinelli, E.; Mussi, M.; Barbieri, M. Fault zone structure and fluid–rock interaction of a high angle normal fault in Carrara marble (NW Tuscany, Italy). *J. Struct. Geol.* **2010**, *32*, 1334–1348. [CrossRef]
37. Molli, G.; Torelli, L.; Storti, F. The 2013 Lunigiana (Central Italy) earthquake: Seismic source analysis from DInSar and seismological data, and geodynamic implications for the northern Apennines. A discussion. *Tectonophysics* **2016**, *668–669*, 108–112. [CrossRef]
38. Carmignani, L.; Giglia, G. Il problema della doppia vergenza delle Alpi Apuane e la struttura del Monte Corchia. *Mem. Soc. Geol. Ital.* **1983**, *26*, 515–525.
39. Carmignani, L.; Conti, P.; Disperati, L.; Fantozzi, P.L.; Giglia, G.; Meccheri, M. *Carta Geologica del Parco delle Alpi Apuane 1: 50,000*; Carte Geologiche: Firenze, Italy, 2000.
40. Fallani, F.; Piccini, L. Pianta generale del Complesso Carsico del Monte Corchia. *Suppl. Talp Riv. Fed. Speleol. Toscana* **1990**, *2*, 10.
41. Molli, G.; Brovarone, A.V.; Beyssac, O.; Cinquini, I. RSCM thermometry in the Alpi Apuane (NW Tuscany, Italy): New constraints for the metamorphic and tectonic history of the inner northern Apennines. *J. Struct. Geol.* **2018**, *113*, 200–216. [CrossRef]
42. Conti, P.; Carmignani, L.; Massa, G.; Meccheri, M.; Patacca, E.; Scandone, P.; Pieruccioni, D. *Note Illustrative della Carta Geologica d'Italia alla scala 1:50,000 "Foglio 249—Massa Carrara"*; ISPRA—Servizio Geologico d'Italia: Roma, Italy, 2019; p. 290.
43. Isola, I.; Mazzarini, F.; Molli, G. Structural analysis in percolating fracture network in karst system: The Antro del Corchia Cave, Alpi Apuane, Italy. In Proceedings of the 28th IUGG Conference on Mathematical Geophysics, Pisa, Italy, 7–11 June 2010.
44. Baldacci, F.; Cecchini, S.; Lopane, G.; Raggi, G. Le risorse idriche del Fiume Serchio ed il loro contributo all'alimentazione dei bacini idrografici adiacenti. *Mem. Soc. Geol. Ital.* **1993**, *49*, 365–391.
45. Dickinson, W.R. Interpreting provenance relations from detrital modes of sandstones. In *NATO ASI Series (Series C: Mathematical and Physical Sciences)*; Provenance of Arenites; Zuffa, G.G., Ed.; Springer: Dordrecht, The Netherlands, 1985; Volume 148, pp. 333–361.
46. Zijderveld, J.D.A. AC Demagnetization of Rocks: Analysis of Results. In *Methods in Palaeomagnetism*; Runcorn, S.K., Creer, K.M., Collinson, D.W., Eds.; Elsevier: Amsterdam, The Netherlands, 1967; pp. 254–286.
47. Butler, R.F. *Paleomagnetism: Magnetic Domains to Geologic Terranes*; Blackwell Scientific Publications: Portland, OR, USA, 1992; p. 319.
48. Fisher, R. Dispersion on a sphere. *Proc. R. Soc. Lond.* **1953**, *A217*, 295–305. [CrossRef]
49. Drysdale, R.N.; Zanchetta, G.; Hellstrom, J.C.; Fallick, A.E.; Zhao, J.X.; Isola, I.; Bruschi, G. Palaeoclimatic implications of the growth history and stable isotope ($\delta^{18}O$ and $\delta^{13}C$) geochemistry of a Middle to Late Pleistocene stalagmite from central-western Italy. *Earth Planet. Sci. Lett.* **2004**, *227*, 215–229. [CrossRef]
50. Drysdale, R.N.; Zanchetta, G.; Hellstrom, J.C.; Fallick, A.E.; Zhao, J.X. Stalagmite evidence for the onset of the Last Interglacial in southern Europe at 129+/−1 ka. *Geophys. Res. Lett.* **2005**, *32*, 1–4. [CrossRef]
51. Drysdale, R.N.; Zanchetta, G.; Hellstrom, J.C.; Fallick, A.E.; McDonald, J.; Cartwright, I. Stalagmite evidence for the precise timing of North Atlantic cold events during the early last glacial. *Geology* **2007**, *35*, 77–80. [CrossRef]
52. Drysdale, R.N.; Zanchetta, G.; Hellstrom, J.C.; Fallick, A.E.; Sanchez-Goni, M.F.; Couchoud, I.; McDonald, J.; Maas, R.; Lohmann, G.; Isola, I. Evidence for obliquity forcing of glacial termination II. *Science* **2009**, *325*, 1527–1531. [CrossRef]
53. Drysdale, R.N.; Zanchetta, G.; Baneschi, I.; Guidi, M.; Isola, I.; Couchoud, I.; Piccini, L.; Greig, A.; Wong, H.; Woodhead, J.; et al. Partitioning of Mg, Sr, Ba and U into a subaqueous calcite speleothem. *Geochim. Cosmochim. Acta* **2019**, *264*, 67–91. [CrossRef]
54. Drysdale, R.; Couchoud, I.; Zanchetta, G.; Isola, I.; Regattieri, E.; Hellstrom, J.; Govin, A.; Tzedakis, P.C.; Ireland, T.; Corrick, E.; et al. Magnesium in subaqueous speleothems as a potential palaeotemperature proxy. *Nat. Commun.* **2020**, *11*, 1–11. [CrossRef] [PubMed]
55. Zanchetta, G.; Drysdale, R.N.; Hellstrom, J.C.; Fallick, A.E.; Isola, I.; Gagan, M.; Pareschi, M.T. Enhanced rainfall in the western Mediterranean during deposition of sapropel S1: Stalagmite evidence from Corchia Cave (Central Italy). *Quat. Sci. Rev.* **2007**, *26*, 279–286. [CrossRef]
56. Zanchetta, G.; Regattieri, E.; Isola, I.; Drysdale, R.N.; Bini, M.; Baneschi, I.; Hellstrom, J.C. The so-called "4.2 event" in the central Mediterranean and its climatic teleconnections. *Alp. Mediterr. Quat.* **2016**, *29*, 5–17.
57. Regattieri, E.; Zanchetta, G.; Drysdale, R.; Isola, I.; Hellstrom, J.; Dallai, L. Late glacial to Holocene trace element record (Ba, Mg, Sr) from Corchia Cave (Apuan Alps, central Italy): Paleoenvironmental implications. *J. Quat. Sci.* **2016**, *29*, 381–392. [CrossRef]
58. Bajo, P.; Borsato, A.; Drysdale, R.; Hua, Q.; Frisia, S.; Zanchetta, G.; Hellstrom, J.; Woodhead, J. Stalagmite carbon isotopes and dead carbon proportion (DCP) in a near-closed-system situation: An interplay between sulphuric and carbonic acid dissolution. *Geochim. Cosmochim. Acta* **2017**, *210*, 208–227. [CrossRef]
59. Bajo, P.; Drysdale, R.N.; Woodhead, J.D.; Hellstrom, J.C.; Hodell, D.; Ferretti, P.; Voelker, A.; Zanchetta, G.; Rodrigues, T.; Wolff, E.; et al. Persistent influence of obliquity on ice age terminations since the Middle Pleistocene transition. *Science* **2020**, *367*, 1235–1239. [CrossRef]

60. Isola, I.; Zanchetta, G.; Drysdale, R.N.; Regattieri, E.; Bini, M.; Bajo, P.; Hellstromm, J.; Baneschi, I.; Lionello, P.; Woodhead, J.; et al. The 4.2 ka event in the central Mediterranean: New data from a Corchia speleothem (Apuan Alps, central Italy). *Clim. Past* **2019**, *15*, 135–151. [CrossRef]
61. Isola, I.; Ribolini, A.; Zanchetta, G.; Bini, M.; Regattieri, E.; Drysdale, R.N.; Hellstrom, J.; Bajo, P.; Montagna, P.; Pons-Branchu, E. Speleothem U/Th age constraints for the Last Glacial conditions in the Apuan Alps, northwestern Italy. *Palaeogeogr. Palaeoclimatol. Palaeoecol.* **2019**, *518*, 62–71. [CrossRef]
62. Hellstrom, J. U–Th dating of speleothems with high initial ^{230}Th using stratigraphical constraint. *Quat. Geochronol.* **2006**, *1*, 289–295. [CrossRef]
63. Woodhead, J.; Hellstrom, J.; Maas, R.; Drysdale, R.; Zanchetta, G.; Devine, P.; Taylor, E. U–Pb geochronology of speleothems by MC-ICPMS. *Quat. Geochronol.* **2006**, *1*, 208–221. [CrossRef]
64. Simon, Q.; Bourlès, D.L.; Thouveny, N.; Horng, C.S.; Valet, J.P.; Bassinot, F.; Choy, S. Cosmogenic signature of geomagnetic reversals and excursions from the Réunion event to the Matuyama–Brunhes transition (0.7–2.14 Ma interval). *Earth Planet. Sci. Lett.* **2018**, *482*, 510–524. [CrossRef]
65. Channell, J.E.T. Magnetic excursions in the late Matuyama Chron (Olduvai to Matuyama-Brunhes boundary) from North Atlantic IODP sites. *J. Geophys. Res. Solid Earth* **2017**, *122*, 773–789. [CrossRef]
66. Frank, M.; Schwarz, B.; Baumann, S.; Kubik, P.W.; Suter, M.; Mangini, A. A 200 kyr record of cosmogenic radionuclide production rate and geomagnetic field intensity from ^{10}Be in globally stacked deep-sea sediments. *Earth Planet. Sci. Lett.* **1997**, *149*, 121–129. [CrossRef]
67. Piccini, L. Caratteri morfologici ed evoluzione dei fenomeni carsici profondi nelle Alpi Apuane (Toscana—Italia). 1994. *Nat. Brescia.* **1996**, *30*, 45–85.
68. Bigazzi, G.; Di Pisa, A.; Gattiglio, M.; Meccheri, M.; Norelli, P. La struttura cataclastico milonitica di foce di Mosceta Alpi Apuane sud-orientali (M. Corchia, Gruppo delle Panie). *Atti Soc. Tosc. Sci. Nat. Mem. Ser. A* **1988**, *95*, 105–116.
69. Conti, P.; Di Pisa, A.; Gattiglio, M.; Meccheri, M. The pre-alpine basement in the Apuan Alps (Northern Apennines, Italy). In *Pre-Mesozoic Geology in the Alps*; Springer Science and Business Media LLC: Berlin/Heidelberg, Germany, 1993; pp. 609–621.
70. Puccinelli, A.; D'Amato Avanzi, C.; Perilli, N. *Note Illustrative della Carta Geologica d'Italia alla Scala 1:50,000 "Foglio 250—Castelnuovo Garfagna"*; ISPRA—Servizio Geologico d'Italia: Roma, Italy, 2019; p. 166.
71. Engel, J.; Woodhead, J.; Hellstrom, J.; Maas, R.; Drysdale, R.; Ford, D. Corrections for initial isotopic disequilibrium in the speleothem U-Pb dating method. *Quat. Geochronol.* **2019**, *54*, 101009. [CrossRef]
72. Bennett, R.A.; Serpelloni, E.; Hreinsdóttir, S.; Brandon, M.T.; Buble, G.; Basic, T.; Casale, C.; Cavaliere, A.; Anzidei, M.; Marjonovic, M.; et al. Syn-convergent extension observed using the RETREAT GPS network, northern Apennines, Italy. *J. Geophys. Res.* **2012**, *117*, B04408. [CrossRef]
73. Piccini, L. Evolution of karst in the Apuan Alps (Italy): Relationships with the morphotectonic history. *Suppl. Geogr. Fis. Din. Quat.* **1998**, *III-4*, 21–31.

geosciences

Review

Active Fault Systems in the Inner Northwest Apennines, Italy: A Reappraisal One Century after the 1920 Mw ~6.5 Fivizzano Earthquake

Giancarlo Molli [1,*], Isabelle Manighetti [2], Rick Bennett [3], Jacques Malavieille [4], Enrico Serpelloni [5], Fabrizio Storti [6], Tiziano Giampietro [2], Aurelien Bigot [4], Gabriele Pinelli [1], Serena Giacomelli [7], Alessio Lucca [6], Luca Angeli [1] and Lorenzo Porta [1]

1. Dipartimento Scienze Della Terra, Università di Pisa, 56126 Pisa, Italy; g.pinelli@marmotest.com (G.P.); lucaangeli12@gmail.com (L.A.); portalorenzo4@gmail.com (L.P.)
2. Université Côte d'Azur, Observatoire de la Côte d'Azur, French Research Institute for Development (IRD), French National Center for Scientific Research (CNRS), Géoazur, Sophia Antipolis, 06905 Valbonne, France; manighetti@geoazur.unice.fr (I.M.); giampietro@geoazur.unice.fr (T.G.)
3. Department of Geosciences, University of Arizona, Tucson, AZ 85721, USA; rb0@email.arizona.edu
4. Géosciences Montpellier, Université de Montpellier, 34095 Montpellier, France; j.malavie@gmail.com (J.M.); aurelien.bigot@hotmail.fr (A.B.)
5. Istituto Nazionale di Geofisica e Vulcanologia, Osservatorio Nazionale Terremoti, 40128 Bologna, Italy; enrico.serpelloni@ingv.it
6. Natural and Experimental Tectonics Research Group, Department of Chemistry, Life Sciences and Environmental Sustainability, University of Parma, 43124 Parma, Italy; fabrizio.storti@unipr.it (F.S.); alessio.lucca@unipr.it (A.L.)
7. Dipartimento di Scienze Biologiche, Geologiche e Ambientali, Università di Bologna, 40126 Bologna, Italy; giacserena@gmail.com

* Correspondence: giancarlo.molli@unipi.it

check for updates

Citation: Molli, G.; Manighetti, I.; Bennett, R.; Malavieille, J.; Serpelloni, E.; Storti, F.; Giampietro, T.; Bigot, A.; Pinelli, G.; Giacomelli, S.; et al. Active Fault Systems in the Inner Northwest Apennines, Italy: A Reappraisal One Century after the 1920 Mw ~6.5 Fivizzano Earthquake. *Geosciences* **2021**, *11*, 139. https://doi.org/10.3390/geosciences11030139

Academic Editors: Jesus Martinez-Frias and Rodolfo Carosi

Received: 31 January 2021
Accepted: 12 March 2021
Published: 18 March 2021

Publisher's Note: MDPI stays neutral with regard to jurisdictional claims in published maps and institutional affiliations.

Copyright: © 2021 by the authors. Licensee MDPI, Basel, Switzerland. This article is an open access article distributed under the terms and conditions of the Creative Commons Attribution (CC BY) license (https://creativecommons.org/licenses/by/4.0/).

Abstract: Based on the review of the available stratigraphic, tectonic, morphological, geodetic, and seismological data, along with new structural observations, we present a reappraisal of the potential seismogenic faults and fault systems in the inner northwest Apennines, Italy, which was the site, one century ago, of the devastating Mw ~6.5, 1920 Fivizzano earthquake. Our updated fault catalog provides the fault locations, as well as the description of their architecture, large-scale segmentation, cumulative displacements, evidence for recent to present activity, and long-term slip rates. Our work documents that a dense network of active faults, and thus potential earthquake fault sources, exists in the region. We discuss the seismogenic potential of these faults, and propose a general tectonic scenario that might account for their development.

Keywords: active faults; earthquakes; inner northwest Apennines; current deformation; satellite geodesy

1. Introduction

Earthquake data, including instrumental and historical earthquakes, are critical information to constrain seismic-hazard assessment [1–4]. However, these data cover periods of time that are generally much shorter than the recurrence times of large earthquakes (i.e., with Mw $\geq$ 6.0) on a given fault and regionally, especially in areas of low deformation rates [5–7], as is the case in Italy. Paleoseismological data [8] may complete and complement the seismic catalogs, but paleoseismological studies are generally difficult to conduct extensively—as is the case in Italy at the present moment—and they generally recover partial information on the past earthquakes. Nevertheless, earthquakes result from the rupture of tectonic faults that are generally long-lived features, and it has been shown that documenting these long-term faults provides critical insights to anticipate some of the properties of the large earthquakes these faults may produce in the future [5,6,9]. On the other hand, identifying where seismogenic faults are in a region is key to anticipate where earthquakes might occur in the future (e.g., [9]).

Faults have intrinsic properties, such as a 3D overall architecture, a lateral segmentation of their principal trace, a certain degree of structural maturity, a given average slip rate, etc., and these properties have been shown to control, at least partly, the earthquake behavior [10–12]. In particular, the fault structural maturity, which relates to the fault slip longevity and rate over geological time, markedly impacts the geometrical and mechanical properties of the fault, which in turn controls the earthquake rupture size (rupture length and slip amplitude) and "energy" (i.e., stress drop); the more immature a fault, the greater its capacity to produce large stress drop, and hence highly damaging earthquakes [11,13,14]. Therefore, identifying seismogenic faults in a given region and documenting their general properties (lateral segmentation, connection to other faults, cumulative displacements, slip rates, etc.) constitutes a fundamental basis to implement and improve the seismotectonic models that are used as input for seismic hazard estimates [9].

In the present study, we conduct this work in the inner northwest Apennines region of Italy. That region was the site, one century ago, of a major devastating earthquake, the 1920 Fivizzano event, the magnitude of which is estimated at Mw 6.5, similar to that of the dramatic main shock of the 2016 Central Italy earthquake sequence [15]. This region also hosted several other moderate to large historical earthquakes over the last few centuries. Despite the occurrence of all these earthquakes, the Italian database of seismogenic structures (DISS-INGV) [16] includes only two seismic sources in the whole region. We document here that many more potential earthquake fault sources exist in the region (some are referred in the Ithaca fault catalog, ISPRA [17]), and should be taken into account in seismic-hazard evaluation.

Based on a comprehensive review of the available geological, tectonic, chronological, morphological, geodetic, and seismological (instrumental and historical) data for the inner northwest Apennines, along with new tectonic data from our own work, we present a reappraisal of the active or possibly active seismogenic faults and fault systems in the internal Apennines north of the Arno river. Our updated fault catalog provides the fault locations, as well as the description of their architecture, large-scale segmentation, cumulative displacements, evidence for recent to present activity, and long-term slip rates. It also includes the offshore domains of the northern Tyrrhenian coast so far ignored in previous compilations [16,17]. Our work eventually allows us to discuss the seismogenic potential of the identified faults, while proposing a general tectonic scenario accounting for these faults.

2. Main Structural Domains in the Northern Apennines

The northern Apennines (Figure 1) are the result of the Neogene subduction of the Adria continental crust and the overlying remnants of a former intraoceanic accretionary wedge [18–20]. The latter is represented by the so-called Ligurian and sub-Ligurian units, which may be observed superimposed to the Tuscan and Romagna–Umbria continental-derived cover and basement of the distal to proximal Adria margin [18–21]. After the inception of the continental subduction during the Oligocene, parts of the Adria continental margin were imbricated and incorporated into the orogenic wedge at a shallow crustal depth (e.g., the Tuscan Nappe), whereas other portions were underplated at deeper crustal levels (e.g., the Tuscan Metamorphic units) and lately exhumed at the surface within tectonic windows in the internal, Tyrrhenian side of the orogen [19–21].

The Neogene-to-recent tectonic evolution of the Apennines has been characterized by contractional tectonic activity in the foreland, accompanied by extension in the internal domain [22–24]. This deformation has produced a great density and variety of fault systems with different kinematics, which may be seen as defining different morphostructural domains (Figure 1), as described below [25–29].

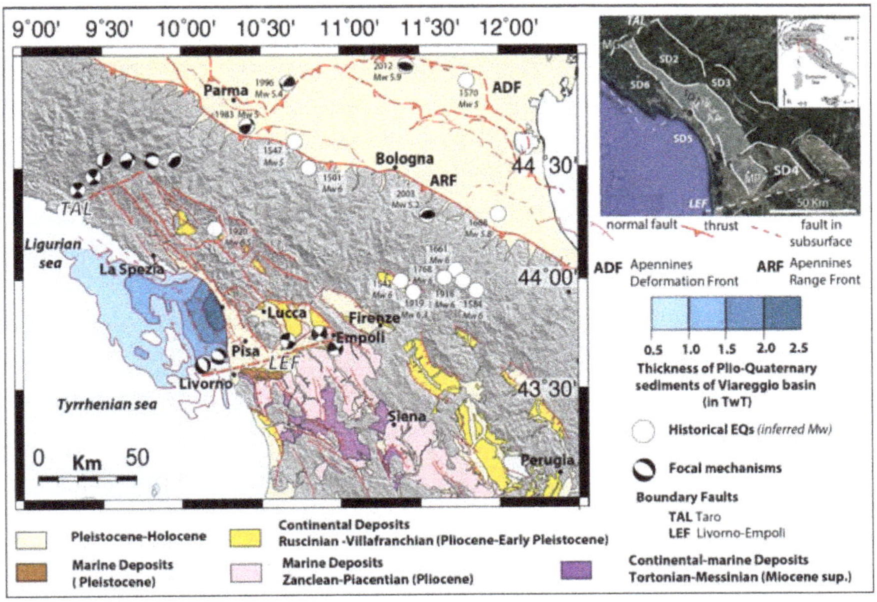

Figure 1. Simplified geological and structural map of the northern Apennines emphasizing the Neogene tectonics in the investigated area. Structural domains (SDs) discussed in the paper are shown in the Google Earth inset. AA: Alpi Apuane; LEF: Livorno–Empoli Fault; MG: Mt. Gottero; MP: Mt. Pisano; TAL: Taro Line. TAL and LEF bound the region of study. Offshore sediment thickness are after [30,31], and earthquakes and focal mechanisms are from the INGV archive [2,32].

The major fault systems in the external domain are the contractional fronts, divided into the northern Apennines Deformation Front (ADF) and the Apennines Range Front (ARF). The northeast Apennines Deformation Front is presently primarily buried beneath the Po Valley and extends offshore in the western Adriatic Sea [23,24,28,29]. Recent seismicity, including the 2012 Emilia Mw 5.9 thrust earthquake [4,33], along with geomorphic studies [29], clearly show that folding and thrusting associated with the ADF are currently active. The Apennines Range Front [25,27–29] lies some 60 km to the southwest of the ADF. The ARF front is marked by a transition from gently dipping alluvial strata in the Po basin to uplifted bedrock in the Apennines, presently standing up to 2000 m above the Po plain near the orographic divide. Different authors have provided evidence that the topography of the range front represents the forward-dipping limb of an anticline overlying a fairly steeply dipping blind reverse fault [29,34]. Similarly to the ADF, the ARF is currently active, as documented by GPS measurements of ongoing crustal deformation [25] and instrumental seismic activity including some recent moderate earthquakes, such as the 1983 Mw 5.0 "Parma" and the 2003 Mw 5.2 "Monghidoro" thrust ruptures [1,2,4,35,36]. Historical earthquakes are also reported on the ARF, such as the 1501 Mw 5 "Modenese", the 1547 Mw 6 "Reggiano", and the 1688 Mw 5.8 "Romagna" earthquakes [37].

The inner Apennines are characterized by a lower frequency of earthquakes, yet of significant magnitude, with most of them showing a normal faulting slip mode. Among the several large earthquakes that have occurred in the region, there are the 1542 and the 1919 "Mugello" events (Mw 6.0 and Mw 6.4, respectively), the 1584, 1661, 1768, and 1918 events in the "Forlivese" region (all ~ Mw 6), and the 1920 ~Mw 6.5 Fivizzano earthquake [35,37,38]. This latter earthquake will be discussed in some detail in the following.

3. Faulting in Main Morphostructural Domains

The characteristics of the faults described in this section are summarized in Table S1. The fault and geological maps described below are derived from the combination of available studies to which we refer, and from our work.

The study area extends over more than 6000 km^2 on the Tyrrhenian side of the northern Apennines (Figure 1). From a morphostructural point of view, it is limited by the main Apenninic orographic divide to the east/northeast, and by the northern Tyrrhenian–Ligurian Sea to the west/southwest. More precisely, the northern boundary of the region coincides with an alignment of reliefs with a transversal, nearly east–west trend bounding the Taro valley, whereas to the South, the region is bounded by the shallow hilly reliefs between Livorno and Empoli (Figure 1).

These northern and southern boundaries represent the surface expression of crustal/lithospheric discontinuities [39,40], corresponding to the westernmost part of the Taro Line (TAL) to the north [18,41–44] and to the westernmost part of the Livorno–Sillaro Line [45–47], called the Livorno–Empoli Fault (LEF in Figure 1), to the south. The Taro Line, already recognized in early studies (refs. in [39,41,42,48]), has been described in different contributions due to its prominent morphological signature [41–43] and seismic activity [4,35,49]. In the most recent interpretations, its westernmost segment, within our region of interest, is associated with a lateral ramp of a basement thrust, named the Taro lateral ramp, that would accommodate a NE–SW crustal shortening [42–44,50]. On the other hand, the Livorno–Empoli Fault coincides with a significant lateral change in the thickness of the crust, with the northern Tuscany crust being ~28 km thick north of the fault and ~20 km thick south of it [47,51,52]. The geology also changes across the LEF, with the base of Late Miocene marine basins exposed at the surface south of the fault (in the southern Tuscany Tectonic Province), while the same horizons are buried in the subsurface north of it (e.g., in the Viareggio basin). Both the Taro Line and the Livorno–Empoli Fault are associated with well-documented deformations within the Neogene–Quaternary deposits [44,46,47]. Their activity is still ongoing, as attested by historical and instrumentally recorded seismicity [49,51,52] (Figure 1).

In between the TAL and LEF, first-order physiographic and structural features allow the definition of six principal structural domains (SDs) (Figure 1). These are related to a set of ranges with a dominant NW to NNW trend (Apenninic trend) separated by intramontane or continental/marine morphotectonic depressions. The division into domains mainly aims to simplify their description.

The structural domain SD1 is bounded toward the Apenninic divide by the upper Lunigiana and Garfagnana grabens, which form the structural domains SD2 and SD3 to the north, and by the Lucca plain and the Montecarlo–Vinci hills (structural domain SD4) to the south. West of SD1 in the southern part of the Alpi Apuane–Mt. Pisano, the structural domain SD5 corresponds to the Versilia and Pisa plains, which are parts of a major submarine half-graben, the Viareggio basin. The latter grades northward to the structural domain SD6, a mixed on-land/offshore domain hosting the Vara Valley–lower Lunigiana tectonic depressions and the western promontory of La Spezia. Plio-Quaternary stratigraphic units form the sedimentary filling of the intramontane or continental/marine domains in all these zones (Figure 2). These Plio-Quaternary sedimentary records will be used to document the neotectonic activity of the principal faults and systems in each of the six structural domains.

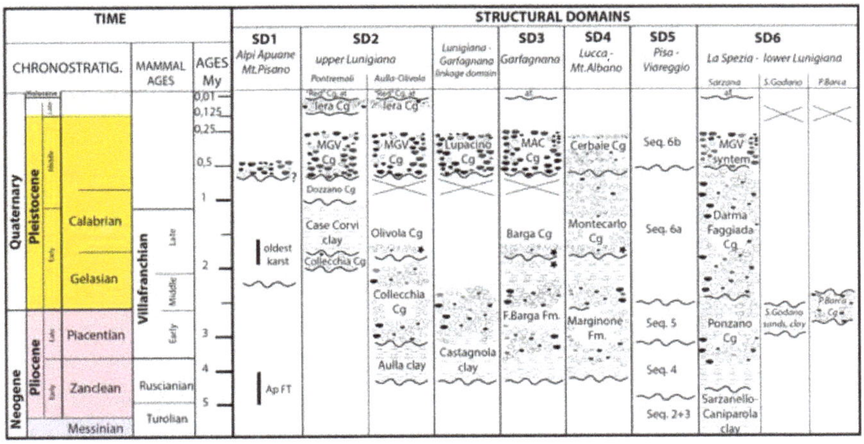

Figure 2. Schematic frame and correlations of continental–marine Plio-Quaternary sedimentary successions in the different structural domains described (MGV and MAC mean Magra Valley and Macigno-rich conglomerates, respectively).

4. Plio-Quaternary Stratigraphic Units, Fault Systems, Time of Activity and Displacements

4.1. Structural Domain SD1: The Mt. Gottero–Alpi Apuane–Mt. Pisano

The present-day large-scale geomorphology of the structural domain SD1 is dominated by the sharp relief of the Alpi Apuane (AA in inset of Figure 1), with its highest summits (~2000 m) only a few kilometers away from the Tyrrhenian coastline. North of the Alpi Apuane, the relief reaches ~1600 m on Mt. Gottero, whereas to the south Mt. Pisano stands at an elevation of ~900 m. The metamorphic units of the Alpi Apuane and the southernmost Mt. Pisano have been the object of low-temperature thermochronological investigations [53,54], which constrained the latest stages of exhumation as occurring in the last 4–5 myrs. This exhumation is taken to have resulted from erosion associated with normal faulting along the steep bounding faults (Figure 1) separating the exposure of the metamorphic rocks from the surroundings [50,54]. The estimated exhumation rates are about 0.5–0.8 mm/yr [54,55]. Within this SD1 domain, few remnants of continental deposits of Pleistocene (or Middle Pleistocene?) age (Figure 2) may be found locally, at an elevation of ~1000 m in the southeast Alpi Apuane [56,57].

4.2. Structural Domain SD2: Upper Lunigiana

Structural domain SD2 includes the upper Lunigiana, a valley hosting the upper course of the Magra River (Figure 3). Continental stratigraphic sequences are observed in the northwest part close to Pontremoli and in the central-southeast area close to Aulla (Figures 2 and 3). The deposits in these two areas show the same sedimentary evolution with a transition from fluvio-lacustrine to fluvial environments, and a diachronous sedimentation (Figure 2).

In the southernmost sector, the Aulla clays, with an estimated thickness of 70–80 m [58,59], mainly consist of fine-grained lithofacies, with medium to thick coarse-grained interbeds of sandstones and conglomerates. The overlaying coarse-grained Collecchia conglomerates, interpreted as alluvial fan deposits formed in a braided fluvial environment, directly cover the bedrock in the marginal zone of the former basin and overlay the Aulla clays in the depocentral area, where they show a thickness of ~150 m [58,59].

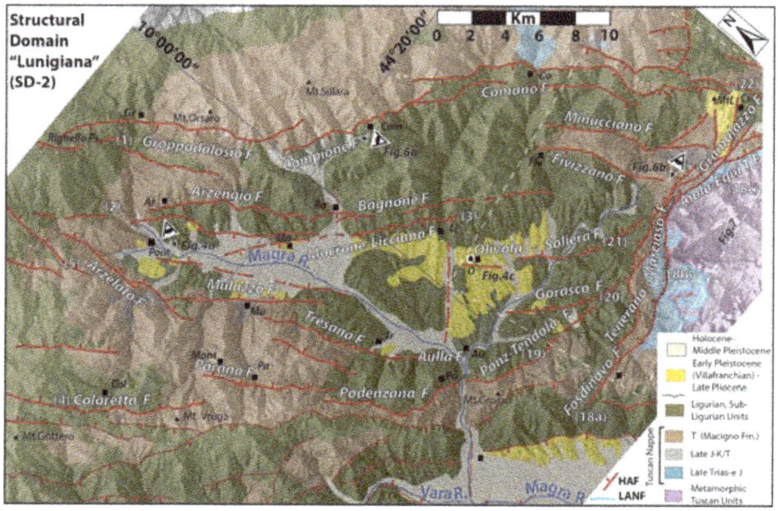

Figure 3. Main fault systems and geology in structural domain SD2 (Lunigiana). The map is derived from [57–60] and our work. Fault identification numbers are as reported in Table S1. HAF: steep normal faults (thicker lines for those described); LANF: low-angle detachment faults; Ar: Arzengio; Au: Aulla; Ba: Bagnone; Co: Comano; Col: Coloretta; Com: Compione; Cs: Castagnola; Fiv. Fivizzano; Gr: Groppodalosio; Li: Licciana; Mont: Montereggio; MtL: Mt. Lupacino; O: Olivola; Pa: Parana; Po: Podenzana; Pont: Pontremoli.

The Olivola conglomerates unconformably overlay the Collecchia conglomerates and are only preserved close to the Olivola village. The sequence starts with clay–sandy deposits, containing the Olivola mammals remains of Late Villafranchian age [59–62], followed by clast-supported and well-cemented conglomerates and sandstones (Figures 2–4). The conglomerates, with a thickness of about 60 m in the type-area, include clasts of metamorphic rocks belonging to the Tuscan metamorphic units.

The sequence exposed close to Pontremoli starts with a ~10 m thick polygenic conglomerate followed by the Case Corvi clays, which mainly consist of blue-grey clays and silty clays with lignite fragments, grading to sandstones and conglomerates, which become more abundant and coarser-grained in the upper portion. The land fossil remains and pollen in the finer-grain deposits allow attributing the unit to the Late Villafranchian [63–65].

The following Dozzano conglomerates mainly consist of coarse-grained sediments and subordinate lenses of sandstones and clays. The sequence, etheropic with the Case Corvi clays, is characterized by channel-shape deposits of well-sorted polygenic conglomerate with poor matrix and with thin sandstone and clay levels. It is followed by a polygenic conglomerate with clay matrix and a thickness of ~20 m. The sedimentary facies suggest debris-flow processes that prograded from the margins basinward. The unit is suggestive of Early Pleistocene, since its deposition occurred between that of the Case Corvi clays and that of the Magra Valley Conglomerates [58,66].

The two continental Villafranchian sequences of the Aulla–Olivola and Pontremoli areas have been related by some authors to independent and distinct basins [59,65–67], whereas other authors have interpreted them as being connected to a single "Lunigiana basin" with a diachronous development and a later segmentation during the post-Villafranchian time [58].

Figure 4. Evidence of Plio-Quaternary faulting and deformation: (**a**) LiDAR-derived DTM showing the fault scarp offsetting a Late Pleistocene–Holocene alluvial fan along the Mulazzo Fault in the structural domain SD2 (Lunigiana). This site has been recently trenched for paleoseismological studies [68,69]; (**b**) fault scarp offsetting Late Pleistocene alluvial fan deposit along the southernmost splay of the Corfino Fault, in structural domain SD3 (Garfagnana); (**c**) normal fault offsetting the Late Villafranchian Olivola Conglomerates, in structural domain SD2 (Lunigiana); (**d**) Ruscinian Sarzanello–Caniparola sands and clay tilted by the Sarzana–Carrara fault, in the northern part of the West Alpi Apuane Fault system in the structural domain SD6 (La Spezia–lower Lunigiana). Locations of these sites are shown in the Figures related to the different SDs.

From a structural point of view, domain SD2 represents a ~40 km long, NW-trending graben with a maximum width of about 15–20 km in the south of the basin and decreasing northwestwards (Figure 3). The graben is bounded by antithetic high-angle normal fault systems overprinting a polydeformed contractional thrust stack and a later system of low-angle extensional detachments [50,68,69].

To the east, the west-dipping faults form 3 subparallel, NW-trending, closely spaced (a few km) fault systems running over a total length of more than 40 km along the entire upper Lunigiana extensional domain [58,70,71]. From east to west these systems are the Groppodalosio–Compione–Comano, the Arzengio–Bagnone, and the Mocrone–Licciana strands (Figure 3). Each of these fault systems is segmented laterally into a few major segments (whose names constitute the system names), most of 10–15 km length. Some of these segments have been studied in detail. The Groppodalosio fault segment runs over ~18 km from the Passo del Righello (west of Cisa Pass) to the southwest of Mt. Sillara. It has a strike from N 130° to N 170° and shows a steep dip and normal slip kinematics with a cumulative displacement of ~400 m near Groppodalosio. The Compione–Comano fault segments [59,66,70–74] form a ~20 km long fairly connected strand dipping 60–70° to the SW. The Groppodalosio and Compione–Comano fault segments show a right-stepping "en échelon" pattern suggestive of a left-lateral component of slip on the whole system, in addition to the dominant normal one. On the Compione–Comano fault zone, a cumulative vertical displacement of more than 2000 m has been inferred from surface and subsurface data [50,59,75]. This suggests that the cumulative vertical slip decreases from south to north along the whole Groppodalosio–Compione–Comano fault system. The Late Pleistocene era conglomerates appear vertically displaced across of the Groppodalosio and the Compione–Comano faults [59,66], attesting for their slip motions in the last 100 kyrs.

The Arzengio–Bagnone and the Mocrone–Licciana faults extend over a total length of ~25 km. Their largest cumulative vertical displacement is estimated at ~600 m [59,72,73]. The Arzengio fault is considered to have been active mainly in the pre-Middle Pleistocene times [59,66,67]. The Mocrone–Licciana Fault, which is closest to the graben axis, has a maximum throw of ~400 m. It is taken to have been active during and after the Middle Pleistocene [66,67]. Two small adjacent faults, Fivizzano and Minucciano, have been suggested to be inactive since the Middle Pleistocene [66].

The antitethic NE-dipping fault system that bounds the graben to the west includes the Coloretta–Parana–Podenzana–Tendola and the Arzelato–Mulazzo–Tresana–Aulla fault zones [50,59,73,74]. The former has a total length of about 35 km, while each segment is about 10–15 km long. The maximum cumulative vertical throw, ~700 m, is observed in the northernmost part of the fault system, where a dip-slip normal kinematics and a Late Pleistocene, possibly Holocene activity are inferred [59,73].

The Arzelato–Mulazzo–Tresana–Aulla fault system bounds the upper Lunigiana graben (Figures 3 and 4a). In its northern part, its normal dip-slip kinematics are attested by various indicators (some close to Mulazzo), while a cumulative vertical displacement of ~1000 m is observed. The fault system has been suggested to be active in the Holocene [59,66,74], which paleoseismological trenching has confirmed recently [68,69].

Both the Coloretta–Parana–Podenzana and the Arzelato–Mulazzo–Tresana–Aulla faults have their strike changing from the NW–SE to W/NW–E/SE toward their southernmost termination, where they seem to connect to the North Apuane Fault System (described below, Fosdinovo–Tenerano–Marciaso–Aiola–Equi Terme Faults in Figure 3).

Recent geomorphic analysis have been locally performed in the upper Lunigiana [66,73] with the aim to estimate the cumulative fault throws and their variations along strike, the fault segmentation pattern, and the possible linkages among the fault segments and systems [59,66,68]. Among others, these studies have derived late Quaternary throw rates in the range of 0.4–0.8 mm/year, in fair agreement with displacement rates suggested in earlier works from geological studies (Table S1).

4.3. Structural Domain SD3: The Garfagnana

Structural domain SD3 includes the Garfagnana, an intramontane depression where the Serchio valley is developed (Figure 6). The Plio-Quaternary stratigraphic units exposed in the area (Figure 2) are characterized by:

(a) Basal, fine-grained lithofacies, mainly clays and silty clays locally interbedded with coarse-grained conglomerates alternating to thin beds and lenses of sands and organic-rich horizons [56,76–78]. Thin dark grey paleosols are also present, at different levels of the clayed–sandy sediments and related to local emersion of the depositional area [56,79–81]. These dominantly fine-grained basal lithofacies, called Fornaci di Barga clays, have a maximum thickness of 170–200 m (inferred in exploration wells near Pieve di Fosciana and Barga [56]). According to the paleontological contents, the formation is attributed to Middle–Late Pliocene; in particular the lower portion could be assigned to the Ruscinian–Early Villafranchian, while the upper portion could reach the Late Middle Villafranchian [56,79–81];

(b) Barga conglomerates, which are dominated by coarse-grained lithofacies with subordinate silty to sandy interbeds mainly distributed into the lowermost portion where organic-rich horizons also are observed. In contrast with the conglomerates interbedded within the Fornaci di Barga clays, the Barga conglomerates are formed more than 50% by Mesozoic carbonate clasts sourced from both the unmetamorphic and the metamorphic Tuscan units. The thickness of the sequence ranges from 70 to ~170 m and can be referred to the Late Villafranchian, based on its correlation with the similar Olivola conglomerates [79–81].

Commonly, the sedimentary record of the Villafranchian sequences is interpreted to attest to a depositional environment with a transition from lacustrine to fluvial [76–78]. More recently, however, some authors [79,80] have related its formation to an alluvial system with NW–SE flow direction developed in a humid subtropical climate regime, followed by the development of a gravel bed-load braided system.

(c) Post-Villafranchian deposits are represented by: (i) Macigno-derived conglomerates; and (ii) terrace alluvial deposits, with decreasing altitude above the present riverbed [66,77,78].

The distribution of the Villafranchian and post-Villafranchian deposits along the present Garfagnana and their internal structural setting document significant syn- and post-Villafranchian [66,81–83] vertical movements and deformations related to the activity of fault systems described hereafter (Figures 5 and 6).

Figure 5. (**a**) Panoramic view (toward N/NW) of the Compione Fault (structural domain SD2). (**b**) Panoramic view (toward W/SW) of the Aiola–Equi Terme fault part of the North Apuane Fault System. The village in the valley is Equi Terme, epicentral area of the 2013 Lunigiana earthquake. Photo taken from the road to Ugliancaldo Village, epicentral area of the 1837 Alpi Apuane earthquake; (**c**) Corfino fault in structural domain SD3. (**d**) Monte Perpoli morphostructural high in SD3. White line indicates the base of the Late Pleistocene deposits in the hanging wall and footwall of the North Mt. Perpoli fault. The latter is taken to have produced the 2013 Garfagnana earthquake. (**e**) Pietrasanta fault (SD5), central southern segment of the West Alpi Apuane fault, the major bounding fault of the Viareggio basin. (**f**) Basin and range landscape forming the western boundary of the Pisa plain (southernmost part of the structural domain SD5). Stereonets show microtectonic measurements we made on the faults (minor fault planes and slickenslides) (Schmid net, lower hemisphere).

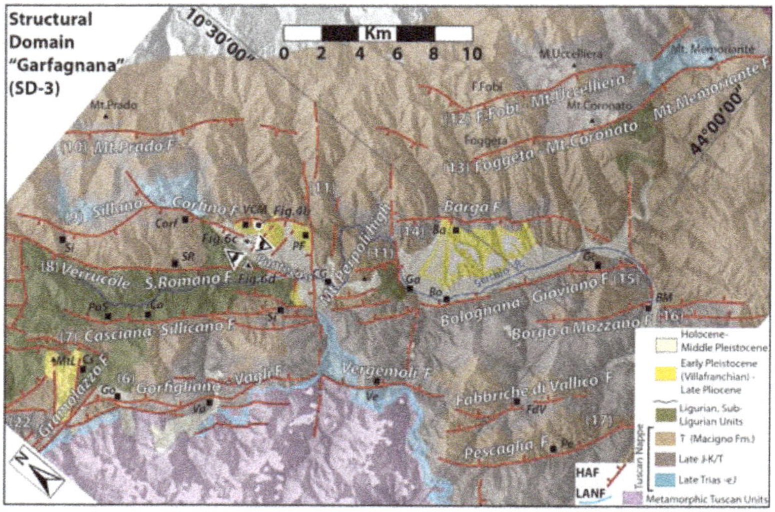

Figure 6. Main fault systems and geology in structural domain SD3 (Garfagnana). The map is derived from [56,57,77,84] and our work. Fault identification numbers are as reported in Table S1. HAF: steep normal faults; LANF: low-angle detachment faults; Ba: Barga; BM: Borgo a Mozzano; Bo: Bolognana; Ca: Casciana; CG: Castelnuovo Garfagnana; Cs: Castagnola; Corf: Corfino; Ga: Gallicano; Gi: Gioviano; Go: Gorfigliano; MtL: Mt. Lupacino; PaS: Piazza al Serchio; Pe: Pescaglia; PF: Pieve Fosciana; Si: Sillicano; SR: San Romano; Va: Vagli; VCM: Villacollemandina; Ve: Vergemoli.

From a structural point of view, domain SD3 represents a ~30 km long, NW-trending graben with a maximum width of about 20 km in the SE, slightly decreasing northwestward (Figure 6). As with SD2, it is bounded by antithetic steep normal fault systems overprinting the contractional thrust stack and a set of low-angle extensional detachments, mainly located at the base and within the Tuscan unmetamorphic unit [50]. SD3 is divided in its center by the Mt. Perpoli high (between Castelnuovo Garfagnana and Barga), a normal fault bounded horst trending subperpendicular to the graben faults (Figure 5d). This transverse relief divides SD3 into two subdomains, Garfagnana North and Garfagnana South.

The Garfagnana North subdomain is bounded to the East by three west-dipping fault systems: the Verrucole–S. Romano, the Sillano–Corfino, and the Mt. Prado systems. Each fault forming any of these systems is 10–15 km long. The Verrucole–S. Romano Fault has a total vertical throw of ~400 m, with offset recorded in the Late Pleistocene Macigno-rich conglomerates [56,66,82]. The fault also offsets by ~18 m more recent alluvial terraces toward its southern prolongation represented by the Pontecosi fault [66,82,83].

The Sillano–Corfino fault (Figures 4b and 5c) shows slickensides that are interpreted to suggest a right-lateral component of slip in addition to the dominant normal one [74]. The total vertical offset of the Corfino fault is ~600 m, with deformation in the Villafranchian and post-Villafranchian deposits. Smaller vertical displacement of ~250 m is measured in its southern part [66,77,82,83], between Villa Collemandina and Castiglione Garfagnana (Figure 4b).

The M. Prado fault, close and at the straddle with the orographic divide, is a ~10 km long structure with a maximum cumulative vertical offset of ~300 m and evidences of post Last Glacial Maximum (LGM) activity [56,66].

To the west, the graben is bounded by two major NE-dipping normal fault systems, the Gorfigliano–Vagli and the Casciana–Sillicano systems. Both seem to continue beyond the transverse Mt. Perpoli high, in the form of the NE-dipping Vergemoli–Fabbriche di Vallico–Pescaglia and the Bolognana–Gioviano–Borgo Mozzano fault zones, respectively.

All these fault systems are segmented laterally, with each identified fault being one of these segments, 10–15 km long.

The Casciana–Sillicano and Bolognana–Gioviano faults bound the present graben axis. Along these faults, the cumulative vertical displacement increases from north to south, up to ~400 m west of Castelnuovo Garfagnana (Casciana–Sillicano Fault), and up to 1.5 km across the Tuscan units around the Gallicano village (Bolognana–Gioviano Fault). There, the Barga conglomerates are vertically offset by about 39 m [56,66].

In the Garfagnana South subdomain, the eastern border faults are represented by the west-dipping Foce a Fobi/Mt. Uccelliera and the Foggeta–Mt.Coronato–Mt.Memoriante normal faults. The former is ~18 km long and shows well-expressed topographic scarps [56,66]. The latter is a ~20 km long structure with more than 300 meters of cumulative vertical displacement that bounds the Mesozoic carbonatic inlier of the Lima Valley to the west [84].

Further west, the Barga fault closely bounds the graben. It is a segmented fault about 10 km long with a NW–SE trend that bends to nearly N–S toward its southern termination close to Borgo a Mozzano, where it abuts against the east-dipping Gioviano fault. The distribution of the Late Pleistocene alluvial deposits has been related to the activity of the Barga Fault [56,66,73,77].

The Mt. Perpoli high that divides SD3 is bounded to the north and south by NE-trending faults with an oblique dextral and normal slip locally well constrained in bedrock minor faults [66,72,83] (Figure 5d). Deformation of Villafranchian and post-Villafranchian deposits has been described across the faults [56,66,82,83], which document a minimum of ~250 m of vertical displacement during Plio-Quaternary and ~150 m in the Middle Pleistocene–Holocene. An intermittent historical activity of the Pieve di Fosciana sinkhole-related lake has been related to the activity of the Mt. Perpoli faults [85].

4.4. The Lunigiana and Garfagnana Linkage Zone

The northern part of domain SD3 includes an area where a structural linkage seems to occur between the major fault systems of domains SD2 and SD3 (Figure 7). In this linkage zone, the oldest Plio-Quaternary continental deposits are largely eroded, although they are observed in three major exposures close to Canova–Ceserano and S. Terenzo in the Lunigiana and between Minucciano and Piazza al Serchio in Garfagnana [57,59,86]. The oldest levels, represented by clays, sands, and mainly polygenic conglomerates [56,57,86], may be compared with the Aulla clay–Collecchia conglomerates in Lunigiana and with the Fornaci di Barga fm. in Garfagnana [56,57,60]. These deposits are unconformably covered by ~200 m thick Mg-dominant conglomerates (called M. Lupacino conglomerates), which are considered of Early Villafranchian age [57], or more likely Late Villafranchian–Middle Pleistocene from correlation with similar deposits in Lunigiana and Garfagnana [58,60] (Figure 2). The original flat erosive base of these Plio-Quaternary deposits is vertically displaced by about +30, +50, and +500 m at the Canova–Ceserano, S. Terenzo, and Castagnola–M. Lupacino exposures, respectively, compared to similar deposits exposed in Lunigiana and Garfagnana.

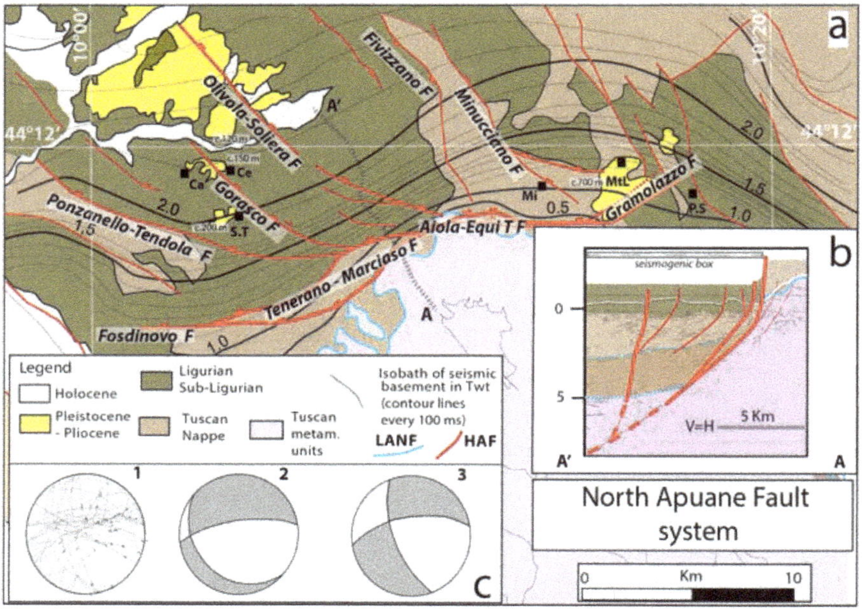

Figure 7. Lunigiana–Garfagnana linkage domain between SD2 and SD3, with the surface and subsurface architecture of the North Apuane Fault System. (**a**) Simplified tectonic and geological map of the Lunigiana–Garfagnana linkage domain (location in Figure 3) with breakthrough surface faults (Fosdinovo–Tenerano–Marciaso–Aiola–Equi T.–Gramolazzo and splays). HAF: steep normal faults; LANF: low-angle detachment faults. The contours are depths (in TWT) of the seismic basement identified in the seismic profiles, down to 5000 m; they image the North Apuane fault plane in the subsurface, as shown in (**b**). Numbers c.000 m indicate the elevation of the base of Plio-Pleistocene deposits. (**b**) Interpreted seismic section across the North Apuane fault (ENI: Equi Terme–Monti migrated line; more details in [50]). (**c**) Fault data (main fault planes and secondary faults with striations) showing kinematics of Aiola–Equi Terme Fault (1) and fault slip data inversion, PT kinematic diagrams (2, 3). All kinematic indicators attest to an oblique slip, both normal and right-lateral. Ca: Canova; Ce: Ceserano; Mi: Minucciano; Mt.L: M. Lupacino; P.S: Piazza al Serchio; S.T: S. Terenzo Monti.

The Lunigiana and Garfagnana linkage domain is related to the North Apuane fault system (Figures 3, 6 and 7). This fault zone, trending about ENE, is the most prominent structure in the morphology (Figure 5b) and the best constrained both at the surface and in the subsurface (Figure 7), thanks to Eni seismic lines [50,66,87]. In the seismic profiles, the North Apuane fault is well imaged down to a ~5 km depth (~2s in TWT, Figure 7b). Dipping by ~50° toward about the north, it separates a metamorphic footwall domain (Apuane metamorphic units in subsurface) from the unmetamorphic cover units of the nappe stack [50,87]. At the surface, the fault extends over more than 30 km long and is divided into three connected, ~10 km long segments: from west to east: the ~E–W trending Fosdinovo fault, the central ~NE-trending Tenerano–Marciaso fault, and the ~E–W trending, Aiola–Equi Terme fault. Its total vertical slip is 2–3 km.

The Aiola–Equi Terme fault shows clear evidence of recent activity, such as bedrock fault scarps, triangular facets, and hydrothermal springs along the fault trace (Figure 5b) [17,66,87–89].

The overall architecture of the North Apuane fault zone and adjacent faults suggests that the Ponzanello–Tendola, Gorasco, and Olivola–Soliera faults are secondary splay faults of the major North Apuane fault. Altogether, these secondary faults form a large horsetail at the western termination of the North Apuane fault zone, in keeping with its right-lateral component of slip (Figure 7c). To the east, the Gramolazzo fault may also be a secondary fault connected to the North Apuane fault, although this has to be better investigated. It displaces Plio-Quaternary deposits and a set of alluvial Late Pleistocene-Holocene terraces [86], which attests to its recent activity.

4.5. Structural Domain SD4: Lucca–Mt.Albano

Structural domain SD4 has been described as the "Montecarlo basin" [90–94] filled with up to ~2000 m of Neogene to Quaternary continental and marine stratigraphic sequences (exposed at surface and revealed in exploration wells; Figure 8).

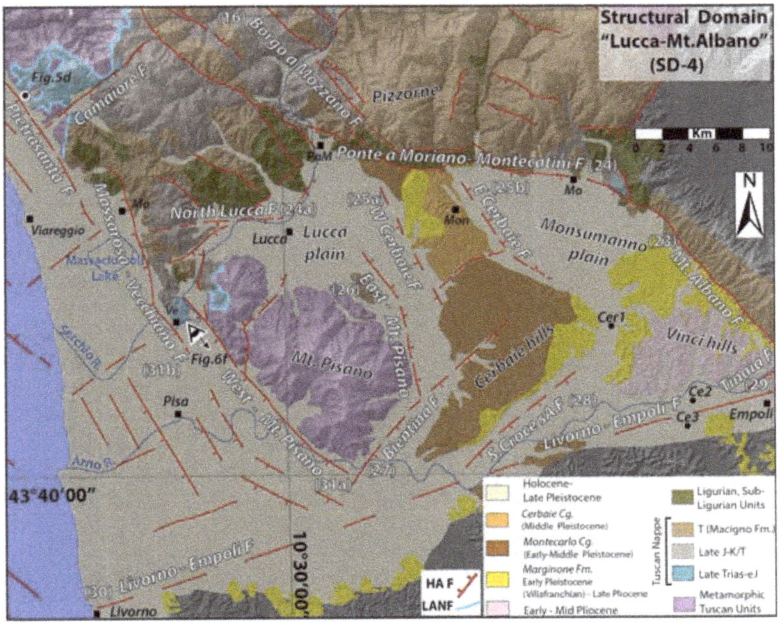

Figure 8. Major faults and geology in structural domain SD4 (Lucca–Mt. Albano). The map is derived from [92–94] and our work. Fault identification numbers as reported in Table S1. HAF: steep normal faults; LANF: low-angle detachment faults; Cer1: Cerbaie well; Ce2, Ce3: Certaldo wells; Ma: Massarosa; Mo: Montecatini; PaM: Ponte a Moriano; Ve: Vecchiano.

More precisely, SD4 includes two sub-basins, the Lucca and the Monsumanno plains, bounded by the hills of Montecarlo–Le Cerbaie and of Vinci. The basal terms of the filling sequence (Figure 2) are only known in subsurface at the Cerbaie 1 and Certaldo 2–3 boreholes, and tentatively referred to Late Miocene [90]. They consist of fluvio-lacustrine marls, clay and silty sands.

Their thickness is about 600 m in the Certaldo 2 and 3 boreholes, while the levels pinch out northward as documented in the Cerbaie 1 borehole. In outcrops, the lowermost terms are represented by the Marginone Fm. [90,91], which is formed by clays, silty clays, and sands locally associated with conglomerates. Fossil remnants and sedimentological features [92,95] document a transitional environment related to an alluvial plain grading southward to a marine environment. With a maximum thickness of ~400 m, the Marginone Fm. is dated to the Late Pliocene to Early Pleistocene (Early to Late Villafranchian) [95,96], although a possible Ruscian age has been suggested based on the presence of mammal remnants (*Alephis lyrix*) in its northermost part [92].

The overlaying Montecarlo conglomerates dominantly consist of a red gravelly deposit composed of quartzites and phyllites mainly sourced from the Triassic Verrucano Formation of the close Mt. Pisano. The unit, with a maximum thickness of 150–200 m, can be attributed to late Early Pleistocene, and is unconformably overlain by the Cerbaie Formation (Figure 2), which displays the same composition as the Montecarlo Formation, although with a finer clast size.

The Cerbaie Formation, with a maximum thickness of 20–30 m, is dated at the Middle Pleistocene through a regional correlation with deposits cropping out on the southern side of the Arno Valley and containing a tuff layer radiometrically dated at about 0.5 Ma [91,93,95–98].

From a structural point of view, SD4 represents a rhomb-like graben (Figure 8). To the south it is bounded by the ENE-trending Livorno–Empoli fault, and to the east by the Mt. Albano west-dipping normal fault. While that fault extends over 20 km with a NNW-strike, it bends anticlockwise at its northern termination so as to form the WNW-trending Ponte a Moriano–Montecatini normal fault. The connection between the two NNW- and WNW-trending normal faults suggest oblique slip on both of them, left-lateral on the Mt. Albano fault and right-lateral on the Ponte a Moriano–Montecatini fault. The Mt. Albano fault seems to extend further south beyond the Livorno–Empoli Fault (Figure 1).

The western boundary of SD4 is marked by the east Mt. Pisano Fault, which is well defined in the subsurface [93] along the eastern side of Mt. Pisano. The interior of the SD4 fault-bounded basin is also dissected by a few faults, mainly the west- and east-dipping Cerbaie normal faults, which bound the Montecarlo–Cerbaie horst in the center of the basin. A few other faults (Bientina, S. Croce sull'Arno, and Tinaia) are inferred in the south of the basin from subsurface data [90,93,94,99], possibly connected to the Livorno–Empoli fault. Although the tectonic studies of the SD4 basin faults are still few, the available works show that the cumulative vertical displacement on the Mt. Albano fault is about 1000 meters in its northern part (around Monsummano–Montecatini), including ~250 m and at least ~150 m of vertical slip since the Early-Middle Pliocene and the Gelasian, respectively [94]. Southeast of the Cerbaie and Vinci hills, "erosional" scarps may represent the expression of the subsurface S. Croce sull'Arno and Tinaia faults [91–93]. Their post-Middle Pleistocene vertical displacement is estimated at ~110 m [91–94].

The WNW-trending Ponte a Moriano–Montecatini normal fault steps to the left in the western part of the basin, resulting in the south-dipping, ~E-W-trending North Lucca normal fault. The left-stepping arrangement of the two normal faults suggests a right lateral component of slip along the whole transverse fault system. The activity of this northern basin-bounding fault system is recorded in the Plio-Quaternary deposits where deformation postdating the Late Pliocene is well documented, as well as in the morphotectonic evolution of the footwall domain represented by the reliefs of Le Pizzorne [78,92,94].

4.6. Structural Domain SD5: The Viareggio Basin

Structural domain SD5 includes the northernmost of the Tuscany and Northern Tyrrhenian Neogene basins [30,31,50,100]. It is made of an onshore part with the coastal plains of Versilia and Pisa and of an offshore domain with a basin-depocenter west of Viareggio (Figure 9). The coastal plains of Versilia and Pisa are formed by Quaternary deposits mainly sourced from the major rivers, Magra, Serchio, and Arno, and their tributaries. The offshore Viareggio basin is known thanks to industrial seismic profiles and exploration wells offshore and onshore, which described a Neogene–Quaternary sedimentary filling up to ~3500 m thick [30,31,46,47]. According to [30,100], the sedimentary sequence starts with 300 m of marine clay and sandstone (Seq. 2+3 of [30,100]) considered to be Late Messinian by [30]. The Pliocene succession shows a thickness of about 1500 m and is subdivided into two sequences (Seq. 4 and 5) from seismic facies and unconformities [47,100,101]. The succession is mainly formed by outer neritic marine clay associated with sandy layers. The following Quaternary deposits include ~600 m of open marine to littoral-brackish alternations of clay and sand (Seq. 6a) with inclined and well-defined reflectors indicating a westward prograding sandy mouth-bar related to the development of the paleo-Arno river. Deposits related to a marine ingression define the bottom of an uppermost, ~100 m thick sequence (Seq. 6b) referred to the Middle Pleistocene–Holocene. This upper sequence is formed by marine to brackish sands associated with gravel layers and covered by 40 m thick clays.

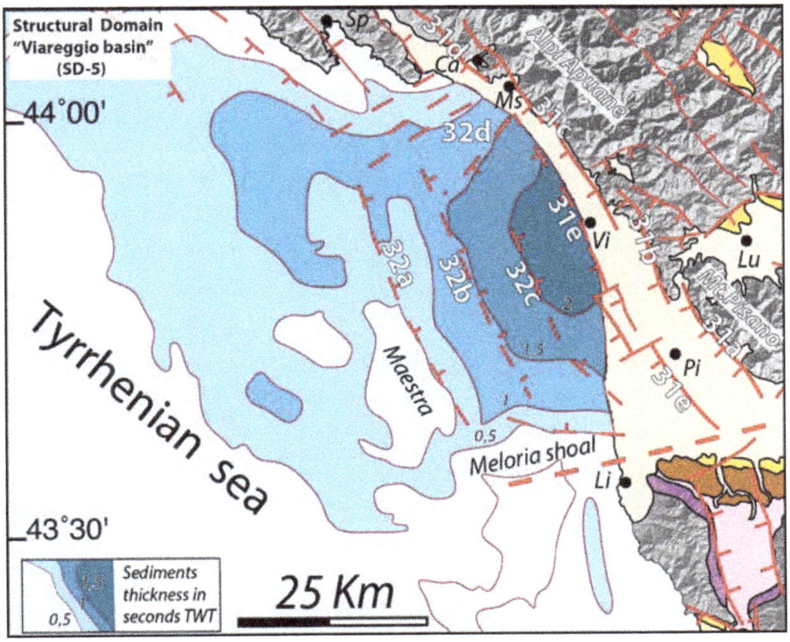

Figure 9. Main fault systems and geology in structural domain SD5 (Viareggio basin). Fault identification numbers are as reported in Table S1. Offshore sediment thickness after [30,31]. Ca: Carrara; Lu: Lucca; Li: Livorno; Ms: Massa; Pi: Pisa; Sp: La Spezia; Vi: Viareggio.

From a structural point of view, the Viareggio basin is an asymmetric, NNW-elongated graben, about 70 km long and 35–40 km wide (Figures 1 and 9). It is bounded to the north by submarine highs shaped by ~E–W trending faults offshore, Marina di Massa and Marina di Carrara, and to the south by the ENE-trending Meloria Shoal fault, which is the western offshore segment of the Livorno-Empoli fault. To the west, a submarine high, Maestra, forms the structural boundary of the basin, whereas to the east, the basin margin is made of a series of closely-spaced, west-dipping, steep normal faults that are part of the West Apuane Fault System (Figure 9). Seismic images and surface geology document the asymmetry of the basin controlled by the southwest-dipping master faults forming the West Apuane Fault System [50,87,99–102]. The relationships between faults and the sedimentary infill document deformation with total vertical offsets of more than 3500 m on the West Apuane system, and of ~200 m since the Middle Pleistocene. They provide evidence of activity of both the west- and east-dipping faults in Early Pliocene and Middle Pleistocene [100,101,103,104]. Although most of the available industrial seismic profiles do not have sufficient resolution to identify fault offsets within the Late Pleistocene–Holocene deposits, a few displacements are locally documented [46,105–107].

The kinematics of basin-margin high-angle faults is well constrained in some of the exhumed segments observable at the surface, as the Pietrasanta fault (Figures 5e and 8).

Close to Pisa, in the Mt. d'Oltre Serchio, a kilometer-wide relay zone between two NNW normal faults shows fault scarps up to 10 m high, likely formed in the Late Pleistocene–Holocene [103–105].

4.7. Structural Domain SD6: La Spezia–Lower Lunigiana

Structural domain SD6 is an on-land/offshore domain shaped by the Vara Valley–lower Lunigiana tectonic depression in which the Vara and the lower course of the Magra flow (Figure 10). The western promontory of La Spezia and the Vezzano–Pta. Bianca reliefs bound the depression. Continental stratigraphic sequences are scatterly observed in this domain, and are especially known in the northwest near Sesta Godano, and along the eastern flank of the lower Lunigiana, from S. Stefano Magra to Sarzana [58,106–109]. Small sedimentary remnants between Sesta Godano and La Spezia [55,105,108] mark a paleohydrographic setting with record of the paleo-Vara now captured by the Magra river [55,108] (Figure 10).

Figure 10. Main fault systems and geology in the structural domain SD6 (La Spezia–lower Lunigiana). The map is derived from [58,107] and our work. Fault identification numbers are as reported in Table S1. HAF: steep normal faults; LANF: low-angle detachment faults; A: Ameglia; BdM: Bocca di Magra; Be: Beverino; BV: Borghetto Vara; PB: Pian di Barca; PBt: Piano Battolla; Po: Ponzano Magra; SG: Sesta Godano; S.St: Santo Stefano Magra; U: Usurana; V: Vezzano.

The deposits of the lower Magra valley are related to a former Neogene–Quaternary "Sarzana basin" [58,107,108] associated with a continental sequence exploited in the past for clays and lignite deposits and containing rich flora and mammal remnants, and the object of investigations since the mid-19th century [107,109]. The sequence includes:

(a) Basal fine-grained lithofacies, mainly formed by clays and silty clays interbedded by thin beds and lenses of sand and lignite-bearing horizons grading and alternating upward with medium- to coarse-grained conglomerates. The dominantly fine-grained basal lithofacies, called Sarzanello–Caniparola clays [58,107–109], has been documented in the old mines around Caniparola and in less visible outcrops along the Albachiara creek south of Sarzanello [106,108,109]. The thickness of this unit is ~40 m and its depositional environment has been considered as related to lacustrian–palustrian in transition to a fluvio-deltaic setting developed in subtropical to temperate conditions. The paleontological contents allowed, with some debate, an attribution to the Early Pliocene, with its lower portion assigned to the Ruscinian age [108,109].

(b) Ponzano Magra clays and conglomerates, which are dominated by coarse-grained lithofacies with subordinate silty to sandy interbeds and fine-grained clay lithofacies mainly distributed in the lowermost portion, where organic-rich horizons are also present. In a quarry close to Ponzano Magra, the fine-grained lithofacies were used for industrial pottery until the mid-1980s. The Ponzano Magra clays and conglomerates, with a thickness of ~120 m [58,106,107], have been related to an alluvial plain environment and are characterized by a paleontological content which allows their dating to the Early–Middle (?) Villafranchian [61,106–109].

(c) Darma–Faggiada conglomerates, which are mainly formed by polygenic coarse-grained conglomerates associated with sands and silty-sand layers locally observable in fining-upward sequences. This unit is interpreted as related to alluvial fans resting unconformably on top of the Ponzano Magra conglomerates. From relative chronology evidence, the unit is suggested of Middle–Late Villafranchian age [58,106–109].

(d) Fan deposits and alluvial terraces. The oldest of these units, collectively called the "lower Val di Magra system" [107], includes alluvial terraces and related fan deposits mainly observable in the eastern flank of the lower Magra valley (between S. Stefano di Magra and Sarzana). The sequences are represented by polygenic unsorted gravels and boulders with minor sands and clays, and are referrable to an evolving braided fluvial system and lateral fans. The oldest of this group of deposits, which may be found at about 50 m above the present Magra river, is well developed between nearby Santo Stefano di Magra and is related to an early stage of development of the present drainage system [107,109]. Collectively these units are attributed to the Middle–Late Pleistocene. The recent alluvial terraces and alluvial deposits are related to the present-day Magra–Vara river system and include two levels of terraces formed by conglomerates, sands, and silts. The oldest of these deposits have been dated to the Late Pleistocene–Holocene, while the youngest are of the Bronze and Iron Ages (due to the presence of a human-made "Statua Stele" found during sand quarrying near Sarzana [107]).

The deposits of Sesta Godano include conglomerates grading to coarse- to fine-grained sands interbedded with clay and silty clays. The sequence has a thickness of about 60 m and is related to a lacustrine–alluvial depositional system [58,107]. Pollen contents allowed its dating to the Early Villafranchian [106,108,109]. Noteworthy are the scattered coarse-grained deposits, mainly conglomerates and sands, observable NW of La Spezia in Pian di Barca surroundings. These deposits are interpreted as related to the confined alluvial systems [58,106] of Pliocene–Early Villafranchian age [58,107,108]. They thus attest to a past drainage of the paleo-Vara river flowing to the proto-La Spezia Gulf before its diversion toward the present-day course track [58].

The central graben—lower Lunigiana—of domain SD6 is bounded by antithetic normal fault systems (Figure 10). Overall, the width of the fault-bounded graben decreases from SE to NW. To the east, the major west-dipping Sarzana–Carrara fault, the northermost part of the West Apuane Fault System, extends over ~25 km from Carrara to S. Stefano di Magra, where it steps to the right to continue through the Mt. Grosso–Mt. Cornoviglio–M. Vruga fault [50,58,60]. The step has developed so that the fault system is intersected by a NE-trending short fault, the S. Stefano Magra fault. From the step toward the northwest, the Sarzana–Carrara fault splays into a network of curved normal faults, forming a sort of horsetail at the northern end of the Sarzana–Carrara fault. This horsetail suggests a left-lateral component of slip on the main NW-trending Sarzana–Carrara fault, in addition to its dominant normal one.

To the west of the graben, the east-dipping, NW-trending bounding normal fault system includes La Spezia fault, about 20 km long, and the Piano Battolla–Vezzano–Ameglia fault on the eastern side of the Punta Bianca promontory. With a total length of about 40 km, the Piano Battolla–Vezzano–Ameglia fault bends counterclockwise in its northern section, likely to connect the La Spezia fault.

The Quaternary vertical displacements on the Sarzana–Carrara fault exceed 700 m, as recorded in tilted lower levels of the "Sarzana Basin" [50,58,72,107,110], and are estimated to ~75 m during Mid-Pleistocene–Holocene. Deformations of Holocene alluvial terraces are also documented across the Ameglia–Piana Battolla fault [72,107,109].

5. Geodetic Data, and Historical and Instrumental Seismicity

5.1. Geodetic Data

We used GPS data recorded in 1998–2020 by continuous Global Navigation Satellite System (GNSS) stations operating in Italy and surroundings, along with GPS data collected in repeated temporary campaign measurements during the RETREAT project [25] in the northern Apennines, to calculate new horizontal velocity and strain-rate fields.

We processed the data following the same procedures as described in [111], to which we refer for more technical details. The main result of our GPS data processing was a set of ground surface velocities, represented in Figure 11 relative to a fixed Eurasia reference frame [25,112]. The discrete velocities at each station, weighted by their uncertainties, were used to estimate a continuous velocity field, and its spatial gradients, using the multiscale method described in [113,114]. Figure 11 also shows the total strain-rate field (i.e., the second invariant of the strain-rate tensor), defined as the square root of the sum of squares of all its components, along with the horizontal principal strain axes (grey arrows). The diverging and converging arrows thus represent the directions of extensional and contractional deformation, respectively.

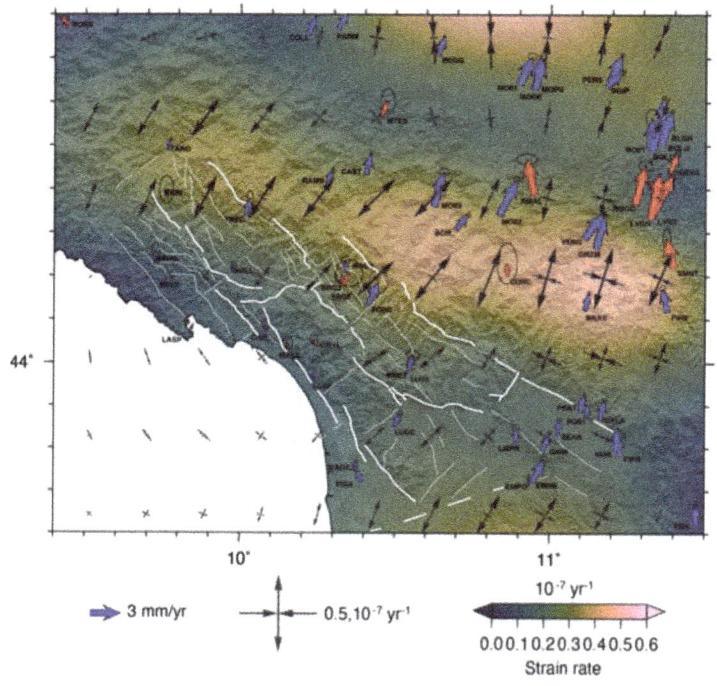

Figure 11. Horizontal GPS velocity field (blue and red vectors are continuous and campaign stations, respectively, with 95% confidence error ellipses) and total strain-rate field (color scale). The diverging and converging grey arrows show the axes of the principal strain rates, indicating extensional and contractional deformation, respectively.

Beyond the general characteristics of the velocity and strain fields, reported in earlier works [25,112], several interesting features were revealed.

First, surface velocities and strain rates abruptly changed across two zones trending about NW–SE: in the southwestmost part of the area, they increased significantly across a curved line that roughly coincided with the westernmost major fault systems of the SD2 and SD3 domains; whereas, further to the northeast, they decreased significantly across a curved line that runs in the Apennines northern flank, about 30–50 km east of the SD2 and SD3 easternmost fault systems. These pronounced changes might result from the current locking of active faults. In the region of concern here, this provides further support for the seismogenic activity of the westernmost SD2 and SD3 normal faults.

Another observation was the existence of a ~E-W narrow zone north of the Firenze graben, where the strain vectors suggest a strike-slip deformation. The zone roughly extends in the eastern prolongation of the ~E-W North Lucca–Ponte a Moriano–Montecatini fault zone that bounds the SD4 basin to the north. This is consistent with this fault zone, which currently has a right-lateral component of slip.

5.2. Historical and Instrumental Seismicity Within the Investigated Areas

The studied area is dominated by shallow earthquakes of low to moderate magnitude and normal to oblique focal mechanisms [1,2,4,27,115–118]. Figure 12 reports instrumental earthquakes with magnitude Mw >1 and historical earthquakes derived from the CPTI15 v.3 catalog [35,37]. Although instrumental and historical seismicity is widespread throughout the study area, only two seismogenic sources (in Lunigiana and Garfagnana, our SD2 and SD3 domains) are included in the national database of seismogenic areas. Figure 12 shows, however, that clusters of recent seismicity and historical earthquakes did occur in all the structural domains in the region of interest. Structural domain SD2 shows shallow crustal earthquakes, with hypocenter depths $\leq$ ~10 km, and focal mechanisms attesting to normal or oblique slips [116,117]. It is noteworthy that historical seismicity in this area includes different earthquakes with an estimated magnitude Mw > 5 (Figure 12c); among them, the "Val di Taro" (1545, Mw ~5.2) and "Bagnone" (1903, Mw ~5.3) events. Similar to upper Lunigiana, SD3 (Garfagnana) is an active seismic region, even though the distribution of instrumental seismicity is quite heterogeneous. Like in Lunigiana, the hypocenters are shallow ($\leq$10 km) and earthquakes have both extensional and oblique-slip kinematics. Leaving aside the 7 September 1920 earthquake (Mw ~6.5), which will be discussed below, at least five historical earthquakes with severe damages have been reported in Garfagnana [35,37] (Figure 12c): the "1740 Piastroso"(Mw ~5.2), the "1746 Barga" (Mw ~5), the "1902 Barga"(Mw ~4.9), the "1919 Piastroso" (Mw ~4.9), and the "1939 Vagli"(Mw ~4.9) earthquakes. The other structural domains described in this study show an instrumental seismicity less dense and of smaller maximum magnitudes (Mw < 3.5) than those recorded in the SD2 and SD3 domains. However, in all domains there is a significant seismicity, including clusters and/or alignments of epicenters, both onland and offshore (Figures 12 and 13). It is worthy to report here the offshore events that occurred along and off the coast of Pisa and Viareggio in 2013. Some occurred in clusters, which superimpose fairly well to the surface traces of some of the major faults we described earlier (Figure 13). Cross-sections across these faults and clusters support that the former may be the sources of the earthquake clusters (Figure 13). This confirms that the bounding faults of the Viareggio basin are currently active, as those north of SD4 and east of SD6. The focal mechanisms are too few to derive any significant information, but they confirm that earthquakes on NNW-trending faults have normal or normal and left-lateral slip.

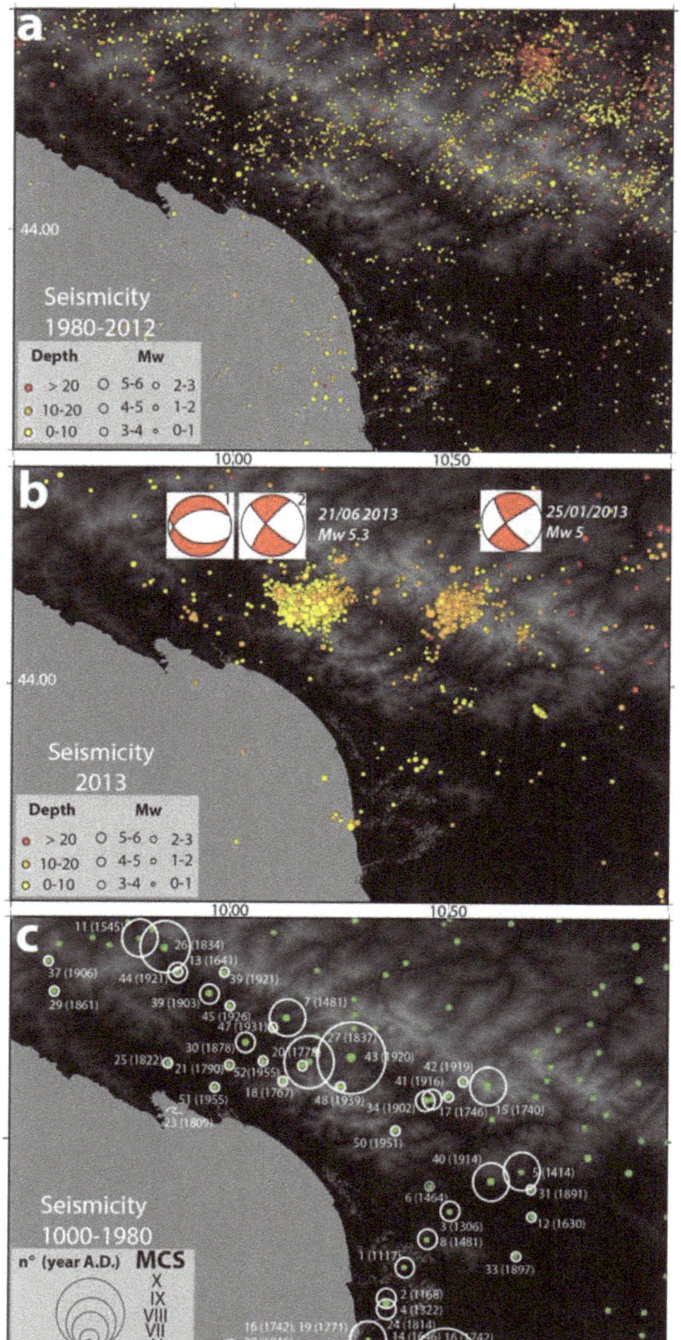

Figure 12. Seismicity in the studied area (instrumental and historical). (**a**) 1982–2012; (**b**) 2013; (**c**) historical moderate to large earthquakes (intensities greater than V) reported from 1000 to 1980 (from CPTI15-DBMI15 catalogs). Identification numbers are the same as those reported in Table S2.

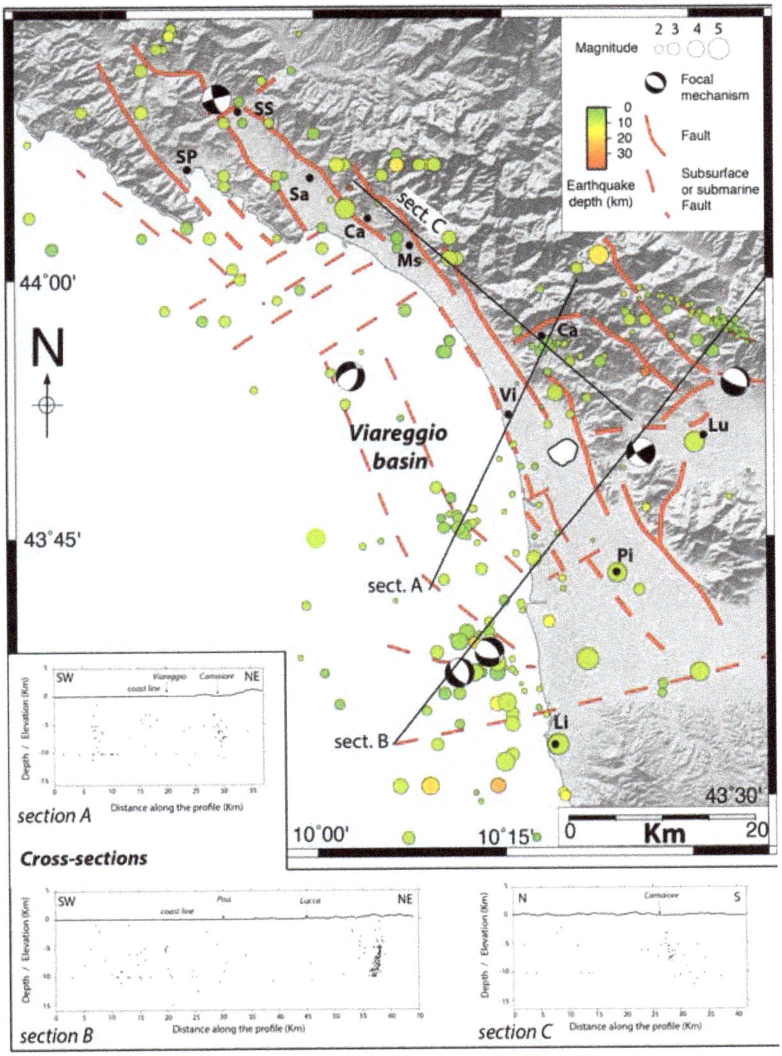

Figure 13. Faults and seismicity (1980–2013) in the Viareggio basin and surroundings. Focal mechanisms are from the INGV archive [2,32].

At least 11 historical moderate earthquakes with significant damages occurred within the structural domains SD4, SD5, and SD6 (Table S2): the 1168 "Pisa" (Mw ~4.5), the 1322 "Pisa"(Mw ~4.5), the 1414 "Villa Basilica" (Mw ~5.2), the 1481 "San Giuliano Terme" (Mw ~4.5), the 1630 "Pescia" (Mw ~4.5), the 1891 "Villa Basilica" (Mw ~4.5), the 1814 "Livorno" (Mw ~5.1), the 1861 "Varese Ligure" (Mw ~4.6), the 1897 "lower Val d'Arno" (Mw ~4.6), the 1914 "Ponte a Moriano" (Mw ~5.8), and the 1955 "Sarzana" (Mw ~4.5) earthquakes. Furthermore, an important earthquake occurred in the Viareggio–Pisa area in 1117 that caused large damages [37]. The significant magnitudes of these earthquakes attest that they ruptured fault lengths of kilometer scales. Altogether, these earthquake records thus confirm that at least several of the faults we have described in the region are currently active and seismogenic.

5.3. The 2013 Seismic Activity

During 2013, a series of seismic events occurred within the investigated area, soon after the end of the Emilia 2012 seismic sequence [4,33,74]. The 2013 events affected domain SD3 (Garfagnana) in January and the transition zone between SD2 and SD3 in June and August. A few other events occurred in the plain of Pisa and in the close offshore area, as well as in the lower Val di Magra (SD6) in April, June, and October (Figures 12 and 13).

On January 25 at 3:48 p.m., an earthquake with Mw 4.8 occurred in mid-Garfagnana at a depth of ~15 km (Figure 12b). The main shock was followed by about 300 aftershocks in the following days, with magnitudes that rarely exceeded Mw 3 [32]. The distribution of the aftershocks roughly draws a NE–SW trend, and the aftershocks migrated away from the main shock toward the NE. The location of the main shock and of the aftershocks appears unrelated to the two seismic sources recognized in the area by the DISS project [16]. The focal mechanisms suggest right-lateral slip on a NE–SW-trending plane or a left-lateral slip on a NW–SE-trending plane. As the cluster of January 2013 geometrically fits the area of the transversal, NE-trending fault zone (Mt. Perpoli high and related faults) that divides SD3 into two parts, the event provides support to the current activity of this fault zone.

On 21 June 2013 (10:33 UTC), an earthquake of Mw 5.2 hit southern Lunigiana, with its epicenter close to Monzone (Figures 1 and 12b). The main event was preceded by a foreshock on June 15 (Mw 3.4), and was followed by more than 2450 aftershocks, 4 of them having a magnitude Mw $\geq$ 4 and 27 with Mw $\geq$ 3 [32]. The data show that most aftershocks extended NE from the epicenter [4] (Figure 12b). INGV calculated two solutions for the focal mechanism of the main shock, with the one noted as (2) in Figure 12b being the best constrained. Both solutions are consistent with a N–S extension, while the best constrained mechanism suggests a normal left-lateral or a normal right-lateral slip on a NW- or NE-trending plane, respectively. The seismic source was initially constrained through the measurement of the coseismic surface deformation field with DInSAR interferometry, which defined a rupture on a normal fault dipping NNW by about 50° [119]. Refined results were later presented based on a combination of DInSAR and seismological data and hypocenter relocation [120]. These identified the source as a ~N 70° E striking normal fault, ~45° NNW-dipping fault with a main dip-slip mechanism, but combined with a small right-lateral strike-slip component [120]. This is consistent with our observations on the transverse faults.

The year 2013 was also marked by a series of small earthquakes throughout western Tuscany and eastern Liguria. While these earthquakes are sparse, some form clusters coinciding with the traces of some of the faults identified in the western flank of the Alpi Apuane, including offshore in the Viareggio basin, but also with the transversal NE–SW-trending faults offshore of SD6 (Figures 10 and 13).

6. The 1920 Mw ~6.5 Fivizzano Earthquake: Hypothesis on its Fault Source

On 7 September 1920 at 05:55, the area between southern Lunigiana and Garfagnana (SD 2 and SD3) was struck by a devastating earthquake with an estimated Mw of 6.4–6.5 [35,37].

The mainshock was preceded by some foreshocks on the previous day, and was followed by aftershocks that lasted until August 1921 [38]. The earthquake induced a vast area of severe damages throughout Lunigiana and Garfagnana and their surroundings, over an area of about 160 km^2 (Figure 14). The mainshock was felt up to the Côte d'Azur in southern France, Aosta in northwestern Italy, the Friuli region in the northeast, and south of Perugia in central Italy [38].

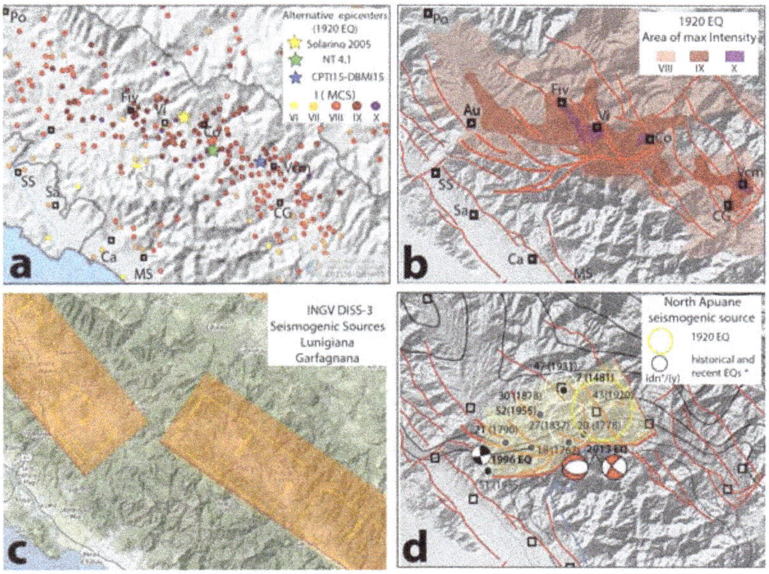

Figure 14. (**a**) Macroseismic field of the 1920 Fivizzano earthquake. MCS intensities are those of the database in [37], while epicentral positions are derived from different works (see legend). (**b**) Area of maximum intensity of the 1920 Fivizzano earthquake [35,37]. (**c**) Recognized seismogenic source according to DISS 3 [16]. (**d**) North Apuane Fault System architecture in surface and subsurface, with location of historical and recent earthquakes *. Light yellow represents the seismogenic source we inferred from Figure 7.

From the interpretations of the macroseismic field (Figure 14a) (which in some zones reached I 10 MCS) [37,38], different causative faults were proposed as the source of the 1920 Fivizzano earthquake. Among them, the most commonly proposed faults are the NW-trending, ~10 km long Minucciano fault in Lunigiana (SD2, Figure 3) and the NW-trending, ~15 km long Casciana–Sillicano fault in Garfagnana (SD3, Figure 6), which are among the seismogenic faults recognized in the national hazard map [16,17,116,117]. As a matter of fact, in the area between Lunigiana (SD2) and Garfagna (SD3) different historical earthquakes with Mw > 4 have been reported [37] (Figure 12c): the 1481 "Fivizzano" (Mw ~5.5), the 1767 "Monzone" (Mw ~5.3), the 1790 "Tendola" (Mw ~4,9), the 1837 "Alpi Apuane" (Mw ~5.8), and the 1878 "Olivola" (Mw ~5.1) earthquakes. Moreover, the epicenter of the 15 October 1995 Mw 4.9 earthquake [116–118,121], as that of the 21 June 2013 earthquake, are located in the same zone. Figure 14d shows the surface projection of the North Apuane Fault System that we described earlier (Figure 7). We suggest that this source may account for some of the recorded historical and instrumental events, and moreover, that its rupture produced the 1920 earthquake. As the North Apuane Fault System forms the northern termination of the NNW-trending, east-dipping westernmost faults of SD3, we hypothesize that the earthquake ruptured both the northern part of the NNW trending faults (Gorfigliano–Vagli or Casciana–Sillicano faults; Figure 6), and its ending oblique North Apuane fault. Coseismic slip might have been greater on this ending oblique fault, as observed elsewhere, such as in the 1999 Mw 7.6 Chichi earthquake [122]. Therefore, while some NNW-trending faults of the SD3 domain likely ruptured in the 1920 earthquake, we suggest that the ~E-W North Apuane fault extending at the northern tip of the SD3 NNW faults also ruptured during the earthquake, and actually likely produced the largest slips and ground accelerations [49,66,74,87]. This hypothesis will of course need to be validated by future fieldwork analyses.

7. Discussion

We have provided a review of most available knowledge on Cenozoic geology/stratigraphy, faulting, seismicity, and current surface strain in the inner northwest Apennines. This compilation allows us to discuss two major issues.

7.1. Fault Distribution and Evolution

Our work shows that the region of study is densely dissected by large faults of 10 to 40 km lengths. The dominant faulting trends NW–NNW, and has normal slip, however with an additional component of left-lateral slip. The faults are steep and commonly combine in conjugate pairs of antithetic faults bounding grabens. Figure 15 presents a simplified map of the principal grabens with their sedimentary Plio-Quaternay filling in yellow. These grabens are disconnected from each other or nested within one another, and their width is narrowing northwestward. Furthermore, collectively, they form an overall area that tapers northwestward. Graben tapering has been described elsewhere and shown to result from the propagation of extensional faulting in the direction of graben narrowing [123]. This architecture might thus suggest the northwestward propagation of extensional faulting in this part of the Apennines. Such a northwestward propagation of extensional faulting has already been proposed in earlier works [5,124]. It is consistent with the northwestward decrease of the total vertical slip along most of the major NW–NNW normal faults, indicative of faults becoming younger in the northwestward direction [125]. Furthermore, we note that the total vertical slips accommodated on the graben-bounding faults decrease overall from west (~3.5 km in the Viareggio basin) to east, where total slip on the easternmost faults is half that on the westernmost faults. This might suggest a migration of extensional strain over time from west to east, in addition to the northwestward propagation.

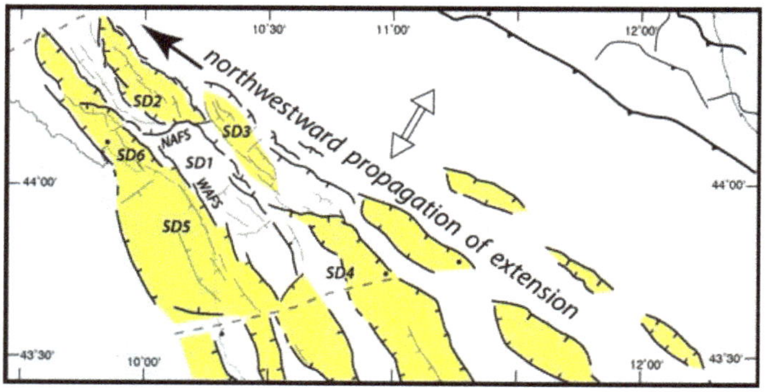

Figure 15. Simplified map of major extensional fault systems in the inner northwest Apennines. Principal grabens are represented, with their sedimentary Plio-Quaternary filling in yellow. The NW–NNW fault systems are suggested to have propagated northwestward in the few last million years. SD: structural domain; NAFS: North Apuane Fault System; WAFS: West Apuane Fault System.

Although most NW–NNW faults are likely to interact (as shown in central Italy, e.g., [5,126]), they are disconnected, sometimes en échelon disposed, so that each principal fault is at most 40 km long, and more generally 10–15 km.

Another fault set exists in the region, trending almost perpendicular to the dominant NW–NNW normal faults. The origin of these transverse faults is not yet fully investigated. As similar ENE faults exit throughout most of Italy, they might likely arise from structural inheritance from earlier phases of deformation [127–130]. In the region of study, the most important of these ENE faults are the TAL and LEF, which may act at depth as major shear zones. They actually have a right-lateral component of slip. Some of these faults might be reactivated by the NW–NNW normal, left-lateral faults. This might be the case for the

North Apuane and the Mt. Perpoli faults. Being strongly oblique to the major NW–NNW west-dipping faults of the SD2 and SD3 domains, and located at their tips, they sustain extension due to the left-lateral component of slip on these principal NW–NNW faults.

More work is needed to confirm the fault organization and evolution suggested in Figure 15, and to validate whether this architecture results from the overall propagation of extensional faulting toward the northwest, along with a possible eastward migration of the strain locus. Whatever it is, the architecture of the faults depicted in this study markedly differs from the seismotectonic model proposed so far in [131], in which faulting would be mainly related to a major low-angle, east-dipping active detachment [47,131,132]. Rather, we show that faults are steep, paired in conjugate sets bounding grabens, and disconnected, although forming a kinematically coherent overall fault system.

7.2. Evidence for Recent Fault Activity and Implications for Seismic Hazard

One of the most important results of this work was to document that the majority of faults described in the study have been active over the Quaternary. Most show displacements post-800 ka, and in some cases clear slips across Late Pleistocene and Holocene deposits [59,66,69,72,83]. In addition, some of these faults show evidence of recent to current activity. Yet, most of these seismogenic faults are not considered in the national and local hazard evaluations. While most of the faults we identified are shorter than 35–40 km, they may have the potential to produce strong earthquakes. Most are indeed immature (as defined in [11]), and it has been shown that, while immature faults are shorter than more mature faults, they produce more energetic earthquakes (i.e., larger stress drops), generating greater displacements at depth and surface [11], and stronger ground motions [14]. Therefore, it is possible that some of the faults we identified can generate one to several meters of slip when they rupture, and large ground accelerations. Furthermore, immature faults tend to rupture in cascade, each of their major segments rupturing individually at successive times [5,11,126]. Such cascading earthquakes have been reported elsewhere in Italy on other similar normal faults (e.g., [5,126]). We have shown that most of the NW–NNW normal faults in northwest Apennines are segmented laterally in generally 3–4 major segments, each 10–15 km length. According to the available earthquake slip-length scaling laws [11], the rupture of one of such segments might produce up to 1–3 m of slip at the ground surface, likely accompanied with strong ground accelerations. As shown in Central Italy, such earthquakes might have a magnitude between 6 and 7, and can follow in close time successions [5,126].

Therefore, our study emphasizes that seismic hazard is significant in the inner northwest Apennines and should be considered [133,134], especially as the presence of thick, soft alluvial deposits in the Viareggio, Pisa, and lower Arno valleys [97,134] may lead to the amplification of ground movements and accelerations. The hazard might even be more complex as some of the seismogenic faults extend offshore where their dip-slip rupture might generate tsunamis, as already recorded in the past in La Spezia and Livorno [135]. Onland, the large vertical slips expected in the faults might trigger significant landsliding.

8. Conclusions

Using the rich information in the literature, along with field observations we made, we propose a reappraisal of the active fault identification and mapping in the inner northwest Apennines. We showed that faults active in Quaternary times are numerous in the region and dominated by NW–NNW antithetic normal faults bounding a series of disconnected grabens. Most of the faults show evidence of Quaternary activity up to the present time. Most of the faults identified in this work are thus likely seismogenic, i.e., potential sources of forthcoming earthquakes. The properties of the faults are such that these earthquakes are expected to be potentially large (Mw $\geq$ 6.0), to produce up to several meters of slip at the ground surface, and to possibly occur in temporal cascades. The seismic hazard is thus significant in the region, and even greater than expected if cascading earthquakes are the norm.

Therefore, our study calls for the need to evaluate in greater detail the recent activity and the earthquake potential of the many faults we identified as potentially seismogenic. We claim that the majority of them should be included in the Italian seismic hazard evaluation. We hope that our work will encourage the national and regional authorities to support the needed geological, tectonic, and seismological research studies in the inner northwest Apennines. These should include tectonic fault analysis with remote sensing and fieldwork, seismological recording and analysis, densification of the GPS network, paleoseismological trenching, and fault-scarp dating, among others. These future studies are critical to assess seismic hazard more accurately in the densely populated areas from La Spezia to Lucca–Pisa–Livorno, which presently have low seismic protection codes.

Supplementary Materials: The following are available online at https://www.mdpi.com/2076-3263/11/3/139/s1, Table S1: Fault systems, timing of activity, displacements, displacement rates, and related references; Table S2: Historical earthquakes in the investigated area based on the CPTI15 and CPTI-DBMI15 databases.

Author Contributions: Conceptualization, G.M., J.M., I.M.; validation, G.M., R.B., J.M., E.S., F.S., I.M.; formal analysis, G.M., R.B., J.M., I.M., E.S., F.S.; investigation, G.M., F.S., A.B., G.P., A.L., T.G., L.A., L.P.; resources, G.M., A.B., G.P., S.G.; data curation, A.B., G.P., S.G.; writing—original draft preparation, G.M., I.M.; writing—review and editing, G.M., I.M.; visualization, S.G.; project administration, G.M.; funding acquisition, G.M., F.S., E.S., R.B., I.M. All authors have read and agreed to the published version of the manuscript.

Funding: This research was funded by Ateneo funds from Pisa University, FIL grants from Parma University, and ANR projects QUAKonSCARPS and FAULTS_R_GEMS. The APC was funded by an Ateneo grant from Pisa University.

Acknowledgments: We thank the two reviewers Paolo Galli and Stefano Mazzoli for their detailed comments that greatly helped us to improve our manuscript.

Conflicts of Interest: The authors declare no conflict of interest.

References

1. Chiarabba, C.; Jovane, L.; Di Stefano, R. A new view of Italian seismicity using 20 years of instrumental recordings. *Tectonophysics* **2005**, *395*, 251–268. [CrossRef]
2. Pondrelli, S.; Salimbeni, S.; Ekström, G.; Morelli, A.; Gasperini, P.; Vannucci, G. The Italian CMT dataset from 1977 to the present. *Phys. Earth Planet. Inter.* **2006**, *159*, 286–303. [CrossRef]
3. Meletti, C.; Galadini, F.; Valensise, G.; Stucchi, M.; Basile, R.; Barba, S.; Vannucci, G.; Boschi, E. A seismic source zone model for the seismic hazard assesment of the Italian territory. *Tectonophysics* **2008**, *450*, 85–108. [CrossRef]
4. Scafidi, D.; Barani, S.; De Ferrari, R.; Ferretti, G.; Pasta, M.; Pavan, M.; Spallarossa, D.; Turino, C. Seismicity of Northwestern Italy during the last 30 years. *J. Seismolog.* **2015**, *19*, 201–218. [CrossRef]
5. Benedetti, L.; Manighetti, I.; Gaudemer, Y.; Finkel, R.; Malavieille, J.; Pou, K.; Arnold, M.; Aumaître, G.; Bourlès, D.; Keddadouche, K. Earthquake synchrony and clustering on Fucino faults (Central Italy) as revealed from in situ 36Cl exposure dating. *J. Geophys. Res.* **2013**, *118*. [CrossRef]
6. Schlagenhauf, A.; Manighetti, I.; Benedetti, L.; Gaudemer, Y.; Finkel, R.; Malavieille, J.; Pou, K. Earthquake supercycles in Central Italy, inferred from 36Cl exposure dating. *Earth Planet. Sci. Lett.* **2011**, *320*, 487–500. [CrossRef]
7. Galli, P. Recurrence times of central-southern Apennine faults (Italy): Hints from paleoseismology. *Terra Nova* **2020**. [CrossRef]
8. Galli, P.; Galadini, F.; Pantosti, D. Twenty years of paleoseismology in Italy. *Earth Sci. Rev.* **2008**, *88*, 89–117. [CrossRef]
9. Field, E.H. UCERF3: A new earthquake forecast for California's complex fault system. *Fact Sheet* **2015**. [CrossRef]
10. Wesnousky, S.G. Seismological and structural evolution of strike-slip faults. *Nature* **1988**, *335*, 340–342. [CrossRef]
11. Manighetti, I.; Campillo, M.; Bouley, S.; Cotton, F. Earthquake scaling, fault segmentation, and structural maturity. *Earth Planet. Sci. Lett.* **2007**, *253*, 429–438. [CrossRef]
12. Perrin, C.; Manighetti, I.; Ampuero, J.-P.; Cappa, F.; Gaudemer, Y. Location of the largest earthquake slip and fast rupture controlled by along-strike change in fault structural maturity due to fault growth. *J. Geophys. Res.* **2016**, *121*, 1–19. [CrossRef]
13. Stirling, M.W.; Wesnousky, S.G.; Shimazaki, K. Fault trace complexity, cumulative slip, and the shape of the magnitude-frequency distribution for strike-slip faults: A global survey. *Geophys. J. Int.* **1996**, *124*, 833–868. [CrossRef]
14. Radiguet, M.; Cotton, F.; Manighetti, I.; Campillo, M.; Douglas, J. Dependency of Near-Field Ground Motions on the Structural Maturity of the Ruptured Faults. *Bull. Seism. Soc. Am.* **2009**, *99*, 2572–2581. [CrossRef]

15. Chiaraluce, L.; Di Stefano, R.; Tinti, E.; Scognamiglio, L.; Michele, M.; Casarotti, E.; Cattaneo, M.; De Gori, P.; Chiarabba, C.; Monachesi, G.; et al. The 2016 Central Italy Seismic Sequence: A First Look at the Mainshocks, Aftershocks, and Source Models. *Seism. Res. Lett.* **2017**, *88*, 757–771. [CrossRef]
16. DISS Working Group. Database of Individual Seismogenic Sources (DISS), Version 3.2.1: A Compilation of Potential Sources for Earthquakes Larger than M 5.5 in Italy and Surrounding Areas. 2018. Available online: http://diss.rm.ingv.it/diss/ (accessed on 15 March 2021).
17. ITHACA Working Group. ITHACA (ITaly HAzard from CApable Faulting), A Database of Active Capable Faults of the Italian Territory. Version December 2019. ISPRA Geological Survey of Italy. Available online: http://sgi2.isprambiente.it/ithacaweb/Mappatura.aspx (accessed on 15 March 2021).
18. Elter, P. Introduction à la géologie de l'Apennin septentrional. *Bullettin Soc. Geol. Fr.* **1975**, *7*, 956–962. [CrossRef]
19. Carmignani, L.; Decandia, F.A.; Fantozzi, P.L.; Lazzarotto, A.; Liotta, D.; Meccheri, M. Tertiary extensional tectonics in Tuscany (Northern Apennines, Italy). *Tectonophysics* **1994**, *238*, 295–315. [CrossRef]
20. Jolivet, L.; Faccenna, C.; Goffé, B.; Mattei, M.; Rossetti, F.; Brunet, C.; Storti, F.; Funiciello, R.; Cadet, J.P.; D'Agostino, N.; et al. Midcrustal shear zones in postorogenic extension: Example from the northern Tyrrhenian Sea. *J. Geophys. Res.* **1998**, *103*, 12123–12160. [CrossRef]
21. Molli, G. Northern Apennines-Corsica orogenic system: An updated overview. Siegesmund, S.; Fügenschuh, B.; Froitzheim, N. Eds.; Tectonic aspects of the Alpine–Dinaride–Carpathian system. *Geol. Soc. Lond. Spec. Publ.* **2008**, *298*, 413–442. [CrossRef]
22. Elter, P.; Giglia, G.; Tongiorgi, M.; Trevisan, L. Tensional and compressional areas in the recent (Tortonian to present) evolution of the northern Apennines. *Boll. Geofis. Teor. Appl.* **1975**, *17*, 3–18.
23. Carminati, E.; Doglioni, C. Alps vs. Apennines: The paradigm of a tectonically asymmetric Earth. *Earth Sci. Rev.* **2012**, *112*, 67–96. [CrossRef]
24. Le Breton, E.; Handy, M.; Molli, G.; Ustaszewski, K. Post-20 Ma motion of the Adriatic plate—new constraints from surrounding orogens and implications for crust-mantle decoupling. *Tectonics* **2017**, *36*, 3135–3154. [CrossRef]
25. Bennett, R.A.; Serpelloni, E.; Hreinsdóttir, S.; Brandon, M.T.; Buble, G.; Basic, T.; Casale, G.; Cavaliere, A.; Anzidei, M.; Marjonovic, M.; et al. Syn-convergent extension observed using the RETREAT GPS network, northern Apennines, Italy. *J. Geophys. Res.* **2012**, *117*, B04408. [CrossRef]
26. Faccenna, C.; Becker, T.W.; Miller, M.S.; Serpelloni, E.; Willett, S. Isostasy, dynamic topography, and the elevation of the Apennines of Italy. *Earth Planet. Sci. Lett.* **2014**, *407*, 163–174. [CrossRef]
27. D'Acquisto, M.; Dal Zilio, L.; Molinari, I.; Kissling, E.; Gerya, T.; van Dinther, Y. Tectonics and seismicity in th Northern Apennines driven by slab retreat and lithospheric delamination. *Tectonophysics* **2020**, *789*. [CrossRef]
28. Mazzoli, S.; Santini, S.; Macchiavelli, C.; Ascione, A. Active tectonics of the outer northern Apennines: Adriatic vs. Po Plain seismicity and stress fields. *J. Geodyn.* **2015**, *84*, 62–76. [CrossRef]
29. Benedetti, L.; Tapponier, P.; Gaudemer, Y.; Manighetti, I.; Van der Woerd, J. Geomorphic evidence for an emergent active thrust along the edge of Po Plain: The Boni-Stradella Fault. *J. Geophys. Res. Space Phys.* **2003**, *108*, 2238. [CrossRef]
30. Mariani, M.; Prato, R. I bacini neogenici costieri del margine tirrenico: Approccio sismico-stratigrafico. *Mem. Soc. Geol. Ital.* **1988**, *41*, 519–531.
31. Bigi, G.; Cosentino, D.; Parotto, M.; Sartori, R.; Scandone, P. Structural model of Italy, Sheets 1–6. Scale 1:500,000. *CNR Quad. Ric. Sci.* **1983**, *114*, 3.
32. Scognamiglio, L.; Tinti, E.; Quintiliani, M. Time Domain Moment Tensor (TDMT). *Ist. Naz. Geofis. Vulcanol. INGV* **2006**. [CrossRef]
33. Martelli, L.; Santulin, M.; Sani, F.; Tamaro, A.; Bonini, M.; Rebez, A.; Corti, G.; Slejko, D. Seismic hazard of the Northern Apennines based on 3D seismic sources. *J. Seism.* **2017**, *21*, 1251–1275. [CrossRef]
34. Picotti, V.; Pazzaglia, F.J. A new active tectonic model for the construction of the northern Apennines mountain front near Bologna (Italy). *J. Geophys. Res. Space Phys.* **2008**, *113*, B08412. [CrossRef]
35. Rovida, A.; Locati, M.; Camassi, R.; Lolli, B.; Gasperini, P. The Italian earthquake catalogue CPTI15. *Bull. Earthq. Eng.* **2020**, *18*, 2953–2984. [CrossRef]
36. Piccinini, D.; Chiarabba, C.; Augliera, P. Compression along the northern Apennines? Evidence from the Mw 5.3 Monghidoro earthquake. *Terra Nova* **2006**, *18*, 89–94. [CrossRef]
37. Guidoboni, E.; Ferrari, G.; Tarabusi, G.; Sgattoni, G.; Comastri, A.; Mariotti, D.; Ciuccarelli, C.; Bianchi, M.G.; Valensise, G. CFTI5Med, the new release of the catalogue of strong earthquakes in Italy and in the Mediterranean area. *Sci. Data* **2019**, *6*, 80. [CrossRef] [PubMed]
38. de Ferrari, R.; Ferretti, G.; Barani, S.; Spallarossa, D. Investigating on the 1920 Garfagnana earthquake (Mw=6.5); Evidences of site effects in Villa Collemandina (Tuscany, Italy). *Soil Dyn. Earthq. Eng.* **2010**, *30*, 1417–1429. [CrossRef]
39. Ghelardoni, R. Osservazioni sulla tettonica trasversale dell'Appennino Settentrionale. *Boll. Della Soc. Geol. Ital.* **1967**, *84*, 1–14.
40. Rosenbaum, G.; Agostinetti, N.P. Crustal and upper mantle responses to lithospheric segmentation in the northern Apennines. *Tectonics* **2015**, *34*, 648–661. [CrossRef]
41. Monteforti, B.; Raggi, G. Lineamenti strutturali fra l'alta Val di Vara e il Passo Cento Croci: Considerazioni sulla linea trasversale Val Taro-Val Parma. *Atti Soc. Toscana Sci. Nat. Mem.* **1980**, *87*, 275–284.
42. Bernini, M.; Vescovi, P.; Zanzucchi, G. Schema strutturale dell'Appennino Nord-Occidentale. *L'ateneo Parm. Acta Nat.* **1997**, *33*, 43–54.

43. Cerrina Feroni, A.; Martelli, L.; Martinelli, P.; Ottria, G.; Catanzariti, R. Carta Geologico-Strutturale dell'Appennino Emiliano-Romagnolo. Note Illustrative. Regione Emilia-Romagna Selca, Firenze. 2002. Available online: http://hdl.handle.net/11380/629938 (accessed on 15 March 2021).
44. Bernini, M.; Papani, G. Alcune considerazioni sulla struttura del margine appenninico emiliano tra lo Stirone e l'Enza (e sue relazioni con il sistema del Taro). *L'ateneo Parm. Acta Nat.* **1987**, *24*, 219–240.
45. Bortolotti, V. La tettonica trasversale dell'Appennino, I—La linea Livorno–Sillaro. *Boll. Soc. Geol. Ital.* **1966**, *85*, 529–540.
46. Bernini, M.; Boccaletti, M.; Moratti, G.; Papani, G.; Sani, F.; Torelli, L. Episodi compressivi neogenico-quaternari nell'area estensionale tirrenica. Dati in mare e a terra. *Mem. Della Soc. Geol. Ital.* **1990**, *45*, 577–589.
47. Argnani, A.; Bernini, M.; Di Dio, G.M.; Papani, G.; Rogledi, S. Stratigraphic record of crustal-scale tectonics in the Quaternary of the Northern Apennines (Italy). *Ital. J. Quat. Sci.* **1997**, *10*, 595–602.
48. Bernini, M.; Bertoldi, R.; Papani, G.; Vescovi, P. Evoluzione in regime compressivo del Bacino villafranchiano di Compiano (Parma). *Atti Ticinensi Sci. Terra* **1994**, *37*, 155–171.
49. Eva, E.; Ferretti, G.; Solarino, S. Superposition of different stress orientations in the western sector of the northern Apennines (Italy). *J. Seismol.* **2005**, *9*, 413–430. [CrossRef]
50. Molli, G.; Carlini, M.; Vescovi, P.; Artoni, A.; Balsamo, F.; Camurri, F.; Clemenzi, L.; Storti, F.; Torelli, L. Neogene 3-D Structural Architecture of the North-West Apennines: The Role of the Low-Angle Normal Faults and basement thrusts. *Tectonics* **2018**, *37*, 2165–2196. [CrossRef]
51. Piana Agostinetti, N. The structure of the Moho in the Northern Apennines: Evidence for an incipient slab tear fault? *Tectonophysics* **2005**, *655*, 88–96. [CrossRef]
52. Chiarabba, C.; Giacomuzzi, G.; Bianchi, I.; Piana Agostinetti, N.; Park, J. From underplating to delamination-retreat in the northern Apennines. *Earth Planet. Sci. Lett.* **2014**, *403*, 108–116. [CrossRef]
53. Abbate, E.; Balestrieri, M.L.; Bigazzi, G.; Norelli, P.; Quercioli, C. Fission-track datings and recent rapid denudation in northern Apennines, Italy. *Mem. Soc. Geol. Ital.* **1994**, *48*, 579–585.
54. Fellin, M.G.; Reiners, P.W.; Brandon, M.T.; Wüthrich, E.; Balestrieri, M.L.; Molli, G. Thermochronologic evidence for the exhumational history of the Alpi Apuane metamorphic core complex, northern Apennines, Italy. *Tectonics* **2007**, *26*, 26. [CrossRef]
55. Isola, I.; Mazzarini, F.; Molli, G.; Piccini, L.; Zanella, E.; Zanchetta, G.; Drysdale, R.; Hellstrom, J.; Woodhead, J.; Roncioni, A.; et al. New chronological constraints from hypogean deposits for late Pliocene to Recent morphotectonic history of the Alpi Apuane (NW Tuscany, Italy). *Geosciences* **2021**, *11*, 65. [CrossRef]
56. Puccinelli, A.; D'Amato Avanzi, G.; Perilli, N. *Note Illustrative Della Carta Geologica d'Italia, Scala 1:50.000, CARG Foglio 250 Castelnuovo Garfagnana ISPRA*; DREA: Pratovecchio, Italy, 2016; p. 167.
57. Conti, P.; Carmignani, L.; Massa, G.; Meccheri, M.; Patacca, E.; Scandone, P.; Pieruccioni, D. *Note Illustrative Della Carta Geologica d'Italia, Scala 1:50.000, CARG Foglio 249 Massa ISPRA*; Regione Toscana: Florence, Italy, 2020; p. 290.
58. Raggi, G. Neotettonica ed evoluzione paleogeografica plio-pleistocenica del bacino del Fiume Magra. *Mem. Della Soc. Geol. Ital.* **1985**, *30*, 35–62.
59. Bernini, M.; Papani, G. La distensione della fossa tettonica della Lunigiana nordoccidentale (con carta geologica alla scala 1:50,000). *Boll. Della Soc. Geol. Ital.* **2002**, *121*, 313–341.
60. Puccinelli, A.; D'Amato Avanzi, G.; Perilli, N. Note Illustrative Della Carta Geologica d'Italia, Scala 1:50.000, CARG Foglio 233 Pontremoli ISPRA. A.T.I.-S.El.CA.srl-L.A.C. Srl-System Cart srl. 2015, p. 127. Available online: https://arpi.unipi.it/handle/11568/781624#.YFIKYdwRW00 (accessed on 15 March 2021).
61. Azzaroli, A. The Villafranchian stage in Italy and the Plio-Pleistocene boundary. *Giorn. Geol.* **1977**, *41*, 61–79.
62. Bertoldi, R. Una sequenza palinologica di eta' rusciniana nei sedimenti lacustri basali del bacino di Aulla-Olivola (Val Magra). *Riv. Ital. Paleontol. Stratigr.* **1988**, *94*, 105–138.
63. Federici, P.R. Nuovi resti di vertebrato nel bacino fluvio-lacustre villafranchiano di Pontremoli (Val di Magra). *Boll. Mus. Stor. Nat. Lunigiana* **1981**, *1*, 71–74.
64. Bertoldi, R. Indagini palinologiche nel deposito fluvio-lacustre villafranchiano di Pontremoli (Val di Magra). *L'ateneo Parm. Acta Nat.* **1984**, *20*, 155–163.
65. Federici, P.R. La tettonica recente dell'Appennino: 2. Il bacino fluvio-lacustre di Pontremoli (alta Val di Magra) e sue implicazioni neotettoniche. *Gruppo Di Studio Del Quat. Padano. Quad.* **1978**, *4*, 121–131.
66. Di Naccio, D. Morphotectonic analysis of the Lunigiana and Garfagnana Grabens (Northern Italy): Implications for Active and Potentially Seismogenic Normal Faulting. Ph.D. Thesis, Università Degli Studi di Chieti-Pescara, Pescara, Italy, 2009; p. 97.
67. Moretti, A. Evoluzione tettonica della Toscana settentrionale tra il Pliocene e l'Olocene. *Boll. Soc. Geol. Ital.* **1992**, *111*, 459–492.
68. Piccardi, L.; Nirta, G.; Montanari, D.; Moratti, G.; Blumetti, A.M.; Di Manna, P.; Vittori, E.; Baglione, M.; Fabbroni, P. Geological Setting and First Paleosismological Data on the Mulazzo Fault (Lunigiana Basin, Northern Tuscany). GNGTS-2019 Volume Abstract. Session-S1, 108–112. 2019. Available online: www3.ogs.trieste.it/gngts/index.php/15-pagina-sessione-2019/75-2019-1-1 (accessed on 15 March 2021.).
69. Nirta, G.; Vittori, E.; Blumetti, A.M.; Di Manna, P.; Benvenuti, D.; Montanari, D.; Perini, M.; Fiera, F.; Moratti, G.; Baglione, M.; et al. Geomorphic and paleo seismological evidence of capable faulting in the Northern Apennines (Italy): Insights into active tectonics and seismic hazard of the Lunigiana basin. *Geomorphology* **2021**, *374*, 107486. [CrossRef]

70. Bernini, M.; Papani, G.; Dall'Asta, M.; Lasagna, S.; Heida, P. The upper Magra valley extensional basin: A cross section between Orsaro Mt. and Zeri (Massa province). *Boll. Della Soc. Geol. Ital.* **1991**, *110*, 451–458.
71. Argnani, A.; Barbacini, G.; Bernini, M.; Camurri, F.; Ghielmi, M.; Papani, G.; Rizzini, F.; Rogledi, S.; Torelli, L. Gravity tectonics driven by Quaternary uplift in the Northern Apennines: Insights from the La Spezia-Reggio Emilia geo-transect. *Quat. Int.* **2003**, *101–102*, 13–26. [CrossRef]
72. Pinelli, G. Tettonica Recente e Attiva nell'Appennino Interno a Nord Dell'arno: Una Revisione Delle Strutture e Delle Problematiche. Diploma Thesis, Università di Pisa, Pisa, Italy, 2013; p. 89.
73. Di Naccio, D.; Boncio, P.; Brozzetti, F.; Pazzaglia, F.J.; Lavecchia, G. Morphotectonic analysis of the Lunigiana and Garfagnana grabens (northern Apennines, Italy): Implications for active normal faulting. *Geomorphology* **2013**, *201*, 293–311. [CrossRef]
74. Bonini, M.; Corti, G.; Delle Donne, D.; Sani, F.; Piccardi, L.; Vannucchi, G.; Genco, R.; Martelli, L.; Ripepe, M. Seismic sources and stress transfer interaction among axial normal faults and external thrust fronts in the Northern Apennines (Italy): A working hypothesis based on the 1916–1920 time–space cluster of earthquakes. *Tectonophysics* **2016**, *680*, 67–89. [CrossRef]
75. Lucca, A.; Storti, F.; Molli, G.; Muchez, P.; Schito, A.; Artoni, A.; Balsamo, F.; Corrado, S.; Mariani, E.S. Seismically enhanced hydrothermal plume advection through the process zone of the Compione extensional Fault, Northern Apennines, Italy. *GSA Bull.* **2019**, *131*, 547–571. [CrossRef]
76. Calistri, M. II-Il Pliocene fluvio-lacustre della conca di Barga. *Mem. Soc. Geol. Ital.* **1974**, *13*, 1–21.
77. Dallan, L.; Nardi, R.; Puccinelli, A.; D'sAmato Avanzi, G.; Trivellini, M. Valutazione del rischio di frana in Garfagnana nella media valle del Serchio (Lucca). Carta geologica e carta della franosità degli elementi Sillano, Corfino, Fosciandora e Coreglia (scala 1:10,000). *Boll. Della Soc. Geol. Ital.* **1991**, *110*, 245–272.
78. Bartolini, C.; Bernini, M.; Carloni, G.C.; Costantini, A.; Federici, P.R.; Gasperi, G.; Lazzarotto, A.; Marchetti, G.; Mazzanti, R.; Papani, G.; et al. Carta neotettonica dell'Appennino settentrionale. *Note illustrative Boll. Della Soc. Geol. Ital.* **1982**, *101*, 523–549.
79. Landi, E.; Ravani, S.; Sarti, G.; Sodini, M. The Villafranchian deposits of the Castelnuovo Garfagnana and the Barga basins (Lucca, Tuscany, Italy): Facies analysis and paleoenvironmental reconstruction. *Atti Della Soc. Toscana Di Sci. Nat.* **2003**, *108*, 81–93.
80. Perilli, N.; Puccinelli, A.; Sarti, G.; D'Amato Avanzi, G. Villafranchian deposits of the Barga and Castelnuovo Garfagnana basins (Tuscany, Italy): Lithostratigraphy and sedimentary features. *IL Quat.* **2004**, *17*, 313–322.
81. Coltorti, M.; Pieruccini, P.; Rustioni, M. The Barga Basin (Tuscany): A record of Plio-Pleistocene mountain building of the Northern Apennines, Italy. *Quat. Int.* **2008**, *189*, 56–70. [CrossRef]
82. Puccinelli, A. Un esempio di tettonica recente nella Val di Serchio: Il sollevamento di Monte Perpoli. *Atti Della Soc. Toscana Di Sci. Nat. Mem. Ser. A* **1987**, *94*, 105–117.
83. Angeli, L. Tettonica Recente-Attiva Nella Media Garfagnana: Revisione Geologico-Strutturale e Problematiche di Microzonazione Sismica. Diploma Thesis, Università di Pisa, Pisa, Italy, 2018; p. 108.
84. Botti, F.; Daniele, G.; Baldacci, F. *Note Illustrative Della Carta Geologica d'Italia, Scala 1:50.000, CARG Foglio 251 Porreta, ISPRA*; System Cart: Roma, Italy, 2017; p. 193.
85. La Rosa, A.; Pagli, C.; Molli, G.; Casu, F.; De Luca, C.; Pieroni, A.; Avanzi, G.D. Growth of a sinkhole in a seismic zone of the northern Apennines (Italy). *Nat. Hazards Earth Syst. Sci.* **2018**, *18*, 2355–2366. [CrossRef]
86. Bartolini, C.; Bortolotti, V. Studi di geomorfologia e neotettonica. I—I depositi continentali dell'Alta Garfagnana in relazione alla tettonica Plio-Pleistocenica. *Mem. Soc. Geol. Ital.* **1971**, *10*, 203–245.
87. Molli, G.; Torelli, L.; Storti, F. The 2013 Lunigiana (Central Italy) earthquake: Seismic source analysis from DInSar and seismological data, and geodynamic implications for the northern Apennines. A discussion. *Tectonophys* **2016**, *668–669*, 108–112. [CrossRef]
88. Brozzetti, F.; Boncio, P.; Tinari, D.P.; Di Naccio, D.; Torelli, L. LANFs attive e relativi meccanismi di trasferimento alla terminazione settentrionale dell'Etrurian Fault System (Lunigiana–Garfagnana, Italia). *Rend. Della Soc. Geol. Ital.* **2007**, *4*, 164–165.
89. Molli, G.; Doveri, M.; Manzella, A.; Bonini, L.; Botti, F.; Menichini, M.; Montanari, D.; Trumpy, E.; Ungari, A.; Vaselli, L. Surface-subsurface structural architecture and groundwater flow of the Equi Terme hydrothermal area, Northern Tuscany Italy. *Ital. J. Geosci.* **2015**, *134*, 442–457. [CrossRef]
90. Ghelardoni, R.; Giannini, E.; Nardi, R. Ricostruzione paleogeografica dei bacini neogenici e quaternari nella bassa valle dell'Arno sulla base dei sondaggi e dei rilievi sismici. *Mem. Soc. Geol. Ital.* **1968**, *7*, 91–106.
91. Federici, P.R.; Mazzanti, R. L'evoluzione della paleogeografia e della rete idrografica del Valdarno Inferiore. *Boll. Soc. Geol. Ital.* **1988**, *5*, 573–615.
92. Dallan, L. Ritrovamento di Alephis lyrix nelle argille della serie lacustre di Montecarlo (Lucca) e considerazioni stratigrafiche sui depositi continentali dell'area tra il Monte Albano e il Monte Pisano. *Atti Soc. Toscana Sci. Nat. Mem. Ser. A* **1988**, *95*, 1–17.
93. Cantini, P.; Testa, G.; Zanchetta, G.; Cavallini, R. The Plio-Pleistocenic evolution of extensional tectonics in northern Tuscany, as constrained by new gravimetric data from the Montecarlo Basin (lower Arno Valley, Italy). *Tectonophysics* **2001**, *330*, 25–43. [CrossRef]
94. Puccinelli, A.; D'Amato Avanzi, G.; Perilli, N.; Verani, M. Note Illustrative Della Carta Geologica d'Italia, Scala 1:50.000, CARG Foglio 262 Pistoia, ISPRA, A.T.I.-S.El.CA.srl-L.A.C.srl-System Cart srl. 2015; p. 157. Available online: https://www.isprambiente.gov.it/Media/carg/262_PISTOIA/Foglio.html (accessed on 15 March 2021).
95. Zanchetta, G. Nuove osservazioni sui depositi esposti sul fianco sud-orientale delle colline delle Cerbaie (Valdarno Inferiore, Toscana). *IL Quat.* **1995**, *8*, 291–304.

96. Sarti, G.; Testa, G.; Zanchetta, G. A new stratigraphic insight of the Upper Pliocene-Lower Pleistocene succession of Lower Valdarno (Tuscany, Italy). *Geoacta* **2008**, *7*, 27–41.
97. Aguzzi, M.; Amorosi, A.; Sarti, G. Stratigraphic architecture of late quaternary deposits in the lower Arno plain (Tuscany, Italy). *Geol. Romana* **2005**, *38*, 1–10.
98. Aguzzi, M.; Amorosi, A.; Castorina, F.; Ricci Lucchi, M.; Sarti, G.; Vaiani, S.C. Stratigraphic architecture and aquifer systems in the eastern Valdarno Basin, Tuscany. *Geoacta* **2006**, *5*, 39–60.
99. Sbrana, A.; Marianelli, P.; Pasquini, G.; Costantini, P.; Palmieri, F.; Ciani, V.; Sbrana, M. The integration of 3D modeling and simulation to determine the energy potential of low-temperature geothermal systems in the Pisa (Italy) sedimentary plain. *Energies* **2018**, *11*, 1591. [CrossRef]
100. Pascucci, V. Neogene evolution of the Viareggio Basin, Northern Tuscany (Italy). *Geoacta* **2005**, *4*, 123–128.
101. Pascucci, V.; Martini, I.P.; Sagri, M.; Sandrelli, F. Effects of transverse structural lineaments on the Neogene-Quaternary basins of Tuscany (inner Northern Apennines, Italy). *Sediment. Process. Environ. Basins Tribut. Peter Friend* **2007**, *38*, 155–182.
102. Bigot, A. Aléa sismique de la regione de Pise (Italie): Etude morphostructural des failles bordières du Bassin de Viareggio. *Master Geosci. Montp.* **2010**, *28*, 108–112.
103. Giampietro, T. Architettura Strutturale e Zone di Danneggiamento in Zone di Trasferimento in Sistemi di Faglie Normali: Il Caso di Studio di Vecchiano (PI) e Modellizzazione Sperimentale. Diploma Thesis, Università di Pisa, Pisa, Italy, 2018; p. 195.
104. Porta, L. Kinematics of Faulting in Extensional Systems: Analogue Modelling and a Comparison with the Viareggio Basin Case Study. Diploma Thesis, Università di Pisa, Pisa, Italy, 2019; p. 144.
105. D'Amato Avanzi, G.; Nardi, I. Indizi di neotettonica nei Monti d'Oltre Serchio: Faglie distensive recenti al bordo della Pianura Pisana e depositi ciottolosi a quota 170 metri. *Boll. Soc. Geol. Ital.* **1993**, *112*, 601–604.
106. Federici, P.R. La tettonica recente dell'Appennino. I. Il bacino villafranchiano di Sarzana ed il suo significato nel quadro dei movimenti distensivi a NW delle Apuane. *Boll. Soc. Geol. Ital.* **1973**, *92*, 287–301.
107. Abbate, E.; Fanucci, F.; Benvenuti, M.; Bruni, P.; Cipriani, N.; Falorni, P.; Fazzuoli, M.; Morelli, D.; Pandeli, E.; Papini, M.; et al. *Note Illustrative della Carta Geologica d'Italia, scala 1:50.000, CARG Foglio 248 La Spezia, ISPRA*; Regione: Liguria, Italy, 2005; p. 204.
108. Bertoldi, R. Lineamenti palino-stratigrafici di depositi continentali del Pliocene–Pleistocene inferiore iniziale dell'Italia nord-occidentale. *Boll. Della Soc. Paleontol. Ital.* **1997**, *36*, 63–73.
109. Raggi, D.; Raggi, G. La bassa Val di Magra ed il sottosuolo della Piana Lunense, da Capellini ai giorni nostri. *Mem. Accad. Lunigianese Sci.* **2016**, *86*, 139–178.
110. Storti, F. Tectonics of Punta Bianca promontory: Insights for the evolution of the Northern Apennines-Northern Tyrrhenian Sea basin. *Tectonics* **1995**, *14*, 832–847. [CrossRef]
111. Serpelloni, E.; Pintori, F.; Gualandi, A.; Scoccimarro, E.; Cavaliere, A.; Anderlini, L.; Belardinelli, M.E.; Todesco, M. Hydrologically-induced karst deformation: Insights from GPS measurements in the Adria-Eurasia plate boundary zone. *J. Geophys. Res. Solid Earth* **2018**, *85*, 457. [CrossRef]
112. Altamimi, Z.; Métivier, L.; Rebischung, P.; Rouby, H.; Collilieux, X. ITRF2014 plate motion model. *Geophys. J. Int.* **2017**, *209*, 1906–1912. [CrossRef]
113. Tape, C.; Musé, P.; Simons, M.; Dong, D.; Webb, F. Multiscale estimation of GPS velocity fields. *Geophys. J. Int.* **2009**, *179*, 945–971. [CrossRef]
114. Serpelloni, E.; Vannucci, G.; Anderlini, L.; Bennett, R.A. Kinematics, seismotectonics and seismic potential of the eastern sector of the European Alps from GPS and seismic deformation data. *Tectonophysics* **2016**, *688*, 157–181. [CrossRef]
115. Chiaraluce, L.; Barchi, M.R.; Carannante, S.; Collettini, C.; Mirabella, F.; Pauselli, C.; Valoroso, L. The role of rheology, crustal structures and lithology in the seismicity distribution of the northern Apennines. *Tectonophysics* **2017**, *694*, 280–291. [CrossRef]
116. Mantovani, E.; Viti, M.; Cenni, M.; Babbucci, D.; Tamburelli, C.; Baglione, M.; D'Intinosante, V. Seismotectonics and present seismic hazard in the Tuscany-Romagna-Marche-Umbria Apennines (Italy). *J. Geodyn.* **2015**, *89*, 1–14. [CrossRef]
117. Basili, R.; Valensise, G.; Vannoli, P.; Burrato, P.; Fracassi, U.; Mariano, S.; Tiberti, M.; Boschi, E. The Database of Individual Seismogenic Sources (DISS), version 3: Summarizing 20 years of research on Italy's earthquakes geology. *Tectonophysics* **2008**, *453*, 20–43. [CrossRef]
118. Ferretti, G.; Massa, M.; Solarino, S. An improved method for recognition of seismic families: Application to the Garfagnana-Lunigiana Area, Italy. *Bull. Seismol. Soc. Am.* **2005**, *96*, 1903–1915. [CrossRef]
119. Stramondo, S.; Vannoli, P.; Cannelli, V.; Polcari, M.; Melini, D.; Samsonov, S.; Moro, M.; Bignami, C.; Saroli, M. X- and C-Band SAR Surface Displacement for the 2013 Lunigiana Earthquake (Northern Italy): A Breached Relay Ramp? *Ieee J. Sel. Top. Appl. Earth Obs. Remote Sens.* **2014**, *7*, 2746–2753. [CrossRef]
120. Pezzo, G.; Boncori, J.P.M.; Atzori, S.; Piccinini, D.; Antonioli, A.; Salvi, S. The 2013 Lunigiana (Central Italy) earthquake: Seismic source analysis from DInSAR and seismological data, and geodynamical implications for the northern Apennines. *Tectonophys* **2014**, *636*, 315–324. [CrossRef]
121. Tertulliani, A.; Maramai, A. Macroseismic evidence and site effects for the Lunigiana (Italy) 1995 Earthquake. *J. Seism.* **1998**, *2*, 209–222. [CrossRef]
122. Dominguez, S.; Avouac, J.-P.; Michel, R. Horizontal coseismic deformation of the 1999 Chi-Chi earthquake measured from SPOT satellite images: Implications for the seismic cycle along the western foothills of central Taiwan. *J. Geophys. Res. Space Phys.* **2003**, 108. [CrossRef]

123. Manighetti, I.; Tapponnier, P.; Gillot, P.Y.; Jacques, E.; Courtillot, V.; Armijo, R.; Ruegg, J.C.; King, G. Propagation of rifting along the Arabia-Somalia Plate Boundary: Into Afar. *J. Geophys. Res. Space Phys.* **1998**, *103*, 4947–4974. [CrossRef]
124. Westway, R.; Gawthorpe, R.; Tozzi, M. Seismological and field observations of the 1984 Lazio-Abruzzo earthquakes: Implications for the active tectonics of Italy. *Geophys. J.* **1989**, *98*, 489–514. [CrossRef]
125. Manighetti, I.; King, G.C.P.; Gaudemer, Y.; Scholz, C.H.; Doubre, C. Slip accumulation and lateral propagation of active normal faults in Afar. *J. Geophys. Res. Solid Earth* **2001**, *106*, 13667–13696. [CrossRef]
126. Tondi, E.; Jablonská, D.; Volatili, T.; Michele, M.; Mazzoli, S.; Pierantoni, P.P. The Campotosto linkage fault zone between the 2009 and 2016 seismic sequences of central Italy: Implications for seismic hazard analysis. *GSA Bull.* **2020**. [CrossRef]
127. Molli, G.; Meccheri, M. Structural inheritance and style of reactivation at mid-crustal levels: A case study from the Alpi Apuane (Tuscany, Italy). *Tectonophysics* **2012**, *579*, 74–87. [CrossRef]
128. Piazza, A.; Artoni, A.; Ogata, K. The Epiligurian wedge-top succession in the Enza Valley (northern Apennines): Evidence of a syn-depositional transpressive system. *Swiss J. Geosci.* **2016**, *109*, 17–36. [CrossRef]
129. Cuffaro, M.; Riguzzi, F.; Scrocca, D.; Antonioli, F.; Carminati, E.; Livani, M.; Doglioni, C. On the geodynamics of the northern Adriatic plate. *Rend. Lincei* **2010**, *21*, 253–279. [CrossRef]
130. Liotta, D.; Brogi, A. Plio-Quaternary fault kinematics in the Larderello geothermal area (Italy): Insights for the interpretation of the present stress field. *Geothermics* **2020**, *83*, 101714. [CrossRef]
131. Boncio, P.; Brozzetti, F.; Lavecchia, G. Architecture and seismotectonics of a regional low-angle normal fault zone in Central Italy. *Tectonics* **2000**, *19*, 1038–1055. [CrossRef]
132. Eva, E.; Solarino, S.; Boncio, P. HypoDD relocated seismicity in northern Apennines (Italy) preceding the 2013 seismic unrest: Seismotectonic implications for the Lunigiana-Garfagnana area. *Boll. Geofis. Teor. Appl.* **2014**, *55*, 739–754.
133. Galadini, F.; Falcucci, E.; Galli, P.; Giacco, B.; Gori, S.; Messina, P.; Moro, M.; Saroli, M.; Scardia, G.; Sposato, A. Time intervals to assess active and capable faults for engineering practices in Italy. *Eng. Geol.* **2012**, *139*, 50–65. [CrossRef]
134. Piccardi, L.; Vittori, E.; Blumetti, A.M.; Comerci, V.; Di Manna, P.; Guerrieri, L.; Baglione, M.; D'Intinosante, V. Mapping capable faulting hazard in a moderate-seismicity, high heat-flow environment: The Tuscia province (southern Tuscany-northern Latium, Italy). *Quat. Int.* **2017**, *451*, 11–36. [CrossRef]
135. Tinti, S.; Maramai, A.; Graziani, L. A new version of the European tsunami catalogue: Updating and revision. *Nat. Hazards Earth Syst. Sci.* **2001**, *1*, 255–262. [CrossRef]

MDPI
St. Alban-Anlage 66
4052 Basel
Switzerland
Tel. +41 61 683 77 34
Fax +41 61 302 89 18
www.mdpi.com

Geosciences Editorial Office
E-mail: geosciences@mdpi.com
www.mdpi.com/journal/geosciences

www.ingramcontent.com/pod-product-compliance
Lightning Source LLC
Chambersburg PA
CBHW041936260926
51850CB00018B/211
* 9 7 8 3 0 3 6 5 2 2 5 1 7 *